Artificial Intelligen
STEM Education

Artificial intelligence (AI) opens new opportunities for STEM education in K-12, higher education, and professional education contexts. This book summarizes AI in education (AIEd) with a particular focus on the research, practice, and technological paradigmatic shifts of AIEd in recent years.

The 23 chapters in this edited collection track the paradigmatic shifts of AIEd in STEM education, discussing *how* and *why* the paradigms have shifted, explaining *how* and *in what* ways AI techniques have ensured the shifts, and envisioning *what* directions next-generation AIEd is heading in the new era. As a whole, the book illuminates the main paradigms of AI in STEM education, summarizes the AI-enhanced techniques and applications used to enable the paradigms, and discusses AI-enhanced teaching, learning, and design in STEM education. It provides an adapted educational policy so that practitioners can better facilitate the application of AI in STEM education.

This book is a must-read for researchers, educators, students, designers, and engineers who are interested in the opportunities and challenges of AI in STEM education.

Chapman & Hall CRC Artificial Intelligence and Robotics Series

Artificial Intelligence in STEM Education: The Paradigmatic Shifts in Research, Education, and Technology
Fan Ouyang, Pengcheng Jiao, Bruce M. McLaren, Amir H. Alavi

Artificial Intelligence in STEM Education

The Paradigmatic Shifts in Research, Education, and Technology

Edited by
Fan Ouyang, Pengcheng Jiao, Bruce M. McLaren and Amir H. Alavi

CRC Press
Taylor & Francis Group
Boca Raton London New York

CRC Press is an imprint of the
Taylor & Francis Group, an **informa** business
A CHAPMAN & HALL BOOK

First edition published 2023
by CRC Press
6000 Broken Sound Parkway NW, Suite 300, Boca Raton, FL 33487-2742

and by CRC Press
4 Park Square, Milton Park, Abingdon, Oxon, OX14 4RN

CRC Press is an imprint of Taylor & Francis Group, LLC

Library of Congress Cataloging-in-Publication Data

Names: Ouyang, Fan, editor.
Title: Artificial intelligence in STEM education : the paradigmatic shifts in research, education, and technology / edited by Fan Ouyang, Pengcheng Jiao, Bruce M. McLaren, Amir H. Alavi.
Description: First edition. | Boca Raton, FL : CRC Press, 2023. | Series: Chapman & Hall/CRC artificial intelligence and robotics series | Includes bibliographical references and index.
Identifiers: LCCN 2022028544 (print) | LCCN 2022028545 (ebook) | ISBN 9781032009216 (hardback) | ISBN 9781032009216 (paperback) | ISBN 9781003181187 (ebook)
Subjects: LCSH: Artificial intelligence--Educational applications. | Science--Study and teaching--Technological innovations.
Classification: LCC LB1028.43 .A795 2023 (print) | LCC LB1028.43 (ebook) | DDC 371.33/4--dc23/eng/20220908
LC record available at https://lccn.loc.gov/2022028544
LC ebook record available at https://lccn.loc.gov/2022028545

ISBN: 9781032009216 (hbk)
ISBN: 9781032019604 (pbk)
ISBN: 9781003181187 (ebk)

DOI: 10.1201/9781003181187

Typeset in Palatino
by Deanta Global Publishing Services, Chennai, India

Contents

List of Contributors vii
Editor Biographies xi

Section I AI-Enhanced Adaptive, Personalized Learning

1. **Artificial Intelligence in STEM Education: Current Developments and Future Considerations** 3
Fan Ouyang, Pengcheng Jiao, Amir H. Alavi, and Bruce M. McLaren

2. **Towards a Deeper Understanding of K-12 Students' CT and Engineering Design Processes** 15
Gautam Biswas and Nicole M. Hutchins

3. **Intelligent Science Stations Bring AI Tutoring into the Physical World** 39
Nesra Yannier, Scott E. Hudson, and Kenneth R. Koedinger

4. **Adaptive Support for Representational Competencies during Technology-Based Problem-Solving in STEM** 51
Martina A. Rau

5. **Teaching STEM Subjects in Non-STEM Degrees: An Adaptive Learning Model for Teaching Statistics** 61
Daniela Pacella, Rosa Fabbricatore, Alfonso Iodice D'Enza, Carla Galluccio, and Francesco Palumbo

6. **Removing Barriers in Self-Paced Online Learning through Designing Intelligent Learning Dashboards** 77
Arta Faramand, Hongxin Yan, M. Ali Akber Dewan, and Fuhua Lin

Section II AI-Enhanced Adaptive Learning Resources

7. **PASTEL: Evidence-based Learning Engineering Methods to Facilitate Creation of Adaptive Online Courseware** 93
Noboru Matsuda, Machi Shimmei, Prithviraj Chaudhuri, Dheeraj Makam, Raj Shrivastava, Jesse Wood, and Peeyush Taneja

8. **A Technology-Enhanced Approach for Locating Timely and Relevant News Articles for Context-Based Science Education** 109
Jinnie Shin and Mark J. Gierl

9. **Adaptive Learning Profiles in the Education Domain** 127
Claudio Giovanni Demartini, Andrea Bosso, Giacomo Ciccarelli, Lorenzo Benussi, and Flavio Renga

Section III AI-Supported Instructor Systems and Assessments for AI and STEM Education

10. **Teacher Orchestration Systems Supported by AI: Theoretical Possibilities and Practical Considerations** 151
Suraj Uttamchandani, Haesol Bae, Chen Feng, Krista Glazewski, Cindy E. Hmelo-Silver, Thomas Brush, Bradford Mott, and James Lester

11. **The Role of AI to Support Teacher Learning and Practice: A Review and Future Directions** 163
Jennifer L. Chiu, James P. Bywater, and Sarah Lilly

12. **Learning Outcome Modeling in Computer-Based Assessments for Learning** 175
Fu Chen and Chang Lu

13. **Designing Automated Writing Evaluation Systems for Ambitious Instruction and Classroom Integration** 195
Lindsay Clare Matsumura, Elaine L. Wang, Richard Correnti, and Diane Litman

Section IV Learning Analytics and Educational Data Mining in AI and STEM Education

14. **Promoting STEM Education through the Use of Learning Analytics: A Paradigm Shift** 211
Shan Li and Susanne P. Lajoie

15. **Using Learning Analytics to Understand Students' Discourse and Behaviors in STEM Education** 225
Gaoxia Zhu, Wanli Xing, Vitaliy Popov, Yaoran Li, Charles Xie, and Paul Horwitz

16. **Understanding the Role of AI and Learning Analytics Techniques in Addressing Task Difficulties in STEM Education** 241
Sadia Nawaz, Emad A. Alghamdi, Namrata Srivastava, Jason Lodge, and Linda Corrin

17. **Learning Analytics in a Web3D Based Inquiry Learning Environment** 259
Guangtao Xu, Yingqian Li, Zhouyang Zhu, Yihui Hu, and Wenting Zhou

18. **On Machine Learning Methods for Propensity Score Matching and Weighting in Educational Data Mining Applications** 277
Juanjuan Fan, Joshua Beemer, Xi Yan, and Richard A. Levine

19. **Situating AI (and Big Data) in the Learning Sciences: Moving toward Large-Scale Learning Sciences** 289
Danielle S. McNamara, Tracy Arner, Reese Butterfuss, Debshila Basu Mallick, Andrew S. Lan, Rod D. Roscoe, Henry L. Roediger III, and Richard G. Baraniuk

20. **Linking Natural Language Use and Science Performance** 309
Scott Crossley, Danielle S. McNamara, Jennifer Dalsen, Craig G. Anderson, and Constance Steinkuehler

Section V Other Topics in AI and STEM Education

21. **Quick Red Fox: An App Supporting a New Paradigm in Qualitative Research on AIED for STEM** 319
Stephen Hutt, Ryan S. Baker, Jaclyn Ocumpaugh, Anabil Munshi, J.M.A.L. Andres, Shamya Karumbaiah, Stefan Slater, Gautam Biswas, Luc Paquette, Nigel Bosch, and Martin van Velsen

22. **A Systematic Review of AI Applications in Computer-Supported Collaborative Learning in STEM Education** 333
Jingwan Tang, Xiaofei Zhou, Xiaoyu Wan, and Fan Ouyang

23. **Inclusion and Equity as a Paradigm Shift for Artificial Intelligence in Education** 359
Rod D. Roscoe, Shima Salehi, Nia Nixon, Marcelo Worsley, Chris Piech, and Rose Luckin

Index 375

List of Contributors

Amir H. Alavi
University of Pittsburgh
Pittsburgh, PA, USA

Emad A. Alghamdi
University of Melbourne
Melbourne, Australia

Craig G. Anderson
University of California
Berkeley, CA, USA

J.M.A.L. Andres
University of Pennsylvania
Philadelphia, PA, USA

Tracy Arner
Arizona State University
Tempe, AZ, USA

Haesol Bae
Indiana University Bloomington
Bloomington, IN, USA

Ryan S. Baker
University of Pennsylvania
Philadelphia, PA, USA

Richard G. Baraniuk
Rice University
Houston, TX, USA

Debshila Basu Mallick
Rice University
Houston, TX, USA

Joshua Beemer
San Diego State University
San Diego, CA, USA

Lorenzo Benussi
Fondazione per la Scuola – Compagnia di San Paolo
Turin, Italy

Gautam Biswas
University of Pennsylvania
Philadelphia, PA, USA

Nigel Bosch
University of Pennsylvania
Philadelphia, PA, USA

Andrea Bosso
Polytechnic University of Turin
Turin, Italy

Thomas Brush
Indiana University Bloomington
Bloomington, IN, USA

Reese Butterfuss
Arizona State University
Tempe, AZ, USA

James P. Bywater
James Madison University
Harrisonburg, VA, USA

Prithviraj Chaudhuri
North Carolina State University
Raleigh, NC, USA

Fu Chen
University of Macau
Macau, China

Jennifer L. Chiu
University of Virginia
Charlottesville, VA, USA

Giacomo Ciccarelli
Polytechnic University of Turin
Turin, Italy

Richard Correnti
University of Pittsburgh
Pittsburgh, PA, USA

Linda Corrin
Swinburne University of Technology
Melbourne, Australia

Scott Crossley
Vanderbilt University
Nashville, TN, USA

Jennifer Dalsen
University of Wisconsin
Madison, WI, USA

Claudio Giovanni Demartini
Polytechnic University of Turin
Turin, Italy

Alfonso Iodice D'Enza
University of Naples Federico II
Naples, Italy

M. Ali Akber Dewan
Athabasca University
Athabasca, Alberta, Canada

Rosa Fabbricatore
University of Naples Federico II
Naples, Italy

Juanjuan Fan
San Diego State University
San Diego, CA, USA

Arta Faramand
Athabasca University
Athabasca, Alberta, Canada

Chen Feng
Indiana University Bloomington
Bloomington, IN, USA

Carla Galluccio
University of Florence
Florence, Italy

Mark J. Gierl
University of Alberta
Edmonton, Alberta, Canada

Krista Glazewski
Indiana University Bloomington
Bloomington, IN, USA

Cindy E. Hmelo-Silver
Indiana University Bloomington
Bloomington, IN, USA

Paul Horwitz
The Concord Consortium
Concord, MA, USA

Scott E. Hudson
Carnegie Mellon University
Pittsburgh, PA, USA

Nicole M. Hutchins
Vanderbilt University
Nashville, TN, USA

Stephen Hutt
University of Pennsylvania
Philadelphia, PA, USA

Pengcheng Jiao
Zhejiang University
Zhejiang, China

Shamya Karumbaiah
University of Pennsylvania
Philadelphia, PA, USA

Kenneth R. Koedinger
Carnegie Mellon University
Pittsburgh, PA, USA

Susanne P. Lajoie
McGill University
Montreal, Quebec, Canada

Andrew S. Lan
University of Massachusetts
Amherst, MA, USA

James Lester
North Carolina State University
Raleigh, NC, USA

Richard A. Levine
San Diego State University
San Diego, CA, USA

Shan Li
Lehigh University
Bethlehem, PA, USA

Yaoran Li
University of San Diego
San Diego, CA, USA

Sarah Lilly
University of Virginia
Charlottesville, VA, USA

Fuhua Lin
Athabasca University
Athabasca, Alberta, Canada

Diane Litman
University of Pittsburgh
Pittsburgh, PA, USA

Jason Lodge
The University of Queensland
Brisbane, Australia

Chang Lu
Shanghai Jiao Tong University
Shanghai, China

Rose Luckin
University College London
London, UK

Dheeraj Makam
North Carolina State University
Raleigh, NC, USA

Noboru Matsuda
North Carolina State University
Raleigh, NC, USA

Lindsay Clare Matsumura
University of Pittsburgh
Pittsburgh, PA, USA

Bruce M. McLaren
Carnegie Mellon University
Pittsburgh, PA, USA

Danielle S. McNamara
Arizona State University
Tempe, AZ, USA

Bradford Mott
North Carolina State University
Raleigh, NC, USA

Anabil Munshi
University of Pennsylvania
Philadelphia, PA, USA

Sadia Nawaz
University of Melbourne
Melbourne, Australia

Nia Nixon
University of California-Irvine
Irvine, CA, USA

Jaclyn Ocumpaugh
University of Pennsylvania
Philadelphia, PA, USA

Fan Ouyang
Zhejiang University
Zhejiang, China

Daniela Pacella
University of Naples Federico II
Naples, Italy

Francesco Palumbo
University of Naples Federico II
Naples, Italy

Luc Paquette
University of Pennsylvania
Philadelphia, PA, USA

Chris Piech
Stanford University
Stanford, CA, USA

Vitaliy Popov
University of Michigan
Ann Arbor, MI, USA

Martina A. Rau
University of Wisconsin–Madison
Madison, WI, USA

Flavio Renga
Fondazione per la Scuola – Compagnia di San Paolo
Turin, Italy

Henry L. Roediger III
Washington University in St. Louis
St. Louis, MO, USA

Rod D. Roscoe
Arizona State University
Tempe, AZ, USA

Shima Salehi
Stanford University
Stanford, CA, USA

Machi Shimmei
North Carolina State University
Raleigh, NC, USA

Jinnie Shin
University of Florida
Gainesville, FL, USA

Raj Shrivastava
North Carolina State University
Raleigh, NC, USA

Stefan Slater
University of Pennsylvania
Philadelphia, PA, USA

Namrata Srivastava
Monash University
Melbourne, Australia

Constance Steinkuehler
University of California
Berkeley, CA, USA

Peeyush Taneja
North Carolina State University
Raleigh, NC, USA

Jingwan Tang
University of Rochester
Rochester, NY, USA

Suraj Uttamchandani
Adelphi University
Garden City, NY, USA

Martin van Velsen
University of Pennsylvania
Philadelphia, PA, USA

Xiaoyu Wan
University of Rochester
Rochester, NY, USA

Elaine L. Wang
RAND Corporation
Santa Monica, CA, USA

Jesse Wood
North Carolina State University
Raleigh, NC, USA

Marcelo Worsley
Northwestern University
Evanston, IL, USA

Charles Xie
Institute for Future Intelligence
Natick, MA, USA

Wanli Xing
University of Florida
Gainesville, FL, USA

Guangtao Xu
Hangzhou Normal University
Hangzhou, China

Hongxin Yan
University of Eastern Finland
Joensuu and Kuopio, Finland

Xi Yan
San Diego State University
San Diego, CA, USA

Nesra Yannier
Carnegie Mellon University
Pittsburgh, PA, USA

Xiaofei Zhou
University of Rochester
Rochester, NY, USA

Gaoxia Zhu
Nanyang Technological University
Singapore

Editor Biographies

Fan Ouyang is Research Professor in the College of Education at Zhejiang University. Dr. Ouyang holds a Ph.D. from the University of Minnesota. Her research interests are computer-supported collaborative learning, learning analytics and educational data mining, online and blended learning, and artificial intelligence in education. Dr. Ouyang has authored/coauthored more than 30 SSCI/SCI/EI papers and conference publications and worked as PI/co-PI on more than 10 research projects, supported by National Science Foundation of China (NSFC), Zhejiang Province Educational Reformation Research Project, Zhejiang Province Educational Science Planning and Research Project, Zhejiang University-UCL Strategic Partner Funds, etc.

Pengcheng Jiao is Research Professor in the Ocean College at the Zhejiang University, China. His multidisciplinary research integrates structures and materials, sensing, computing, networking, and robotics to create and enhance the smart ocean. His research interests include mechanical functional metamaterials, SHM and energy harvesting, marine soft robotics, and AIEd. In recent years, he has authored/co-authored more than 100 peer-reviewed journals and conference publications and worked as PI/co-PI on more than 10 research projects.

Bruce M. McLaren is Associate Research Professor at Carnegie Mellon University, current Secretary and Treasurer, and ex-President of the International Artificial Intelligence in Education Society (2017–2019). McLaren is passionate about how technology can support education and has dedicated his work and research to projects that explore how students can learn with educational games, intelligent tutoring systems, e-learning principles, and collaborative learning. He holds a Ph.D. and M.S. in Intelligent Systems from the University of Pittsburgh, an M.S. in Computer Science from the University of Pittsburgh, and a B.S. in Computer Science (cum laude) from Millersville University.

Amir H. Alavi is Assistant Professor in the Department of Civil and Environmental Engineering and Department of Bioengineering at the University of Pittsburgh. He holds a Ph.D. in Civil Engineering from the Michigan State University. His original and seminal contributions to developing and deploying advanced machine learning and bioinspired computation techniques have established a road map for their broad applications in various engineering domains. He is among the Web of Science ESI's World Top 1% Scientific Minds in 2018, and the Stanford University list of Top 1% Scientists in the World in 2019 and 2020.

Section I

AI-Enhanced Adaptive, Personalized Learning

1

Artificial Intelligence in STEM Education: Current Developments and Future Considerations

Fan Ouyang, Pengcheng Jiao, Amir H. Alavi, and Bruce M. McLaren

CONTENTS

1.1 Introduction ... 3
1.2 Paradigmatic Shifts of AI in STEM Education ... 4
1.2.1 Paradigm One: AI-Directed STEM Education ... 5
1.2.2 Paradigm Two: AI-Supported STEM Education ... 6
1.2.3 Paradigm Three: AI-Empowered STEM Education ... 7
1.3 Discussion and Future Considerations ... 8
1.4 Structure of the Book ... 9
1.5 Conclusions ... 11
Acknowledgment ... 11
References ... 11

1.1 Introduction

During the last decade, applications of artificial intelligence (AI) methods in various academic fields have significantly increased due to the rapid development of data processing and computing technologies. Artificial Intelligence in Education (AIEd) is a well-established, interdisciplinary field that uses AI methods to facilitate instruction, learning, and decision-making processes (Hwang et al., 2020; Holmes et al., 2019; Roll & Wylie, 2016; O'Shea & Self, 1986; Self, 2016). AIEd can assist instructors in various instructional processes, such as automatically evaluating students' performance (Smith et al., 2019), providing recommendations and feedback to students (Bywater et al., 2019), or identifying at-risk students (Holstein et al., 2018; Hung et al., 2017). AIEd can also support student learning processes, such as tutoring students (VanLehn, 2006, 2011), providing learning materials based on students' need (Chen, et al., 2020), diagnosing students' strengths, weaknesses, and knowledge gaps (Liu et al., 2017), supporting student self-regulated learning (Aleven et al, 2016; Azevedo et al, 2008), or promoting collaboration between learners (Aluthman, 2016; Walker et al., 2009). AIEd can help administrators and managers monitor attrition patterns across colleges or departments and make decisions about their program developments (Hwang et al., 2020). Different AI techniques (e.g., artificial neural networks, ANN; deep learning, DL) have been successfully deployed to provide intelligent learning–teaching environments for building prediction models, learning recommendation, detecting behavior, etc. (Chen et al., 2020; Scruggs et al., 2020).

The emergence and continued work of AIEd has provided extensive opportunities for innovations in the field of science, technology, engineering, and mathematics (STEM) education (Xu & Ouyang, 2022). STEM education focuses on the integration of the subjects in STEM to improve students' interdisciplinary domain knowledge and understanding, as well as higher-order thinking and problem-solving skills (Kennedy & Odell, 2014; McLaren et al., 2010). STEM education usually faces challenges such as generating STEM problems, tracking students' learning, and evaluating their performance. The implementation of AI within instructional systems has the potential to solve developmental challenges in STEM education through creating active, interactive, or adaptive learning environments, automatically generating STEM problems and exercises, and evaluating or predicting students' performances (Alabdulhadi & Faisal, 2021; Jeong et al., 2019; Walker et al., 2014). For example, Yannier et al. (2020) introduced a mixed-reality AI system supported with computer vision algorithms to create and follow children's active learning behaviors in STEM education. In this book, Chapter 3 by Yannier et al. further introduces a new genre of Intelligent Science Stations,

DOI: 10.1201/9781003181187-2

a mixed-reality system that bridges the physical and virtual worlds to improve children's inquiry-based STEM learning. In addition, intelligence tutoring systems (ITSs) equipped with machine learning (ML) techniques have been used to predict students' learning preferences and time to complete specific tasks, and categorize them into clusters of similar properties to form learning groups (Alabdulhadi & Faisal, 2021). Yağci and Çevik (2019) proposed automatic AI-based algorithm models to predict the academic students' achievements in science courses (physics, chemistry, and biology) and put forward suggestions to facilitate students' successful learning.

Currently, AI-directed STEM education, AI-supported STEM education, and AI-empowered STEM education are known as three main paradigm shifts transforming AI in STEM. This opening chapter discusses various aspects of these paradigm shifts supported by the AIEd frameworks. Their capacity to design AI-based STEM educational methods is highlighted. The chapter provides further insight into the advantages, disadvantages, and future trends of AI applications in STEM education.

1.2 Paradigmatic Shifts of AI in STEM Education

AIEd has undergone several research, practice, and technological paradigmatic shifts in its brief history (Ouyang & Jiao, 2021). The first major shift is *AI-directed* (i.e., learner-as-recipient) education, which is based on behaviorism theory (Skinner, 1953, 1958). In this paradigm shift, the primary role of AI technology is to present STEM knowledge and/or course content to students, who receive the service of knowledge representations and learning pathways provided by AI systems. For example, Stat Lady Intelligent Tutoring System (Shute, 1995), Cognitive Tutors (Koedinger et al., 1997), and ASSISTment Builder (Razzaq et al., 2009) are categorized within this paradigm. The theoretical underpinning of the second paradigm called *AI-supported* (i.e., learner-as-collaborator) education is cognitive and social constructivism, in which the AI provides learning supports as the core component and students act as active collaborators to learn and progress. For example, dialogue-based tutoring systems (DTSs) (Pai et al., 2021) and exploratory learning environments (ELEs) (Rosé et al., 2019) are categorized in this paradigm. The theoretical underpinning of the third paradigm called *AI-empowered* (i.e., learner-as-leader) education is complex adaptive system theory, in which AI serves as a dynamic agent for empowering students' active learning. Students in this paradigm can be effective leaders who actively interact with AI systems and dynamically adjust self-directed learning. Emerging concepts such as human-centered AI systems (Riedl, 2019), human–AI collaboration (Hwang et al., 2020), or human-centered artificial intelligence in education (Yang et al., 2021) can be categorized into this paradigm.

Figure 1.1 shows the number of AIEd publications between 2010 and 2020. As seen, there has been a growing interest in studies related to all three paradigmatic shifts. Interestingly, a large portion of these studies are dealing with AI-supported and AI-empowered paradigms (Figure 1.1).

Although these paradigmatic shifts are general educational frameworks, they can also be applied to STEM education more generally. This process involves gradual reshaping of STEM education from the teacher-directed instruction mode to the student-centered

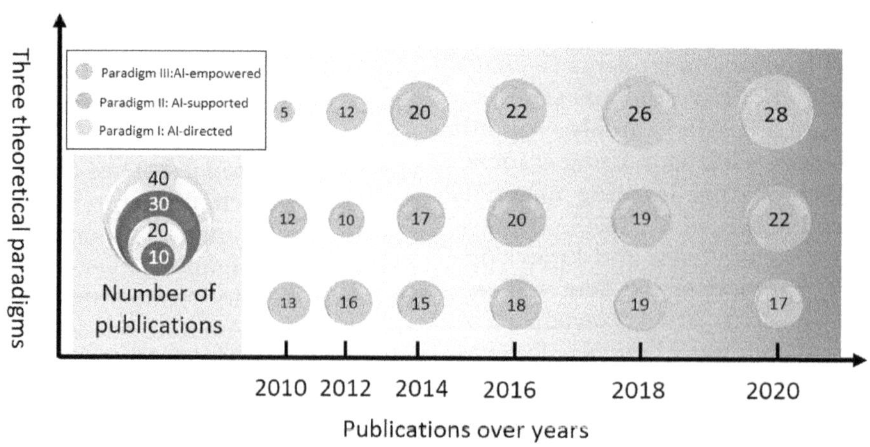

FIGURE 1.1
Number of publications related to AIEd under three theoretical paradigms.

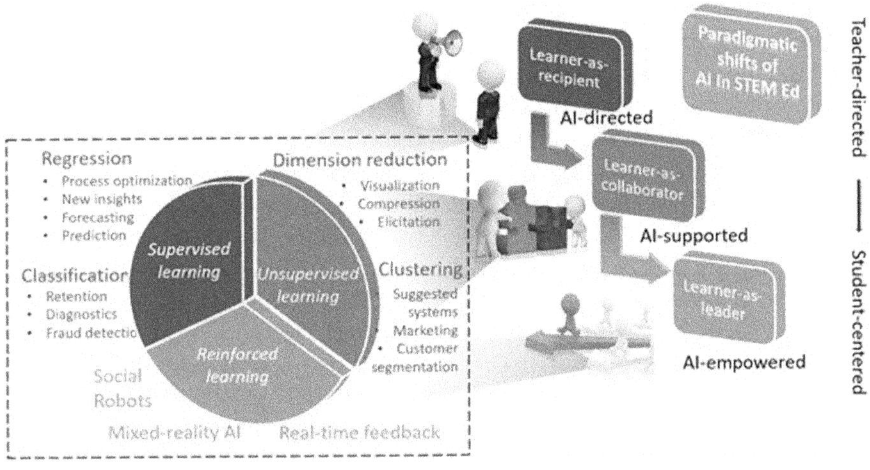

FIGURE 1.2
The shifts of three paradigms of AIEd in STEM education.

learning mode (see Figure 1.2). The following sections discuss the three paradigms of AIEd in STEM education in more detail – about the applications of the three paradigms in STEM education, the current AI applications for STEM education under these three paradigms, and related design and research.

1.2.1 Paradigm One: AI-Directed STEM Education

In the AI-directed paradigm, AI is equipped with the subject knowledge and guides the whole learning process in which the learner acts as a recipient to follow the AI-enabled learning path (Ouyang & Jiao, 2021). Behaviorism emphasizes prior systematic sequencing of learning content for the learner and guiding the learner to achieve an expected outcome (Skinner, 1953). For the AI-directed paradigm, learning is about helping learners reinforce their knowledge through a programmed instruction pattern or learning path. For instance, when learners are learning new concepts, the pattern requires immediate feedback for incorrect responses and the presentation of stimuli to guide students to mastery (Greeno et al., 1996; Schommer, 1990; Skinner, 1958). The student is required to respond to preprepared knowledge by following predetermined learning procedures and pathways and by continuously performing intended learning activities set by the AI until the desired goal is achieved (Burton et al., 2004; Holmes et al., 2019; Koschmann, 2009). Under the AI-directed paradigm, AI systems share similar characteristics to Skinner's 'teaching machine' (Skinner, 1958), which presents learners with logical subject matters and different learning pathways, that require learners to respond overtly, show they know the correct knowledge immediately, and move on to personalized learning paths (Burton et al., 2004).

For example, Shute (1995) introduces the Stat Lady Intelligent Tutoring System, which presents statistics content in a predefined order and requires learners to follow the learning sequence. Koedinger et al. (1997) use Cognitive Tutors in practical algebra curriculums to help students understand and use multiple representations of information. Mitrovic (2003) presents a Web-enabled Intelligent SQL Tutor in an introductory database course to observe students' actions and adapts to their knowledge and learning abilities. Moreover, Chin et al. (2010) and Biswas et al. (2016) use teachable agents that enable students to draw on the social metaphor of teaching to help them learn. Razzaq et al. (2009) utilize ASSISTment Builder as a tool to support teachers to effectively create, edit, test, and deploy tutor content in mathematics courses. McLaren et al. (2011) introduce a web-based intelligent tutor, namely, the Stoichiometry Tutor, to support students in chemistry learning. Overall, in the AI-directed paradigm, the AI directs the whole learning process, while the learner solves problems, engages in cognitive inquiry, and ultimately achieves learning goals by receiving AI services.

From the perspective of the AI-directed paradigm, the application of AI technologies in STEM education is a process of AI-reinforced instruction and learning. On the one hand, the instructor's teaching in STEM education is released or replaced by AI techniques. In particular, AI usually serves as a tutoring platform or a pedagogical agent to help instructors deliver teaching materials and resources, impart information and knowledge to students, and carry out teaching activities (Anderson et al., 1990; Shute, 1995). For instance, the ACT Programming Tutor system generates a production rules database for programming knowledge and presents students with a personalized learning sequence by calculating the probability of them learning these rules (Anderson et al., 1990). On the other hand, a student's learning in STEM education is reinforced through behaviorism-oriented AI techniques. For example, intelligent tutors train and supervise students mastering knowledge, completing homework, and passing examinations (Chin et al., 2010; Koedinger & Corbett, 2006; McLaren et al., 2011; Mitrovic et al., 2001; VanLehn, 2011). Consequently, the AI-directed paradigm mainly utilizes AI technology to represent knowledge in a certain instructional pattern or learning pathway, in order to impart course materials during instruction and learning in STEM education.

1.2.2 Paradigm Two: AI-Supported STEM Education

The AI-supported paradigm indicates that the AI system loosens its control and acts as a learning support system, while the role of the learner changes to that of a collaborator working with the system, focusing on the individual self-directed learning (Ouyang & Jiao, 2021). This paradigm assumes that learning occurs when learners interact with people, resources, and technology in the social environment in light of social constructivism learning (Bandura, 1986; Liu & Matthews, 2005; Vygotsky, 1978). According to this paradigm, the active and bidirectional interaction between the learner and the AI system should be formed, optimizing the learner-centered learning context. In other words, the AI system continuously collects data from the learner during the learning process as incremental input to optimize the student model, while the learner achieves better or more effective learning as a result of the interaction with the AI system (Baker & Smith, 2019; du Boulay, 2019; Rose et al., 2019). Overall, the AI-supported paradigm promotes learner-centered learning through effective interaction and ongoing collaboration between learners and the AI systems.

In STEM education, various AI implementations have been established to enable effective interaction between AI systems and learners, representative of which are dialogue-based tutoring systems (Pai et al., 2021) and exploratory learning environments (Rosé et al., 2019). On the one hand, these AI systems accurately understand the learner's learning situation by tracking their learning process and collecting and analyzing multimodal data about the learner. For example, Gerard et al. (2019) present a natural language processing tool embedded in student scientific explanations in learning. This tool can automatically score students' responses based on human-designed rubrics, adaptively guide students' learning based on the scoring results, and provide real-time feedback to the teacher on learning status. On the other hand, sustained interaction between learners and the AI systems can improve understanding of the system's decision-making process and appropriate adjustments for the upcoming learning activities. For example, Caballé et al. (2014) introduce a learning resource, the Collaborative Complex Learning Resource (CCLR), in a software engineering course. CCLR virtualizes the collaborative learning process, enabling students to observe how avatars discuss and collaborate, how discussion threads grow, and how knowledge is constructed, refined, and consolidated. Berland et al. (2015) use a tool AMOEBA, with real-time analyses of students' programming progressions, to support collaboration in a programming classroom setting in real time among middle and high school students. In summary, the learner follows the predefined learning path of the AI systems in the AI-directed paradigm, while in the AI-supported paradigm, the learner and the AI systems form a continuous mutual interaction, thus facilitating the development of learner-centered learning (Ouyang & Jiao, 2021).

From the perspective of Paradigm Two, major educational subjects (i.e., instructor, student) collaborate with the AI technologies to enhance the instruction and learning process. On the one hand, the instructional process in STEM education can be understood as a complementarity process between AI and the instructor. As an assistant, AI in STEM education can help instructors carry out instructional activities through automated question generation, assessment, feedback, and monitoring. For instance, Smith et al. (2019) propose a multimodal computational model that enables a more accurate portrait of learners by automatically analyzing students' writing and drawing in science learning. Bywater et al. (2019) describe a teacher-responding tool based on natural language processing technology that automatically generates response suggestions to assist teachers in providing personalized feedback to students in a mathematics course. On the other hand, student's learning in STEM education can be understood as a collaborative process between AI and students. The AI system acts as a support tool that does not dominate the learning process, while the student works with the system and thus focuses more on the individual student's learning process. In this case, a collaborative relationship is established between AI techniques and students during STEM education. For example, Howard et al. (2017) created an intelligent dialogue agent that helps college students learn Computer Science concepts. This agent tracks the student's learning behavior and tries to guide them toward more productive behavior. Di Mitri et al. (2021) introduced CPR Tutor, a real-time multimodal feedback system for cardiopulmonary resuscitation (CPR) training, to help students correct mistakes and improve their learning performance. In this book, Chapter 15 by Zhu et al uses various machine learning methods (e.g., text classification, transition rate analysis and sequential pattern mining, network analysis, and multilevel modeling) to understand the relationships between students' learning outcomes and processes in terms of students' discourse, multifaceted engagement, self-regulation, as well as evaluation behaviors during collaborative inquiry learning. In summary, the AI-supported paradigm focuses on using AI technologies to improve learners' engaging and collaborative roles to support individualized learning in STEM education.

1.2.3 Paradigm Three: AI-Empowered STEM Education

Driven by learner agency and instructor agency, the AI-empowered paradigm brings together multiple learners and instructors using AI as a support engine to empower quality instruction and learning. The complexity theory as the theoretical foundation of the AI-empowered paradigm holds education as a complicated intelligent system (Mason, 2008), which enhances learner intelligence through a collaborative approach between multiple agents. Moreover, stakeholders in this system should realize that AI technology is part of a complex system that consists of teachers, students, and other humans from the point of view of system design and application (Riedl, 2019). Numerous emerging concepts are proposed to build synergistic collaboration in the complex system by considering human conditions, expectations, and contexts. These typical concepts include human–machine cooperation (Hoc, 2000), human-centered AI systems (Riedl, 2019), human–AI collaboration (Hwang et al., 2020), human-centered artificial intelligence in education (Yang et al., 2021), etc. In the AI-empowered paradigm, AI enables augmented intelligence by providing learners and teachers with higher transparency of the learning process, more accurate feedback, and more practical advice (Riedl, 2019; Yang et al., 2021). AI systems support teachers in improving their understanding of the teaching and learning process, interpretation and personalized learning-oriented support, further enhancing student-centered learning activities (Baker & Smith, 2019; Holmes et al., 2019; Roll & Wylie, 2016). The learners, with empowerment of AI, lead their learning processes, hedge the risks of AI-automated decisions, and develop more effective learning (Gartner, 2019). Overall, the trends in the AI-empowered paradigm reflect the ultimate goal of AI applications in education, which is to enhance human intelligence, capability and potential (Gartner, 2019; Law, 2019; Tegmark, 2017).

Human–machine cooperation systems can achieve the AI-empowered goal by integrating AI technologies and human decision-making in STEM education. On the one hand, emerging intelligent technologies (e.g., deep learning, brain–computer interfaces, etc.) facilitate the collection and analysis of multimodal data, while ensuring transparency and accuracy. For example, Arguedas et al. (2016) use a fuzzy logic model to provide emotional feedback in an online technology course. In this way, students' emotional data is collected by AI technologies to make a more accurate representation of their emotions, enabling students to be aware of their own emotions, assess these emotions, and provide appropriate affective feedback. In turn, the role of AI has changed as human–artificial cognition has evolved (Hwang et al., 2020). On the other hand, humans can dynamically optimize the decision-making process for teaching and learning through the AI's intelligent, personalized feedback. For example, Yağci and Çevik (2019) use artificial neural networks to predict students' academic achievements in science courses (physics, chemistry, and biology) and put forward suggestions to support students. Holstein et al. (2019) use Lumilo, wearable, and real-time learning analytics glasses, to help teachers support students' learning in AI-enhanced physical classrooms. In this book, Chapter 3 by Yannier et al. introduces a new genre of Intelligent Science Stations, a mixed-reality systems that bridge the physical and virtual worlds to improve children's inquiry-based STEM learning. Chapter 21 by Hutt et al designs a new app that leverages user modeling techniques (e.g., behavior and affect-sensing) to direct interviewers to learners at critical, theory-driven moments as they learn with AIEd technologies in the classroom. The research uses machine learned models to gain a deeper insight into students' behaviors and their motivations in a qualitative way, thus furthering AIEd research. In summary, in the AI-empowered paradigm, emphasis is placed on generating adaptive, personalized learning through a synergistic interaction, integration, and collaboration between artificial intelligence systems and human intelligence (Arguedas et al., 2016; Yağci & Çevik, 2019; Yannier et al., 2020).

From the perspective of Paradigm Three, the applications of AI in STEM education will transform into a new level, namely, AI-enhanced co-design processes. Educational subjects (i.e., instructor, student) take agency and decide how to use AI technologies to enhance their instruction or learning processes and qualities (Bower, 2019). On the one hand, instructors take advantage of AI technologies to predict students' performance, identify the potential risk students, and analyze students' engagement, thereby improving STEM education (Hussain et al., 2018; Yağci & Çevik,

2019; Yannier et al., 2020). For instance, Hussain et al. (2018) use machine learning techniques to analyze students' engagement in an online social science course, and instructors can use the analysis results to adjust their teaching and thus promote students' engagement. On the other hand, in STEM education, students can use the right AI technology to avoid decision-making risks and become the owners of their own learning. In this case, AI provides personalized learning path recommendations and corresponding knowledge graphs to support student learning, based on the understanding of a student's knowledge structure and learning preferences (Arguedas et al., 2016; Chi et al., 2018; Wang et al., 2017). For example, a knowledge graph presents connections between knowledge points and concepts through graph structure, enabling students to differentiate and master complex concepts in STEM education (Chi et al., 2018; Wang et al., 2017). Based on the learner-centered principle, the AI-empowered paradigm uses AI technologies to make learners the center leader in STEM education, where learners become active participants rather than passive receivers.

1.3 Discussion and Future Considerations

AI systems can truly revolutionize STEM education via reducing teacher burdens, personalizing learning experiences for students, and transforming the roles of teacher and student. Furthermore, AI technologies can free teachers and students from redundant, elementary tasks and focus on more advanced, complicated tasks (Holstein et al., 2019; Hwang et al., 2020; VanLehn, 2011). For example, automatic evaluation techniques help reduce instructors' assessment tasks, while automatic translation tools improve the efficiency of students' language learning (Xu & Ouyang, 2021). AI applications can help instructors create and convey course content and materials (Razzaq et al., 2009) and provide students with tailored learning experiences, such as personalized tutoring (Mitrovic, 2003; Yang & Zhang, 2019). These techniques can also work as a supplementary assistant to serve STEM education and help teachers and students understand the teaching and learning process and the environment by continuously tracking data on the learning process (Figaredo, 2020; Papamitsiou & Economides, 2014). For example, AI-enabled algorithms and modeling can be used to predict students' learning performance (Yağci & Çevik, 2019), while wearable devices can be utilized to track students' learning behaviors (Holstein et al., 2019). AI in STEM education can potentially transform the teacher–student relationships from teacher-directed to student-centered learning. In traditional STEM courses, instruction refers to the effective transfer of teachers' knowledge and skills to students, which is characterized by teacher-directed, performance-oriented, and highly structured teaching model (DynaGloss, 1998). In contrast to the traditional modes where the teacher plays a substantial leadership role, designing and leading the teaching and learning process, the role of the teacher may shift to that of a supporter, collaborator, and facilitator in AI-enabled STEM education (Xu & Ouyang, 2021).

However, there are major challenges ahead of AI applications in STEM education, including the ambiguity of the responsibility of AI, overreliance on AI technologies, AI bias, and invasion of data privacy. When AI functions as a part in STEM education, it can partially take on human responsibilities but can hardly replace humans, as AI-empowered agents lack social competence and self-reflection even if they possess human-like intelligence. In STEM education, instructors are expected to impart knowledge to students, proactively reflect on instructional strategies, and adapt teaching appropriately based on the understanding of student learning (Collinson, 1996; Turner-Bisset, 2001). Peers can collaborate and communicate with students in different learning situations and establish interactions in social life (Muhisn et al., 2019). Therefore, relevant questions are raised: *Can AI replace instructor responsibilities in STEM education?* Whether and how does *the use of AI technology improve the quality of STEM education and improve teaching?*(Xu & Ouyang, 2021). Although AI can free teachers and students from redundant tasks in STEM education, it still lacks the ability to solve critical problems (Gary, 2019; Selwyn, 2016). For example, how to develop learners' interest and motivation and foster a desire to learn. Hence, when using AI techniques in STEM education, teachers need to consider why and how they should use AI. Are the reasons for using AI to reduce their workload or improve efficiency? Will the use of AI technology lead to decreased students' performance or other ethical issues? Third, privacy is a key challenge in applying AI in STEM education. To some extent, AI technologies such as educational data mining and learning analytics have the potential to enhance the teaching and learning process in STEM education. However, in the complex process of collecting, storing, transmitting, and using data, it may easily cause the disclosure of personal privacy or improper use of data (Zawacki-Richter et al., 2019). Before applying AI in STEM education, instructors should consider the risks of using technologies and pay attention to protecting students' privacy.

AI applications and research need to address the complexity of STEM education. The challenge is how to match the complexity of learning processes with the complexity of AI systems and the complexity of educational contexts (Ouyang & Jiao, 2021). AI technology should be designed to offer constant communications with and updates to instructors and students, to align AI models with learners' learning values, and to support the emergent, changing learning processes (Segal, 2019). Furthermore, AI applications should also consider how to empower stakeholders in reflecting on teaching and learning processes and goals, and accordingly how to inform AI systems to adapt and lead an iterative cycle of design, instruction, and development.

Development of AIEd in STEM education has experienced paradigm shifts from the traditional teacher-centered approach to the AI-enabled, learner-centered strategy (Ouyang & Jiao, 2021). The AI techniques have been thoroughly involved in AIEd to ensure such instructional changes, from designing teaching strategy using machine learning to predict learners' performance, and from capturing learners' responses using natural language processing (NLP) to analyzing learners' reactions using pattern recognition (PR) in teaching (Chen et al., 2020; Ouyang & Jiao, 2021; Xu & Ouyang, 2021). According to the AI technique perspective, the characteristics of AI technologies and algorithms contain automaticity, intelligence, and self-adapting, which might prompt another paradigmatic shift of AIEd in STEM education. In the future, AIEd can be developed for achievements in three main directions. The first direction is to apply AI techniques to analyze multimodal data from the instruction and learning process. It has potential to eliminate misunderstandings between learners and instructors, eventually improving students' learning quality and performance in STEM education (Belpaeme et al., 2018). The second direction is to build an AI-empowered virtual learning environment for the STEM education to better represent and convey knowledge that is difficult to understand or practice in real-world environments (Mystakidis et al., 2021). AI techniques not only can support the design and implementation of virtual environments, but also can provide learners with real-time feedback, personalized learning materials, and suggestions of learning paths. For example, the application of augmented reality (AR) in STEM education can spatially merge virtual and physical worlds with the support of digital devices (e.g., handheld devices, portable, glasses) (Riegler et al., 2019). In AR-based STEM education, AI techniques have the potential to improve students' understandings of abstract concepts and knowledge through visible and touchable artifacts (Ke & Hsu, 2015; Mystakidis et al., 2021). Third, collaboration, inclusion, and equity are involved as a paradigm shift for AI in STEM education. For example, Roscoe et al. propose Chapter 23 that AI algorithm models need to be disaggregated to include more nuanced variables and effects related to participants' social identities. Tang et al. in Chapter 22 argued that AI applications designed and implemented to support collaborative learning should be further strengthened, such as how AI supports group formation and students' interactions.

1.4 Structure of the Book

This opening chapter of the book presents an overview of recent advances in the area of AIEd. Underpinned by the AIEd paradigm frameworks, the chapter illustrates how AI in STEM can be observed through the three paradigmatic shifts of AI-directed STEM education (learner-as-recipient), AI-supported STEM education (learner-as-collaborator), and AI-empowered STEM education (learner-as-leader). The book is structured into five sections:

- Section I: AI-Enhanced Adaptive, Personalized Learning
- Section II: AI-Enhanced Adaptive Learning Resources
- Section III: AI-Supported Instructor Systems and Assessments for AI and STEM Education
- Section IV: Learning Analytics and Educational Data Mining in AI and STEM Education
- Section V: Other Topics in AI and STEM Education

Section I: AI-Enhanced Adaptive, Personalized Learning, includes four chapters. The present chapter introduces Intelligent Science Stations. Experiments indicate that the automated reactive guidance, made possible by a specialized AI computer vision algorithm, can provide personalized interactive feedback to children. Chapter 2 by Biswas and Hutchins combines AI and machine learning methods to support curriculum and learning environment design in an earth sciences module, and then developed analytics to analyze middle school students' learning performance and behaviors in the environment. The integrated methods provide an understanding of students' learning pathways as they transition from applying their conceptual knowledge to constructing computational models to solve an engineering design problem. Chapter 3 et al. by Yannier introduces a new genre of Intelligent Science Stations, a mixed-reality system that bridges the physical and virtual worlds to improve children's inquiry-based STEM learning. Chapter 4 by Rau presents a series of studies of sense-making and perceptual fluency in problem-solving activities that enhances students' learning of STEM content knowledge and describes how learning analytics reveal that prior knowledge affects students' interaction with representational competency supports. Chapter 5 by Pacella et al. develops an Adaptive LEArning for Statistics (ALEAS) app to provide an adaptive learning environment that allows students to assess their own knowledge in statistics. Learners are assessed by two multivariate methods: (i) for each topic, a multidimensional latent-class IRT model is defined, in which dimensions corresponding to the students' ability are measured; (ii) within each area, archetypal analysis allows integrating and comparing the students' performances. Chapter 6 by Faramand et al. proposes a methodology of intelligent learning dashboard focusing on SPOL and discusses how to construct mechanisms for adaptive formative assessment and student engagement detection with the state-of-the-art AI techniques, how to design and integrate these technologies in intelligent learning dashboards, and how to include these mechanisms in the course learning design loop to ensure data collection and pedagogical connection.

Section II: AI-Enhanced Adaptive Learning Resources includes three chapters. Chapter 7 by Matsuda et al. proposes PASTEL, a pragmatic method to develop adaptive and scalable technologies for next-generation e-learning. PASTEL is a collection of methods to assist courseware developers to build adaptive online courseware. The chapter provides details about the PASTEL technology and results from its early-stage evaluations. Chapter 8 by Shin and Gierl introduces a technology-enhanced framework based on machine learning and natural language processing techniques to understand and evaluate science articles. The chapter reveals that the best model can identify an interpretable topic to accurately classify the science articles based on their curriculum standards. Chapter 9 by Demartini et al. enhances the comprehension of teaching and learning within the educational domain by leveraging data gathered along the student learning life cycle. The integrated data mining and machine learning techniques make this conceptual platform an adaptative and innovative tool to develop reinforcement and personalization of educational experiences.

Section III: AI-Supported Instructor Systems and Assessments for AI and STEM Education includes four chapters. Chapter 10 by Uttamchandani et al. proposes considerations that emerge in the design of orchestration assistant, an AI-supported teacher orchestration system. The theoretical possibilities are proposed for supporting pedagogy with AI. And possibilities are proposed when the teacher orchestration system is enacted, complicated, or transformed in the context of real classroom activity. Relevant design considerations are proposed for designing such AI-supported systems for teachers. Chapter 11 by Chiu et al. reviews current approaches in education that use AI technologies to provide targeted learning opportunities for teachers. This chapter leverages the ICAP framework to discuss current and future directions for AI-based tools that put teachers in-the-loop, which provides automated feedback on teachers' practices and improve students' knowledge construction. Chapter 12 by Chen and Lu proposes an overview of the mainstream learner models that are commonly used in computer-based assessments for learning as well as recent advances in learning outcome modeling. Chapter 13 by Matsumura et al. develops an automated writing evaluation system (eRevise) to support argument writing. The chapter proposes that the AWE systems communicate the features of authentic tasks, provide information that is transparent, actionable, and fair, and open up avenues for student-centered classroom collaborations.

Section IV: Learning Analytics and Educational Data Mining in AI and STEM Education includes seven chapters. Chapter 14 by Li and Lajoie introduces a theory-driven learning analytics model, which has the potential to promote the evolution of STEM education and research. This chapter presents an example study to illustrate how theory-driven learning analytics can be applied into practice in a STEM learning context. Chapter 15 by Zhu et al. discusses how learning analytics can be used to analyze students' discourse and behaviors in technology-enhanced STEM learning environments. Machine learning methods such as text classification, transition rate analysis and sequential pattern mining, network analysis, and multilevel modeling are adopted to understand the learning outcomes and processes. Chapter 16 by Nawaz et al. discusses the notion of task difficulty, how it is defined and operationalized in digital learning environments. This work further highlights how artificial intelligence and learning analytics offer opportunities to provide timely support to students when they experience task difficulties. Chapter 17 by Xu integrates inquiry learning and Web3D technology into virtual experiments in order to improve learner experiences and learning quality. A general framework is proposed, which includes three application branches: data collection and processing, learner modeling, and learning recommendation. Chapter 18 by Fan et al. reviews recent developments of ensemble learning machinery for propensity score matching and weighting. This work extends and improves the use of learning analytics for estimating treatment in the personalized medicine observational studies literature. Chapter 19 by McNamara et al. proposes that AI (and data science) can reveal nuanced patterns of student retention, persistence, and performance, but expertise in learning theory and psychological sciences is needed to suggest mechanisms and explanations for these patterns. Chapter 20 by Crossley et al. uses a widely used educational data mining technique – natural language processing – to extract linguistic attributes of students' collaborative problem-solving and links it to their final science performance.

Section V: Other Topics in AI and STEM Education deals with qualitative research and collaborative learning practice in AI and STEM education. Chapter 21 by Hutt et al. designs a new app that leverages user modeling techniques (e.g., behavior and affect-sensing) to direct interviewers to learners at critical, theory-driven moments as they learn with AIEd technologies in the classroom. The research indicates that using machine learning models to optimize researcher time helps researchers gain a deeper insight into students' behaviors and their motivations, thus furthering AIEd research. Chapter 22 by Tang et al. conducts a systematic literature review to understand the development of AI to support computer-supported collaborative learning (CSCL) in STEM education from 2011 to 2021. This review examines the overall trend of AI applications designed and implemented to support CSCL and evaluates the effects of proposed AI techniques and applications in supporting group formation and students' interaction. Chapter 23 by Roscoe et al. stresses the importance of AIEd to include more nuanced variables and effects related to demographic factors and social identities. This work also proposes that intersectional approaches are needed to represent learners' multiple identities, associated power, or privilege and to interpret observed effects.

1.5 Conclusions

This opening chapter presented an overview of recent advances in the area of AIEd and STEM education underpinned by the AIEd paradigm frameworks. We showed how AI in STEM can be observed through the three paradigmatic shifts of AI-directed STEM education (learner-as-recipient), AI-supported STEM education (learner-as-collaborator), and AI-empowered STEM education (learner-as-leader). We examined how AI applications are connected to existing educational and learning theories, the extent of which AI technologies influence teaching, and the different roles of AI in education. The capacity of the three shifts to transform the AI-based STEM educational methods was further highlighted. We discussed the future AIEd practices and research in STEM education from teacher-directed education to learner-centered learning, where learner agency, initiations, and lifelong learning are valued. Finally, summaries of the chapters included in this book are provided.

Acknowledgment

Fan Ouyang acknowledges the financial support from the National Natural Science Foundation of China, No. 62177041.

References

Alabdulhadi, A., & Faisal, M. (2021). Systematic literature review of STEM self-study related ITSs. *Education and Information Technologies*, *26*(2), 1549–1588. https://doi.org/10.1007/s10639-020-10315-z

Aleven, V., Roll, I., McLaren, B. M., & Koedinger, K. R. (2016). Help helps, but only so much: Research on help seeking with intelligent tutoring systems. *International Journal of Artificial Intelligence in Education*, *26*(1), 205–223. https://doi.org/10.1007/s40593-015-0089-1

Aluthman, E. (2016). The effect of using automated essay evaluation on ESL undergraduate students' writing skill. *International Journal of English Linguistics*, *6*(5), 54. https://doi.org/10.5539/IJEL.V6N5P54

Anderson, J. R., Boyle, C. F., Corbett, A. T., & Lewis, M. W. (1990). Cognitive modeling and intelligent tutoring. *Artificial Intelligence*, *42*(1), 7–49.

Arguedas, M., Xhafa, F., Casillas, L., Daradoumis, T., Peña, A., & Caballé, S. (2016). A model for providing emotion awareness and feedback using fuzzy logic in online learning. *Soft Computing*, *22*(3), 963–977. https://doi.org/10.1007/S00500-016-2399-0Azevedo

Azevedo, R., Moos, D. C., Greene, J. A., Winters, F. I., & Cromley, J. G. (2008). Why is externally-facilitated regulated learning more effective than self-regulated learning with hypermedia? *Educational Technology Research and Development*, *56*(1), 45–72. https://doi.org/10.1007/s11423-007-9067-0

Baker, T., & Smith, L. (2019). Educ-AI-tion rebooted? Exploring the future of artificial intelligence in schools and colleges. https://media.nesta.org.uk/documents/Future_of_AI_and_education_v5_WEB.Pdf

Bandura, A. (1986). *Social foundations of thought and action: A social cognitive theory*. Prentice-Hall.

Belpaeme, T., Kennedy, J., Ramachandran, A., Scassellati, B., & Tanaka, F. (2018). Social robots for education: A review. *Science Robotics*, *3*(21), eaat5954. https://doi.org/10.1126/scirobotics.aat5954

Berland, M., Davis, D., & Smith, C. P. (2015). AMOEBA: Designing for collaboration in computer science classrooms through live learning analytics. *International Journal of Computer-Supported Collaborative Learning*, *10*(4), 425–447. https://doi.org/10.1007/S11412-015-9217-Z

Biswas, G., Segedy, J. R., & Bunchongchit, K. (2016). From design to implementation to practice – A learning by teaching system: Betty's Brain. *International Journal of Artificial Intelligence in Education*, *26*(1), 350–364. https://doi.org/10.1007/s40593-015-0057-9

Bower, M. (2019). Technology-mediated learning theory. *British Journal of Educational Technology*, *50*(3), 1035–1048. https://doi.org/10.1111/BJET.12771

Burton, J. K., Moore, D. M. M., & Magliaro, S. G. (2004). *Behaviorism and instructional technology*. Lawrence Erlbaum Associates Publishers.

Bywater, J. P., Chiu, J. L., Hong, J., & Sankaranarayanan, V. (2019). The teacher responding tool: Scaffolding the teacher practice of responding to student ideas in mathematics classrooms. *Computers and Education*, *139*, 16–30. https://doi.org/10.1016/J.COMPEDU.2019.05.004

Caballé, S., Mora, N., Feidakis, M., Gañán, D., Conesa, J., Daradoumis, T., & Prieto, J. (2014). CC-LR: Providing interactive, challenging and attractive collaborative complex learning resources. *Journal of Computer Assisted Learning*, *30*(1), 51–67. https://doi.org/10.1111/JCAL.12021

Chen, L., Chen, P., & Lin, Z. (2020). Artificial intelligence in education: A review. *IEEE Access*, *8*, 75264–75278. https://doi.org/10.1109/ACCESS.2020.2988510

Chin, D. B., Dohmen, I. M., Cheng, B. H., Oppezzo, M. A., Chase, C. C., & Schwartz, D. L. (2010). *Preparing students for future learning with teachable agents*. Society for Research on Educational Effectiveness.

Chi, Y., Qin, Y., Song, R., & Xu, H. (2018). Knowledge graph in smart education: A case study of entrepreneurship scientific publication management. *Sustainability*, *10*(4), 995. https://doi.org/10.3390/su10040995

Collinson, V. (1996). *Reaching students: Teachers ways of knowing*. Corwin Press.

Di Mitri, D., Schneider, J., & Drachsler, H. (2021). Keep me in the loop: Real-time feedback with multimodal data. *International Journal of Artificial Intelligence in Education*, 1–26. https://doi.org/10.1007/s40593-021-00281-z

du Boulay, B. (2019). Escape from the skinner box: The case for contemporary intelligent learning environments. *British Journal of Educational Technology*, *50*(6), 2902–2919. https://doi.org/10.1111/bjet.12860

DynaGloss. (1998). Instructionism. http://Seed.cs.colorado.edu/dynagloss.makeGlossPage.fcgi$URLinc=1&Term=Instructionism

Figaredo, D. D. (2020). Data-driven educational algorithms pedagogical framing. *Revista Iberoamericana de Educación a Distancia*, *23*(2), 65–84. http://doi.org/10.5944/ried.23.2.26470

Gartner. (2019). Hype cycle for emerging technologies, 2019. https://www.gartner.com/en/documents/3956015/hype-cycle-for-emerging-technologies-2019

Gary, K. (2019). Pragmatic standards versus saturated phenomenon: Cultivating a love of learning. *Journal of Philosophy of Education*, *53*(3), 477–490. https://doi.org/10.1111/1467-9752.12377

Gerard, L., Kidron, A., & Linn, M. C. (2019). Guiding collaborative revision of science explanations. *International Journal of Computer-Supported Collaborative Learning*, *14*(3), 291–324. https://doi.org/10.1007/s11412-019-09298-y

Greeno, J. G., Collins, A. M., & Resnick, L. B. (1996). Cognition and learning. In D. C. Berliner & R. C. Calfee (Eds.), *Handbook of educational psychology* (pp. 15–46). Lawrence Erlbaum Associates.

Hoc, J. M. (2000). From human–machine interaction to human–machine cooperation. *Ergonomics*, *43*(7), 833–843.

Holmes, W., Bialik, M., & Fadel, C. (2019). *Artificial intelligence in education: Promises and implications for teaching and learning*. Center for Curriculum Redesign.

Holstein, K., McLaren, B. M., & Aleven, V. (2018). Student learning benefits of a mixed-reality teacher awareness tool in AI-enhanced classrooms. In C. Rosé, R. Martínez-Maldonado, H. U. Hoppe, R. Luckin, M. Mavrikis, K. Porayska-Pomsta, B. McLaren, & B. du Boulay (Eds.), *Proceedings of the 19th international conference on artificial intelligence in education (AIED 2018)*. LNAI 10947 (pp. 154–168). Springer.

Holstein, K., McLaren, B. M., & Aleven, V. (2019). Co-designing a real-time classroom orchestration tool to support teacher–AI complementarity. *Journal of Learning Analytics*, *6*(2), 27–52. https://doi.org/10.18608/jla.2019.62.3

Howard, C., Jordan, P., Di Eugenio, B., & Katz, S. (2017). Shifting the load: A peer dialogue agent that encourages its human collaborator to contribute more to problem solving. *International Journal of Artificial Intelligence in Education*, *27*(1), 101–129. https://doi.org/10.1007/S40593-015-0071-Y

Hung, J. L., Wang, M. C., Wang, S., Abdelrasoul, M., Li, Y., & He, W. (2017). Identifying at-risk students for early interventions – A time-series clustering approach. *IEEE Transactions on Emerging Topics in Computing*, *5*(1), 45–55. https://doi.org/10.1109/TETC.2015.2504239

Hussain, M., Zhu, W., Zhang, W., & Abidi, S. M. R. (2018). Student engagement predictions in an e-learning system and their impact on student course assessment scores. *Computational Intelligence and Neuroscience*, *2018*, 6347186. https://doi.org/10.1155/2018/6347186

Hwang, G. J., Xie, H., Wah, B. W., & Gašević, D. (2020). Vision, challenges, roles and research issues of artificial intelligence in education. *Computers and Education: Artificial Intelligence*, *1*, 100001. https://doi.org/10.1016/j.caeai.2020.100001

Jeong, H., Hmelo-Silver, C. E., & Jo, K. (2019). Ten years of computer-supported collaborative learning: A meta-analysis of CSCL in STEM education during 2005–2014. *Educational Research Review*, *28*, 100284. https://doi.org/10.1016/j.edurev.2019.100284

Ke, F., & Hsu, Y.-C. (2015). Mobile augmented-reality artifact creation as a component of mobile computer-supported collaborative learning. *The Internet and Higher Education, 26*, 33–41. https://doi.org/10.1016/j.iheduc.2015.04.003

Kennedy, T. J., & Odell, M. R. L. (2014). Engaging students in STEM education. *Science Education International, 25*(3), 246–258.

Koedinger, K. R., Anderson, J. R., Hadley, W. H., & Mark, M. A. (1997). Intelligent tutoring goes to school in the big city. *International Journal of Artificial Intelligence in Education, 8*(1), 30–43.

Koedinger, K. R., & Corbett, A. T. (2006). Cognitive tutors: Technology bringing learning sciences to the classroom. In R. K. Sawyer (Ed.), *The Cambridge handbook of the learning sciences* (pp. 61–78). Cambridge University Press.

Koschmann, T. (2009). Chapter 1. Paradigm shifts and instructional technology: An introduction. In T. Koschmann (Ed.), *CSCL: Theory and practice of an emerging paradigm* (pp. 1–23). Routledge.

Law, N. W. Y. (2019). *Human development and augmented intelligence*. The 20th International Conference on Artificial Intelligence in Education (AIED 2019). Chicago, IL, USA.

Liu, C. H., & Matthews, R. (2005). Vygotsky's philosophy: Constructivism and its criticisms examined. *International Education Journal, 6*(3), 386–399.

Liu, M., Li, Y., Xu, W., & Liu, L. (2017). Automated essay feedback generation and its impact on revision. *IEEE Transactions on Learning Technologies, 10*(4), 502–513. https://doi.org/10.1109/TLT.2016.2612659

Mason, M. (2008). What is complexity theory and what are its implications for educational change? *Educational Philosophy and Theory, 40*(1), 35–49.

McLaren, B. M., Deleeuw, K. E., & Mayer, R. E. (2011). Polite web-based intelligent tutors: Can they improve learning in classrooms? *Computers and Education, 56*(3), 574–584. https://doi.org/10.1016/J.COMPEDU.2010.09.019

McLaren, B. M., Scheuer, O., & Mikšátko, J. (2010). Supporting collaborative learning and e-Discussions using artificial intelligence techniques. *International Journal of Artificial Intelligence in Education, 20*(1), 1–46. https://doi.org/10.3233/JAI-2010-0001

Mitrovic, A. (2003). An intelligent SQL tutor on the web. *International Journal of Artificial Intelligence in Education, 13*(2–4), 173–197.

Mitrovic, A., Mayo, M., Suraweera, P., & Martin, B. (2001). Constraint-based tutors: A success story. Budapest, Hungary: Engineering of intelligent systems. *Proceedings of the 14th international conference on industrial and engineering applications of artificial intelligence and expert systems (IEA/AIE 2001), 4–7 June 2001. Lecture notes in computer science, 2070*, 931–940.

Muhisn, Z. A. A., Ahmad, M., Omar, M., & Muhisn, S. A. (2019). The impact of socialization on collaborative learning method in e-learning management system (eLMS). *International Journal of Emerging Technologies in Learning, 14*(20), 137–148. https://www.online-journals.org/index.php/i-jet/article/view/10992

Mystakidis, S., Christopoulos, A., & Pellas, N. (2021). A systematic mapping review of augmented reality applications to support STEM learning in higher education. *Education and Information Technologies*, 1–45. https://doi.org/10.1007/S10639-021-10682-1/FIGURES/10

O'Shea, T., & Self, J. (1986). *Learning and teaching with computers: The artificial intelligence revolution*. Prentice Hall Professional Technical Reference.

Ouyang, F., & Jiao, P. (2021). Artificial intelligence in education: The three paradigms. *Computers and Education: Artificial Intelligence, 2*, 100020. https://doi.org/10.1016/J.CAEAI.2021.100020

Pai, K. C., Kuo, B. C., Liao, C. H., & Liu, Y. M. (2021). An application of Chinese dialogue-based intelligent tutoring system in remedial instruction for mathematics learning. *Educational Psychology, 41*(2), 137–152. https://doi.org/10.1080/01443410.2020.1731427

Papamitsiou, Z., & Economides, A. (2014). Learning analytics and educational data mining in practice: A systematic literature review of empirical evidence. *Educational Technology & Society, 17*(4), 49–64.

Razzaq, L., Patvarczki, J., Almeida, S. F., Vartak, M., Feng, M., Heffernan, N. T., & Koedinger, K. R. (2009). The assistment builder: Supporting the life cycle of tutoring system content creation. *IEEE Transactions on Learning Technologies, 2*(2), 157–166. https://doi.org/10.1109/TLT.2009.23

Riedl, M. O. (2019). Human-centered artificial intelligence and machine learning. *Human Behavior and Emerging Technologies, 1*(1), 33–36.

Riegler, A., Wintersberger, P., Riener, A., & Holzmann, C. (2019). Augmented reality windshield displays and their potential to enhance user experience in automated driving. *I-Com, 18*(2), 127–149. https://doi.org/10.1515/icom-2018-0033

Roll, I., & Wylie, R. (2016). Evolution and revolution in artificial intelligence in education. *International Journal of Artificial Intelligence in Education, 26*(2), 582–599.

Rosé, C. P., McLaughlin, E. A., Liu, R., & Koedinger, K. R. (2019). Explanatory learner models: Why machine learning (alone) is not the answer. *British Journal of Educational Technology, 50*(6), 2943–2958. https://doi.org/10.1111/bjet.12858

Schommer, M. (1990). Effects of beliefs about the nature of knowledge on comprehension. *Journal of Educational Psychology, 82*(3), 498–504.

Scruggs, R., Baker, R. S., & McLaren, B. M. (2020). Extending deep knowledge tracing: Inferring interpretable knowledge and predicting post-system performance. In H. J. So, M. M. Rodrigo, J. Mason, & A. Mitrovic (Eds.), *Proceedings of the 28th international conference on computers in education (ICCE 2020)* (pp. 195–204). Asia-Pacific Society for Computers in Education.

Self, J. (2016). The birth of IJAIED. *International Journal of Artificial Intelligence in Education, 26*(1), 4–12. https://doi.org/10.1007/s40593-015-0040-5

Segal, M. (2019). A more human approach to artificial intelligence. *Nature, 571*(7766), S18–S18.

Selwyn, N. (2016). *Is technology good for education?* Polity Press.

Shute, V. J. (1995). SMART: Student modeling approach for responsive tutoring. *User Modeling and User-Adapted Interaction, 5*(1), 1–44.

Skinner, B. F. (1953). *Science and human behavior.* Macmillan.

Skinner, B. F. (1958). Teaching machines. *Science, 128*(3330), 969–977.

Smith, A., Leeman-Munk, S., Shelton, A., Mott, B., Wiebe, E., & Lester, J. (2019). A multimodal assessment framework for integrating student writing and drawing in elementary science learning. *IEEE Transactions on Learning Technologies, 12*(1), 3–15. https://doi.org/10.1109/TLT.2018.2799871

Tegmark, M. (2017). *Life 3.0: Being human in the age of artificial intelligence.* Knopf.

Turner-Bisset, R. (2001). *Expert teaching: Knowledge and pedagogy to lead the profession.* David Fulton Publishers.

VanLehn, K. (2006). The behavior of tutoring systems. *International Journal of Artificial Intelligence in Education, 16*(3), 227–265. https://content.iospress.com/articles/international-journal-of-artificial-intelligence-in-education/jai16-3-02

VanLehn, K. (2011). The relative effectiveness of human tutoring, intelligent tutoring systems, and other tutoring systems. *Educational Psychologist, 46*(4), 197–221. https://doi.org/10.1080/00461520.2011.611369

Vygotsky, L. S. (1978). *Mind in society.* Harvard University Press.

Walker, E., Rummel, N., & Koedinger, K. R. (2009). Integrating collaboration and intelligent tutoring data in the evaluation of a reciprocal peer tutoring environment. *Research and Practice in Technology Enhanced Learning, 4*(3), 221–251. https://doi.org/10.1142/S1793206809000074X

Walker, E., Rummel, N., & Koedinger, K. R. (2014). Adaptive intelligent support to improve peer tutoring in algebra. *International Journal of Artificial Intelligence in Education, 24*(1), 33–61.

Wang, Q., Mao, Z., Wang, B., & Guo, L. (2017). Knowledge graph embedding: A survey of approaches and applications. *IEEE Transactions on Knowledge and Data Engineering, 29*(12), 2724–2743. https://doi.org/10.1109/TKDE.2017.2754499

Xu, W., & Ouyang, F. (2021). A systematic review of AI role in the educational system based on a proposed conceptual framework. *Education and Information Technologies.* https://doi.org/10.1007/s10639-021-10774-y

Xu, W., & Ouyang, F. (2022). The application of AI technologies in STEM education: a systematic review from 2011 to 2021. *International Journal of STEM Education, 9* (59), 1–20. https://doi.org/10.1186/s40594-022-00377-5

Yağci, A., & Çevik, M. (2019). Prediction of academic achievements of vocational and technical high school (VTS) students in science courses through artificial neural networks (comparison of Turkey and Malaysia). *Education and Information Technologies, 24*(5), 2741–2761. https://doi.org/10.1007/s10639-019-09885-4

Yang, J., & Zhang, B. (2019). Artificial intelligence in intelligent tutoring robots: A systematic review and design guidelines. *Applied Sciences, 9*(10), 2078. https://doi.org/10.3390/APP9102078

Yang, S. J., Ogata, H., Matsui, T., & Chen, N. S. (2021). Human-centered artificial intelligence in education: Seeing the invisible through the visible. *Computers and Education. Artificial Intelligence, 2,* 100008.

Yannier, N., Hudson, S. E., & Koedinger, K. R. (2020). Active learning is about more than hands-on: A mixed-reality AI system to support STEM education. *International Journal of Artificial Intelligence in Education, 30*(1), 74–96. https://doi.org/10.1007/s40593-020-00194-3

Zawacki-Richter, O., Marín, V. I., Bond, M., & Gouverneur, F. (2019). Systematic review of research on artificial intelligence applications in higher education – Where are the educators? *International Journal of Educational Technology in Higher Education, 16*(39), 1–27. https://doi.org/10.1186/s41239-019-0171-0

2

Towards a Deeper Understanding of K-12 Students' CT and Engineering Design Processes

Gautam Biswas and Nicole M. Hutchins

CONTENTS

2.1 Introduction ... 15
2.2 Background and Motivation ... 17
2.2.1 21st-Century K-12 Classrooms ... 17
2.2.2 Understanding Students' STEM Learning Strategies ... 17
2.2.2.1 Computational Modeling Strategies ... 18
2.2.2.2 Engineering Design Strategies ... 18
2.3 SPICE Curriculum and Learning Environment ... 18
2.3.1 System Design Perspectives ... 18
2.3.1.1 Evidence-Centered Design ... 19
2.3.1.2 Coherence across Modeling Representations ... 19
2.3.1.3 Domain-Specific Modeling Languages ... 19
2.3.2 SPICE Learning Trajectory and Curriculum ... 20
2.4 Methods ... 22
2.4.1 Implementation ... 22
2.4.2 Data Sources ... 22
2.4.3 Analysis Methods ... 23
2.5 Results ... 26
2.5.1 The Role of Computational Thinking ... 26
2.5.1.1 Effects on Computational Modeling ... 26
2.5.1.2 Effects on Engineering Design ... 27
2.5.1.3 Effects on Posttest Scores ... 27
2.5.2 Learning with Multiple Representations ... 28
2.5.3 Impact of Strategy Use on Learning ... 30
2.5.3.1 Computational Modeling Strategies ... 30
2.5.3.2 Engineering Design Strategies ... 33
2.6 Discussion and Future Directions ... 35
Notes ... 36
Bibliography ... 36

2.1 Introduction

The Framework for K-12 Science Education and the Next Generation Science Standards (NGSS) calls for the integration of science and engineering content in K-12 classrooms. This provides opportunities for authentic learning experiences that can motivate students and better prepare them for 21st-century STEM careers [1]. The environments we have designed to support such integrated activities are open-ended in nature, i.e., students are given a specific problem to solve (e.g., design a playground that meets specified constraints), but they are free to explore the environment to learn their science, computation, and engineering concepts by constructing conceptual and computational models and performing an engineering design task [2, 3]. This approach has been successful in improving students' *synergistic* learning and understanding of science, engineering, and computational thinking (CT) concepts [4, 5, 6]. However, students sometimes have difficulties in learning their science and engineering concepts and applying them

DOI: 10.1201/9781003181187-3

to solve problems [7, 5]. These students may need support and guidance to help them succeed and learn from the activities in the environment [8]. Our overall goal has been to design open-ended educational systems that leverage artificial intelligence (AI) representations and support the design of assessments and learning analytics methods that are sufficiently rich to capture students' evolving knowledge and problem-solving processes in STEM domains and CT to better support the learners, the learning designers, and educators in classroom environments.

In this chapter, we develop two relevant research directions that leverage AI, analytics, and machine learning methods:

1. **Creating multiple, linked representations that support students' learning processes** [9, 10]: To do so, we leverage representational and reasoning schemes developed in AI. This allows students to create executable model representations online using visual interfaces. As a result, students have opportunities to construct, execute, and analyze their developing models, which gives them deeper insights into the scientific modeling processes and interpreting and applying the behaviors generated by these processes to problem-solving tasks. We discuss a computational modeling environment for scientific processes that is linked to an engineering design challenge [11, 12].
2. **Using learning analytics and machine learning techniques to analyze students' learning and problem-solving tasks**: Our goal is to use learning analytics and machine learning methodologies that capture students' learning progressions and their problem-solving processes grounded in learning sciences theory. To accomplish this, we develop analytics that capture and help us interpret students' understanding of the primary concepts and practices in each representation and across representations [10]. In this chapter, we introduce *path analysis* methods [13] as a simplified form of Bayesian causal networks [14] for linking students' performance across multiple representations. In addition, we have developed coherence analysis measures [15] and sequence mining algorithms [16] in machine learning to understand students' learning behaviors and their use of strategies as they go about their model building and engineering design tasks. In this chapter, we develop Markov chain (MC)–based methods to study and interpret students' learning and problem-solving behaviors in the computational modeling engineering design tasks [17].

Our research is motivated by currents trends that call for integrating the learning of science and engineering in middle school classrooms [1]. In this project, we introduce CT concepts and practices along with computational modeling activities to serve as the platform for integrated engineering and science learning. Our environment, SPICE (Science Projects Integrating Computation and Engineering), implements a water runoff curriculum, where students build computational models of water runoff for different ground surface materials after heavy rainfall [11]. Students use their developed models to solve an engineering design challenge that meets runoff, cost, and accessibility constraints. They need to generate and test multiple designs to find the "optimal" design solution that meets all of the constraints. This design task is challenging for young learners. Typically, the more absorbent and accessible materials also tend to have high costs, so students need to analyze the trade-offs between cost, absorption, and accessibility as they search for their engineering design solutions. A nonsystematic trial-and-error approach may overwhelm students as they search for their optimal solution.

Our research targets a core research question: *How does a learning sequence, which adopts multiple linked representations and combines conceptual and computational modeling and engineering design help students develop critical thinking skills in STEM and CT?* In more detail, we investigate:

1. What is the *role of* CT in facilitating science learning and engineering design?
2. How do a sequence of linked representations, anchored in CT, support students learning of science and engineering concepts and practices?
3. How do students *develop and utilize strategies* to facilitate their learning and problem-solving processes (i.e., the construction of computational models and generating engineering designs)?

We adopt a learning analytics framework combined with machine learning methods to answer these questions, and report the results of a study conducted in four middle school classrooms with 99 students. We discuss the implications of using AI and machine learning methods in supporting and assessing STEM learning, and then discuss extensions of our work to support adaptive scaffolding in the SPICE learning environment.

2.2 Background and Motivation

This section reviews previous work on the integration of computational modeling and engineering design into middle school classrooms, and on the use of learning analytics to better understand students' learning behaviors and strategies as they implement such curricula.

2.2.1 21st-Century K-12 Classrooms

While technology is changing K-12 science classrooms, state and national standards are also emphasizing the need to integrate computing and engineering problem-solving into science curricula to better prepare our students for future success [1]. Initial work on building *open-ended learning environments* (OELEs) to support such initiatives has been successful in engaging students in scientific inquiry and engineering problem-solving, while providing learning and engagement across multiple domains [11, 18]. These approaches have enabled *synergistic learning* opportunities, integrating science and computation through computational modeling [5, 3], by

1. lowering the learning threshold for science concepts by reorganizing them around intuitive computational representations that introduce discrete and qualitative forms of fundamental laws, which are simpler to understand than equation-based continuous forms [19];
2. studying a phenomenon as a discrete time process, where dynamic behavior evolves in a step-by-step fashion which is easier for students to comprehend compared to continuous dynamics [20];
3. representing programming and computational modeling as core scientific practices, such as modeling, verification, and explanation [21];
4. contextualizing computational constructs in order to make it easier to learn programming [22].

For example, in terms of computational representations of science phenomena, learning to decompose a modeling task, e.g., separating upward and downward motion for a projectile, simplifies the modeling and reasoning tasks, while helping make assumptions more explicit, and students' conceptions more visible. In addition, *visualizations*, such as animations and graphs, afforded by simulating computational models make it easier for learners to judge the legitimacy and correctness of their evolving models [19].

Engineering design curricula have also emphasized engaging students in scientific investigation and design activities that improve their knowledge, reasoning, and problem-solving skills. In K-12 settings, this translates to students engaging in integrated science and engineering activities that include question posing, design testing, and solution generation. Curricular implementations of engineering design support students conceptual science and engineering learning [11]. Moreover, recent reports also indicate that the learning of engineering and CT concepts and practices can be synergistic, empowering learning in each domain. Engineering and CT complement each other in problem-solving and system design, and their conceptual underpinnings may make engineering a productive discipline for extending CT learning and applications [23, 24].

However, teachers and their students often face difficulties in implementing and learning such integrated curricula. Research has demonstrated that students have difficulties in

1. translating their developing scientific knowledge into computational forms [3];
2. understanding the mathematical and causal relations between variables [25];
3. applying key computational practices, such as problem decomposition and debugging [7];
4. deriving links between multiple representations to help reinforce the underlying conceptual phenomena [26, 9].

More work is needed to improve instruction and support students as they integrate their developing scientific ideas during computational modeling and engineering design.

2.2.2 Understanding Students' STEM Learning Strategies

In our research, we have developed methods that identify the strategies students use to learn and complete their computational modeling and engineering design tasks. We focus on understanding how multiple representations influence their use of strategies to support their learning and problem-solving and also developing a deeper understanding of students' productive behaviors and their difficulties.

In the context of educational research, strategies generally represent systematic plans or subplans students develop for achieving goals [27]. In their simplest representational form, strategies are *conditional* constructs that can be expressed as *if-then-else* rules [28]. In our research, we define strategies as "students' conscious and controllable sequences of actions to facilitate and enhance task performance" ([29], p.13).

2.2.2.1 Computational Modeling Strategies

The ability to analyze students' activities in the context of the tasks they are performing using learning analytics has provided us with systematic methods to study the processes and strategies students apply during their computational modeling tasks [29]. Early efforts, such as the work of Brennan and Resnick, used frequency analysis of students' log data to identify students' model-building behaviors. However, this approach did not provide enough information on the relations between students' strategy use and their understanding of the problem domain. Blikstein et al. [30] used log data to identify program states and assess the likelihood of students reaching a solution state or facing a "sink" state in which they were likely to get stuck.

In other work, Basu, Biswas, and Kinnebrew [4] characterized students' modeling progress by calculating and tracking its distance to an expert model at each model revision. Others have implemented clustering methods to evaluate students' learning based on action data [31, 15, 32]. More recently, we have established a generalizable framework using task modeling and coherence analysis [33] for identifying productive and unproductive strategies in OELEs [29]. We used differential sequence mining [16] to compare strategy use by high- and low-performing groups (determined by their performance on a pre–post summative assessment). Applications of this framework in multiple domains have allowed us to identify the difficulties students face by tracking and characterizing their activity data during model construction tasks. However, very little research has targeted context-preserving strategy analysis to analyze students' understanding of domain concepts, and the impact this has on their model-building and debugging processes.

2.2.2.2 Engineering Design Strategies

Recent work has also leveraged log data to evaluate engineering testing strategies. This includes characterizing differences in students' design processes using time series analysis [34], analytics methods to evaluate systematic and unsystematic experimentation processes [35], and Bayesian network models to automatically evaluate students' engineering design performance [36]. These approaches have demonstrated how differences in students' design processes can be evaluated using log data.

Zhang et al. [12] used analytic measures, such as the number of tests conducted, the number of satisfying design solutions generated, the best score achieved for the design solutions generated, and the score of their submitted solution to characterize the approaches middle school students' used to generate their design solutions. Correlations between these measures and students' science and engineering pre–post test scores showed interesting links between their science, engineering, and CT proficiency and the quality of their design solution. Bywater et al. [37] developed a new sequence segmentation method known as the Differential Segmentation of Categorical Sequences (DiSCS) to identify meaningful periods of design activity as students implemented engineering design tasks in OELEs.

However, research that leverages log data and learning analytics to evaluate engineering design practices at the K-12 level is still in its infancy. For example, current methods lack contextual information about design testing strategies that may be beneficial for a deeper understanding of students' domain-specific successes and difficulties as they integrate science, CT, and engineering.

2.3 SPICE Curriculum and Learning Environment

SPICE implements a NGSS-aligned Water Runoff Challenge (WRC) that emphasizes the movement of surface water in a system after heavy rainfall and the human impact of this runoff on the environment [11]. WRC is a three-week middle school curricular unit that challenges students to redesign their schoolyard using multiple constraints: (1) use surface materials that minimize the amount of water runoff after a storm, (2) reduce overall costs, and (3) ensure sufficient accessibility while providing for required functionalities for the schoolyard. Overall, this curriculum integrates computation and engineering into science curricula by using a CT framework to model science phenomena and designing engineering solutions. In addition, the computational modeling task is embedded in a real-world design problem.

Given this structure, computational modeling serves as the bridge that supports understanding of a scientific process (water runoff) via inquiry and modeling, and then developing an engineering solution for a system (schoolyard). This bridge is especially important because water runoff phenomena for different materials cannot be easily investigated in classrooms.

2.3.1 System Design Perspectives

We discuss three system design perspectives that we applied to develop the integrated STEM unit.

2.3.1.1 Evidence-Centered Design

We have used evidence-centered design (ECD) [38] as an overarching framework to systematically integrate the STEM and CT disciplines and to align the design of curricular activities and assessment tasks. ECD promotes design coherence by explicitly linking claims about students' learning, evidence from students' work products, and design features of instructional and assessment tasks that elicit the desired evidence. This approach extends previous work on developing science assessments [39] and backward design to develop instructional materials.

2.3.1.2 Coherence across Modeling Representations

As discussed in Section 2.2.1, technology-enhanced environments can support students' inquiry and problem-solving using computational modeling processes [3, 18]. Computational modeling experiences need to be anchored to underlying conceptual models of phenomena. Together, conceptual and computational representations provide a more complete framework for developing a systematic understanding of phenomena among novice learners by providing links between model representations [40].

We adopt AI representations and reasoning methods to enable links between conceptual and computational models. Besides, our computational models are executable, so students can simulate their models as they are building them to gain a deeper understanding of the science phenomena that they are modeling. Executable computational models also facilitate debugging practices that may further help students to encounter and overcome their difficulties with domain and computational concepts during their model-building activities. Moreover, our approach builds on work that investigates how students use computational models to solve problems [41], specifically problems in engineering design.

In extending previous system design perspectives, it becomes important to maintain coherence across system representations. However, novice students may have difficulties in connecting and integrating multiple linked representations, therefore, the representations need to be conceptually coherent [26, 9]. For example, in our curriculum, each system representation is explicit about the principle of conservation of matter, which captures the relationship between rainfall and a surface material's absorption limit. Some of the rainwater is absorbed by surface materials and the rest produces water runoff. Connected with ECD, ensuring coherence across conceptual and computational representations helps students develop a deeper understanding of science, CT, and engineering concepts and practices anchored in the problem-solving scenarios [10]. The AI representations we have developed facilitate this coherence.

2.3.1.3 Domain-Specific Modeling Languages

Section 2.2.1 outlined a number of challenges that arise in implementing an integrative, learning-by-modeling curricula; for example, the difficulties students experienced in their introduction to programming [42] and the challenges they face translating their science knowledge to computational form [3]. To support the learning of science with understanding, the designed modeling environments must incorporate appropriate and intuitive representational mechanisms linked to interpretable computational (i.e., language) constructs that can easily be adopted by classroom teachers and students, and are situated in the relevant domain concepts and practices [43]. Designing modeling languages whose component structures clearly imply their associated semantics provides a method for accomplishing these links [44].

Domain-specific modeling languages (DSMLs) provide the needed framework to establish the simultaneous "situating" in domain modeling and CT constructs seamlessly to better support the synergistic learning of STEM and CT through computational modeling [5]. DSMLs are categorized by their "focused, expressive power in the problem domain" ([45], p. 26). In other words, DSMLs are analogous to knowledge representation (KR) schemes in AI with their corresponding reasoning mechanisms [46]. Much like the formulation of DSMLs, KR research addresses the problem of finding compact and expressive languages to express domain knowledge in a form that computer algorithms can be designed to make inferences with the knowledge structures created.

In the context of K-12 science classrooms, a DSML plays an important role in focusing learner attention towards the relevant STEM concepts and practices when building models and solving problems in specific domains. In addition to providing the mechanisms to develop solutions at an appropriate level of abstraction in the target domain (e.g., using variables relevant to the specific STEM domain), a DSML allows for the building of programs that are concise and self-documenting (e.g., by associating variables with properties of objects, and working with templates that let students set and update variable values in discrete time steps). DSMLs also allow for the verification and validation of the visual representation and data generated by the models (e.g., are my object's velocity values changing as they should?) [43]. This affordance in terms of visual feedback of the model relates to the self-documenting nature of DSMLs, e.g., by providing visual information in support of a physical phenomenon. The DSML we have designed for our earth science curriculum helps students understand and interpret the computational constructs that constitute these models in a domain-specific way. We have successfully implemented educational DSMLs in additional domains, including physics and marine biology [43].

Overall, we hypothesize that the three design perspectives will support students as they work through the linked modeling representations from conceptual to computational, thus reducing the difficulties they face in integrating science and CT modeling concepts, and then using their computational model to solve an engineering design problem.

2.3.2 SPICE Learning Trajectory and Curriculum

The SPICE WRC curriculum consists of three core units: physical experiments and conceptual modeling, computational modeling of the water runoff phenomenon, and engineering design. Figure 2.1 illustrates our designed learning trajectory from conceptual to computational model to finding engineering design solutions. Students learn to apply the conservation of matter principle (representing a scientific principle or law) through the relation: *Total rainfall = total absorption + total runoff.* After acquiring a basic understanding of the runoff concept, students develop conceptual, pictorial representations that express the amount of water runoff as the difference (if any) between the total rainfall and water absorbed by surface materials. For example, students are tasked with predicting the amounts of absorption and runoff for 3 inches of rainfall and a 1-inch absorption limit of the surface material.

As a next step, students work on a paper-and-pencil rule creation task (Figure 2.2b). This task elicits a more general representation of the water runoff phenomenon that enables students to express the relation between total rainfall, total absorption, absorption limit, and total runoff. These relations model the conservation law, and use conditional logic expressions to specify the different situations (e.g., no runoff versus a certain amount of runoff). Subsequent modeling forms require application of additional CT concepts to (1) specify the computational runoff model in a more general form (see Figure 2.1) and (2) conduct fair tests to support the engineering design process. Translating the rules to the computational model form is facilitated by our DSML blocks (Figure 2.3a), and requires additional knowledge of variables and mathematical and relational operators, as seen in Figure 2.1.

Next, students use their computational models to solve the engineering design problem, i.e., redesigning their schoolyard to provide multiple functionalities (e.g., soccer field, basketball court) while adhering to the set of design constraints. A visual interface (Figure 2.4) allows students to populate individual playground squares with 1 of 6 available schoolyard material (defined in Table 2.1) so that they can test their design solutions for different material options. An extended version of the computational runoff model, implemented under the hood, supports the 4 × 4 schoolyard design visual interface. Figure 2.5 (on the right) illustrates the visual interface for the engineering design task. A photo of the school terrain corresponding to 4 × 4 schoolyard area is shown on the left of Figure 2.5. Students create their design solutions by implementing a search process to find the "optimal" solution. To do so, students conduct fair tests as they explore the design space to choose appropriate materials, so that

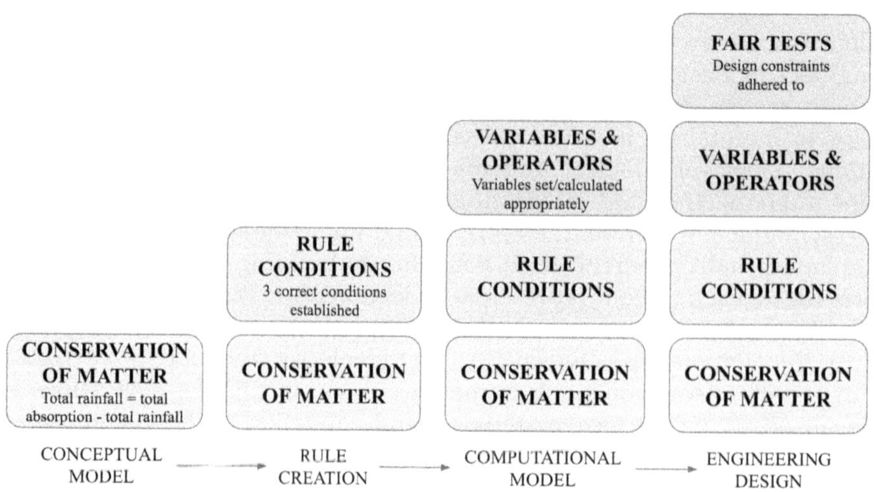

FIGURE 2.1
Learning trajectory from conceptual to computational model tasks. (*Source:* Adapted from [10]).

(a)

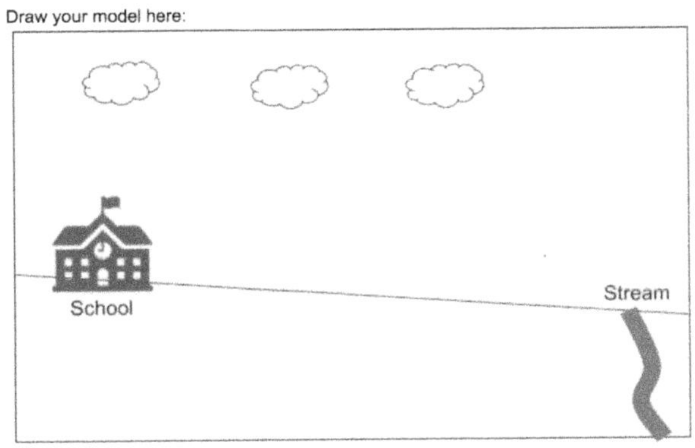

(b)

WRITE PSEUDOCODE FOR YOUR COMPUTER MODEL

Before you program your computer model, think about the instructions you want to give to the computer. In English language, write out your inputs, outputs, and rules for calculating the outputs using the inputs.

Input variables: ______________________________

Output variables: ______________________________

Rules for calculating the outputs given the inputs:

Rule #1:

Rule #2:

Rule #3:

FIGURE 2.2
Task examples for paper-and-pencil conceptual modeling (a) and rule creation (b).

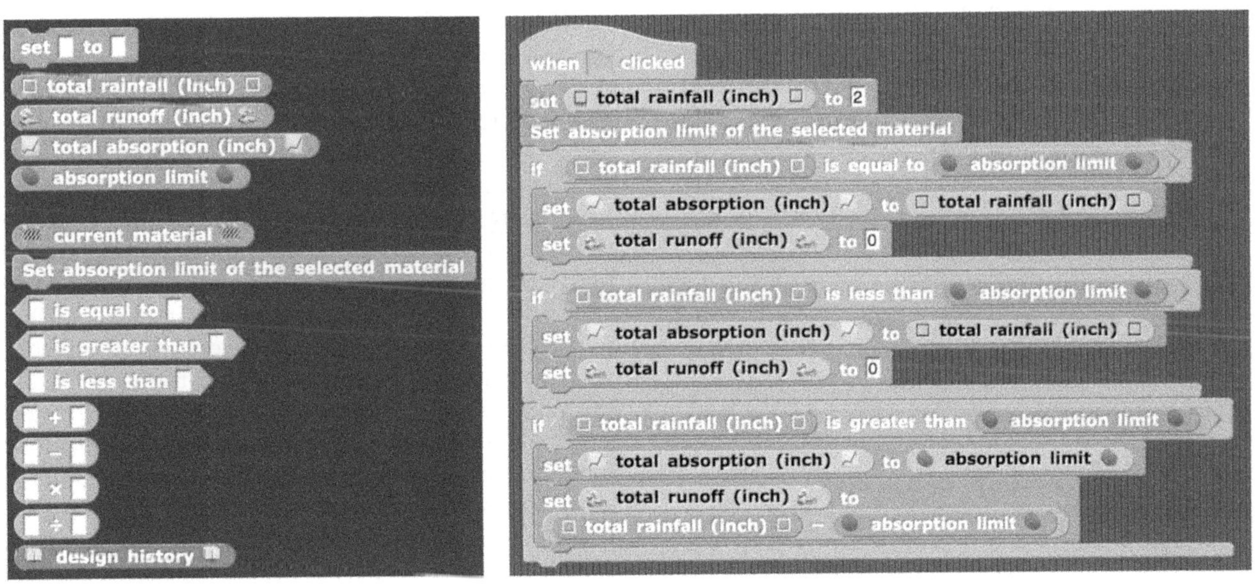

(a) (b)

FIGURE 2.3
Earth Science DSML (a) and example computational model (b) [43].

they may generate an "optimal" solution for their playground design. Students can document chosen design solutions throughout this process.

In designing each of the model-building activities, we maintained coherence across the representations, and gradually introduced students to CT concepts and practices. The design of these representations combined visual, abstraction, synergy, and reasoning mechanisms derived from past work in AI Knowledge Representation and reasoning schemes to facilitate the

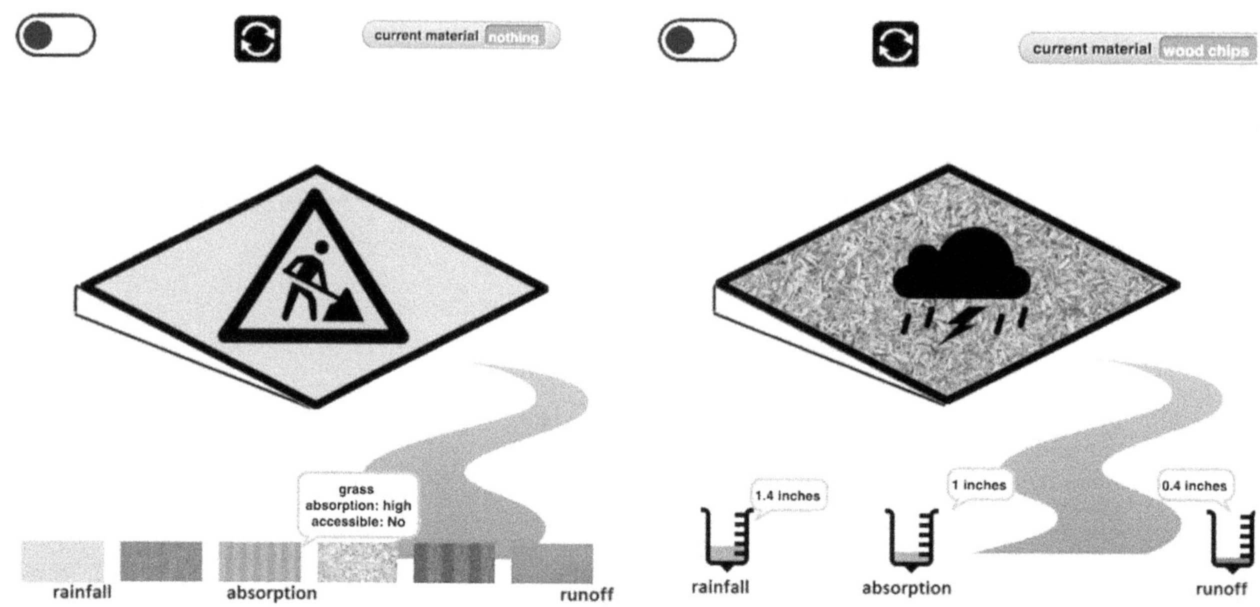

(a) (b)

FIGURE 2.4
The computational modeling visual interface for selecting material (a) and the resulting calculations (b) [24].

TABLE 2.1
Schoolyard Material Information

Material	Absorption Limit	Cost	Accessible
Concrete	0.1 inches	$37,500	Yes
Permeable concrete	1.3 inches	$93,750	Yes
Natural grass	1.2 inches	$18,750	No
Wood chips	1.0	$37,500	No
Artificial turf	0.6	$112,500	Yes
Poured rubber	1.2	$187,500	Yes

development of the SPICE environment [47]. In addition, this approach provides a framework for evaluating students' modeling artifacts across different representations and understanding how these representations support students' learning trajectories.

2.4 Methods

This section outlines our study, the participants, and the data sources and analysis techniques used to answer our research questions. We also discuss our analysis methods, and emphasize the use of analytics and machine learning approaches in analyzing students' learning performance and behaviors.

2.4.1 Implementation

The data was collected from a classroom study with 99 students in a 6th-grade classroom in the southeastern United States. The study, led by two experienced science teachers, was run in the Fall of 2019. Three Vanderbilt researchers provided additional support in the classroom but mostly acted as observers during the study. The two teachers participated in four days of professional development conducted by the research team (Vanderbilt researchers and collaborators from the University of Virginia) during the summer session before the study. Elementary programming classes are part of the middle school curriculum in this school, and all participating students had varying amounts of prior programming experience with Scratch (https://scratch.mit.edu/about).

The WRC curriculum was covered by the teachers with intervening student work on the system for 45 minutes per day, three days a week in their regular science classes, and 75 minutes, twice a week that included additional personalized-learning time. The curriculum was covered in 15 school days, with the pre–post tests administered during two additional 45-minute class sessions. Table 2.2 lists a breakdown of the topics of the day-by-day learning activities.

2.4.2 Data Sources

The data analyzed in this study came in multiple formats and consisted of (1) the NGSS-aligned science and engineering and CT pre–post assessments; (2) formative assessments in science, engineering, and CT administered as homework; (3) students' paper-based responses on the conceptual modeling and rule creation tasks; and (4) system logs of students'

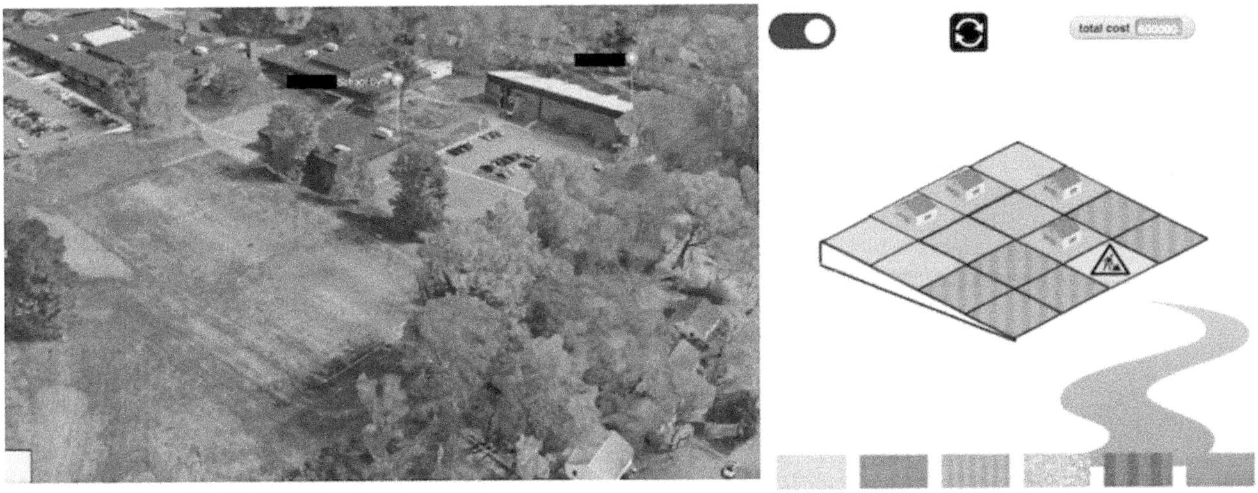

FIGURE 2.5
SPICE engineering design task [24].

TABLE 2.2
Themes of the Learning Activities of the 15-Day WRC Implementation

Lesson	Theme
1	What problem are we trying to solve?
2	Where does the rain go?
3	How much does it rain at the school?
4	Why do some materials soak in more water?
5	How do we calculate water runoff?
6	How can we design the schoolyard to reduce water runoff?
7	What language does a computer understand?
8	Build a computer model (part 1)
9	Build a computer model (part 2)
10	Build a computer model (part 3)
11	How can we test if our schoolyard design meets the project criteria?
12	How do you know what design will be the best?
13	How can you use the model to improve your design?
14	How can you convince the Principal of the school to use your design?
15	Class presentations

computational model-building activities as well as their engineering design and testing activities. The data collected in (1)–(3) was mostly in textual form, where students filled in the blanks or generated short answers to questions. Data for (4) was collected online in the form of formatted log files as part of the SPICE environment.

The learning environment logged individual learners' actions during their *computational model-building* activities. In addition, using *post hoc* analyses, we labeled computational blocks as being in one of three states (*not present* in the students' model-building work space, present but *disconnected* from other blocks in the work space, and *connected*) to other blocks in the work space. This allowed us to track students' sequences of model-building actions, i.e., sequences of AddBlock, ConnectBlock, DisconnectBlock, and RemoveBlock in more detail using Figure 2.6, the computational blocks life cycle.

We logged two other actions in the computational model-building activity: FillField and StartSimulation. As shown in Figure 2.3, any block with formal argument fields can use numeric or character string literals as the actual argument. FillField represents such an action. In addition, StartSimulation indicates when a student started a simulation to test the computational model.[1]

The learning environment also logged students' actions specific to their playground engineering design process. These actions included AddMaterial (add new material to square in design), RemoveMaterial (remove the material from a square in the design), RunSimulation (clicking of the Green Flag to test the current design), ViewDesignHistory (opening the Design History table to view past designs), and Reset (clearing or removing all squares in a design).

2.4.3 Analysis Methods

In line with our research questions repeated below, our analysis was directed toward understanding:

1. What is the role of CT in supporting synergistic learning of science and engineering concepts and practices?
2. How does a sequence of linked representations supported students' synergistic learning?
3. What strategies did students adopt to facilitate their synergistic learning?

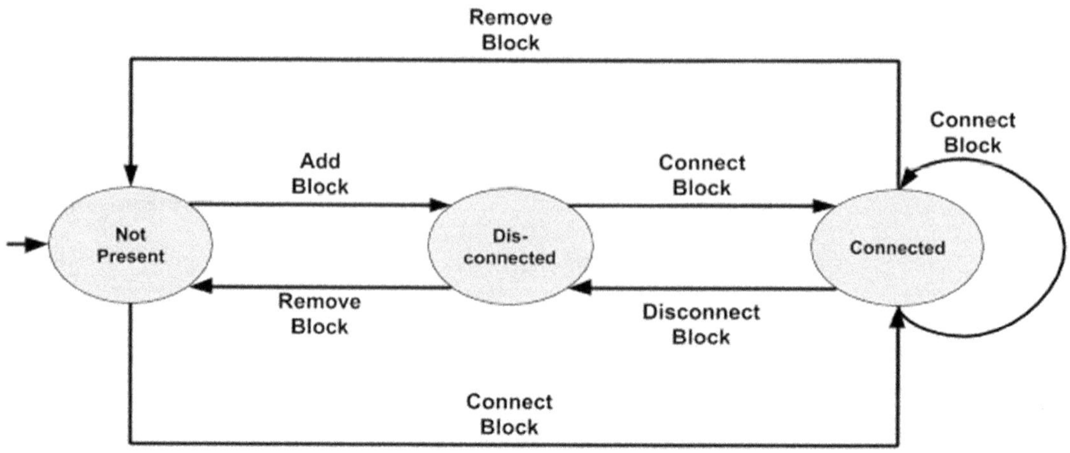

FIGURE 2.6
Computational block life cycle.

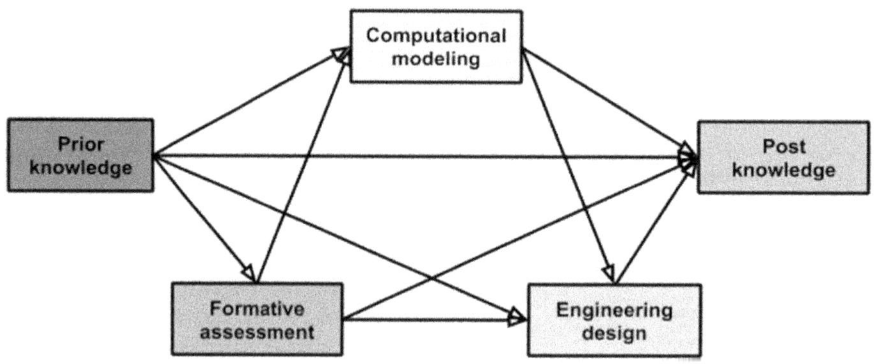

FIGURE 2.7
A hypothesized path model of the direct effects on the different categories of learning behavior and performance variables.

To answer the first research question, we used *path analysis* to study the relationship between science, engineering, and CT learning in the WRC. We hypothesized that students' knowledge gains, the behaviors they developed, and their performance in the tasks they worked on would influence their learning, behaviors, and performance in subsequent tasks and assessments in the WRC curriculum. Path analysis can be seen as a variation of Structural Equation Modeling without the latent variables. In many ways, the methods used for path analysis can be related to work in deriving Bayesian causal models from data [48, 49]. We hypothesized the causal paths shown in Figure 2.7 among the different tasks students performed in the WRC. Each arrow in that figure indicates a direct effect of the variable on the left of the arrow to the corresponding variable on the right of the arrow. The horizontal positions of the variables also corresponded to their temporal order in the WRC curriculum (pretests $\prec$ formative assessments $\preceq$ computational modeling $\prec$ engineering design $\prec$ posttests).

Research question 2 took a deeper dive into students' transitions from conceptual to computational modeling. For this analysis, we combined analysis of students pre–post test results in science, CT, and engineering along with our coherence framework (Figure 2.1), which we applied to two vignettes to demonstrate how our approach provides a domain-specific understanding of students' learning trajectories. We coded students' paper-based Conceptual models to categorize their representation of the conservation of matter principle (see Figure 2.2) as (1) mechanistic (and correct), (2) numeric (and correct), and (3) developing. To demonstrate mechanistic proficiency, students could use pictorial representations and/or text to express the causal relations between total rainfall, absorption of water by the surface material, and the amount of runoff. Many students expressed this relation using algebraic and numerical expressions in addition to providing explanatory text. A numeric categorization implied that the students' representations showed all of the required elements and correct values, but did not describe the causal relations. The developing categorization reflected misunderstandings or errors in students' application of the conservation of matter principle. Students' numerical, pictorial, and written descriptions were coded separately, and their understanding was represented by the representation that was the most expressive and correct.

For the rule creation task, each rule that students developed was scored separately. For rule conditions, students received scores for expressing the correct conditional relation between total rainfall and material absorption limit (e.g., if total rainfall is greater than the material absorption limit). The conservation relation was scored for a correct expression of the values for each required output: total absorption and total runoff.

Students' computational models were scored using a predefined rubric targeting the application of the conservation of matter rules, the conditional statements for the different rules, and the variable assignments (for rainfall, absorption, runoff). The maximum score attainable was 15. To achieve this score, students had to assign appropriate values to total rainfall and absorption limit, generate the code for the three conditional statements based on a comparison of variables (total rainfall and absorption limit), and update the absorption and runoff variables for each of the three conditions. Students were given points for generalizability only if they used expressions comprising variables and operators to express variable values (e.g., setting total runoff to "total rainfall – absorption limit" in the overflow condition) as opposed to just assigning numeric values to variables.

We answered research question 3 by evaluating students' learning strategies as they completed their computational modeling and engineering design tasks. Previous work has shown that students' strategy use can vary substantially, and these differences are indicative of differences in their model-building performance and learning [50, 33]. We implemented a pattern mining approach that used Markov chain modeling. Students' action sequences logged in the WRC were modeled as *Markov chains* with actions representing the model states, and the transitions between actions captured as conditional probabilities. For example, a transition

$$\text{Add Block} \xrightarrow{0.54} \text{Connect Block}$$

during model-building can be associated with a conditional probability:

$$\Pr(\text{action}_{t+1} = \text{Connect Block} \mid \text{action}_t = \text{Add Block}) = \frac{\#\text{Add Block} \rightarrow \text{Connect Block}}{\#\text{Add Block}} = 0.54$$

Taking into account the learning activities of a group of students working on the WRC, we derived a MC model that represented the students' dynamic learning processes. For example, Figure 2.8 provides a visualization of the aggregated first-order Markov chain model of the computational model-building activities of all participants in the study ($N = 99$).

In addition, we applied a lift measurement [51] that defines the *interestingness* and importance of the relations between actions implied by the links in the MC model. The value of the lift is the odds of the confidence of a sequence of actions and the expected confidence of the sequence, calculated as follows:

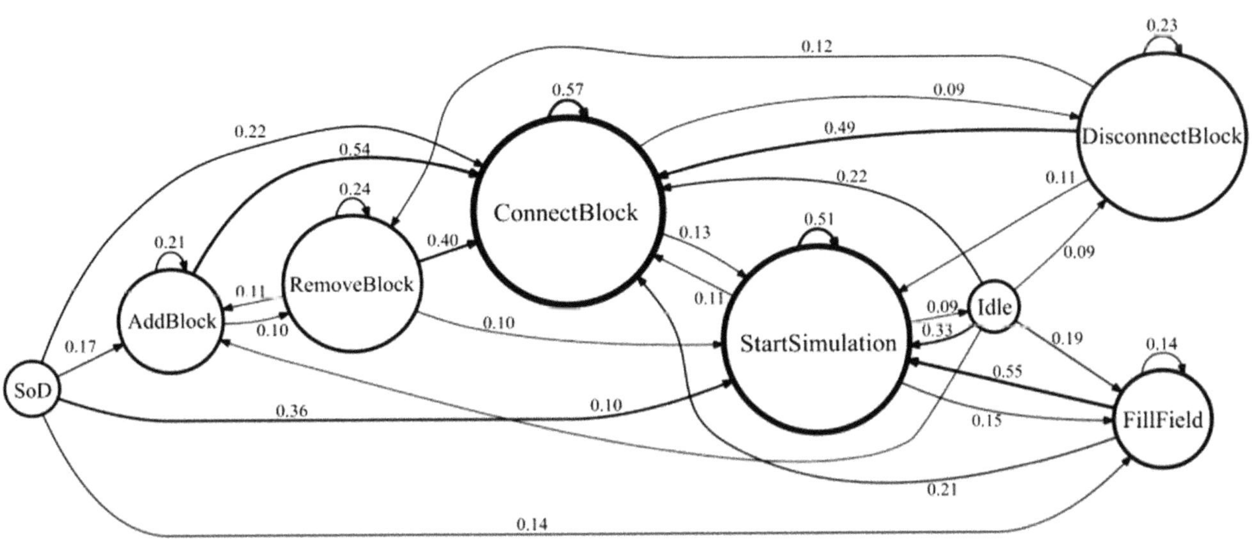

FIGURE 2.8
Sample Markov chain model of students' computational model-building processes.

$$\text{lift} = \frac{\text{Confidence}(A \to B)}{P(B)} = \frac{P(A \wedge B)}{P(A)P(B)}$$

$$\text{lift} = \frac{\text{Confidence}(A \to B)}{P(B)} = \frac{P(A \wedge B)}{P(A)P(B)}$$

2.5 Results

Students successfully completed the WRC as intended. Pre–post assessment scores were determined to be normally distributed and a paired t-test analysis showed significant learning gains in science and engineering ($p < 0.0001$, Cohen's $d = 0.82$) as well as CT ($p < 0.0001$, Cohen's $d = 0.83$). This provides aggregated evidence that the intervention helped students learn science, engineering, and CT concepts. In the rest of this section, we address each one of our research questions in greater detail.

2.5.1 The Role of Computational Thinking

Using the hypothesized structure shown in Figure 2.7, we created a path diagram to further study the relations between the measured variables using the IBM® SPSS® Amos 26 software. We modeled a total of 47 direct effects from the 15 variables in the path diagram. Using a pre-analysis suggested by [52], we evaluated the assumptions of multivariate normality and then removed four outliers from subsequent analyses, leaving a sample size of 95 for the path analysis. Bootstrap samples ($N = 1{,}000$) were generated to estimate the standard errors and calculate the 95% confidence intervals. The standard errors and their critical ratios were later used to evaluate the statistical significance of the modeled causal effects while reducing the variance in the observed variables.

We also calculated the model-fitting statistics of the path model as compared to the saturated model that includes pairwise associations among all variables [52]: $\chi^2 = 40.89$ (DF = 54, p-value = 0.91); the goodness of fit (GFI) was 0.95 (≥ 0.95 threshold); the comparative fit index (CFI) was 0.99 (>0.9 threshold); and the root mean square error of approximation (RMSEA) was 0.01 (<0.06 threshold). These statistics indicate that the path model fit the measurements well. All of the hypothesized paths in Figure 2.7 were confirmed as direct or indirect effects. Figure 2.9 shows the statistically significant causal paths that had a direct standardized effect ($|\beta| > 0.1$). The bolder lines represent the stronger causal relations. We report our results on the direct and indirect effects for the following three categories of variables.

2.5.1.1 *Effects on Computational Modeling*

The students' learning behaviors and performance in the computational modeling activity (Figure 2.10) were directly affected by variables in the same category (i.e., the number of computational tests performed, *comp_test*, and the number of editing actions between computational tests, *edit_btw_test*). Figure 2.10 shows the subsection of the discovered causal paths terminating in the computational model-building variables, number of computational edits, *comp_edit*, and the computational model score achieved by the student *comp_model_score*.

The formative assessment scores (*formative*) had significant effects on three of the four of these variables directly. The direct and total effects from *formative* to *comp_test*, *edit_btw_ test*, *comp_edit*, and *comp_model_ score* were 0.3, 0.1, 0.18, and 0.54, respectively. These results, especially with the high η^2 of 0.29 between *formative* and *comp_model_score*, showed that the formative assessment tasks closely mirror the curricular activities in the WRC. Furthermore, the formative assessments

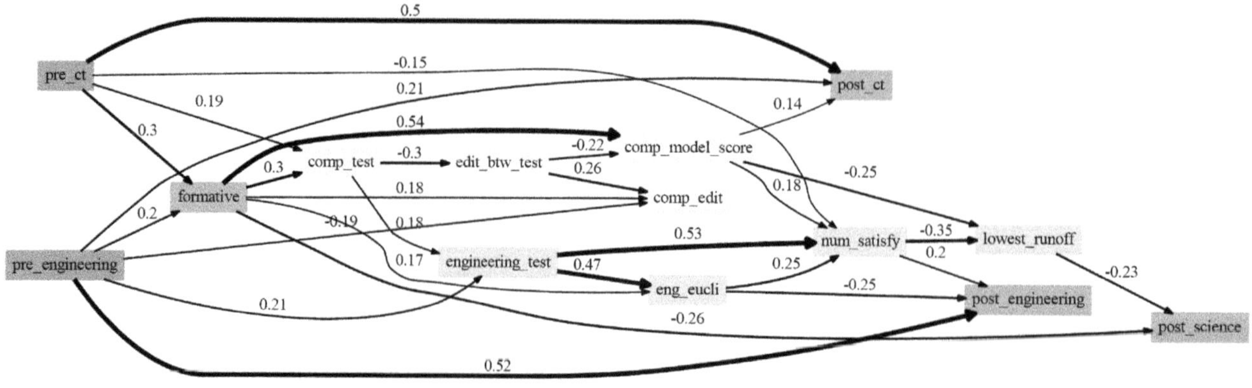

FIGURE 2.9
Discovered causal paths with statistically significant direct effects.

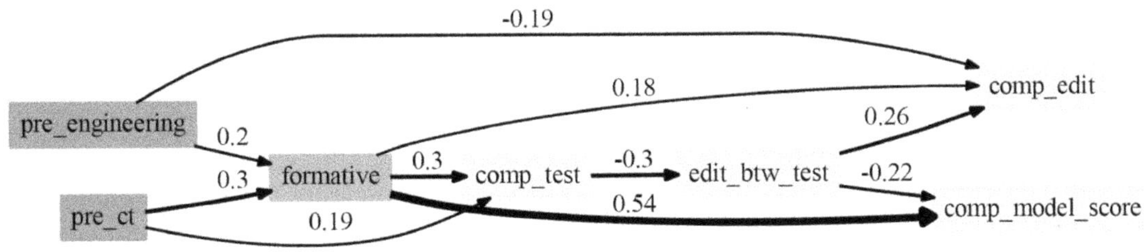

FIGURE 2.10
Effects on the computational model-building measures.

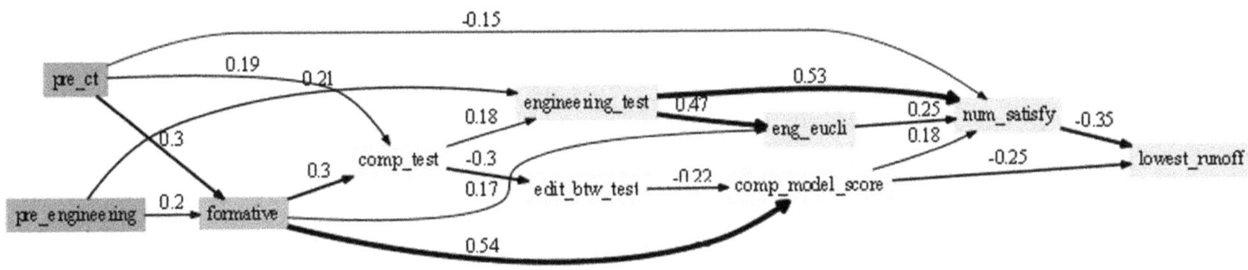

FIGURE 2.11
Effects on the engineering design measures.

were a strong indicator of students' learning outcomes, and contributed to nearly 30% of the variance in the computational model score. In addition, the CT pretest score also directly or indirectly affected the *comp_model_score* and *comp_edits* (via *formative*, *comp_test*, and *edit_btw_tests*) with $\Sigma\beta$'s of 0.28 for both.

As one of the main learning outcomes, the students' *comp_model_score* was affected by the students' engineering pretest score *pre_engineering* indirectly, with $\Sigma\beta = 0.12$. The median number of model edits between tests, *edit_btw_test*, also had a strong indirect effect on *comp_model_score* ($\Sigma\beta = -0.24$), indicating that students who edited and tested their model in *smaller* chunks performed better in the computational model-building task. Similar results of smaller edit chunks being associated with better modeling performance have been reported in [4].

2.5.1.2 Effects on Engineering Design

For the engineering design activity (Figure 2.11), the number of unique satisfying designs (*num_satisfy*) and the lowest amount of runoff values from satisfying designs (*lowest_runoff*) were the two variables we used for evaluating the quality of students' designs. These two variables were significantly affected by the number of exogenous variables.

For *num_satisfy*, the strongest direct effects came from the number of tests conducted on the design solutions students generated (*engineering_test*, $\beta = 0.53$) and the total standardized Euclidean distance between the tested designs (*eng_euclid*, $\beta = 0.25$). The *lowest_runoff* was most strongly affected by *num_satisfy* ($\beta = -0.35$), *comp_model_score* ($\beta = -0.25$), and *formative* (mainly indirectly, $\Sigma\beta = -0.15$). In addition, *engineering_test* had a moderate total effect on *lowest_runoff* ($\Sigma\beta$'s $= -0.32$), and this effect mainly came via *num_satisfy*.

In summary, students who explored a larger portion of the problem space systematically were more likely to generate better engineering design solutions [12]. This also matched the *scientific discovery as dual search* theory [53], which states that successful learners connect the *hypothesis space* and the *experiment space* by making inferences with data drawn from their investigations. More importantly, these results also suggest a strong connection between computational modeling (*comp_model_score*) and engineering design (*lowest_runoff*) with a total standardized effect of −0.32 ($\beta = -0.25$, the total indirect effect is −0.07). The negative value indicates that students generating better computational models on their own produced better design (i.e., lower cost and lower runoff) solutions, even though all students were given the correct implementation of the computational model before the engineering design activity. It also indicates the benefits of having students develop their own computational model to use for designing and testing, as opposed to providing students with a model that has been developed by experts.

2.5.1.3 Effects on Posttest Scores

The posttest scores (Figure 2.9) were affected by all other categories of variables in the WRC learning activities and most significantly by the pretest scores as exogenous variables. The *science* posttest scores (*post_sci*) were significantly influenced by *lowest_runoff* (directly, $\beta = -0.23$), *num_satisfy* (indirectly, $\Sigma\beta = 0.08$), *engineering_test* (indirectly, $\Sigma\beta = 0.08$), *comp_ model_score* (indirectly, $\Sigma\beta = 0.04$), and *formative* (directly, $\beta = -0.26$). The effect of the engineering design activity (*lowest_runoff*) on science learning and understanding (*post_sci*) again demonstrated students' synergistic learning and the effectiveness of the design principle of integrating engineering with science learning that has been called out in the NGSS framework.

The *engineering* posttest scores were affected by *pre_eng* (directly, $\beta = 0.52$), *eng_euclid* (directly, $\beta = -0.25$), and *num_satisfy* (directly, $\beta = 0.2$). The effects of both the learning behaviors (*eng_euclid*) and the performance (*num_satisfy*) indicate that students' success in solving the engineering design problem by searching for the optimal combinations of surface materials for the schoolyard contributed to higher learning outcomes. This finding again demonstrated the benefit of the WRC as a curricular unit for young learners.

As for the *CT* posttest score, it was only significantly affected by the related pretest scores. The variable *comp_model_score* had a relatively large total effect of 0.14 on *post_ct*, yet the effect was not statistically significant. Nevertheless, this effect shows a positive trend that students' performance in the computational model-building activity contributed to their CT learning.

Overall, our positive results suggest that the students' success with the engineering design activities can be linked to their science and engineering proficiency, providing evidence for the benefit of integrating engineering with science learning [1]. With regard to RQ 1, we found clear evidence of significant links between CT pretest scores and formative assessments, the computational modeling scores, and success in engineering design measured by the ability to generate designs with low runoff values. In other words, students who started with high CT knowledge, or learned their CT concepts from the formative assessments also constructed better computational models and generated better engineering designs.

2.5.2 Learning with Multiple Representations

We classified 32 students' conceptual models as mechanistic, 59 as numerical, and 7 as developing (1 student packet missing). For the rules creation task's rule condition component, 57 students correctly described the three conditions while 35 had errors in at least one rule condition, and the remaining responses were either illegible or missing. For the conservation of matter component, 35 students correctly calculated both absorption and runoff for each rule and 63 students had incorrect and missing elements. One student's packet was missing. A number of students had errors in applying the conservation of matter principle for rule creation task. The primary errors were linked to missing variable assignments for either runoff or total absorption in the three conditions. For example, students would only describe the resulting runoff when describing the situation where total rainfall was greater than the absorption limit. The computational model mean score for the class (we could retrieve 62 student models) was 13.75, Std dev = 2.42. We used this as a reference for our case studies discussed below. Spearman's ρ correlation analysis indicated a moderate but significant correlation between the conceptual model representation and rule-based model scores ($r = 0.35, p = 0.0007, n = 90$), but a small non-significant correlation between rule-based model and computational model scores ($r = 0.19, p = 0.13, n = 62$), implying the relation may not be linear. We observed that 68% of the students had correctly working computational models and 92% implemented the three conditions correctly. These results suggest that constructing of computational models improved students' understanding of the runoff system. As discussed earlier, some students received help from the teachers or the researchers when constructing their models.

We used comparative case analysis to further investigate students' transitions from conceptual modeling to the intermediate rule creation task, and then to their computational model-building tasks in the WRC. We investigated two students, Alex's and Taylor's (anonymized, gender-neutral names) trajectories (see Figure 2.1), analyzing their science and CT knowledge as they worked through the three modeling tasks. Figure 2.12 shows the two students' final computational models.

Alex performed well on the pre–post assessment and assigned curricular assessments. On the pre–post assessment, Alex's score improved from 17.5 to 20 (out of 23) on the science and engineering assessment and from 10 to 13 (out of 13) on the CT assessment.

During the conceptual modeling task, Alex correctly modeled each rule and demonstrated mechanistic understanding of conservation of matter. In response to the prompt to model water flow for 3 inches of rainfall and an 1 inch absorption limit, Alex wrote: "1 inch of that gets absorbed into the ground and since that's the absorption limit, the rest of the two inches becomes runoff." This response indicated that Alex correctly calculated the runoff and total absorption using the absorption limit and the total rainfall, while providing a correct causal description of the process.

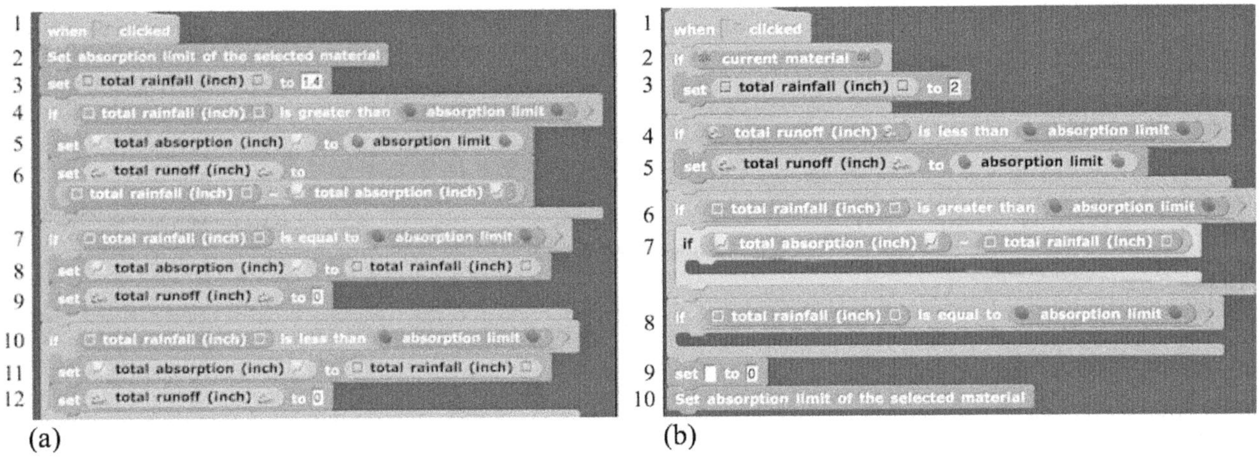

FIGURE 2.12
Final computational model code for (a) Alex and (b) Taylor.

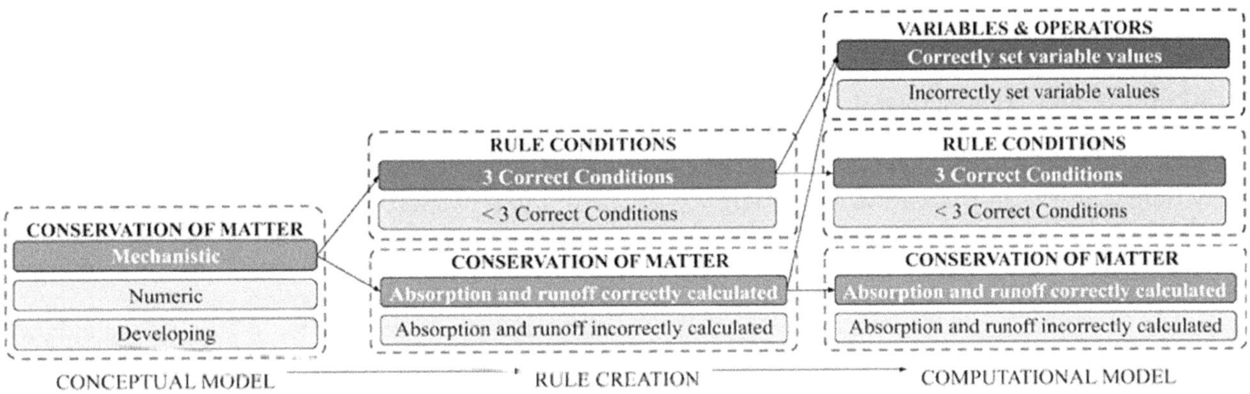

FIGURE 2.13
Alex's learning trajectory.

Alex's successfully applied the conservation of matter principle to the rule creation task by correctly writing the conditions and the expressions for the variable values for each rule. For example, in describing the "greater than" rule, Alex wrote: "If the total rainfall is greater than absorption limit, set absorption to absorption limit, set total runoff to total rainfall-absorption." This translated to building the correct computational model (see Figure 2.12a) for which Alex received the maximum score of 15. Alex's learning trajectory is shown in Figure 2.13.

Taylor had a relatively low score on the pre-assessment, but did achieve pre–post gains in the science and engineering (improving from 11.5 to 14.5 out of 23) and CT (improving from 5 to 8 out of 13) test. For the conceptual modeling task, Taylor demonstrated a numerical understanding of the conservation of matter rules. Taylor correctly determined the total absorption and calculated the runoff based on a prompt of a total rainfall of 3 inches and an absorption limit of 1 inch. For the description, Taylor wrote "the cloud rained 3 inches and the ground absorbed 1 and others become runoff." This description includes the needed variables, but does not describe the relations in a mechanistic form. However, Taylor correctly defined each rule condition for the rule creation task, and successfully calculated absorption and runoff for each rule, demonstrating an improved ability to apply the conservation of matter relation.

Taylor had difficulties with the computational modeling task, implying that he was unable to convert his science knowledge into the computational model form, As illustrated in Figure 2.12b, Taylor tried multiple arrangements of conditional blocks, including an if-block in which the expression was set to a "total absorption – total rainfall" expression (line 7) inside of the if-block for the greater than condition (line 6). Taylor's computational model received a score of 5 out of 15. Taylor demonstrated some knowledge of the if-blocks and stated the correct conditional statements. However, they had difficulty translating their science domain knowledge to develop correct expressions

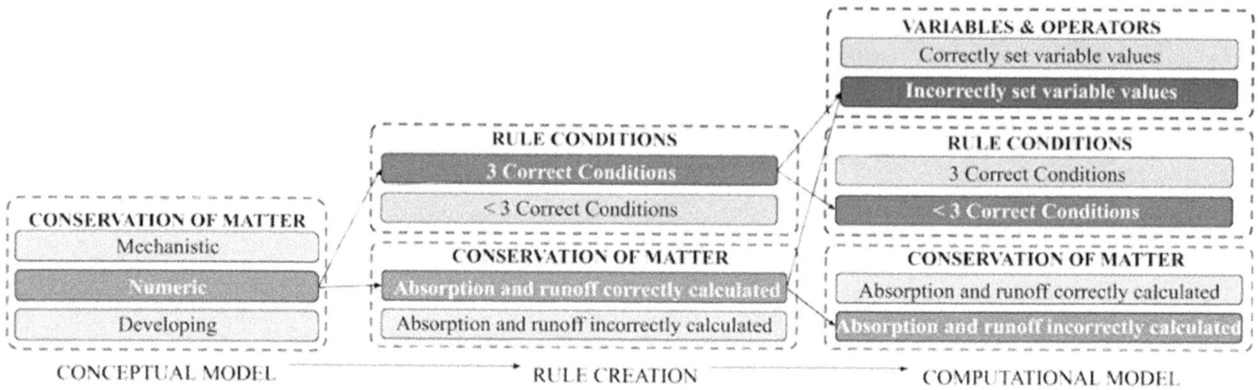

FIGURE 2.14
Taylor's learning trajectory.

for the total absorption and runoff and showed poor debugging skills. Taylor's final learning trajectory is shown in Figure 2.14.

In summary, with respect to RQ 2, our analyses from this section and the last, show clear relations between the conceptual and computational modeling tasks, and their impact on students' abilities to generate good design solutions. In more detail, our case studies demonstrate that we can track students' progressions through the linked representations, and performance in one representation is predictive of how students are likely to perform in subsequent tasks.

2.5.3 Impact of Strategy Use on Learning

When students work on complex problems, they develop and apply a variety of strategies to support their learning and problem-solving tasks. The use of strategies is latent; thus, they cannot be observed or measured directly. However, the use of strategies manifests students' declarative, procedural, and conditional knowledge [54], and we hypothesize that the use of strategies is reflected in students' observable behaviors that can be derived from their logged activity data. We have developed an exploratory learning approach, based on Markov chain models that support finer-grained analyses of student behaviors.

2.5.3.1 Computational Modeling Strategies

To study students' strategy use in computational model-building activities, we analyze sequences of students' actions collected in log files. These actions include AddBlock, ConnectBlock, DisconnectBlock, RemoveBlock, FillField, and StartSimulation. A total of 23,805 actions were logged for the 99 students participating in the computational model-building processes. When forming action sequences, we removed the outlier actions, whose time interval between succeeding actions was either too short (less than 10 milliseconds, indicating a batch operation such as undo or redo in the computational modeling environment) or too long (greater than 40 minutes, indicating the end of a class period for the students). We then used the 95th percentile of the remaining 23,319 actions to determine an estimated max time students spent on an action (80 seconds). We hypothesized that a gap of more than 80 seconds between actions implied that the student was idling between the two actions. We used this heuristic to add an implicit Idle action to students' action sequences if they did not perform any activity for at least 80 seconds when working on the system.

Table 2.3 presents the number of instances of the actions and their proportions. At the highest level, building (ConnectBlock) and testing (StartSimulation) the computational model were the most frequent types of actions. The interpretation of Idle actions is context-dependent. The action as described could represent students reflecting on their problem-solving processes and interpreting simulation results, or simply becoming disengaged from their tasks for a period of time.

TABLE 2.3
Count and Proportion of Actions in the Computational Model-building Activity

Action	Number of Instances	Proportion
Add block	1,233	0.050
Connect block	8,876	0.361
Disconnect block	1,932	0.079
Remove block	1,629	0.067
Fill field	2,453	0.100
Start simulation	6,970	0.284
Idle	1,144	0.047
Disengaged	118	0.005

We derived a first-order Markov chain model that included sequence data for all participants ($n = 99$) by (1) using the proportions of actions listed in Table 2.3 as *prior probabilities* of each state in the MC model and (2) the number of instances of any consecutive pairs of actions as the conditional probability, i.e.,

$$\frac{\text{Count}(\text{Action}_t = A \wedge \text{Action}_{t+1} = B)}{\text{Count}(\text{Action}_t = A)}, \forall t \in \{T\},$$

$$\forall A, B \in \{\text{Model-building actions}\}$$

Figure 2.15 shows the MC model for the computational model-building activities (the GraphViz software [55] was used to generate this figure). Each state in the model represents one of the actions discussed earlier. The size of the states is proportional to the number of occurrences of the particular action, as listed in Table 2.3. An arrow between two states represent a transition between the states, and the number on the arrow represents the first-order Markov transition probability for that transition. Arrows with transition probabilities less than 0.09 were not included in Figure 2.15 to reduce clutter. The topology of the states is also determined by the transition probabilities. Figure 2.16 shows the transition matrix of the Markov chain model for all students ($n = 99$). The transitions with a probability >0.2 are displayed in the bold format in Figure 2.16.

As shown in Table 2.3, ConnectBlock is a very frequent model-building activity and it made up 36.1% of the total amount of actions. In Figure 2.15, ConnectBlock with all incoming transitions (and a self-loop) represents a *sink* state in the Markov chain model. In particular, this was because it was a follow-up action to AddBlock, DisconnectBlock, and RemoveBlock. All of these represent expected model-building actions.

StartSimulation was another frequent action that accounted for 28.4% of all the actions. This action provides key information for the learner to estimate the correctness of their models. After running a simulation, students can examine the values of the output variables (the amounts of absorption and runoff in Figure 2.4). Therefore, it is a central step in *debugging* the computational models. As shown in Figure 2.15, actions frequently preceded StartSimulation included FillField, Idle, Disengaged, and the action itself (implying a self-loop).

The FillField action was the third most frequent action during the computational modeling activity. In the modeling process, it mainly appeared in two contexts: (1) as part of the initial process building the model and (2) as part of tweaking parameters to test the model. The transitions between FillField, ConnectBlock, and StartSimulation correspond to these two contexts.

To understand how students' use of strategies facilitated their learning processes, we hypothesized that the differences in certain action sequences would be indicative of the students' learning performance. Using the same method to fit a Markov chain model, we derived MC models for each individual student. A Laplace smoothing was applied to avoid likelihoods of zero in the empirical data if a transition did not occur for certain students (for example, some students never idled).

We used the Markov chains as generative models to estimate the likelihoods of sequences of actions (e.g., Start Simulation $\rightarrow$ Idle $\rightarrow$ Remove Block) generated

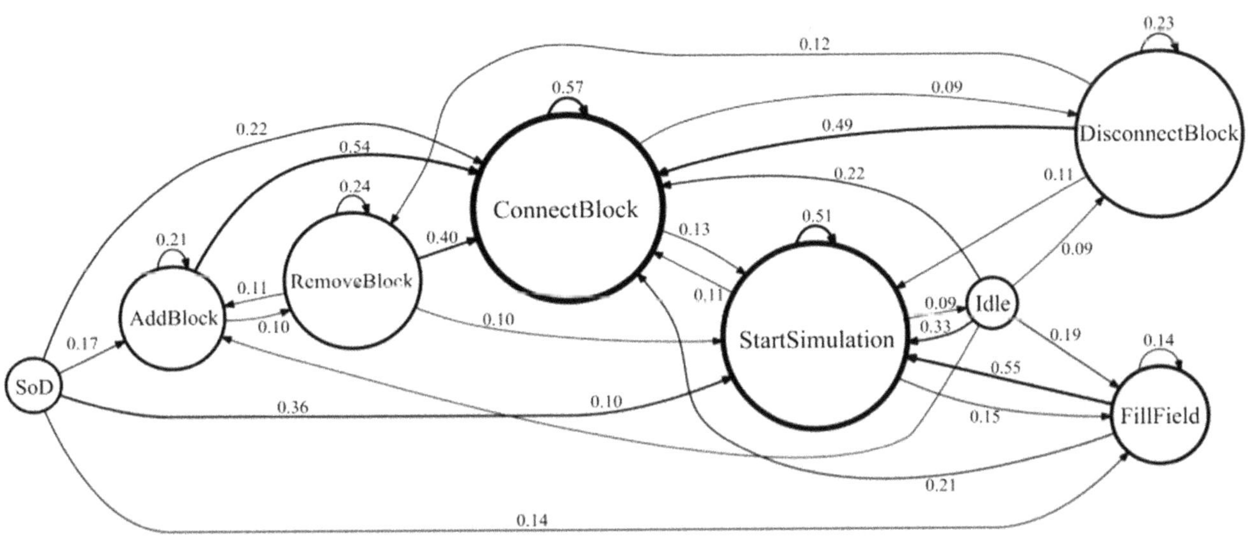

FIGURE 2.15
Discrete-time Markov chain model of students' computational model-building processes.

	AddBlk	ConnectBlk	DisconnectBlk	FillField	RemoveBlk	StartSim	Idle	Disengaged
AddBlk	**0.208**	**0.537**	0.039	0.009	0.102	0.081	0.021	0.002
ConnectBlk	0.035	**0.567**	0.091	0.084	0.055	0.132	0.031	0.003
DisconnectBlk	0.021	**0.488**	**0.234**	0.017	0.115	0.109	0.013	0.000
FillField	0.024	**0.206**	0.025	0.137	0.020	**0.546**	0.033	0.002
RemoveBlk	0.112	**0.401**	0.044	0.033	**0.242**	0.104	0.053	0.004
StartSim	0.021	0.115	0.054	0.149	0.039	**0.508**	0.094	0.011
Idle	0.101	**0.221**	0.089	0.190	0.065	**0.335**	n/a	n/a
Disengaged	0.169	**0.220**	0.085	0.136	0.025	**0.364**	n/a	n/a

FIGURE 2.16
Transition probabilities of students' computational model-building processes.

TABLE 2.4
Statistically Significant Predictors (Likelihoods of Pairs of Actions) of the Learning Gains

	Action Sequence	β	Std. Error	p-Value	Lift
1	Disconnect block → fill field	0.3952	0.142	0.008	0.17
2	Idle → connect block	0.3426	0.168	0.048	0.61
3	Start simulation → disconnect block	0.3326	0.147	0.029	0.68
4	Connect block → idle	0.3045	0.137	0.032	0.65
5	Remove block → add block	–0.2736	0.133	0.046	2.21
6	Fill field → remove block	–0.2962	0.124	0.022	0.30
7	Idle → remove block	–0.3093	0.120	0.014	0.97
8	Fill field → disconnect block	–0.4093	0.118	0.001	0.31
9	Start simulation → disengaged	–0.6423	0.269	0.022	2.34

by each individual Markov chain model. The generated likelihoods were then used to predict students' learning performance using regression methods. As the first step, we calculated the likelihoods of pairs and triplets of actions generated by individual MC models for all 99 students.[2] We then applied multivariate regression analyses to check if their likelihoods predicted the students' learning gains in the summative assessments. Table 2.4 lists the statistically significant predictors (likelihoods of action pairs) of the normalized learning gains. The regression coefficients were standardized (denoted as β) and ranked in descending order. Therefore, both the sign and the magnitude of the coefficients provide information on the link between learning behaviors and the learning outcomes. In addition, we calculated the lift measures of the patterns as the ratio between the expected probability of an action relative to the probability of the action pair [51]. Sequences #1, #5, #6, #8, and #9 in Table 2.4 had either large lift measures greater than 1 or small lift measures close to 0; thus, they were considered to be *interesting* behaviors for further discussion. The value of the lift is the odds of the confidence of the pattern and the expected confidence of the pattern. It is calculated as follows:

$$\text{lift} = \frac{\text{Confidence}(A \to B)}{P(B)} = \frac{P(A \wedge B)}{P(A)P(B)}$$

Table 2.4 shows that some action sequences emerged as significant predictors of students' learning gains. We observed four pairs of actions as positive predictors and five pairs of action as negative predictors of learning gains. One characteristic of the action pairs related to the role of the Idle action combined with the ConnectBlock action – the action with the highest number of instances among all students (Table 2.3). Both sequences #2 and #4 are positive predictors of the learning gains. In both situations, students seemed to pause and revisit the task at hand (indicated by the idle action), and for the students who demonstrated these behaviors, we saw higher learning gains.

Second, Figure 2.16 shows that the probability of sequence #3 among all students (Start Simulation → DisconnectBlock) was small (0.05); therefore, the corresponding edge was not visualized in Figure 2.17. Instead, a more likely sequence observed included an intermediate idle state (i.e., StartSimulation → Idle → DisconnectBlock). The positive link between the likelihood of sequence #3 and the learning gains suggests that students who disconnected a block from the computational model after running a simulation, presumably to replace it by a more correct block (disconnecting a block may be considered to be part of a debugging action), showed a better understanding of the computational modeling task.

As for the negative predictors, i.e., sequences #5, #6, and #7, they all include the RemoveBlock action, one of the least frequent actions during the computational modeling activity. On the one hand, the RemoveBlock action itself appeared to be an indicator of an ineffective model-building approach: if the block that is removed is correct or needed for the computational model, it implies an *ineffective model edit*. In other situations, the removal of an incorrect or unnecessary block clearly demonstrates an effective action. In addition, the more likely actions following Idle include StartSimulation, ConnectBlock, FillField, and DisconnectBlock, which all appeared as part of *debugging* or *tinkering* strategies students used for learning during the model-building activity.

Similarly, in the computational modeling activity, the FillField actions mainly appeared in two contexts: (1) as part of the initial process of building the model and (2) as part of tweaking parameters to test the model. The high transition probabilities between the FillField action, ConnectBlock, and StartSimulation actions in the Markov chain (including a self-loop for FillField) in Figure 2.15 confirm the two interpretations. On the other hand, the transition FillField → DisconnectBlock has a very low probability (0.025), and such an unnatural flow of actions suggests suboptimal learning strategy use.

Finally, the negative link between the learning gains and the likelihood of the Start Simulation → Disengaged sequence suggests the undermining effect of being *disengaged* during the computational modeling activity. In this case, instead of using the simulation results to make an informed modification to the model (as part of the *debugging* or *tinkering* strategies), the student tested the model without doing anything until the end of the class.

2.5.3.2 Engineering Design Strategies

We logged the following actions in the engineering design learning activity: AddMaterial, RemoveMaterial, ResetDesign, RunSimulation, and ViewDesignHistory. Table 2.5 lists the number of instances and the proportion of these actions. The types of actions are fewer than the computational modeling activity but the number of instances of each action was much greater.

We used the same method to determine the threshold of Idle and Disengaged actions in the engineering design activity. Figure 2.17 shows the first-order Markov chain model derived from all students' activities with interpolated Idle and SoD (start of day) actions. Figure 2.18 shows the transition probabilities between the state space defined as the engineering design actions.

TABLE 2.5

Count and Proportion of Actions in the Engineering Design Activity

Action	Number of Instances	Proportion
Add material	17,613	0.491
Remove material	12,327	0.344
Run simulation	3,460	0.096
View design history	120	0.003
Reset	501	0.014
Idle	1,648	0.046
Disengaged	213	0.006

As the two most frequent actions, AddMaterial (49%) and RemoveMaterial (34%) were *sink* states (see Figure 2.17), and most of the other states had high transition probabilities to these two actions. This reflects the learning activity in the engineering design of the WRC, i.e., students needed to search for satisfying schoolyard designs from a vast problem space of 6^{12} possible solutions.

Similar to the computational modeling activity, students had short gaps in time between actions in the engineering design activity. In other words, students did not seem to spend much time on applying cognitive and metacognitive strategies, such as reflection and planning in the engineering design activity.

Using the same method for analyzing students' strategy use in computational modeling, we analyzed the students' learning behaviors in the engineering design activity to link strategy use and performance. Table 2.6 lists the significant predictors and their standardized regression coefficients (β) ranked in descending order. All of the eight action sequences had high lifts.

Sequence #1 involved the Reset action, which only accounted for 1.4% of the total number of actions performed by the students. This sequence indicated the starting of a schoolyard design instance. Sequences #2 and #6 include the ViewDesignHistory action. Sequence #2 (SoD → ViewDesignHistory) had a positive association with the learning performance and a very high lift measure. On the other hand, the likelihood of sequence #6 (ViewDesignHistory → RunSimulation) was the strongest negative predictor of learning performance. ViewDesignHistory is an action with the lowest total number of instances ($n = 120$) and used by very few students. Note that clicking on the *DesignHistory* block pops out a well-formatted table that lists the entire history of all unique schoolyard designs a student generated, including the timestamp, the cost, the composition of the materials, and the values of the input and output variables (total rainfall, total absorption, and total runoff). This feature was designed specifically

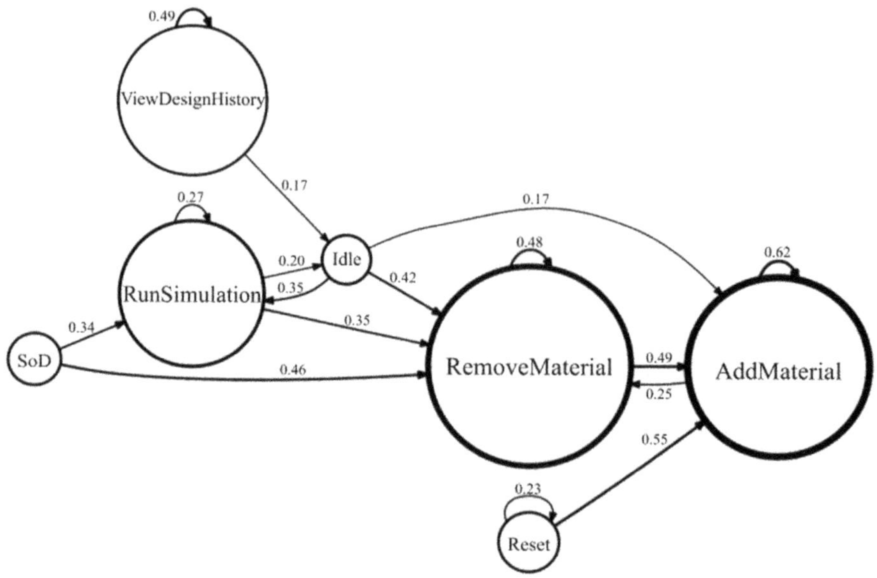

FIGURE 2.17
Discrete-time Markov chain model of students' engineering design processes.

	AddMaterial	RemoveMaterial	Reset	RunSimulation	ViewHistory	Idle	SoD
AddMaterial	**0.62**	**0.25**	0.00	0.10	0.00	0.02	0.01
RemoveMaterial	**0.49**	**0.48**	0.01	0.01	0.00	0.01	0.00
Reset	**0.55**	0.00	**0.23**	0.06	0.01	0.08	0.05
RunSimulation	0.04	**0.35**	0.03	**0.27**	0.00	0.20	0.09
ViewDesignHistory	0.13	0.00	0.03	0.11	**0.49**	0.17	0.08
Idle	0.17	**0.42**	0.05	**0.35**	0.01	n/a	n/a
SoD	0.12	**0.46**	0.06	**0.34**	0.02	n/a	n/a

FIGURE 2.18
Transition probabilities of students' engineering design actions.

TABLE 2.6
Statistically Significant Predictors (Likelihoods of Pairs of Actions) of the Learning Performance

	Action Sequence	β	Std. Error	p-Value	Lift
1	Reset → add material	0.45	0.20	0.030	1.13
2	SoD → view design history	0.42	0.20	0.047	6.06
3	Idle → run simulation	0.37	0.18	0.048	3.58
4	Add material → add material	−0.31	0.15	0.035	1.25
5	Run simulation → disengaged	−0.36	0.18	0.049	6.11
6	View design history → run simulation	−0.57	0.24	0.021	1.12

to support students' search for optimal engineering design solutions. Some students' in the study used the feature strategically to optimize their schoolyard designs. For example, some high-performing students used Sequence #2 as a reflection strategy to review the design solutions they had generated on previous days. On the other hand, the action sequence ViewDesignHistory→RunSimulation suggests a non-effective use of ViewDesignHistory. Viewing the table of previously generated solutions does not result in a schoolyard design, and, therefore, the *RunSimulation* action provides no new information on the student's generated design solutions.

Sequence #3 (Idle→RunSimulation) was positively correlated with the learning performance, and indicates a useful *metacognitive strategy*: a more *systematic approach to generating and testing* design solutions as opposed to using a *trial-and-error* strategy.

Sequence #4 involves multiple *AddMaterial* actions and it accounted for 49% of the student actions overall. This sequence is indicative of design solution construction [29]. It had a negative association with learning performance ($\beta = -0.31$), implying that students spent a lot of their time and effort constructing design solutions, but did not put as much effort into comparing design solutions, and reflecting on how to generate better ones.

Last, Sequence #5 (RunSimulation→Disengaged) revealed a behavior where students, instead of using the simulation results to improve the engineering design, became disengaged (we defined disengagement as a state, where students were idle [did nothing] for at least seven minutes). The likelihood of this behavior was a significant predictor of the learning performance during the computational model-building activity as well, and in both cases, our analyses showed a negative link between students' disengagement and the learning performance.

In summary, our MC analyses showed that students used a number of strategies (sequence of actions) for their computational modeling and engineering design tasks. By correlating the use of strategies with their performance on these two tasks, we successfully characterized strategies as being productive and unproductive. These analyses also provide us with answers to RQ3.

2.6 Discussion and Future Directions

The work presented in this chapter has clearly demonstrated how AI-based representation schemes can support the development of open-ended STEM learning environments directed toward gaining a deep understanding of scientific processes and solving engineering design problems. Two innovative characteristics of our work derived from AI representations and reasoning schemes are (1) using domain-specific modeling languages that create a terminology and constructs to adopt a computational thinking (CT) framework to support the integrated learning of science and engineering; and (2) developing multiple linked representations that help students develop, reason about, and connect important science and engineering concepts, starting from intuitive conceptual models to computational simulation models and engineering design solution representations through a sequence of representations that include conceptual and computational modeling.

In addition, we have used learning analytics and machine learning–based algorithms, such as path analysis and Markov chain modeling to characterize students' learning progressions (path analysis) and understand their learning behaviors and strategies (Markov chain modeling). Using additional analytics measures, we have linked students' strategy use to their performance, thus developing a good understanding of a set of productive and unproductive strategies that students apply to their learning and problem-solving tasks.

Our overall approach to curriculum and assessment design has adopted an ECD approach, and this has enabled us to develop a set of analytics measures that directly link to the domain-specific concepts and practices (e.g., debugging strategies that students employ when building their computational models) that we expect the students to learn. As noted in the literature review, learning analytics efforts targeting learning behaviors and strategy analysis often lack contextual information necessary to link students' behaviors to domain-specific performance and knowledge construction. We presented an initial, qualitative approach leveraging results from our ECD-created tasks and assessments that demonstrate the potential of our scheme in providing such context. We are currently working on additional AI methods, for example, natural language analysis methods to assess students' short answer questions in formative assessments (including the conceptual model and rule creation task discussed in Section 2.5.2), so that we can directly support the learners and the classroom teachers as the curriculum is being taught in classroom settings.

In the future, we will further leverage AI and machine learning techniques in two primary directions:

1. Developing more *automated online assessments* of students' conceptual and computational models, and their engineering design solutions. This will require developing natural language processing mechanisms and machine learning–based analyses that are closely tied to instructor- and researcher-generated rubrics so that the assessment results are reliable and validated. Automated online grading will provide students and their teachers timely feedback on their learning progressions, and serve as formative assessments to aid their learning processes. Using visualizations and other explanatory mechanisms developed in AI to present relevant information on students' learning performance and behaviors in the form of dashboards can aid teachers in being more responsive to students' needs and improving their instructional practices [56, 57].

2. Designing and implementing *adaptive scaffolding frameworks* that can monitor students' progress and difficulties as they work on their model-building and problem-solving tasks. In previous work, we have successfully developed adaptive scaffolding systems [58, 4, 59], but these have been handcrafted systems that were designed to address a small number of difficulties that students faced. Machine learning models like the ones presented in this chapter are likely to be more comprehensive and complete.

Furthermore, from the analyses presented in this chapter and many others conducted in the AI and education and the learning sciences domains, in-depth analyses of students' learning performance and behaviors have been conducted on small amounts of data (say between 50 and 200 students). Therefore, it is not clear how the results presented in these approaches will scale robustly to much larger and diverse populations of students. Future work in this area needs to simultaneously address the data impoverishment problem, while also developing robust and explainable deep learning methods to analyze student data.

Notes

1. The engineering design activity also has a start simulation action, but they are logged differently to differentiate the type of learning activity, as an extension to the generic pressStart action of NetsBlox.
2. Longer action sequences can be seen as the combination of shorter sequences and thus provide limited added information. Therefore, we focused on short sequences.

Bibliography

1. National Research Council. *Next Generation Science Standards: For States, by States*. The National Academies Press, Washington, DC, 2013.
2. Chia-Jung Chang, Chen-Chung Liu, Cai-Ting Wen, Li-Wen Tseng, Hsin-Yi Chang, Ming-Hua Chang, Shih-Hsun Fan Chiang, Fu-Kwun Hwang, and Chih-Wei Yang. The impact of light-weight inquiry with computer simulations on science learning in classrooms. *Computers and Education*, 146:103770, 2020.
3. Pratim Sengupta, John S Kinnebrew, Satabdi Basu, Gautam Biswas, and Douglas Clark. Integrating computational thinking with k-12 science education using agent-based computation: A theoretical framework. *Education and Information Technologies*, 18(2):351–380, 2013.
4. Satabdi Basu, Gautam Biswas, and John S Kinnebrew. Learner modeling for adaptive scaffolding in a computational thinking-based science learning environment. *User Modeling and User-Adapted Interaction*, 27(1):5–53, 2017.
5. Nicole M Hutchins, Gautam Biswas, Miklós Maróti, Ákos Lédeczi, Shuchi Grover, Rachel Wolf, Kristen Pilner Blair, Doris Chin, Luke Conlin, Satabdi Basu, et al. C2stem: A system for synergistic learning of physics and computational thinking. *Journal of Science Education and Technology*, 29(1):83–100, 2020.
6. Ningyu Zhang, Gautam Biswas, Kevin W McElhaney, Satabdi Basu, Elizabeth McBride, and Jennifer L Chiu. Studying the interactions between science, engineering, and computational thinking in a learning-by-modeling environment. In Ig Ibert Bittencourt, Mutlu Cukurova, Kasia Muldner, Rose Luckin, and Eva Millán, editors. *Artificial Intelligence in Education*, 598–609. Springer International Publishing, Cham, 2020.
7. Satabdi Basu, Gautam Biswas, Pratim Sengupta, Amanda Dickes, John S Kinnebrew, and Douglas Clark. Identifying middle school students' challenges in computational thinking-based science learning. *Research and Practice in Technology Enhanced Learning*, 11(1):13, 2016.
8. James P Bywater, Jennifer L Chiu, James Hong, and Vidhya Sankaranarayanan. The teacher responding tool: Scaffolding the teacher practice of responding to student ideas in mathematics classrooms. *Computers and Education*, 139:16–30, 2019.
9. Satabdi Basu, Gautam Biswas, and John S Kinnebrew. Using multiple representations to simultaneously learn computational thinking and middle school science. In D. Schuurmans and Michael Wellman (Chairs), *Thirtieth AAAI Conference on Artificial Intelligence*, 3705–3711, 2016.
10. Nicole M Hutchins, Satabdi Basu, Kevin McElhaney, Jennifer Chiu, Sarah Fick, Ningyu Zhang, and Gautam Biswas. Coherence across conceptual and computational representations of students' scientific models. In *The International Society of the Learning Sciences Annual Meeting 2021*. International Society of the Learning Sciences (ISLS), 2021.
11. Kevin W McElhaney, Ningyu Zhang, Satabdi Basu, Elizabeth McBride, Gautam Biswas, and Jennifer L Chiu. Using computational modeling to integrate science and engineering curricular activities. In *The Interdisciplinarity of the Learning Sciences, 14th International Conference of the Learning Sciences (ICLS) 2020*, 1357–1364. International Society of the Learning Sciences (ISLS), 2020.
12. Ningyu Zhang, Gautam Biswas, Jennifer L Chiu, and Kevin W McElhaney. Analyzing students' design solutions in an NGSS-aligned earth sciences curriculum. In Seiji Isotani, Eva Millán, Amy Ogan, Peter Hastings, Bruce McLaren, and Rose Luckin, editors. *Artificial Intelligence in Education*, 532–543. Springer International Publishing, Cham, 2019.

13. Peter Zeegers*. Student learning in higher education: A path analysis of academic achievement in science. *Higher Education Research and Development*, 23(1):35–56, 2004.
14. Wei Wen Wu. Linking Bayesian networks and pls path modeling for causal analysis. *Expert Systems with Applications*, 37(1):134–139, 2010.
15. James R Segedy, John S Kinnebrew, and Gautam Biswas. Using coherence analysis to characterize self-regulated learning behaviours in open-ended learning environments. *Journal of Learning Analytics*, 2(1):13–48, 2015.
16. John S Kinnebrew, Kirk M Loretz, and Gautam Biswas. A contextualized, differential sequence mining method to derive students' learning behavior patterns. *JEDM – Journal of Educational Data Mining*, 5(1):190–219, 2013.
17. Aphrodite Galata, Neil Johnson, and David Hogg. Learning variable-length Markov models of behavior. *Computer Vision and Image Understanding*, 81(3):398–413, 2001.
18. David Weintrop, Elham Beheshti, Michael Horn, Kai Orton, Jona Kemi, Laura Trouille, and Uri Wilensky. Defining computational thinking for mathematics and science classrooms. *Journal of Science Education and Technology*, 25(1):127–147, 2016.
19. Bruce L Sherin. How students understand physics equations. *Cognition and Instruction*, 19(4):479–541, 2001.
20. Andrea diSessa. *Changing Minds: Computers, Learning, and Literacy*. MIT Press, Cambridge, MA, 2001.
21. Elliot Soloway. Should we teach students to program? *Communications of the ACM*, 36(10):21–24, 1993.
22. Seymour Papert. Situating constructionism. In I Harel, and S Papert, editors. *Constructionism*, 193–206. Ablex Publishing, Westport, CT, 1991.
23. Valerie J Shute, Chen Sun, and Jodi Asbell-Clarke. Demystifying computational thinking. *Educational Research Review*, 22:142–158, 2017.
24. Ningyu Zhang. *Supporting the Integrated Learning of Science, Engineering, and Computational Thinking in an Open-Ended Learning Environment*. PhD thesis. Vanderbilt University, 2020.
25. Pratim Sengupta, and Amy Voss Farris. Learning kinematics in elementary grades using agent-based computational modeling: A visual programming-based approach. In *Proceedings of the 11th International Conference on Interaction Design and Children*, IDC '12, 78–87. Association for Computing Machinery, New York, 2012.
26. Shaaron Ainsworth. Deft: A conceptual framework for considering learning with multiple representations. *Learning and Instruction*, 16(3):183–198, 2006.
27. Rebecca L Oxford. Strategies for learning a second or foreign language. *Language Teaching*, 44(2):167, 2011.
28. Philip H Winne, Dianne Jamieson-Noel, and Krista Muis. Methodological issues and advances in researching tactics, strategies, and self-regulated learning. *Advances in Motivation and Achievement: New Directions in Measures and Methods*, 12:121–155, 2002.
29. Ningyu Zhang, Gautam Biswas, and Nicole Hutchins. Measuring and analyzing students' strategic learning behaviors in open-ended learning environments. *Manuscript*, Submitted for Publication, nd.
30. Paulo Blikstein, Marcelo Worsley, Chris Piech, Mehran Sahami, Steven Cooper, and Daphne Koller. Programming pluralism: Using learning analytics to detect patterns in the learning of computer programming. *Journal of the Learning Sciences*, 23(4):561–599, 2014.
31. Nicole Hutchins, Gautam Biswas, Shuchi Grover, Satabdi Basu, and Caitlin Snyder. A systematic approach for analyzing students' computational modeling processes in c2stem. In Seiji Isotani, Eva Millán, Amy Ogan, Peter Hastings, Bruce McLaren, and Rose Luckin, editors. *Artificial Intelligence in Education*, 116–121. Springer International Publishing, Cham, 2019.
32. Ningyu Zhang, Gautam Biswas, and Yi Dong. Characterizing students' learning behaviors using unsupervised learning methods. In *International Conference on Artificial Intelligence in Education*, 430–441, Springer, 2017.
33. J S Kinnebrew, J R Segedy, and G Biswas. Integrating model-driven and data-driven techniques for analyzing learning behaviors in open-ended learning environments. *IEEE Transactions on Learning Technologies*, 10(2):140–153, April 2017.
34. Charles Xie, Zhihui Zhang, Saeid Nourian, Amy Pallant, and Edmund Hazzard. Time series analysis method for assessing engineering design processes using a cad tool. *International Journal of Engineering Education*, 30:218–230, 2014.
35. Camilo Vieira, Molly Hathaway Goldstein, Şenay Purzer, and Alejandra J Magana. Using learning analytics to characterize student experimentation strategies in the context of engineering design. *Journal of Learning Analytics*, 3(3):291–317, December 2016.
36. Wanli Xing, Chenglu Li, Guanhua Chen, Xudong Huang, Jie Chao, Joyce Massicotte, and Charles Xie. Automatic assessment of students' engineering design performance using a Bayesian network model. *Journal of Educational Computing Research*, 59(2):230–256, 2021.
37. James P Bywater, Mark Floryan, and Jennifer L Chiu. Discs: A new sequence segmentation method for open-ended learning environments. In *International Conference on Artificial Intelligence in Education*. Springer, 2021.
38. Robert J Mislevy, and Geneva D Haertel. Implications of evidence-centered design for educational testing. *Educational Measurement: Issues and Practice*, 25(4):6–20, 2006.
39. Christopher J Harris, Joseph S Krajcik, James W Pellegrino, and Kevin W McElhaney. *Constructing Assessment Tasks That Blend Disciplinary Core Ideas, Crosscutting Concepts, and Science Practices for Classroom Formative Applications*. SRI International, Menlo Park, CA, 2016.
40. John R Frederiksen, Barbara Y White, and Joshua Gutwill. Dynamic mental models in learning science: The importance of constructing derivational linkages among models. *Journal of Research in Science Teaching*, 36(7):806–836, 1999.

41. Charles Xie, Corey Schimpf, Jie Chao, Saeid Nourian, and Joyce Massicotte. Learning and teaching engineering design through modeling and simulation on a cad platform. *Computer Applications in Engineering Education*, 26(4):824–840, 2018.
42. Shuchi Grover, and Satabdi Basu. Measuring student learning in introductory block-based programming: Examining misconceptions of loops, variables, and Boolean logic. In *Proceedings of the 2017 ACM SIGCSE Technical Symposium on Computer Science Education*, SIGCSE '17, 267–272. Association for Computing Machinery, New York, 2017.
43. Nicole M Hutchins, Gautam Biswas, Ningyu Zhang, Caitlin Snyder, Ákos Lédeczi, and Miklós Maróti. Domain-specific modeling languages in computer-based learning environments: A systematic approach to support science learning through computational modeling. *International Journal of Artificial Intelligence in Education*, 30(4):537–580, 2020.
44. Kurt VanLehn. Model construction as a learning activity: A design space and review. *Interactive Learning Environments*, 21(4):371–413, 2013.
45. Arie van Deursen, Paul Klint, and Joost Visser. Domain-specific languages: An annotated bibliography. *ACM SIGPLAN Notices*, 35(6):26–36, June 2000.
46. Hector J Levesque. Knowledge representation and reasoning. *Annual Review of Computer Science*, 1(1):255–287, 1986.
47. Rivka Oxman. The mind in design: A conceptual framework for cognition in design education. In *Design Knowing and Learning: Cognition in Design Education*, 269–295. Elsevier, 2001.
48. Byron Ellis and Wing Hung Wong. Learning causal Bayesian network structures from experimental data. *Journal of the American Statistical Association*, 103(482):778–789, 2008.
49. Martin S Hagger, and Kyra Hamilton. Motivational predictors of students' participation in out-of-school learning activities and academic attainment in science: An application of the trans-contextual model using Bayesian path analysis. *Learning and Individual Differences*, 67:232–244, 2018.
50. John S Kinnebrew, James R Segedy, and Gautam Biswas. Analyzing the temporal evolution of students' behaviors in open-ended learning environments. *Metacognition and Learning*, 9(2):187–215, 2014.
51. Agathe Merceron, and Kalina Yacef. Interestingness measures for association rules in educational data. In *Educational Data Mining 2008*, 2008.
52. James B Schreiber, Amaury Nora, Frances K Stage, Elizabeth A Barlow, and Jamie King. Reporting structural equation modeling and confirmatory factor analysis results: A review. *The Journal of Educational Research*, 99(6):323–338, 2006.
53. David Klahr, and Kevin Dunbar. Dual space search during scientific reasoning. *Cognitive Science*, 12(1):1–48, 1988.
54. Gregory Schraw, Kent J Crippen, and Kendall Hartley. Promoting self-regulation in science education: Metacognition as part of a broader perspective on learning. *Research in Science Education*, 36(1–2):111–139, 2006.
55. Emden R Gansner. *Drawing Graphs with Graphviz*. Technical Report, 2009.
56. Vanessa Echeverria, Roberto Martinez-Maldonado, Simon Buckingham Shum, Katherine Chiluiza, Roger Granda, and Cristina Conati. Exploratory versus explanatory visual learning analytics: Driving teachers' attention through educational data storytelling. *Journal of Learning Analytics*, 5(3):72–97, November 2018.
57. Ioana Jivet, Maren Scheffel, Marcus Specht, and Hendrik Drachsler. License to evaluate: Preparing learning analytics dashboards for educational practice. In *Proceedings of the 8th International Conference on Learning Analytics and Knowledge*, 31–40, 2018.
58. Satabdi Basu, Pratim Sengupta, and Gautam Biswas. A scaffolding framework to support learning of emergent phenomena using multiagent-based simulation environments. *Research in Science Education*, 45(2):293–324, 2015.
59. James R Segedy, John S Kinnebrew, and Gautam Biswas. The effect of contextualized conversational feedback in a complex open-ended learning environment. *Educational Technology Research and Development*, 61(1):71–89, 2013.

3

Intelligent Science Stations Bring AI Tutoring into the Physical World

Nesra Yannier, Scott E. Hudson, and Kenneth R. Koedinger

CONTENTS

3.1 Introduction 39
3.2 Intelligent Science Stations: A Guided AI Mixed-Reality Interactive Experience 40
3.2.1 Physical Setup and AI Computer Vision Technology 40
3.2.2 Scenario 41
3.3 AI in the 3D Physical World versus on Flat Screen 42
3.4 AI Guidance Matters: Making Science Exhibits Yield Better Learning and More Engagement 44
3.5 The Kind of AI Guidance Matters: Scaffolded Science Inquiry Produces Better Engineering 45
3.6 Discussion and Next Steps 46
Note 48
References 48

3.1 Introduction

Intelligent tutoring systems have been one quite successful application of artificial intelligence (AI) in education (e.g., Aleven et al., 2017; Akbulut & Cardak, 2012; Aleven et al., 2006; Anderson et al., 1995). There the emphasis was on using AI (1) to create an expert model of domain competence, (2) to use this model and plan-recognition to track and adapt to students' progress in step-by-step problem-solving, and (3) to use statistical inference to track and adapt to students' knowledge growth (Koedinger et al., 2013; Aleven et al., 2017). The efforts we describe use AI to bring the power of intelligent tutoring into the world of 3D physical interaction, building on recent advances in educational technology, learning science, and tangible interfaces (Mayer & Alexander, 2016; Papavlasopoulou et al., 2017; Baykal et al., 2018; Khan et al., 2019). The key here is to use sensing technology and AI vision algorithms to watch what students are doing as they engage in physical activities, like running an experiment on stability or rolling speed, building and testing a tower or model car. We present a new genre of Intelligent Science Stations and a sequence of experiments that justify their key features.

In Intelligent Science Stations, children do experiments and make discoveries in the real world accompanied by interactive feedback and guidance from a gorilla character (AI agent) that becomes part of their activities and works with them. Depth camera sensing and AI computer vision (CV) algorithms track what children are doing as they act for themselves with physical objects, so that the system can apply guidance, structure, and pedagogy. For example, in our first Intelligent Science Station, EarthShake, children make predictions about which of the given towers will fall first when the table shakes, with interactive feedback and guidance. Section 3.2 provides more detail.

One key feature of Intelligent Science Stations is that students are able to work, and get feedback on their work, in the 3D physical world. AI vision is used to make such feedback possible. But is it necessary? What if students learn just as much (and have as much fun) with a flat-screen implementation where the same interactive guidance is provided, but students watch videos of towers falling on the earthquake table? This question is important and non-obvious given past experimental results showing mixed results in learning outcomes from physical versus virtual interactions (Klahr et al., 2007; Olympiou & Zacharias, 2012). Section 3.3 reports on random assignment experiments that rigorously demonstrate that students learn much more and report greater enjoyment from the 3D physical interaction than from the matched flat screen interaction.

DOI: 10.1201/9781003181187-4

Another key feature of Intelligent Science Stations is that students get interactive guidance and, again, the AI vision makes that possible. However, for years children have engaged in science lab experiences without such guidance. And they have had fun with, and perhaps learned from, many museum exhibits where no interactive guidance is readily available. Furthermore, there are reasonable concerns about too much guidance in instruction (Chi & Wylie, 2014; Papert & Harel, 1991; Kafai & Resnick, 1996; Resnick, 2014; Jeffery-Clay, 1998) and even evidence that unstructured play may be better for some learning outcomes (Barker et al., 2014). Thus, we performed a random assignment experiment comparing a typical unfacilitated museum exhibit structure with an Intelligent Science Station exhibit and found that students learned much more and were engaged longer with the AI-based exhibit. These results are described in Section 3.4.

A third key feature of Intelligent Science Stations is a pedagogical approach that focuses on *deliberate practice* toward *use and understanding underlying scientific principles*. Deliberate practice involves carefully designed tasks toward key learning objectives, repeated opportunities to demonstrate successful actions and explanations, and interactive feedback that is responsive to students' learning-by-doing attempts. Focusing on the underlying scientific principles involves, for example, including contrasting cases that isolate the differences of targeted learning principles (Chase et al., 2010), a predict–observe–explain cycle (Chi et al., 1989; Aleven & Koedinger, 2000), or self-explanation menus that prompt the student to explain underlying reasons of physical phenomena they have observed (White & Gunstone, 1992).

Another approach, consistent with an interpretation of constructivist pedagogy (Applefield et al., 2000; Cakir, 2008) and common in other mixed-reality learning efforts (Tseng et al., 2011; Resnick et al., 1998; Ryokai & Cassell, 1999), is to present interesting project activities with open-ended engineering/construction challenges and with natural feedback from the physical world. The emphasis here is on building – constructing a physical artifact as referenced in this quote: 'Constructionism (Papert & Harol, 1991; Kafai & Resnick, 1996; Papert, 1980) suggested that learners engage in design challenges and that they have a personally meaningful physical artifact to take home with them' (Kolodner et al., 2003). These can be thought of as engineering activities and the idea is that probing underlying scientific principles is contextualized and motivated with such constructive engineering tasks. In Section 3.5, we summarize experiments showing that a focus on *deliberate practice* toward *use and understanding underlying scientific principles* yields better learning than a focus on engineering construction not only on assessments of science use and explanation, but also on engineering construction tasks.

3.2 Intelligent Science Stations: A Guided AI Mixed-Reality Interactive Experience

3.2.1 Physical Setup and AI Computer Vision Technology

Our base Intelligent Science Station platform consists of an earthquake table, physical towers, a depth camera facing the objects, an LCD or a projector screen, and a display screen with the interactive game. The depth camera and our specialized computer vision algorithm detect if and when an object is placed on the table giving feedback if the placed towers match the correct ones or not (based on the shape and position of the object). It is critical for students to place the correct contrasting case towers on the table so that they can isolate target variables. The computer vision algorithm detects when an action happens on the table (e.g., a tower falls) so that it can give interactive feedback in sync with what is happening in the real world. The CV algorithm can also detect any tower that it builds, using different types of materials such as lego blocks, cardboard, etc. When it shakes the earthquake table by pressing the shake button on the screen, it can tell how many seconds it took for the tower to fall down. The system gives different challenges (e.g., with differing height and base widths) as it progresses and ensures that the tower built in the physical world satisfies the challenges that the system gives (e.g., the tower they build is higher than the Eiffel tower on the screen). The computer game that is displayed on a display screen provides personalized interactive feedback to the users, telling if their prediction was correct or not and prompting them to explain why. The AI computer vision technology is critical to provide personalized immediate feedback to individual children as they do physical experiments using scientific apparatus, extending intelligent tutoring technology into the real world. The earthquake table consists of a computer-controlled relay, a small motor, a mechanism for converting from rotary to reciprocating linear motion, and rails to support the moving platform. When the switch or the relay is activated through the game, it activates the motor, which in turn moves the platform back and forth.

The computer vision algorithm uses depth information that is extracted from a Kinect depth camera. We decided to use depth information instead of color information since the depth information does not change according to lighting and is more reliable in real-world settings. First, we do filtering on the image using background segmentation to segregate the blocks from the background and create a clean image. Then we use blob detection (Wang and Ju, 2008) where a blob is a contiguous group of similar pixels (e.g., with color, brightness, or depth values distinguished from surrounding pixels). A tower is identified as a blob with a particular shape. The shape of a blob is determined by a Moment of Inertia metric (see Yannier et al., 2020) that is calculated along multiple dimensions, for example, a horizontal (x) and vertical direction (y). This x–y pair (or larger vector) is compared to the known x–y moment vectors of all the towers computing the Euclidean distance in each case. The tower is identified when the distance is both the smallest and below a threshold – there can be images in the display, like hands, that are not identified. The system must also detect when a tower falls. A fall is detected when the height of a blob in the same x-position as an original tower is lower than the height of that original tower.

This computer vision algorithm allows us to both determine if the correct towers that were specified (contrasting case towers) were placed on the table to impose pedagogical structure and to track the actions in the real world to provide interactive real-time feedback accordingly. The technical challenge is in creating tangible interfaces and mixed-reality platforms that can adapt to the changes in the physical environment, while also providing children with room for exploration. Such structure and interactivity is comparatively easy in purely virtual settings, but not intuitive in a real-world physical setting. The AI vision algorithm is highly reliable. While it is possible that AI might incorrectly recognize a tower or the moment at which a tower has fallen, these errors have not occurred in practice, once full algorithm debugging was completed.

The mixed-reality platform we have created can be extended to many different content areas. For example, our second Intelligent Science Station is called SmartRamps to teach children the scientific concepts of force and motion. In this Intelligent Science Station, we use the same base platform, including the depth camera, display screen, tablet input screen, and the additional physical objects/apparatus (in this case cars running down ramps) that can be placed on top of the modular base platform. The AI vision algorithm described above can be adapted to different content areas, tracking different types of experiments, actions, and physical apparatus on the base platform.

3.2.2 Scenario

Our first Intelligent Science Station, EarthShake, consists of two different modes: *Guided-discovery* mode and *Explore-construct* mode.

The **Guided-discovery mode** implements an inquiry cycle of deliberate practice, drawing on specific evidence-based techniques that have been shown to be effective in intelligent tutoring systems and in the learning science literature: predict–explain–observe–explain (White and Gunstone, 1992), contrasting cases (Chase et al., 2010), self-explanation (Chi et al., 1989; Aleven and Koedinger, 2000), and real-time interactive feedback (Corbett and Anderson, 2001), developing critical thinking skills by making decisions about comparisons (Holmes et al., 2015).

With the guidance of a gorilla character, the users are first asked to place the matching towers shown on the screen on the physical earthquake table (see Figure 3.1). These contrasting case towers have only one difference between them, targeting physics

FIGURE 3.1
Guided-discovery mode of Intelligent Science Stations. Children are asked to make a prediction about which of the towers will fall first when the earthquake table shakes and then explain their observations, all with interactive feedback from an AI agent.

principles of stability (height, base width, symmetry, center of mass) so that the students can focus on isolated principles. The system detects if the correct towers are placed on the table or not and gives feedback accordingly. If they place the correct towers matching the ones on the screen, a green check appears above the tower on the screen and they can continue. Otherwise, if the tower they place does not match the tower on the screen that they were asked to place, the computer vision system detects that there was a mismatch and a red cross appears on the tower on the screen, asking them to place the correct tower.

Once the correct towers have been placed, the gorilla character asks them to make a prediction about which of the two towers placed on the table will fall first when the table shakes. They can choose either 1, 2 or the same by clicking on one of the towers on the screen. After they make a prediction, the gorilla character prompts them to explain why they think this tower will fall first and discuss with others. After they discuss their prediction, they can click the 'Shake' button on the touch screen which in turn starts shaking the physical earthquake table (the relay that is connected to the motor of the earthquake table is triggered).

When one of the towers falls down with the table shaking, the depth camera and the system detects that there was a fall and stops the earthquake table. The gorilla character gives feedback about whether their prediction was correct or not based on the computer vision algorithm that detects whether the left or right tower fell and if the outcome matches with the prediction of the user. If the user's prediction was correct, then the gorilla character says: 'Good job! Your hypothesis was correct! Why do you think this tower fell first?', and starts to jump and dance on the screen giving them positive feedback. If the user's prediction was wrong, then the gorilla character says: 'Uh oh! Your prediction was not quite right! Why do you think this tower fell first?' This time the users are asked to explain the outcomes of the experiment using an explanation menu with four different choices that they can choose from: 'It is taller', 'It has a thinner base', 'It has more weight', 'It is not symmetrical'. These explanations correspond with the four different principles of stability and balance, which are primary science content learning objectives for the Intelligent Science Station. This explanation menu comes up with both if their prediction was correct or wrong. In the rare cases where the unexpected tower falls first in the physical world (e.g., short tower falls instead of the tall tower), the system notices and provides feedback to the student: 'Uh oh, that's not what happens usually, please try again'. When this rare event does occur, it can be a good opportunity for students and teachers to talk about uncertainty in science.

When the users click on one of the choices in the explanation menu, the gorilla character tells them if their explanation was correct or wrong and shows a visualization overlaid on the images of the towers on the screen to explain why the tower actually fell (e.g., if the tower fell because it was not symmetrical, the visualization shows a line dividing down the middle of the tower to demonstrate that the tower is not the same on both sides of the line). If their explanation was not correct, the gorilla says: 'Actually it fell first because it was not symmetrical. Good try. Click CONTINUE to play again'. Or if it fell because it was taller, the ruler visualizations this time highlight the height of each tower. This scenario is repeated for different contrasting cases (Yannier et al., 2015).

In the **Explore-construct** mode of the Intelligent Science Station, users can build any tower they want using different types of blocks (e.g., legos, magnet blocks, wooden blocks). The gorilla character asks them to build a tower that they think will withstand the earthquake. When they have built their tower and click the SHAKE button on the touch screen, the table starts to shake. If their tower falls down in less than 5 seconds, the system detects the fall and stops the table, displaying how many seconds it took for the tower to fall (see Figure 3.2). The gorilla character then gives feedback saying: 'Uh oh! Your tower fell down! Press CONTINUE to make another tower'. If the tower does not fall down in 5 seconds, the table stops shaking and the gorilla character starts jumping/dancing, and says: 'Good job! Your tower stayed up! Press CONTINUE to make another tower'. When they press continue, they are asked to make another tower this time with a challenge (e.g., 'Can you make a tower that is taller than the house on the screen?'). As they proceed, the challenges get more difficult (e.g., taller than the Eiffel tower or with a smaller base than the ruler on the screen).

3.3 AI in the 3D Physical World versus on Flat Screen

Our first experiments explored whether it is worth developing AI to track and support learning in the physical world. No AI sensing and perception is needed to provide similar guidance in a simulated environment within a typical flat-screen computer interaction. Do students learn just as much from analogous flat-screen guidance (cf. Klahr et al., 2007) or does providing for physical interaction without sacrificing any guidance better for learning? To address this question, we compared the mixed-reality version

FIGURE 3.2
Explore-construct mode of Intelligent Science Stations. Children are asked to build a tower that will withstand an earthquake using different materials. The system displays the time it took for their tower to fall. If their tower stays up, they start getting different challenges.

of EarthShake with an equivalent virtual-only version. These versions differ only in the medium of presentation. In the mixed-reality condition, students interact with the mixed-reality Intelligent Science Station where they are asked to make predictions about which of the physical towers will fall first with interactive feedback on an earthquake table. In the virtual condition, the video of the towers shaking and falling are incorporated into the same interaction on the flat screen. All other important variables are tightly controlled – the within-game and assessment questions, the game scenario, and the interactive feedback are kept the same. Only the medium of presentation is varied between conditions: virtual-only or mixed-reality (physical with interactive feedback).

The results of two experiments clearly demonstrated the benefits of the guided physical interaction made possible by the AI sensing and perception. In the *first experiment* (Yannier et al., 2016), we had a 2 x 2 comparison, with children interacting with mixed-reality or virtual-only versions of EarthShake, in pairs or alone. The results revealed that children in the mixed-reality group scored significantly and substantially higher on a challenging set of post-assessments than children in the virtual-only group, with post-test means of 64% versus 48%, respectively. Controlling for pretest scores in an ANCOVA analysis, these post-test group differences were highly significant ($F(1,66) = 23.3$, $p < 0.0001$) and the effect size estimate ($d = 0.78$) indicates a large effect. An interesting secondary finding of this experiment was that no positive effect of collaboration was found: learning outcomes were not different based on whether children were (through random assignment) working alone or in pairs.

The post-test assessed children's ability both to make accurate predictions when presented with novel towers and to provide accurate scientific explanations (in kid language) of these predictions. The overall benefits of the mixed-reality condition held for both prediction and explanation items in the pre/post-tests. The average post-test score for the mixed-reality condition was 76%, whereas the average score for the virtual condition was 70% ($F(1,66) = 3.1$, $p < 0.0035$, $d = 0.39$). These results demonstrate that the mixed-reality system improved prediction skills more than the virtual control. The results on explanation items were similar. A two-way ANCOVA with post-test explanation as the outcome variable, pre-test explanation score as the covariate, and media type and collaboration as fixed factors revealed that there was a significant main effect of media type. The average post-test score for the mixed-reality condition was 52%, while the average post-test score of the virtual-only condition was 26% ($F(1,66) = 18.6$, $p < 0.0001$, $d = 0.87$). The large effect size (0.87) showed a big difference in learning to provide a scientific explanation, showing deeper understanding of the scientific concepts of stability.

In addition to the prediction and explanation pre- and post-tests to measure understanding of scientific concepts, we also gave students a tower-building task before and after interacting with the game, to see if their learning could transfer to a constructive building/problem-solving task as well. We scored each student's towers according to three principles: height, symmetry, and center of mass (they would get a 1 on a specific principle if their pre- to post-tower improved on that principle; 0 if there was no difference, –1 if their tower became worse on that principle). The results of the tower tests revealed that the towers of the mixed-reality condition improved significantly more than those of the virtual-only condition ($F(1,66) = 6.9$, $p = 0.01$, $d = 0.48$). Thus, children were not only learning the scientific concepts better but they could also transfer them to a real-world constructive problem-solving task.

We also gave students a survey based on Likert scale to measure their enjoyment. The results of the survey demonstrated that the mixed-reality condition had higher mean ratings for enjoyment compared to the virtual-only condition, and an ANOVA showed that this difference was significant (F(1,66) = 6.9, p = 0.01, d = 0.48). On the other hand, there was no significant difference between the solo and pair groups for enjoyment.

In a *follow-up experiment*, we compared the mixed-reality system with a virtual-only condition, but this time we also added a simple physical control to see if adding a simple physical control such as shaking a tablet would increase learning or enjoyment, or if observing physical phenomena in the real world in the context of a mixed-reality system was more critical for learning. The results of the experiment showed a significant effect of media type (F(1,91) = 8.2, p <0.01, d = 0.37), with benefits for mixed-reality. The average score on the post-tests was 45% across the mixed-reality conditions and 39% across the virtual conditions. The overall improvement from pre- to post-test was 11.3% in the mixed-reality conditions and 2.4% in the virtual-only conditions (see Figure 3.3). Thus, the mixed-reality game improved learning almost five times more than the screen-only alternatives. On the other hand, there was no significant effect for control type or no significant interaction effects. Thus, this result indicated that observing physical phenomena in a mixed-reality setting was critical for children's learning; however surprisingly, having a simple physical trigger such as shaking the tablet did not increase learning or enjoyment.

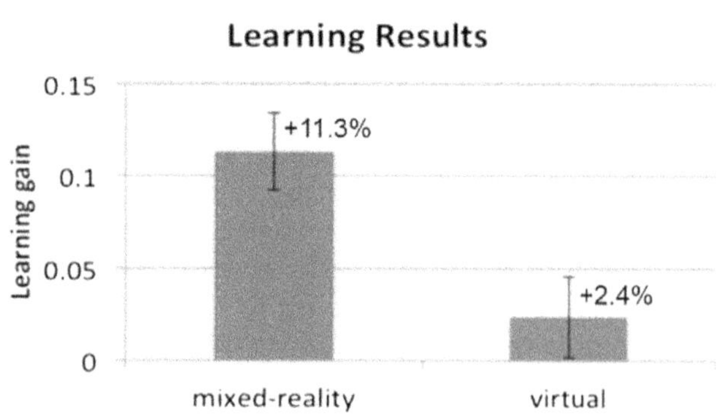

FIGURE 3.3
The results of the experiment comparing mixed-reality and virtual versions of EarthShake revealed that having mixed-reality interaction in the Intelligent Science Stations improved children's learning by *five times* compared to equivalent virtual-only conditions (the overall improvement from pre- to post-test was 11.3% in the mixed-reality conditions and 2.4% in the virtual-only conditions).

The results of this follow-up experiment were aligned with the first experiment, replicating the results for the learning benefits of mixed-reality and physical interaction compared to only screen-based interaction. Children's scores were significantly higher for both prediction and explanation items in mixed-reality conditions compared to virtual-only conditions (Yannier et al., 2015). Similarly, a two-way ANCOVA showed that there was a significant effect of media type for the tower scores, in favor of mixed-reality (F(1,91) = 6.9, p = 0.01, d = 0.64), while there was no significant effect for control type and no interaction effect of media type and control type. Students in the mixed-reality conditions improved their towers more than students in the virtual conditions.

In addition to the learning results, an ANOVA on the survey results showed a significant difference in enjoyment by media type, with the mixed-reality condition indicating more enjoyment (F(1,92) = 6.7, p = 0.01, d = 0.55), whereas there was no significant effect of control type for enjoyment.

An analysis of the videos of children's interactions revealed that children were using more meaningful gestures while describing the towers in the mixed-reality version where they were observing physical phenomena with interactive feedback compared to the screen-only version (p = 0.001, d = 0.72). For example, they would mimic the towers with their hands, with gestures indicating the shape of a base as they explained, 'Because number one has a sturdier bottom'. We found that there was a significant correlation between these meaningful gestures and overall learning gains (R = 0.21, p <0.05). This result suggests that children's spontaneous gestures might be reflecting their mental simulation and processes (Hostetter et al., 2008), thus leading to deeper learning.

These experiments show the promise of Intelligent Science Stations, integrating AI into the physical world using computer vision techniques, for improving learning and enjoyment, above and beyond equivalent purely screen-based alternatives.

3.4 AI Guidance Matters: Making Science Exhibits Yield Better Learning and More Engagement

Another key research question is whether the AI guidance added on top of the physical world is critical and necessary. Does it improve learning and engagement outcomes in informal and formal learning settings compared to physical exploration alone? This question is especially important as currently most of the maker spaces and museum exhibits rely mainly on hands-on open-ended exploration and do not usually provide AI guidance. Especially in museums, there have been some concerns raised by museum professionals that adding guidance may decrease enjoyment. Might intelligent guidance reduce student agency and subsequent engagement (Schwartz et al., 2016)?

To test these hypotheses, we conducted a controlled experiment comparing our Intelligent Science Exhibit with a matching standard exhibit in a Science Center (with hands-on exploration without the AI layer). This matching standard exhibit represents current practice at museums (mimicking the earthquake exhibit at the Science Center where we conducted our experiment). It uses the same materials as the Intelligent Science Exhibit, with only the intelligent guidance turned off.

3.5 The Kind of AI Guidance Matters: Scaffolded Science Inquiry Produces Better Engineering

We have seen from the experiments described above that an AI-based mixed-reality system combining physical and virtual worlds can improve children's learning and engagement, compared to both virtual-only and physical-only versions. However, it's not clear from previous literature how the mixed-reality systems should be designed to maximize learning and enjoyment, and what kinds of interactivity features are critical (e.g., is it better to have guided discovery activities with deliberate practice or exploration/construction activities with problem-solving or a combination of both?).

Many researchers advocate for tangible interfaces and mixed-reality environments that are purely exploratory and support learning through discovery (Marshall et al., 2010). Also, with the Maker Movement becoming more popular, there is a lot of focus on exploration with physical objects. Constructivist learning theory, where open-ended discovery learning and hands-on exploration is the primary focus, is believed to lead to better learning (Papert, 1980).

Other researchers criticize the discovery learning approach, suggesting that it defies cognitive learning theory and that learners are not able to retain the amount of information that is needed to process the content (Kirschner et al., 2006). In contrast to open-ended discovery learning, guided discovery provides scaffolding that recognizes the boundaries of cognitive load while also encouraging exploration. Students engage in inquiry learning, where deliberate practice and reflection opportunities are offered (Harvel, 2010).

Intelligent Science Stations provide the opportunity to incorporate guided discovery with personalized interactive feedback as children experiment in the physical world, while also supporting exploration and construction (using different modes described in Section 3.2). We have conducted an experiment to investigate whether guided discovery with scientific thinking practice, exploration with engineering practice, or a combination of both leads to better learning outcomes in mixed-reality environments.

In this experiment, children interact with different versions of the Intelligent Science Stations: *Guided-discovery mode* (described in Section 3.2), where they make predictions on contrasting cases with interactive feedback; *Explore-construct mode* (described in Section 3.2), where they are given an engineering challenge and asked to build a structure that will withstand an earthquake; and a *Combined* condition, where they go between Guided-discovery and Explore-construct modes (Yannier et al., 2020). The timing was matched in all conditions. Similar to the previous experiments, we used prediction and explanation as well as building tests to measure scientific and engineering outcomes.

Based on an ANOVA testing the effect of the three conditions on the engineering outcomes (improvement in tower construction), we found a statistically significant effect of condition ($F(2, 72) = 4.24$, $p < 0.02$). Most surprisingly, even though children in the Explore-Construct condition practiced more building, they showed the least improvement on tower-building. The Combined condition, where children did some Guided- discovery in addition to Explore-Construct, produced nearly a standard deviation effect ($d = .92$; $M = 2.31$) over the Explore-Construct condition ($F(1,48) = 9.38$, $p < 0.01$) on tower-building, even though children were doing less tower-building overall (see Figure 3.4b). Learning of building/engineering skills, as measured by tower-building improvement, was more than ten times larger for the Combined condition compared to the Explore-Construct condition (Yannier et al., 2020). This result suggests that some scientific inquiry guidance enhances engineering outcomes better than less-guided constructive exploration, especially when some construction is integrated with the inquiry guidance.

In addition to the tower-building measures, the explanation assessments also showed clear and consistent evidence of the benefit of guided discovery over constructive exploration. An ANCOVA test with post-test prediction explanation score as the outcome variable and pre-test prediction explanation score as a covariate showed that the Combined ($F(1,48) = 7.72$, $p < 0.01$, $d = 0.66$) and Guided-discovery ($F(1,47) = 9.74$, $p < 0.01$, $d = 0.75$) conditions learned to provide better

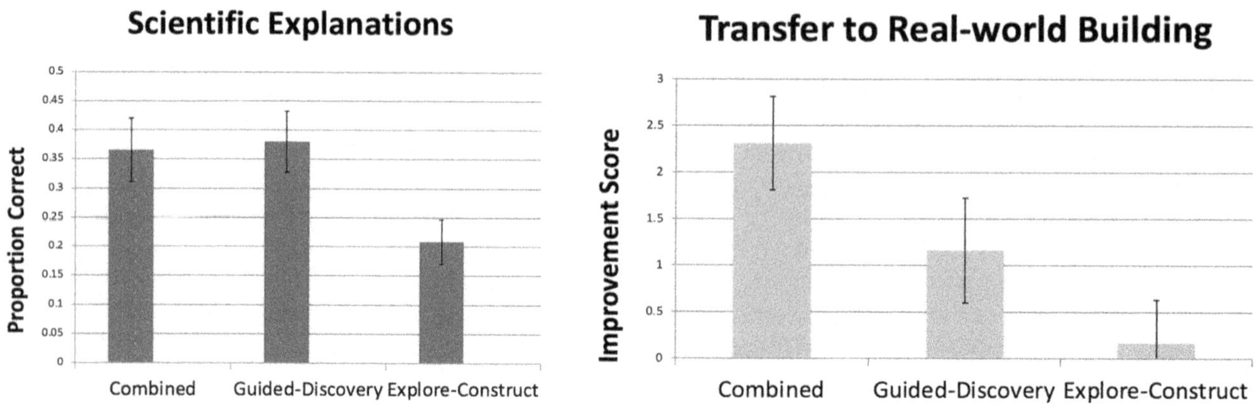

FIGURE 3.4
(a) Students showed significantly better learning gains (as measured by scientific explanations) in the AI-enhanced Guided-discovery and Combined conditions compared to Explore-construct condition. (b) The Combined condition where children practiced AI-guided inquiry in addition to building/construction with interactive feedback showed better transfer to real-world building, compared to the Explore-construct condition even though children practiced more building in this condition.

explanations of the scientific phenomena compared to the Explore-construct condition. Their improvement on the explanation tests was more than four times than for the Guided-discovery condition (M = 0.29 mean improvement) and the Combined condition (M = 0.22) compared to the Explore-construct condition (M = 0.06) (see Figure 3.4a).

Thus, AI-facilitated inquiry guidance not only improved children's understanding of the scientific principles, but also produced more transfer to a real-world constructive problem-solving task. Our results revealed that the guided discovery facilitated by the AI agent in our Intelligent Science Station helped children formulate better, more scientific theories of the physical phenomena they experience. We also found that children receiving guidance during inquiry are better able to learn to *apply* science in engineering tasks, particularly when guided discovery is interleaved with construction. On the other hand, exploration and construction through a minimally guided interactive experience resulted in relatively poor learning in general.

Constructive activities are believed to engage the students more through playful experimentation or tinkering with materials (Resnick et al., 2014). The results of our experiment suggest that children enjoyed the Guided-discovery activities with deliberate practice as much as the Explore-construct activities. Based on an enjoyment survey, there was no significant difference between average enjoyment scores (94% for Combined condition, 90% for the Guided-discovery condition, and 91% for the Explore-construct condition). Many children also had excited comments: 'I never thought something we do at school could be so much fun', 'I wish all my science classes were fun like this', 'Can you make one for my birthday? I can trade my toys', 'This is the best day of my life!', or 'I'm gonna be a builder when I grow up because a lot of these didn't fall!'

These results suggest that AI-supported Guided-discovery activities supporting deliberate practice have the potential to improve learning in mixed-reality environments without sacrificing enjoyment, especially when combined with construction and exploration activities.

3.6 Discussion and Next Steps

Work on Intelligent Science Stations has nicely demonstrated how Intelligent Tutoring System technology can be brought into the 3D world, helping children learn science by doing science. AI vision technology makes it possible to track students' progress as they engage in scientific experimentation and give them personalized feedback when needed. Random assignment experiments have demonstrated both that learning from 3D physical interactions is more effective than matched 2D flat-screen interactions and that learning with AI guidance on physical interactions is more effective than physical interactions without guidance.

In ongoing efforts, we are getting Intelligent Science Stations in use in some 20 school, museum, and after-school settings (and increasing), which reach hundreds of thousands of children and families yearly. The potential for broad dissemination is evidenced by the fact that many schools are ready to pay for R&D costs so as to have an Intelligent Science Station installation at their schools. At the time of writing, there are Intelligent Science Stations (aka Intelligent Science Exhibits) at public museums in two different large cities (see Figure 3.5). In addition to EarthShake, we now have Intelligent Science Stations working for two new scientific domains and associated physical apparatuses: the science of physical motion as revealed in experiments using rolling objects on ramps (called 'SmartRamps' – see Figure 3.6) and the science of torque as revealed in experiments with weights at different distances on a balance scale (called 'BalanceScale'). Our goal is to generalize Intelligent Science Stations to many different content areas.

These new applications not only provide for more interactive learning experiences, which schools and students highly desire, but also provide testbeds for advancing scientific investigations addressing innovation in AI in education technology and in advancing principles of STEM learning. One important AI research question is how best to develop new vision and perception algorithms in each new content domain. For SmartRamps and BalanceScale, the AI must recognize new object variations (e.g., rolling disks with smooth versus rough surfaces or hollow or solid cores; model cars with more weight in front versus back; different numbers of weights at different distances) and new actions (e.g., when an object reaches the end of a ramp before another; when a balance scale tips right, left, or balances). So far, the AI vision algorithms have been modified by hand, but we envision future research that enables the authoring of Intelligent Science Stations in new domains by demonstrating domain objects and important actions to a machine learning system. We suspect about ten examples per object will be sufficient, but we will need to test that in future work. Another important AI research question is adding more adaptivity to the

FIGURE 3.5
Intelligent Science Exhibits at the Children's Museum of Atlanta (left and center) and Carnegie Science Center in Pittsburgh (rightmost).

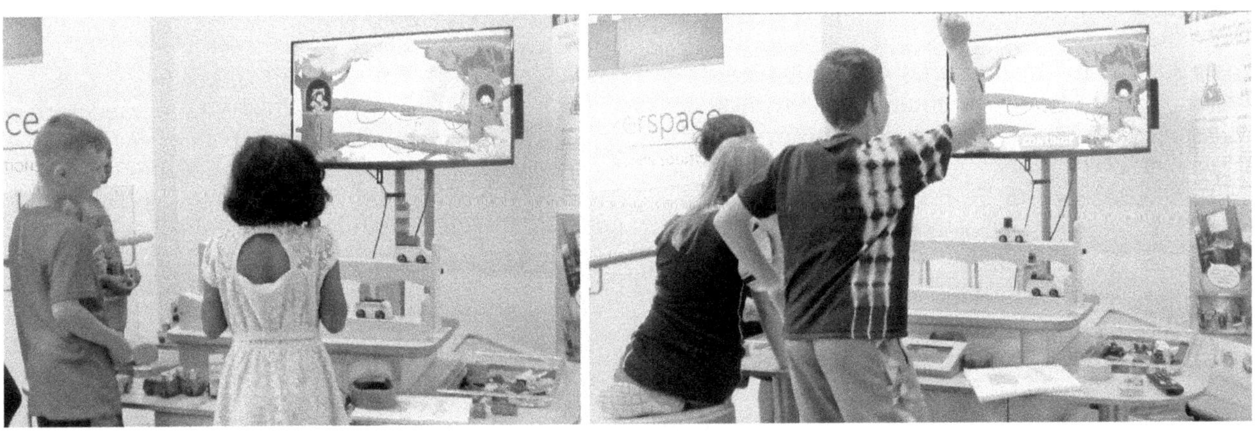

FIGURE 3.6
SmartRamps (our second Intelligent Science Station) to teach children concepts of force and motion.

system (adapting to their background) and extending the tracking of students' activity within well-structured experimental activities to tracking students' progress as they work on more open-ended engineering challenges (e.g., building the highest stable tower or the fastest car) and providing feedback and advice adapted to particulars of a child's distinctive solution. What kinds of adaptivity work best in a mixed-reality setting?

Intelligent Science Stations also provide a great context in which to explore learning science questions. For example, we have theoretical reasons and case study evidence to suggest the general STEM learning hypothesis that guided scientific inquiry (e.g., using the predict-observe-explain approach) in combination with engineering activities are best pursued together and in that order. The theory here is that more learning occurs during engineering activities when students have some fundamental scientific knowledge to try to apply and think about. We plan to investigate further what combinations of scientific inquiry and construction/problem-solving activities work better in a mixed-reality setting and in what ordering.

Another interesting open question is what is the underlying explanation for our replicated results that learning is much better with the 3D physical interactions in EarthShake than it is with tightly matched interactions on a 2D flat screen? One hypothesis is that children believe the physical 3D results, whereas they have some skepticism about videos on the flat screen. This hypothesis would suggest, for example, that 3D interaction with a physical but scale-model simulation of the solar system would not show a benefit over matched 2D interactions. Another hypothesis is that young children's perceptual systems are more efficient in processing 3D physical input than 2D virtual input. This hypothesis predicts the effect would go away for adults or even older children. It provides one possible explanation for why Klahr et al. (2007), whose studies were with grade 7 and 8 students rather than grade K-3 in ours, did not find a benefit for physical over virtual interaction (they did not have mixed-reality condition or a combination of physical and virtual as in our studies).[1] A third hypothesis is that children (or even adults) gain benefits of 'embodied cognition' (Glenberg, 2015) from the physical interaction. This hypothesis is consistent with our observation that children produce more relevant gestures in the physical/mixed-reality than virtual condition (Yannier et al., 2016), showing signs of mental visualization and deeper understanding (Hostetter and Alibali, 2008).

A final set of open questions involves how far or narrowly does learning with Intelligent Science Station transfer. Our results so far indicate much better learning of target scientific principles and use of them in more effective engineering and problem-solving. In the ongoing work, we investigate whether students may be acquiring more general scientific inquiry skills or gaining a greater interest in learning more science. These are promising possibilities worthy of future investigation.

As demonstrated by studies described above, Intelligent Science Stations have great potential to improve inquiry-based STEM learning as well as 21st-century skills like critical thinking, persistence, and collaboration by integrating AI into the physical world.

Note

1. Another possible explanation is that the Klahr, Triona, and Williams (2007) conditions did not have AI feedback for students as in our studies. They only had physical-only or virtual-only conditions but not a combination of both as in our mixed-reality system. Another study also showed that a combination of physical and virtual works better than physical and virtual alone for college students (Olympiou & Zacharias, 2012).

References

Akbulut, Y., & Cardak, C. S. (2012). Adaptive educational hypermedia accommodating learning styles: A content analysis of publications from 2000 to 2011. *Computers & Education, 58*(2), 835–842. https://doi.org/10.1016/j.compedu.2011.10.008.

Aleven, V., & Koedinger, K. R. (2000). The need for tutorial dialog to support self-explanation. Building dialogue systems for tutorial applications, papers of the 2000 AAAI fall symposium.

Aleven, V., McLaren, B. M., Roll, I., & Koedinger, K. R. (2006). Toward meta-cognitive tutoring: A model of help seeking with a cognitive tutor. *International Journal of Artificial Intelligence in Education, 16*, 101–128.

Aleven, V., McLaughlin, E. A., Glenn, R. A., & Koedinger, K. R. (2017). Instruction based on adaptive learning technologies. In R. E. Mayer & P. Alexander (Eds.), *Handbook of research on learning and instruction* (2nd ed., pp. 522–560). Routledge.

Anderson, J. R., Corbett, A. T., Koedinger, K. R., & Pelletier, R. (1995). Cognitive tutors: Lessons learned. *Journal of the Learning Sciences, 4*(2), 167–207.

Applefield, J. M., Huber, R., & Moallem, M. (2000). Constructivism in theory and practice: Toward a better understanding. *The High School Journal, 84*(2), 35–53.

Barker, J. E., Semenov, A. D., Michaelson, L., Provan, L. S., Snyder, H. R., & Munakata, Y. (2014). Less-structured time in children's daily lives predicts self-directed executive functioning. *Frontiers in Psychology, 5*, 593.

Baykal, G. E., Alaca, I. V., Yantaç, A. E., & Göksun, T. (2018). A review on complementary natures of tangible user interfaces (TUIs) and early spatial learning. *International Journal of Child-Computer Interaction, 16*, 104–113.

Cakir, M. (2008). Constructivist approaches to learning in science and their implications for science pedagogy: A literature review. *International Journal of Environmental & Science Education, 3*(4), 193–206.

Chase, C. C., Shemwell, J. T., & Schwartz, D. L. (2010). Explaining across contrasting cases for deep understanding in science: An example using interactive simulations. In S. Goldman & J. Pellegrino (Eds.), *Proceedings of the 9th International Conference of the Learning Sciences* (pp. 153–160). International Society of the Learning Sciences.

Chi, M., & Wylie, R. (2014). The ICAP framework: Linking cognitive engagement to active learning outcomes. *Educational Psychologist, 49*(4), 219–243.

Chi, M. T., Bassok, M., Lewis, M. W., Reimann, P., & Glaser, R. (1989). Self-explanations: How students study and use examples in learning to solve problems. *Cognitive Science, 13*(2), 145–182.

Corbett, A. T., & Anderson, J. R. (2001). Locus of feedback control in computer-based tutoring: Impact on learning rate, achievement and attitudes. In J. Jacko, A. Sears, M. Beaudouin-Lafon, & R. Jacob (Eds.), *CHI'2001 conference on human factors in computing systems* (pp. 245–252). ACM.

Glenberg, A. M. (2015). Few believe the world is flat: How embodiment is changing the scientific understanding of cognition. *Canadian Journal of Experimental Psychology/Revue canadienne de psychologie expérimentale, 69*(2), 165.

Harvel, C. (2010). *Guided discovery learning: Faith-based education that constructs: A creative dialogue between constructivism and faith-based education* (pp. 169–172). Wipf and Stock.

Holmes, N. G., Wieman, C. E., & Bonn, D. A. (2015). Teaching critical thinking. *Proceedings of the National Academy of Sciences of the United States of America, 112*(36), 11199–11204.

Hostetter, A. B., & Alibali, M. W. (2008). Visible embodiment: Gestures as simulated action. *Psychonomic Bulletin & Review, 15*(3), 495–514.

Jeffery-Clay, K. R. (1998). Constructivism in museums: How museums create meaningful learning environments. *Journal of Museum Education, 23*(1), 3–7.

Kafai, Y. B., & Resnick, M. (1996). *Constructionism in practice: Designing, thinking, and learning in a digital world*. Routledge.

Khan, T., Johnston, K., & Ophoff, J. (2019). The impact of an augmented reality application on learning motivation of students. *Advances in Human–Computer Interaction, 2019*, 1–14.

Kirschner, P., Sweller, J., & Clark, R. E. (2006). Why unguided learning does not work: An analysis of the failure of discovery learning, problem-based learning, experiential learning and inquiry-based learning. *Educational Psychologist, 41*(2), 75–86.

Klahr, D., Triona, L. M., & Williams, C. (2007). Hands on what? The relative effectiveness of physical versus virtual materials in an engineering design project by middle school children. *Journal of Research in Science Teaching, 44*(1), 183–203.

Koedinger, K. R., Brunskill, E., Baker, R. S., McLaughlin, E. A., & Stamper, J. (2013). New potentials for data-driven intelligent tutoring system development and optimization. *AI Magazine, 34*(3), 27–41.

Kolodner, J. L., Camp, P. J., Crismond, D., Fasse, B., Gray, J., Holbrook, J., Puntambekar, S. & Ryan, M. (2003). Problem-based learning meets case-based reasoning in the middle-school science classroom: Putting learning by design (tm) into practice. *Journal of the Learning Sciences, 12*(4), 495–547.

Marshall, P., Cheng, P. C. H., & Luckin, R. (2010, January). Tangibles in the balance: A discovery learning task with physical or graphical materials. In *Proceedings of the fourth international conference on tangible, embedded, and embodied interaction* (pp. 153–160).

Mayer, R. E., & Alexander, P. A. (Eds.). (2016). *Handbook of research on learning and instruction*. Taylor & Francis.

Olympiou, G., & Zacharia, Z. C. (2012). Blending physical and virtual manipulatives: An effort to improve students' conceptual understanding through science laboratory experimentation. *Science Education, 96*(1), 21–47.

Papavlasopoulou, S., Giannakos, M. N., & Jaccheri, L. (2017). Empirical studies on the maker movement, a promising approach to learning: A literature review. *Entertainment Computing, 18*, 57–78.

Papert, S. (1980). *Mindstorms: Children, computers, and powerful ideas*. Basic Books, Inc.

Papert, S., & Harel, I. (1991). Situating constructionism. *Constructionism, 36*(2), 1–11.

Resnick, M. et al. (1998). Digital manipulatives. *CHI '98: Proceedings of the SIGCHI Conference on Human Factors in Computing Systems*, 281–287. http://doi.org/10.1145/274644.274684

Resnick, M. (2014). Give P's a chance: Projects, peers, passion, play. Constructionism and creativity: Proceedings of the third international constructionism conference. Austrian Computer Society, Vienna.

Ryokai, K., & Cassell, J. (1999). StoryMat: A play space for collaborative storytelling. In *CHI '99 extended abstracts on human factors in computing systems* (pp. 272–273). CHI EA '99. ACM.

Schwartz, D. L., Tsang, J. M., & Blair, K. P. (2016). *The ABCs of how we learn: 26 scientifically proven approaches, how they work, and when to use them*. WW Norton & Company.

Tseng, T., Bryant, C., & Blikstein, P. (2011). Collaboration through documentation: Automated capturing of tangible constructions to support engineering design. In *Proceedings of the 10th international conference on interaction design and children* (pp. 118–126). IDC '11. ACM.

Wang, L., & Ju, H. (2008, December). A robust blob detection and delineation method. In *2008 international workshop on education technology and training & 2008 international workshop on geoscience and remote sensing* (Vol. 1, pp. 827–830). IEEE.

White, R., & Gunstone, R. (1992). Prediction-observation-explanation. *Probing Understanding*, *4*, 44–64.

Yannier, N., Hudson, S. E., Wiese, E. S., & Koedinger, K. R. (2016). Adding physical objects to an interactive game improves learning and enjoyment: Evidence from EarthShake. *ACM Transactions on Computer-Human Interaction (TOCHI)*, *23*(4), 1–31.

Yannier, N., Hudson, S. E., & Koedinger, K. R. (2020). Active learning is about more than hands-on: A mixed-reality AI system to support STEM education. *International Journal of Artificial Intelligence in Education*, *30*(1), 74–96.

Yannier, N., Koedinger, K. R., & Hudson, S. E. (2015, April). Learning from mixed-reality games: Is shaking a tablet as effective as physical observation? In *Proceedings of the 33rd annual ACM conference on human factors in computing systems* (pp. 1045–1054).

4

Adaptive Support for Representational Competencies during Technology-Based Problem-Solving in STEM

Martina A. Rau

CONTENTS

4.1 Introduction 51
4.2 Literature Review 52
4.2.1 Sense-Making Competencies 52
4.2.2 Perceptual Fluency 53
4.3 Empirical Studies 54
4.3.1 Combining Sense-Making Support and Perceptual-Fluency Support 54
4.3.2 How Should Sense-Making and Perceptual-Induction Activities Be Sequenced? 55
4.3.3 Should Representational-Competency Supports Adapt to Students' Current Knowledge Level? 56
4.4 Discussion 57
Note 58
References 58

4.1 Introduction

In most STEM domains, visual representations play an important role in learning and problem-solving (Ainsworth, 2008; NRC, 2006). For example, chemistry students typically encounter the visual representations shown in Figure 4.1. Visual representations have the potential to enhance students' learning because they can make visuospatial concepts accessible to students (Ainsworth, 2006). However, visual representations can also pose an obstacle to students' learning because many students have difficulties interpreting them (Ainsworth et al., 1998; Rau et al., 2014). Instructors often fail to realize that their students have difficulties working with visual representations (Dreher & Kuntze, 2015).

To support students' learning with visual representations, prior research has investigated which competencies students need to use representations in a way that supports their learning (Ainsworth, 2006; Gilbert, 2008; Rau, 2017a). Helping students acquire representational competencies is an important educational goal in most STEM domains because the ability to use representations for problem-solving is part of scientific and professional practices (Fan, 2015; Kozma & Russell, 2005; Latour, 1986; NRC, 2006).

However, providing instructional supports for representational competencies is not straightforward. As detailed below, separate lines of research have given rise to different types of representational-competency supports. This leads to the question of whether students need multiple types of supports, and – if so – how to sequence these supports to maximize students' learning.

This chapter provides an overview of a research program that addressed these questions. Throughout, the goal was to support students' representational competencies while students were using the representations to learn domain knowledge through problem-solving. For example, while students were solving chemistry problems with the visual representations in Figure 4.1, the goal was to provide support for students to understand these representations in the context of the given problem. This approach meant to ensure that representational-competency supports did not require additional instructional time but instead could be integrated with existing instructional problem-solving activities. Further, the ultimate goal was to enhance students' learning of domain knowledge. The acquisition of representational competencies was considered a means to an end, namely, to enable students to use the representations to learn the domain-relevant ideas they visualize.

In the following section, an overview of prior research on representational competencies and representational-competency supports is provided.

DOI: 10.1201/9781003181187-5

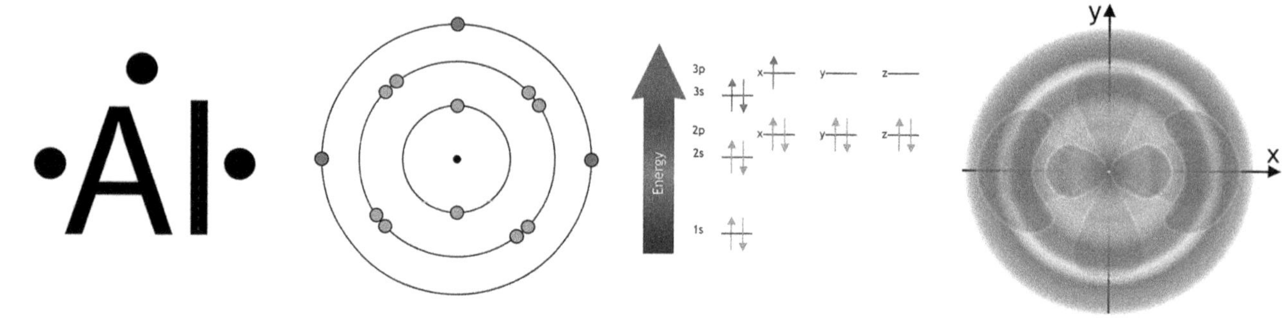

FIGURE 4.1
Visual representations of atoms: Lewis structure, shell model, atomic orbital energy diagram, atomic orbital diagram.

Then, a series of empirical studies is presented that investigated (1) whether different supports for representational competencies should be combined to enhance students' learning, (2) how these supports should be sequenced, and – building on findings that effects depended on students' prior knowledge, and (3) whether representational-competency supports should be adaptive.

4.2 Literature Review

Two mostly separate lines of research have identified two broad types of representational competencies (Rau, 2017a).

4.2.1 Sense-Making Competencies

Most prior research has focused on students' ability to conceptually make sense of visual representations (Ainsworth, 2006; Schnotz, 2014). Sense-making competencies describe students' ability to explain how visual features of the representations depict domain-relevant concepts (Ainsworth, 2006; Rau, 2017a). For example, a student should be able to explain that the dots in the Lewis structure in Figure 4.1 show electrons, which is a subatomic particle with a negative charge. Further, sense-making competencies involve the ability to explain connections among multiple representations based on visual features that show corresponding concepts (Ainsworth, 2006; Rau, 2017a). For example, students should be able to explain that the dots in the Lewis structure and the dots on the outermost shell of the shell model (see Figure 4.1) both correspond to valence electrons, which can participate in chemical bonding. Making sense of connections also means that students should be able to explain differences between the representations (Ainsworth, 2006; Rau, 2017a). For example, students should be able to explain that the Lewis structure does not show core electrons, whereas the shell model does (see Figure 4.1).

Students acquire sense-making competencies through explanation-based sense-making processes (Chi et al., 1989; Koedinger et al., 2012). These processes are verbally mediated and require willful cognitive effort (Chi et al., 1989; Chi et al., 1994). Hence, to support students' sense-making competencies, instructional activities ask students to verbally explain how visual features map to concepts (Berthold & Renkl, 2009; Seufert & Brünken, 2006), for instance through self-explanation prompts (Berthold et al., 2008). Further, students should become active in constructing such mappings (Bodemer & Faust, 2006; Chi, 2009). Finally, because students often rely on surface features (e.g., color that is irrelevant for establishing mappings between representations) rather than structural features (i.e., features that carry conceptual meaning), students should receive assistance in establishing such mappings (Ainsworth et al., 1998; Rau et al., 2014).

Figure 4.2 provides an example from Chem Tutor, an intelligent tutoring system for undergraduate chemistry (Rau, 2017b), which provides instructional activity with sense-making support (henceforth sense-making activity). Students are given a visual representation of an atom and asked to construct a second representation of the same atom. Then, students complete fill-in-the-gap sentences that serve as prompts to self-explain how the two representations show conceptual information about atomic structure. These prompts ask students to explain similarities between the representations (e.g., both the energy diagram and the orbital diagram show orbitals), and differences (e.g., the energy diagram shows how many electrons occupy the orbitals, but the orbital diagram does not). Students receive conceptual feedback on their responses, such that if they make a mistake, they are given a detailed explanation about why their response is incorrect. Further, students receive guidance on each step with conceptual information about the given question.

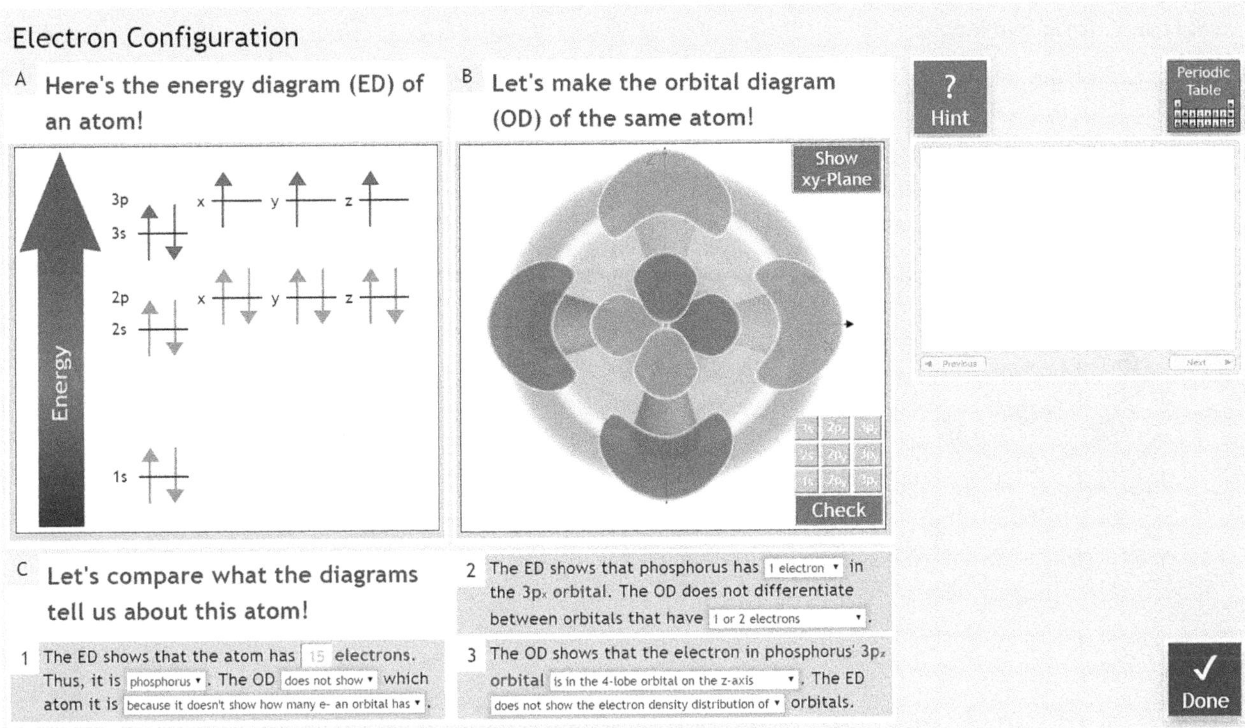

FIGURE 4.2
Example sense-making activity. Students are given one visual representation (an energy diagram, top left) and construct a second visual representation of the same atom (an orbital diagram, top right). Then, they complete fill-in-the-gap prompts that ask them to explain similarities and differences between the two representations.

4.2.2 Perceptual Fluency

A second type of representational competency describes students' ability to fluently perceive information in visual representations (Kellman & Massey, 2013; Rau, 2017a). Perceptual fluency describes a student's ability to efficiently extract meaningful information from visual representations based on perceptual cues (Kellman & Massey, 2013; Massey et al., 2011). For example, students should be able to see at a glance that the Lewis structure in Figure 4.1 shows aluminum. Further, perceptual fluency involves the ability to quickly translate among multiple representations, without cognitive effort, as fluently as if they were bilinguals translating between languages (Massey et al., 2011; Rau, 2017a). For example, students should be able to establish at a glance that the shell model and the energy diagram in Figure 4.1 show the same element. Perceptual fluency plays an important role in students' learning of domain knowledge because if frees cognitive resources that students can invest in higher-order thinking (Goldstone & Barsalou, 1998; Richman et al., 1996).

Students acquire perceptual fluency through implicit, inductive, nonverbal processes (Gibson, 2000; Goldstone & Barsalou, 1998). These perceptual-induction processes are often unintentional and unconscious (Frensch & Rünger, 2003; Shanks, 2005). Students inductively learn to recognize visual patterns through experience with a variety of visual representations (Kellman & Massey, 2013). Perceptual processing cannot easily be explained; instead, verbal explanations have been shown to interfere with perceptual processing (Chin & Schooler, 2008; Schooler & Engstler-Schooler, 1990). Therefore, instructional activities that support perceptual fluency are designed very differently than sense-making activities (see Kellman et al., 2008; Kellman et al., 2010). They expose students to numerous example visual representations. The example representations vary irrelevant features and repeat relevant features. Students are asked to solve quick classification tasks that require them to pay attention to the relevant features. Throughout, students are encouraged not to explain their reasoning but to rely on their perceptual intuitions. They receive immediate, nonverbal feedback on their task performance.

Figure 4.3 shows an example Chem Tutor activity that provides support for perceptual fluency (henceforth perceptual-induction activity). Students are prompted to work on the tasks quickly and without overthinking their answer. Given one visual representation, the task is to select one of four other representations that shows the same element. The four choice

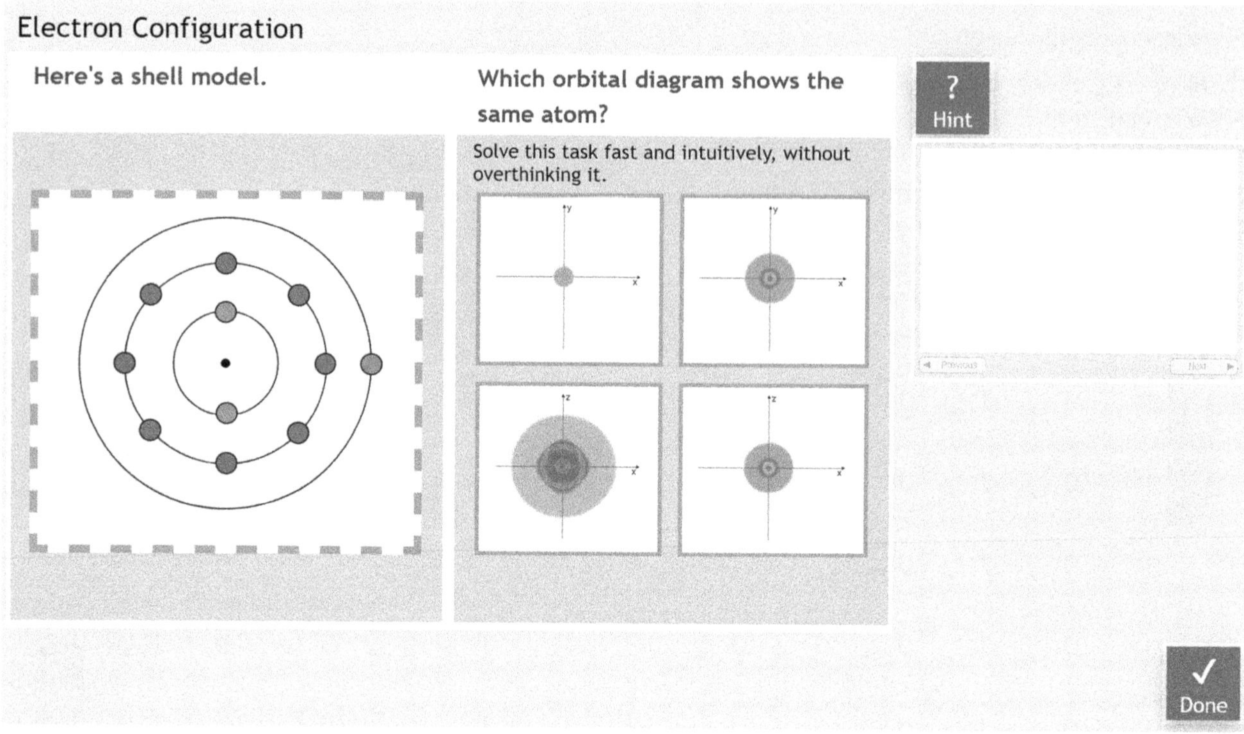

FIGURE 4.3
Example perceptual-induction activity. Students are given one visual representation (a shell model, left) and are asked to choose one of four other visual representations (four orbital diagrams, right) while making their selection fast and intuitively, without overthinking it.

options present contrasting cases by varying visual features that students should learn to use to identify elements. Students receive only correctness feedback but no explanation about why their answer is correct or incorrect. If they provide an incorrect answer, they can try again until they are correct.

4.3 Empirical Studies

The brief literature review shows that it established two different types of representational competencies that are supported through different types of instructional activities. This yields the question: To optimally support students' ability to use visual representations to learn domain knowledge, whether they need support for both sense-making competencies *and* perceptual fluency, or whether one type of instructional support is sufficient? The goal of the first empirical study, presented in the following, was to address this question.

4.3.1 Combining Sense-Making Support and Perceptual-Fluency Support

Rau and Wu (2018) investigated the question: Is instruction most effective if it combines support for sense-making competencies and perceptual fluency? To address this question, $N = 117$ undergraduate students without prior knowledge about the targeted chemistry content were recruited. The experiment used a 2 × 2 design to compare effects of providing students with sense-making activities (yes/no) and of providing perceptual-induction activities (yes/no). Hence, students received (1) regular instructional activities without support for representational competencies, (2) regular activities and sense-making activities, (3) regular activities and perceptual-induction activities, or (4) regular activities, sense-making activities, and perceptual-induction activities. Time on task was held constant by adjusting the number of activities such that they took the same amount of time across conditions (e.g., the sense-making and perceptual-induction condition would receive sense-making activities than the sense-making only condition, and fewer perceptual-induction activities than the perceptual-induction only condition).

Students worked with a Chem Tutor unit that introduced them to atomic structure. Four different versions of Chem Tutor were created that corresponded to the four experimental conditions. Students were then randomly assigned to one of the four conditions. Prior to working on the Chem Tutor activities, students received a pretest that assessed their knowledge of chemical bonding. After finishing the Chem Tutor activities, they took an equivalent posttest. The experiment was conducted in a research lab.

Results showed that students in the combined sense-making *and* perceptual-induction condition had higher learning gains than students in the other conditions. The sense-making only condition was no more effective than the control condition. Students in the perceptual-induction only condition had significantly lower learning gains than students in the control condition that received no representational-competency supports.

Additional analyses using eye-tracking showed that both types of representational-competency supports increased students' switching between visual representations, which has been associated with integrative mental processes. Cued retrospective reports revealed that switching between representations was associated with different cognitive processes, depending on which type of representational-competency support students had received. Students who received sense-making activities showed higher-quality reasoning about differences between the visual representations. By contrast, students who received perceptual-induction activities were better able to reason about similarities between the representations.

Rau (2018a) attempted to replicate these findings in the field. Two experiments were conducted with N = 95 students and N = 113 students, each in an introductory chemistry course. Students used Chem Tutor for a homework assignment and were randomly assigned to a version of Chem Tutor that corresponded to their condition. Students' learning of domain knowledge was assessed as before. Results differed somewhat from the Rau and Wu (2018) findings and showed that students' benefit from sense-making and perceptual-induction activities depended on their prior knowledge: students needed only a preliminary amount of knowledge about the targeted concepts before they could benefit from sense-making activities. Students needed considerable knowledge about the target concepts to benefit from perceptual-induction activities.

In sum, Rau and Wu (2018) found that students' learning of domain knowledge is enhanced *only* by a combination of support for sense-making competencies *and* perceptual fluency. It appears that the different types of support enhance complementary cognitive processes, both of which were necessary to allow students to successfully learn with multiple visual representations. Rau (2018a) found that students' needs for these different types of support depend on their prior domain knowledge.

If sense-making and perceptual-induction activities need to be combined to enhance students' learning with visual representations (Rau & Wu, 2018), then this yields the question of how these activities should be sequenced. Further, if students with low prior knowledge benefit from sense-making activities and students with high prior knowledge benefit from perceptual-induction activities (Rau, 2018a), this yields the hypothesis that a sequence that provides sense-making activities followed by perceptual-induction activities may be most effective.

4.3.2 How Should Sense-Making and Perceptual-Induction Activities Be Sequenced?

Rau (2018b) conducted three experiments that investigated the question: Which sequence of sense-making and perceptual-induction activities is most effective at enhancing students' learning of domain knowledge? Experiment 1 recruited students from a chemistry course for non-science majors (N = 48), and Experiments 2 and 3 recruited students from a chemistry course for science majors (N = 61 in Experiment 2; N = 607 in Experiment 3). All experiments were conducted with Chem Tutor (using a unit on atomic structure in Experiments 1 and 2, and a unit on chemical bonding in Experiment 3). Students first received regular activities, and then were randomly assigned to receive sense-making activities followed by perceptual-induction activities, or perceptual-induction activities followed by sense-making activities.[1] Students received pretests and posttests assessing their learning of the targeted domain knowledge. Additionally, causal path analysis tested whether the errors students made while working on the Chem Tutor activities mediated condition effects on learning gains.

The causal path analysis revealed multiple mechanisms that explained a complex interplay among the sequences of activities and students' prior knowledge. Sense-making activities enhanced students' benefit from subsequent perceptual-induction activities (sense-enhancement mechanism); but the reverse was also true: perceptual-induction activities helped students learn from subsequent sense-making activities (perceptual-enhancement mechanism). Students' prior knowledge moderated the strength of these mechanisms. For students with high prior knowledge, the sense-enhancement mechanism was stronger than the perceptual-enhancement mechanism. Consequently, these students showed higher learning gains if they received sense-making activities followed by perceptual-induction activities.

By contrast, students with low prior knowledge showed higher learning gains if they received perceptual-induction activities followed by sense-making activities. However, the causal path analysis revealed that this effect could not be attributed to the perceptual-enhancement mechanism. Rather, causal path analysis also identified two mechanisms that described costs of switching between the different types of activities. Switching from sense-making activities to perceptual-induction activities interfered students' benefit from the perceptual-induction activities – likely because students had to shift gears and engage in a different type of learning process, which is cognitively demanding (sense-interference mechanism). But the reverse was also true: switching from perceptual-induction activities to sense-making activities interfered with students' benefit from the sense-making activities (perceptual-interference mechanism). For students with low prior knowledge, the sense-interference mechanism was stronger than the perceptual-interference mechanism. Consequently, these students showed higher learning gains if they received perceptual-induction activities followed by sense-making activities.

This latter finding aligns with the results from the previously mentioned Experiment 2 by Rau (2018a), which had compared the sequence in which sense-making and perceptual-induction activities were combined to conditions that received only regular activities, regular and sense-making activities, or regular and perceptual-induction activities. If students' benefit from sense-making and perceptual-induction activities varies with their prior knowledge, then this explains why the combined conditions were not most effective. It seems that over the course of a longer learning period, the combination of sense-making and perceptual-induction activities is effective because the different activities provide complementary benefits (Rau & Wu, 2018). But during a shorter intervention, which type of activity is most effective depends on students' knowledge level at that particular point in time (Rau, 2018a, 2018b).

However, the finding that students with low prior knowledge benefited from receiving perceptual-fluency activities first is somewhat contradictory compared to Experiment 2 by Rau (2018a), which found that students with high prior knowledge benefited from receiving perceptual-fluency activities first. Note that Experiment 2 in Rau (2018a) was conducted at the end of the semester of the respective chemistry course, whereas the experiments in Rau (2018b) were conducted at the beginning of the semester. Therefore, a possible explanation is that experienced students may benefit from receiving perceptual-induction activities because they already have sense-making competencies that allow them to benefit from perceptual-induction activities. Indeed, secondary analyses of the log data from Rau (2018b)'s Experiment 2 confirmed this interpretation. Students with high prior knowledge exhibited a perceptual-enhancement mechanism that was stronger than the sense-enhancement mechanism.

Altogether, it is difficult to compare findings across these experiments because they were cross-sectional; that is, different students had different levels of prior knowledge in each experiment. To address this limitation, a longitudinal experiment by Rau and Zahn (2018) compared sequences of sense-making and perceptual-induction activities across six weeks. N = 71 undergraduate chemistry students participated in the experiment, using an expanded Chem Tutor curriculum on atomic structure as part of their chemistry course. They were randomly assigned to receive sense-making activities followed by perceptual-induction activities or vice versa in the Chem Tutor assignment for the given week of their course. Results suggested a learning progression whereby the perceptual-sense sequence is most effective at the beginning (in line with Rau, 2018b's findings for students with low prior knowledge). In an intermediate phase, the sense-perceptual sequence is most effective (in line with Rau, 2018b's findings for students with high prior knowledge). In a final phase, the perceptual-sense sequence is most effective (in line with Rau, 2018a's Experiment 2 findings for students with high prior knowledge).

Taken together, findings by Rau (2018a, 2018b) show that students' prior knowledge predicts which type of representational-competency support they benefit from. Findings by Rau and Zahn (2018) show that students need different types of representational-competency supports as their knowledge level changes during a longer learning period. Given that all students learn at different rates, it seems reasonable to hypothesize that adapting representational-competency supports to the individual student's current knowledge level might be most effective at enhancing their learning of domain knowledge.

4.3.3 Should Representational-Competency Supports Adapt to Students' Current Knowledge Level?

Rau and colleagues (2021) conducted an experiment that investigated the question: Does adaptive support for sense-making competencies and perceptual fluency increase students' learning of domain knowledge compared to static support for these representational competencies? The first step in addressing this question was to develop an algorithm that predicts, based on an assessment of a student's current knowledge level, whether she would benefit from regular activities, sense-making activities, or perceptual-induction activities. To this end, data was collected from N = 129 undergraduate students who were enrolled in a chemistry course. Over the course of ten weeks, they were randomly assigned to one of five versions of Chem Tutor, which provided (1) only regular activities, (2) regular activities followed by sense-making activities, (3) regular activities followed by perceptual-induction activities, (4) regular activities followed by sense-making and then perceptual-induction activities, or (5) regular activities followed by perceptual-induction and then sense-making activities. For all conditions, the first two activities were identical regular activities. This experimental design was implemented for each week of the course. Each week, students received a pretest and a posttest of the content covered in the Chem Tutor unit assigned for that week. Data analysis then used linear regression to predict, based on students' performance on the problem-solving steps in the first two regular activities that were identical across conditions, whether students would benefit from sense-making activities and/or perceptual-induction activities. This analysis yielded a set of rules. For example, IF students make a mistake on a step that asks about the filling order of orbitals (see Figure 4.4), THEN they should receive sense-making activities for this unit.

The next step was to compare an adaptive version of Chem Tutor, which assigned students to sense-making and perceptual-induction activities based on the identified rules and which was more effective than a static version of Chem Tutor. To this end, $N = 44$ undergraduate students who were enrolled in a chemistry course were randomly assigned to the adaptive or the static version of Chem Tutor for ten weeks. The static version provided sense-making and perceptual-induction activities for each unit, following the results by Rau and Wu (2018). In each week, students took a pretest and posttest of the content covered in the Chem Tutor unit assigned for that week.

Results showed a significant advantage of the adaptive over the static version. Additional analyses of student reports showed that the adaptive version reduced students' confusion about the visual representations they encountered in Chem Tutor. Further analyses of the Chem Tutor log data revealed that students in the adaptive version made fewer errors during problem-solving, particularly on steps that involved interactions with visual representations. Finally, an inspection of how the adaptive version of Chem Tutor assigned students to sense-making and perceptual-induction activities revealed patterns similar to the learning progression identified by Rau and Zahn (2018): students typically received sense-making activities or only regular activities in the early units and received perceptual-induction activities only after having received sense-making activities. Further, most students received perceptual-induction activities at two points during the ten-week sequence. With one exception, all students received both sense-making and perceptual-induction activities over the course of the ten units, in alignment with Rau and Wu (2018). However, in alignment with Rau (2018a), no student received sense-making and perceptual-induction activities for the same unit.

4.4 Discussion

Taken together, this research demonstrates that support for representational competencies can

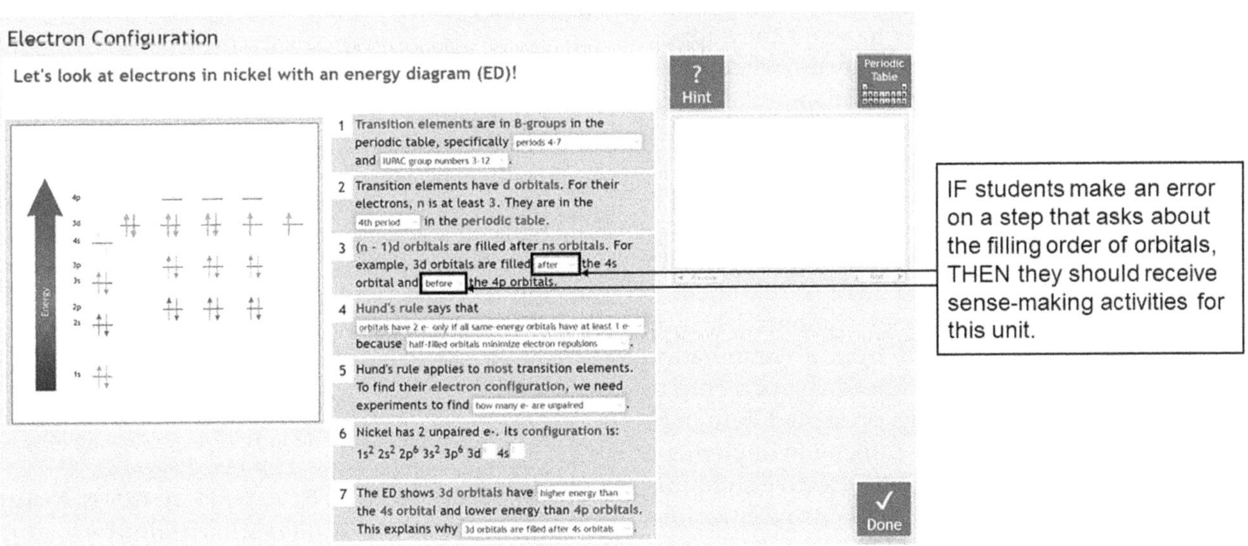

FIGURE 4.4
Example of an IF-THEN rule that specifies a condition for students receiving sense-making activities for the given unit.

significantly enhance students' learning of domain knowledge. However, supporting representational competencies is not straightforward. Students need different types of support for different types of representational competencies, and which type of support is most beneficial for their learning of domain knowledge depends on their current knowledge level. The final experiment presented here shows that adaptively assigning students to representational-competency supports based on their current knowledge level is effective.

These findings expand prior research in multiple ways. First, prior research had focused either only on sense-making competencies or on perceptual fluency, but had not investigated whether students need support for both types of competencies. This research shows that students need support for both types of competencies. Second, prior research had not investigated the interplay between sense-making competencies and perceptual fluency. This research shows that these two types of competencies mutually enhance one another. However, switching between activities that support sense-making competencies and activities that support perceptual fluency is cognitively demanding, which is problematic for students with low prior knowledge. Third, prior research had not investigated the relationship between students' prior knowledge and their benefit from representational-competency supports. This research shows that decisions about how to support representational competencies should depend on students' prior knowledge level. Finally, while prior research has demonstrated the effectiveness of adaptive tutoring, these studies adapted students' assignment of different problem-solving tasks that had different difficulty levels (e.g., Koedinger & Corbett, 2006; VanLehn, 2011). The novel approach in the research presented here was to adapt the support for specific representational competencies, whereas the content students learned remained the same.

It is important to note that research on the interplay between sense-making competencies and perceptual fluency is nascent, and that consequently it is too early to make generalizable claims about student populations, visual representations, or content not investigated in the research presented here. While there is evidence from elementary-school students learning fractions that they also benefit from a combination of sense-making activities followed by perceptual-induction activities, the findings presented here indicate that such combinations are difficult to implement. Future research should investigate effective combinations for students from various domains with various levels of background knowledge to establish domain-general principles about how to combine sense-making and perceptual-induction activities. Further, future research should unpack dependencies on prior knowledge. Specifically, in the domain of chemistry, knowledge about the visual representations is difficult to separate from domain knowledge because visual representations are part of the disciplinary discourse that is used to communicate about domain-relevant ideas. However, it may be possible to distinguish domain knowledge from knowledge about visual representations in other domains to investigate which aspects of prior knowledge representational-competency supports should adapt to.

Given the prevalence of visual representations in STEM and that students who lack representational competencies are at a severe disadvantage in these fields, the finding that adaptive support for representational competencies enhances students' learning of domain knowledge will likely have high impact on STEM instruction.

Note

1. Experiment 3 additionally varied how frequently students switched between sense-making and perceptual-induction activities, as reported in Rau, M.A. (2018b). Sequencing support for sense-making and perceptual induction of connections among multiple visual representations. *Journal of Educational Psychology, 110*(6), 811–833. https://doi.org/http://dx.doi.org/10.1037/edu0000229.

References

Ainsworth, S. (2006). DeFT: A conceptual framework for considering learning with multiple representations. *Learning and Instruction, 16*(3), 183–198. https://doi.org/10.1016/j.learninstruc.2006.03.001

Ainsworth, S. (2008). The educational value of multiple-representations when learning complex scientific concepts. In J. K. Gilbert, M. Reiner, & A. Nakama (Eds.), *Visualization: Theory and practice in science education* (pp. 191–208). Springer. https://doi.org/10.1007/978-1-4020-5267-5_9

Ainsworth, S., Bibby, P., & Wood, D. (1998). Analysing the costs and benefits of multi-representational learning environments. In M. W. van Someren, P. Reimann, H. P. A. Boshuizen, & T. de Jong (Eds.), *Learning with multiple representations* (pp. 120–134). Pergamon.

Berthold, K., Eysink, T. H. S., & Renkl, A. (2008). Assisting self-explanation prompts are more effective than open prompts when learning with multiple representations. *Instructional Science, 27*(4), 345–363. https://doi.org/10.1007/s11251-008-9051-z

Berthold, K., & Renkl, A. (2009). Instructional aids to support a conceptual understanding of multiple representations. *Journal of Educational Research, 101*(1), 70–87. https://doi.org/dx.doi.org/10.1037/a0013247

Bodemer, D., & Faust, U. (2006). External and mental referencing of multiple representations. *Computers in Human Behavior, 22*(1), 27–42. https://doi.org/10.1016/j.chb.2005.01.005

Chi, M. T. (2009). Active-constructive-interactive: A conceptual framework for differentiating learning activities. *Topics in Cognitive Science, 1*(1), 73–105.

Chi, M. T., Bassok, M., Lewis, M. W., Reimann, P., & Glaser, R. (1989). Self-explanations: How students study and use examples in learning to solve problems. *Cognitive Science, 13*(2), 145–182. https://doi.org/10.1016/0364-0213(89)90002-5

Chi, M. T., de Leeuw, N., Chiu, M. H., & Lavancher, C. (1994). Eliciting self-explanations improves understanding. *Cognitive Science, 18*(3), 439–477. https://doi.org/10.1016/0364-0213(94)90016-7

Chin, J., & Schooler, J. (2008). Why do words hurt? Content, process, and criterion shift accounts of verbal overshadowing. *European Journal of Cognitive Psychology, 20*(3), 396–413.

Dreher, A., & Kuntze, S. (2015). Teachers facing the dilemma of multiple representations being aid and obstacle for learning: Evaluations of tasks and theme-specific noticing. *Journal für Mathematik-Didaktik, 36*(1), 23–44. https://doi.org/10.1007/s13138-014-0068-3

Fan, J. E. (2015). Drawing to learn: How producing graphical representations enhances scientific thinking. *Translational Issues in Psychological Science, 1*(2), 170–181. https://doi.org/10.1037/tps0000037

Frensch, R., & Rünger, D. (2003). Implicit learning. *Current Directions in Psychological Science, 12*(1), 13–18. https://doi.org/10.1111/1467-8721.01213

Gibson, E. J. (2000). Perceptual learning in development: Some basic concepts. *Ecological Psychology, 12*(4), 295–302. https://doi.org/10.1207/S15326969ECO1204_04

Gilbert, J. K. (2008). Visualization: An emergent field of practice and inquiry in science education. In J. K. Gilbert, M. Reiner, & M. B. Nakhleh (Eds.), *Visualization: Theory and practice in science education* (Vol. 3, pp. 3–24). Springer.

Goldstone, R. L., & Barsalou, L. (1998). Reuniting perception and conception. *Cognition, 65*(2), 231–262. https://doi.org/10.1016/S0010-0277(97)00047-4

http://www.redi-bw.de/db/ebsco.php/search.ebscohost.com/login.aspx?direct=true&db=psyh&AN=2006-07157-005&site=ehost-live

Kellman, P. J., & Massey, C. M. (2013). Perceptual learning, cognition, and expertise. In B. H. Ross (Ed.), *The psychology of learning and motivation* (Vol. 558, pp. 117–165). Elsevier Academic Press.

Kellman, P. J., Massey, C. M., Roth, Z., Burke, T., Zucker, J., Saw, A., Aguero, K. E., & Wise, J. (2008). Perceptual learning and the technology of expertise: Studies in fraction learning and algebra. *Pragmatics and Cognition, 16*(2), 356–405. https://doi.org/10.1075/pc.16.2.07kel

Kellman, P. J., Massey, C. M., & Son, J. Y. (2010). Perceptual learning modules in mathematics: Enhancing students' pattern recognition, structure extraction, and fluency. *Topics in Cognitive Science, 2*(2), 285–305. https://doi.org/10.1111/j.1756-8765.2009.01053.x

Koedinger, K. R., & Corbett, A. (2006). Cognitive tutors: Technology bringing learning sciences to the classroom. In R. K. Sawyer (Ed.), *The Cambridge handbook of the learning sciences* (1st ed., pp. 61–77). Cambridge University Press. koedinger@cmu.edu

Koedinger, K. R., Corbett, A. T., & Perfetti, C. (2012). The knowledge-learning-instruction Framework: Bridging the science-practice chasm to enhance robust student learning. *Cognitive Science, 36*(5), 757–798. https://doi.org/10.1111/j.1551-6709.2012.01245.x

Kozma, R., & Russell, J. (2005). Students becoming chemists: Developing representational competence. In J. Gilbert (Ed.), *Visualization in science education* (pp. 121–145). Springer.

Latour, B. (1986). Visualization and cognition: Thinking with eyes and hands. In H. Kuklick (Ed.), *Knowledge and society: Studies in the sociology of culture past and present* (Vol. 6, pp. 1–40). Jai Press.

Massey, C. M., Kellman, P. J., Roth, Z., & Burke, T. (2011). Perceptual learning and adaptive learning technology - Developing new approaches to mathematics learning in the classroom. In N. L. Stein & S. W. Raudenbush (Eds.), *Developmental cognitive science goes to school* (pp. 235–249). Routledge.

NRC. (2006). *Learning to think spatially*. National Academies Press.

Rau, M. A. (2017a). Conditions for the effectiveness of multiple visual representations in enhancing STEM learning. *Educational Psychology Review, 29*(4), 717–761. https://doi.org/10.1007/s10648-016-9365-3

Rau, M. A. (2017b). A framework for discipline-specific grounding of educational technologies with multiple visual representations. *IEEE Transactions on Learning Technologies, 10*(3), 290–305. https://doi.org/10.1109/TLT.2016.2623303

Rau, M. A. (2018a). Making connections among multiple visual representations: How do sense-making competencies and perceptual fluency relate to learning of chemistry knowledge? *Instructional Science, 46*(2), 209–243. https://doi.org/10.1007/s11251-017-9431-3

Rau, M. A. (2018b). Sequencing support for sense making and perceptual induction of connections among multiple visual representations. *Journal of Educational Psychology, 110*(6), 811–833. https://doi.org/10.1037/edu0000229

Rau, M. A., Aleven, V., Rummel, N., & Pardos, Z. (2014). How should intelligent tutoring systems sequence multiple graphical representations of fractions? A multi-methods study. *International Journal of Artificial Intelligence in Education, 24*(2), 125–161. https://doi.org/10.1007/s40593-013-0011-7

Rau, M. A., & Wu, S. P. W. (2018). Support for sense-making processes and inductive processes in connection-making among multiple visual representations. *Cognition and Instruction, 36*(4), 361–395. https://doi.org/10.1080/07370008.2018.1494179

Rau, M. A., & Zahn, M. (2018). Sequencing support for sense making and perceptual fluency with visual representations: Is there a learning progression? In J. Kay & R. Luckin (Eds.), *Rethinking learning in the digital age: Making the learning sciences count (ICLS) 2018* (Vol. 1, pp. 264–271). International Society of the Learning Sciences.

Rau, M. A., Zahn, M., Misback, E., Herder, T., & Burstyn, J. (2021). Adaptive support for representational competencies during technology-based problem solving in chemistry. *Journal of the Learning Sciences, 30*(2), 163–203. https://doi.org/10.1080/10508406.2021.1888733

Richman, H. B., Gobet, F., Staszewski, J. J., & Simon, H. A. (1996). Perceptual and memory processes in the acquisition of expert performance: The EPAM model. In K. A. Ericsson (Ed.), *The road to excellence? The acquisition of expert performance in the arts and sciences, sports and games* (pp. 167–187). Erlbaum Associates.

Schnotz, W. (2014). An integrated model of text and picture comprehension. In R. E. Mayer (Ed.), *The Cambridge handbook of multimedia learning* (2nd ed., pp. 72–103). Cambridge University Press.

Schooler, J. W., & Engstler-Schooler, T. Y. (1990). Verbal overshadowing of visual memories: Some things are better left unsaid. *Cognitive Psychology, 22*(1), 36–71.

Seufert, T., & Brünken, R. (2006). Cognitive load and the format of instructional aids for coherence formation. *Applied Cognitive Psychology, 20*(3), 321–331. https://doi.org/10.1002/acp.1248

Shanks, D. (2005). Implicit learning. In K. Lamberts & R. Goldstone (Eds.), *Handbook of cognition* (pp. 202–220). Sage.

VanLehn, K. (2011). The relative effectiveness of human tutoring, intelligent tutoring systems and other tutoring systems. *Educational Psychologist, 46*(4), 197–221. https://doi.org/10.1080/00461520.2011.611369

5

Teaching STEM Subjects in Non-STEM Degrees: An Adaptive Learning Model for Teaching Statistics

Daniela Pacella, Rosa Fabbricatore, Alfonso Iodice D'Enza, Carla Galluccio, and Francesco Palumbo

CONTENTS

5.1 Introduction 61
5.2 State of the Art 63
 5.2.1 A Hierarchical Structure for Statistical Knowledge 63
 5.2.2 Knowledge Space Theory 64
 5.2.3 Dublin Descriptors 65
5.3 Methodology 65
 5.3.1 Multidimensional IRT Model 65
 5.3.2 Archetypal Analysis 66
5.4 Implementation 67
 5.4.1 App Interface and Workflow 67
 5.4.2 User Model and Data Processing 67
5.5 Simulation Data Results 69
 5.5.1 Design of the Study 69
 5.5.2 Results 69
 5.5.3 Preliminary Application on Real-world Data 71
5.6 Discussion 71
5.7 Limitations and Future Work 71
5.8 Conclusion 73
References 74

5.1 Introduction

Statistical knowledge is a transversal competency that helps develop not only traditional hard skills such as the ability to infer causal relationships and understand probability and statistical significance (Utts, 2003), but also important soft skills such as critical thinking (Vanhoof et al., 2011) and methodological and quantitative reasoning (Lehman & Nisbett, 1990). These skills vastly contribute to a more responsible and critical approach to fruiting information daily delivered by all media. Thus, statistical literacy is a fundamental component of undergraduate education in STEM and non-STEM courses such as Medicine and all healthcare professions, Psychology, Sociology, and Political Science. It is a matter of fact that today the curricula of most of these disciplines require statistical knowledge, at least at the introductory level, as part of their knowledge base. However, during their statistical learning experience, undergraduate students in these disciplines often experience anxiety, negative emotions, and low self-efficacy that significantly alter their performance and outcome (Walker & Brakke, 2017). *Statistics anxiety* refers to *"the feeling of anxiety encountered when attending a statistics course or doing statistical analyses"* (Cruise et al., 1985), while *self-efficacy* can be considered as the confidence to learn or do statistics (Baloğlu et al., 2017; Finney & Schraw, 2003). The much scientific work and the many articles in the specialized journals that focus on statistics anxiety show great interest about that. According to the most accepted literature, many among the students enrolled in non-STEM degree curricula, regardless of ethnicity or age, experience statistics anxiety when approaching the study of the discipline (Bui & Alfaro, 2011; Moradi et al., 2021); gender differences are still under debate (Baloğlu et al., 2011; Galli et al., 2011; Lin & Tang, 2017).

Along with anxiety, attitude and motivation also affect statistical learning and performance; in other

DOI: 10.1201/9781003181187-6

words, attitude toward statistics predicts students' achievement (Paechter et al., 2017), while self-efficacy has a direct relation with motivation and significantly affects learning (Ramos Salazar, 2018). It is evident that approaching statistical teaching in an innovative manner that incorporates each student's ability, needs, and pedagogical and psychological characteristics is a necessity.

The continuous innovations in advanced learning technologies have provided opportunities for enhancing traditional teaching methods, adding the possibility to design systems that revolve around the learners and adapt to them, and also allow gaining immediate, reliable, and personalized assessment using portable and globally accessible technologies. In fact, most recent progress in information technology allows reaching rapidly and effectively the geographically sparse learners; moreover, artificial intelligence (AI)- and machine-learning-based technologies promote new means of peer-to-peer comparison and evaluation. In general, adaptive learning systems that employ AI algorithms for helping students' learning, providing customized feedback and teaching material, are referred to as Intelligent tutoring systems (ITS). An ITS is "an adaptive system able to personalize learning according to individual characteristics as the user knowledge, mood and emotion" (Anderson et al., 1985). Generally, the adoption of Intelligent tutoring systems leads to improved learning outcomes in comparison with other teaching methods (Akyuz, 2020; Mousavinasab et al., 2021).

The design of a modern and performing ITS should integrate cross-sectional and longitudinal data to adapt to the user and personalize the learning experience and provide accurate feedback. In particular, the student model should be designed to incorporate two levels of information: on the one hand, an assessment of the student in comparison with the student population should be provided, establishing the knowledge depth achieved by each student in comparison with their peers; on the other hand, an individual longitudinal assessment should be carried out to assess the students' performance changes over time – and therefore the individual learning outcomes. In literature, two different approaches – stemming from either psychometric or computer science – to solve these research questions are usually considered when building tutoring systems; however, they have rarely been combined. Nevertheless, a more integrated design would be extremely beneficial for the effectiveness of the learning process, especially to obtain an accurate user model tailored to the learner.

We here describe an integrated methodological framework for teaching and assessing statistical knowledge with a focus on university students enrolled in non-scientific degree programs that is developed and built in the ALEAS (Adaptive LEArning system for Statistics) App.

In particular, the user model and intelligent tutoring system designed within ALEAS allow fulfilling the following aims:

- Dynamically organizing the learning outcome and teaching material according to the learner's skills: This is achieved through Knowledge Space Theory (KST), which is widely used in the field of expert systems to organize, by a directed acyclic graph in which each node represents a competency, the full knowledge required to master a specific subject.
- Detecting the learners' overall ability level and consequently adapting its behavior: It is worth noting that several approaches exist to assess the ability level. In ALEAS, it is achieved through Item Response Theory (IRT) approach. IRT refers to a quite wide class of parametric and semiparametric psychometric models that exploit the statistical inference theory to get the "best" models' parameter estimations (Rasch, 1993). In more detail, IRT allows both modeling the probability that a student answers correctly to a specific item given his or her other responses and providing the estimates of the difficulty of each item, given the set of answers. ALEAS exploits the capabilities of a particular class of IRT model that also allows discretizing the user's ability distribution into a given number of latent classes. The integration of KST and IRT models allows establishing the ability thresholds used to track the students' progress and evaluate the knowledge state.
- Assessing, for each subtopic, the student's ability level according to the item difficulty level and three of five learning dimensions known as Dublin descriptors of learning outcomes.
- Giving tailored feedback, a motivational reinforcement, and the assessment selecting the most appropriate set of topics, questions, and items to present to each student according to their knowledge progress: This is done via archetypal analysis, a machine learning algorithm that allows classifying students according to their ability profile into homogeneous "categories." Compared to the most known model-based and geometric clustering algorithms, which aim to identify the group

prototype as the most "central," observed or unobserved, statistical unit, archetypal analysis identifies the group prototypes among the most extreme points lying on the convex hull. Such kind of strategy allows finding groups that are mainly characterized between the prototypes rather than the homogeneity within groups.

The following sections will describe the ALEAS App model and architecture in detail and present results on simulated data and student preliminary data collected during the interaction with the platform.

5.2 State of the Art

Statistical competency education through online and interactive platforms has seen a high rise of applications, such as in the case of the Shinyapp online platform developed by Potter et al. (2016). Gamification generally proved to improve the students' performance and learning of statistics (Legaki et al., 2020). Along with a variety of MOOCs (see Albert et al., 2020, for a review), several apps that make use of fictional characters, cartoons, or game-like interactions have been developed. One example is the Multimedia Probability and Statistics System (MMPASS) (Krishnasamy et al., 2020), which makes use of a tutoring character, animations and other media to deliver statistical learning material. However, it does not integrate any assessment system.

Although several learning approaches and tools exist to teach and assess statistics knowledge at the high school and university level, there are no applications specifically designed for undergraduate students in non-scientific courses. In fact, statistics in humanistic courses is often overlooked, and there is a lack of undergraduate programs and courses to prepare teachers in statistics education, that is why research on teaching and learning statistics is fragmented (Zieffler et al., 2008). Recently, Lopez et al. demonstrated the advantages of using a virtual environment for teaching statistics in Medicine degree courses, although not grounding their learning assessment upon specific educational theories (López Lamezón et al., 2018) Another example aimed at non-STEM courses is the app described in Ramos-Galarza et al. (2020), which delivers the subject of descriptive statistics. However, this app also lacks an intelligent tutoring system and a student model.

Concerning the construction of the knowledge structure, several frameworks for teaching statistics have been proposed, among which the formalization called "technological pedagogical statistical knowledge" (TPSK) is used mostly as a pedagogical path to guide students from descriptive statistics to inference (Lee & Hollebrands, 2011), but it has rarely been used for quantifying the learner's knowledge and providing an adaptive assessment.

Among the few applications addressed to non-STEM students and that integrate an intelligent tutoring system with a formalized statistical domain theorization, there are ALEKS, an adaptive learning framework for statistics and other subjects that allows creating a model of the user's competency (for a recent meta-analysis, see Fang et al., 2019), and Stat-Knowlab, a web app that allows monitoring adaptively the students' progress and providing tailored learning material according to their ability (de Chiusole et al., 2020). These frameworks, however, do not integrate adaptive feedback, dynamic student modeling, and student knowledge update based on individual item responses, do not allow student profiling in comparison with their peers, and do not make use of any gamification or game-like features.

5.2.1 A Hierarchical Structure for Statistical Knowledge

The statistical learning domain in ALEAS is arranged into a hierarchical structure defined by Areas, Topics, and Units. In particular, an Area consists of one or more Topics, and each Topic contains one or more Units, which are the smallest educational modules to learn (Davino, et al., 2020b). Within each Unit, students can verify their level of ability by answering three different sets of questions or short exercises that aim to assess their knowledge, application ability, and judgment (critical) skill. These three pedagogical aspects refer to three of the five learning skills, also known as Dublin descriptors. ALEAS combines into one synthetic but analytical feedback the skill level reached by each student. Then ALEAS provides students with a reinforcement that aims to motivate them in case of satisfying or unsatisfying performance. To summarize, students are assessed as follows:

- At Topic level, for each statistics topic, a multidimensional IRT user model estimates each item's difficulty and discriminating power and categorizes each student into a latent class based on his or her ability to answer correctly.
- At Area level, students are clustered into homogeneous groups by the archetypal analysis on their own average ability levels according to the latent class IRT models. Each

student is classified into one of four academic performance profiles, and an equal number of feedback structures are defined accordingly, adopting traditional principles of educational psychology teleXbe.

Such scalable and modular structure allows a dynamic implementation of more Units, Topics, and Areas without changing the underlying model and is potentially applicable to other STEM subjects.

5.2.2 Knowledge Space Theory

KST (Doignon & Falmagne, 1985, 2016) is a methodological framework that analyzes the performance of individual students at each of their knowledge states K. The knowledge state is the current student ability and is a subset of the domain Q, representing the collection of all nodes in the knowledge graph so that $K \subseteq Q$. Each domain Q contains a set of items q. KST aims to analytically compute, among all the $\{q_1, q_2, q_3 ... q_n\} \not\subseteq K$ in the skill map, those that the student is ready to learn since the prerequisite knowledge is already in K. Thus, KST is particularly useful not only for the assessment of the current knowledge of the student, but also to adaptively select the most tailored items to present to the students, given their competencies. Therefore, KST aims to build a structure of a domain into nodes of a directed acyclic graph that allows assessing each learner according to the subsets of the domain they master. The challenges of utilizing this framework lie in two different aspects:

- To construct the structure of the knowledge domain investigating the relationship among the different subsets of the domain;
- To select the most appropriate criteria to assess the current knowledge subsets the student has mastered.

Constructing the structure and dependencies of the knowledge domain can be conducted with several methods: analyzing response patterns, querying experts, and/or analyzing components or skills.

In order to select the next node the student can progress to, that is the set of questions to ask the student, it is necessary to assess first what are the domains the student has or has not mastered, and then propose an item that falls outside of the subsets that the student has mastered. There are several implementations for the knowledge state progression algorithm, one of them being the probabilistic model, which is based on each knowledge state having a likelihood function whose parameters are updated according to the students' answers following the basic local independence model (BLIM). For the ALEAS App, we adopt a structure built using expert knowledge and subdivided statistical knowledge into Topics contained in Areas (Davino, et al., 2020a; Fabbricatore et al., 2019). The knowledge graph is displayed in Figure 5.1. To progress from one Topic to the next, mastery – considered as the highest level of ability classified by the multidimensional IRT models based on Dublin descriptors – must be obtained in all the prerequisite topics. The level of ability is evaluated through the corresponding Topic's multidimensional IRT model.

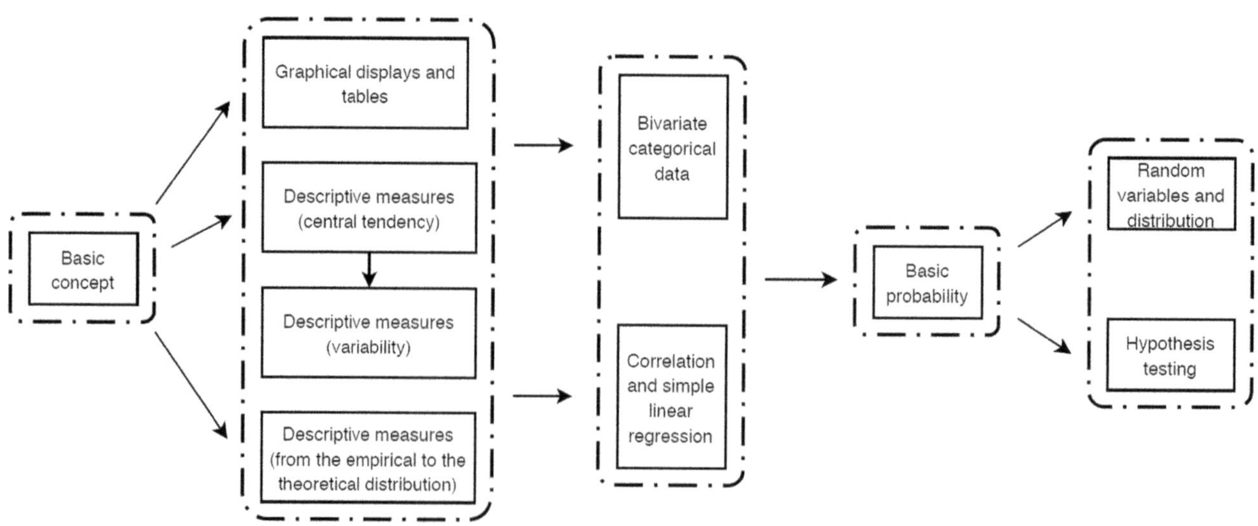

FIGURE 5.1
Knowledge graph of the ALEAS App. Solid lines define Topics, while dotted lines encompass Areas. Arrows declare pre-requisite knowledge.

5.2.3 Dublin Descriptors

Within each KST node or knowledge state, the knowledge depth should be evaluated to establish when every student is ready to progress in their curriculum. For this purpose, the Dublin descriptors that served as one of the bases for the Framework for Qualifications of the European Higher Education Area are vastly used to assess the students' learning outcomes. In particular, the Dublin descriptors are general statements that are used to assess the qualifications obtained after the completion of each specific cycle of higher education studies (Gudeva et al., 2012) and are built on the following objectives:

- *Knowledge and understanding* – the ability to demonstrate knowledge and understanding, including a theoretical, practical and critical perspective on the topic;
- *Applying knowledge and understanding* – the ability to apply the knowledge for identifying, analyzing, and solving problems sustaining an argument;
- *Making judgments* – the ability to gather, evaluate, and present information exercising appropriate judgment;
- *Communications skills* – the ability to communicate ideas, problems, and solutions effectively and disseminate to non-specialist audience;
- *Learning skills* – the ability to identify learning needs and fill the knowledge gaps.

Since we aim at assessing introductory statistical ability via a technological platform, only the first three descriptors – knowledge and understanding (K), applying knowledge and understanding (A), and making judgments (J) are integrated into our student model. As described in the following methodological sections, these dimensions constitute the latent variables of the multidimensional IRT model.

5.3 Methodology

5.3.1 Multidimensional IRT Model

In the educational field, IRT models are widely used to assess students' ability, typically assumed as a normally distributed latent construct (Rasch, 1960). One of the IRT model peculiarities lies in estimating the probability of a correct answer as a function of both student's ability and item characteristics (i.e., discriminating power, item difficulty, guessing, and ceiling parameters). This item-specific probability follows a logistic curve called Item Characteristic Curve (ICC), whose shape depends on the item parameters (see Bartolucci et al., 2019, for a review on the most used IRT models).

In the ALEAS system, item behaviors are described according to the two-parameter logistic (2PL) formalization (Birnbaum, 1968), accounting only for discrimination and difficulty parameters. Thus, the parameters of both guessing and ceiling are consequently set to 0, considering that each item has at least four different possible answers that lowered the presence of guessed responses.

Since traditional IRT models ground on the three main assumptions of unidimensionality, monotonicity, and local independence, extensions releasing some model constraints proved to be best suited for addressing ALEAS project goals. In particular, multidimensional IRT models releasing the unidimensionality assumption allow accounting for the multidimensional nature of students' ability (Bartolucci, 2007), defined in ALEAS by the Dublin descriptors. Furthermore, the latent class IRT models releasing the constraint about the continuous nature of the latent trait follow a semi-parametric formulation and allow detecting homogeneous groups of students according to their performance.

Hence, at the Topic level, the ALEAS system exploits the multidimensional latent class IRT models simultaneously considering more students' ability dimensions (Dublin descriptors), each represented by a discrete distribution allowing for learners' classification. It is worth noting that we adopted the between-item multidimensional formulation; thus, each item is related only to one learning dimension as defined by the Dublin descriptors.

More formally, the vector $\Theta = (\Theta_1, \Theta_2, \dots, \Theta_D)'$ of the D latent variables (herein $D = 3$ according to the Dublin descriptors) is assumed to follow a discrete distribution with $\xi_1, \xi_2, \dots, \xi_k$ support points defining k latent classes. Thus, students belonging to the same latent class share the same profile according to the Dublin dimensions defining students' ability in statistics. The prior probabilities of belonging to latent classes determine the class weights $\pi_1, \dots, \pi_k$, where $\pi_c = P(\Theta = \xi_c)$ with $c = 1, \dots, k$, $\sum_{c=1}^{k} \pi_c = 1$ and $\pi_c > 0$.

Given $\theta = (\theta_1, \theta_2, \dots, \theta_D)$, the possible realization of Θ with $\xi_c = (\theta_{c1}, \theta_{c2}, \dots, \theta_{cD})$, the probability of correct answer to a dichotomously scored item i (with $i = 1, \dots, I$) can be formalized as follows:

$$g[P(X_i = 1 \mid \Theta = \xi_c)] = \log \frac{P(X_i = 1 \mid \Theta = \xi_c)}{P(X_i = 0 \mid \Theta = \xi_c)} = a_i \left(\sum_{d=1}^{D} \delta_{id} \theta_{cd} - b_i \right), \quad (5.1)$$

where g(·) is the logit link function; X_i is the response at item i with realization $x_i \in [0;1]$; and δ_{id} is a dummy variable equal to 1 if the item i measures the latent trait d. Moreover, according to the 2PL parametrization, only the item discrimination a_i and the item difficulty b_i affect response probability.

Due to the assumption of *local independence*, the manifest distribution of the entire response vector $\mathbf{X} = (X_1, \dots, X_I)'$ can be expressed as follows:

$$P(\mathbf{X} = \boldsymbol{x}) = \sum_{c=1}^{k} P(\mathbf{X} = \boldsymbol{x} \mid \Theta = \xi_c)\pi_c, \tag{5.2}$$

where

$$P(\mathbf{X} = \boldsymbol{x} \mid \Theta = \xi_c) = \prod_{d=1}^{D} \prod_{i \in I_d} P(X_i = x_i \mid \Theta_d = \theta_{cd}). \tag{5.3}$$

The estimation of the model parameters is performed through the maximum marginal likelihood (MML) approach (Thissen, 1982). In particular, the following log-likelihood is maximized:

$$\ell(\eta) = \sum_{\boldsymbol{x}} n_{\boldsymbol{x}} \log[P(\mathbf{X} = \boldsymbol{x})], \tag{5.4}$$

where the vector η contains all the free model parameters, n_x is the frequency of the response vector x, and $P(\mathbf{X} = x)$ is defined according to Equation 5.2. Then, the maximization of $\ell(\eta)$ is obtained using the expectation-maximization (EM; A. P. Dempster et al., 1977) algorithm that alternates two steps – E-step and M-step – until convergence.

In the E-step, the model estimates the individual's conditional probability of belonging to one of the latent classes given the response configuration. Then, the M-step maximizes the expected value of the complete data log-likelihood based on the posterior probabilities computed in the E-step.

In sum, the model parameters to be estimated are the matrix of ability levels where each entry is equal to the corresponding estimated support point θ_{cd}, the class weights, the item discriminant, and difficulty parameters. It is worth noting that to ensure the identifiability of the model, for each latent trait it is required that one discriminating index is equal to 1 (usually $a_1 = 1$) and one difficulty parameter is equal to 0 (usually $b_1 = 0$). In addition, the number of latent classes can be either theoretically assumed or selected comparing models with different number of classes through the BIC index (Schwarz, 1978). The estimation process is performed through the R package MultiLCIRT (Bartolucci et al., 2014).

Once the model parameters are estimated, each student is assigned to the class corresponding to the highest posterior probabilities of belonging.

Moreover, the expected a posteriori (EAP) estimates of the ability are computed with respect to the three Dublin descriptor learning dimensions. More in depth, we obtain the ability level of student s (with $s = 1, \dots, n$) in the ability dimension d as follows:

$$\hat{\theta}_{sd} = \sum_{c=1}^{k} \hat{z}_{sc} \hat{\theta}_{cd} \tag{5.5}$$

where $\hat{z}_{sc} = P(\Theta_s = \xi_c \mid \mathbf{X}_s = \boldsymbol{x}_s)$.

Estimated parameters are stored in the ALEAS App database and are used for the a posteriori categorization of each new individual. Therefore, when a new individual t logs into the system, him or her response vector $X_t = (X_{t1}, X_{t2}, \dots, X_{tI})'$ is used to compute the a posteriori classification as follows (Bartolucci et al., 2015):

$$P(\Theta_t = \xi_c \mid \boldsymbol{X}_t = \boldsymbol{x}_t) = \frac{\pi_c \prod_{d=1}^{D} \prod_{i \in I_d} P(X_{ti} = x_{ti} \mid \Theta_{td} = \theta_{cd})}{P(\boldsymbol{x}_t)}. \tag{5.6}$$

Subject t is assigned to latent class c if

$$c = \arg\max_{g=1,\dots,k} P(\Theta_t = \xi_g \mid X_t = x_t). \tag{5.7}$$

Finally, the EAP estimate of the ability is computed according to Equation 5.5.

5.3.2 Archetypal Analysis

An archetype is a reference model such that objects of the same kind are copied from it or based on it. The concept of archetypes appeared in statistics, thanks to Cutler and Breiman (1994) who presented the archetypal analysis (AA) as a novel clustering method. AA is an unsupervised learning technique seeking to synthesize multivariate observations using a reduced number of special vectors, that is, the *archetypes*.

Let $\mathbf{X}$ be a $n \times p$ data matrix with observations on rows and attributes on columns, and let $z_1, z_2, \dots, z_k$, $k \ll n$, be a reduced set of convex combinations of the observations, such that $z_j = \sum_{i=1}^{n} b_{ij} x_i$, $i = 1, \dots, n$ and $j = 1, \dots, k$. The aim is to approximate each observation x_i via the convex combination $\sum_{j=1}^{k} a_{ji} z_j = \sum_{j=1}^{k} a_{ji} \left(\sum_{i=1}^{n} b_{ij} x_i \right)$. In algebraic notation:

$$\min_{A,B} \| \mathbf{X}^T - \mathbf{ZA} \|^2 = \min_{\mathbf{A},\mathbf{B}} \| \mathbf{X}^T - \mathbf{X}^T\mathbf{BA} \|^2 \tag{5.8}$$

where $\mathbf{A}$ and $\mathbf{B}$ are respectively $k \times n$ and $n \times k$ column stochastic matrices, that is, both $\mathbf{A}$ and $\mathbf{B}$ have non-negative elements, and each of their columns add up to 1.

The problem does not admit solution in closed form; therefore, Cutler and Breiman (1994) proposed an iterative procedure to optimize the loss function in Formula 5.8 alternately with respect to **A** for fixed **B**, and the other way round. The convergence of the procedure is guaranteed, yet multiple random starts are needed to reduce the risk of local optima.

For optimal **A** and **B**, the k convex combinations z_j are the archetypes. In Mørup and Hansen (2012), it is proven that they lie on the convex hull generated by **X**. Moreover, if the convex hull of the data has $q \leq n$ vertices and $k = q$, then the archetypes are exactly the vertices of the convex hull; while for $k < q$, they can be thought of as extreme observations defining the vertices of the principal convex hull (PCH), the dominant approximation of the convex hull of **X**.

5.4 Implementation

5.4.1 App Interface and Workflow

The ALEAS App is a mobile and freely available web-based application. A diagram of the app flow is presented in Figure 5.2. The front end is constituted by the user interface, which displays personal user information, the exercises and/or the learning material that are being viewed, and the tutoring character, Ronny McStat, resembling the famous statistician Ronald Fisher, who welcomes the students and follows them during the learning process, giving feedback on the performance. After an introductory login and registration screen where the user is asked for username, password, e-mail, language, and demographic data such as gender and nationality, the terms of service and privacy policy is displayed. If the user agrees to them, he/she is redirected to the main menu. The main menu contains a list of all the currently unlocked Topics – i.e., the knowledge nodes judged as requiring the same mastery level of the student by the KST structure – that the user is free to explore and view. When selecting a Topic within an Area, the student is presented with the choice of examining the learning material first or of proceeding directly to the exercises. In particular:

- The learning material consists of animated 3D cartoons, short stories, and dynamically compiled Latex files that are specific to each knowledge domain (an example is shown in Figure 5.3);
- Exercises are organized into sessions of 15 exercises, 5 investigating the learning outcomes described by each of the Dublin descriptors. From the database of all exercises in the Topic, the set of 15 exercises are randomly drawn and their numerical and textual content is dynamically generated from a Latex template containing R code. Exercises can require an open or a multiple choice answer. After completing each exercise, in case the answer provided is wrong, the student is provided with the correct solution and a conceptual explanation (as shown in Figure 5.4).

5.4.2 User Model and Data Processing

After the student completes the session, he/she is categorized according to the multidimensional IRT model and a level of ability is computed and stored in the database. Additionally, the Area-level student model based on archetype analysis is updated (Adabbo et al., 2021) integrating the information gathered with the latest student–app interaction. After all the Topics within an Area have been examined and completed, the student is provided with a motivational feedback tailored to his/her level of ability among the four

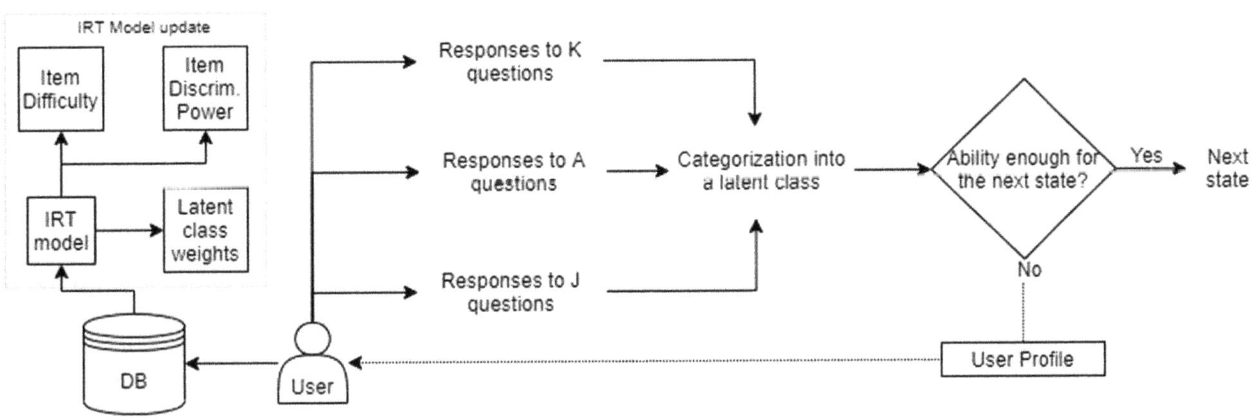

FIGURE 5.2
Workflow diagram of the ALEAS App within each KST node.

FIGURE 5.3
On the left, the tutoring character Ronny McStat with his nephew, a fictional learning character. On the right, an example of learning material content.

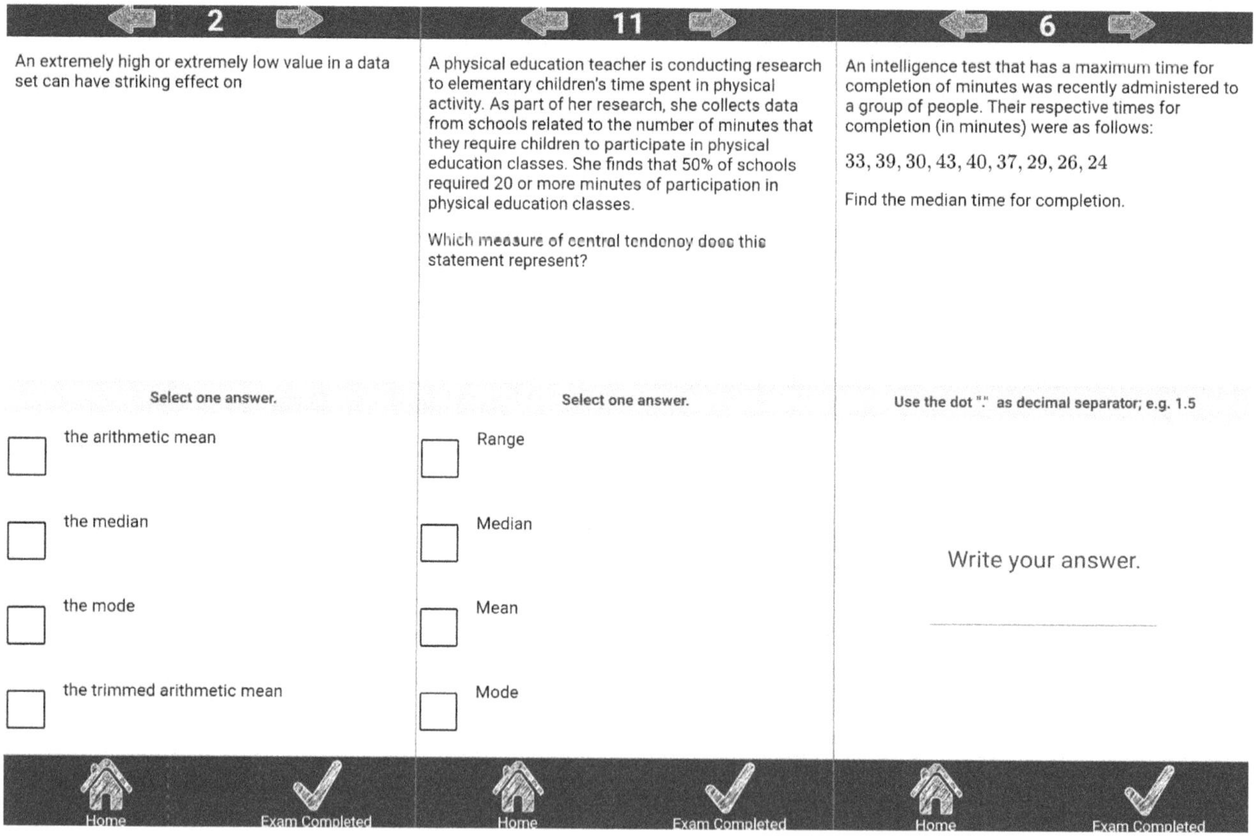

FIGURE 5.4
Screenshots of user interface of three exercises that are part of the learning material of the ALEAS App.

preset academic profiles. The tutoring character also displays a facial expression that is consistent with the student evaluation.

While the client-side is written in C# and XML, the app back end is implemented using the SmartFox server technology enriched with extensions developed using the Java programming language. Data is stored in JSON format using a MYSQL database. The server manages the communications with clients, providing access to the information contained in the database and providing tools to get statistics about all users' learning progress, to be eventually accessed by class professors. The server also runs the R statistical framework and the R/Exams package, taking care of the dynamic generation of all content and thus allowing the App to be lighter on the end-user side. The server communicates with the client-side using the RESTful Web API.

5.5 Simulation Data Results

This section describes the simulation study used to depict how the model detects groups of students with different ability levels. To this aim, we consider three Topics of the same learning Area.

5.5.1 Design of the Study

The simulation design is organized as follows:

- *Definition of the item bank:* We generated a database of item parameters for each Topic according to the two-parameter logistic (2PL) formulation expressed in Equation 5.1. The item difficulty parameters were randomly drawn from a standard Gaussian distribution, whereas the discrimination parameters were generated from a standard log-normal distribution. The entire database included 45 items, 15 for each Topic, equally distributed in Knowledge, Application, and Judgment domains.
- *Generation of item responses:* For the response patterns generation, we considered different ability levels according to the Dublin descriptors. In this regard, several experts in the subject of statistics involved in the ALEAS project suggested using the following learning outcome combinations:
  1. Poor performance in all the three domains;
  2. Good performance in all the three domains;
  3. Good performance in Knowledge and poor performance in both Application and Judgment;
  4. Average performance in all the three domains.

For each combination, $n = 200$ patterns of item responses were randomly generated using the R package MAT (Choi & King, 2014). The latent trait was assumed to be normally distributed with $\sigma = 1$, and $\mu = 1$ for good performers, $\mu = 0$ for average performers, and $\mu = -1$ for poor performers. Moreover, we set the correlation between dimensions equal to 0.5.

- *Multidimensional latent class IRT model:* We carried out the multidimensional latent class IRT model on the set of $N = 800$ (200×4 outcome combinations) simulated response patterns to obtain students' Topic-level classification. First, we assumed $k = 4$ latent classes, according to the number of simulated learning scenarios. Then, we estimated students' ability levels employing the EAP estimation according to the three Dublin descriptors (see Equation 5.5). Each student was assigned to the class that corresponds to the highest probability of belonging.
- *Archetypal analysis:* We computed the average ability levels in each learning dimension and carried out the AA to obtain the Area-level classification. For each Area, students were clustered around $k-1$ archetypes and one more barycentric point of the archetypal space. The space around the barycenter embraces a residual group that includes average performers.

This approach allowed defining a set of reference students with specific learning outcomes, which were used to build a recommendation system.

5.5.2 Results

Results from the simulation study are of particular significance, enabling the reader to appreciate the functioning of the ALEAS system. In this vein, the detailed description of the Topic-level and the Area-level results gives some interesting hints.

Regarding the Topic-level classification, the results of Topic 1 (see Table 5.1) show that Class 1 includes subjects with poor performance in all the three domains; Class 2 encompasses subjects with good performance in Knowledge, average performance in Application, and poor performance in Judgment; Class 3 regards subjects with average performance in all the three dimensions (slightly worse in Knowledge and Judgment); finally, Class 4 consists of the subjects with good performance in all Dublin descriptor dimensions. It is worth noting that the Judgment domain reports the lowest score for all the latent classes.

Moreover, looking at the differences between the maximum and the minimum value estimated for the support points in each dimension, we notice that

TABLE 5.1

Estimated Support Points for Each Dimension and Class Weights π_c for the Multidimensional Latent Class IRT Model Applied on Topic 1

	Knowledge	Application	Judgment	π_c
Class 1	−1.10	−1.60	−3.02	0.32
Class 2	0.91	0.12	−2.16	0.30
Class 3	−0.25	0.14	−0.42	0.16
Class 4	1.25	2.71	0.50	0.22

Application is the dimension that better discriminates among subjects, reporting a difference equal to 4.31.

As regards class weights π_c, results indicate that Class 1 and Class 2 are the largest ones, accounting for 62% of the subjects. Otherwise, Class 3 is the smallest one, with a weight $\pi_3 = 0.16$.

About Topic 2, the results depicted in Table 5.2 show that Class 1 and Class 4 include subjects with poor and good performance in all the three dimensions, respectively; on the other hand, Class 2 regards subjects with medium to high performance in Knowledge and Application and medium to low performance in Judgment; finally, Class 3 encompasses subjects with poor performance in Knowledge, average performance in Application, and good performance in Judgment. However, in this case, Knowledge is the dimension that better discriminates among subjects (difference equal to 3.45).

Finally, the results about Topic 3 (see Table 5.3) show that Class 1 includes subjects with poor performance in all the three dimensions, especially in Judgment; Class 2 defines subjects with good performance in Knowledge and poor performance in Application and Judgment; Class 3 regards subjects with poor performance in Knowledge and Application, and medium to low performance in Judgment; whereas Class 4 encompasses subjects with good performance in Knowledge and Judgment and a medium to high performance in Application. Judgment reveals to be the dimension that better discriminates among subjects (difference equal to 6.43). As regards the weights π_c, it is worth noting that Class 1, Class 2, and Class 4 show quite similar values, whereas Class 3 reports a value lower than the others.

TABLE 5.2

Estimated Support Points for Each Dimension and Class Weights π_c for the Multidimensional Latent Class IRT Model Applied on Topic 2

	Knowledge	Application	Judgment	π_c
Class 1	−2.12	−1.96	−0.43	0.29
Class 2	0.56	−0.53	0.79	0.30
Class 3	−1.34	0.44	1.67	0.16
Class 4	1.33	1.55	2.71	0.25

TABLE 5.3

Estimated Support Points for Each Dimension and Class Weights π_c for the Multidimensional Latent Class IRT Model Applied on Topic 3

	Knowledge	Application	Judgment	π_c
Class 1	−1.67	−3.00	−4.08	0.28
Class 2	1.11	−1.33	−1.74	0.27
Class 3	−1.25	−0.85	−0.35	0.19
Class 4	1.25	0.31	2.35	0.26

Afterward, the ability levels for each simulated response pattern were estimated exploiting the EAP estimation computed accounting for the ability matrix and the individual vector of belonging probabilities to the latent classes. The resulting matrix included for all subjects three ability values for each dimension, corresponding to Topic 1, Topic 2, and Topic 3 estimates, respectively.

In order to obtain the Area-level classification of the subjects, we performed the mean by rows of Topic-level abilities according to the Dublin descriptors. An example of the transition from the Topic-level to the Area-level ability scores based on the simulated data is provided in Table 5.4.

Then, we performed the archetypal analysis to identify a set of reference students with a specific learning outcome. Here, we make use of three archetypes and an additional barycentric point of the archetypal space. We made this choice to resemble the number of latent classes in the Topic-level classification.

Table 5.5 shows the archetype profiles. In particular, the first archetype represents students with poor performance in all the three Dublin descriptors considered; the second one identifies students with a good performance in Knowledge, and slightly poor performance in Application and Judgment; the third archetype corresponds to students with good performance in all the three dimensions. Finally, the barycenter refers to an average performance.

Students are assigned to the "closest" archetype, as shown in Figure 5.5. The figure depicts the ternary map, where both students and archetypes are represented in a two-dimensional map. Indeed, in the ternary plot, the original coordinates of the ith student $\{a_{1i}, a_{2i}, a_{3i}\}$ turned into $x_{1i} = a_{2i} + a_{3i}/2$ and $x_{2i} = a_{3i}\sqrt{3}/2$ coordinates.

Thus, to customize the students' experience based on their ability level on the different dimensions, we defined a recommendation for each of the four reference students. Hence, for example, a student that belongs to the second archetype will receive the following recommendation:

> You seem to have understood the theoretical perspective and learned quite well the formal definition of the concepts. On the other hand, your ability to apply knowledge for solving problems and make judgments is lacking. I strongly advise you to do more exercises to improve your calculation skills and your ability to evaluate information to reach an appropriate judgment in statistical matters.

TABLE 5.4

Area-level Ability Scores (on the Right) Obtained from the Topic 1, Topic 2, and Topic 3 Ability Estimates (on the Left)

	Topic 1			Topic 2			Topic 3				Area			
Id	K	A	J	K	A	J	K	A	J		Id	K	A	J
1	−1.10	−1.60	−3.02	−2.11	−1.95	−0.42	−1.66	−2.93	−3.97	⇒	1	−1.62	−2.16	−2.47
2	−1.09	−1.60	−3.02	−1.99	−1.84	−0.33	−1.14	−0.92	−0.49		2	−1.40	−1.45	−1.28
3	−0.75	−1.25	−2.77	−2.11	−1.95	−0.43	−1.29	−1.09	−0.77		3	−1.39	−1.43	−1.32
⋮	⋮	⋮	⋮	⋮	⋮	⋮	⋮	⋮	⋮		⋮	⋮	⋮	⋮

TABLE 5.5

Archetypes Defining the Reference Students in the Simulated Data

	Knowledge	Application	Judgment
Archetype 1	−1.61	−2.10	−2.37
Archetype 2	0.89	−0.45	−0.91
Archetype 3	1.01	1.41	1.75
Barycenter	0.33	0.33	0.33

Students with the same ability level are assigned to the same archetype; therefore, they will receive the same recommendation.

Results obtained from the simulation study represented the reference for the categorization of the first App users, as illustrated in the following section.

5.5.3 Preliminary Application on Real-world Data

A preliminary application on real-world data was performed considering students enrolled in the first year of the Bachelor's Degree in Psychological Sciences and Techniques at the University of Naples Federico II. Data included the first 38 students who completed the learning Area on probability, consisting of the following three Topics: sampling space, axioms of probability, and conditional probability.

Given the students' response pattern for each Topic, the a posteriori probabilities of belonging to latent classes were computed, as showed in Section 5.3. Consequently, the expected a posteriori estimate of the students' ability in Knowledge, Application, and Judgment was obtained for each Topic.

Once the Area-level ability was calculated, students were projected in the archetypal space and classified according to their smallest Euclidean distance from the reference archetypes and barycenter (see Figure 5.6). It is worth noting that no students were classified in the class defined by the first archetype; however, 4 students were assigned to the second archetype, 19 students to the third archetype, and 15 students to the fourth archetype.

In particular, in the figure, we highlighted one student for each archetype as an example: student 18 for the first archetype, student 29 for the third one, and student 34 for the fourth archetype. Table 5.6 are shown the Area-level and the Topic-level ability for the three students considered. Looking at the table, it is possible to notice that the students profile is very similar to the archetype they were assigned: indeed, student 18 shows a good performance in Knowledge and slightly poor performance in Application and Judgment; student 29 reports a medium to high performance in Knowledge and Application, and a good performance in Judgment; while student 34 presents an average performance in Knowledge and Judgment and a medium to low performance in Application.

5.6 Discussion

ALEAS allows both to (1) provide a tailored learning experience to students, integrating their learning style and performance to compute a personalized curriculum and (2) adaptively assess students using cross-sectional and longitudinal data in comparison with their peers and with a standardized learning framework. Additionally, the App incorporates a game-like interface, a 3D tutoring character, and comics-like visual style to help reduce the educational burden and improve the attitude toward statistics. The innovation of the ALEAS architecture stands in the possibility not only to consider a student ability discrimination at item level and at Topic level, thanks to the adoption of the multidimensional IRT model, but also to provide a dynamic feedback generation at Area-level that helps the students feel motivated and engaged with the platform and with the subject.

5.7 Limitations and Future Work

We are aware of the current study limitations due to the constrained sample size of students that

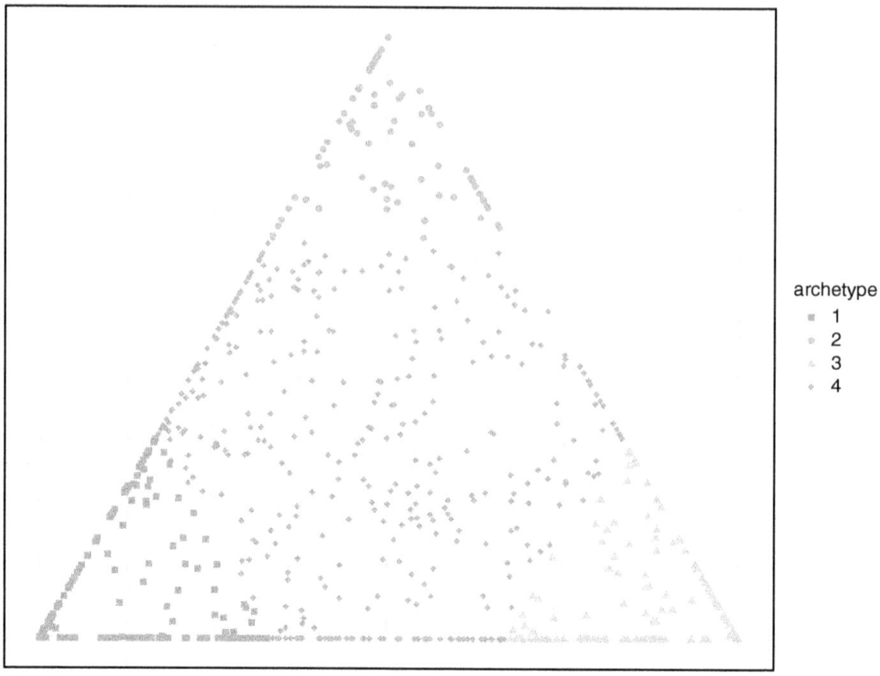

FIGURE 5.5
Ternary map obtained from the simulated data: three archetypes on the vertices and a fourth barycentric one.

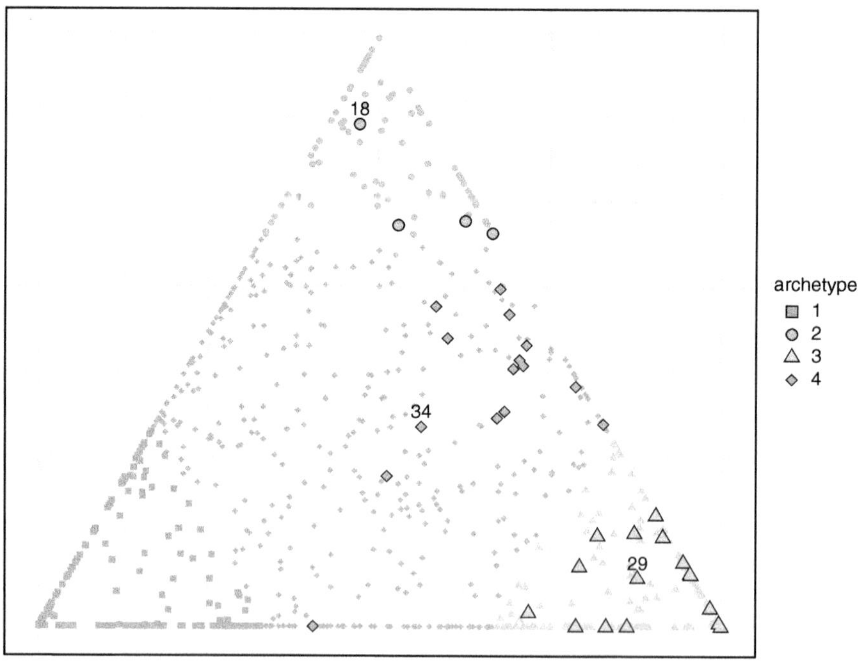

FIGURE 5.6
Projection of the students on the simulated data ternary map (marked with black borders). Each individual is projected in the archetypal space and classified according to their smaller distance from archetypes and the barycenter.

interacted with the app, although simulated data used for the model evaluation and initialization showed promising and interesting results. In particular, the whole model seems to be able to correctly identify strengths and weaknesses of each student, not only by Topic, but more importantly by learning outcome, providing indications as to how to improve the learning method and on which aspect each student should focus to increase their ability and reach higher performance in the context of the Dublin descriptors, may it be Knowledge (K), Application (A), and/or Judgment (J).

TABLE 5.6

Area-level and Topic-level Ability for the Students 18, 29, and 34

	Sampling Space			Axioms of Probability			Conditional Probability				Probability Area			
Id	K	A	J	K	A	J	K	A	J		Id	K	A	J
18	0.83	0.14	−2.02	0.54	−0.53	0.79	0.58	−1.28	−1.52	⇒	18	0.65	−0.56	−0.92
29	1.19	2.55	0.39	−0.02	−0.17	1.12	1.25	0.30	2.33		29	0.80	0.89	1.28
34	−0.82	−1.33	−2.84	0.57	−0.51	0.81	1.25	0.31	2.35		34	0.33	−0.51	0.10

In general, data suggests that the Knowledge graph defined in the context of KST is consistent and that, analogously, the four identified archetypes satisfyingly reflect the differences in the students within their population, allowing for an easier but reliable peer-to-peer comparison.

Data computed by the model is also accessible to examine from the perspective of a lecturer or professor, which, glancing at the ability distribution of students by domain and learning outcome, can quickly realize the key aspects on which to focus the next lectures.

As regards to the machine learning model behind the Area-level classification, although Archetypal analysis has been chosen to satisfy the need to find "typical" learners' performances and classify users on the basis of how close they perform to those archetypes, there are other possible alternatives for the user model classification using data output from the multidimensional IRT model that we plan to explore for the improvement of the ALEAS model. Recently, several approaches based on deep learning for IRT parameter estimates have been proposed, such as the Deep Knowledge Tracing approach (Piech et al., 2015), the deep-IRT approach (Tsutsumi et al., 2021), and the even more recent Knowledge Interaction-enhanced Knowledge Tracing model (KIKT) (Gan et al., 2020). These models, however, are still highly debated in terms of interpretability and usually require a larger amount of data to converge.

Future work and applications of ALEAS will focus on

- increasing the students' sample size so as to also increase the precision and accuracy of the IRT and AA models. A larger data collection is already in progress, and the app is kept free to download on the mobile app stores;
- evaluating the user-experience with the app as well as the usability. Although preliminary research has been conducted on this aspect and leads to promising expectations, more data needs to be collected for an exhaustive evaluation;
- comparing the effectiveness of multiple assessments among different statistical learning platforms and ITS; and
- extending the current app model and architecture to other STEM subjects.

5.8 Conclusion

This chapter presented the theoretical framework, implementation, and application of the Adaptive LEArning system for Statistics (ALEAS) App. ALEAS aims to provide an adaptive assessment of non-STEM degrees undergraduate students' statistical abilities integrating the advantages of technology and accessibility, gamification strategies, dynamic user modeling, and a multilevel intelligent tutoring system. In order to develop a user model of the students who use the ALEAS App, we constructed with the help of experts a Knowledge Structure for basic statistics knowledge and built a directed acyclic graph consisting of different statistical modules, in particular Areas, Topics, and Units. Then, within each knowledge state – or node – we structured statistics questions that can assess the student ability on three of the Dublin descriptors related to knowledge of the Topic, Concept Application, and Judgment. A multidimensional IRT user model estimates each of these exercises' difficulty and discriminating power taking into account the three Dublin descriptors and uses the computed parameters to estimate each student's level of ability classifying them into a latent class. As the individual students' ability increases and the frequency of correct answers rises, they will be assigned to the next latent class, and he or she will be allowed to progress to the other knowledge node and switch to a new statistics Topic in the knowledge structure. Additionally, students receive a feedback on their performance that takes into account their progress in all the Topics belonging to each Area with the use of archetypal analysis, which categorizes the students' achievements into one of four learning profiles. Through a personalized curriculum and features such as dynamic feedback, 3D animations, videos, and humorous comics, ALEAS also helps students cope with statistical anxiety and lack of motivation. Formative and motivational feedback is generated to increase the students' engagement and improve their learning strategies, rewarding their efforts. In conclusion, the environment provided by ALEAS is an innovative and integrated tool to improve statistical learning strategies, as a stand-alone training and learning ground or as a complement to traditional academic courses.

References

Adabbo, B., Fabbricatore, R., Iodice D'Enza, A., & Palumbo, F. (2021). Statistics knowledge assessment: An archetypal analysis approach. In C. Perna, N. Salvati, & F. Schirripa Spagnolo (Eds.), *Bosp sis 2021* (pp. 1388–1393). Pearson.

Akyuz, Y. (2020). Effects of intelligent tutoring systems (ITS) on personalized learning (PL). *Creative Education, 11*(6), 953–978.

Albert, J., Cetinkaya-Rundel, M., & Hu, J. (2020). Online statistics teaching and learning. In J. P. Howard II & J. F. Beyers (Eds.), *Handbook of teaching and learning mathematics online* (pp. 99–116). Chapman & Hall/CRC.

Anderson, J. R., Boyle, C. F., & Reiser, B. J. (1985). Intelligent tutoring systems. *Science, 228*(4698), 456–462.

Baloglu, M., Abbassi, A., & Kesici, S. (2017). Multivariate relationships between statistics anxiety and motivational beliefs. *Education, 137*(4), 430–444.

Baloğlu, M., Deniz, M. E., & Kesici, Ş. (2011). A descriptive study of individual and cross-cultural differences in statistics anxiety. *Learning and Individual Differences, 21*(4), 387–391.

Bartolucci, F. (2007). A class of multidimensional IRT models for testing unidimensionality and clustering items. *Psychometrika, 72*(2), 141–157.

Bartolucci, F., Bacci, S., & Gnaldi, M. (2014). Multilcirt: An R package for multidimensional latent class item response models. *Computational Statistics and Data Analysis, 71*, 971–985.

Bartolucci, F., Bacci, S., & Gnaldi, M. (2019). *Statistical analysis of questionnaires: A unified approach based on R and Stata*. Chapman & Hall/CRC.

Bartolucci, F., Dardanoni, V., & Peracchi, F. (2015). Ranking scientific journals via latent class models for polytomous item response data. *Journal of the Royal Statistical Society: Series A (Statistics in Society), 178*(4), 1025–1049.

Birnbaum, A. (1968). Some latent trait models and their use in inferring an examinee's ability. In F. M. Lord & M. R. Novick (Eds.), *Statistical theories of mental test scores* (pp. 395–479). Addison-Wesley.

Bui, N. H., & Alfaro, M. A. (2011). Statistics anxiety and science attitudes: Age, gender, and ethnicity factors. *College Student Journal, 45*(3), 573–586.

Choi, S., & King, D. (2014). Mat: Multidimensional adaptive testing. *R package version 2.2*. Retrieved from https://CRAN.R-project.org/package=MAT

Cruise, R. J., Cash, R. W., & Bolton, D. L. (1985). Development and validation of an instrument to measure statistical anxiety. *American Statistical Association Proceedings of the Section on Statistical Education, 4*(3), 92–97.

Cutler, A., & Breiman, L. (1994). Archetypal analysis. *Technometrics, 36*(4), 338–347. Retrieved from http://www.jstor.org/stable/1269949

Davino, C., Fabbricatore, R., Galluccio, C., Pacella, D., Vistocco, D., & Palumbo, F. (2020a). Teaching statistics: An assessment framework on multidimensional IRT and knowledge space theory. In A. Pollice, N. Salvati, & F. Schirripa Spagnolo (Eds.), *Bosp sis 2020* (pp. 1093–1098). Pearson.

Davino, C., Fabbricatore, R., Pacella, D., Vistocco, D., & Palumbo, F. (2020b). ALEAS: A tutoring system for teaching and assessing statistical knowledge. *CEUR Workshop Proceedings, 2730*.

de Chiusole, D., Stefanutti, L., Anselmi, P., & Robusto, E. (2020). Stat-knowlab. Assessment and learning of statistics with competence-based knowledge space theory. *International Journal of Artificial Intelligence in Education, 30*(4), 668–700.

Dempster, A. P., Laird, N. M., & Rubin, D. B. (1977). Maximum likelihood from incomplete data via the EM algorithm. *Journal of the Royal Statistical Society: Series B (Methodological), 39*(1), 1–22.

Doignon, J.-P., & Falmagne, J.-C. (1985). Spaces for the assessment of knowledge. *International Journal of Man-Machine Studies, 23*(2), 175–196.

Doignon, J.-P., & Falmagne, J.-C. (2016). Knowledge spaces and learning spaces. In *New handbook of mathematical psychology* (Vol. 2, pp. 274–321).

Fabbricatore, R., Galluccio, C., Davino, C., Pacella, D., Vistocco, D., & Palumbo, F. (2019). The effects of attitude towards statistics and math knowledge on statistical anxiety: A path model approach. In M. Carpita & L. Fabbris (Eds.), *ASA conference 2019 "statistics for health and well-being" BoSP* (pp. 97–100). CLEUP sc.

Fang, Y., Ren, Z., Hu, X., & Graesser, A. C. (2019). A meta-analysis of the effectiveness of ALEKS on learning. *Educational Psychology, 39*(10), 1278–1292.

Finney, S. J., & Schraw, G. (2003). Self-efficacy beliefs in college statistics courses. *Contemporary Educational Psychology, 28*(2), 161–186.

Galli, S., Chiesi, F., & Primi, C. (2011). Measuring mathematical ability needed for "non-mathematical" majors: The construction of a scale applying irt and differential item functioning across educational contexts. *Learning and Individual Differences, 21*(4), 392–402.

Gan, W., Sun, Y., & Sun, Y. (2020). Knowledge interaction enhanced knowledge tracing for learner performance prediction. In *2020 7th international conference on behavioural and social computing (BESC)* (pp. 1–6).

Gudeva, L. K., Dimova, V., Daskalovska, N., & Trajkova, F. (2012). Designing descriptors of learning outcomes for higher education qualification. *Procedia-Social and Behavioral Sciences, 46*, 1306–1311.

Krishnasamy, S., Ling, L., & Kim, T. (2020). Improving learning experience of probability and statistics using multimedia system. *International Journal of Emerging Technologies in Learning (iJET), 15*(1), 77–87.

Lee, H. S., & Hollebrands, K. F. (2011). Characterising and developing teachers' knowledge for teaching statistics with technology. In C. Batanero, G. Burrill, & C. Reading (Eds.), *Teaching statistics in school mathematics-challenges for teaching and teacher education* (pp. 359–369). Springer.

Legaki, N.-Z., Xi, N., Hamari, J., Karpouzis, K., & Assimakopoulos, V. (2020). The effect of challenge-based gamification on learning: An experiment in the context of statistics education. *International Journal of Human-Computer Studies, 144*, 102496.

Lehman, D. R., & Nisbett, R. E. (1990). A longitudinal study of the effects of undergraduate training on reasoning. *Developmental Psychology, 26*(6), 952.

Lin, Y.-J., & Tang, H. (2017). Exploring student perceptions of the use of open educational resources to reduce statistics anxiety. *Journal of Formative Design in Learning, 1*(2), 110–125.

López Lamezón, S., Rodríguez López, R., Amador Aguilar, L. M., & Azcuy Lorenz, L. M. (2018). Social significance of a virtual environment for the teaching and learning of descriptive statistics in medicine degree course. *Humanidades Médicas, 18*(1), 50–63.

Moradi, S., Maraghi, E., Babaahmadi, A., & Younespour, S. (2021). Application of pop quiz method in teaching biostatistics to postgraduate midwifery students and its effect on their statistics anxiety, test anxiety and academic achievement: A quasiexperimental study with control group. *Journal of Biostatistics and Epidemiology, 7*(2), 181–188.

Mørup, M., & Hansen, L. K. (2012). Archetypal analysis for machine learning and data mining [special issue on machine learning for signal processing 2010]. *Neurocomputing, 80*, 54–63. Retrieved from https://doi.org/10.1016/j.neucom.2011.06.033.

Mousavinasab, E., Zarifsanaiey, N., Niakan Kalhori, S. R., Rakhshan, M., Keikha, L., & Ghazi Saeedi, M. (2021). Intelligent tutoring systems: A systematic review of characteristics, applications, and evaluation methods. *Interactive Learning Environments, 29*(1), 142–163.

Paechter, M., Macher, D., Martskvishvili, K., Wimmer, S., & Papousek, I. (2017). Mathematics anxiety and statistics anxiety. Shared but also unshared components and antagonistic contributions to performance in statistics. *Frontiers in Psychology, 8*, 1196.

Piech, C., Spencer, J., Huang, J., Ganguli, S., Sahami, M., Guibas, L., & Sohl-Dickstein, J. (2015). Deep knowledge tracing. arXiv preprint arXiv:1506.05908.

Potter, G., Wong, J., Alcaraz, I., Chi, P., et al. (2016). Web application teaching tools for statistics using R and shiny. *Technology Innovations in Statistics Education, 9*(1).

Ramos Salazar, L. (2018). Examining the relationship between math self-efficacy and statistics motivation in the introductory business statistics course: Self-compassion as a mediator. *Decision Sciences Journal of Innovative Education, 16*(2), 140–160.

Ramos-Galarza, C., Acosta-Rodas, M., Bolaños-Pasquel, M., & Cruz-Cárdenas, J. (2020). Mobile app for psychostatistics learning. In *International conference on applied human factors and ergonomics* (pp. 170–174).

Rasch, G. (1960). *Studies in mathematical psychology: I. Probabilistic models for some intelligence and attainment tests.* Nielsen & Lydiche.

Rasch, G. (1993). *Probabilistic models for some intelligence and attainment tests.* ERIC.

Schwarz, G. (1978). Estimating the dimension of a model. *Annals of Statistics, 6*(2), 461–464.

Thissen, D. (1982). Marginal maximum likelihood estimation for the one-parameter logistic model. *Psychometrika, 47*(2), 175–186.

Tsutsumi, E., Kinoshita, R., & Ueno, M. (2021). Deep item response theory as a novel test theory based on deep learning. *Electronics, 10*(9), 1020.

Utts, J. (2003). What educated citizens should know about statistics and probability. *American Statistician, 57*(2), 74–79.

Vanhoof, S., Kuppens, S., Castro Sotos, A. E., Verschaffel, L., & Onghena, P. (2011). Measuring statistical attitudes: Structure of the survey of attitudes toward statistics. *Statistics Education Research Journal, 10*(1).

Walker, E. R., & Brakke, K. E. (2017). Undergraduate psychology students' efficacy and attitudes across introductory and advanced statistics courses. *Scholarship of Teaching and Learning in Psychology, 3*(2), 132.

Zieffler, A., Garfield, J., Alt, S., Dupuis, D., Holleque, K., & Chang, B. (2008). What does research suggest about the teaching and learning of introductory statistics at the college level? A review of the literature. *Journal of Statistics Education, 16*(2).

6

Removing Barriers in Self-Paced Online Learning through Designing Intelligent Learning Dashboards

Arta Faramand, Hongxin Yan, M. Ali Akber Dewan, and Fuhua Lin

CONTENTS

6.1 Introduction....77
6.1.1 Intelligent Learning Dashboards....77
6.1.2 Automatic Generation of Formative Assessments....78
6.1.3 Automatic Analysis of Students' Engagement....78
6.2 Overview of the Intelligent Learning Dashboard Framework....79
6.3 Adaptive Formative Assessments....80
6.4 Automatic Detection of Students' Engagement....82
6.5 Learning Analytic Dashboards....84
6.6 Conclusions and Future Work....86
Acknowledgment....87
References....88

6.1 Introduction

Self-paced online learning (SPOL) is also called individualized online study in which students set their own schedule within a course contract period to complete the course. In SPOL, students can learn anywhere, anytime, at any pace, and via any pathway to meet individual's learning needs. Thus, SPOL offers considerable flexibility for learners, especially adult learners who usually have work and family commitments. Moreover, the popularity of science, technology, engineering, and mathematics (STEM) fields has been increased in postsecondary education to meet the global demand [1]. Due to the nature of openness and flexibility of the SPOL model and the high demand for STEM courses, the STEM courses in SPOL have attracted a wide range of learners. The STEM learners have seen SPOL as an opportunity to learn from anywhere at their own pace. However, this flexibility of SPOL also imposes some barriers to learning in terms of social connection, collaboration, learning awareness, and academic intervention.

In SPOL, instructors and students usually communicate asynchronously where direct student–instructor interaction is not available. Also, students mostly interact with learning content rather than with instructors and peers, especially in STEM disciplines. As a result, meaningful social interaction is often lacking in SPOL, and many students feel isolated. The STEM courses require a higher degree of collaborative multidisciplinary activities, specialized software and tools, and personalized educational paths for students with diverse backgrounds [1]. The lack of collaborative sessions and synchronous or face-to-face interaction among the students and the instructor often result in the lower self-awareness of learning in students, the difficulties of tracking students' learning, and the challenges for the instructor to provide effective academic intervention. Thus, these learning barriers greatly hinder students' success in SPOL STEM courses. Various AI technologies have been used to deal with such challenges in SPOL with little success. More research effort is needed to make these systems effective, especially for the STEM courses in the SPOL model.

6.1.1 Intelligent Learning Dashboards

Research shows that learning dashboards can have cognitive and metacognitive benefits for the students [2]. Learning dashboards can assist students by providing information about their progress in a course and flag if they are at risk of failing a course [2]. Furthermore, learning dashboards can provide personalized recommendations in relation to reading material and learning activities, which could

DOI: 10.1201/9781003181187-7

greatly improve students' engagement in SPOL STEM courses [3]. Intelligent learning dashboards can also tailor the course offering based on students' needs [2]. We believe that effective learning dashboard design to support SPOL students is a critical research topic in online learning. An effective learning dashboard design requires a good understanding of the learning sciences and the effective usage of learning pedagogies. Suthers and Verbert [4] stated that the learning analytics research should investigate which educational concepts constitute the theoretical foundation for developing student-facing learning dashboards. Existing learning dashboards also need to be evaluated for the effectiveness of the design and level of engagement with the learners [5]. Designing a learning dashboard that draws from previous studies allows us to identify the features that are most important to learners as well as the dashboards' limitations. For example, Jo et al. [6] identify that students' log-in frequency, total log-in time, log-in regularity, and visits on the board and repository are some useful indicators to predict students' learning performance. Hodges [7] identifies the correlation between students' motivation and self-awareness with their performance. Therefore, the dashboard functionalities are designed to visualize variables, which can influence online learners' organizational skills, self-awareness, and motivation [6]. In this chapter, we discuss further how the educational concepts can be integrated into designing effective intelligent dashboards to support students of SPOL STEM courses. Especially, we focused on how the techniques for adaptive formative assessment and students' engagement detection can be integrated into an intelligent learning dashboard.

6.1.2 Automatic Generation of Formative Assessments

Automatic generation of formative assessment is an intelligent system that has a great potential to support students in SPOL STEM courses. Formative assessment refers to evaluation that can generate formative feedback on performance and a systematic process to continuously gather evidence about learning [8, 9]. Formative assessment allows students to reflect upon their learning, increase self-awareness of learning, and identify their learning gaps. Also, by formative assessment designed across the entire course, students can be engaged in or initiate meaningful discussions with peers and instructors. Thus, it can be an effective way of promoting social connection in SPOL.

However, traditional fixed formative assessment has some limitations. For example, it does not customize questions to individuals' knowledge and skill level and does not accommodate students' knowledge change during the assessment. Therefore, it may not estimate students' knowledge levels accurately and reliably. Also, fixed formative assessment does not consider shortening the testing time in identifying students' knowledge levels. As warned by Vie et al. [10],

> The ratio of the amount of the evidence to the breadth of the assessment is particularly critical for systems that cover a large array of skills, as it would be unacceptable to ask hours of questions before making a usable assessment.

In addition, some forms of formative assessments are difficult to implement in SPOL. For example, peer assessments that allow students to use one another as learning resources are hard to be done in SPOL as the students may be learning different concepts or skills at a point in time.

To better estimate students' knowledge and skill proficiency using formative assessment, an adaptive mechanism using an appropriate knowledge estimation algorithm can be embedded in the assessment [11]. Adaptive assessment or computer adaptive testing (CAT) automatically generates a subsequent question from a question bank based on their responses to previous questions and adapts to individual students' abilities [12]. Compared with the traditional fixed-form assessment, adaptive assessment can estimate a student's knowledge and skill proficiency in a more accurate, reliable, efficient, and flexible way. A few algorithms have been researched and used in adaptive assessment, such as Item Response Theory (IRT) [13], Bayesian Knowledge Tracing (BKT) [14], Deep Knowledge Tracing (DKT) [15], and so on. These traditional models, however, face some challenges in the context of quiz-based formative assessment. For example, they assume students' knowledge level is static during the test [10] and require considerable amount of training data [16]. Lastly, most research on adaptive formative assessment (AFA) focuses on classroom teaching or traditional paced online learning. This chapter will focus on SPOL and discuss possible AI algorithms that can consider knowledge changing and eliminate the need for training data to generate adaptive formative assessment automatically.

6.1.3 Automatic Analysis of Students' Engagement

Automatic analysis of students' engagement is another AI application that can help to improve learners' engagement in SPOL STEM courses with their educational activities. Learners' engagement has been a key topic in the education literature [17], and this interest may be driven by the concerns about high dropout rates in

SPOL STEM courses [18]. It is widely acknowledged that engagement and affect are linked to increased productivity and learning gain. Some research shows that engagement is malleable, and proper pedagogical interventions, learning designs, and feedback can enhance learners' engagement [19]. To provide personalized pedagogical support through intervention to online learners, detecting learners' engagement can play an important role in online education [20].

Several facets of learners' engagements have been discussed in the literature, including *behavioral, cognitive, emotional,* and *psychological* [21–23]. *Behavioral* engagement draws on the idea of participation, including participation in the classroom and extracurricular activities, stay focused, submit assigned tasks, and follow the instructors' dictation [24]. *Cognitive* engagement refers to the thoughtfulness and willingness to exert the effort necessary to comprehend complex ideas and master difficult skills (e.g., focused attention, memory, and creative thinking [23]). *Emotional* engagement encompasses positive and negative reactions to teachers, classmates, and academics [22]. Finally, *psychological* engagement refers to the sense of belonging and relationships with teachers and peers [25]. Different types of engagements in learning are helpful to know for personalized intervention design to improve learners' experience. However, studies that focus on learners' engagement need a way of measuring it [26]. Such measuring can be done with one of the two types of data identified by engagement theorists: internal to the individual (cognitive) and external observable factors (perceptible facial features, postures, speech, and actions) [21]. Some research studies also emphasized that measuring engagement requires bringing together observational data with the data internal to the individual (e.g., self-reports) [17]. This chapter discusses the existing automatic engagement detection methods, their research challenges, and how these methods can be more effective in improving students' engagement in the SPOL STEM courses.

The rest of the chapter is organized as follows. Section 6.2 presents a framework for intelligent learning dashboards integrating different AI technologies. Section 6.3 discusses the methods used for automatic formative assessments to support intelligent learning dashboards. Section 6.4 discusses the systems for automatic analysis of the students' engagement and how these can be used for pedagogical intervention design for SPOL STEM courses. Section 6.5 discusses the effective design and development methodologies of the intelligent learning dashboards targeting SPOL STEM courses. Finally, Section 6.6 concludes the chapter with discussions and recommendations for further research.

6.2 Overview of the Intelligent Learning Dashboard Framework

SPOL courses are growing in demand and popularity in online education [11]. A major advantage of SPOL courses is that the learners can learn from anywhere, anytime, and at any pace via any pathway based on their individual learning needs. SPOL courses give considerable flexibility to online learners because of their asynchronous and individualized learning characteristics. However, this flexibility also creates several barriers or challenges for online learners. For example, direct student–student and student–instructor interactions are not available, and students mostly learn in an individualized mode. This learning distance causes students to feel isolated and disengaged from their courses. In addition to the solitude, students lack information on how they are performing compared to their peers. Another major limitation with the online courses is that most online courses are designed using a one-size-fits-all model. Thus, a course could be too difficult or too easy for different student groups [11]. Also, many courses are designed in such a way that students can decide when they submit their assessments. Thus, students who submit their assessments toward the end of the course may receive minimal feedback, which could affect their course performance. Without having social interaction and timely feedback, students usually lack self-awareness of learning and often become disengaged from their online courses. In such situations, students often fail to decide when to seek help, how to motivate themselves by improving their engagement, and how to regulate learning [11]. Increasing students' self-regulation is vital because research shows that students who start SPOL courses exhibit poor self-regulation skills, negatively impacting their academic performance, awareness, and persistence [27].

Intelligent learning dashboards have a great potential to mitigate these issues and improve students' self-regulation by visualizing their engagement status and performance in their online courses and providing personalized feedback. The improved self-regulation can help to facilitate students' learning experience in the SPOL courses. An architecture of an intelligent learning dashboard is shown in Figure 6.1. There are three primary design considerations for developing an intelligent learning dashboard system – *adaptive formative assessment, engagement detection,* and *learning analytics dashboard* – with the state-of-the-art methods that are discussed in detail in Sections 6.3–6.5. Developing an intelligent student-facing learning dashboard is important because modern education systems are transitioning

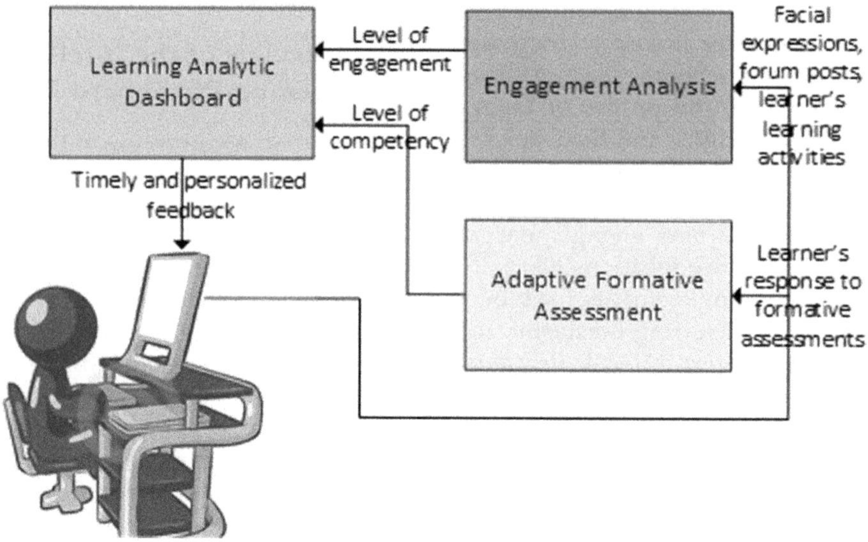

FIGURE 6.1
Architecture for an intelligent student-facing learning dashboard.

from teacher-centered to student-centered learning and instruction. With the greater emphasis on learner-centered education, students need to become more personally responsible, self-aware, and self-directed in their learning. When students have meta-cognitive, motivational, and behavioral control in their learning process, they will be able to better achieve learning objectives (LOs) [28]

In the dashboard, adaptive formative assessments could provide students personalized feedback based on their course progress. Students can use the feedback to adjust their learning tactics and time management strategies to improve self-regulation, academic performance, awareness, and persistence. In addition, the dashboard can also support students' engagement in the courses. The automatic detection of engagement can help detect students' academic effort, persistence, and attention to detail [29]. This can provide students feedback about the emotional dimension of engagement, which describes students' willingness to work, feelings of belonging, level of interest, enthusiasm, absence of anger, anxiety, and boredom [29]. The behavioral, cognitive, and emotional dimensions detected by the automatic detection of students' engagement can be used to predict students' ability to achieve learning goals [29]. The personalized feedback can positively impact aspects of students' self-regulation and adjustment of learning strategies. The dashboard can generate feedbacks timely and personalized to the students' progress and targeted at developing students' self-awareness and self-regulation [27]. The advantage of personalized feedback is to enable students to use comprehensive time managements strategies that involve revisiting and enhancing planning, which is associated with higher engagement as well as performance to the courses.

6.3 Adaptive Formative Assessments

Adaptive formative assessment has a great potential to support students in SPOL for STEM courses. Through the formative feedback on performance and the evidence gathered about learning, formative assessment can increase students' learning awareness and identify their learning gap. To customize questions to students' knowledge state and skill level and accommodate students' knowledge change during the assessment, an adaptive mechanism can be embedded in the formative assessment. Compared with the traditional fixed-form assessment, adaptive assessment can estimate a student's knowledge and skill level in a more accurate, reliable, efficient, and flexible way. The following sections will discuss some possible adaptive algorithms to be embedded in formative assessment.

Adaptive formative assessment can be used for different purposes in different contexts. In this chapter, we argue that there are two primary modes of AFA – Diagnostic and Training. In the diagnostic mode, AFA is used to detect students' knowledge and skills level through a limited number of questions. For example, the diagnostic AFA can quickly identify a student's learning gap or learning weakness. While in a classroom or online synchronous teaching, teachers or instructors often want to know students' prior knowledge to teach a class accordingly, in self-paced online learning where no direct teaching activities are available, students need to be aware of their own knowledge proficiency of a learning topic and of their readiness to proceed with the next learning topic. So, the diagnostic mode is often used at the beginning or at the end of a topic or course. In such cases, the adaptive mechanism embedded in the formative assessment aims to shorten the question number or reduce the testing time while maintaining the diagnose accuracy.

In the training mode, AFA is used to promote students' learning by optimizing the sequence of the exercise questions based on students' knowledge and skills status and change. Skills development is critical in STEM courses, and it often needs practicing. As a type of formative assessment, these exercises usually provide feedback and cover different skills at different difficulty levels for a learning topic. One goal of sequence optimization is to avoid too difficult or too easy questions for a student so that she/he can learn more effectively. The Zone of Proximal Development (ZPD) [30] is such an area where students do not feel bored or frustrated. With the questions at a student's ZPD area, the student can gain knowledge or skills most efficiently. Another goal of such sequence optimization is to provide the proper order of skills for students to learn. Different skills across a learning unit may have some dependencies. If skill-B depends on skill-A, students should master skill-A first. So, the training mode is often used during the process of learning a topic or course.

In the diagnostic mode of AFA, students are expected to check their knowledge or skill levels with a limited number of questions or as quickly as possible. Their knowledge level can be assumed static. Therefore, some traditional adaptive algorithms can meet the needs, such as IRT, BKT, etc. Although the knowledge change is not a concern of these traditional algorithms, they must use historical data to train their models. In the training mode of AFA, students can use as much time as needed to do the exercise and check learning materials during the assessment for answers or solutions. Therefore, their knowledge and skills may change during the assessment. This mode of AFA design aims to find the best sequence of exercises to improve students' learning most effectively. Because of the possible change in skills during the assessment and their dependency, such formative assessment becomes a dynamic and uncertain environment. Reinforcement learning (RL) is an intelligent approach that finds the best sequence of actions to generate the optimal outcome through interaction with a dynamic and uncertain environment. Therefore, RL algorithms are explored to be the adaptive agency in AFA in this chapter, especially for the training mode of the AFA.

Among the reinforcement learning algorithms, multi-armed bandit (MAB) algorithms [31] have attracted significant attention in the educational field. MAB algorithms allow a truly personalized learning experience relying on little domain knowledge [32]. For example, Clement et al. [32] used MAB-based models to maximize students' learning progress in intelligent tutoring systems. The MAB algorithm makes use of online learning, by which the algorithm analyzes data in real time and returns results on the fly. Therefore, the behavior of the algorithm depends on the data it sees, but the data the algorithm sees depends on the behavior of the algorithm [33]. Thus, MAB algorithms have great potential in designing adaptive formative assessment [34].

For SPOL formative assessment, using MAB algorithms can optimize the sequence of questions to (a) promote students' learning and (b) determine students' knowledge and skill level. Of the many MAB algorithms, the Upper Confidence Bound (UCB) [35] and Thompson Sampling [36] are very popular. Researchers started to realize its effectiveness in solving educational sequencing problems, such as questions sequencing, learning activities sequencing, etc. Two key factors to be determined in MAB for sequencing is the arms and rewarding mechanism.

We argue that the learning outcomes of a course can be regarded as arms. Creating learning outcomes is usually the starting point of a course design. Assessment is designed to evaluate if students have achieved the expected learning outcomes. Questions in AFA are associated with a series of learning outcomes of a learning topic or course. Learning outcomes are also created based on a specific framework or model, such as Bloom's taxonomy. Bloom's taxonomy consists of six hierarchical cognitive levels: remembering, understanding, applying, analyzing, evaluating, and creating [37]. For a concept or skill, students are tested at one of the six difficulty levels. Therefore, if the arms of MAB in the context of formative assessment mean where the questions should draw from, the learning outcomes can be regarded as the arms.

Deciding on rewards is another key task for a real-world MAB problem. To trade off between exploration and exploitation, the rewarding mechanism of an MAB algorithm will determine the arms (options) that are exploited the most. That means such mechanism will lead the convergency direction regardless of the MAB algorithms adopted (e.g., the Upper Confidence Bounds or weighted least squares Thompson Sampling [38]). Depending on the purpose of the adaptive assessment, a different rewarding mechanism can be created. The following are two examples of MAB rewarding mechanisms: one is to quickly detect a student's weakest points in the diagnose mode, and the other is to keep students in the ZPD [39] when they are self-practicing in the training mode.

To quickly detect a student's weakest skills or concepts among a set of targeted learning outcomes, the algorithm should converge at the most challenging learning outcomes (skills). That means if a student answers a question incorrectly, a reward should be given to encourage more questions to be drawn from this learning outcome (or this arm). In contrast, if the student answers a question correctly, no reward should be given to discourage further questions from this learning outcome (or this arm).

In the case of the training or self-practicing, adaptive assessment aims to engage students with customized exercises, the ZPD, where students can gain the knowledge or skills most efficiently when practicing. Suppose the formative assessment is used for training or self-practicing. In that case, students have no time pressure to finish the exercises and are encouraged to check the learning materials or ask for help during the assessment. Therefore, during self-practicing, students' knowledge or skill level is expected to improve, resulting in a changing ZPD. It will be ideal if the adaptive mechanism is able to detect such changing ZPD and customize the exercises accordingly. In this way, the adaptive mechanism can generate exercises that keep matching up with students' updated knowledge or skill levels. The following example shows how to use the concept of ZPD to create a simple version of the rewarding mechanism.

ZPD area should contain questions that students do not feel too easy or too difficult to answer. So, if only considering the answer correctness (not the answer time in this case), we can argue that a ZPD area is where a student can only answer a portion of the questions correctly. If the student can answer questions 50% correctly, we can say the student is at the center of the ZPD for that area. The following rewarding mechanism can be suggested for the MAB algorithm to keep students practicing or answering questions at their changing ZPD areas: $R = \sin\left(\pi C^{\alpha}\right)$. In this equation, R stands for the Reward ($R \in [0,1]$), C stands for the correct answer percentage of an LO ($C \in [0,100\%]$). The parameter α determines the skew of the rewarding curve (see Figure 6.2).

In Figure 6.2, when $\alpha = 1$, the four points in Table 6.1 illustrate how the rewarding mechanism works.

At the end of the assessment, the correct answer percentage (C) of each LO can be regarded as the mastery level of this LO. Therefore, the LO with the lowest

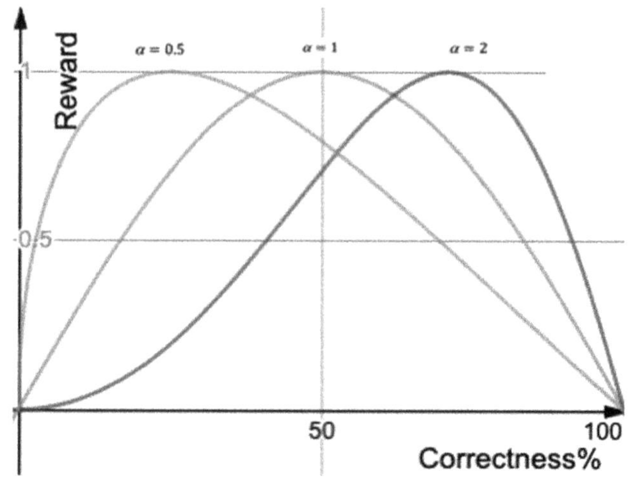

FIGURE 6.2
A ZPD-based MAB rewarding mechanism.

C is their weakest LO, the LO with the highest C is their strongest LO, and the LOs with medium C are at the ZPD level. By using the AFA, not only can students advance their learning by practicing exercises at their ZPD areas most of the time, but they will also be aware of where they are stuck if it ever happens. For example, if their ZPD is not moving forward after a long-time of practicing, the students may get stuck at that point. As the α will affect the curve of rewarding plot, it is necessary to test different values of α against the effectiveness of the MAB in the training mode of AFA and find the optimal value for a specific context. Because of the complex and dynamic nature of the reinforcement learning, simulation is usually crucial for testing before the algorithms are experimented with in real-world cases.

6.4 Automatic Detection of Students' Engagement

Automatic detection of students' engagement has been a challenging problem especially in eLearning platforms. Lack of participation and effort, disaffection, and failure to invest deeply in the academic content and activities are the key indicators of disengagement [22]. Although online learning has gained immense popularity and attention, online courses have significantly higher dropout rates than conventional courses mainly because of lack of engagement [22, 40]. Students' engagement is a complex but achievable goal if this can be detected in time and provided appropriate pedagogical intervention [22]. In a traditional classroom setting, the instructor can observe and intervene with students as needed to improve their engagement over time. However, it is quite impossible in an eLearning platform, especially with the SPOL model as students rarely get chances to meet with their instructors face-to-face or in live sessions. Most communication happens between the instructors and the students offline via email, phone, and discussion forums. Thus, the automatic identification of students' engagement in a SPOL model can greatly benefit the instructors in identifying students with lower engagement levels and the students in improving learning performance.

As mentioned earlier, students' engagements are multidimensional, including *behavioral*, *emotional*, and *cognitive* [22]. A good number of methods have been proposed to detect students' *emotional* and *behavioral* engagement automatically. However, the automatic detection of students' *cognitive* engagement is still rare in educational information systems. Facial expression is an important and widely used cue for automatically detecting students' *behavioral* and *emotional* engagements. These methods are popular because facial expressions can be captured and analyzed non-intrusively over time and these can be done using low-cost webcams or mobile devices [41, 42]. Facial action units (AUs) are more commonly used for facial expression recognition. AUs are the different muscle movements of the faces that represent specific discrete emotions [43, 44]. Psychologists and neuroscientists have extensively used AUs on various aspects of facial expression analysis. The AUs can occur either singly or in combination. *Computer Expression Recognition Toolbox* (CERT) is commonly used to track AUs for emotion analysis. Graafsgaard et al. [45] uses CERT to track AUs consisting of eyebrow raising (inner and outer), brow lowering, eyelid tightening, and mouth dimpling for predicting students' engagement and frustration. Bosch et al. [46] use CERT to track AUs and identify educational emotions, such as boredom, confusion, delight, frustration, and engaged concentration using C4.5 trees and Bayesian classifiers. In another study, Bosch et al. [47] identify positive and negative mind wondering. Six levels of granularity (including upper-body movement, head pose, facial texture patches, facial action units, co-occurring AU pairs, and temporal dynamics of AUs) have been used in this study where the support vector machines and deep neural networks have been used for classification. Manseras et al. [48] track the AUs consisting of lid tighter and lip corner puller, upper lid raiser, lip's part, and jaw drop, and identify the students as 'engaged' or 'not engaged' using support vector machine, naive Bayes, and random forest classifies. Alkabbany et al. [49] track AUs to identify the engagement as one of four levels: no face detected, behaviorally not engaged, look engaged, and emotionally engaged using convolutional neural network and support vector machine. Most authors in their study indicate that the engagement can be detected at levels above chance, though these are still far from perfect recognition.

Holistic features along with AU's have also been used in several studies for students' engagement detection. For example, Whitehill et al. [17] use GentleBoost with Box Filter features, support vector machines with Gabor filter, and Multinominal logistic regression with AU's for students' engagement detection. This study recognizes two-level (engaged and not-engaged) and four-level (not engaged, nominally engaged, engaged in tasks, and very engaged) emotional states. Authors indicate that the engagement levels of 10-second video clips can be reliably predicted from the average labels of their constituent frames. The authors also reported that more granularity in engagement levels can degrade the performance of the engagement detection methods.

Various cues from different sources have also been used with facial expression for educational emotion detection. For example, Farrell et al. [50] use head pose with facial expression to address the issues with variable lighting and when glasses were worn during engagement detection. Zhang et al. [51] use two aspects of students' behavior data: face data (using adaptive weighted Local Gray Code Patterns for facial expression recognition) and mouse interaction from the Learning Management System (LMS) to improve the accuracy of learning engagement detection. Kerdawy et al. [52] use EEG and facial expressions to build models that detect two cognitive states

TABLE 6.1
Reward Mechanism in MAB Algorithms

(0, 0)	When C is 0 (the correct answer percentage of an LO is 0%), it means the questions from this LO are too hard. In this case, the reward is 0, which discourages further questions from this LO
(1, 0)	When C is 1 (the correct answer percentage of an LO is 100%), it means the questions from this LO are too easy. In this case, the reward is 0, which also discourages further questions from this LO
(0.5, 1)	When C is 0.5 (the correct answer percentage of an LO is 50%), it means the questions from this LO are at the student's ZPD level. In this case, the reward is 1, which encourages further questions from this LO
Other cases	The reward is between 0 and 1 by the equation

(engagement and instantaneous attention) and three cognitive skills (focused attention, planning, and shifting).

6.5 Learning Analytic Dashboards

Student-facing intelligent dashboards can improve the learning experience for students in SPOL courses. Learning dashboards can be used for various dimensions, including monitoring, predicting, tutoring, feedback, recommendation, and reflection [11]. Learning dashboards can be defined as a specific class of 'personal informatics' applications, a class of tools that support users in collecting personal information about various aspects of their life, behavior, habits, and thoughts [53]. They help students to improve self-regulation by providing tools for review and analysis of their learning history. Increasing self-regulation has several benefits such as promoting insight and increasing self-control [53].

Learning dashboards can be adopted in online learning as a supporting tool to track students' activities and performance in the course. This objective can be achieved by using data mining to analyze data collected from log files to discover meaning, followed by visualizing the results to bring visibility into students' performance [54]. Evidence shows that learning dashboards can have several cognitive and metacognitive benefits for students – for example, by providing information about their progress – flag if they are at risk of failing a course, provide customized recommendations in relation to reading material and learning activities, and measure students' improvement [2]. Moreover, interactive learning dashboards can tailor the course offering based on students' needs [2]. Some of the benefits of a well-designed learning dashboard from students' perspective include (i) capturing information in real time, which can give students the ability to measure their own progress more accurately; (ii) allowing students to map their individual objectives to their learning path; (iii) giving students an overview of their personal learning preferences and style, which could help them to make better decisions for the next steps in their learning path, and (iv) making incremental changes during their learning process and adjusting learning strategies to mitigate various learning challenges [54, 55].

Different machine learning models have been used in educational data mining and these models have great potential to use in intelligent dashboard design [56]. Decision trees, K-means clustering, artificial neural network, linear discriminant analysis, naive Bayes classifier, support vector machines, and various deep learning models could be used in the dashboards for students' performance prediction, progress analysis, automated feedback generation, and educational decision-making. For real-time analysis of data and visualizing the results, these methods need to be computationally efficient and optimized by tuning different parameters [57–59]. As a proof of concept, to verify the effectiveness of the machine learning models in intelligent dashboard design, we designed a simple prototype and used a students' performance and engagement dataset [60] with 485 student records to empirically evaluate the dashboard. We used the K-means clustering model to group students into *persistent learners* (characterized by low-level of self-regulation and engagement), *regular learners* (characterized by high emotional self-control and high appropriate cognitive regulation strategies), and *irregular learners* (characterized by low motivation and strong procrastination) [56]. We examined 20 different variables on students' progress and performance. After clustering, among 485 students, 58.22% of the students are found regular, 45.68% persistent, and 1.44% irregular [32]. Figure 6.3 is K-mean clustering model results, representing three distinct categories of students (persistent, regular, and irregular) [56]. Using this model, any new learner can be compared with the other learners based on his/her performance and identified in one of the groups in the intelligent learning dashboard. This could be used as a key performance indicator (KPI) for a student for the intelligent learning dashboard that can support their cognitive, metacognitive, behavioral, and emotional competencies while students are enrolled in the course.

A key goal of designing a student-facing learning dashboard is to promote self-regulation and increase motivation in students enrolled in SPOL courses [61]. To accomplish this task, learning dashboards need to remove learning barriers identified for SPOL courses. Three critical learning dashboard applications need to be considered when designing an intelligent learning dashboard: (i) increase learning awareness; (ii) identify struggling students; and (iii) provide academic intervention [53]. In SPOL, students' self-awareness of knowledge state and learning gaps can help them to reflect on learning and decide when to ask for help. Learning dashboards can achieve this objective through social comparison, goal achievement, and monitoring learning progress [3]. Venant et al. [62] found that social comparison with other students in online courses can be a motivating factor and could cause students to work harder, increase their course engagement, and simulate a feeling of connectedness to peers.

Learning dashboards need to identify struggling students in a fashion that give them enough time to seek

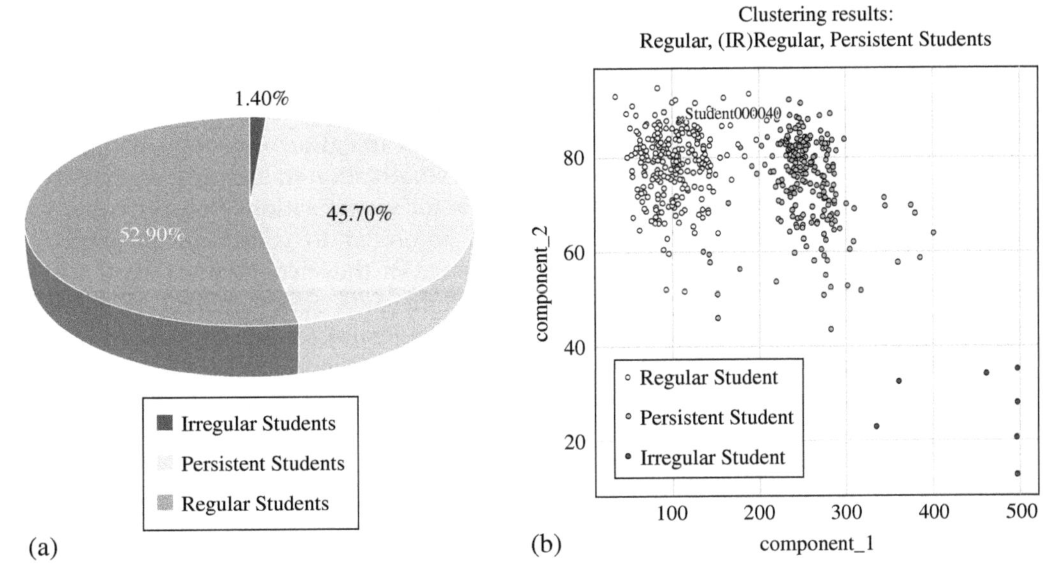

FIGURE 6.3
K-means clustering results. (a) Pie chart showing the percentage of students categorized as regular, persistent, and irregular. (b) Clustering result with a particular student highlighted in the clusters [56].

extra support. Students get the detailed information regarding the topics and concepts that are important when seeking help. This information also helps instructors to provide effective intervention [11]. The dashboard needs to support students' sense-making from learning and teaching data. This will allow students to monitor their course progress in real time and help them to better understand their engagement and motivation level [63]. To accomplish this, four main categories of indicators need to be considered: *key performance indicators*, *data hierarchy*, *dashboard design*, and *filters* [63]:

- *Key performance indicators* can be defined as quantifiable measures that are used to determine to what extent students set goals and objectives are achieved. For example, for students the goal could be to get over cognitive thresholds or the temporal progress in the course.
- *Data hierarchy* can be defined as both the structure (level) of the data and users' permission (privileges). For example, learning managers may need to have access to a dataset to enable them to explain the overall institution system performance. However, students should only be able to see their personal data such as progress in the course and the totality of their activities.
- *Dashboard design* can be defined as the appropriate way for an effective presentation and consolidation of the information. For example, is the relevant information displayed effectively on a student's mobile device?
- *Filters* can be defined as the methodology used to effectively display the relevant information. For example, showing an instructor the students' ID, or highlighting specific information in a quiz or assessment result [63].

To design an effective learning dashboard, reviewing several prior studies is essentials. The review process allows developers to better understand the key performance indicators that students find useful and not useful. The dashboard design process starts with data ingestion from the LMS. Learners enrolled in SPOL courses naturally generate a considerable amount of data as they progress through different course activities. For example, clicking a hyperlink to open web resources is data about a learner's cognition and motivation based on the context of the link [64].

Modern LMSs also collect data about the number of times students access resources, date and time of access, the number of discussion board posts generated, the number of discussion boards read, and the type of resources accessed. Aggregated trace data can support inferences about (i) learner's preferred work schedule that mildly support inferences about procrastination, (ii) which resources are judged more relevant and appealing, (iii) motivation to calibrate judgement of learning and efficacy, and (iv) value attributed to contributing, acquiring, or clarifying by exchanging information with peers [64]. Learning dashboards use data mining to draw insight and discover patterns from the data captured from the LMS. This information can be used to predict students' behavior, study patterns, students' planning and organization ability, motivation, and engagement level in a course. The information can even be used to identify different learner types, as well as the characteristics of their interactions with the LMS [65]. By mining student data from both LMSs and instructor records, researchers can accurately predict students' performance in relation to the course material. This information is both descriptive and predictive and can lead to prescriptive recommendations. From the descriptive to perspective, data mining may help answer the questions such as 'what happened?', 'where was the problem?', and 'what actions should be taken?' [66].

Language has a fundamental importance in the field of education as it is a conduit for the communication and understanding of information [67]. LMSs can generate a considerable amount of information about language generated by the students within discussion board forums and emails. This data can be mined using Natural Language Processing (NLP) to better understand the depth of students' grasp of course topics. In addition, the analysis of discussion forum posts using NLP can provide insights into students' social interaction with colleagues and instructor, which is a good indicator of their course engagement [67]. As a result, NLP analysis can be summarized in the dashboard to present students with information about their level of understanding for various course topics, which topics they found challenging, and how they interact with the instructor and other students in the course [68]. The theory is that by showing this information to students, we could potentially help students enrolled in SPOL courses to feel less isolated, and increase their self-awareness, which could affect their motivation and engagement level in the course.

Once the data captured from the LMS is analyzed, using data mining and NLP, the next step is selecting the KPIs displayed to the students. KPIs can be defined as quantifiable measures that can be used to determine to what extent the set of objectives are achieved. For example, in an intelligent student-facing dashboard, we can give students information about their *course progress, sentiment/level of understanding of the course topics presented in the discussion board posts*, and *if they are active or passive learners.* This information can help students to overcome certain learning barriers or improve their temporal progress in the course. [69].

Modern businesses have been using static visualizations to support storytelling, usually in the form of diagrams and charts embedded in the body of text. In this format, the text conveys the story and the image typically provides supporting evidence [70]. Data storytelling can be defined as an 'information compression' technique to help an audience focus on what is most important to communicate via data visualizations [71]. The goal of data storytelling is to communicate insights through the combination of data, visuals, and narrative [72]. In online education, data storytelling and visualizations can be used to communicate insights more effectively to students [73]. The following four principles should be used when communicating key insights to students using data storytelling and visualization in learning dashboards [71]: (i) have a title for visualizations that provides a succinct message intended to communicate explicitly the main message of the visualization explicitly; (ii) highlight a specific data series related to a specific KPI; (iii) declutter visualizations by removing grids, eliminating indirect legend, and reducing the number of colors; (iv) use narrative texts and labels in visualization to add context to interesting insights and minimize distraction.

Visualizations can provide an overview of a students' learning activities and how they relate to their peers, which could help improve the learning experience [74]. Visualizations can also convert abstract and complex data into concrete and simple to understand information by amplifying human cognition. Thus, the proper visualization can lead to effective communication and correct decision-making. To effectively communicate the trends and important information, several design principals must be adopted. First, the visualization should highlight important data and trends in a way that stands out from the rest. Second, the information must be organized logically, support situated awareness, and help rapid perception by using diverse visualization techniques. Third, the information should be deployed so that the elements of the information support viewers' immediate goal for decision-making [54].

As mentioned in Section 6.2, the three elements discussed in this chapter (adaptive formative assessment, automatic detection of engagement, and visualization) connect together under the intelligent student-facing dashboard to provide students with personalized feedback and detect their academic effort, persistence, and attention to detail based on behavioral, cognitive, and emotional dimensions to help increase their awareness, persistence, self-regulation, and academic performance.

6.6 Conclusions and Future Work

This chapter discussed a framework for an intelligent learning dashboard and its design considerations that can support online learners in the SPOL STEM courses. Specifically, we discussed the state-of-the-art adaptive formative assessment and students' engagement detection approaches and how to integrate them in a framework of intelligent learning dashboard. With the increasing pressure to access higher education, many institutions are focusing on increasing the number of SPOL courses they offer [75]. Some key challenges that institutions face are low learning awareness and engagement, lack of academic intervention, etc. These barriers often affect learners' progress and academic performance in the SPOL model, especially in STEM courses, where it requires a more collaborative, supportive, and engaging environment for learners to become successful [75]. To address this problem, academic institutions are evaluating strategies to get students more engaged [75].

As part of this effort, higher education institutions are exploring how information technology can be used to evaluate teaching, learning strategies, methods, activities, and the way students engage with their studies [76]. In this context, students' engagement can be thought of as the level of effort that students place on learning the content for a specific course [77]. Previous research on students' engagement in SPOL has focused on exploring the relationship between self-regulation and students' engagement [78]. However, most of these studies looked at the impact of the instructor on students' engagement [78]. In more recent studies, there has been a shift in focus from instructors' impact on students' engagement to the students [78]. To examine the effects of the instructor on students' engagement, most of these studies used questionnaires or surveys as their primary tool. Few studies looked at using big data to analyze students' engagement in SPOL courses [78]. However, a study by the authors of [79] identified language as a conduit for the communicating and understanding information. Modern LMSs collect an abundance of information about students, including information about students' discussions with other students and the instructor [79]. This data can be mined using NLP to better understand the depth of students' grasp of course topics. Also, the analysis of discussion forum posts using NLP can provide insights into students' social interactions with other students and the instructor, which is a good indicator of their course engagement [79]. By studying students' social processes, researchers can discover patterns about students' cognition and behavior in SPOL courses, which could have gone unaccounted for [80].

SPOL courses rely on tutorials and activities to transmit knowledge. As a result, students' engagement is important in determining students' motivation and self-regulation [81]. A study by Shulman [82] found that first, students' engagement can be used to measure individuals' intrinsic involvement with learning. Second, the study also found that students' engagement could be used as an indirect measure to determine if students will be successful in a course [83]. Third, engagement could be used as a direct measure of students' involvement in the educational processes [84]. Research has shown that students' engagement is important because a lack of engagement can affect students' final grades, retention of material, and course dropout rates. Additionally, students who engage in discussion board posts and other course activities are less likely to withdraw from a course [77]. As a result, students' engagement should be measured from the perspective of their interaction with other students, using tools such as NLP and more traditional methodologies. The analysis results will then be presented to students in learning dashboards, which can provide students with indicators about their performance, learning process, learning activities, and time spent studying. LDs help with showing students factors that influence their engagement in the learning environment [78]. Such information could help students make micro-adjustments in their learning strategies, which could result in finding ways to improve their self-regulation leading to better course performance.

Another part of the solution of mitigating the SPOL learning barriers is to design adaptive formative assessment across a course. We suggested some possible reinforcement learning algorithms that can be used for the adaptive agent. Some rationale and advantages of those algorithms have been provided. But the effectiveness of those algorithms needs to be tested out. Because of the complexity of the reinforcement learning, we suggested that simulation of the algorithms is conducted before experimenting with real students' data in a real learning environment. One important criticism or concern is that AI in education often separates from the human teaching process. To prevent this, we think that subject matter experts should validate if the questions sequenced by the algorithms are reasonable and explainable. To verify if adaptive formative assessment can improve students' learning awareness and help instructors provide intervention, the learning behavior data obtained from the learning system should be exposed to students. The messages delivered to students can make students aware of their learning progression, strengths and weaknesses, and the areas that need further help. Such messages can be delivered through a well-designed dashboard.

Acknowledgment

This research was supported by NSERC Discovery Grant and Athabasca University.

References

1. I. Chirikov, T. Semenova, N. Maloshonok, E. Bettinger and R. F. Kizilcec, "Online education platforms scale college STEM instruction with equivalent learning outcomes at lower cost," *Science Advances*, 6(15), pp. 1–10, 2020.
2. E. Durall and B. Gros, "Learning analytics as a metacognitive tool," *6th International Conference on Computer Supported Education*, pp. 380–384, 2014.
3. R. Bodily, J. Kay, V. Aleven, I. Jivet, D. Davis, F. Xhakaj and K. Verbert, "Open learner models and learning analytics dashboards: A systematic review," in *Proceedings of the 8th International Conference on Learning Analytics and Knowledge*, 2018.
4. D. Suthers and K. Verbert, "Learning analytics as a middle space," in *International Conference on Learning Analytics and Knowledge*, New York, 2013.
5. I. Jivet, M. Scheffel, H. Drachsler and M. Specht, "Awareness is not enough: Pitfalls of learning analytics dashboards in the educational practice," in *Data Driven Approaches in Digital Education: European Conference on Technology Enhanced Learning*, 2017.
6. I.-H. Jo, D. Kim and M. Yoon, "Analyzing the log patterns of adult learners in LMS using learning analytics," in *International Conference on Learning Analytics and Knowledge*, Indiana, 2014.
7. C. B. Hodges, "Self-efficacy in the context of online learning environments," *Performance Improvement Quarterly*, 20(3–4), pp. 7–25, 2008.
8. R. Sadler, "Formative assessment: Revisiting the territory," *Assessment in Education. Principles, Policy and Practice*, pp. 77–84, 1998.
9. M. Heritage, J. Kim, T. Vendlinski and J. Herman, "From evidence to action: A seamless process in formative assessment?," *Educational Measurement: Issues and Practice*, 28(3), pp. 24–31, 2009.
10. J. J. Vie, F. Popineau, E. Bruillard and Y. Bourda, "A review of recent advances in adaptive assessment," *Learning Analytics: Fundamentals, Applications, and Trends*, 94, pp. 113–142, 2017.
11. H. Yan, F. Lin and Kinshuk, "Including learning analytics in the loop of self-paced online course learning design," *International Journal of Artificial Intelligence in Education*, pp. 1–18, 2020.
12. M. Rezaie and M. Golshan, "Computer adaptive test (CAT): Advantages and limitations," *International Journal of Educational Investigations*, 2(5), pp. 128–137, 2015.
13. R. J. de Ayala, *The theory and practice of item response theory*. New York: The Guilford Press, 2009.
14. A. T. Corbett and J. R. Anderson, "Knowledge tracing: Modeling the acquisition of procedural knowledge," *User Modeling and User-Adapted Interaction*, pp. 253–278, 1994.
15. C. Piech, J. Bassen, J. Huang, S. Ganguli, M. Sahami, L. Guibas and J. Sohl-Dickstein, "Deep knowledge tracing," in *Advances in Neural Information Processing Systems*, 2015.
16. A. S. Lan and R. G. Baraniuk, "A contextual bandits framework for personalized learning action selection," in *International Conference on Educational Data Mining*, 2016.
17. J. Whitehill, Z. Serpell, Y.-C. Lin, A. Foster and J. R. Movel, "The faces of engagement: Automatic recognition of student engagement from facial expressions," *IEEE Transactions on Affective Computing*, 5(1), pp. 86–98, January–March 2014.
18. L. Rothkrantz, "Dropout rates of regular courses and MOOCs," in *International Conference on Computer Supported Education*, Rome, 2016.
19. H. Monkaresi, N. Bosch, R. A. Calvo and S. K. D'Mello, "Automated detection of engagement using video-based estimation of facial expressions and heart rate," *IEEE Transactions on Affective Computing*, 8(1), pp. 15–28, March 2017.
20. S. Karumbaiah, R. Lizarralde, D. Allessio, B. Woolf, I. Arroyo and N. Wixon, "Addressing student behavior and affect with empathy and growth mindset," in *International Conference on Educational Data Mining*, Wuhan, June 2017.
21. N. Bosch, "Detecting student engagement: Human versus machine," in *Conference on User Modeling Adaptation and Personalization*, Halifax, 2016.
22. J. A. Fredricks, *Eight myths of student disengagement – Creating classrooms of deep learning*. CA: Corwin, 2014.
23. A. R. Anderson, S. L. Christenson, M. F. Sinclair and C. A. Lehr, "Check & Connect: The importance of relationships for promoting engagement with school," *Journal of School Psychology*, 42(2), pp. 95–113, 2004.
24. S. L. Christenson, A. L. Reschly and C. Wylie, *Handbook of research on student engagement*. New York: Springer, 2012.
25. S. L. Christenson and A. R. Anderson, "The centrality of the learning context for students' academic enabler skills," *School Psychological Review*, 31(3), pp. 378–393, 2002.
26. L. R. Harris, "A phenomenographic investigation of teacher conceptions of student engagement in learning," *The Australian Educational Researcher*, pp. 57–79, 2008.
27. L.-A. Lim, D. Gasevic, W. Matcha, N. A. Uzir and S. Dawson, "Impact of learning analytics feedback on self-regulated learning: Triangulating behavioural logs with students' recall," in *International Learning Analytics and Knowledge Conference*.
28. E. Delen and J. Liew, "The use of interactive environments to promote self-regulation in online learning: A literature review," *European Journal of Contemporary Education*, 15(1), pp. 24–33, 2016.
29. A. Silvola, P. Naykki, A. Kaveri and H. Muukkonen, "Expectations for supporting student engagement with learning analytics: An academic path perspective," *Computers and Education*, 168, p. 104192, 2021.
30. L. Vygotsky, *Mind in society: The development of higher psychological processes*. Cambridge, MA: Harvard University Press, 1978.
31. D. Berry and B. Fristedt, *Bandit problems: Sequential allocation of experiments*. London: Chapman and Hall, 2013.

32. B. Clement, D. Roy, P. Y. Oudeyer and M. Lopes, "Multi-armed bandits for intelligent tutoring systems," *Journal of Educational Data Mining*, 7(2), pp. 20–48, 2015.
33. J. White, *Bandit algorithms for website optimization*. O'Reilly Media, Inc, 2012.
34. J. Mui, F. Lin and M. A. A. Dewam, "Multiarmed bandit algorithms for online adaptive learning: A survey," in *Artificial Intelligence in Education (AIED 2021)*, Utrecht, I. Roll et al. (Eds.), AIED 2021, LNAI 12749, pp. 273–274, 2021.
35. R. Agrawal, "Sample mean based index policies with O (log n) regret for the multi-armed bandit problem," *Advances in Applied Probability*, 27, pp. 1554–1078, 1995.
36. B. Clement, D. Roy, P. Y. Oudeyer and M. Lopes, "On the likelihood that one unknown probability exceeds another in view of the evidence of two samples," *Biometrika*, 25(3–4), pp. 285–294, 1933.
37. E. J. Furst, *Bloom's taxonomy: Philosophical and educational issues*. Chicago, IL: University of Chicago Press, 1994.
38. A. Slivkins, Introduction to multi-armed bandits, 2019. https://arxiv.org/pdf/1904.07272.pdf
39. H. Kuusisaari, "Teachers at the zone of proximal development: Collaboration promoting or hindering the development process," *Teaching and Teacher Education*, pp. 46–57, 2014.
40. P. Bawa, "Retention in online courses: Exploring issues and solutions – A literature review," SAGE Open, pp. 1–11, 2016. https://doi.org/10.1177/2158244015621777
41. A. Dewan, M. Murshed and F. Lin, "Engagement detection in online learning: A review," *Smart Learning Environments*, pp. 1–17, 2019.
42. A. Dewan, F. Lin, D. Wen, M. Murshed and Z. Uddin, "A deep learning approach to detecting engagement of online learners," in *IEEE International Conference on Internet of People*, Guangzhou, pp. 1895–1902, October 2018.
43. P. Ekman and W. Friesen, *Facial action coding system: A technique for the measurement of facial movement*. Consulting Psychologists Press, 1978.
44. P. Ekman, W. Friesen and J. C. Hager, *"Facial action coding system," a human face*, 2002.
45. J. F. Grafsgaard, J. B. Wiggins, K. E. Boyer, E. N. Wiebe and J. C. Lester, "Automatically recognizing facial indicators of frustration: A learning-centric analysis," in *International Conference on Affective Computing & Intelligent Interaction*, Geneva, pp. 159–165, September 2013.
46. N. Bosch, S. K. D'Mello, R. S. Baker, J. Ocumpaugh, V. Shute, M. Ventura, L. Wang and W. Zhao, "Detecting student emotions in computer-enabled classrooms," in *International Joint Conference on Artificial Intelligence*, New York, 2016.
47. N. Bosch and S. K. D'Mello, "Automatic detection of mind wandering from video in the lab and in the classroom," *IEEE Transactions on Affective Computing*, pp. 1–16, 2019.
48. R. Manseras, T. Palaoag and A. Malicdem, "Class engagement analyzer using facial feature classification," in *Future Technologies Conference*, Vancouver, 2017.
49. I. Alkabbany, A. Ali, A. Farag, I. Bennett, M. Ghanoum and A. Farag, "Measuring student engagement level using facial information," in *IEEE International Conference on Image Processing*, Taipei, 2019.
50. C. C. Farrell, C. Markham and C. Deegan, "Real time detection and analysis of facial features to measure student engagement with learning objects," in *Irish Machine Vision and Image Processing*, Dublin, 2019.
51. Z. Zhang, Z. Li, H. Liu, T. Cao and S. Liu, "Data-driven online learning engagement detection via facial expression and mouse behavior recognition technology," *Journal of Educational Computing Research*, 58(I), pp. 63–86, 2019.
52. M. E. Kerdawy, M. E. Halaby, A. Hassan, M. Maher, H. Fayed, D. Shawky and A. Badawi, "The automatic detection of cogniion using EEG and facial expression," *Sensors*, pp. 1–32, 2020.
53. K. Verbert, E. Duval, J. Klerkx, S. Govaerts and J. L. Santos, "Learning analytics dashboard applications," *American Behavioral Scientist*, 57(10), pp. 1500–1509, 2013.
54. Y. Park and I.-H. Jo, "Development of the learning analytics dashboard to support students' learning performance," *Journal of Universal Computer Science*, pp. 111–133, 2015.
55. C. Howlin and D. Lynch, "Learning and academic analytics in the realizeit system," in *World Conference on E-Learning in Corporate, Government, Healthcare, and Higher Education*, pp. 1–11, October 2014.
56. A. Farahmand, M. A. A. Dewan and F. Lin, "Constructing intelligent learning dashboard for online learners," in *IEEE Cyberscience and Technology Conference 2021*, 2021.
57. O. Y. Al-Jarrah, P. D. Yoo, S. Muhaidat, G. K. Karagiannidis and K. Taha, "Efficient machine learning for big data: A review," *Big Data Research*, 2(3), pp. 87–93, 2015.
58. IGI Global, "What is computational cost," [Online]. Available: https://www.igi-global.com/dictionary/computational-cost/82637. [Accessed 03 October 2021].
59. G. Wang, J. Xu and B. He, "A novel method for tuning configuration parameters of spark based on machine learning," in *IEEE International Conference on High Performance Computing and Communications*, pp. 586–593, 2016.
60. A. Moubayed, M. Injadat, A. Shami, A. BouNassif and H. Lutfiyya, "Student performance and engagement prediction in e-learning datasets," *IEEE Dataport*, 2020.
61. A. Farahmand, M. A. A. Dewan and F. Lin, "Student-facing educational dashboard design for online learners," in *IEEE Cyber Science and Technology Congress*, Calgary, pp. 345–349, 2020.
62. R. Venant, P. Vidal and J. Broisin, "Evaluation of learner performance during practical activities: An experimentation in computer education," in *International Conference on Advanced Learning Technologies*, Austin, pp. 237–241, 2016.
63. B. Fazlagic, A. Dipace and T. Minerva, "The design of a learning analytics dashboard," *Journal of E-Learning and Knowledge Society*, 15(3), pp. 30–47, 2019.

64. P. H. Winne, "Chapter 21: Learning analytics for self-regulated learning," *Handbook of Learning Analytics – First Edition*, pp. 241–249, 2017.
65. M. Bienkowski, M. Feng and B. Means, *Enhancing teaching and learning through educational data mining and learning analytics*. U.S. Department of Education, pp. 1–26, 2012.
66. B. Dietz-Uhler and J. E. Hurn, "Using learning analytics to predict (and improve) student success: A faculty perspective," *Journal of Interactive Online Learning*, 12(1), pp. 17–26, 2013.
67. D. S. McNamara, L. K. Allen, S. A. Crossley, M. Dascalu and C. A. Perret, "Chapter 8: Natural language processing and learning analytics," *Handbook of Learning Analytics*, pp. 93–100, 2017.
68. A. S. Alblawi and A. A. Alhamed, "Big data and learning analytics in higher education demistifying variety, acquisition, storage, NLP, and analytics," in *2017 IEEE Conference on Big Data and Analytics*, 2017.
69. A. Dipace, B. Fazlagic and T. Minerva, "The design of a learning analytics dashboard: Eduopen MOOC platform redefinition procedures," *Journal of e-Learning and Knowledge Society*, 15(3), pp. 30–47, 2019.
70. E. Segel and J. Heer, "Narrative visualization: Telling stories with data," *IEEE Transactions on Visualization and Computer Graphics*, 15(6), pp. 1139–1148, 2010.
71. V. Echeverria, R. Martinez-Maldonado, S. B. Shum, K. Chiluiza, R. Granda and C. Conati, "Exploratory versus explanatory visual learning analytics: Driving teachers' attention through educational data storytelling," *Journal of Learning Analytics*, 5(3), pp. 72–97, 2018.
72. B. Dykes, "Data storytelling: What it is and how it can be used to effectively communicate analysis results," *Applied Marketing Analytics*, 1(4), pp. 299–315, 2015.
73. R. Martinez-Maldonado, V. Echeverria, G. F. Nieto and S. B. Shum, "From data to insights: A layered storytelling approach for multimodal learning analytics," in *Proceedings of the 2020 Chi Conference on Human Factors in Computing Systems*, 2020.
74. E. Duval, "Attention please!: Learning analytics for visualization and recommendation," in *Proceedings of the International Conference on Learning Analytics and Knowledge*, 2011.
75. K. A. Meyer, "Student engagement in online learning: What works and why," in *ASHE Higher Education Report*, 2014.
76. C. Beer, K. Clark and D. Jones, "Indicators of engagement," in *Proceedings of the Ascilite*, Sydney, 2010.
77. M. Hussain, W. Zhu, W. Zhang and S. M. R. Abidi, "Student engagement predictions in an e-learning system and their impact on student course assessment score," *Computational Intelligence and Neuroscience*, pp. 1–22, 2018.
78. J. Ma, X. Han, J. Yang and J. Cheng, "Examining the necessary condition for engagement in an online learning environment based on learning analytics approach: The role of the instructor," *Internet and Higher Education*, 24, pp. 26–34, 2015.
79. D. S. McNamara, L. K. Allen, S. A. Crossley, M. Dascalu and C. A. Perret, "Natural language processing and learning analytics," in *Handbook of Learning Analytics*, 2017.
80. B. Chen, Y.-H. Chang, F. Ouyang and W. Zhou, "Fostering student engagement in online discussion through social learning analytics," *The Internet and Higher Education*, 37, pp. 1–10, 2018.
81. H. Coates, "A model of online and general campus-based student engagement," *Assessment and Evaluation in Higher Education*, 32(2), pp. 121–141, 2007.
82. L. S. Shulman, "Making differences: A table of learning," *Change: The Magazine of Higher Learning*, 34(6), pp. 36–44, 2002.
83. P. T. Ewell and D. P. Jones, "Actions matter: The case for inndirect measure in assessing higher education's progress on the national education goal," *The Journal of General Education*, 42, pp. 123–148, 1993.
84. G. D. Kuh, C. Pace and N. Vesper, "The development of process indicators to estimate student gains associated with good practices in undergraduate education," *Research in Higher Education*, 38(4), pp. 435–454, 1997.

Section II

AI-Enhanced Adaptive Learning Resources

7

PASTEL: Evidence-based Learning Engineering Methods to Facilitate Creation of Adaptive Online Courseware

Noboru Matsuda, Machi Shimmei, Prithviraj Chaudhuri, Dheeraj Makam, Raj Shrivastava, Jesse Wood, and Peeyush Taneja

CONTENTS

7.1 Introduction 93
7.2 CyberBook 94
7.3 PASTEL 95
7.3.1 WATSON 95
7.3.2 SMART 96
7.3.3 QUADL 97
7.3.4 RAFINE 98
7.3.5 RADARS 99
7.4 Evaluation 100
7.4.1 SMART 101
7.4.2 QUADL 101
7.4.3 RAFINE 103
7.4.4 RADARS 105
7.5 Conclusion 106
Notes 107
References 107

7.1 Introduction

Massive Online Open Course (MOOC) has been growing rapidly with accomplished impact on hundreds of millions of students all over the world. Current online-course technology has a great potential for providing effective instructions with rich multimedia contents and multi-modal activities. It is therefore expected that online courses, when appropriately engineered, can promote deep *conceptual understanding of complex subjects*. Cognitive tutors, on the other hand, offer an adaptive instructional technology that promotes the *mastery learning of problem-solving skills*. While these approaches have been remarkably successful in supporting learners within these specifics, there continues to be a lack of educational technologies that promote robust learning with *synergetic competency* across procedural and conceptual understanding.

There is ample evidence that adaptive and personalized learning technologies can dramatically improve learning outcomes (Hattie, 2008). However, most online learning platforms offer personalization only at minimal levels – e.g., video lectures and simple multiple choice activities combined with unmanaged discussion forums. One example of this problem can be seen in the popular online courses – MOOCs – that often provide a limited and fixed set of instructions for all students. The current online-course technologies are deficient in the practice for skill mastery.

Cognitive tutors, on the other hand, provide highly adaptive instructions for skill acquisition (Ritter et al., 2007; VanLehn, 2011). Cognitive tutors themselves, however, might not necessarily convey conceptual understanding. We therefore hypothesize that effectively integrating cognitive tutors into well-designed online courses will create a new type of online learning environment that promotes the synergetic competency.

There has been intensive research on effective instructions, e.g., learning by doing (Anzai & Simon, 1979), learning from examples (Chi & Bassok, 1989), self-explanation (Chi, 1996), spaced practice (Pavlik Jr. & Anderson, 2008), assistance dilemma (Koedinger & Aleven, 2007), etc. However, there is a critical lack of

DOI: 10.1201/9781003181187-9

theory development surrounding active online learning. In general, active online learning has the potential to provide students with richer learning activities than any other research-based interventions – multimedia instructions, multi-modal hands-on (e.g., problem-solving, simulation), homework assignments, and formative and summative assessments. In the current work, we further add cognitive tutors to the mix to promote student's synergetic competency.

There are some issues to achieve this goal. First, there is a lack of efficient and effective authoring technology for cognitive tutors and their integration into online courses. Cognitive tutors are notoriously expensive to build (Murray, 1998, 1999, 2003) and hence nearly impossible to truly integrate into a larger adaptive online course. Second, ensuring that these courses are authored in an effective, evidence-based way is often reliant on expensive and inconsistent human support.

We therefore propose to develop a next-generation, scalable, cyberlearning architecture for adaptive online courses, called CyberBook, that provide students with adaptive online instruction and researchers with sharable big learning data as the basis for theory development on how students learn in the adaptive online-course learning environment. We also propose to develop a suite of artificial intelligence techniques to assist courseware developers to build adaptive online courseware, called PASTEL (*P*ragmatic methods to develop *A*daptive and *S*calable *T*echnologies for next-generation *E*-*L*earning).

7.2 CyberBook

The online course content implemented on CyberBook is organized as a hierarchical collection of 'chapters' that contains a sequence of 'sections' that consist of 'units'. This parallels the format of a traditional textbook. CyberBook is designed to facilitate acquisition of *interconnected knowledge*. We hypothesize that to acquire interconnected knowledge, students must be given opportunities to navigate through the knowledge network *and* revisit nodes to strengthen their understanding of individual concepts (encoded as nodes in the knowledge network) and the connections between them. To facilitate acquisition of interconnected knowledge, CyberBook provides students with adaptive support.

CyberBook carries out four types of 'adaptive' supports among many different kinds of adaptive instructions implemented on online courses: (1) timely and contextualized scaffolding for learning by doing (aka tutored problem-solving), (2) optimal problem selection, (3) the impasse-driven adaptive remediation supported by dynamic linkage between a learning-by-doing and online course material, and (4) the need-based adaptive assessment.

The first two adaptive supports are realized by adaptive instructions provided by cognitive tutors (Corbett, 2001; Ritter et al., 2007). The cognitive tutor provides the double-looped adaptive instruction (VanLehn, 2006) where the outer loop is to compute an optimal problem sequence driven by the knowledge-tracing technique, and the inner loop is to provide immediate feedback and just-in-time hint driven by the model-tracing technique. The embedded cognitive tutors provide the adaptive problem selection (within individual cognitive tutors) and adaptive problem-solving scaffolding. Integration of cognitive tutors to online courseware is done by the WATSON technology, as described in Section 7.3.1.

The third adaptive support, the *impasse-driven adaptive remediation*, is realized by the wheel-spinning detector, called RADARS, as described in Section 7.3.5. *Wheel-spinning*, by definition, occurs when cognitive tutor fails to stop posing problems in the face of a student's inability to meet the predefined level of mastery. The wheel-spinning detector classifies a sequence of student's attempts on cognitive tutors into two categories: a sequence of attempts that will eventually meet the mastery level and the one that will never meet the mastery level. Once wheel-spinning is detected on a particular skill, the corresponding instructional material on the online courseware (often written material, but potentially richer activities) is identified and presented to the student. This can be done by the embedded skill model that provides a three-way connection among didactic instructions (i.e., paragraph texts), cognitive tutors, and traditional assessment items. The skill model is automatically identified by analyzing pedagogical texts appearing on the courseware by the SMART technology, as described in Section 7.3.2.

The fourth adaptive support, the *need-based adaptive assessment*, is realized by an extension of knowledge-tracing technique. The excessive training is a known issue that occurs when a student reaches a predefined mastery level, but the system inadequately provides further formative assessment (Koedinger et al., 2012; Martin et al., 2011). The excessive training decreases student's motivation and causes the serious early course termination problem (Gates et al., 2012; Goodman, 2002; Seymour & Hewitt, 1997; Watkins & Mazur, 2013). Addressing this challenge requires a mechanism to dynamically compute the optimal activities for students to reach mastery while maintaining their motivation and engagement. In the current implementation of CyberBook, when a cognitive tutor detects a mastery on a particular skill, the system identifies assessment items that are most closely related to the mastered skill based on the embedded skill model. Cognitive tutors in this design arguably function as formative assessments (with adaptive scaffolding) just like other types of formative assessments. In other words, we design the mastery-based skill learning distributed across multiple cognitive tutors and formative assessments. Since cognitive tutors on an adaptive online course can be designed to share skills, cognitive tutors can be dynamically dropped just like other formal assessment items.

The concept of CyberBook is an independent platform, hence it can be implemented on any online learning platform. At the time of writing, a prototype of CyberBook has been implemented using the Open edX platform.[1] We have implemented several customized xBlocks to seamlessly integrate adaptive scaffolding techniques.

7.3 PASTEL

This section briefly introduces the current collection of PASTEL techniques. The results from their primary evaluations are given in Section 7.4. Table 7.1 shows a list of current PASTEL technologies.

7.3.1 WATSON

WATSON (*W*eb-browser-based *A*uthoring *T*echnique for adaptive tutoring *S*ystem on *On*line courseware) is a web-browser-based authoring environment where course developers can create cognitive tutors by demonstrating solutions. We aim to provide an authoring aid for broader content creators who are not necessarily experienced developers for cognitive tutors, which otherwise is very costly to build (Murray, 2016). The primary design objective of WATSON is therefore to carry out an authoring environment where authors can build cognitive tutors without heavy programming and seamlessly integrate them into online courseware.

The cognitive tutor provides students with guided problem-solving that consists of adaptive problem selection and scaffolding on solution steps (Ritter et al., 2007; VanLehn, 2006). These two types of adaptation are driven by a *domain expert model* that represents a set of skills necessary to solve problems on a particular *tutoring interface* reified on the tutor.

The observation above implies that, to build a cognitive tutor, the author only needs to create a tutoring interface and a domain expert model. The algorithms to provide adaptive problem selection and scaffolding are subject agnostic, hence they can be implemented as a generic library and automatically embedded into cognitive tutors. Indeed, two known techniques, knowledge tracing (Corbett & Anderson, 1995) and

TABLE 7.1

The Current PASTEL Technologies

Technology	Description
WATSON (3.1)	A web-browser-based authoring tool that allows course developers to create cognitive tutors by demonstrating solutions. It consists of a WYSIWYG HTML interface builder (CTAT) and an interactive machine learning agent (SimStudent). Resulted cognitive tutors are HTML5 based, hence can be easily integrated into online courseware given that task generic pedagogy (i.e., model tracing and knowledge tracing) is available as an external service
SMART (3.2)	A text analysis tool to automatically extract latent skills from didactic and assessment texts on the given courseware content. The given texts will be first clustered based on their semantic similarity and then each cluster will be given a label that best represents the texts in the cluster
QUADL (3.3)	A deep-neural network model to generate questions from didactic texts and learning objectives. The questions generated are aimed to facilitate learning to attain the given learning objectives. To achieve this goal, QUADL consists of the Answer Prediction model (AP) and the Question Conversion model (QC). The QC model converts a given sentence into a question for which a particular token specified in the given sentence becomes an answer. The AP model determines whether a given sentence could be converted into a question to attain the learning objective. The output from the AP model is the given sentence (i.e., the input) with the pair of indices indicating the start and end tokens to be the prospective answer
RAFINE (3.4)	An application of reinforcement learning to identify instructional components on the given online courseware – e.g., assessment questions, hint messages, videos, etc. RAFINE requires actual students' performance and learning outcome data taken from the target online courseware. The basic idea is to convert the learning data into a Markov Decision Process (MDP) where states represent a history of learning activities, edges represent a learning activity taken, and rewards reflect the increase of predicted mastery level computed by Additive Factor Model
RADARS (3.5)	A regression model to make a prediction for at-risk students who, by definition, are unlikely to achieve a predefined masterly level within the predefined amount of time, aka wheel-spinning detection. One of the criteria for RADARS performance is the speed of detection – the earlier, the better to make a prediction with high accuracy

Numbers in the parentheses show sections in which the detailed explanations are given.

model tracing (Anderson & Pelletier, 1991), have been implemented and combined into WATSON.

To assist building a tutoring interface, WATSON uses an existing interactive WYSIWYG interface builder, Cognitive Tutor Authoring Tools, or CTAT for short (Aleven et al., 2006; Aleven et al., 2016; Koedinger et al., 2003). CTAT runs on a standard web-browser CTAT and allows authors to *draw* a tutoring interface on the web-browser. The tutoring interface will then be automatically converted into an HTML5 code that is rendered on a standard web-browser.

To assist creating a domain expert model, WATSON allows authors to *tutor* an interactive machine-learning agent, called SimStudent, how to solve problems using the tutoring interface just built on the web-browser (Matsuda et al., 2015). Figure 7.1 shows an example screenshot of WATSON. Through guided problem-solving, SimStudent not only learns cognitive skills that are sufficient to solve the exact problems tutored but also generalizes those skills so that they can be used to solve similar problems – e.g., those with different values or those that require a different order of operations (Li et al., 2015). The learned skills are represented in the form of production rules written in the Jess language (Friedman-Hill, 2003) that will be used as a domain expert model for the target cognitive tutor (Matsuda et al., 2015). The problems used for tutoring SimStudent will be saved for candidate problems given to students once the cognitive tutor is deployed through CyberBook.

The cognitive tutor is deployed on CyberBook using a custom xBlock that uses the iframe technology to render the cognitive tutor. The actual instance of cognitive tutor is hosted a dedicated server machine that also hosts model-tracing and knowledge-tracing services.

7.3.2 SMART

Written pedagogical texts for instructions and assessments presumably reflect 'skills' that are a collection of knowledge that students need to learn, aka, *knowledge*

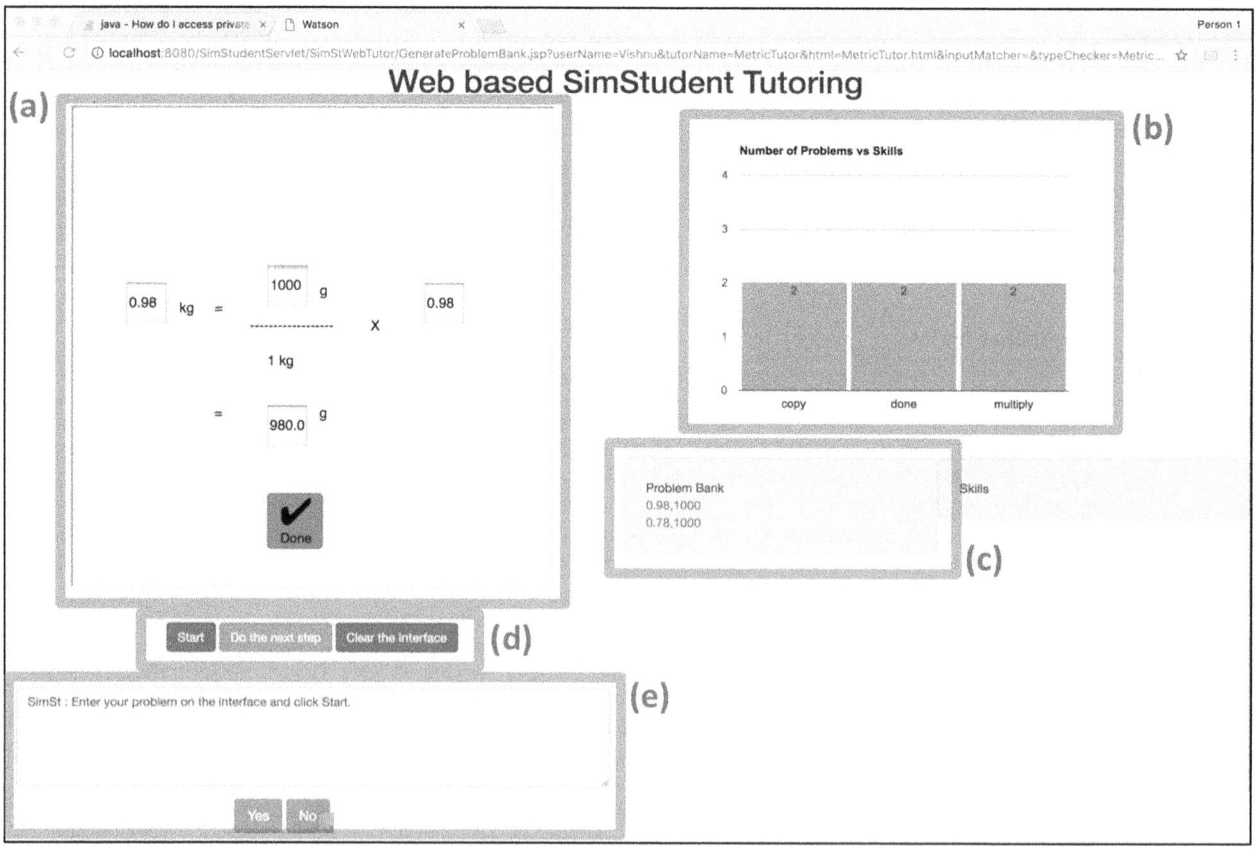

FIGURE 7.1
An example screenshot of WATSON – a web-browser-based cognitive tutor authoring tool. The author is teaching SimStudent how to solve problems using the tutoring interface (a) created by CTAT HTML editor. (b) Name of skills that has been tutored and the number of problems on which the corresponding skills were tutored. (c) A list of problems tutored. (d) Buttons to enter a new problem, perform a next step for SimStudent, and clear the interface. (e) A 'chat' box showing message from SimStudent – when SimStudent performs a step, it asks the author the correctness of the step, and the author responds by clicking the (yes/no) button underneath. When SimStudent gets stuck on a problem, it asks the author to demonstrate the next step.

components (Koedinger et al., 2012). We therefore hypothesize that those latent skills can be automatically extracted by analyzing texts used in the written instruction and assessments. To test this hypothesis, we have developed the SMART method (*S*kill *M*odel mining with *A*utomated detection of *R*esemblance among *T*exts) as a combination of a text clustering technique for skill discovery and a text ranking technique for skill labeling.

In the current context, the *pedagogical text* includes didactic instructions (paragraph text), formative assessment items (a few sentences of question for multiple-choice and fill-in-the-blank items), and hint messages used in cognitive tutors (a few sentences).

The basic idea of automated skill discovery is as follows: (1) First, a set of skills (called a skill model) is identified by clustering the assessment items. (2) Next, each cluster of text is then given a label that becomes a name of the corresponding skill. (3) Finally, each didactic instruction and cognitive tutor's hint message is assigned to one of the clusters based on their linguistic similarity to make an association between assessments and didactic instructions/hint messages. That way *a three-way association among the assessments, didactic instructions, and cognitive tutors are made through the skill model.*

To cluster the text of assessment items, we apply the K-means clustering technique (MacQueen, 1967) with an assumption that each cluster of text represents a unique skill. Once the cluster is formed, we apply the TextRanking algorithm (Mihalcea & Tarau, 2004) to each cluster of text to compute a keyword that hypothesizes/summarizes the cluster, hence represents a name of the skill.

After skill clusters are identified and labeled, the association among didactic instructions, the hint messages of cognitive tutors, and the skill clusters is computed using the cosine-similarity measure (Salton & McGill, 1983). The document embeddings are used to compute the cosine similarity. First, the document embeddings for each individual cluster of skills are computed. Then the document embeddings for individual didactic instructions and cognitive tutor's hint messages are computed. Finally, for each skill cluster, a didactic instruction and a cognitive tutor's hint message with the closest cosine similarity are declared as the associated entities.

The knowledge on skill association among instructional components (assessments, paragraphs, and cognitive tutors) gained through analyzing text done by SMART is used as a clue to provide adaptive instruction implemented on CyberBook (e.g., problem-hiding) as mentioned above – e.g., assessment hiding. Videos could be transcribed, and the same technique could apply to associate videos with the skill cluster, but this feature would be a future extension.

7.3.3 QUADL

One of the challenges for courseware developers when creating online course is to produce formative assessments. Automation of question generation therefore becomes a critical assistive technology for learning engineering. For the pedagogical point of view, the 'quality' of the questions generated is our primary concern. We are also interested in investigating the effect of test-enhanced learning that conjectures that answering formative assessment facilitate students' learning (Binks, 2018; McConnell et al., 2015; Roediger et al., 2011). Although there are a number of studies on question generation in the field of AI in education (Kurdi et al., 2020; Pan et al., 2019), little has been studied about the pedagogical value of the generated questions.

To fill this gap, we have developed an artificial intelligence technology for generating questions that supposedly inquire about the key concepts that the students ought to grasp to attain the learning objectives. We shall call the proposed technique for question generation QUADL (*QU*estion generation with an *A*pplication of *D*eep *L*earning). As far as we are aware of at the time this chapter is written, there have been no studies conducted that aim to generate questions that align with the learning objectives.

As the initial effort toward generating pedagogically valuable questions, the current version of QUADL focuses on *verbatim questions* in the form of short-answer questions. An answer for a verbatim question is a word or a phrase within a source sentence from which the question is generated. For example, consider the following learning objective:

> **Learning Objective**: Describe metabolic pathways as stepwise chemical transformations either requiring or releasing energy; and recognize conserved themes in these pathways.

For this learning objective, the following sentence can be converted into a question for which the underlined tokens will be the answer, and the resulted question will well support students' learning to attain the learning objective:

> **Sentence**: Among the main pathways of the cell are photosynthesis and *cellular respiration*, although there are a variety of alternative pathways such as fermentation.
>
> **Question**: Along with photosynthesis, what are the main pathways of the cell?

The current literature suggests that verbatim questions prompt memory retrieval that facilitates learning the factual and conceptual knowledge (Bjork et al., 2014; Pan & Rickard, 2017; Smith & Karpicke, 2014).

QUADL consists of two parts: the Answer Prediction Model (APM) and the Question Conversion Model (QCM). The APM predicts whether a given sentence is suitable to generate a verbatim question for a given learning objective. The output from APM is a *token index* <*Is*, *Ie*> indicating a start and end of the *target tokens* (one or more consecutive words) that represent key concepts in the given sentence. The target tokens become the answer for a question converted from the given sentence. The conversion is done by the QCM. The input to QCM are the sentence and the target tokens (that are the output from APM).

The APM for the current version of QUADL is implemented using BERT, Bidirectional Encoder Representation from Transformers (Devlin et al., 2018). BERT was trained for the answer prediction task as follows: the learning objective (*LO*) and sentence (*S*) were combined as a single input to the model. The final hidden state of the BERT model was fed to a single-layer classifier that outputs logit for the start index (*Is*) and another single-layer classifier that outputs logit for the end index (*Ie*) for each token in the sentence *S*. The final score is calculated by taking the softmax of the sum of the start logit and end logit for every possible span (*Is* < *Ie*) in the sentence. The score is also calculated for <*Is* = 0, *Ie* = 0> indicating that the given sentence is not suitable to generate a question for the given learning objective. The index <*Is*, *Ie*> with the largest score became the final prediction.

For the QCM, an existing pretrained model, QG-Net (Wang et al., 2018) was used. QG-Net has already been trained on the SQuAD dataset, which consists of question–answer pairs regarding the Wikipedia articles, and tested on educational textbooks. The current study provides an opportunity to test the generality of QG-Net.

7.3.4 RAFINE

One of the most challenging issues for online courseware engineering is to maintain the quality of instructional components (e.g., written text, video, and assessments). Learning engineers would like to know how individual instructional components contributed to students' learning. However, identifying effectiveness of each instructional components relative to students' learning is a hard task, because it requires significant expertise in learning science, learning technology, and subject matter pedagogy. To address this challenge, we developed an innovative application of reinforcement learning as an assessor of instructional components implemented in given online courseware. The proposed technology is called RAFINE (*R*einforcement learning *A*pplication *F*or *IN*cremental courseware *E*ngineering).

Figure 7.2 shows an overview of RAFINE. To apply RAFINE, the initial version of the target online courseware needs to be used by students while their activities being logged. These activity data consist of standard clickstream data, including students' responses for formative assessments and their correctness. We call

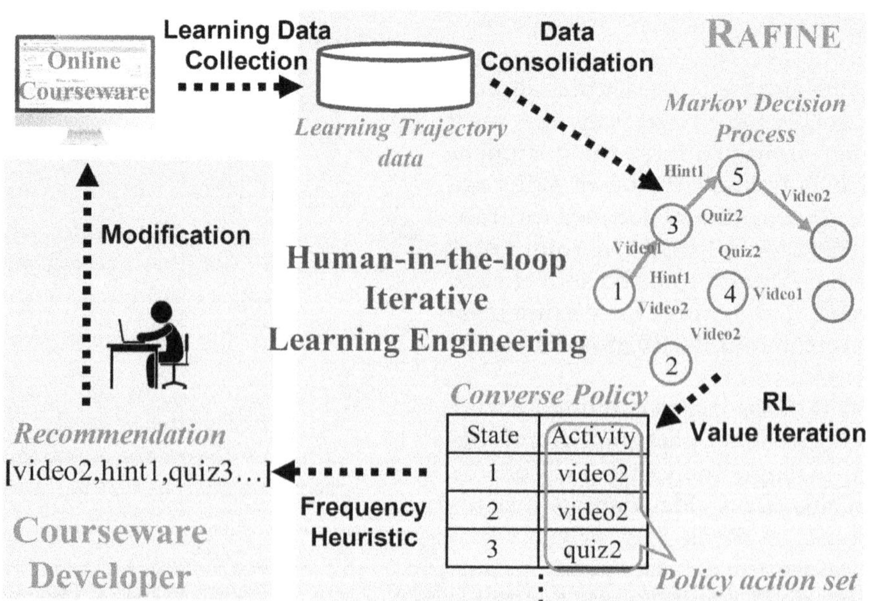

FIGURE 7.2
Overview of RAFINE.

these activity data the *learning trajectory data*. The learning trajectory data from all students are first consolidated into a single-state transition graph called *learning trajectory graph* (LTG).

The LTG is a Markov decision process (MDP) where states represent students' intermediate learning status and actions represent instructional components taken. A learning status consists of a pair of *action history* and *mastery level*; $<\boldsymbol{ah}_{i,T}, p_{i,T}(\varphi)>$. Action history $\boldsymbol{ah}_{i,T}$ is a binary vector $<ah_i^1, \ldots, ah_i^K>$ where ah_i^m shows whether student i has taken the mth instructional component in Φ^φ by time T (assuming the instructional components are ordered and $|\Phi^\varphi| = K$). Mastery level $p_{i,T}(\varphi)$ is a scalar value showing a predicted probability of student i applying skill φ at time T correctly. The value of mastery level is rounded to the nearest multiple of 0.05 (e.g., 0.12 becomes 0.10) to reduce the number of states in the LTG (which will be otherwise intractable). Mastery level is computed based on the history of learning activities. An underlying assumption is that commitment to a particular type of learning activity would increase the mastery level by a specific amount. There are several known techniques available to achieve this goal, including Bayesian models and regression models. As long as masterly level is monotonically updated, any student-modeling technique would work for the RAFINE method.

LTG is annotated with predefined *rewards* that represent quantitative benefits of the learning activity that causes transition from one state to another in the LTG. Finally, a value iteration technique is applied to compute a *converse policy* that shows the worst action to be taken at each state to achieve the expected learning outcome (represented as a table in Figure 7.1). As a consequence, a collection of actions suggested by a converse policy corresponds to a set of instructional components that have the least likelihood at each state to contribute to the ideal learning outcome. We call this collection of actions the *policy action set*.

To create a recommendation for refinement of the courseware contents, RAFINE *interprets the policy as a whole*. That is, all actions in a policy action set are holistically analyzed. Note that in most cases, all instructional components are likely to be included in a converse policy action set, since the number of states in the LTG gets larger than the number of instructional components available on the given online courseware. Therefore, whether the instructional component has been selected as a converse policy action is not a sufficient criterion to decide that the component should be included in a recommendation for refinement. The relative effectiveness of individual instructional components must be analyzed based on the overall frequency. We call this heuristic the *frequency heuristic*. In other words, we hypothesize that actions that frequently appear in the converse policy action set are likely to be the culprit for poor performance. The empirical question is then 'how frequent is frequent?' We examined two frequency cutoffs through an evaluation study as described in Section 7.4.3.

Given the recommendation for refinement, courseware developers revise the courseware. The RAFINE method can be iteratively applied to the revised courseware by collecting a new batch of learning trajectory data to further improve the courseware.

7.3.5 RADARS

Cognitive tutors provide mastery learning on cognitive skills (Corbett, Koedinger, & Hadley, 2001; Ritter et al., 2007). Mastery learning is controlled by a student-modeling technique called knowledge tracing (Corbett & Anderson, 1995) that computes the likelihood of mastering individual cognitive skills to be learned. The output from the knowledge tracer is used to compute an optimal sequence of training problems for individual students to achieve the mastery for desired cognitive skills (VanLehn, 2006).

One of the challenges under this paradigm of model-tracing-based mastery learning happens when students cannot achieve the mastery within a reasonable amount of time. From the students' point of view, this means that they are continuously posed problems one after another even it is unlikely that they eventually achieve the mastery. This phenomenon is called *wheel-spinning* (Beck & Gong, 2013; Beck & Rodrigo, 2014).

There are several reasons for wheel-spinning. We are particularly interested in situations where students indeed never reach the mastery, or they require notable efforts to reach the mastery. In either case, students are faced to unproductive practice activities that may be rather harmful. It is therefore quite important to detect wheel-spinning as soon as possible. A reliable student-modeling technique to predict wheel-spinning is therefore required.

To resolve this issue, we have developed RADARS (*RA*pid *D*etection *A*nd *R*ecovery from Wheel-*S*pinning), an application of a neural-network model to detect potential wheel-spinning at an early phase of learning in the context of cognitive tutoring. The goal of RADARS is to *predict* wheel-spinning *as early as possible*.

The essential system design of RADARS is inspired by the theory of Bayesian knowledge tracing (aka, BKT) that conjectures that student's mastery level can be estimated with a history of answering formative assessments (Corbett & Anderson, 1995). Consequently, the input to RADARS is a response sequence, i.e., a series of 0's and 1's showing the correctness of application of a particular skill performed by a particular student. In particular, we aim to integrate the wheel-spinning detector into the adaptive online course where cognitive tutors (which provide student with adaptive scaffolding) and traditional formative assessments (e.g., multiple choice questions and fill-in-the-blank questions) are integrated into online courseware. We then detect wheel-spinning while students work on cognitive tutors.

Cognitive tutors implement the double-loop pedagogy (VanLehn, 2006). The outer-loop adaptively selects problems to be worked on, whereas the inner-loop provides students with adaptive scaffolding for individual *steps* while solving a problem. Each step is associated with a single skill. The underlying student model (e.g., BKT) predicts the probability of a particular student correctly perform the next step. The wheel-spinning detector predicts a failure to achieve the predefined threshold.

We hypothesized that wheel-spinning can be detected based on the chronological record of the correctness of applying a target skill. While solving a problem, students may make multiple *attempts* per step. The correctness of apply a skill (correct versus incorrect) is measured at the first attempt on a step.

RADARS is implemented as a multilayer linear regression, as shown in Figure 7.3. The input to RADARS is the *L*-parameter taken from BKT. The first layer computes the change in the sequence of *L*-parameters as a slope parameter in a linear regression given the first j opportunities of applying a particular skill k. That is, *L*-parameter for the skill k at the jth opportunity, L^k_j, is regressed as a function of time i $(i = 1, \ldots, j) - L^k_j = a_j * i + b_j$. The second layer computes a change in the slope at time $j - d^k_j = L_{j+1} - L_j$. The last layer is a logistic regression that regressed the probability of wheel-spinning at time j.

Note that the prediction is made per skill per student. In other words, we need to build as many RADARS detectors as the number of unique skill-by-student pair appeared in the data. Each individual kill-specific detector is trained on the first j opportunities in a student-skill response sequence. An important research question is then how small j can be. When we tugged the examples as Wheel-Spinning or Not Wheel-Spinning, the coders looked at the whole sequence of 0's and 1's before making a decision. After training, the detector outputs a classification of Wheel-Spinning or Not Wheel-Spinning based on the jth opportunity of the skill application. That is, the RADARS detector computes the likelihood of wheel-spinning *in the future* after observing the correctness of skill applications for the first j times.

7.4 Evaluation

The PASTEL project is an actively ongoing project at the time this chapter is written. We have conducted preliminary evaluation studies for each PASTEL technology mentioned in Section 7.3 except WATSON. Most of the studies were conducted using data provided by Open Learning Initiative (OLI) at Carnegie Mellon University. OLI provided both students' learning data and the course contents data. This section provides results of the evaluation studies.

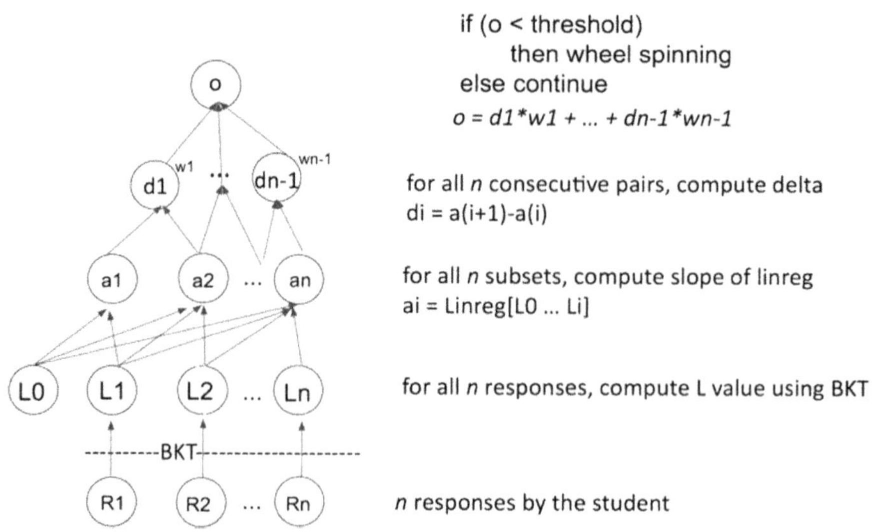

FIGURE 7.3
The structure of the RADARS wheel-spinning detector.

7.4.1 SMART

SMART was applied to two existing OLI courses: Biology[2] and Chemistry.[3] SMART was applied to the entire courseware contents for both courses. The performance of SMART is assessed as the goodness of a model fit of the knowledge component (KC) model discovered by SMART. To compute the model fit, we obtained students' learning data that showed their performance on formative assessments on the corresponding courseware. The learning data were obtained from DataShop.[4] Table 7.2 shows an overview of the learning data obtained. The last column (Number of KCs on DataShop) shows the number of human-crafted KCs uploaded by OLI.

The model fit was measured as an accuracy in predicting the probability of students' answering assessment questions correctly using a slight variation of the additive factor model, or AFM for short (Draney et al., 1995). Relative to the original AFM, we entered the intercept term unique for the skill by student with an assumption that the individual difference exists for the initial difficulty for different skills. As a result, the following logistic regression model was used:

$$\log\left(\frac{p_{ij}}{1-p_{ij}}\right)=\sum_k Q_{kj}\left(\alpha_k T_{ik}\right)+\sum_k Q_{kj}\left(\beta_k+\gamma_{ik}\right)+\theta_i$$

p_{ij}: A probability of student i answering item j correctly.

Q_{kj}: A binary matrix with the cell (k, j) being 1 if item j involves skill k, aka Q-matrix.

T_{ik}: The number of opportunities that student i applied skill k.

α .: The slope coefficient

β ., θ ., γ ..: The intercept for skill, student, and skill by student.

The K-means clustering technique requires the number of clusters, i.e., k, to be given as a hyperparameter. With a lack of empirical and theoretical justification, we compared four levels of the K-value – 10, 50, 100, 150, 200 – using the model fit with RMSE. Figure 7.4 shows the result.

The current data show that the more the number of skills, the better the fit to students' learning data. DataShop contains skills models defined by the expert course instructors. The number of skills for the best expert-crafted skill models are 185 and 186 for Chemistry and Biology, respectively. The RMSE values for those expert-crafted skill models are 0.4049 and 0.3875 for Chemistry and Biology, respectively. Figure 7.4 shows that *SMART can automatically find skill models from existing courseware content data and resulted skill models fit to actual students' learning reliably well relative to the expert-crafted models.*

It must be advised that this observation is likely to be due to the nature of the OLI courseware that we analyzed this time. That is, the OLI courseware that we analyzed might actually contain a large number of skills. Since SMART relies on unsupervised learning techniques to mine skills by clustering the assessment items and subsequently map the paragraphs to the skills, the underlying assumption is that the data include a set of paragraphs and assessment items for each skill emphasized in the course. Therefore, data from courses that require a larger set of skills must include a proportionally larger set of paragraphs and assessment items for SMART to effectively mine the data. On average, a growth in the size of the data results in approximately a linear increase in run time for SMART.

7.4.2 QUADL

To train the Answer Prediction model, we created training data using text data from four existing online courses at Open Learning Initiative (OLI) – Anatomy and Physiology; Modern-Biology; Biochemistry; and Statistics. The course content data consist of modules. Each section in a module contains a *learning objective*, didactic instruction *sentences*, and assessment questions with their *answers*.

The total of 557 target sentences (i.e., the token index was not <0, 0>, which suggests that the sentence might be suitable to generate a question) and 29,529 non-target sentences were extracted from the four OLI courses. Since the data were extremally imbalanced (98:2), the non-target sentences were undersampled so that the data were balanced (i.e., 50:50) when we trained the Answer Prediction model.

TABLE 7.2

Course Content Data Taken from OLI Biology and Chemistry Courses

	Number of Assessments	Number of Paragraphs	Mean Number of Tokens Per Sentence		Number of Students	Number of Student-step Transactions	Number of KCs on DataShop
			Assessments	Paragraphs			
Biology	1,095	1,275	163.7	216.1	480	268,822	186
Chemistry	1,608	2,860	217.1	158.3	616	200,327	209

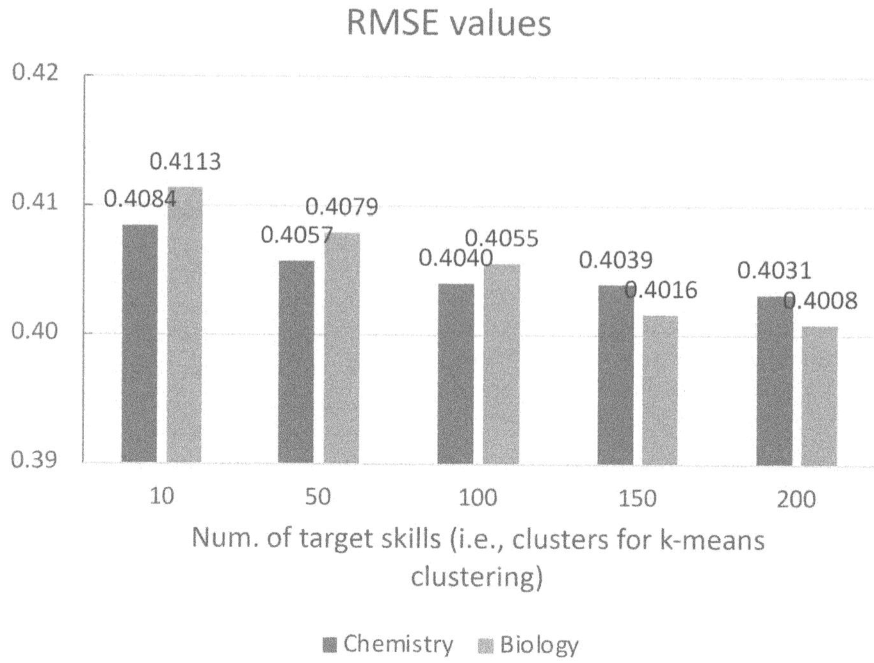

FIGURE 7.4
The model fit for the skill models discovered by SMART.

Since the training data created from the OLI course did not capture all appropriate target tokens, we speculated that the ordinal confusion matrix-based measures did not work well. We also did not use commonly used metrics such as BLUE. This is because those metrics require reference (ground truth) questions to compare the generated questions with, and such data were not available in the current study. Therefore, to evaluate QUADL, we conducted a survey on Amazon Mechanical Turk.

In the survey, the participants were shown four triplets of <*LO*, *S*<*Is*, *Ie*>, *Q*> where *LO* was a learning objective, *S* was a sentence with the target token at the index <*Is*, *Ie*> highlighted, and *Q* was a question generated from *S*. For the sentence that was predicted as a non-target by the Answer Prediction model (i.e., the token index was <0, 0>), a target token was randomly chosen (by excluding stop words) and a question was generated accordingly. For each triplet <*LO*, *S*<*Is*, *Ie*>, *Q*> shown, the participants were asked if they agreed or disagreed with the following two statements: (1) To get a question that helps attain the learning objective *LO*, it is adequate to convert the sentence *S* into a question whose answer is the token <*Is*, *Ie*> highlighted. (2) The question *Q* is suitable for attaining the learning objectives *LO*. Participants were allowed to choose the option 'The text is nonsensical' if a sentence or a question is not linguistically sense-making.

There were 342 survey items with more than one response received. Among 342 survey items, 178 items had a non-zero target token index, and 164 items had a zero <0, 0> index. The total number of participants was 245. The overall mean number of responses per survey item was 2.6. We adopted majority votes to finalize the evaluation from the participants. For each of the two questions in a survey item, if over half of responses were 'agree', then we concluded that the corresponding target token (<*Is*, *Ie*>) or question (*Q*) were accepted as suitable for the learning objective by the participants.

Table 7.3 summarizes the results showing how well the Answer Prediction model identified target tokens for the given sentences relative to specific learning objectives. The data showed that 49% (166/342) of the total predictions about the target token index from the Answer Prediction model were accepted by the Amazon Mechanical Turk (AMT) participants. For the predictions with a non-zero target index, 70% (123/178) of the predictions *were accepted*. As for the non-target sentence predictions (i.e., the Answer Prediction model output the zero <0,0> index), only 26% (43/164) were accepted. That is, 55% (90/164) of the predicted non-target sentences were considered to be target sentences by the AMT participants.

These results show that *the Answer Prediction model is conservative. When it outputs 'positive' predictions (i.e., treating a given sentence as a target sentence), 70% of such predictions are appropriate. However, there is a large number of sentences that should have been predicted as a target*

TABLE 7.3

The Evaluation of the Predicted Target Tokens by the Answer Prediction Model

	Model Prediction		
	Non-zero Target Index <*Is* ≠ 0, *Ie* ≠ 0>	**Zero Target Index <0, 0>**	**Total**
Accepted	**123 (70%)**	**43 (26%)**	166 (49%)
Tie	32 (18%)	25(15%)	57 (17%)
Not accepted	22 (12%)	90 (55%)	112 (33%)
Nonsensical	1 (<1%)	6 (4%)	7 (2%)
Total	178 (100%)	164 (100%)	342 (100%)

There were 178 sentences that the Answer Prediction model predicted target tokens (non-zero index) and 164 sentences that the model predicted non-target (zero index <0, 0>). The table shows how many of them were accepted/not accepted by the majority vote by Amazon Mechanical Turk (AMT) participants.

sentence but missed. We argue that for the educational purposes, these results are good and pragmatic.

One of the challenges in training the Answer Prediction model is the notable skewness in the data – in the training data we used, there were about 560 target sentences and 30,000 non-target sentences. The current study shows that about 1,000 data give us a fair result on detecting target sentences. However, looking into the reports about the amount of data used for various tasks to fine-tune the BERT model, we supposedly need at least more than 2,000 to get higher accuracy of detecting both target and non-target sentences. If we need 2,000 datapoints, then (since we have only 560 target sentences) we would need to deal with data imbalance problem. Applying a data augmentation technique or semisupervised learning might be a practical option.

The output from the Answer Prediction model (i.e., a pair of <sentence, answer index>) was then fed to GQ-net, an existing question conversion model. The AMT participants mentioned above were then asked to assess the quality of the generated questions. To investigate the capability of QG-Net separately from the performance of the Answer Prediction model, we analyzed the performance of QG-Net only for the inputs that the AMT participants judges as 'appropriate' relative to the learning objectives. There were 123 sentences that satisfied this criterion. The result showed that 62% (76/123) of questions generated from 'appropriate' sentences were considered to be suitable for achieving the associated learning objective. This indicates that *the pretrained QG-Net can generate a fair number of suitable questions for domains other than the one it was originally trained.*

7.4.3 RAFINE

To evaluate the RAFINE method, it is necessary to conduct a study with learning data collected from students working on the online courseware that is structured with a skill model tagged to individual instructional components. However, when we conducted the evaluation study, no such online courseware was available. Therefore, we conducted a simulation study as a proof of concept toward an evaluation with actual students. The current evaluation study uses hypothetical learning trajectories in mock online courseware.

We created mock online courseware where there was only one skill involved. When there were multiple skills involved in the given courseware, RAFINE would simply need to be applied separately to each skill. Therefore, this assumption does not harm the generality of the study.

All instructional components in mock online courseware were tagged as either *effective* or *ineffective*. In the current simulation study, we included three types of instructional components: (1) videos, (2) formative assessments (aka quizzes), and (3) hint messages associated with formative assessments. Learning trajectories were generated by simulating students' learning activities.

In the real world, the growth of mastery level depends on the learning activities actually taken and students' latent traits of learning. In the current simulation, the masterly level shows a probability of answering a quiz correctly and the simulated students' performance on a quiz was determined by the masterly level. The growth of the mastery level, $p_{i,T}$, was simulated using a logistic regression model as shown below:

$$p_{i,T} = \left[\frac{1}{1+e^{-Z_{i,T}}}\right]$$

$$Z_{i,T} = Z_{i,T-1} + \delta_1\left(c, e\left(a_{i,T-1}\right)\right) + \delta_2\left(rspns\left(a_{i,T-1}\right)\right)$$

The [X] operator is to round the value X to the nearest multiple of 0.05 and $a_{i,T-1}$ is an instructional component that a simulated student i took at time $T-1$. Logit $Z_{i,T}$ was directly increased with δ_1 and δ_2 that model learning gain obtained by taking an action

$a_{i,T-1}$. $\delta_1(\mu_1, \sigma_1)$ is a rectified random variable that follows a normal distribution with mean μ_1 and standard deviations σ_1. μ_1 and σ_1 are given a priori based on c and $e(a_{i,T-1})$. The parameter c represents the *contrast* in the increase of logit between effective and ineffective instructional elements – large versus moderate versus small. The function $e(a_{i,T-1})$ models the *effectiveness* of the instructional element $a_{i,T-1}$ – effective versus ineffective.

$\delta_2(rspns(a_{i,T-1}))$ is also a rectified random variable that follows a normal distribution with mean $_{\mu 2}$ and standard deviations σ_2. The variable δ_2 was set to be zero if $a_{i,T-1}$ was not a quiz. Otherwise, $_{\mu 2}$ and σ_2 were determined a priori based on a student's response rate, *rspns* = correct/incorrect. We assume that when a student was able to answer the quiz correctly, logit $Z_{i,T}$ increases more than when the student was not able to answer it.

Two instances of mock courseware with two different number of pages (2 versus 3) were created for three levels of quality (low versus mid versus high). Those six instances of courseware were crossed with three levels of contrast, resulting in 18 different simulated-learning scenarios. For each of the 18 learning scenarios, 100 course offerings were created, each with 1,000 hypothetical students. In other words, this simulation study modeled a large-scale field trial as if 1,800 instances of online course offerings were tested each with 1,000 student participants.

Using the frequency heuristic, the effectiveness of each instructional component was computed – those that appeared as converse policy more than the frequency threshold were coded as ineffective. So, the question was what the threshold should be. We would argue that the best threshold value is an empirical call. We therefore compared two threshold values using mean (M) and standard deviation (SD) of the normalized frequency: M + SD versus M – SD.

To evaluate the accuracy in identifying the effectiveness of individual instructional components, we computed precision and recall as follows:

$$\text{Precision} = \frac{\left|\Phi^{R}_{\text{ineff}}\right|}{\left|\Phi^{R}\right|}$$

$$\text{Recall} = \frac{\left|\Phi^{R}_{\text{ineff}}\right|}{\left|\Phi_{\text{ineff}}\right|}$$

$$F1 = 2 * \frac{\text{Precision} \cdot \text{Recall}}{\text{Precision} + \text{Recall}}$$

$\left|\Phi^{R}_{\text{ineff}}\right|$: Number of *ineffective* instructional components included in a recommendation

$\left|\Phi^{R}\right|$: Number of total instructional components included in a recommendation

$\left|\Phi_{\text{ineff}}\right|$: Number of *ineffective* instructional components in courseware

We investigated how precision and recall scores vary depending on the threshold and the condition of the learning data (contrast, quality). Figure 7.5 shows precision and recall scores comparing M – SD and M + SD thresholds for each quality of the courseware. For each data point, three levels of contrasts are aggregated, because there was no notable difference among them. The figure shows that when the quality of courseware is low to medium, the M – SD threshold had better recall and precision scores than M + SD. F1 score for M – SD was 0.99 and 0.92 for low and medium qualities, respectively. On the other hand, when the quality is high, the M + SD threshold outperformed M – SD. F1 scores of M + SD for high-quality courseware was 0.88.

In sum, the frequency heuristic adequately works to determine which instructional components must be taken into a recommendation for courseware refinement. In the current simulation study, over 90% of ineffective instructional components were correctly taken into a recommendation when an appropriate threshold was used based on the maturity of the courseware.

When the courseware is newly built (which is usually in a low to medium quality), the M – SD threshold should be used, whereas the M + SD threshold should be used for matured (high-quality) courseware. In the current study, even with the high-quality courseware where only 10–20% of all instructional components in the courseware are ineffective, RAFINE was able to

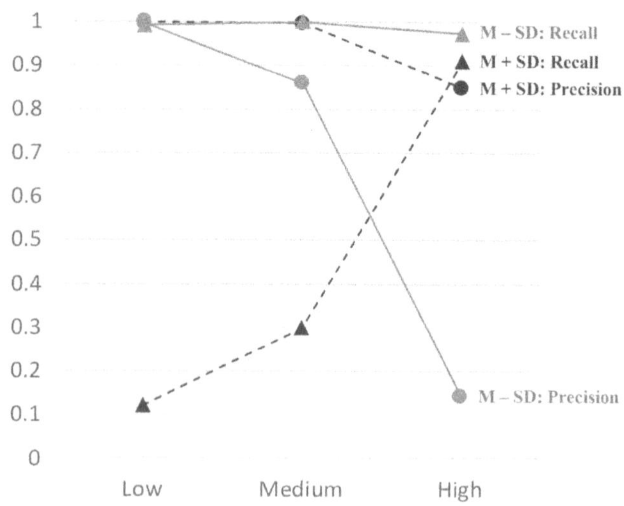

FIGURE 7.5
Precision and recall of a recommendation. The *x*-axis represents the quality of the courseware. Dashed lines show results from M + SD and solid lines show M – SD.

correctly include ineffective components in the recommendation with the M + SD threshold.

7.4.4 RADARS

We used cognitive tutor data taken from DataShop. The name of the dataset is 'Cog Model Discovery Experiment Spring 2010' in the study called 'Geometry Cognitive Model Discovery Closing-the-Loop' (Stamper & Ritter, 2010). In the data, there are 123 unique students and 51 unique skills. There are 5,385 student–skill response sequences where each student–skill response sequence shows a chronological record of the correctness of applying the corresponding skill. Among the 5,385 student–skill response sequences, there are 2,883 response sequences that have more than and equal to 5 responses. We filtered out response sequences with less than 5, because there would not be enough attempts to capture wheel-spinning. Out of 2,883, there are 842 response sequences that do not reach to the mastery according to BKT (hence potentially wheel-spinning). In these 842 response sequences, there are 122 unique students and 44 unique skills included – all but one student had at least one skill for which the BKT's mastery criteria were not met, and students do not achieve mastery on 86% (44 out of 51) of the skills. For those 842 student–skill response sequences, the mean number of opportunities is 7.7 (SD = 3.7). It is those 842 student–skill response sequences that we used for the study mentioned below.

Two coders codified the student–skill response sequences into two categories: will-be-wheel-spinning versus will-not-be-wheel-spinning. The coding manual was iteratively developed during initial coding phase. This iteration happened three times. At the time coders meet a satisfactory agreement on the coding manual, the inter-coder reliability, the Cohen's kappa, showed 0.82. The two coders then coded the rest of the student–skill response sequences ($N = 842$).

To understand the impact of detecting wheel-spinning cases, we analyzed how often students wheel-spun on the current data according to human coding. Figure 7.6 shows the number of students who wheel-spun on each skill, inclusive of skills on which at least one student wheel-spun. Out of the 842 student–skill response sequences that did not achieve mastery, there are 18 skills (41% out of 44) on which at least one student wheel-spun. *On average, there are 4.6 students who wheel-spun on each of those 18 skills.* Table 7.4 shows the number of skills each student wheel-spun, inclusive of students who wheel-spun on at least one skill. There are 56 students (46% out of 122) who wheel-spun on at least one skill. *On average, each of those 56 students*

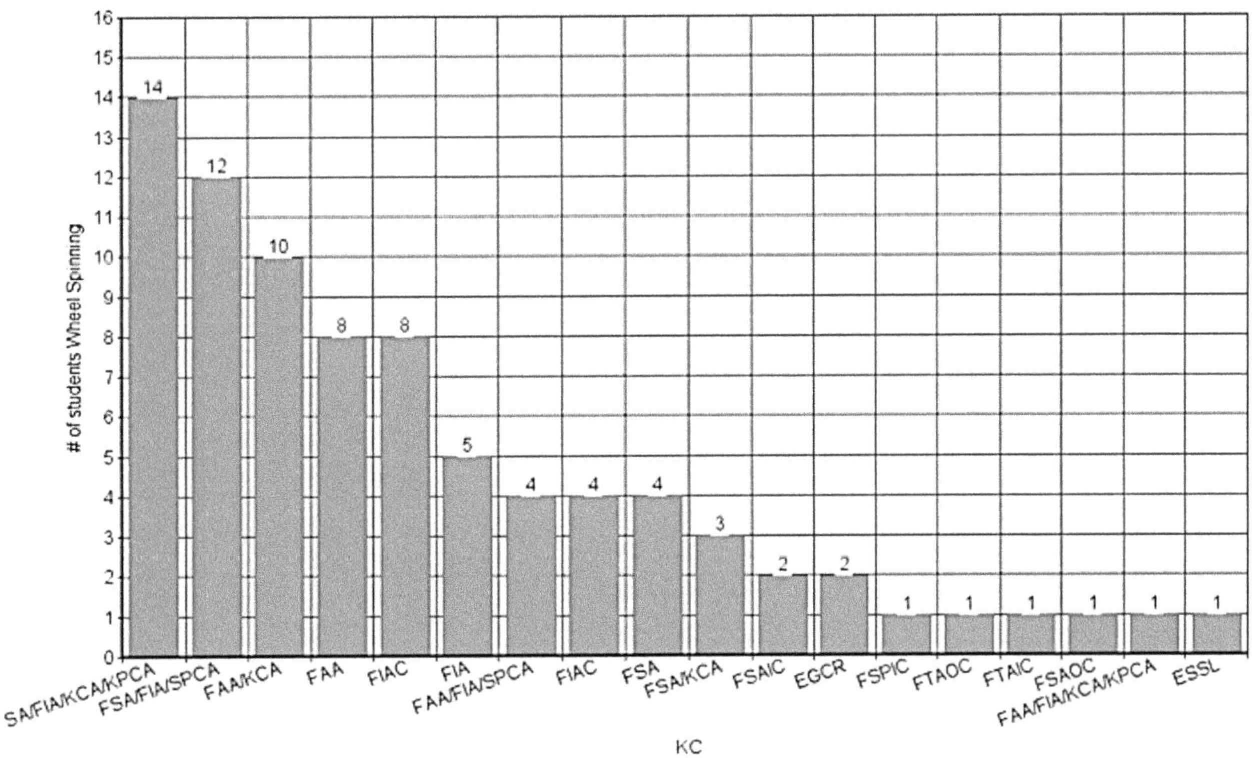

FIGURE 7.6
Number of students who wheel-spun on each skill.

TABLE 7.4

Number of Skills on Which Each Student Wheel-Spun

Number of wheel-spinning skills	0	1	2	3	4
Number of students	66	37	13	5	1

wheel-spun on 1.5 skills. These results show that wheel-spinning is a noteworthy phenomenon, at least on this cognitive tutor, hence detecting the moment of wheel-spinning early on is important.

To evaluate how quickly RADARS can detect wheel-spinning, we trained the detector using the first N (N = 5–10) responses (or 'opportunities' if you will) and validated it by only using the first N responses. Figure 7.7 shows precision, recall, and F1 (which is $2*P*R/(P + R)$, where P and R show precision and recall, respectively) scores for N = 5–10. The x-axis shows the number of observations (i.e., N). As the graph shows, the model with the first five response observations makes relatively better prediction than other models, but the difference in the accuracy of observations is rather small. RADARS detects wheel-spinning with notably high recall, while the precision is still low. This study shows that *RADARS can detect most of the actual wheel-spinning cases as quick as only observing five responses, but it also classifies non-wheel-spinning cases (by definition according to human coding) as wheel-spinning.*

7.5 Conclusion

This chapter introduced the evidence-based learning engineering technique, called PASTEL, which is a collection of artificial intelligence technologies to assist a creation of adaptive online courseware called CyberBook. A key driving force for adaptive instruction is a skill model. SMART is a technology to automatically discover a skill model from pedagogical text on the courseware. CyberBook has a built-in capacity, called RADARS, to detect at-risk students by building a model to predict individual students' growth of skill mastery for each skill. RAFINE is a technology to detect ineffective instructional components that evidently do not contribute to achieve a mastery for certain skill. QUADL generates questions from given pedagogical texts that align with particular learning objectives that are closely related to latent skills taught on the courseware. Another unique characteristic of CyberBook is the integration of cognitive tutors, which is assisted by the WATSON technology. We hypothesize that by integrating cognitive tutors into the traditional online courseware, CyberBook can facilitate learning the synergetic knowledge that, by definition, is the complex web of knowledge, including both conceptual and procedural knowledge.

The demand for open online learning is rapidly growing. Providing effective and reliable online learning technology as well as the assistive technology for efficiently building online courseware is therefore utmost important research agenda. We believe that PASTEL and CyberBook technologies have promising potential. A large-scale empirical evaluation study needs to be conducted to measure the effectiveness of CyberBook and the PASTEL technologies.

Ethics of AI becoming essential as the AI plays an important role in technology-assisted learning and teaching. Education is one of the most severely

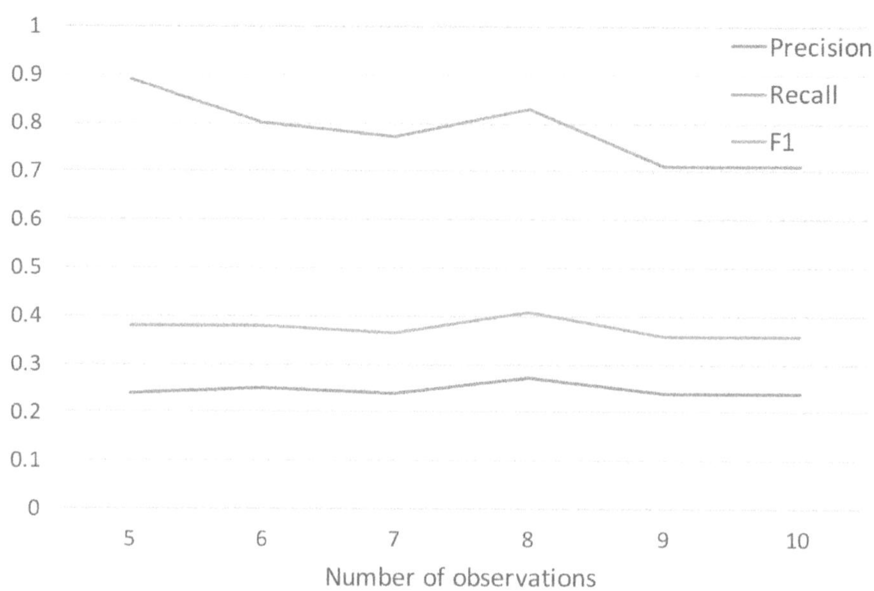

FIGURE 7.7
The precision, recall, and F1 scores computed on the first N response observations.

impacted application domains of ethical computing. Any uncertainly and potential risk in predictions and model output must be proactively reviewed and treated. As an exploratory phase of the project, we have not explicitly paid attention to the ethical issues for PASTE and CyberBook thus far yet. Some obvious concerns include (but not limited to) the correctness of an expert model generated by WATSON that influences the soundness of the cognitive tutor, the validity of the assessment questions generated by QUADL, the accuracy of the recommendations made by RAFINE, and the accuracy of the prediction on the likelihood of failure made by RADARS. In addition to the human-in-the-loop approach (as mentioned in Section 7.3.4 for RAFINE), developing proactive and plausible solutions for each of these issues is one of the important goals for future research on the PASTEL project.

Notes

1. https://open.edx.org
2. https://pslcdatashop.web.cmu.edu/DatasetInfo?datasetId=1934
3. https://pslcdatashop.web.cmu.edu/DatasetInfo?datasetId=4650
4. https://pslcdatashop.web.cmu.edu

References

Aleven, V., McLaren, B. M., Sewall, J., & Koedinger, K. R. (2006). The cognitive tutor authoring tools (CTAT): Preliminary evaluation of efficiency gains. In M. Ikeda, K. D. Ashley, & T. W. Chan (Eds.), *Proceedings of the 8th international conference on intelligent tutoring systems* (pp. 61–70). Springer Verlag.

Aleven, V., McLaren, B. M., Sewall, J., Velsen, M., Popescu, O., Demi, S., Ringenberg, M., & Koedinger, K. R. (2016). Example-tracing tutors: Intelligent tutor development for non-programmers. *International Journal of Artificial Intelligence in Education*, 1–46. https://doi.org/10.1007/s40593-015-0088-2

Anderson, J. R., & Pelletier, R. (1991). A development system for model-tracing tutors. In *Proceedings of the of the international conference on the learning sciences* (pp. 1–8).

Anzai, Y., & Simon, H. A. (1979). The theory of learning by doing. *Psychological Review, 86*(2), 124–140.

Beck, J. E., & Gong, Y. (2013). Wheel-spinning: Students who fail to master a skill. In H. C. Lane, K. Yacef, J. Mostow, & P. Pavlik (Eds.), *Artificial intelligence in education* (Vol. 7926, pp. 431–440). Springer Berlin Heidelberg.

Beck, J. E., & Rodrigo, M. M. (2014). Understanding wheel spinning in the context of affective factors. In S. Trausan-Matu, K. Boyer, M. Crosby, & K. Panourgia (Eds.), *Intelligent tutoring systems* (Vol. 8474, pp. 162–167). Springer International Publishing.

Binks, S. (2018). Testing enhances learning: A review of the literature. *Journal of Professional Nursing, 34*(3), 205–210. https://doi.org/10.1016/j.profnurs.2017.08.008

Bjork, E. L., Little, J. L., & Storm, B. C. (2014). Multiple-choice testing as a desirable difficulty in the classroom. *Journal of Applied Research in Memory and Cognition, 3*(3), 165–170. https://doi.org/10.1016/j.jarmac.2014.03.002

Chi, M. T. H. (1996). Constructing self-explanations and scaffolded explanations in tutoring. *Applied Cognitive Psychology, 10*(Special Issue), S33–S49.

Chi, M. T. H., & Bassok, M. (1989). Learning from examples via self-explanations. In L. B. Resnick (Ed.), *Knowing, learning, and instruction: Essays in honor of Robert Glaser* (pp. 251–282). Erlbaum.

Corbett, A. T. (2001). Cognitive computer tutors: Solving the two-sigma problem. In M. Bauer, P. Gmytrasiewicz, & J. Vassileva (Eds.), *User modeling 2001: 8th international conference* (pp. 137–147).

Corbett, A. T., & Anderson, J. R. (1995). Knowledge tracing: Modeling the acquisition of procedural knowledge. *User Modeling and User Adapted Interaction, 4*(4), 253–278.

Corbett, A. T., Koedinger, K. R., & Hadley, W. S. (2001). Cognitive tutors: From the research classroom to all classrooms. In P. S. Goodman (Ed.), *Technology enhanced learning: Opportunities for change* (pp. 235–263). Erlbaum.

Devlin, J., Chang, M.-W., Lee, K., & Toutanova, K. (2018). Bert: Pre-training of deep bidirectional transformers for language understanding. arXiv Preprint ArXiv:1810.04805.

Draney, K. L., Pirolli, P., & Wilson, M. (1995). A measurement model for a complex cognitive skill. In P. D. Nichols, S. F. Chipman, & R. L. Brennan (Eds.), *Cognitively diagnostic assessment* (pp. 103–125). Erlbaum.

Friedman-Hill, E. (2003). *Jess in action: Java rule-based systems*. Manning.

Gates, S. J., Handelsman, J., Lepage, G. P., & Mirkin, C. (Eds.). (2012). *Engage to excel: Producing one million additional college graduates with degrees in science, technology, engineering, and mathematics*. Office of the President, DC: PCAST STEM Undergraduate Working Group.

Goodman, I. F. (2002). *Final report of the women's experiences in college engineering (WECE) project*. Goodman Research Group.

Hattie, J. (2008). *Visible learning: A synthesis of over 800 meta-analyses relating to achievement*. Routledge.

Koedinger, K. R., & Aleven, V. (2007). Exploring the assistance dilemma in experiments with cognitive tutors. *Educational Psychology Review, 19*(3), 239–264.

Koedinger, K. R., Aleven, V., & Heffernan, N. (2003). Toward a rapid development environment for cognitive tutors. In U. Hoppe, F. Verdejo, & J. Kay (Eds.), *Proceedings of the International Conference on Artificial Intelligence in Education* (pp. 455–457). IOS Press.

Koedinger, K. R., Corbett, A. T., & Perfetti, C. (2012). The knowledge-learning-instruction framework: Bridging the science-practice chasm to enhance robust student learning. *Cognitive Science, 36*(5), 757–798. https://doi.org/10.1111/j.1551-6709.2012.01245.x

Kurdi, G., Leo, J., Parsia, B., Sattler, U., & Al-Emari, S. (2020). A systematic review of automatic question generation for educational purposes. *International Journal of Artificial Intelligence in Education, 30*(1), 121–204. https://doi.org/10.1007/s40593-019-00186-y

Li, N., Matsuda, N., Cohen, W. W., & Koedinger, K. R. (2015). Integrating representation learning and skill learning in a human-like intelligent agent. *Artificial Intelligence, 219*, 67–91. https://doi.org/10.1016/j.artint.2014.11.002

MacQueen, J. (1967). Some methods for classification and analysis of multivariate observations. In *Proceedings of the fifth Berkeley symposium on mathematical statistics and probability* (Vol. 1, Statistics, pp. 281–297). University of California Press.

Martin, B., Mitrovic, A., Koedinger, K. R., & Mathan, S. (2011). Evaluating and improving adaptive educational systems with learning curves. *User Modeling and User-Adapted Interaction, 21*(3), 249–283. https://doi.org/10.1007/s11257-010-9084-2

Matsuda, N., Cohen, W. W., & Koedinger, K. R. (2015). Teaching the teacher: Tutoring SimStudent leads to more effective cognitive tutor authoring. *International Journal of Artificial Intelligence in Education, 25*(1), 1–34. https://doi.org/10.1007/s40593-014-0020-1

McConnell, M. M., St-Onge, C., & Young, M. E. (2015). The benefits of testing for learning on later performance. *Advances in Health Sciences Education, 20*(2), 305–320. https://doi.org/10.1007/s10459-014-9529-1

Mihalcea, R., & Tarau, P. (2004). Textrank: Bringing order into texts. In D. Lin & D. Wu (Eds.), *Proceedings of the EMNLP* (pp. 404–411).

Murray, T. (1998). Authoring knowledge-based tutors: Tools for content, instructional strategy, student model and interface design. *Journal of the Learning Sciences, 7*(1), 5–64.

Murray, T. (1999). Authoring intelligent tutoring systems: An analysis of the state of the art. *International Journal of Artificial Intelligence in Education, 10*, 98–129.

Murray, T. (2003). An overview of intelligent tutoring system authoring tools. In T. Murray, S. Ainsworth, & S. B. Blessing (Eds.), *Authoring tools for advanced technology learning environment* (pp. 491–544). Kluwer Academic.

Murray, T. (2016). Coordinating the complexity of tools, tasks, and users: On theory-based approaches to authoring tool usability. *International Journal of Artificial Intelligence in Education, 26*(1), 37–71. https://doi.org/10.1007/s40593-015-0076-6

Pan, L., Lei, W., Chua, T.-S., & Kan, M.-Y. (2019). Recent advances in neural question generation. arXiv preprint arXiv:1905.08949.

Pan, S. C., & Rickard, T. C. (2017). Does retrieval practice enhance learning and transfer relative to restudy for term-definition facts? *Journal of Experimental Psychology: Applied, 23*(3), 278–292. https://doi.org/10.1037/xap0000124

Pavlik Jr., P. I., & Anderson, J. R. (2008). Using a model to compute the optimal schedule of practice. *Journal of Experimental Psychology: Applied, 14*(2), 101–117.

Ritter, S., Anderson, J. R., Koedinger, K. R., & Corbett, A. (2007). Cognitive tutor: Applied research in mathematics education. *Psychonomic Bulletin and Review, 14*(2), 249–255.

Roediger, H. L., III, Agarwal, P. K., McDaniel, M. A., & McDermott, K. B. (2011). Test-enhanced learning in the classroom: Long-term improvements from quizzing. *Journal of Experimental Psychology: Applied, 17*(4), 382–395. https://doi.org/10.1037/a0026252

Salton, G., & McGill, M. J. (1983). *Introduction to modern information retrieval*. McGraw-Hill.

Seymour, E., & Hewitt, N. M. (1997). *Talking about leaving: Why undergraduates leave the sciences*. Westview Press.

Smith, M. A., & Karpicke, J. D. (2014). Retrieval practice with short-answer, multiple-choice, and hybrid tests. *Memory, 22*(7), 784–802. https://doi.org/10.1080/09658211.2013.831454

Stamper, J., & Ritter, S. (2010). Cog model discovery experiment spring 2010. Dataset 392 In *Data Shop*. https://pslcdatashop.web.cmu.edu/DatasetInfo?datasetId=392

VanLehn, K. (2006). The behavior of tutoring systems. *International Journal of Artificial Intelligence in Education, 16*.

VanLehn, K. (2011). The relative effectiveness of human tutoring, intelligent tutoring systems, and other tutoring systems. *Educational Psychologist, 46*(4), 197–221. https://doi.org/10.1080/00461520.2011.611369

Wang, Z., Lan, A. S., Nie, W., Waters, A. E., Grimaldi, P. J., & Baraniuk, R. G. (2018). *QG-Net: A data-driven question generation model for educational content*. Paper presented at the Proceedings of the Fifth Annual ACM Conference on Learning at Scale.

Watkins, J., & Mazur, E. (2013). Retaining students in science, technology, engineering, and mathematics (STEM) majors. *Journal of College Science Teaching, 42*(5), 36–41.

8

A Technology-Enhanced Approach for Locating Timely and Relevant News Articles for Context-Based Science Education

Jinnie Shin and Mark J. Gierl

CONTENTS

8.1 Literature Review 110
8.1.1 Science News Articles and Assessment in Context-Based Science Education 110
8.1.2 Automated Curriculum Alignment of Learning Resources in Science Education 111
8.1.3 Topic Modeling and Latent Dirichlet Allocation 112
8.1.4 Present Study 113
8.2 Methods 113
8.2.1 Data 113
8.2.1.1 Stage 1: Science Article Preprocessing and Vectorization 114
8.2.1.2 Stage 2: Science Article Topic Modeling 115
8.2.1.3 Stage 3: Topic Structure Prediction and Evaluation of the Units of Study 115
8.2.1.4 Stage 4: Science Article Recommendation Based on Topic Similarity 116
8.3 Results 116
8.3.1 Science Article Topic Analysis 116
8.3.2 Topic Prediction and Topic Evaluation Units of Study 117
8.3.3 Science Article Recommendation and Item Generation 118
8.4 Conclusions and Discussion 120
Appendix A 122
Appendix B 123
Appendix C 124
Appendix D 124
Notes 125
References 125

Applying classroom lessons to real-life contexts has long been considered one of the most important aspects of science learning (Miller, 2011; Fensham, 2009; Kuhn & Müller, 2014). The fact that the world is evolving more rapidly than ever with new scientific findings and discoveries provides students with authentic opportunities to apply and evaluate real-world problems using the knowledge and skills they have acquired from their classroom experiences. To keep up with this rapidly changing scientific information, students are encouraged to relate their learning within the classroom to novel contexts outside of the classroom. The purpose of context-based science education is to motivate students by presenting them with accurate scientific issues that are relevant and important in order to enhance and promote more effective learning. Therefore, providing a seamless instructional experience in the context-based science classroom is dependent on the teacher's ability to introduce novel and appropriate science resources to students in a timely manner.

Advocates for context-based curriculum argue that providing meaningful and interesting science stories is a promising way of expanding students' scientific knowledge and skills beyond school (Kuhn & Müller, 2014). More specifically, instructions using 'interesting' and 'relatable' science stories, such as newspaper articles, can enhance students' learning motivation and positive engagement (Mysliwiec et al., 2003). In addition, students' understanding of science stories can be used to measure their competency of applying the information in novel settings (Gayford, 2002; Elliott, 2006; McClune & Jarman, 2010; Oliveras et al., 2013). To underscore the importance of context-based science education, PISA includes context-based items that accompany real-world events to evaluate science

DOI: 10.1201/9781003181187-10

literacy (OECD, 2019). In other words, providing students with relevant and meaningful scientific content could play an important role in engaging (Mysliwiec et al., 2003), instructing (Kuhn & Müller, 2014), and assessing students (Oliveras et al., 2013).

Unfortunately, locating and providing relevant and appropriate news required for science instruction and assessment is a challenging task for most teachers. This is due, in part, to their inability to locate, evaluate, and utilize resources wisely and effectively (Anderman et al., 2012). To add to this challenge, the recent introduction of technologies and the prevalent use of Internet resources in education have produced an overflow of 'somewhat-relevant resources' in science education that further complicate the process of locating appropriate science contexts for teachers to efficiently provide novel and practical assessment items (Bracewell et al., 1998). Science articles include diverse scientific constructs that could be used for classroom instruction and assessment. But this diversity only serves to increase the challenge of evaluating the appropriateness of the resources based on specific curriculum standards.

Traditionally, locating relevant science news for classroom instruction and assessment was manually conducted by teachers and content specialists with little or no guidelines (So et al., 2011; Kuhn & Müller, 2014). Also, the evaluation of the resources and their relatedness to the curricular concepts and expected learning outcomes was commonly measured using teachers' judgments. Moreover, the unstructured text format of the curriculum standards and resources made aligning the two sources complex and challenging. Only abstract guidelines were available for content specialists who were instructed to use science resources that 'support and be consistent with curriculum outcomes', 'appropriate for the subject area', and 'present high stands of quality in factual content and presentation' (e.g., Prince Edward Island Department of Education, 2008).

In order to address these challenges, previous studies have used various linguistic features and dimensions to provide systematic decisions in mapping science learning resources with curriculum standards. However, the lack of performance accuracy and the supervised nature of the proposed frameworks which require extensive content expert intervention hinder systems from becoming widely adopted. Hence, the purpose of the current study is to introduce and demonstrate a machine learning framework that can be used to increase and advance the automaticity of the traditional processes for locating science resources in context-based science education. Our system can be used to locate and evaluate vast amounts of current and relevant scientific information from science articles that cover course-related materials according to the curriculum standards. Various text mining techniques and a topic modeling algorithm called latent Dirichlet allocation (LDA) were used to identify the topic structure of recent science articles that were openly published for students in Grades K to 12. Then, their curricular relatedness and appropriateness was evaluated based on the expected curriculum outcomes described in the programs of study using a topic structure prediction analysis. Finally, the benefits of this application for locating science resources were demonstrated by generating test items that can be used to evaluate students' scientific literacy. Our system contributes significantly to enhancing the traditional processes of locating relevant science articles while addressing the previous challenges that machine-learning approaches encountered when used to create content for science assessments.

8.1 Literature Review

8.1.1 Science News Articles and Assessment in Context-Based Science Education

Context-based science education refers to a stream of philosophies that emphasize the context of learning science. Pilot and Bulte (2006) claimed that context-based science learning should be able to bring the learning of science closer to students' life and interests while showing students how the use of context increases their interest in science, thereby improve their understanding. In context-based science education, learning science concepts in schools cannot be isolated from understanding the context in which they appear (Holbrook, 2014). Students are often required to locate and identify scientific concepts from the resources that were initially used to introduce various applications in order to demonstrate their scientific understanding, application knowledge, and higher-order thinking in the classroom (Avargil et al., 2012). This requirement indicates how much educators value scientific application as the origin and development of scientific concepts and ideas (Bennett et al., 2007).

To provide students with opportunities to demonstrate their learning in various contexts and real-life scenarios, researchers and practitioners strive to use resources to successfully implement assessment in context-based science classrooms (Vos et al., 2011; Demuth et al., 2006; Nentwig & Waddington, 2006; Shwartz et al., 2013; Prins et al., 2018). One particular form of assessment is a science story from a newspaper or online article that can be widely adopted to create source-based test items (Kuhn & Müller, 2014).

Newspaper and online science articles feature real-life contexts on contemporary scientific issues with relatively formal language. Kuhn and Müller (2014) argued that the value of using science news articles that are professionally prepared by journalists and experts is to introduce science stories that are both high quality and interesting. They emphasized that good science articles could positively affect students' motivation and engagement (Kuhn & Müller, 2014). On the strong assumptions that teachers could carefully select articles that consider important science concepts in an unbiased manner and evaluate these articles using curriculum standards, science articles promote effective assessment experiences for students because it allows them to use and demonstrate their higher-order thinking skills (Grant et al., 2009).

Researchers have investigated the benefits of using newspaper articles in various context-based science learning environments to promote assessment. Levinson et al. (2001) claimed that science teachers showed relatively coordinated perspectives of using newspaper articles in science lessons and assessment. They also reported that teachers found the articles to be the most useful in teaching and assessing socio-scientific concepts and issues, such as climate change, animal testing, and genetic engineering. Similarly, Gayford (2002) and Elliott (2006) reported that science teachers commonly identified teaching and assessing students' scientific literacy using controversial issues as more beneficial than using more straightforward science news articles. This outcome occurred, in part, because science teachers strived to provide more critical and creative instructions when introducing students to relevant resources and materials (McClune & Jarman, 2010). Using scientific articles for classroom assessment also helps students become more engaged and motivated, thereby improving their science literacy achievement (McClune & Jarman, 2010). Mysliwiec et al. (2003), for example, used science news articles in three different biology courses with the goal of encouraging students to take control of their learning and enhance their interest in science. After a semester of numerous course assignments and assessments using news articles, students consistently reported that these resources helped facilitate more effective learning.

8.1.2 Automated Curriculum Alignment of Learning Resources in Science Education

In our fast-changing world with scientific discoveries and findings being updated each day, locating and maintaining current real-life resources for classroom instruction can be challenging. A small number of studies have been conducted to create methods that can be used to align and recommend various learning resources (e.g., formative assessment questions, reading resources) based on curriculum standards.

Devaul et al. (2011) described cataloging tools that used various natural language processing techniques to understand whether the tool could make comparable decisions to content experts when assigning curriculum standards to various online learning resources. They adopted a cataloging method, the Content Assignment Tool (CAT), to process the curriculum standards and the learning resource together to appropriately map linguistic and grammatical features, such as the entity tagging and parts-of-speech tagging, with expert-identified rule-based systems. The evaluation of their system was compared with human judgments using inter-rater agreement. The human-raters produced relatively poor results with agreement of only 32% when tested using 29 learning resources with the curriculum standards extracted from the Grades 5–8 National Science Education Standards. The CAT system produced higher but still poor performance accuracy with results ranging from 31% to 51%.

In sum, while research has been conducted to develop systematic and automated methods for identifying and assigning learning resources, these methods could not produce high accuracy or agreement with the content expert's classification decisions. Moreover, the previous methods focused on identifying linguistic patterns (e.g., part-of-speech tagging, grammatical structure) and word overlap (e.g., term-frequency and inverse document-frequency), rather than focusing on the overarching topic of the learning resources to find the most appropriate standard classification. Finally, the resources that were provided for the studies we cited often came from the pre-collected science learning resources rather than the open-access resources, such as online articles. Unlike pre-collected learning resources, online resources have important advantages in providing the most up-to-date information about recent events. However, online resources often require careful selection and filtering to ensure their readability based on grade level. Hence, the previous methods and systems were inherently limited because they could not be used to ensure the quality (e.g., readability level) of the learning resources when directly applied to open-access online resources. This limitation hinders practitioners and educators from finding recent and up-to-date resources that are available in open-access platforms.

Access to scientific news for students is proliferating with the introduction of commercial websites such as *Science News for Students*,[1] *Science Journal for Kids*,[2] and *DOGO news*.[3] These websites provide recent news articles as resources for science teachers. While the websites feature numerous public resources, science teachers cannot use the resources effectively without implementing systematic methods to manage and evaluate this abundant and constantly changing information. Moreover, the absence of proper methods to make a systematic connection between the resources and curriculum outcomes limits the capacity of teachers to identify whether specific outcomes of interest could be assessed effectively using specific resources.

8.1.3 Topic Modeling and Latent Dirichlet Allocation

Traditionally, locating documents based on their common topics has been considered a tedious manual task. Having science teachers and content specialists review thousands of science articles – which are updated daily – to search for appropriate resources is a time-consuming and inefficient exercise. By way of contrast, topic modeling is a machine learning– and natural language processing–based method that can automatically uncover hidden topics found in large numbers of documents. Topic modeling provides a method to automatically organize, understand, search, and summarize large text data without tedious manual review (Blei, 2003). A number of topic modeling approach have been introduced, such as the latent semantic analysis (LSA), probabilistic latent semantic analysis (pLSA), and non-negative matrix factorization (NMF) approaches (Steyver & Griffiths, 2007). Both LSA and NMF use a dimensionality reduction approach in order to extract the latent topic structure from a set of documents, by representing them as a word-document co-occurrence matrix. The pLSA, on the other hand, configures the document-topic structure over a set of probabilities, such as the topic distribution, distribution of topics for a particular document, and the probability distribution over words given the topic. This probabilistic topic model is used to infer the documents' topic structure by reversing the generation process to identify the topic assignment of a given set of documents. Steyvers and Griffiths (2007) provide a detailed walkthrough regarding how the different variations of topic models operate. Surging number of recent studies in education introduced the variations of topic models to understand the educational text data (Daenekindt & Huisman, 2020; Liu et al., 2017; Wu et al., 2018) to support education systems. In this study, we adopted one of the most commonly adopted generative probabilistic topic models, Latent Dirichet Allocation approach.

Latent Dirichlet Allocation (LDA; Blei et al., 2003) is a generative probabilistic topic modeling algorithm in which a document is assumed to consist of a mixture of several topics. These topic structures are referred to as the latent structure of a document which can be identified based on sets of vocabularies that frequently occur together. To discover the topic structures by locating groups of vocabularies that tend to appear frequently together, LDA utilizes the two major distributions – word-topic and document-topic distributions – to mimic a document-generation process. That is, LDA assumes that a topic representation of each document comes from a document-topic distribution, which specifies how each document includes a mixture of different topics, $P(\text{topic} \mid \text{document})$. Then, a set of vocabularies are generated to represent each topic in a document, $P(\text{word} \mid \text{topic})$. Taken together, the generation process of a document can be defined as the probability of a word to be introduced in one document by multiplying the two distributions, $P(\text{word} \mid \text{topic}) * P(\text{topic} \mid \text{document})$.

This process can also be described using a graphical representation of LDA (see Figure 8.1). For example, word-topic distribution (φ) is drawn from a Dirichlet

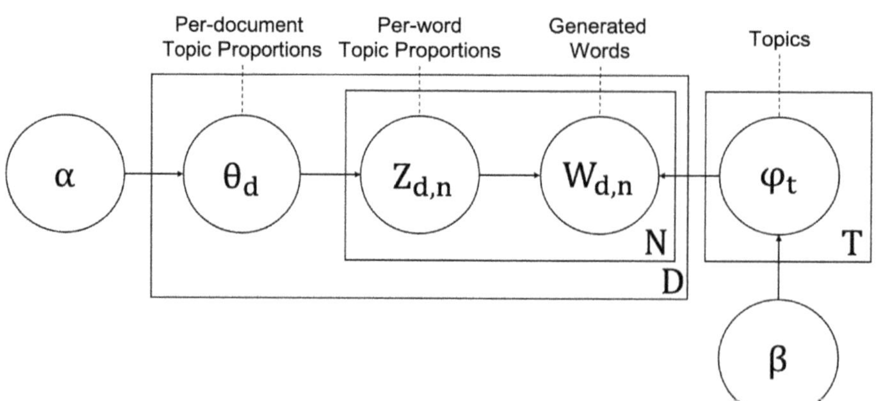

FIGURE 8.1
A graphical representation of LDA.

distribution with a hyper-parameter (β). This can be denoted as $P(\varphi \mid \beta)$. For each document, document-topic distributions are drawn from another Dirichlet distribution with a parameter (α , which can be noted as $P(\theta \mid \alpha)$. For each N word, a word-topic distribution is chosen as $P(Z \mid \theta)$ and a word is generated from the word-topic and document-topic assignment as $P(W \mid \varphi, Z)$. We then identify a join distribution of a document-topic proportion (θ), word-topic distributions (Z), and the number of words (N) as follows:

$$P(\theta, Z, W \mid \alpha, \beta) = P(\theta \mid \alpha) \prod_{n=1}^{N} P(Z \mid \theta) P(W \mid \varphi, Z).$$

8.1.4 Present Study

Researchers have demonstrated that novel contexts can be used to evaluate students' higher-order thinking skills in science (Gayford, 2002; Elliott, 2006; McClune & Jarman, 2010; Avargil et al., 2012; Oliveras et al., 2013). However, providing appropriate science resources has been a daunting task because overly general guidelines exist for evaluating science resources that must be located using a manual process. Moreover, the abundance of online resources has complicated the task of identifying appropriate science resources. To overcome these challenges, we introduce a new machine learning framework that can be used to locate and evaluate vast amounts of current and relevant science articles that are closely associated with specific curriculum standards and learning outcomes. Within this framework, a topic modeling algorithm was used to identify topics from a vast number of science articles and evaluate their relatedness to curriculum standards. Finally, the applicability of our system is demonstrated by generating constructed-response test items that can be used to evaluate science literacy.

8.2 Methods

8.2.1 Data

A set of 1,025 recently published science articles were collected to train our topic modeling system using Webhose.io. Webhose.io allows researchers to access web resources using specific keywords and categories. Using the system, science articles were gathered for K to 12 students that were published within the same month. Then, the Flesch–Kincaid reading-ease test was used to identify the appropriate grade level of the provided resources. In the Flesch–Kincaid reading-ease test, the readability of the resources is measured using the basic syntactic structure of the text, such as the number of syllables, words, and sentences. The reading-ease scores range from 0 to 100 with higher scores indicating that the material is easier to read. The conventional score categorization suggests that reading materials with scores between 80 and 60 are most appropriate for students from Grade 7 to Grade 9. The results indicated that all the extracted science articles met the requirement, with an average score of 66.56. After the initial filtering, these articles served as a training dataset to develop initial topic-model structures. On average, the articles had 383 words with the longest and the shortest articles containing 800 and 50 tokens, respectively.

Next, a Programs of Study (POS) document was used as our testing dataset to evaluate and rank the articles based on their relatedness to the curriculum standards. A POS provides comprehensive descriptions of curriculum frameworks using a list of topics and expected student learning outcomes. In the study, we used the Science POS for Grades 7–9 from a province in western Canada. The POS is publicly available in all subjects for educators to help design their lesson plans by providing guidelines for the topic, key concepts, focusing questions, and the expected learning objectives. Fifteen units of study were selected to evaluate the initial topic structures using the POS. Table 8.1 presents specific topics for each grade. In this example, each grade introduces five units of study that cover distinct topics encompassing various areas of science, such as biology (e.g., interactions and ecosystems, biological diversity), chemistry (e.g., matter and chemical change, environmental chemistry), physics (e.g., mechanical systems, electrical principles, and technologies), and earth sciences (e.g., planet earth, space exploration).

TABLE 8.1

Science Grades 7–9 Topics of the Units of Study

Unit	Grade 7	Grade 8	Grade 9
A	Interactions and Ecosystems	Mix and Flow of Matter	Biological Diversity
B	Plants for Food and Fiber	Cells and Systems	Matter and Chemical Change
C	Heat and Temperature	Light and Optical Systems	Environmental Chemistry
D	Structures and Forces	Mechanical Systems	Electrical Principles and Technologies
E	Planet Earth	Fresh and Saltwater Systems	Space Exploration

Figure 8.2 provides a conceptual representation of the framework of general units of study in science education. A unit of study consists of three major components: focusing questions, key concepts, and expected outcomes. Focusing questions refer to central questions that encompass the overall themes to guide teachers and students throughout the unit. Therefore, to successfully answer the focusing questions, students are required to understand the key concepts and achieve the skills and knowledge described as the expected outcomes of the unit. Key concepts are a list of scientific concepts that specify the scope of the topic for a particular unit. Key concepts are often repeatedly addressed across the grades while the intended depth of understanding varies based on the grade level. For example, Unit 'Interactions and Ecosystems' in Grade 7 introduced key concepts, such as species distribution and extinction. This narrows down the focus of the topic, while the 'expected outcomes' specify the depth and scope of the required knowledge and skills.

8.2.1.1 Stage 1: Science Article Preprocessing and Vectorization

Data processing was first used to locate more distinguishable and interpretable topics by converting words into stem words and original forms (i.e., lemmas). This allowed us to reduce the derivational forms of words into their lemmas and stems. Stemming is a process of reducing inflectional forms and related forms of a word into its base form by removing the ends of words to locate its stem words, such as converting the word 'forcing' to 'force'. Lemmatization also attempts to group words so that they are categorized and analyzed based on their dictionary or original forms. For example, the words 'environmental' and 'environments' are categorized together based on their lemma, 'environment'. Both stemming and lemmatization were conducted using the porter stemmer and wordnet lemmatizer imported from the Python NLTK library (Bird et al., 2009). Lemmatization is a process of reinforcing and complementing a rather crude and heuristic nature of stemming. We also removed the stopwords from the text using the

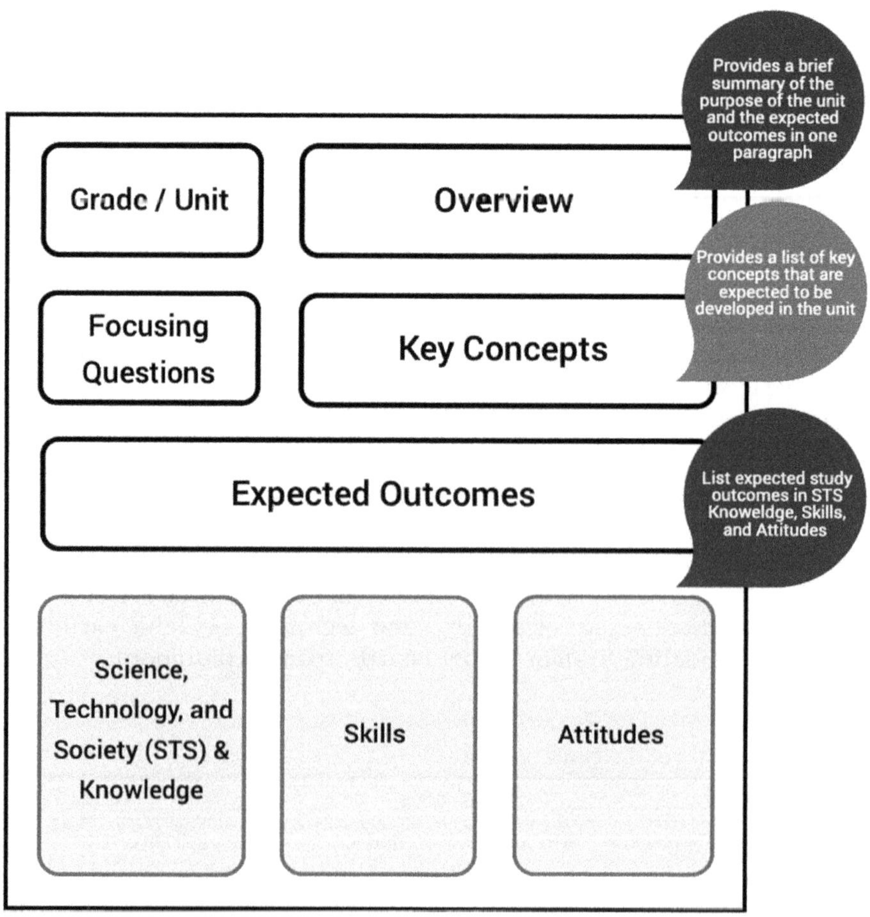

FIGURE 8.2
A conceptual representation of the framework.

available list of English stopwords provided by the NLTK library (Bird et al., 2009). After the data cleaning stage was completed, the science articles were vectorized by counting the frequency of each vocabulary to create a word-count vector to represent each text. In the vectorization process, no vocabulary with an extreme number of occurrences were discarded. Instead, every unique occurrence of vocabularies in the document was taken into consideration as part of our analysis.

8.2.1.2 Stage 2: Science Article Topic Modeling

Using the vectorized representation of science articles, the topic structure of the science articles was identified using LDA (Blei et al., 2003). The Python library Gensim was used to implement our online LDA model. This LDA model provided two major latent structures (i.e., document-topic and word-topic distributions) using a collapsed Gibbs sampling method while optimizing two hyper-parameters: the number of topics (K) and the smoothing parameter (α). The model was initiated with the default setting of the smoothing parameter ($\alpha = 0.10$) and set to iterate every 300 documents for each update until the model converged. For each class of prior, the LDA models were evaluated by varying the number of topics (K) up to a relatively large number, 30. The performance of the initial model was evaluated using the perplexity measure and the topic coherence score. Perplexity is a commonly used topic-model measure that is calculated by dividing a negative log-likelihood by the number of words (Blei et al., 2003). As the name suggests, perplexity provides the degree of uncertainty or confusion the model has in assigning probabilities to text. Therefore, the optimal number of topics could be determined by locating the model with the lowest perplexity. Conversely, the coherence score assesses the quality of the learned topic to optimize the number of topics in the model. The CV coherence measure was used. It ranges from 0 to 1, a higher score indicates more coherence. We identified models with the score of 0.40 or higher as an accepting model (Röder et al., 2015). Once the best model was identified with the optimal number of topic clusters, the model was saved to identify the topic structure in the unit of study document.

8.2.1.3 Stage 3: Topic Structure Prediction and Evaluation of the Units of Study

To evaluate the relatedness between the science articles and the curriculum standards, the topic structures for each unit of study contained in the POS was predicted using the best-performing model from the previous stage. Each unit of study consisted of several statements representing the expected learning outcomes in the domains of Science, Technology, and Society (STS) and Knowledge, Skills, and Attitudes. Each domain consisted of several statements defining and specifying the requirement that students were expected to demonstrate in order to achieve the learning outcomes. Hence, each unit of study was parsed into learning-outcome texts based on their domain. Then, the trained topic model was used from the previous stage to investigate whether the important context and topic of the learning outcomes in the units of study could be understood using the preidentified topic structure from the science articles. In other words, the topic structures of the expected outcomes in the major categories (e.g., STS and knowledge, skills, and attitudes) in the units of study were predicted using the initial trained topic model using the science articles.

To begin with, the same preprocessing was conducted on our three categories of learning outcomes extracted from the units of study to ensure the model prediction is clear and interpretable. Then, a 'bag-of-words' representation of our training dataset was obtained. Bag-of-words is a simplified representation of a document that disregards the grammatical structure and word-sequence of the original text. It is important to notice that Gensim only considers words that were introduced in the original training dataset (or science articles) to build bag-of-words representations of the unseen dataset. While this approach significantly reduces the potential errors that could arise from handling unknown or unintroduced vocabularies, the learning outcomes could be underrepresented due to the intentional removal of unintroduced words. To mitigate this problem, we ensured that our training dataset came from various science disciplines to cover different topics in science education. Finally, the topic structure of each unit was visualized by normalizing the topic contribution to understanding the prediction results. In summary, the outcome of the topic structure predictions of the three categories of learning outcomes was represented as a mixture of topics which were identified from the science articles.

The topic prediction results were evaluated by introducing human judgment as a point of reference and by comparing agreement with our system. This approach was used to understand whether the topic structures identified from the science articles could sufficiently cover the contexts and topics presented in different dimensions of learning outcomes in the units of study. Hence, two content specialists who were familiar with the science programs of study were asked to locate appropriate topics for each study outcome (see Appendices A and B for more information about the evaluation tool). Cohen's Kappa was used to measure interrater agreement for topic assignment between human-raters and our system. Kappa ranges from 0 (i.e., no agreement) to 1 (i.e., perfect agreement) and it provides a chance-corrected agreement score between the raters. It is computed by taking the ratio of the exact agreement over the perfect agreement (or 1) while penalizing the chance agreement. A kappa score between 0.60 and 0.80 is considered as an acceptable level of agreement.

8.2.1.4 Stage 4: Science Article Recommendation Based on Topic Similarity

After a thorough evaluation, training dataset of science articles was located and ranked based on their topic structure similarity to the topic structure of the three categories of learning outcomes (i.e., STS and knowledge, skills, and attitudes) from the units of study. This process locates and recommends the most appropriate science articles by identifying the highest content-relatedness with the curriculum standards and learning outcomes depicted in the units of study. Hence, the outcome could directly map and identify the most appropriate science articles that could be assigned to help students demonstrate the specific learning outcomes from the units of study.

To evaluate the topic structure similarity or the content-relatedness, the Jensen–Shannon distance was used. This measure is a symmetric distributional index that compares the divergence of the two topic distributions (Dagan et al., 1999). That is, the Jensen–Shannon distance measure is used to evaluate which documents have a statistically closer relationships using the Kullback–Leibler (KL) divergence. For example, to produce the distance between the two distributions (*A* and *B*), the measure computes the average distance between *A* and *B* with the average of 2 (*M*) using the equation:

$$\mathrm{JSD}(A\,||\,B)=\frac{1}{2}D(A\,||\,M)+\frac{1}{2}D(B\,||\,M),$$

where $M=\frac{1}{2}(A+B)$.

Therefore, the science articles with the least Jensen–Shannon distance are expected to be the most closely related to the units of study based on their topic structures.

To demonstrate benefits of our system and to effectively utilize the learning resources (i.e., science articles) that were identified to have appropriate content matching with the learning outcomes, constructed-response items for science assessments were also generated to evaluate students' understanding of the given resource and their demonstration of the mapped learning outcome. This analysis was conducted to demonstrate an effective usage of science resources that are systematically and automatically aligned with the curriculum standards in our framework. To develop corresponding test items using the science articles evaluated in Stage 2, the focusing questions were used from the units of study to create constructed-response prompts. Focusing questions refer to central questions that encompass the overall themes of the unit that can be used as guides for teachers and students. In other words, a constructed-response item was created that could measure students' application skills in real-life contexts using the focusing questions from the science articles. For example, the unit 'Interaction and Ecosystems' introduces two focusing questions: 'How do human activities affect ecosystems?' and 'What methods can we use to observe and monitor changes in ecosystems, and assess the impacts of our actions?' (see Table 8.2). To make sure that the modified question could include the science reading article as a reading resource, the questions were modified to 'What human activities does the article introduce that affect ecosystem?' and 'What methods do the articles introduce that can be used to observe and monitor changes in ecosystems and assess the impacts of our actions?' The modified questions allowed students to directly apply the knowledge and skills depicted in the original questions to a broader context in the science articles. In addition, to visualize the connections between the reading prompt and the question, our system was used to highlight corresponding parts of the text that represent the correct solution or answers for the focusing questions based on our topic weights and distributions. This allowed us to, in effect, create the key for each constructed-response test item. Figure 8.3 contains an overview of the conceptual representation of our three-stage system.

8.3 Results

8.3.1 Science Article Topic Analysis

The topic structures located with the keywords and their corresponding weights are presented in Table 8.3. After preprocessing, 647 unique vocabularies were identified from the bag-of-words representations of the original science articles. The best-performing model included 12 topic structures ($K=12$ and $\alpha=0.10$) with a perplexity of –21.09 and a coherence score of 0.46. All topic structures were easily interpretable and readily distinguishable. The title of each topic cluster was identified based on the keywords with

TABLE 8.2

Example Focusing Questions in the Units of Study

Unit	Grade 7	Focusing Questions
A	Interactions and Ecosystems	• How do human activities affect ecosystems? • What methods can we use to observe and monitor changes in ecosystems, and assess the impacts of our actions?
B	Plants for Food and Fiber	• How do we produce useful plant products? • What techniques do we use, what knowledge are these techniques based on, and how do we apply these techniques in a sustainable way?
C	Heat and Temperature	• What heat-related technologies do we use to meet human needs? • Upon what scientific principles are these technologies based? • What implications do these technologies have for sustainable use of resources?
D	Structures and Forces	• How does structure stand up under load? • What forces act on structure, and what materials and design characteristics contribute to structural strength and stability?
E	Planet Earth	• What do we know about Earth – about its surface and what lies below? • What evidence do we have, and how do we use this evidence in developing and understanding of Earth and its change?

the highest corresponding weights and the lowest common hypernyms among the keywords. The corresponding weights represented normalized weights that explain the degree of contribution for each word to explain the topic structure. For example, Topic 4 included keywords such as 'water', 'use', 'star', 'earth', 'ocean', 'heat', and 'universe' which represent the topic 'planet and resources'. Another example is Topic 5, which consisted of keywords such as 'forest', 'tree', 'insect', 'wind', 'fire', 'bird', 'plant', 'farm', and 'agriculture', representing 'nature and agriculture'. Article 892, 'What are we able to see moving objects against moving background', showed Topic 12 (Research and New Technology, $w=0.38$) and Topic 4 (Ageism and Health, $w = 0.23$) as the two highest contributing topics with the highest weights. This article described a new cellular technology that could enhance the observance of moving objects regardless of human age.

8.3.2 Topic Prediction and Topic Evaluation Units of Study

Topic prediction results for the units of study produced reliable outcomes using the best-performing model from Stage 1. For example in Table 8.4, the overall topic structure of the unit 'Interactions and Ecosystems' showed a mixture of four contributing topics, 'Research and New Technology' (Topic 12, $w=0.26$), 'Data and Information' (Topic 1, $w=0.17$), 'Planet and Resource' (Topic 4, $w=0.15$), and 'Nature and Agriculture' (Topic 5, $w=0.10$).

The system also produced a comparable agreement score to the human judgments. First, we evaluated the agreement between the two raters to assess the human–human agreement. The ratings between two content specialists produced a kappa score of 0.78, which indicates that 78% of the score agreement occurred between the two human-raters after correcting for chance agreement. Then, we evaluated the agreement between the two human-raters and our system separately to assess the human–system agreement. The content specialists ratings produced kappa scores of 0.77 and 0.89, respectively, with the results from the system. This agreement indicates that, on average, 83% of the agreement over the two raters was made between the current system and science content specialists in identifying topics for expected outcomes in the units of study. It is also noteworthy that our system achieved a higher agreement kappa score with the two human-raters compared to human–human agreement. The human raters commented that topic labeling of study outcomes is an extremely difficult and time-consuming task. On the other hand, using our system, the raters claimed that they could predict and label the topic structures more accurately and quickly. Tables 8.5 and 8.6 includes a list of STS and knowledge, science skills, and attitudes that showed the highest associating weights for each topic. For example, Topic 1, 'Data and Information', was highly associated with students' scientific knowledge to interpret examples of scientific investigations that could inform environmental decision-making. For Topic 5, 'Nature and Agriculture', and Topic 4, 'Planet and Resources', the most highly associated study outcomes were related to content-specific knowledge in ecosystems regarding 'energy', 'cycle', 'plant', 'species', 'habitats', and 'food'. For scientific skills and attitudes, demonstrating one's understanding by identifying science-related issues and showing interest in such issues produced the highest corresponding weights to Topic 12. Similarly, 'data', 'information', 'methods', and 'quantitative and qualitative research' generated

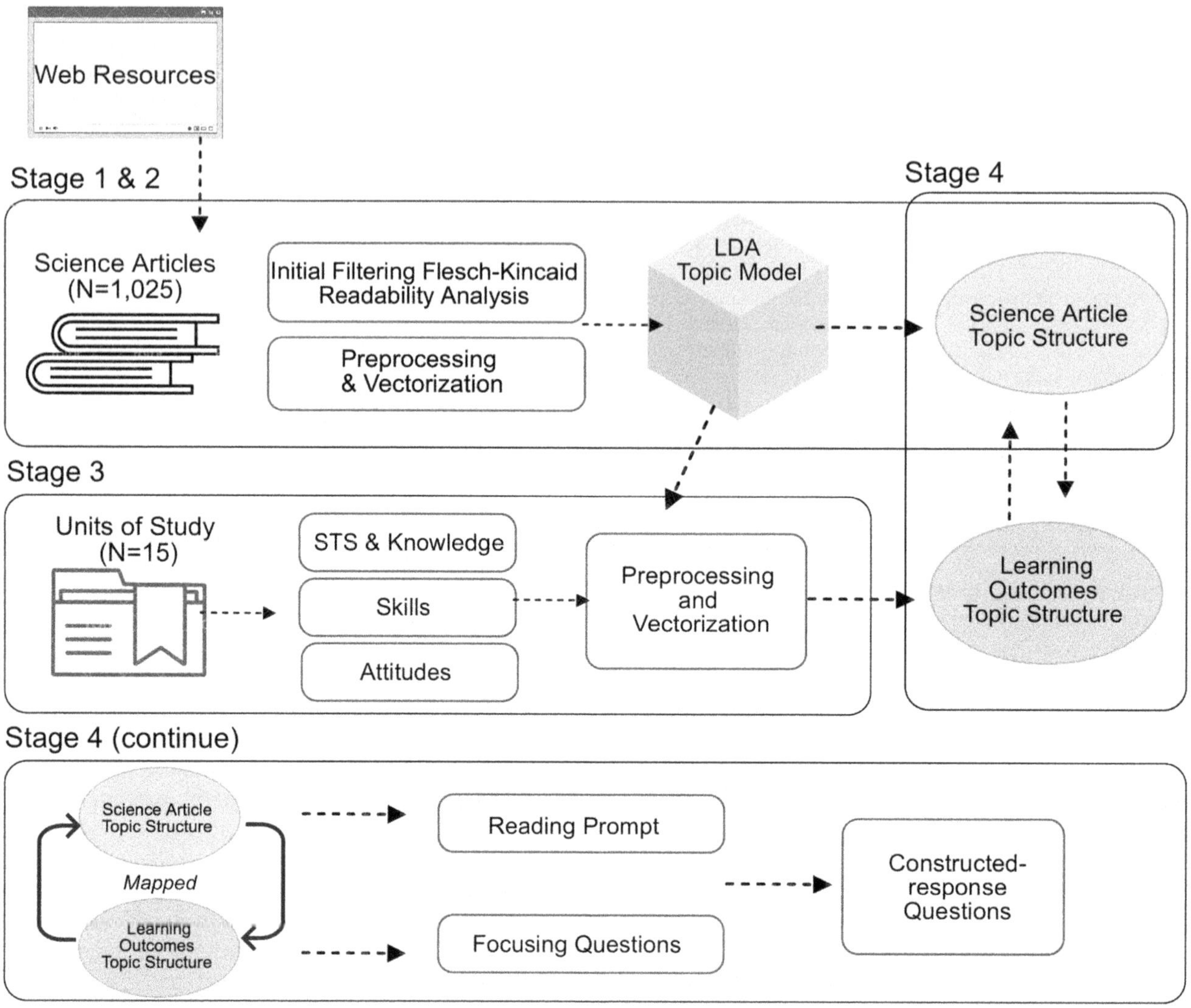

FIGURE 8.3
An overview of the conceptual representation's items.

the highest weights to represent Topic 1, 'Data and Information'. Therefore, the prediction results at the outcome level could be used to effectively understand and identify specific scientific knowledge, skills, and attitudes that were represented in the science articles.

8.3.3 Science Article Recommendation and Item Generation

Based on the Jensen–Shannon distance, eight relevant articles were located for each unit that represented similar topic structures with the units of study. Table 8.7 describes the eight most relevant articles for 'Interactions and Ecosystems'. The articles introduce new scientific innovations and technologies that were invented to overcome environmental issues. For example, Article 284, 'New forest treatment helps trees adapt better to climatic change', describes a new technique that could help recover and maintain forest ecosystems that are dealing with extreme climate change. The article briefly introduces the human impact on climate change that severely affected the forest ecosystems while providing a novel suggestion to overcome the current environmental challenges (see Table 8.5 for more information).

Based on the previous outcomes, the content of the relevant articles was clearly associated with outcomes in STS and knowledge, science skills and attitudes. Using Article 284, 'New forest treatment helps trees adapt better to climate changes', for instance, could be used to generate a constructed-response item to evaluate students' scientific literacy. The modified focusing questions of the unit was used: 'What human activities does the article introduce that affect ecosystem?' and 'What methods does the article introduce that can be used to observe and monitor changes in

TABLE 8.3
The Science Topic Modeling Results

ID	Topic Title	Keywords and Weights
1	Data and Information	0.015 * **Research** + 0.013 * **Student** + 0.011 * **Data** + 0.011 * **Use** + 0.009 * **University** +0.009*System + 0.007*Education + 0.007 * **Inform** + 0.007 * **Policy** + 0.007 * **Model**
2	Geology	0.012 * **Material** + 0.010 * **Use** + 0.009 * **Methane** + 0.009 * **Electron** + 0.009 * **Graph** + 0.008 * **Energy** + 0.007 * **Temperature** + 0.007 * **Emission**
3	Genetics	0.024 * **Genome** + 0.010 * **DNA** + 0.023 * **Genet** + 0.021 * **Mutant** + 0.018 * **Sequence** + 0.015 * **Gene** + 0.015 * **Species** + 0.014 * **Plasma** + 0.010 * **Virus**
4	Planet and Resources	0.018 * **Water** + 0.010 * **Use** + 0.007 * **Star** + 0.006 * **Research** + 0.006 * **Earth** + 0.005 * **One** + 0.005 * **Ocean** + 0.005 * **Surface** + 0.005 * **Heat** + 0.005 * **Universe**
5	Nature and Agriculture	0.036 * **Forest** + 0.026 * **Tree** + 0.020 * **Insect** + 0.018 * **Wind** + 0.017 * **Fire** + 0.014 * **Bird** + 0.014 * **Plant** + 0.013 * **Farm** + 0.013 * **Habitat** + 0.011 * **Agriculture**
6	Age and Health	0.013 ***Age** + 0.010 * **Year** + 0.008 * **People** + 0.008 * **Health** + 0.008 * **Women** + 0.007 * **Study** + 0.007 * **Veteran** + 0.007 * **Robot** + 0.007 * **Use**
7	Bacteria and Infection	0.029 * **Bacteria** + 0.027 * **Infect** + 0.025* **Antibiotic** + 0.024 * **Gut** + 0.016 * **Resist** + 0.012 * **Prescribe** + 0.011 * **Dietarian**
8	Physics and Computing	0.114* **Quantum** + 0.031***Computers** + 0.030***Cluster** + 0.019***Photon** + 0.017***System** + 0.016***Cloud** +0.011***Physicist**
9	Diseases and Health	0.023***Study** + 0.022***Children** + 0.017***Health** + 0.013***Research** + 0.009***Diabetic** + 0.008***Risk** + 0.008***Increase** + 0.008***Mental** + 0.007***Disorder** + 0.007***Depress**
10	Galaxy and Universe	0.015 * **Galaxy** + 0.012 * **Science**+ 0.010 * **Said** + 0.008 * **Manage** + 0.007 * **Universe** + 0.007 * **Research** + 0.007 * **Inform** + 0.007 * **Ecosystem** + 0.007 * **Study**
11	Cancer and Health	0.028 * **Cancer** + 0.019 * **Patient** + 0.014 * **Research** + 0.013 * **Health** + 0.013 * **Study** + 0.010 * **Treatment** + 0.010 * **Medicine** + 0.008 * **Drug** + 0.008 * **Use**
12	Research and Technology	0.023***Research** + 0.017***University** + 0.011***Science** + 0.009***New** + 0.009***Use** + 0.008***Institute** + 0.007***Develop** + 0.006***Technology** + 0.006***Work**

TABLE 8.4
POS Topic Prediction Results for 'Interaction and Ecosystem'

Topic	1	2	3	4	5	6	7	8	9	10	11	12
Weight	**0.17**	0.01	0.00	**0.15**	**0.10**	0.00	0.00	0.03	0.05	0.05	0.00	**0.26**

TABLE 8.5
Topic Prediction Evidence in STS and Knowledge Outcomes

Topic	STS and Knowledge Outcomes
12B	• Students will describe and interpret examples of scientific investigations that serve to inform environmental decision-making
1A	• Students will analyze personal and public decisions that involve consideration of environmental impacts, and identify needs for scientific knowledge that can inform those decisions
4	• Students will describe the process of cycling carbon and water through an ecosystem
5	• Students will identify examples of human impacts on ecosystems and investigate and analyze the link between these impacts and the human wants and needs that give rise to them • Students will analyze ecosystems to identify producers, consumers, and decomposers; and describe how energy is supplied to and flows through a food web, by describing and giving examples of energy and nutrient storage in plants and animals. • Students will analyze ecosystems to identify producers, consumers, and decomposers; and describe how energy is supplied to and flows through a food web, by describing how matter is recycled in an ecosystem through interactions among plants, animals, fungi, bacteria, and other microorganisms • Students will investigate a variety of habitats and describe and interpret distribution patterns of living things found in those habitats. • **Students will identify intended and unintended consequences of human activities within local and global environments** • Students will illustrate how life-supporting environments meet the needs of living things for nutrients, energy sources, moisture, suitable habitat, and exchange of gases

TABLE 8.6

Topic Prediction Evidence in Skills and Attitudes Outcomes

Topic	Skill and Attitude Outcomes
12B	• Student will identify science-related issues (e.g., identify a specific issue regarding human impacts on environments) • Students will show interest in science-related questions and issues and pursue personal interests and career possibilities within science-related fields
1A	• Students will state a prediction and a hypothesis based on background information or an observed pattern of events • Students will select appropriate methods and tools for collecting data and information • Students will conduct investigations into the relationships between and among observations and gather and record qualitative and quantitative data • Students will research information relevant to a given problem or issue • Students will select and integrate information from various print and electronic sources or from several parts of the same source • Students will use tools and apparatus effectively and accurately for collecting data • Students will take an interest in media reports on environmental issues, and seek out further information

TABLE 8.7

Relevant Science Articles to 'Interactions and Ecosystems'

Article ID	Title
284	New forest treatment helps trees adapt better to climatic change
375	Study suggests crash location of MH370 near 25S, north of underwater search area
577	NIST presents first real-world test of new smokestack emissions sensor designs
715	USDA awards $1.8M to Cornell for packaging, beverage concentrate research
769	Story tips from the Department of Energy's Oak Ridge National Laboratory, July 2019
784	Building up an appetite for a new kind of grub
812	Analysis finds US ecosystems shifting hundreds of miles north
981	UCI, UC Merced: California forest die-off caused by depletion of deep-soil water

ecosystems and assess the impacts of our actions?'. These two questions evaluated students' application skills and scientific literacy required in Unit A Grade 7. Appendices C and D contain two example questions that were developed using Article 284 and 375. The two articles showed the highest relatedness to the unit 'Interactions and Ecosystems'. It is important to note that the same modified questions were used to generate two distinct example test items. This demonstrates the generalizability and the universal nature of the focusing questions while emphasizing that the focus of our system was in generating and recommending appropriate science reading prompts rather than developing a variety of test item stems.

To exemplify the rationale behind the key for each question, our system was also used to *highlight* the corresponding parts of the text in the reading prompt most related to each focusing question. This highlighted text serves as the key for each question. For example, yellow highlights (A1 and A2 of Q1s) of the texts in the example questions were located based on their corresponding weights. More specifically, lines 2 and 3 of the first prompt 'the thinning techniques is effective as an alternative approach in the … extreme climates' showed the highest corresponding topic weights to Q1, which discusses human activities related to ecosystems. Similarly, lines 15 and 16 of the second prompt 'The data-based Markov-chain model … pollutants in the ocean' had the highest corresponding topic weights to Q2, which discusses methods to observe and monitor changes in ecosystems. Therefore, the highlighted lines were used to support the decision-making process of our topic modeling system while presenting them as exemplary explanations for correct student responses.

8.4 Conclusions and Discussion

Introducing appropriate science articles for assessment in context-based science classrooms provides benefits and opportunities for evaluating higher-order thinking skills. Students apply their scientific knowledge and understanding acquired from classroom instruction to real-life contexts, thereby demonstrating their scientific literacy (Kuhn & Müller,

2014). Moreover, teachers can motivate and engage students by providing them with interesting and challenging real-world tasks (Mysliwiec et al., 2003; 16Gayford, 2002; Elliott, 2006). Hence, providing students with appropriate science articles that feature relevant and up-to-date information in a timely manner has long been recognized as an important task for context-based science education. In the past, locating and evaluating scientific resources was handled by teachers with little or no systematic search strategies or guidelines. Despite the abundance of online public resources, it is an exceedingly time-consuming process for teachers to identify and evaluate this material. To overcome this problem, we introduced and demonstrated a machine learning framework that can be used to locate and evaluate large numbers of science articles which are directly linked to outcomes in curriculum standards. These materials, in turn, can be used to generate source-based constructed-response test items using the articles as reading prompts.

In the first stage, the best-performing LDA model was used to predict 12 distinct and interpretable topics from the science articles. Each topic was represented with several keywords and their corresponding weights that could describe the topic structure. Then, in the second stage, our system predicted the topic structure of each unit of study showing a high agreement score with content specialist. The predicted topic results provide a comprehensive representation of the units of study in terms of the specific contents featured (e.g., STS and knowledge) as well as the learning activities and cognitive processes (e.g., skills and attitudes) students were expected to demonstrate during the lessons. Based on the topic structure prediction results, eight relevant science articles were identified using the Jensen–Shannon distance measure. In the third stage, source-based constructed-response test items were generated using the articles as a reading source from each unit as a stem. The generated items demonstrated the scope and value of our system for locating relevant and up-to-date science reading resources. Our system was also used to highlight the parts of the text in the reading source that directly supported an exemplary student responses for the test item. The highlighted text serves as the keyed responses in addition to explicitly connecting the reading sources (i.e., science articles) with the constructed-response prompt (e.g., focusing questions).

Traditional methods for locating science resources required content experts to manually search for relevant science articles and evaluate the resources based on curriculum standards. This approach is time-consuming and labor-intensive. Using the current system, relevant science resources can be automatically identified, updated, and managed in a timely and cost-effective manner while significantly reducing the need for human judgments to complete the search process. Our solution helps eliminate the need for manual search. In addition, the system yields results that are interpretable. Machine learning applications in science assessment often yield uninterpretable results (Conati et al., 2018; Krajcik & Mun, 2014). However, our system includes features that enhanced the interpretability of the machine learning solution by including visualization that yields direct evidence supporting the decision-making processes of our LDA algorithm. For example, our system could provide a list of study outcomes that were clearly associated with each topic to increase the interpretability and the validity of the topic prediction decision. Moreover, our system could highlight the parts of text to provide direct evidence which explains the strong connection between science articles and focusing questions.

While the primary purpose of this study did not include evaluating the validity of the source-based constructed-response items generated using our system, we acknowledge the importance of future research regarding the validity and reliability of the items generated using the current system. The source-based items generated in the study served two purposes. First, it demonstrated the importance of locating science articles that could be used for content-based science assessment. Second, we demonstrated how a natural language system could yield clearly interpretable results that, in turn, could support science assessment. For any assessment task, providing transparent decision-making processes to evaluate the system is as valued as demonstrating high prediction and classification accuracies. By directly connecting the supporting parts of the text with the assessment prompt, we provided visible evidence to support the performance of the system. But future research is still required to understand the psychometric properties of the generated items to further evaluate the capacity and capabilities of this type of machine learning system. In addition, we encourage future research to investigate and compare the performance of various topic modeling approaches to improve the performance of the proposed system architecture. For instance, few recent studies have reported that the non-negative matrix factorization (NMF) approach could outperform LDA when discovering topics from shorter texts (e.g., Mifrah & Benlahmar, 2020; Chen et al., 2016, 2019) when less linguistic ambiguity and noise are involved in the data (e.g., Chen et al., 2016). Hence, we encourage future research to understand how such variations could improve the performance of the current system.

Appendix A

Programs of Study Topic Evaluation Tool

Grade 7 Unit A: Interactions and Ecosystems (Social and Environmental Emphasis)

Please select the **best topic** that could be used to represent and categorize each student outcome based on its content.

- **Topic A**: Using data and information to inform systems and policy
- **Topic B:** Research and new findings in science
- **Topic C:** Research about natural resources (e.g., water, ocean, heat)
- **Topic D**: Research about forest (e.g., tree, plants), animals (e.g., insect, bird, habitat), and agriculture (e.g. farm)

Outcomes for Science, Technology and Society (STS) and Knowledge	**Topic**			
	A	**B**	**C**	**D**
Students will identify intended and unintended consequences of human activities within local and global environments (e.g., changes resulting from habitat loss, pest control or from introduction of new species; changes leading to species extinction)				
Students will analyze personal and public decisions that involve consideration of environmental impacts, and identify needs for scientific knowledge that can inform those decisions				
Students will investigate a variety of habitats, and describe and interpret distribution patterns of living things found in those habitats (e.g., describe and compare two areas within the school grounds—a relatively undisturbed site and a site that has been affected by heavy use; describe and compare a wetland and a dryland area in a local parkland)				
Students will identify examples of human impacts on ecosystems, and investigate and analyze the link between these impacts and the human wants and needs that give rise to them (e.g., identify impacts of the use of plants and animals as sources of food, fibre and other materials; identify potential impacts of waste products on environments)				
Students will describe the process of cycling carbon and water through an ecosystem.				
Students will analyze ecosystems to identify producers, consumers, and decomposers; and describe how energy is supplied to and flows through a food web, by describing how matter is recycled in an ecosystem through interactions among plants, animals, fungi, bacteria and other microorganisms				
Students will illustrate how life-supporting environments meet the needs of living things for nutrients, energy sources, moisture, suitable habitat, and exchange of gases				
Student will describe and interpret examples of scientific investigations that serve to inform environmental decision making.				
Students will analyze ecosystems to identify producers, consumers and decomposers; and describe how energy is supplied to and flows through a food web, by describing and giving examples of energy and nutrient storage in plants and animals				

Appendix B

Programs of Study Topic Evaluation Tool

Grade 7 Unit A: Interactions and Ecosystems (Social and Environmental Emphasis)

Please select the **best topic** that could be used to represent and categorize each student outcome based on its content.

- **Topic A**: Using data and information to inform systems and policy
- **Topic B:** Research and new findings in science
- **Topic C:** Research about natural resources (e.g., water, ocean, heat)
- **Topic D**: Research about forest (e.g., tree, plants), animals (e.g., insect, bird, habitat), and agriculture (e.g. farm)

Outcomes for Skills and Attitudes	**Topic**			
	A	**B**	**C**	**D**
Students will show interest in science-related questions and issues, and pursue personal interests and career possibilities within science-related fields.				
Students will take an interest in media reports on environmental issues, and seek out further information.				
Students will select appropriate methods and tools for collecting data and information.				
Students will identify intended and unintended consequences of human activities within local and global environments (e.g., changes resulting from habitat loss, pest control or from introduction of new species; changes leading to species extinction).				
Students will research information relevant to a given problem or issue.				
Students will select and integrate information from various print and electronic sources or from several parts of the same source.				
Students will use tools and apparatus effectively and accurately for collecting data.				
Students will conduct investigations into the relationships between and among observations, and gather and record qualitative and quantitative data.				
Students will describe the process of cycling carbon and water through an ecosystem.				
Student will identify science-related issues (e.g., identify a specific issue regarding human impacts on environments).				
Students will state a prediction and a hypothesis based on background information or an observed pattern of events.				

Appendix C

Example Question (1) Generated from the Current System

New forest treatment helps trees adapt better to climatic change

Researchers from the University of Granada, the Andalusian Institute of Agricultural Research and Training, Fishing, Food and Organic Production (IFAPA), and the Pyrenean Institute of Ecology of the Spanish National Research Council (CSIC) have verified that the thinning technique is effective as an alternative approach in the recovery and maintenance of forest ecosystems dealing with extreme climates. The thinning method consists of reducing the number of trees in a given area, so that those remaining are able to access more resources. The researchers validated this technique using a novel method based on taking high-resolution measurements in the variations of the diameters of tree trunks. The method contributes to a better understanding of the short-term relationships between climatic changes and tree stem growth. To date, the technique commonly used has been to measure the width of the trunk's rings, but this is ineffective for discerning growth over a short time scale.

Forest thinning consists of reducing the density of trees per hectare, to thus decrease the competition for available resources and improve the growth of the remaining trees, rendering them less vulnerable to water stress. The researchers have set out this technique in a recent paper entitled "Using stem diameter variations to detect and quantify growth and relationships with climatic variables on a gradient of thinned Aleppo pines" published in the journal *Forest Ecology and Management*. They position it as a viable alternative in the fight against climate change in certain Mediterranean areas, demonstrating that the growth of forests improves when there is less competition between individual trees.

These experts analyzed the evolution of the trees over time. IFAPA researcher Francisco Bruno Navarro Reyes, co-author of the article, explains: "We have monitored the daily contractions and dilations of the trunks over the course of three years, to assess whether this technique triggers earlier growth in Spring and lengthens the period through to the autumn, effectively prolonging the period during which the trees develop."

Q1. What **human activities** does the article introduce that affect ecosystems?

A1: The thinning technique is effective as an alternative approach in the recovery and maintenance of the forest ecosystems dealing with extreme climates (**line 2 – 3**)

Q2. What **methods** does the article introduce that can be used to observe and monitor changes in ecosystems, and assess the impacts of our actions?

A2: The thinning method consists of reducing the number of trees in a given area, so that those remaining are able to access more resources (**line 3 – 4**)

Appendix D

Example Question (1) Generated from the Current System

Study suggests crash location of MH370 near 25°S, north of underwater search area

MIAMI--A new analysis of Malaysian Airlines flight MH370 found that the most probable crash site in the Indian Ocean is near 25°S, north of the area where most of the underwater search missions were performed.

The study, led by oceanographers at the University of Miami Rosenstiel School of Marine and Atmospheric Science (UM), applied probabilistic tools from nonlinear dynamics on all information that was known from the plane disappearance, including; data on the trajectories of unanchored drifting buoys from NOAA's Global Drifter Database, known oceanographic conditions during the time and the date and location of plane debris found along the coast of Reunion Island, Madagascar, Mauritius and coastal East African countries, to reverse model the most likely crash site in the southern Indian Ocean.

"Monsoons play an important role in the dynamics of the Indian Ocean," said the study's lead author Philippe Miron, a postdoctoral associate at the UM Rosenstiel School. "It's a crucial piece of the puzzle to locate the most probable crash site since its influence on the dispersion of floating debris is quite significant."

The disappearance of Malaysian Airlines flight MH370 in the southeastern Indian Ocean on March 8, 2014 in route from Kuala Lumpur to Beijing remains one of the biggest aviation mysteries. With the loss of all 227 passengers and 12 crew members on board, flight MH370 is the second-deadliest incident involving a Boeing 777 aircraft. At a cost nearing $155 million, its search is the most expensive in aviation history. In January 2017, almost three years after the airplane disappearance, the Australian Government's Joint Agency Coordination Centre halted the search after failing to locate the airplane across more than 120,000 square kilometers in the eastern Indian Ocean. In 2018, Ocean Infinity, an ocean exploration company was unsuccessful in locating the aircraft during a several-month cruise. The data-based Markov-chain model developed by the researchers in this new study could also help scientists track oil spills, and other types of marine debris and pollutants in the ocean.

Q1. What **human activities** does the article introduce that affect ecosystems?

A1: MIAMI--A new analysis of Malaysian Airlines flight MH370 found that the most probable crash site in the Indian Ocean is near 25°S, north of the area where most of the underwater search missions were performed (**line 1 – 2**)

A2: "It's a crucial piece of the puzzle to locate the most probable crash site since its influence on the dispersion of floating debris is quite significant." (**line 8 – 9**)

Q2. What **methods** does the article introduce that can be used to observe and monitor changes in ecosystems, and assess the impacts of our actions?

A1: The data-based Markov-chain model developed by the researchers in this new study could also help scientists track oil spills, and other types of marine debris and pollutants in the ocean. (**line 15 – 16**)

Notes

1. https://www.sciencenewsforstudents.org/
2. https://www.sciencejournalforkids.org/
3. https://www.dogonews.com/category/science

References

Anderman, E. M., Sinatra, G. M., & Gray, D. L. (2012). The challenges of teaching and learning about science in the twenty-first century: Exploring the abilities and constraints of adolescent learners. *Studies in Science Education, 48*(1), 89–117.

Avargil, S., Herscovitz, O., & Dori, Y. J. (2012). Teaching thinking skills in context-based learning: Teachers' challenges and assessment knowledge. *Journal of Science Education & Technology, 21*(2), 207–225.

Bennett, J., Lubben, F., & Hogarth, S. (2007). Bringing science to life: A synthesis of the research evidence on the effects of context-based and STS approaches to science teaching. *Science Education, 91*(3), 347–370.

Bird, S., Klein, E., & Loper, E. (2009). *Natural language processing with Python: Analyzing text with the natural language toolkit*. O'Reilly Media, Inc.

Blei. (2003). https://www.slideshare.net/hustwj/probabilistic-topic-models

Blei, D. M., Ng, A. Y., & Jordan, M. I. (2003). Latent Dirichlet allocation. *Journal of Machine Learning Research, 3*(January), 993–1022.

Bracewell, R., Breuleux, A., Laferrière, T., Benoit, J., & Abdous, M. H. (1998). The emerging contribution of online resources and tools to classroom learning and teaching. *Report Submitted to SchoolNet/Rescol by TeleLearning Network Inc. En ligne*. http://www.tact.fse.ulaval.ca/fr/html/apportnt.html.

Chen, Y., Bordes, J. B., & Filliat, D. (2016, September). An experimental comparison between NMF and LDA for active cross-situational object-word learning. In *2016 joint IEEE international conference on development and learning and epigenetic robotics (ICDL-EpiRob)* (pp. 217–222). IEEE.

Chen, Y., Zhang, H., Liu, R., Ye, Z., & Lin, J. (2019). Experimental explorations on short text topic mining between LDA and NMF based Schemes. *Knowledge-Based Systems, 163*, 1–13.

Conati, C., Porayska-Pomsta, K., & Mavrikis, M. (2018). AI in education needs interpretable machine learning: Lessons from open learner modelling. arXiv preprint arXiv:1807.00154.

Daenekindt, S., & Huisman, J. (2020). Mapping the scattered field of research on higher education: A correlated topic model of 17,000 articles, 1991–2018. *Higher Education, 80*(3), 571–587.

Dagan, I., Lee, L., & Pereira, F. C. (1999). Similarity-based models of word co-occurrence probabilities. *Machine Learning, 34*(1–3), 43–69.

Demuth, R., Parchmann, I., & Ralle, B. (Eds.). (2006). *Chemie im Kontext – Sekundarstufe II* ['Chemistry in Context' – Secondary education II]. Cornelsen.

Elliott, P. (2006). Reviewing newspaper articles as a technique for enhancing the scientific literacy of student-teachers. *International Journal of Science Education, 28*(11), 1245–1265.

Fensham, P. J. (2009). Real world contexts in PISA science: Implications for context-based science education. *Journal of Research in Science Teaching: The Official Journal of the National Association for Research in Science Teaching, 46*(8), 884–896.

Gayford, C. G. (2002). Environmental literacy: Towards a shared understanding for science teachers. *Research in Science & Technological Education, 20*(1), 99–110.

Grant, E., Gardner, M., Jones, G., & Ferzli, M. (2009). Popular media in the biology classroom: Viewing popular science sceptically. *The American Biology Teacher, 71*(6), 332–335.

Holbrook, J. (2014). A context-based approach to science teaching. *Journal of Baltic Science Education, 13*(2), 152–154.

Krajcik, J. S., & Mun, K. (2014). Promises and challenges of using learning technologies to promote student learning of science. In N. G. Lederman & S. K. Abell (Eds.), *Handbook of research on science education* (Vol. II, pp. 337–360). Routledge.

Kuhn, J., & Müller, A. (2014). Context-based science education by newspaper story problems: A study on motivation and learning effects. *Perspectives in Science, 2*(1–4), 5–21.

Levinson, R., Douglas, A., Evans, J. E., Kirton, A., Koulouris, P., Turner, S., & Finegold, P. (2001). *Valuable lessons: Engaging with the social context of science in schools*. Wellcome Trust.

Liu, S., Ni, C., Liu, Z., Peng, X., & Cheng, H. N. (2017). Mining individual learning topics in course reviews based on author topic model. *International Journal of Distance Education Technologies (IJDET), 15*(3), 1–14.

McClune, B., & Jarman, R. (2010). Critical reading of science-based news reports: Establishing a knowledge, skills and attitudes framework. *International Journal of Science Education, 32*(6), 727–752.

Mifrah, S., & Benlahmar, E. H. (2020). Topic modeling coherence: A comparative study between LDA and NMF models using COVID'19 corpus. *International Journal of Advanced Trends in Computer Science & Engineering*, 5756–5761.

Miller, A. (2011). The use of current events as assessment tools. *Journal of Microbiology & Biology Education: JMBE, 12*(1), 59.

Mysliwiec, T. H., Shibley Jr, I., & Dunbar, M. E. (2003). Using newspapers to facilitate learning: Learning activities designed to include current events. *Journal of College Science Teaching, 33*(3), 24–28.

Nentwig, P., & Waddington, D. (Eds.). (2006). *Making it relevant: Context based learning of science*. Waxmann Verlag.

OECD. (2019). *PISA 2018 technical report*. OECD Publishing.

Oliveras, B., Márquez, C., & Sanmartí, N. (2013). The use of newspaper articles as a tool to develop critical thinking in science classes. *International Journal of Science Education, 35*(6), 885–905.

Pilot, A., & Bulte, A. M. (2006). Why do you "need to know"? Context-based education. PISA 2006. OECD.

Prince Edward Island Department of Education. (2008). *Evaluation and selection of learning resources: A guide.* Department of Education.

Prins, G. T., Bulte, A. M., & Pilot, A. (2018). Designing context-based teaching materials by transforming authentic scientific modelling practices in chemistry. *International Journal of Science Education, 40*(10), 1108–1135.

Röder, M., Both, A., & Hinneburg, A. (2015, February). Exploring the space of topic coherence measures. In *Proceedings of the eighth ACM international conference on Web search and data mining* (pp. 399–408). ACM. Routledge.

Shwartz, Y., Dori, Y. J., & Treagust, D. F. (2013). How to outline objectives for chemistry education and how to assess them. In *Teaching chemistry–A studybook* (pp. 37–65). Brill Sense.

So, W. M. W., Ching, N. Y., Kong, S. C., & Cheng, M. H. M. (2011). *Teacher's selection and use of internet-based resources and tools to facilitate learning in primary classrooms.* Formatex Research Center.

Steyvers, M., & Griffiths, T. (2007). Probabilistic topic models. In *Handbook of latent semantic analysis* (pp. 439–460). Psychology Press.

Vos, M. A. J., Taconis, R., Jochems, W. M., & Pilot, A. (2011). Classroom implementation of context-based chemistry education by teachers: The relation between experiences of teachers and the design of materials. *International Journal of Science Education, 33*(10), 1407–1432.

Wu, P., Yu, S., & Wang, D. (2018). Using a learner-topic model for mining learner interests in open learning environments. *Journal of Educational Technology & Society, 21*(2), 192–204.

9

Adaptive Learning Profiles in the Education Domain

Claudio Giovanni Demartini, Andrea Bosso, Giacomo Ciccarelli, Lorenzo Benussi, and Flavio Renga

CONTENTS

9.1 Introduction 128
9.2 Academic Analytics 128
9.2.1 Decision Support in Education: Loyalty and Dropout 128
9.2.2 Academic Analytics at Politecnico di Torino 129
9.2.3 Research Questions 130
9.2.4 The Information System Course 131
9.2.4.1 Framework 132
9.2.4.2 Classroom 132
9.2.4.3 Course Delivery Organization 132
9.2.4.4 Assessment 132
9.2.4.5 Course Management: Student Behavior and Assessment 133
9.3 Academic Analytics Platform 134
9.4 Information Systems Course Data 134
9.4.1 Dataset 134
9.4.2 Dataset Initial Exploration 135
9.4.3 Clustering Algorithm Overview and Benchmark 136
9.4.3.1 DBSCAN 137
9.4.3.2 k-Means and k-Medoids 138
9.4.4 Performance-Based Clustering Visualization 139
9.4.5 Association Rules 140
9.4.5.1 Cluster 0: Exam Passed – 'Advanced' Python/UML Section, 'Intermediate' MC Section, and 'Intermediate' Project Work 141
9.4.5.2 Cluster 1: Exam Passed – 'Advanced' Python/UML Section, 'Expert' MC Section, and 'Intermediate' Project Work 141
9.4.5.3 Cluster 2: Exam Failed –'Basic' MC Section, 'Being Developed' Python/UML Section, and No Project Work 141
9.4.5.4 Cluster 3: Exam Passed/Failed –'Intermediate' Python/UML Section, 'Advanced' MC Section, and 'Intermediate' Project Work 142
9.4.5.5 Cluster 4: Exam Failed – 'Being Developed' Python/UML and MC Sections and 'Intermediate' Project Work 142
9.4.5.6 Cluster 5: Exam Failed – 'Being Developed' Python/UML and MC Sections and No Project Work 142
9.4.5.7 Cluster 6: Exam Passed/Failed – All 'Intermediate' Sections 143
9.4.5.8 Cluster 7: Exam Passed – 'Expert' Python/UML and MC Sections and 'Intermediate'/'Advanced' Project Work 143
9.4.5.9 Cluster 8: Exam Failed – 'Being Developed' Python/UML Section, 'Basic'/'Intermediate' MC Section, and 'Intermediate' Project Work 143
9.4.6 Discussion on the Results 143
9.4.7 Research Questions and Answers 145
9.5 Conclusions and Future Work 146
References 147

DOI: 10.1201/9781003181187-11

9.1 Introduction

This study aimed to gain experience in academies and analogous educational organizations practicing methodologies related to the academic analytics domain. More specifically, it proposes a sustainable survey scenario for teacher and student communities who deliver and attend courses in any educational context. This investigation is rooted in the master's degree in engineering and management program of Politecnico di Torino and focuses on the information systems (IS) course, spanning the three years 2017–2019 and becoming a reference to set up the testbed. The master's degree study plan carries out a route that strengthens students' knowledge and awareness on private and public companies' and organizations' life cycles. It primarily stakes economic, legal, and financial perspectives and deals with new product development, including strategies, organizational planning, production system, and quality management. The final purpose of the course, within the master's degree, is to create a highly professional profile that can act effectively in strategic and technical operative decisions, such as shaping business models, organizational structures, development projects for new products/services, and systems.

This multifaceted profile drives the competitiveness of companies and other complex organizations operating in technological domains featuring cutting-edge innovation and showing extensive market entanglement.

The course examined in this analysis was designed for students interested in the business domain who are close to becoming business professionals. The main goal is to help them learn how to use and manage information and information technologies to improve the business process, support decision-making, and gain a competitive advantage. Therefore, a significant emphasis is placed on up-to-date coverage of Internet technologies from the perspective of an adaptive platform for business, commerce, and collaboration processes, established among all business stakeholders.

In this study, academic analytics techniques, including learning analytics (LA) and educational data mining (EDM), were used to explore data collection, mainly expressing the performance achievements accomplished by the students attending the information systems course.

9.2 Academic Analytics

Academic analytics (Costa, 2019) applies statistical, predictive modeling, data mining, and artificial intelligence (AI) techniques to analyze, evaluate, and summarize various organizational, educational, and bibliographic data. This information derives from higher education and research institutions' investigations to provide numerical results that can drive strategic planning and decision-making practices in these contexts. It is becoming increasingly used for student and faculty assessment, deciding the allocation of funding and evaluating the standing and productivity of individual academic departments and entire universities.

Examples of such analyses include the degree of cross-institutional and international authorship of scholarly publications to indicate the importance of research project outputs (Wong, 2016). In addition, the correlation of grades with students' interactions with university services such as libraries and virtual learning environments can improve the detection accuracy of the learning performance of individual students. Moreover, the dropout rates and degree distributions of different universities can be employed to evaluate the quality of teaching. Beneficiaries of these investigations include either university administrators and individual academics, at first, or, in the case of learning analytics, the students themselves and their parents, who are becoming increasingly more involved.

Universities often collect student data concerning exam results or information about their school background, but this information is rarely exploited to its full potential. Leveraging this information can significantly impact higher education by allowing universities to improve the quality of services they provide to students, increasing the success rate to complete their careers (Jones, 2019).

As more profound knowledge of students' profiles increases, a better prediction of their possible failures ensues. This result may have a significant influence on the various stakeholders involved in academic life, acting in the following main domains:

- Course management, involving teachers, assistants, and tutors;
- Student behavior and performance, affecting students and teachers;
- Decision support in higher education, involving the administrative directors of the university;
- Retention and dropout of studies, primarily affecting students and, indirectly the academic bodies at a local and national level.

9.2.1 Decision Support in Education: Loyalty and Dropout

There has been much debate in the literature about how important it is to have meaningful information available to decision-makers within high and higher education institutions. However, it is a challenge to obtain the information they need quickly and efficiently. Within this domain, the methodologies of academic analytics have found fertile ground.

Many institutions have implemented this type of analysis to improve, among different contexts, recruitment management, as Campbell highlights in his work (Campbell, 2007). Institutional researchers working with admissions staff have created complex formulas based on standardized exam scores, high school courses, and other information to determine which candidates have the right profile to access courses. The 'usable intelligence' generated by statistical analyses of the many data sources can drive more efficient use of admission budgets and staff schedules. For some institutions, establishing this analysis means that they can provide applicants with an immediate answer to their admissions questions. Analytical models and decision frameworks have been refined over the years to produce reasonably predictable enrollment rates and a balance in areas such as current/off-course students, program enrollments, and other demographics. Based on the current data, models improve every year to enhance and make more effective registration decisions (Robin Pappas, 2021).

Besides applications management, analytics supports fundraising by building a data warehouse containing information on alumni and affiliates. Institutions can use predictive models to identify and select those donors who would most likely donate. In addition to academic and curricular history, the information may include an individual's response to past requests or interest in particular college and university initiatives. Other parameters are employment, contributions, and any awards received and participation in institutional events.

A fundamental aspect for university organizations is undoubtedly student loyalty and the reduction of school dropouts (Ang et al., 2020). For both professors and administrative managers, the dispersion of students who do not complete a course of study is still one of the central problems in terms of the performance indices with which institutions are evaluated globally, besides economic and financial assessment. In this case, data mining serves as a tool for predicting which types of students are at risk of dropping out. Luan applied both quantitative and qualitative research techniques to uncover student success factors. This research marks a milestone because it has demonstrated the success of applying educational data mining tools to retain students (Luan, 2002). Engaging research was carried out by Gray and Perkins (Gray and Perkins, 2019) in which they applied machine learning techniques to identify the earliest signs of struggling students, being able to make those predictions after the first four weeks of the fall semester.

Furthermore, in another study (Daud et al., 2017), researchers developed models to predict whether a student can complete his degree or not. In their models, features like students' personal information and family expenditures are used besides the 'classic' variables related to academic performances.

9.2.2 Academic Analytics at Politecnico di Torino

For the reasons discussed above, the present situation in academia is urgent and requires further engagement following all emerging perspectives. According to Politecnico's Board of Directors, the academy should take control of the data analytics infrastructures it may practice, which need to be kept open for this purpose, with transparent governance to ensure the healthy functioning of the academic community. While the existing scholarly publishing infrastructure is well established and hard to change quickly, the use of data analytics and AI in academia is still nascent and in flux. Hence, it should be relatively easy to prevent ceding complete control of these activities to commercial vendors, who, of course, are merely doing what they exist to do, namely, to maximize profits for their owners and shareholders.

The idea that was conceived suggests a solution able to design statistics on the historical data concerning students and teachers observed over several academic time slots, without directly involving institutional data repositories, pursuing some collection of the information that directors, teachers, and even students need for making their own organizational and career choices.

The model architecture was validated with information taken from students and teachers during the last three years. The findings provide teachers and deans the awareness to assume systemic decisions on using data collected and made available in the information system repositories of any academic institution or school.

After the due political and technical validation of the model, it was stated that it is a right of any student to know the reports on their academic performance to carry out their self-assessment process.

The same holds for teachers who can be aware of those results to assess the learning processes and shape a continuous adaptation of their course content to follow the learning dynamics emerging in the classroom. In turn, the head of the study program, taking care to support the whole teaching community, can make decisions based on the same data processing findings.

Hence, applying analysis techniques lets any educational organization collect, measure, and process information on each student, linked to its context, showing how that link interferes with any learning processes experienced by the same student. Further enhancements also include describing the student's strengths and weaknesses to shape the quality and effectiveness of his production and performance by representing individual or/and group profiles. As there is a new generation of students and teachers who need new solutions to improve education, the use and implementation of this methodology allows for collecting data produced by both the learner and the teacher. Together with the analysis models, management of these data helps discover useful information to shape the best response to the requirements from the education community.

9.2.3 Research Questions

The project currently in progress, *adaptive learning*, is depicted in Figure 9.1. It focuses on enhancing the comprehension of the teaching and educational domain by exploiting a data-driven approach. The integration of data mining and machine learning techniques makes the developed platform an adaptative and smart tool able to produce a relevant impact on the learning process in terms of reinforcement and personalization of educational experiences (Agasisti and Bowers, 2017).

In particular, the automatic generation and assignment of learners' profiles enable a deeper understanding of the dynamics behind the currently-in-use educational and teaching activities (Brancaccio, 2015). The process of content adaptation and support to teaching activities heavily relies on data-driven methodologies and AI algorithms to increase the personalization and contextualization of both the learning content and the learners' profiles, as depicted in Figure 9.1. In such a way, learning outcomes can be mapped onto the learner profile according to competencies, skills, and knowledge achieved (ESG, 2015).

The extracted knowledge is relevant to feed tools and platforms built around data and centered on people. The main purpose is to affect the education community's processes (Benussi, 2017), which is usually expressed in the formal domain of the organization but often represented in the informality of its perceived image.

Attention is paid to the student profile detector concerning this specific work, as depicted in Figure 9.1. It senses the student *learning outcomes profile* through personal data repositories joined to other historical and current performance data. This profile is compared to the *reference job profile*, built on the learning outcomes profile suggested by available standards and market assessment (e.g., ISO 15288: Technical Standard in Systems Engineering). The computed distance or error determines the regulator/control actions driven by the stated course project work syllabus (PWS) (Demartini, 2017) in terms of the establishment and subsequent executions of flipped classes, in the presence of (or in virtual) living labs,

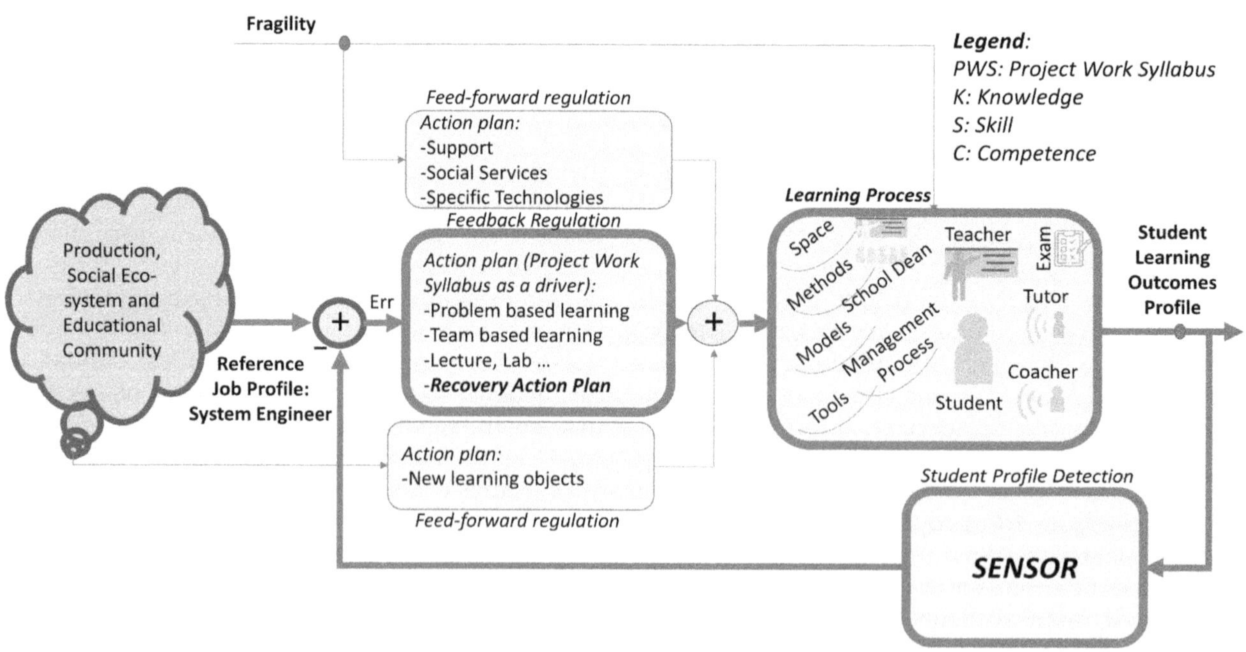

FIGURE 9.1
Adaptive learning at Politecnico di Torino, master's degree in engineering and management: 'Information systems' course.

problem-based learning, lectures, and other specific learning actions or environments. The focus within the *regulator* is here expressed as a *recovery action plan* to be established according to measurements carried out by the sensor. The blocks relevant for this investigation are highlighted with bold lines traced on the image.

Hence, the research questions mainly deal with the following two items, the *sensor* to detect students' *learning outcomes profile* and the *actions* to be planned inside the *regulator* to compensate the *error*:

(RQ1) What is the *sensor*? How does it work? What are the tools used?

(RQ2) What are the *actions* to be planned in the *regulator*? How have they been identified? How are they executed?

9.2.4 The Information System Course

In the *information systems* course (Paravati et al., 2015), the learning process follows a six-stage life cycle, as shown in Figure 9.2. In the first stage, the problem is depicted as unfolded by the company/organization. It often calls for catching images perceived through the lens of the innovation trends, forcing companies to reconsider their value chain based on the flow, as molded by the technology advancement and the practices imposed by competitors. The second stage concerns problem-posing, where a selection of methods and tools applied to the specific domain to understand a broader picture is arranged. The problem is understood, and its impact on systems and the environment is grasped and assessed. The third stage aims at developing suitable strategies for solving the problem by fostering algorithmic approaches based on results and views extracted from the previous step and suitably formalized using practices and standards. After choosing a strategy, a prototype takes shape in the fourth stage, where hardware platforms, software, and languages are assembled. The rapid prototyping phase exploits the 'reuse' principle, making previously developed components easily assembled to obtain a sustainable performance level, improving cost reduction and time to market. The following stage, which includes deployment and dissemination, explores how marketing occurs using suitable communication channels, pointing to different stakeholders, and pursuing appropriate funding means. In the end, according to the training perspective, the assessment is performed on the base of the learning objectives and outcomes specified for both the course and Master of Science plan. The enterprise/organization perspective has a fundamental role in the assessment process. Self-assessment is also promoted to force team components to account for efforts and to estimate costs sustained during the development of the whole life cycle.

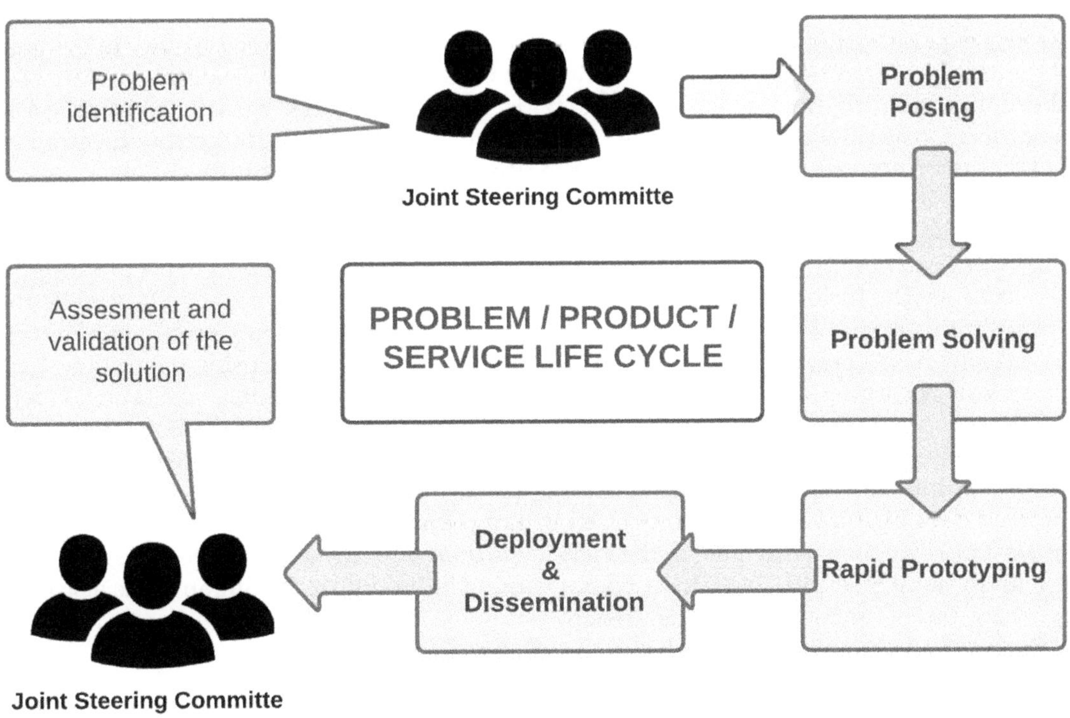

FIGURE 9.2
'Information systems' learning process.

9.2.4.1 Framework

The case study runs in the context of the information systems (IS) course, which today gathers more than 350 freshmen yearly, taking the first position among all courses afferent to Politecnico's Master of Science paths. Students attend other disciplines in parallel to the IS course, such as project management, object-oriented programming, business planning, and quality management.

9.2.4.2 Classroom

The *constructivist* classroom gathers a whole population of more than 900 students, spanning the academic years 2017–2018, 2018–2019, and 2019–2020, permanently organized in three lab teams. Groups, composed of six people each, are self-established into a set of cooperating individuals facing complex problems related to specific projects, shaped according to brainstorming results often carried out with the support of enterprises. Cooperation within each group occurs through collaborative tools such as, among others, Dropbox and Google Drive to store data and Skype, Teams, or Zoom to communicate synchronously. The lab features different isles, where students can gather and sit in circles in contrast with traditional, teacher-centered schema, where students sit in rows listening to an endless stream of lectures.

9.2.4.3 Course Delivery Organization

The *active learning* experience (Paravati et al., 2015) was developed by mixing the traditional and constructivist approaches into a learning framework that molds the course organization as a living lab, where the project life cycle takes place according to the project work syllabus (Demartini, 2017). Part of the weekly schedule is devoted to the project's development (60%), while the remaining part (40%) focuses on traditionally delivered lectures. The fusion of these trajectories answers a twofold dilemma. On one side, there is a university's company-based organization, where time is scanned regularly according to labor coordination and passive interactions. On the other side, this structure copes with requirements expressed by creativity-based processes, mainly rooted in inspiring student activities. Students are also directly engaged in the course organization, since they participate in the kick-off meeting with the Joint Steering Committee, ad hoc established, where the main basic questions arise to disclose all the facets of the problem at hand.

The course spans 13 calendar weeks. While the first week focuses on initial activities such as introducing the course schedule and organization, the second week deals with kickoff issues concerning problems that companies intend to face. Furthermore, students have time to self-organize their teams, gathering complementary skills, knowledge, and experiences. In the end, teams are finalized after having inferred the introductory remarks concerning the problems that companies brought out. Between the second and fourth week, the problem-posing domain is tackled, where projects take their shapes, applying a top-down deductive approach. Hence, at the beginning of the project life cycle management, the existing framework is recognized and assumed as a starting point for subsequent new proposal development. Questions are the primary means, within the 'problem-posing' domain, to shape a scene inspection and gather as much information as possible about the reference arrangement for the product/service to be conceived as an appropriate return to the problem made known by the organization. Therefore, students access into this domain to achieve a clear comprehension of the issue within its context, mainly through logical framework analysis (LFA) and quality functional deployment (QFD).

From the fifth to the seventh week, the problem-solving approach is pursued, focusing on formal and informal specification development, where algorithmic mechanisms predominate. Profiting from the problem analysis carried out previously and also from the established corresponding process planning, students become familiar with the integrated computer-aided manufacturing definition for function modeling (IDEF0) and the unified modeling language (UML) notation for specification processing to prepare a new picture, 'to be', in comparison with eventual benchmarks assumed as state of the art, 'as is'. Weeks 8 through 10 include building a sustainable prototype to meet the targets and restraints stated by the Joint Steering Committee.

From the 11th to the 13th week, the students engage in deployment and dissemination activities: for the last three weeks, first, they try out the prototype to be delivered, working on an appropriate test-bed; then, each team plans and sets up proper communication programs to design the closing exposition shaped for the Joint Steering Committee, including videos, reports, and a complete technical demo for the final discussion. Intermediate release dates stress the delivery schedule for the provision of LFA, QFD, UML specifications, and a final tentative prototype implementation. Furthermore, an appropriate time plan for programming individual skills alignment is also established.

9.2.4.4 Assessment

Students' assessment (Paravati et al., 2015) is divided into four steps, as reported in Table 9.1. The project work has the most significant impact and is collectively

TABLE 9.1
Students' Assessment

Section	Description	Weight
A	Project work	66%
B	Test bank	16%
C	Reverse engineering	16%
D	Self-assessment	2%

expressed by the Steering Committee after each team has jointly discussed on the project development. The closing discussion is sustained through a detailed report with its summary slide sequence, coherent with the project work syllabus framework, a close to ten-minute technical video to illustrate the prototype functional behavior, accompanied by another three-minute emotional Kickstarter-like video, the coding software of the prototype, its testing, and the toolkit for its management and development. The *project work syllabus* (Demartini, 2017) is the reference point at the base of the planning role played by the regulation unit in Figure 9.1. It depicts the job profile interface and its corresponding description, which specifies the worker's primary activities in the enterprise/organization and the link to the learning outcomes profile, which states skills, attitudes, competencies, and knowledge elements (ESG, 2015). The second item listed in the assessment table concerns a test bank that comprises the course reference text, *Information Systems for Business and Beyond*, an OER dealing with information systems at large. Students answer a multiple-choice test based on this course text. Furthermore, a reverse engineering section is also settled on to allow students to process and draw functional and system diagram interpretations working on made available Python code segments. Figure 9.3 depicts the relationship between the project syllabus, the course text, the reverse engineering process, and corresponding assessment tools.

Moreover, a self-assessment mechanism is also exposed to differentiate, individually, the project work assessment based on each group member's abilities and participation. It consists of virtually dispensing a certain number of credits to everyone, who should redeploy them among the team members according to his/her awareness of the practical cooperation each colleague has engaged for the prototype development.

9.2.4.5 Course Management: Student Behavior and Assessment

Managing a course and simultaneously improving the learning process is often a vast and unintuitive task for teachers, professionals, and their assistants. The issues and challenges concerning the enhancement of teaching and learning are known. However, they can also be easily detected through questionnaires

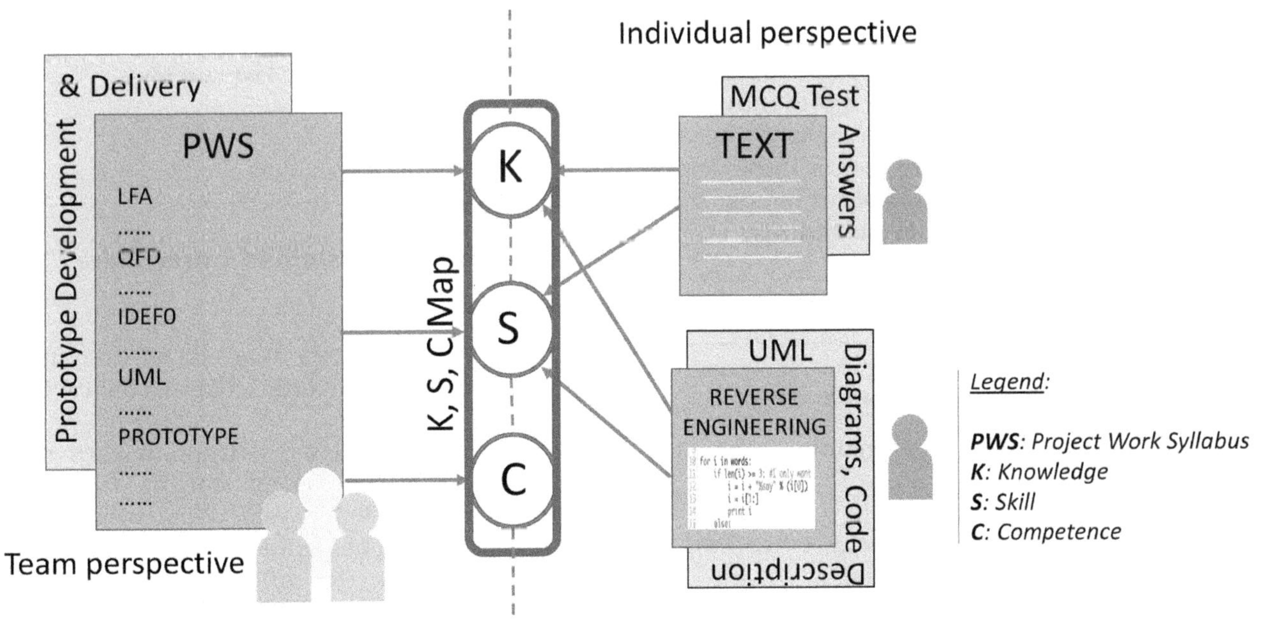

FIGURE 9.3
Assessment scenario: project work syllabus, course text, reverse engineering, and the knowledge, skills, and competency map, together with the corresponding assessment tools and targets (teams and individuals).

submitted to both teachers and students. They may express their perceptions and assess the various facets of course delivery through semi-quantitative scales. Educational data mining and learning analytics (EDM/LA) focuses on bringing out hidden knowledge from educational data. The datasets may be composed of the corresponding data collected during course delivery periods by the teachers themselves from the university's information system and the digital learning frameworks. Instructors can assess the structure of the course content and its effectiveness in the learning process. Tools may classify students based on statements and monitoring perspectives. In some cases, it is also possible to determine students' regular and atypical patterns so as to identify their most frequent errors and develop more practical activities.

After describing the course management domain, it is also necessary to consider the other side of the coin, taking a more granular perspective than that of the university course, the one addressed to the individual student. The two perspectives share part of the knowledge generated by the examples described above, and both teachers and students can benefit from it, since the improvement in teaching also has repercussions on the objectives of those who study. Regarding the EDM/LA applications that focus mainly on modeling behavior and evaluating students' learning performance, there are documents in the literature where authors theorize and implement systems that generate feedback for teachers and students. E-learning courses and those delivered face-to-face detect students' atypical learning behavior and finally notify possible problems in a preventive way. They are student-oriented methodologies for recommending potentially useful activities or resources, or suggesting curriculum curtailment, or simply links that would foster and enhance their learning.

9.3 Academic Analytics Platform

Investigation was carried out mainly using two instruments:

- Microsoft Excel for initial dataset handling and data integration processing;
- RapidMiner for pursuing the data mining process, clustering algorithm execution, and association rule establishment.

Both tools are well-known in the educational data mining field (Slater et al., 2017), and RapidMiner is a visual tool that allows developing data mining analysis and creating models without knowing any specific coding language. The main advantage offered by the tool is accessibility and intuitiveness of its functionalities, though they cannot offer the same flexibility as an actual coding language such as Python or R. Users can create workflows named 'Process' and add 'Operators' to them. Each 'Operator' handles a single task, and RapidMiner, natively, offers a broad array of 'Operators' to blend, clean, model, and score data. The specific tasks of 'Operators' can include regression analysis, classification algorithms, clustering algorithms, and association rules algorithms. Nevertheless, if more flexibility should be required, RapidMiner allows users to execute Python or Java scripts in its workflow. These workflows can be created intuitively using a 'drag & drop' mechanism or writing XML code if required. The software also integrates visualization tools allowing users to see and explore the results of their analysis quickly. In addition, a rich portfolio of tutorials is available to learn how to use the software and develop data mining techniques most efficiently. RapidMiner offers a free Educational License to universities and a free limited tier to everyone. Both tools introduced above constitute the principal sensor component, described in Section 9.4.7, entirely devoted to the analysis phase, to be carried out after execution of the assessment phase. The latter requires sample collection, accessing student personal data, and assessing the performance on both the historical window and the last profile picture taken in the last run test.

9.4 Information Systems Course Data

9.4.1 Dataset

The dataset used in this analysis, as described in Table 9.2, was derived mainly from two data sources:

- *Course Annual Registers:* Contain each student's exam assessment starting from the 2016–2017 academic year. This information links the student identifier to any assessment undertaken in the different exam sections and sessions, including the final exam outcomes and score (i.e., *passed or failed*).
- *Students' Personal Data:* Made available from Politecnico's information system, which includes references to universities where the student previously

TABLE 9.2

Description of the Dataset

Attribute	Description
Student I.D. Code	The student identification code
Multiple-Choice Question Section	Exam assessment: theoretical section score; range, [0,60]
Python/UML Section	Exam assessment: practical section score; range, [0,60]
Project Work Grade	Exam assessment: project work score; range, [0,25]
Final Grade	Weighted sum of the three-partial assessments; range, [0,30]
Passed/Not Passed	Exam outcome: *passed* in the case of success, *not passed* in the case of failure
Sex	Student genre: F for females and M for males
Cultural Area	Student's geocultural macro-area of origin: Europe, Middle East, South American, Asian, and African
Bachelor's Degree University	University where the student got their bachelor's degree. Three values allowed: POLI (for Politecnico), FOREIGN (for students who got their degree abroad), and ITALIAN NOT POLI (for students graduated in Italy, but outside Politecnico)
Erasmus	Student's participation in an Erasmus Program – Yes if true, No otherwise
Italy Zone	Student's origin. It has four values: North, Center, South/Islands, and Abroad; the first three options feature Italian students' birth region
Residence in Turin	Specifies Turin residence – Yes, if true, No otherwise
Bachelor's degree Mark	The student bachelor's degree mark; range, [0,110]
Cat_istms_1	Type of high school attended
High School Exam Mark	High school final exam score
Full Time/Part Time	Type of engagement
Bachelor's Degree	Student bachelor's degree
English Level	Student English knowledge level, assuming the Cambridge Certification levels: B1, B2, C1, and C2
Enrollment Number	Academic year tracking, when the student attended the course
Master's Degree Mark Average	Student master's degree average score

achieved their undergraduate degree, the type of enrollment, the high school of origin, and other details potentially useful for profile-building.

The student assessment covers three sections, each one described by an attribute specified in the dataset. The objective of this analysis is to group students according to their exam scores. Hence, the clustering process is rooted in the following three relevant assessment sections:

- *Python/UML:* Measures students' competency in Python coding and UML modeling;
- *Multiple-choice Questions:* Measures theoretical knowledge acquired during class lectures and reading the correspondent textbook chapters;
- *Project Work:* Based on the autonomous project work development made by the student team. This project addresses the real-world problems usually proposed by business actors who benefit from being part of the game, taking advantage of the potential outcomes of projects shaped in prototyped solutions.

Table 9.3 associates the three evaluation sections mentioned earlier with an abbreviated form used in the subsequent investigation.

The following paragraphs focus on data scrutiny, taking a step-by-step approach, and assuming related activities as the main phases:

- Initial dataset exploration;
- Overview and benchmark of clustering algorithms;
- Performance-based clustering visualization;
- Cluster characterization based on association rules.

9.4.2 Dataset Initial Exploration

The first step of the process drove an opening exploration of the dataset mainly to clean it and explore potential correlations among attributes.

The dataset gathered 975 rows and 20 columns and brought out interesting findings. Some attributes contained different types and levels of missing values, so a twofold approach was chosen to deal with them. Some mitigation techniques were applied to attributes showing a lower number of missing values by considering each attribute's nature and composition. Instead,

TABLE 9.3
Abbreviations of the Assessment Sections

Evaluation Section	Attribute	Attribute Abbreviation
Reverse Engineering	Python/UML section	Python s#
Test Bank	Multiple-choice questions section	MC section
Project Work	Project work section	Project work grade

Attributes	TEST N...	TEST P...	MALE	FEMALE	ERASM...	ERASM...	MC sect...	Python...	Project...	RESIDE...	BACHEL...	HIGH SC...	FULL TI...	MASTE...
TEST NOT PASSED	1	-1	0.057	-0.057	0.130	-0.130	-0.708	-0.865	-0.616	-0.115	-0.305	-0.233	0.030	-0.103
TEST PASSED	-1	1	-0.057	0.057	-0.130	0.130	0.708	0.865	0.616	0.115	0.305	0.233	-0.030	0.103
MALE	0.057	-0.057	1	-1	0.037	-0.037	-0.032	-0.074	-0.032	-0.004	-0.124	-0.121	0.057	-0.081
FEMALE	-0.057	0.057	-1	1	-0.037	0.037	0.032	0.074	0.032	0.004	0.124	0.121	-0.057	0.081
ERASMUS = NO	0.130	-0.130	0.037	-0.037	1	-1	-0.114	-0.166	-0.091	0.021	-0.306	0.037	0.031	-0.136
ERASMUS = YES	-0.130	0.130	-0.037	0.037	-1	1	0.114	0.166	0.091	-0.021	0.306	-0.037	-0.031	0.136
MC section	-0.708	0.708	-0.032	0.032	-0.114	0.114	1	0.775	0.715	0.112	0.268	0.272	0.024	0.347
Python s#	-0.865	0.865	-0.074	0.074	-0.166	0.166	0.775	1	0.661	0.151	0.384	0.236	-0.042	0.292
Project Work grade	-0.616	0.616	-0.032	0.032	-0.091	0.091	0.715	0.661	1	0.103	0.258	0.293	0.022	0.215
RESIDENCE = Turin	-0.115	0.115	-0.004	0.004	0.021	-0.021	0.112	0.151	0.103	1	0.317	-0.225	-0.199	0.400
BACHELOR DEGREE GRADE	-0.305	0.305	-0.124	0.124	-0.306	0.306	0.268	0.384	0.258	0.317	1	0.186	0.002	0.594
HIGH SCHOOL EXAM MARK	-0.233	0.233	-0.121	0.121	0.037	-0.037	0.272	0.236	0.293	-0.225	0.186	1	0.149	-0.171
FULL TIME PART TIME	0.030	-0.030	0.057	-0.057	0.031	-0.031	0.024	-0.042	0.022	-0.199	0.002	0.149	1	-0.107
MASTER'S DEGREE GRADE	-0.103	0.103	-0.081	0.081	-0.136	0.136	0.347	0.292	0.215	0.400	0.594	-0.171	-0.107	1

FIGURE 9.4
Correlation matrix.

attributes with a higher rate of missing values were canceled to avoid the mitigation techniques introducing noise in the dataset, leading to wrong analysis. Actions to discover any possible anomaly or inconsistency in some attributes that could lead to erroneous or misleading outcomes were also carried out.

An example of this is the 'Turin Residence' attribute. The evidence showed that students from the center or south of Italy or Islands were prevalently non-domiciled in Turin. This outcome was unexpected, and an explanation for this information concerns the time students provided their residence data to the university: In fact, this information usually is relevant in the student enrollment process, when students declare their origin town domicile, since they have yet to settle in Turin.

To analyze the correlation between attributes, we used the RapidMiner operator, named 'Correlation Matrix', which, for each pair of attributes, calculates a correlation coefficient (ϱ) from –1 to +1 (if $\varrho = 1$, the two attributes correlate perfectly and positively; if $\varrho = 0$, the attributes are not correlated; and if $\varrho = -1$, they are perfectly but negatively correlated). In Figure 9.4, some relevant correlations are depicted, where, for example, the students that failed the exam ('TEST NOT PASSED') were substantially negatively correlated with the 'Python s#' attribute and slightly less strongly correlated with the property 'MC section'. The project work achieved a lower correlation coefficient because of its nature: it is a one-size-fits-all evaluation for all the students in the team, apart from the self-assessment modulation mechanism, so differences between them tend to flatten out.

9.4.3 Clustering Algorithm Overview and Benchmark

Three different algorithms belonging to two distinct types were tested to identify the best cluster composition. The first algorithm was the DBSCAN (density-based spatial clustering of applications with noise), belonging to the density-based model, which operates by creating clusters and gathering the points closest to one another. The two other algorithms tested in this investigation were k-means and k-medoids, which base their actions on the related concepts of the centroid and medoid, and both aggregate points in a space according to the distance computed from a central point in the cluster. All these algorithms are widely known and used in the Educational Data Mining domain, as highlighted by Dutt in his work (Dutt, 2017).

These three machine-learning clustering algorithms ran on a single high-end performance computer with no timing issues, thanks to the small dataset – just 975 rows – containing information about a three-year course. For a larger dataset, a cloud-based solution with scalable high-performance computing (HPC) can be used to accelerate the process.

9.4.3.1 DBSCAN

Density-based clustering refers to an unsupervised learning method that identifies distinctive groups/clusters in the data, endowed on the concept that a cluster in the data space is a contiguous region of high point density separated from other such clusters through contiguous regions of low point density.

DBSCAN (Ester, 1996) is a reference algorithm for density-based clustering. It looks for and uncovers clusters of different shapes and sizes from a large amount of data containing noise and outliers.

The DBSCAN algorithm works on two parameters:

- *Minimum points:* The minimum number of points to consider a region 'dense';
- *Epsilon (ε):* The maximum distance to consider two points close to one another.

These two parameters were optimized through an iterative approach evaluating the better combination through a k-distance plot and the cluster composition, as shown in Figure 9.5.

In the k-distance plot, the following variables are represented:

- *On Y-axis:* k-distance, the average distance between each point in the dataset and its k-nearest neighbors;
- *On X-axis:* All dataset points sorted by descending k-distance.

Based on an iterative approach, a Minimum Points number 55 was chosen – as reported in Figure 9.5 – to shape the corresponding k-distance plot, used to determine the optimal ε value.

Figure 9.5 also exposes a gap around an ε of 0.1, meaning the distance has grown and a new cluster has emerged. Furthermore, Table 9.4 reports the average within-cluster distance, showing clusters are denser with an $\varepsilon = 0.1$, the appropriate value chosen for the analysis.

Based on this configuration, the algorithm identified three clusters, shown in Figure 9.6:

- Cluster_0: Students who failed the exam, detected as outliers;
- Cluster_1: Students who approached the exam without the required preparation and therefore failed or withdrew;
- Cluster_2: Students who passed the exam.

Despite the exciting suggestions they offer, these results are not relevant when focusing on this research target. All of the attendees who passed the exam were grouped in Cluster_2 without providing more detailed information on the students' performance. Considering the scarce efficacy of DBSCAN to pursue

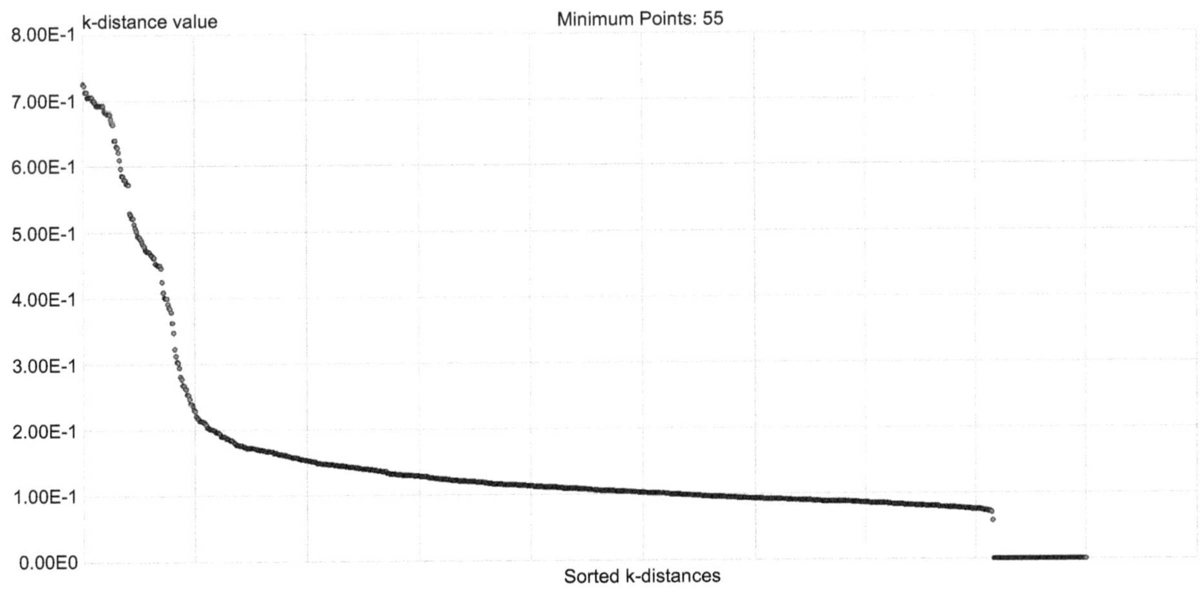

FIGURE 9.5
k-Distance plot.

TABLE 9.4
Average Distance within Cluster (DBSCAN)

Epsilon (ε)	Average Within-cluster Distance
0.09	127.325
0.1	122.323
0.160	158.960
0.220	164.049
0.280	172.191
0.340	192.871
0.400	219.354

the objectives established for this work, the efforts changed the direction, addressing the other family of algorithms, the partitioning algorithms.

9.4.3.2 k-Means and k-Medoids

k-Means is a simple unsupervised machine learning algorithm, conceived in 1967 by James MacQueen (MacQueen, 1967), well-known for its effectiveness in clustering data. This algorithm aims to assign each dataset point to a cluster, trying to minimize the distance between each point and the respective cluster centroid to which it has been assigned.

The other algorithm tested was k-medoids (Leonard Kaufman, 1990), which is similar to k-means but makes some substantial changes. Centroids are always chosen from the dataset points, while in k-means, they can also be points not belonging to the dataset.

Two metrics were used to evaluate the performance of these algorithms and to choose the one that fitted best the dataset:

- *Average within Centroid Distance:* This measures the distance between cluster points and the respective centroid. It is an intra-cluster measure, and the lower the value, the better the cluster compactness. This measure is used to assess the best parameters for the algorithms.
- *Davies–Bouldin Index (DBI)* (Bouldin, 1979): This measures the ratio of intra-cluster distances to inter-cluster distances. It compares the performance of various algorithms. The lower the value, the higher the overall quality of the clustering process.

These algorithms require the cluster number K as input to identify the best K value. An 'elbow graph' was used, where the metric 'average within centroid distance' was represented as the number of clusters K varies. K was set to 9 in both algorithms, being the value where the curve generated an 'elbow'. It is also the point beyond which an increase in the number of clusters k does not lead to a significant reduction

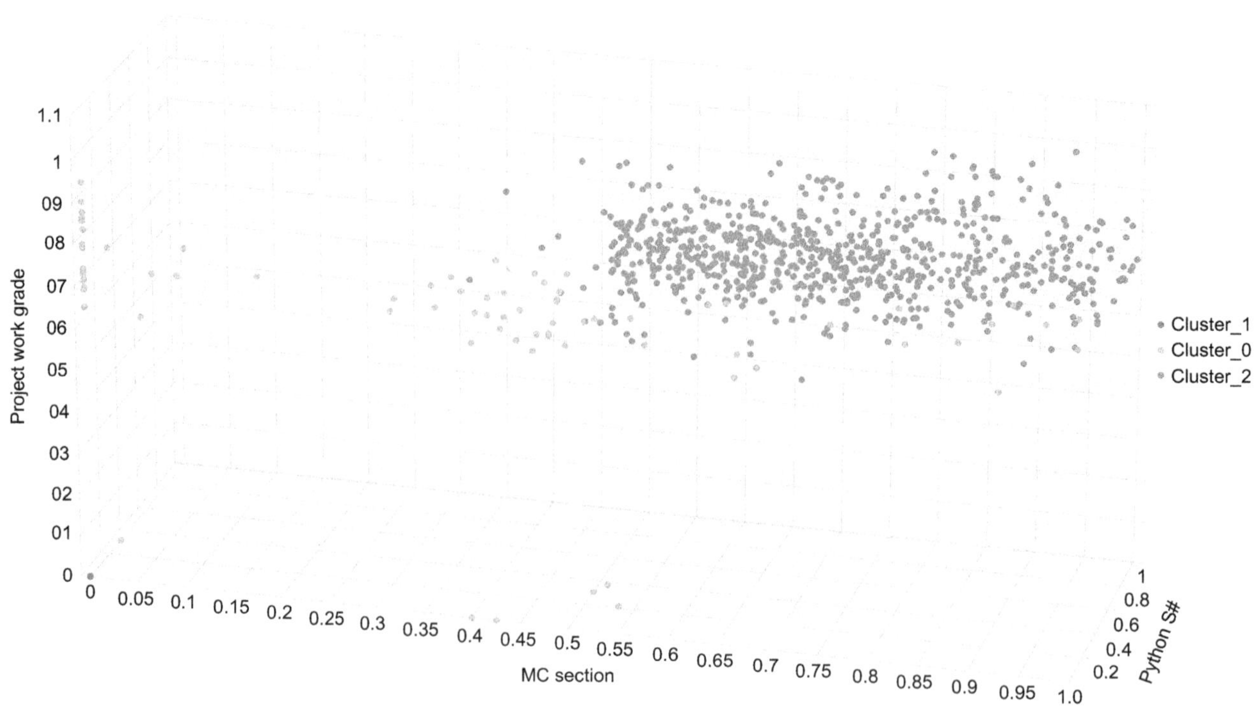

FIGURE 9.6
DBSCAN clusters.

in the intra-cluster distance and consequently in the algorithm performance as shown in Figure 9.7.

To conclude this phase, the choice of the algorithm most suitable for the clustering process was determined by comparing the performance of the k-means and the k-medoids using the Davies–Bouldin Index. The choice fell on the k-means, thanks to its lower index using the selected k of 9, as shown in Table 9.5.

9.4.4 Performance-Based Clustering Visualization

Figure 9.8 depicts nine clusters identified by the k-means algorithm. On the x-axis is the attribute MC section (which measures theoretical knowledge). On the z-axis, the project work grade, and on the y-axis, the Python s# attribute (which measures students' competencies in Python coding language and UML models) are reported.

The dataset exposed different densities, and as shown in Figure 9.9, the highest density was among students who passed the exam. A lower density is associated with students who failed the exam, and, in many cases, they are grouped directly on the axis, since they ultimately failed the exam or did not participate. The 'pass area' borders are visible and determined by the minimum number of points, a threshold required in the MC section and the Python section (with a minimum threshold of 30 out of 60 points). A decision was made not to consider these points as outliers, but to use the assigned cluster to understand better the students who struggle to pass the exam.

To adequately compare the assessment grades, they were mapped onto the 'Career Framework' (University of British Columbia) competency levels. More specifically, Table 9.6 shows the Python s# and MC section competency levels, while Table 9.7 shows the project work levels:

Figure 9.10 shows a bidimensional view of the clusters in which the attributes Python section and MC section are considered. The minimum thresholds for the MC and Python section attributes determined four quadrants divided into exam success and exam failure. The latter concerns three quadrants described in terms of competency achievement as follows:

- Q_2: MC section 'being developed'/'basic'
- Q_3: MC and Python/UML sections 'being developed'/'basic'
- Q_4: Python/UML section 'being developed'/' basic'

Looking at the quadrants where failures prevail, the most populated one is linked to poor performance in the Python section. Furthermore, the project work is not shown in this picture, having a lower variance in the grade within the assessment process, since teams complete the final delivery attending a reserved meeting with tutors and coaches.

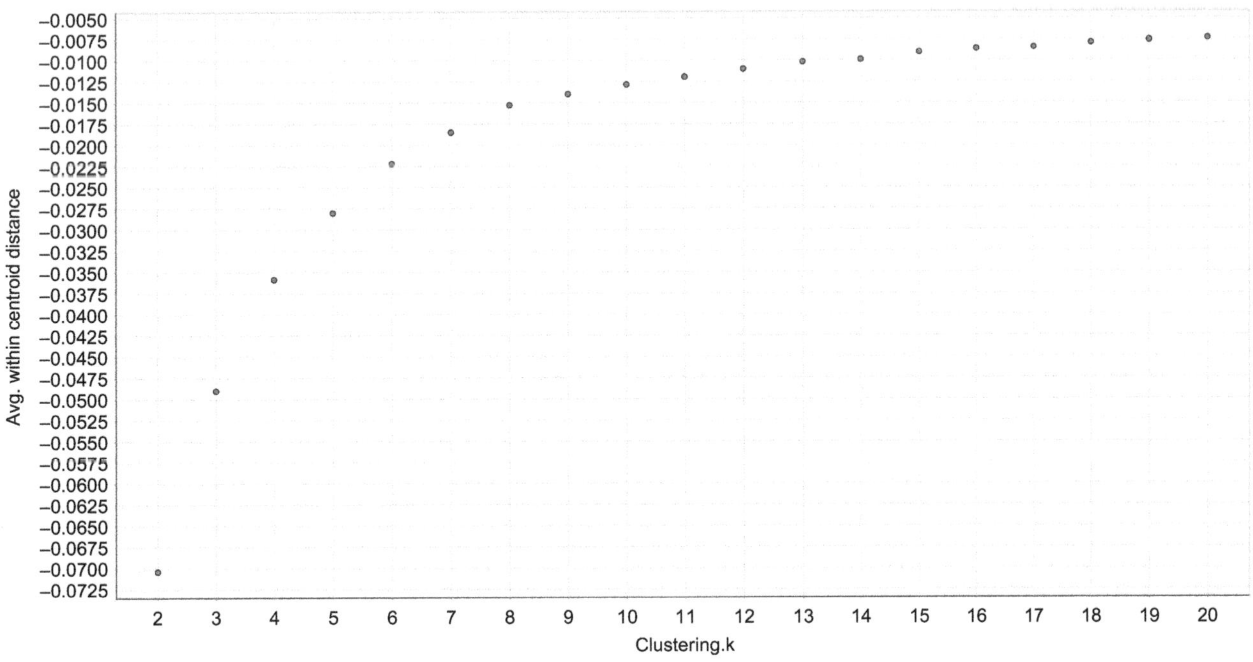

FIGURE 9.7
k-Means Elbow Method graph.

TABLE 9.5
Comparison of Clustering Algorithms

k-Value	k-Means	k-Medoids
2	0.583	1.402
4	0.902	0.798
6	0.786	0.991
7	0.714	1.198
9	0.845	1.136
11	0.920	1.203
13	0.876	1.102
15	0.928	1.085

Figure 9.11 displays a bidimensional view of the clusters considering the Python s# and project work grade attributes. The data show how the 'project work grade' evaluations are distributed almost evenly among the clusters. The differences among them tend to be small because of the nature of the project work. Students work together within a group, and each one obtains the same grade shared with all other members. Usually, when working together as a group, students can overcome individual difficulties by creating a mix of skills, experience, and knowledge. Thus, for these dynamics, though featuring different performances and engagements, students can achieve the same assessment grade, apart from the self-assessment modulation mechanism, which introduces a small component in the direction of individual assessment.

9.4.5 Association Rules

The association rules (Swami, 1993) are one of the most used methodologies in data mining to extract the relationship between data without having a priori knowledge of the correlations sought. An association rule is an expression based on the schema {A, B}->{C}, that is, the presence of attributes A and B (antecedents) implies, to a certain degree, the frequency of the attribute C (consequent). This methodology has its origins in databases of large retail supermarkets, where these techniques help understand which products customers buy together more frequently. With this knowledge, the organization of the spatial arrangement of products within their store improves.

There are three key metrics to evaluate the association rule {A}->{B}:

- *Support:* Represents the fraction of transactions containing both attributes A and B;
- *Confidence:* Measures the rule 'strength', computing the ratio between the support of the rule and the support of its premise;

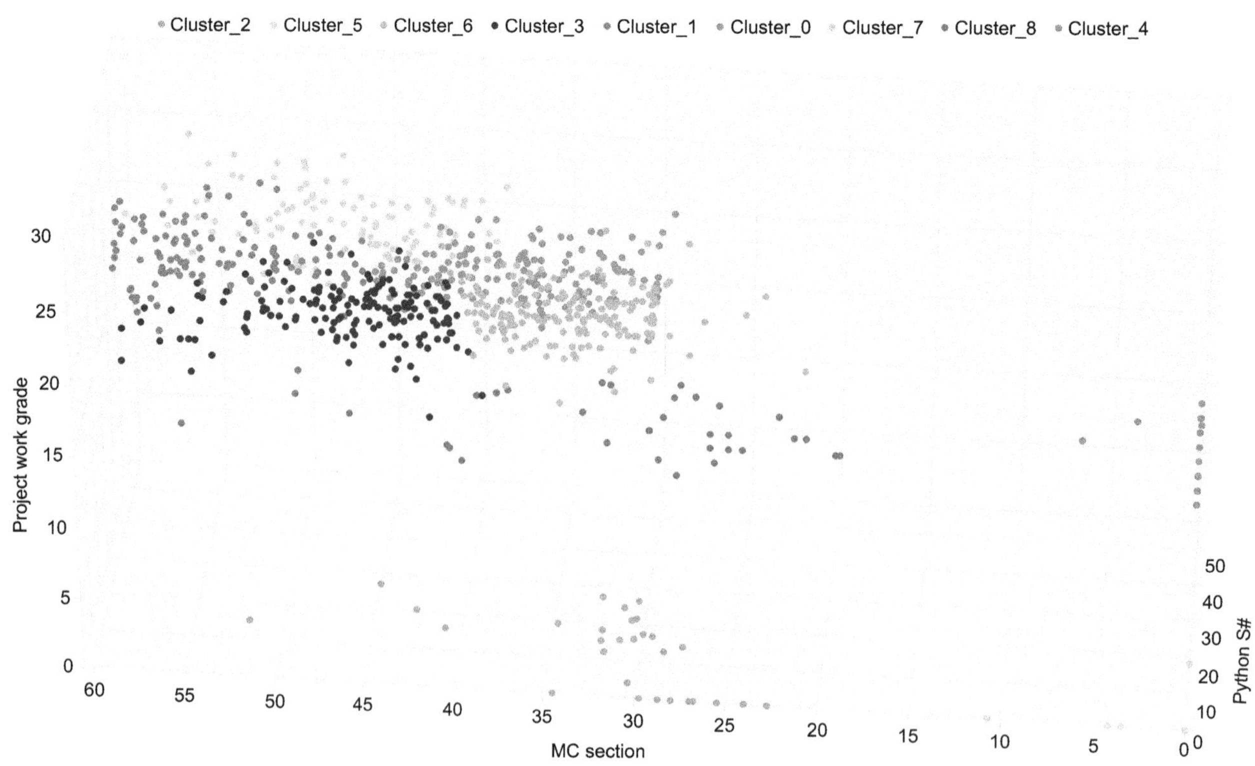

FIGURE 9.8
k-Means clusters.

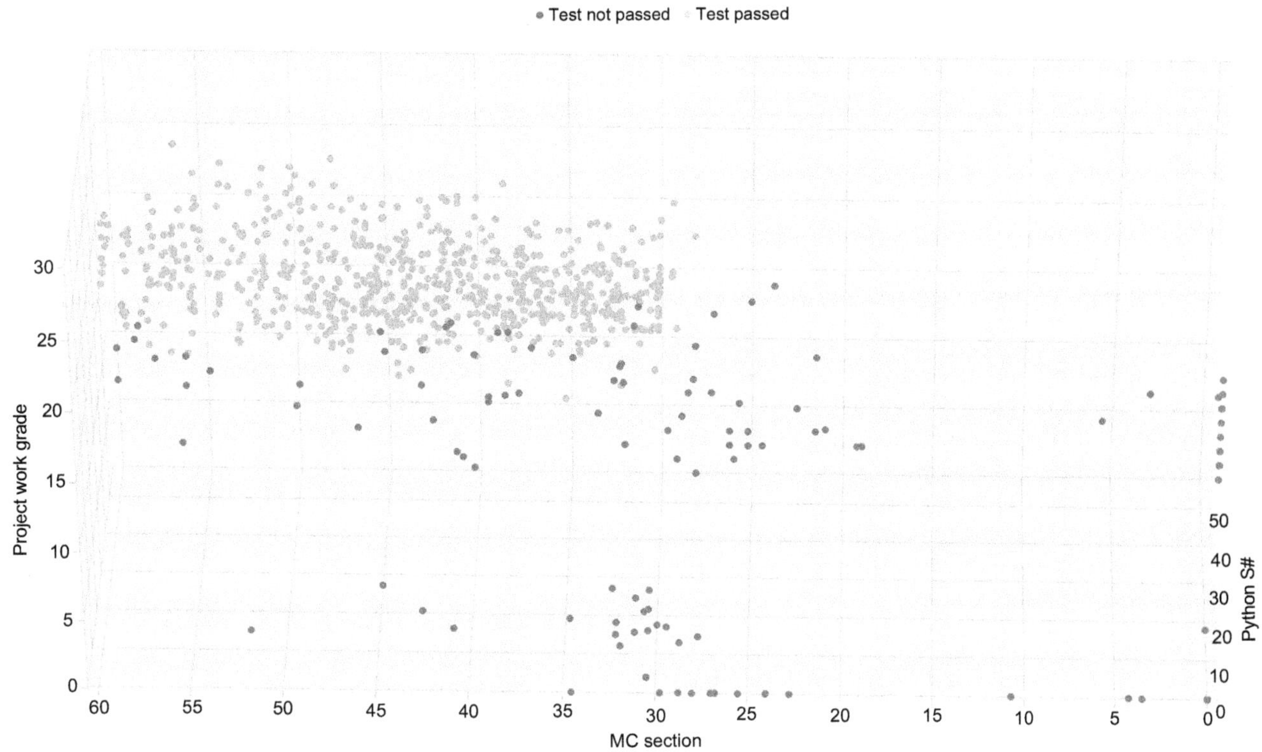

FIGURE 9.9
Students by test passed/not passed.

- *Lift:* Measures the statistical dependence between A and B. If the lift is equal to 1, the two items in the rule are statistically independent, and, therefore, the rule has little meaning.

All identified clusters were investigated here using the association rules, focusing on their composition and characteristics.

9.4.5.1 Cluster 0: Exam Passed – 'Advanced' Python/UML Section, 'Intermediate' MC Section, and 'Intermediate' Project Work

This cluster comprises 142 students who passed the exam. The average final mark is aligned to all of those who passed (26.03 versus 26.08). The project work assessment was slightly higher, but still comparable to its general average (18.6 versus 16.15). The theoretical MC section showed an 'intermediate' competency level, while the Python/UML one was 'advanced'.

TABLE 9.6
Description of the Competence Level of MC and Python/UML Sections

Score	Competency Level	Section Outcome
[0; 15]	Being developed	Section failed
(15; 30)	Basic	Section failed
[30; 40]	Intermediate	Section passed
(40; 50]	Advanced	Section passed
(50; 60]	Expert	Section passed

9.4.5.2 Cluster 1: Exam Passed – 'Advanced' Python/UML Section, 'Expert' MC Section, and 'Intermediate' Project Work

One hundred and thirteen (113) students passed the exam with a final mark above the average of all of those who passed it (27.63 versus 26.08). Their Python/UML performance exhibited an 'advanced' competency level (average 41/60). On the contrary, the theoretical MC section performance achieved the 'expert' level. This profile indicates a significant association rule that links most students from the mechanical engineering bachelor's degree to a Python/UML 'intermediate' competency, lower than the 'advanced' level achieved by the cluster.

9.4.5.3 Cluster 2: Exam Failed –'Basic' MC Section, 'Being Developed' Python/UML Section, and No Project Work

This cluster comprises 30 students who failed the exam primarily because they did not complete the project work. Moreover, they also had troubles in

TABLE 9.7
Description of the Project Work Competence Level

Score	Competency Level	Section Outcome
[0; 13)	Project work missing	Section failed
[13; 16]	Basic	Section failed
(16; 21]	Intermediate	Section passed
(21; 23]	Advanced	Section passed
(23; 25]	Expert	Section passed

other sections. Specifically, in the theoretical MC segment, they achieved an assessment just above the 'basic' level (31.56 versus 36.8 global average), and in the Python/UML section, their competency level remained at the 'being developed' level (12.01 versus 31.8 global average). Exploring this cluster revealed that these students repeatedly registered in more academic years, which signals this cluster as one with more struggling students.

9.4.5.4 Cluster 3: Exam Passed/Failed –'Intermediate' Python/UML Section, 'Advanced' MC Section, and 'Intermediate' Project Work

This cluster contains 164 students, whose final mark average was slightly lower than that achieved by those who passed the exam (25.13 versus 26.08). Close to 10% of the students of the cluster failed the exam, and the association rules unveiled that the cause was the Python/UML assessment, being under the minimum threshold. Most other students in the cluster had an 'Intermediate' competency level in this section, so this cluster is associated with some difficulties in the Python segment. In contrast, the performance achieved in the MC section was 'advanced'.

9.4.5.5 Cluster 4: Exam Failed – 'Being Developed' Python/UML and MC Sections and 'Intermediate' Project Work

Twenty-five (25) members of this cluster achieved the 'being developed' competency level in both the Python/UML and MC sections, even though their project work grade is comparable to that of the other students. A large proportion of the grades in Python/UML and MC were null, so it is likely they withdrew during the exam or did not show up at all.

9.4.5.6 Cluster 5: Exam Failed – 'Being Developed' Python/UML and MC Sections and No Project Work

This cluster gathers 89 elements and is similar to cluster 4. Again, the competency level in both the MC and Python/UML sections was 'being developed', below the minimum threshold needed to pass the exam, suggesting that the students attempted the exam without adequate skills, withdrawing soon after reading the text. The student portrait features a lack of engagement in the project work. It is worth noting that the

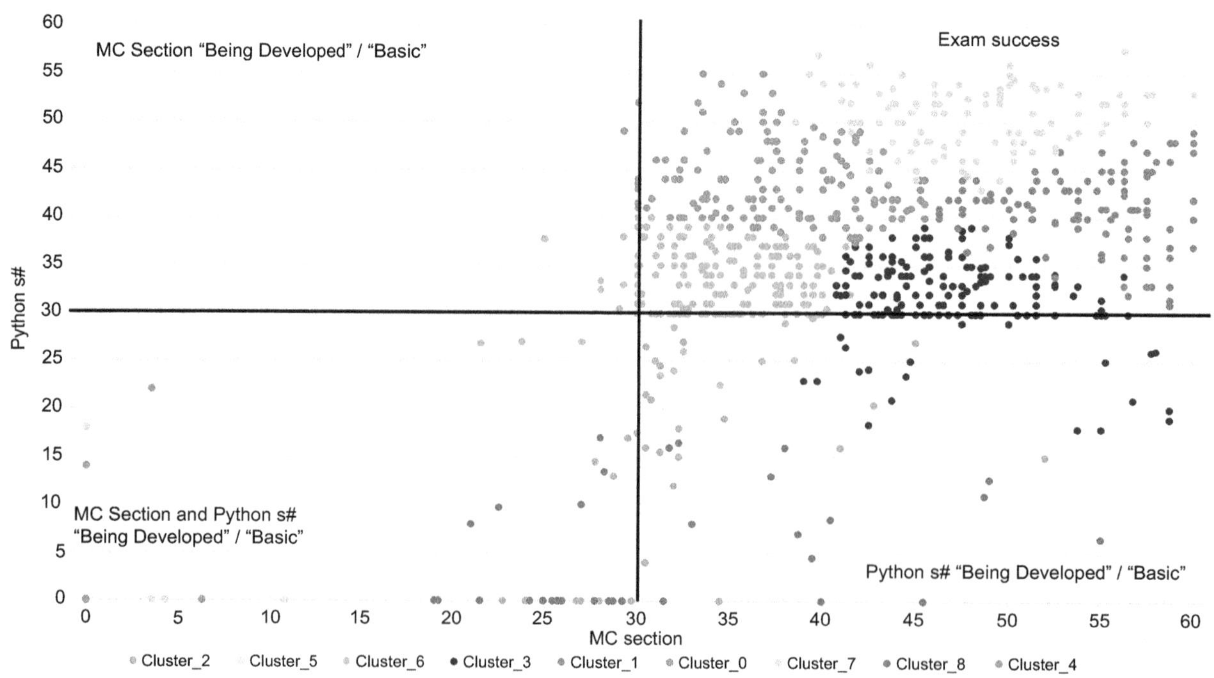

FIGURE 9.10
Clusters by Python s#, MC section, and exam outcome.

percentage of part-time students in this cluster was slightly higher than average.

9.4.5.7 *Cluster 6: Exam Passed/Failed – All 'Intermediate' Sections*

This cluster has the highest number of attendees (202 students) characterized by 'intermediate' performance in all three exam sections. The average final grade was lower than that of the other students who passed the exam. In Figure 9.10, some points are clustered on lower edges bounding the success area of the exam, with some points in the failure zone. However, the few failures are mainly accountable to a 'basic' competency level in the practical part (Python/UML) and, to a lesser extent, the theoretical component (MC section). A particular feature is represented by the students' distribution in universities undertaking a bachelor's degree: most students came from abroad, having achieved their bachelor's degree in a university of their own country.

9.4.5.8 *Cluster 7: Exam Passed – 'Expert' Python/UML and MC Sections and 'Intermediate'/'Advanced' Project Work*

This cluster rounds up the 102 best students. The average grade was 28.57, remarkably higher than all students who passed the exam (26.08). These performances were mainly due to the theoretical (MC section) and practical parts (Python/UML), while the project work showed grades only slightly higher than those of the other clusters, but this is probably due to the teams' internal dynamics. Furthermore, 94% of the members of this cluster came from bachelor's degree courses in engineering and management delivered not only at Politecnico di Torino.

9.4.5.9 *Cluster 8: Exam Failed – 'Being Developed' Python/UML Section, 'Basic'/ 'Intermediate' MC Section, and 'Intermediate' Project Work*

This cluster contains the 34 attendees who did not pass the exam. Their competency level was around the minimum threshold ('basic'/'intermediate') in the MC section and project work sections, but 'being developed' in the practical part (Python/UML). The latter is the more relevant cause of their failure.

9.4.6 Discussion on the Results

The aim was to identify some learning profiles based on students' performance using clustering techniques to describe them. Finally, identifying common patterns and exposing non-trivial evidence by exploiting the potential of association rules became an added value for the whole investigation process.

Beyond the specific results obtained, this study aimed to provide evidence, even though circumscribed, of how it is possible to employ the

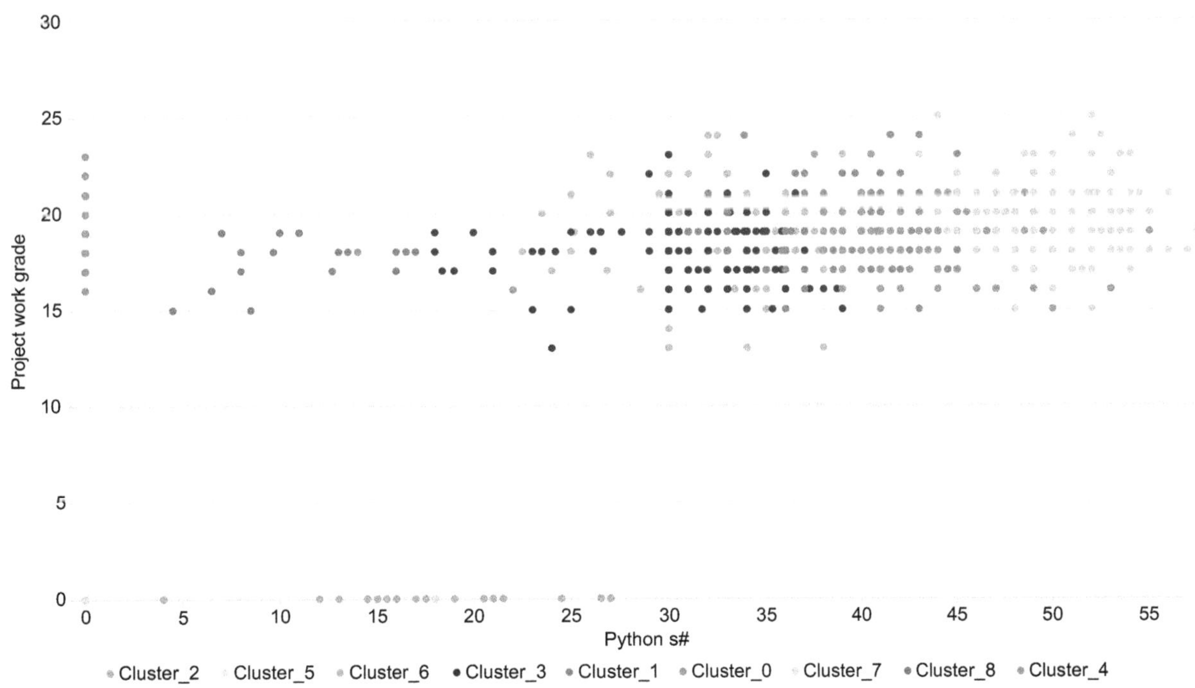

FIGURE 9.11
Clusters by Python s# and project work grade.

methodologies of inquiry and visualization of educational data.

This scenario ensures that teachers and assistants, who lecture high school or university courses, can receive proper support, shaped in a more general effort, to improve teaching, on the one hand, and student performance, on the other. In fact, by understanding more in-depth students' characteristics and detecting their behaviors, it is possible to implement teaching strategies that align study performance toward specific desired learning objectives.

Improving the course statistics provides an advantage to the students, who obtain more vital skills and knowledge. In addition, it improves the image and reputation of the university that graduates students more quickly and effectively, making them available sooner and prepared for the job market or scientific research.

In the following, the main findings of this study are stressed in more detail.

According to both the students' performance and their approach to engaging with the exam, the clusters identified in Table 9.8 provide a detailed representation of the different learning profiles shown by the course attendees.

Cluster 7 students gained 'expert' competence levels, making this cluster the reference learning outcomes profile. Hence, all other profiles of the remaining clusters exposed lower performance compared to the former.

Attractive also are clusters 2 and 5, where the students did not complete the project work, compromising their chances of passing the exam. Regarding cluster 5, the circumstance is even more complicated because the students, other than not having completed the project work, also showed inadequate training for the other two evaluation sections (Python/UML and MC sections). These weaknesses may derive from insecurity in their skills and knowledge or relevant weight perceived about the course content. This information helps to activate further analysis to investigate the causes of this behavior and to eventually make some changes to the course structure to minimize this issue.

From the analysis carried out thus far, several aspects emerged that should help pursue preventive actions and improve teaching and learning. More interesting is a pronounced relationship between failing the exam and the poor or weak performance in the Python/UML section. The latter includes understanding Python code and drawing UML diagrams according to a reverse engineering approach. This evaluation section proves to be more complex than the MC section and is one of the predominant causes of exam failure for at least 20% of students within the academic year.

Analyzing the clusters labeled as having 'intermediate' or lower competence levels in the Python/UML section, that is, clusters 2, 3, 5, 6, and 8 shown in Figure 9.12, it was revealed that there was a higher concentration of students from abroad, especially Asia (China, India, and Pakistan) and the Middle East (Uzbekistan, Iran, and Azerbaijan). Based on this evidence, a further investigation could deepen the details to understand the needs and requirements of these students and to help them manage the difficulties they are experiencing in this exam.

Another exciting consideration emerging from the association rule learning application concerns a plausible link between students who obtained a bachelor's degree in mechanical engineering and the same students' poor performance in the Python/UML section. It is likely that students from that area encounter some stiffness in Python coding, a topic they did not deepen during their bachelor's degree.

TABLE 9.8

Summary of the Clusters

Cluster	Exam Outcome	Description	Numerosity %
Cluster 0	Exam passed	'Advanced' Python/UML section, 'intermediate' MC section, and 'intermediate' project work	16%
Cluster 1	Exam passed	'Advanced' Python/UML section, 'Expert' MC section, 'Intermediate' project work	13%
Cluster 2	Exam failed	'Basic' MC section, 'being developed' Python/UML section, and no project work	3%
Cluster 3	Exam passed/failed	'Intermediate' Python/UML section, 'advanced' MC section, and 'Expert' project work	18%
Cluster 4	Exam failed	'Being developed' Python/UML and MC sections and 'intermediate' project work	3%
Cluster 5	Exam failed	'Being developed' Python/UML and MC sections and no project work	10%
Cluster 6	Exam passed/failed	All sections 'intermediate'	22%
Cluster 7	Exam passed	'Expert' Python/UML and MC sections and 'intermediate'/'advanced' project work	11%
Cluster 8	Exam failed	'Being developed' Python/UML section, 'basic'/'intermediate' MC section, and 'intermediate' project work	4%

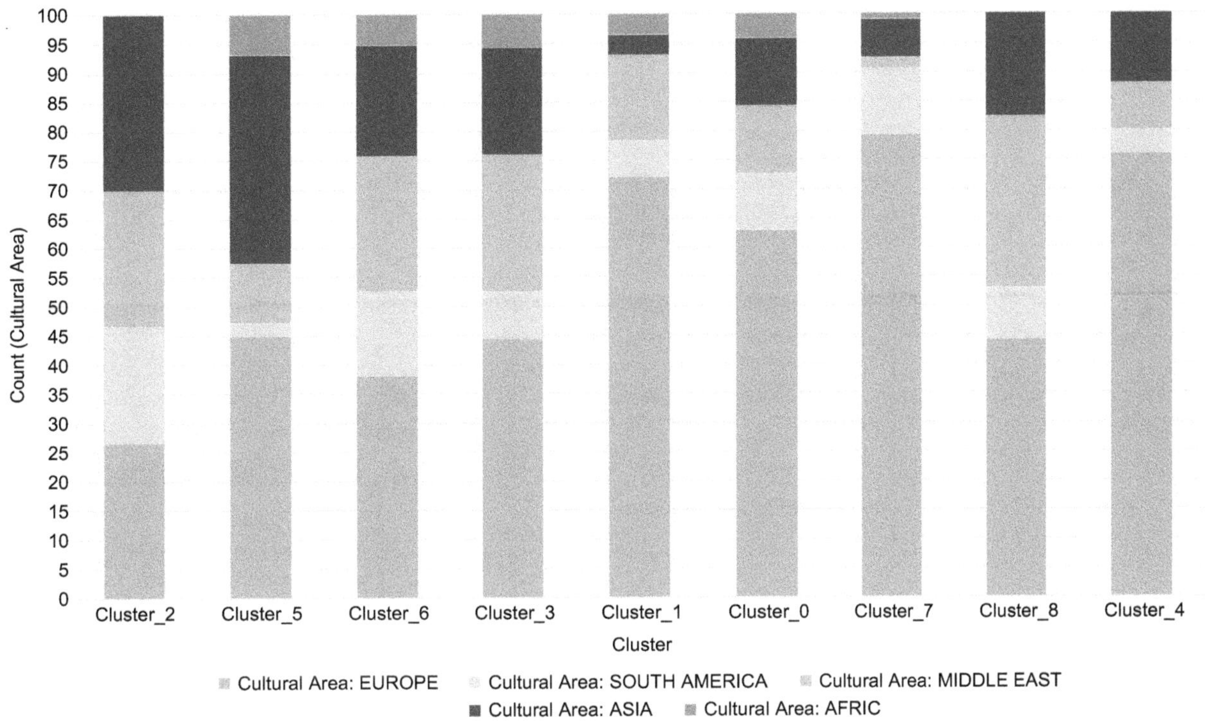

FIGURE 9.12
Clusters by cultural area.

However, while several students' learning profiles showed some weaknesses in the Python/UML section, the excellent performance shown in the same field was closely related to an excellent final mark. Hence, introducing an intermediate test focused on Python and UML might help achieve a self-appraisal concerning how the class perceives the topic, also suggesting remedies for the critical issues that the class encountered.

One of the expectations in the pre-analysis phase assumed that students enrolled in the part-time mode showed lower results than those enrolled in the full-time mode, particularly regarding the project work, because of the higher extra-university commitments of these students. However, the data showed no statistically meaningful differences among the averages of the students enrolled in the two modes.

9.4.7 Research Questions and Answers

Answering the research questions involves deepening the sensor and recovery action roles as highlighted since the beginning and recorded explicitly in Figure 9.13.

Concerning the research questions, 'What is the *sensor*? How does it work? What are the tools used?', Figure 9.13 depicts the 'device' and its main components, including the student performance assessment process, dealing with the learning outcomes description outlined in terms of knowledge, skills, and competency (ESG, 2015). The same 'device' also includes an analytics tool used to detect and validate performance and to shape individual and collective profiles. The tool (RapidMiner) refers directly to the process of harvesting evidence to sustain the appropriateness of interpretations, uses, and decisions based on the assessment results. Furthermore, analytics provides an adequate description of individual performance projected in a collective representation through appropriate assessment clustering mechanisms. The primary driver for planning and executing the learning process, which includes the assessment phase, is provided by the project work syllabus, shaped according to the Bologna process framework.

Hence, in summary, the sensing device components are as follows:

- Assessment and validation of the deliverables:
  - The prototype developed in the project work, together with evidence of the whole development process;
  - The multiple-choice question answers;
  - The Python coding description and UML diagrams drafted within the reverse engineering trial.

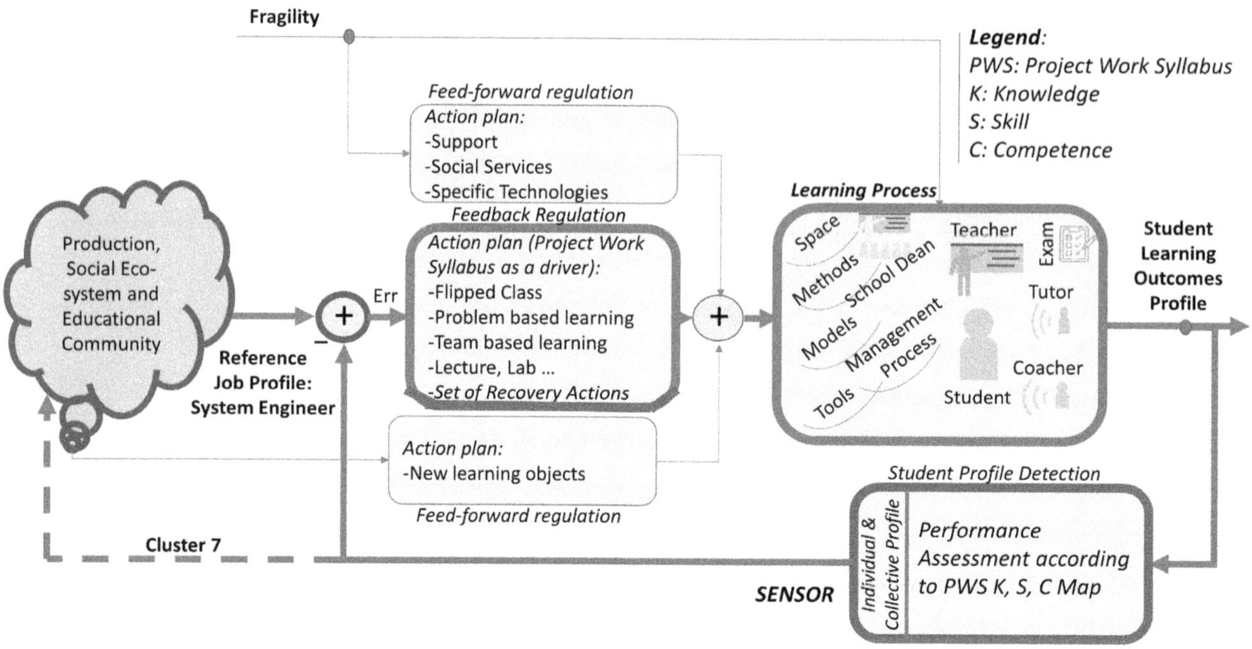

FIGURE 9.13.
Adaptive learning: the role of the 'sensor' within closed-loop regulation, working on the comparison of a reference profile to the one the learning process builds for students, also addressing the set of recovery actions to be planned consequently.

- Analytics applied on the collected data;
- Integrated tools: RapidMiner.

Regarding the regulation unit, the research questions 'What are the actions to be planned in the regulator? How have those actions been identified? How will they be put in execution?' address the planned recovery actions and their subsequent execution within the learning process. These actions come into view as a result achieved by applying analytics to performance assessment, as Section 9.4.6 describes in detail.

According to these results, which point to reverse engineering as the most critical assessment domain, an appropriate recovery action suggests that the open education resources should be a further extension of class lecturers dealing with the UML formal/informal specification domain. Moreover, newly added coaches to the course team can also spread and deepen coding much more to support engineering and reverse engineering techniques. Another initiative concerns enforcing close links between this academy and the universities/colleges from where students enter Politecnico. This more vital link helps empower the clarity of prerequisite specifications. Students planning to submit applications may know long before what the minimum requirements are to complete the academy path successfully. In short, the following aspects should be organized and planned for accomplishment starting from the next academic year:

- Make available OER (open education resources) for the Python coding and UML topics to enforce reverse engineering;
- Increase the number of coaches to support 'learning by doing', addressing the specific student profiles detected by the investigation;
- Update and spread more broadly the course prerequisites in terms of the minimum learning outcomes that students must possess when applying for the course.

9.5 Conclusions and Future Work

This study's main aim was to focus on the capabilities and potential of educational data mining to sustain both teaching on the faculty side and learning from the students' perspective. This work addressed the performance achieved by more than 900 students attending the master's degree in engineering and management at Politecnico di Torino. More specifically, the investigation converged on the 'information systems' course. The most relevant suggestions to drive future work include the following:

- Collect and produce additional data regarding students to extend the attributes set and to bring new knowledge to increase the dataset's size; interesting attributes concern class attendance, knowledge, skills, and competence obtained by students in their training path;

- Pay attention to data quality to exploit the full potential of mining approaches pursuing high-quality data shaped at different granularities;
- Make data analysis systemic within high school and higher education, addressing both the institution as a whole and single courses to implement continuous improvement.

Thus far, the work carried out is only the first step in a roadmap of trials aimed at an increasingly focused investigation on unseen information and unexpected behavior dynamics emerging within primary school, high school, and university paths.

Following an explorative approach, this work focused on applying educational data mining techniques to a university course. This experiment will further develop by shifting the focus to the master's degree in engineering and management or even all university master's degree programs.

The analysis carried out in this investigation outlined how lower coding skills are one of the primary causes of failure within the 'information systems' course. The slow spread of digital and computing competence is a widespread problem to counteract. A systemic approach is needed that involves imparting these types of skills to children as early as in primary schools. In Turin, since 2017, the Riconnessioni Project of the Foundation for the School, an instrumental body of the Compagnia di San Paolo Foundation (Demartini et al., 2020), has been active. The project aims to improve the Internet connection quality for the schools and adopt innovative approaches to teaching by disseminating even among children's skills such as programming from the earliest levels of education. Riconnessioni proposes a pedagogical model that takes advantage of the Digital Revolution to reaffirm the values, methods, and techniques of progressive, cooperative, popular education (Dewey, Freinet, Lodi). In this vision, the school becomes a community and the classroom a laboratory to develop passion in students, creative collaboration, and knowledge-sharing and to shape the personalization of teaching.

Focusing on the schools subset involved in the Riconnessioni project could be an opportunity to integrate data made available both by schools and by the Ministry of Education, University, and Research (MIUR) with information gathered and processed during the same project experience.

References

Agasisti, T., & Bowers, A. J. (2017). Data Analytics and Decision-Making in Education: Towards the Educational Data Scientist as a Key Actor in Schools and Higher Education Institutions. In *Handbook of Contemporary Education Economics*. Edward Elgar.

Ang, K. L., Ge, F. L., & Seng, K. P. (2020). Big Educational Data & Analytics: Survey, Architecture and Challenges. *IEEE Access*, *8*, 116392–116414.

Benussi, C. D. (2017). Do Web 4.0 and Industry 4.0 Imply Education X.0? *IT Professional*, *19*(3), 4–7.

Bouldin, D. L. (1979). A Cluster Separation Measure. *IEEE Transactions on Pattern Analysis and Machine Intelligence*, vol. PAMI-1, 2, 224–227.

Brancaccio, M. M. (2015). Problem Posing and Solving: Strategic Italian Key Action to Enhance Teaching and Learning Mathematics and Informatics in the High School. In *2015 IEEE 39th Annual Computer Software and Applications Conference*, pp. 845–850.

Campbell, J. (2007). Academic Analytics: A New Tool for a New Era. *Educause Review*, *42*.

Costa, L. S. (2019). Evaluation of academic performance based on learning analytics and ontology: A systematic mapping study. In *Proceedings of the frontiers in education conference* (pp. 1–5). IEEE.

Daud, A., Aljohani, N. R., Abbasi, R. A., Lytras, M. D., Abbas, F., & Alowibdi, J. S. (2017). Predicting student performance using advanced learning analytics. In *Proceedings of the 26th international conference on world wide web companion* (pp. 415–421). International World Wide Web Conferences Steering Committee.

Demartini, C. (2017). *Information Systems: Project Syllabus, Its Framework and Delivery Schedule*. Internal Report, Politecnico di Torino.

Demartini, G., Benussi, L., Gatteschi, V., & Renga, F. (2020). Education and Digital Transformation: The "Riconnessioni" Project. *IEEE Access*, *8*.

Dutt, M. A. (2017). A Systematic Review on Educational Data Mining. *IEEE Access*.

ESG. (2015). *Standards and Guidelines for Quality Assurance in the European Higher Education Area*. Brussels, Belgium.

Ester, M. K. (1996). *A Density-Based Algorithm for Discovering Clusters in Large Spatial Databases with Noise*.

Gray, C. C., & Perkins, D. (2019). Utilizing Early Engagement and Machine Learning to Predict Student Outcomes. In *Computers and Education*, *131*, 22–32.

Jones, K. (2019). Learning Analytics and Higher Education: A Proposed Model for Establishing Informed Consent Mechanisms to Promote Student Privacy and Autonomy. *International Journal of Educational Technology in Higher Education 16*, 24(1).

Leonard Kaufman, P. J. (1990). *Partitioning Around Medoids (Program PAM)*.

Luan, J. (2002). Data Mining and Its Applications in Higher Education. *New Directions for Institutional Research*, 17–36.

MacQueen, J. (1967). Some methods for classification and analysis of multivariate observations. In *Proceedings of 5th Berkeley symposium on mathematical statistics and probability* (pp. 281–297). University of California Press.

Paravati, G., Lamberti, F., & Gatteschi, V. (2015). Joint Traditional and Company-Based Organization of Information Systems and Product Development Courses. *2015 IEEE 39th Annual Computer Software and Applications Conference*, pp. 858–867.

Robin Pappas, B. R. (2021). Navigating Learning Analytics in Higher Education. IDEA Paper #82.

Slater, S. S., Joksimović, S., Kovanovic, V., Baker, R. S., & Gasevic, D. (2017). Tools for Educational Data Mining: A Review. *Journal of Educational and Behavioral Statistics, 42*(1), 85–106.

Swami, R. A. (1993). Mining Association Rules between Sets of Items in Large Databases. *ACM Sigmod Record, 22*(2), 207–216.

Wong, Y. (2016). Academic Analytics: A Meta-analysis of Its Applications in Higher Education. *International Journal of Services and Standards, 11*(2).

Section III

AI-Supported Instructor Systems and Assessments for AI and STEM Education

10

Teacher Orchestration Systems Supported by AI: Theoretical Possibilities and Practical Considerations

Suraj Uttamchandani, Haesol Bae, Chen Feng, Krista Glazewski, Cindy E. Hmelo-Silver, Thomas Brush, Bradford Mott, and James Lester

CONTENTS

10.1 Classroom Orchestration 151
10.2 Artificial Intelligence for Classroom Orchestration 152
10.3 Research Context 153
10.4 The Role of AI in Orchestration Assistant Design 153
10.5 Classroom Context Interviews 154
10.6 Orchestration Assistant Design 155
  10.6.1 During Class versus Beyond Class Assistance 156
  10.6.2 Automated Systems versus Teacher Agency 157
  10.6.3 Detailed Information versus Actionable Information 158
  10.6.4 Classroom Management versus Ambitious Learning Practices 158
10.7 Discussion and Future Research 159
References 160

As artificial intelligence (AI) technologies develop and even become ubiquitous, new possibilities emerge for how they can enhance teaching and learning. To support teachers and teaching, AI technologies can be designed in ways that are human-centered, amplifying the teacher's cognitive capacity and expertise. In particular, AI technologies have immense potential to support classroom orchestration – how teachers organize classroom activity across individual, small group, and whole-class scales. Effective classroom orchestration is key to supporting STEM inquiry, productive disciplinary engagement, and collaboration, but given the practical challenges of classroom activity, effective orchestration can be difficult to achieve (Dillenbourg et al., 2018). Teacher orchestration systems supported by AI can assist teachers in facilitating classroom activity by offering them information about pedagogical tools (e.g., curriculum materials, technologies) and suggestions for strategies, tracking student activity in real time, and seamlessly completing managerial tasks to 'free up' the teacher for supporting authentic inquiry. In this chapter, we outline design considerations that emerged as we designed such an *orchestration assistant*, an AI-supported teacher orchestration system. We first consider the theoretical possibilities currently available for supporting pedagogy with AI. Next, we discuss how those possibilities might be enacted, complicated, or transformed in the context of real-classroom activity. We ground our discussion in design considerations from our orchestration assistant project. We conclude with some open questions about such AI-supported systems for teachers.

10.1 Classroom Orchestration

Given the diverse ways researchers and designers have leveraged AI to support learning, we focus our discussion on how AI can support teachers in classroom orchestration. Classroom orchestration is a concept from the learning sciences and computer-supported collaborative learning (CSCL) communities that Dillenbourg et al. (2018) define as 'the real-time management of multi-plane scenarios under multiple constraints' (p. 181). By 'plane', the authors refer to three levels of activity: the individual, the team or small group, and the whole class or large group. Classroom orchestration thus refers to teachers' practices in facilitating activities at each of these levels as well as managing transitions between activities.

To a small degree, classroom orchestration involves classroom management, that is, ensuring that each

DOI: 10.1201/9781003181187-13

group has the directions, worksheets, technologies, or other materials needed to complete the activity at hand. However, orchestration more broadly involves coordinating students' collaborative activity and disciplinary engagement. As an example of orchestrating collaborative activity, consider a situation in which a teacher supports a student who says their group members are not adequately participating in the activity, or how a teacher might re-engage a disengaged student in contributing to their group. As an example of orchestrating disciplinary engagement, consider how a teacher would ask prompting and facilitating questions that help a student engage in the disciplinary practice of computer programming in the context of designing an app for public stakeholders. Classroom orchestration may invoke the metaphor of an orchestra in which a teacher plays the part of a conductor, ensuring that learners are able to play together in harmony.

In schools, we assert that successful classroom orchestration is fundamental for deep learning, but it is also challenging. It is especially important in the context of learning environments that seek to support deep cognitive, sociocultural, and sociopolitical engagement (rather than simply rote memorization), such as environments based on project-based, problem-based, and inquiry learning (Savery, 2015), productive disciplinary engagement (Engle & Conant, 2002), consequential learning (Hall & Jurow, 2015), or other ambitious learning practices (Glazewski & Hmelo-Silver, 2019; Uttamchandani et al., 2020).

The amount of effort needed to support students through effective classroom orchestration may be referred to as a facilitator's or teacher's *orchestration load* (Prieto et al., 2015). Some elements of orchestration load are intrinsic to the learning environment, content, lesson plan, and audience. For example, if a teacher has a 40-minute class period and wants to include a 15-minutes whole-class introduction, 10 minutes of small-group activity, 10 minutes of whole-class discussion, and then 5 minutes of individual exit ticket work with their 13-year-old students, this will be a significantly higher orchestration load than if a college professor planned to deliver a two-hour lecture to his Ancient World History class. In addition to these issues, orchestration load is a measure at the interaction of the particular teacher and the learning environment; a relatively novice teacher will have a higher orchestration load than a relatively experienced teacher even in the same learning environment. Prieto et al. (2018) suggest that orchestration load is influenced by a variety of factors. These include factors related to classroom activity, such as the teacher's current coordination activity, the social plane for this coordination, and the classroom resources the students are using. Other factors include the teacher's level of expertise and familiarity with the classroom situation as well as any other external resources (e.g., the help of a second teacher or teaching assistant).

Classrooms featuring collaborative inquiry require especially complex orchestration from teachers. Collaborative inquiry, initially derived from science education, refers to the process when students learn by working in groups to perform steps of inquiry similar to scientists (Bell et al., 2010). For example, problem-based learning (PBL) is one of the widely adopted pedagogies involving collaborative inquiry. In PBL, learning is no longer passing on information but a complex, dynamic, and context-sensitive process requiring students' deep engagement and delicate facilitation from the teachers (Hmelo-Silver & Barrows, 2006). It often involves multiple learning activities occurring at multiple social levels – individual, group, and classroom – which increases orchestration load. The challenge for teachers' orchestration is expanded when scaling up collaborative inquiry to larger classrooms, as teachers need to monitor the inquiry progress and interactions of multiple groups of students at the same time (Dobber et al., 2017). Therefore, collaborative inquiry especially calls for the development of new tools and new technologies to manage orchestration load.

Taken together, the concepts of classroom orchestration and orchestration load provide a valuable framework for considering how AI technologies might support learning in the complex classroom environment by decreasing orchestration load and supporting teachers in successful classroom orchestration.

10.2 Artificial Intelligence for Classroom Orchestration

Scholars are increasingly asserting the potential for AI to support classroom orchestration (Roschelle et al., 2020). AI can support successful classroom orchestration in at least two ways. First, AI can reduce orchestration load by managing issues such as attendance and information distribution, allowing the teacher to have more resources available to facilitate dialogical tasks that can only be human-led. Second, AI can help process information for a wide variety of learners in real time, learning what is and is not usual learning activity, and thus direct the teacher's energy and attention to those who could benefit most or are most in need of their mentorship, apprenticeship, or intervention.

These AI systems, however, are not without risk. Important questions around ethics, privacy, and social justice are raised when AI is brought into the classroom. For example, some learners may not want their facial expressions or eye movements to be tracked for privacy reasons, whereas many AI systems rely on this information for decision-making (Rubel & Jones, 2016). There are also risks around research methodology, as the presence of AI technology could actually increase orchestration load by introducing a new technology into the classroom and thus adding to the collection of things a teacher needs to keep track of. Finally, there are technical limitations. For example, some data cannot be processed and interpreted in real time, or require more significant Wi-Fi or hardware capacities than are available in a typical classroom.

In designing AI to support classroom orchestration, therefore, we take a 'human-centered' perspective, in which the effectiveness of an AI system is evaluated through its potential to augment the intelligence of teachers and learners in ambitious dialogue-based pedagogical activity. We picture an 'orchestration assistant', a technological interface powered by AI that can support teachers in getting the various information needed to effectively support classroom activity. We draw inspiration from promising work around teacher dashboards and other technology support. In a review of research around such tools, van Leeuwen and Rummel (2019) classified teacher dashboards as having three main functions: mirroring, alerting, and advising:

> By mirroring, we mean systems that provide information but do not aid in the interpretation thereof. By alerting, we mean systems that in some way alert the teacher to important events during collaboration. By advising, we mean systems that advise the teacher about the status of the current situation or about possible ways to act to support students.
>
> **(p. 148)**

These dashboards sometimes but do not always rely on AI, and are often but not always used only during real-time classroom activity (see van Leeuwen et al., 2019, for more on teacher dashboards in computer-supported collaborative learning contexts).

In sum, effective classroom orchestration is crucial to successful collaborative inquiry, but challenging for teachers to accomplish effectively. Advances in AI create new possibilities for delivering timely and actionable information about student learning to teachers in ways that support such orchestration. Research on teacher dashboards suggests an array of possibilities to explore (e.g., mirroring, alerting, and advising) regarding how an AI-enhanced tool might support orchestration and thus support collaborative inquiry. Next, we discuss our design of an AI-supported orchestration assistant in the context of a game-based learning environment, in which we encountered a number of tensions around AI-supported classroom orchestration tools.

10.3 Research Context

We build on prior work (Saleh et al., 2019) to design an orchestration assistant in the context of middle school science. The technological context is a game-based learning environment, Crystal Island: EcoJourneys. In the Crystal Island: EcoJourneys learning environment, students arrive on an island in the Philippines as part of a school field trip. On this island, locals depend on tilapia fish farming for a living. While interacting with the in-game characters, students discover that the fish technician, Jasmine, has a problem: her tilapia fish are falling sick at alarming rates. Thus, students work in groups of four to resolve the aquatic ecosystems problem while exploring the game environment and talking to the characters. The game is based on PBL research (Glazewski & Hmelo-Silver, 2019; Hmelo-Silver et al., 2018), and explicitly seeks to engage students in deep reasoning, occasional disagreement, and authentic scientific practices around ecosystems.

Throughout the game, teachers are equipped with a dashboard that informs them about students' progress and participation. This dashboard forms the basis for the orchestration assistant we discuss in the remainder of this chapter. As students play, the game collects large amounts of trace data, that is, information about each learner's gameplay state (Serrano-Laguna et al., 2017). This information, as well as the relatively bounded disciplinary content of the Crystal Island: EcoJourneys curriculum, makes the game-based learning environment an ideal context to investigate the potential of AI technology in supporting classroom orchestration.

10.4 The Role of AI in Orchestration Assistant Design

Several AI technologies are relevant to the design of an orchestration assistant. Machine learning, which refers to automated model building, can create models of student learning. These models can be the basis of understanding how a variety of students engage with the learning environment and what kinds of information should be presented to a teacher for further analysis and possible action (for example, because student activity is unusual or seems exemplary). Several families of machine learning techniques, such as Bayesian regression and classification, probabilistic graphical models, and deep neural networks, are particularly useful for contexts with relatively small base datasets, such as classrooms. In addition, computer vision techniques, which can similarly help AI make meaning of visual data, are relevant to processing data such as students' gaze, facial expression, or body posture. While deploying these technologies at scale in real time requires significant computational power, machines continue to become faster, smaller, and more affordable with each passing year. This in turn leads to increasing efficiency of training models and decreased runtime for relevant AI algorithms, constantly increasing the reasonable technological possibilities of an orchestration assistant in a classroom context.

Importantly, an orchestration assistant is a *human-in-the-loop* technology, that is, an AI support system in which human actors ultimately make consequential decisions. All AI technologies operate with some degree of uncertainty, with confidence intervals around particular inferences, predictions, classifications, and so on. However, incorrect conclusions and predictions have different consequences. In the context of an orchestration assistant, for example, failing to note that a student has made repeated content-related errors can prevent a teacher from offering helpful guidance. More egregious, however, is flagging a student as incorrect when they are not, which can cause frustration for students and teachers and dramatically reduce their confidence in AI-supported inferences. Group-level, class-level, and other data used in aggregate can decrease uncertainty. The impact of uncertainty can also be minimized by consistently ensuring that highly consequential decisions about learners (e.g., course grades) are ultimately made by humans, even as an orchestration assistant can provide relevant information. Further, the user's experience of an orchestration assistant should be designed so that alerts convey information as worth further investigation by a human, rather than as a known example of something gone wrong (e.g., 'The assistant suggests you visit Group 2 as they may be having trouble collaborating' versus 'Group 2 is having trouble collaborating').

Finally, a challenge with classroom-based AI technologies is that what may be a typical or even large dataset from the perspective of education research can be a very small one in comparison to the significantly larger datasets on which AI technologies in other contexts are trained. Fortunately, significant research is emerging around machine learning methods for smaller datasets, such as transfer learning (Ng et al., 2015; Shu, 2019). As these methods develop, education researchers can increasingly conduct design work in contexts with relatively small datasets.

10.5 Classroom Context Interviews

To inform our design, we conducted 14 semistructured interviews with middle school science teachers about their teaching goals within problem-based learning (e.g., 'What were some of your teaching goals for the students this year?') and about their ideas for dashboards that could support those goals (e.g., 'How might a dashboard support and recommendation system assist you with monitoring and increasing student learning in relation to standards? Can you provide an example of what this might look like before, during, and after teaching a PBL lesson?'). Each interview lasted approximately one hour. The data were transcribed verbatim, partitioned into extracts, and then analyzed using thematic analysis (Braun & Clark, 2006). The analysis formed the basis of our designs.

During the thematic analysis of the 14 interviews, we developed 19 codes in total under the following six thematic categories: (1) engagement and motivation; (2) individual learning; (3) group collaboration; (4) group progress; (5) aims of PBL; and (6) teaching improvement. These six thematic categories capture the areas in which teachers would find a dashboard support most valuable. Despite teachers' common interest in having the orchestration assistant in their classroom, teachers expressed different needs aligning with their own teaching styles and context. For example, one teacher shared that his teaching goal for that year was to help students understand that failure is part of growth, with a special attention to students' learning while they were dealing with mental health issues. In contrast, another teacher expressed that her biggest goal was to 'spark that interest in science and academics in general in students', a more discipline-based outcome. The findings thus suggested that teachers had varied visions for how an orchestration assistant could support them in achieving content, process, and affective outcomes for students.

In the interviews, researchers inquired about three potential functions of an orchestration assistant: (1) making classroom management easier; (2) supporting teachers to facilitate students' collaborative inquiry; and (3) scaffolding teachers adaptively to grow into PBL experts. Although at the beginning of their interviews some teachers were most concerned about classroom management, as the interview progressed they indicated a wider variety of teaching goals (e.g., having their students become better collaborators). Some teachers also hoped experiences with the game and the orchestration assistant could help them become more familiar with PBL and more comfortable facilitating a PBL classroom without the assistance of researchers.

Our findings enabled us to compare and contrast what teachers saw as important dashboard features with what our design team believed to be necessary features to support classroom orchestration. Teachers saw as essential the dashboard's ability to (1) inform the teacher of students' current progress or location (in a virtual learning environment) and (2) identify or 'flag' students who are struggling or in need of assistance. These aligned with our goals as designers to support classroom orchestration of collaborative inquiry across individual, small-group, and whole-class planes of activity. Teachers also saw as essential the ability of a dashboard to (3) assess and deliver information to the teacher about students' progress in meeting curricular standards, and (4) grant teachers control over students' devices. This raised several tensions, as we worried such features would compromise the potential of a dashboard to create opportunities for student disagreement, dialogue, and inquiry. Whereas designers saw it as essential that the dashboard (5) provide scaffolds to teachers, such as sample scripts and questions, that could support such inquiry, teachers did not perceive this feature to be fundamental. Neither group saw it as essential for the dashboard to (6) group students by ability.

The interview analysis raised several important issues around the utility of AI-based dashboards in real-classroom contexts. Most notably, there was some tension around the role of an orchestration assistant as monitoring students to limit the ways they might participate versus as providing teachers with information that could support dialogue. These interviews, alongside our prior experiences implementing preliminary versions of the Crystal Island: EcoJourneys game-based learning environment, led us to several design conjectures for the creation of an AI-supported orchestration assistant.

10.6 Orchestration Assistant Design

As discussed above, effective classroom orchestration is an exceptionally complex task for designers to design for in ways that will reduce load and enhance learning experiences (Dillenbourg et al., 2018; Prieto et al., 2018; Slotta et al., 2013). This is especially the case with inquiry-based learning environments (e.g., PBL contexts) where multiple groups of students engage in complex inquiry-related activities, such as disciplinary reasoning and problem-solving processes (e.g., problem-finding, deliberation, argumentation; Glazewski et al., 2014; Hmelo-Silver et al., 2019). Such complex environments require real-time management to productively coordinate and provide supportive interventions (Dillenbourg et al., 2009).

Providing support that is adaptive and sensitive to the current status of student learning and collaborative inquiry requires a teacher to continuously diagnose the status of understanding of multiple groups and immediately make decisions of how and when to provide support. For example, a teacher might further elicit students' ideas through questioning in order to engage students in more sophisticated reasoning discourse when students demonstrate accurate answers during disciplinary reasoning (Bae et al., 2021). However, judging when and how to provide appropriate support in an authentic classroom where various activities are occurring simultaneously with high unpredictability and immediacy imposes a great challenge for teachers. Thus, we discuss how an orchestration assistant that employs AI-based multimodal learning analytics may provide just-in-time guidance.

Employing AI-based learning analytics affords teachers opportunities to better track students' inquiry learning activities and extend their instructional capacity to provide contingent and adaptive support. Given limited time and attention during ongoing teaching, a teacher can benefit from an orchestration assistant by offloading redundant and simple monitoring tasks to the system and utilizing the reclaimed teaching time for facilitating deeper discussions. In particular, an orchestration assistant can help teachers with prioritizing which students or small groups may be most in need of a teacher's immediate attention with what type of support, based on real-time analytics about groups' progress and collaboration performance. However, it is important to determine the right mix of a teacher and the AI-based system for specific learning environments (Alkhatib & Bernstein, 2019; Heer, 2019;

Holstein et al., 2019; Olsen et al., 2018) and design how the collected information will be interpreted and presented to the teachers through the system. Given the risk of suggesting inaccurate or contextually-irrelevant recommendations to the teacher, it is still critical to take advantage of the rich contextual information and experienced pedagogical knowledge that the teachers have access to but the system may not (Alkhatib & Bernstein, 2019). In our orchestration assistant design, we tried to judge the right balance between benefits and risks of using the AI-based system (An et al., 2020) in a PBL environment. Next, we present four design tensions we encountered as we build the prototype for the orchestration assistant.

10.6.1 During Class versus Beyond Class Assistance

From the teacher interviews, we recognized the importance of having the ability to monitor the real-time progress of individuals and groups, to scaffold the learning process, and to conduct assessment. However, we also recognized that support of real-time orchestration requires additional guidance that happens before and after the class, such as planning and reflection. Because PBL is not a single-day activity but rather unfolds over several days or even a whole semester, it was important to acknowledge the continuity of the learning environment and design a way to support it.

In order to meet these needs, we designed three different phases of guidance from the orchestration assistant: (1) prospective guidance, (2) concurrent guidance, and (3) retrospective guidance (Bae et al., 2019). Prospectively, before class, the orchestration assistant will proactively recommend orchestration planning as 'forward guidance' for teachers to anticipate pre-identified obstacles, to manage orchestration load, and to implement successful classes (see Figure 10.1). Next, during class, the orchestration assistant will provide concurrent guidance to support teachers to determine who or which group needs the most help, what approaches are going to be the most effective, and what kinds of prompts can be useful (see Figure 10.2). The primary goal of the concurrent guidance is to help teachers develop facilitation skills, such as initiating inquiry, scaffolding problem-solving processes, and pushing for deep knowledge construction. Lastly, after each class session, the orchestration assistant will provide guidance for reflection on their real-time orchestration moves and facilitation strategies and suggestions to help them consider future moves. At this stage, more detailed information on the class implementation will be shared, such as student groups' chat messages, artifacts, and history of

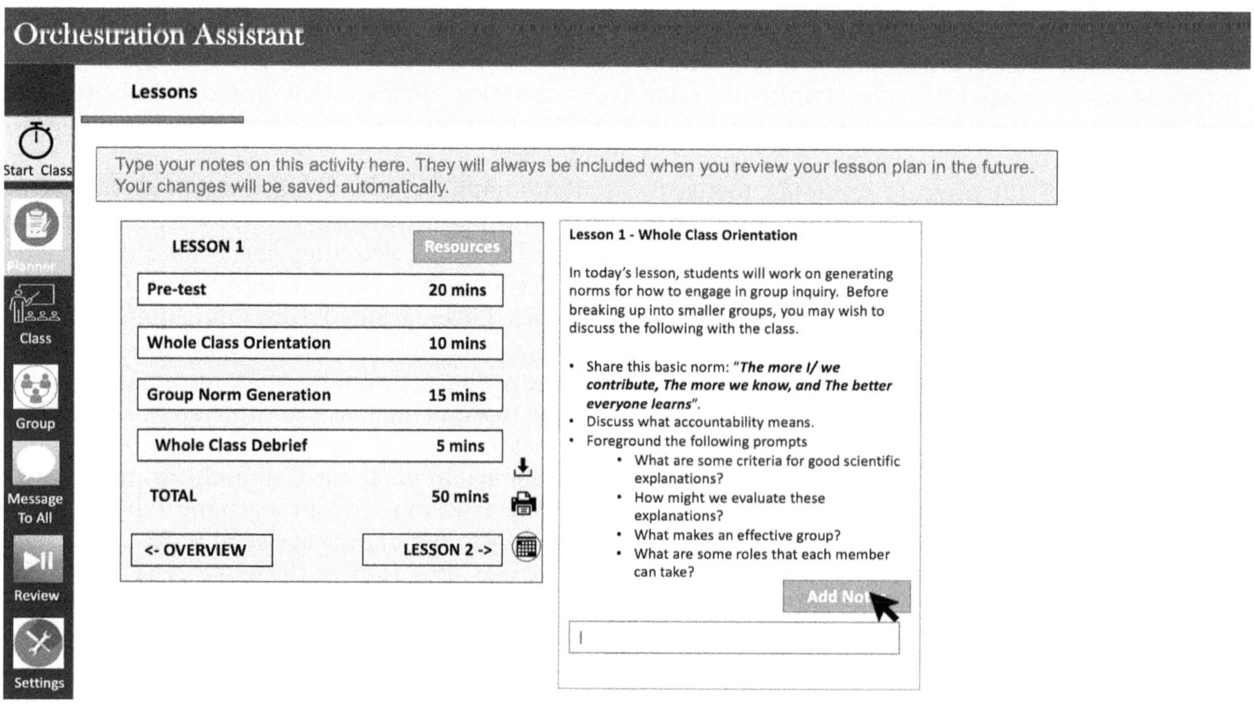

FIGURE 10.1
Planning page at prospective guidance.

prompts. This stage is considered as a critical step to offer useful reflection space to recognize what went well, what did not, and why, and to revise plans for the next class. Further, AI-based prospective, concurrent, and retrospective guidance will be improved in a tight feedback loop with the same teacher's data that is collected during class time.

10.6.2 Automated Systems versus Teacher Agency

A crucial challenge in orchestration assistant design was creating automation that could reduce teachers' workload while not automating decisions in which the teacher should participate (An et al., 2020). In other words, it was crucial to design the orchestration assistant so that it decreased teachers' orchestration load without decreasing their agency over classroom activity.

In order to support teachers' autonomy, we first placed a user survey in the beginning of the setting up stage of orchestration assistant to collect information about how experienced teachers were and their specific teaching approach for the PBL class. Depending on the level of teaching experience, the orchestration assistant's forms of support might be perceived differently. For instance, certain forms of support could be perceived as helpful by novice PBL teachers but could be perceived as intrusive by teachers with greater PBL experience. By identifying the level of teaching experience and their teaching style at the outset, the orchestration assistant can eventually calibrate the level of support for the specific teacher who will actually use the system.

Next, we laid out which specific teaching-related tasks (e.g., monitoring group progress, providing just-in-time feedback, assessing student understanding, and managing frustration in groups) could be automated by the orchestration assistant versus those that should be directly facilitated by a teacher. For example, spamming or bullying in the chat could be simply handled through an automatically generated warning message by the orchestration assistant, but supporting a group where students were stuck at a specific stage in the inquiry activity could be executed by a teacher through sending an eliciting question that is content-specific to the moment. The orchestration assistant might be able to suggest recommended prompts for specific groups so that the teacher can save time to diagnose any challenges in the group and determine the appropriate support during ongoing class. Yet, in order to leave space for supporting teacher autonomy, we designed ways that the teacher can modify settings for system-generated prompts, such as less automated support or more automated support to the teacher for specific dimensions after each class during retrospective guidance (Figure 10.3).

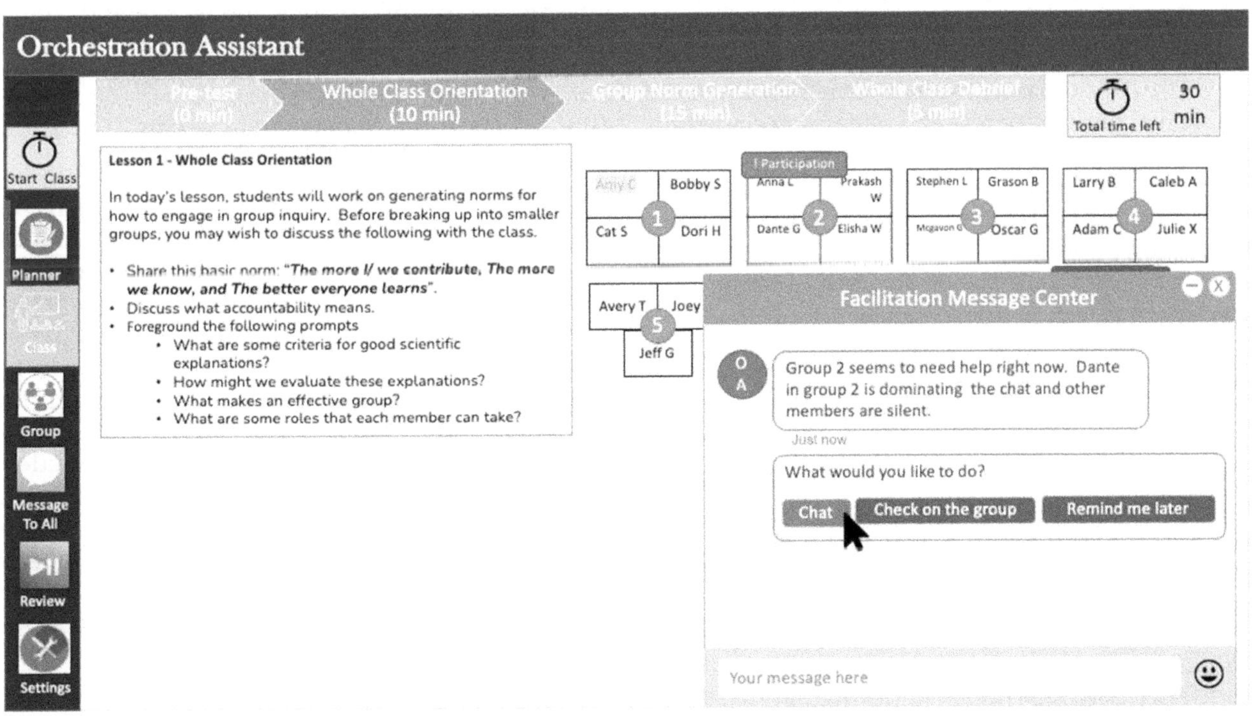

FIGURE 10.2
Receiving alert message during concurrent guidance.

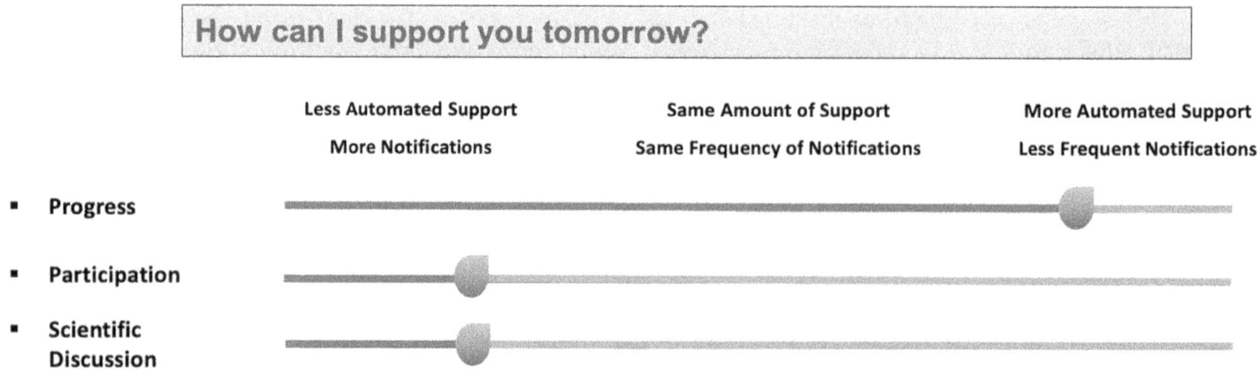

FIGURE 10.3
End-of-day retrospective settings for teacher modification.

10.6.3 Detailed Information versus Actionable Information

While the information presented by the orchestration assistant should be informative enough to support teacher orchestration, it also needs to be unobtrusive and seamless enough to reduce orchestration load without adding another layer of teaching demands. It was challenging to judge what kinds of information were needed and how much information was needed to be presented during ongoing class time and to decide how the task of interpreting classroom information would be divided between the teacher and orchestration assistant. No matter how simple and direct the system-generated information may be, the way information is displayed will always represent a particular framing of the concurrent classroom situations (An et al., 2020). For example, certain information about the class, such as group progress and remaining class time, can be available to the teacher at all times, so that the teacher can have a basic awareness of the class without intentionally clicking or asking for it. On the other hand, certain information might best be displayed only after the orchestration assistant determines certain conditions are met (e.g., a group is silent for a certain predetermined amount of time). This information, which may require the teacher's immediate attention or direct intervention, should not be displayed at all times so as to not overwhelm the teacher.

Furthermore, we tried to design a display that presents more consumable and actionable information to better support teachers in prioritizing their teaching tasks. The ways of delivering information could be designed by either presenting lower-level, less processed/interpreted information, such as simple text or cumulative number of chat lines, or higher-level, more processed/interpreted information, such as a visualized warning icon for groups who were not on task. Less interpreted information could provide greater potential to support teacher agency over when and how to intervene on relevant but not time-sensitive situations, but could require more time to make sense of the information presented and make the best decision for the specific situation. On the other hand, using deliberate, simple, and informative visualizations for the mass amounts of trace data of students could provide more immediately meaningful – and thus actionable – information. Such visualizations are necessary to effectively translate the current state of the classroom on behalf of the teacher (Holstein et al., 2019). We designed the concurrent class stage in orchestration assistant by selectively combining the higher-level information with the lower-level information to reduce teachers' attentional threshold for interpreting the information and to extend the immediate usefulness of the given information.

10.6.4 Classroom Management versus Ambitious Learning Practices

One of the most challenging design tensions that was identified in our previous study (Bae et al., 2020) centered on including features that teachers identified for handling basic classroom management, such as monitoring student participation, controlling students' screen, and administering assessment. In particular, teachers who participated in the interviews emphasized the importance of having the quantified information from assessments to check students' understanding. It was important to them to ascertain if the students' acquired knowledge met the targeted standards. However, our design goal of the orchestration assistant was not simply to develop classroom management tools with quantified information from student participation or standards-based assessment data, but instead to create an actionable cognitive assistant that can provide contingent support for teacher orchestration in a PBL classroom. Our intention was not solely focused on building a classroom management tool that will be used to compare student participation and progress and rank them. Rather, our intention was to build a support environment for teachers that can supply the most valuable information so that they can make the most helpful decisions for their students.

As we consider the fundamental teacher needs, in our prioritization, we chose to surface thinking processes in students' collaborative problem-solving process and provide alerts and recommended prompts while displaying the basic classroom management information, such as absent students, dominating or silent students in chat, and group progress. In particular, orchestration assistants can supply the most pertinent information that can afford a teacher's ability to choose the most beneficial support at the moment. For instance, the teacher can have better information about who to talk to, such as an individual student, a specific group, or the whole class without disturbing the group workflow. Additionally, we designed a reflective space for teachers to put their input that presents a summary of prompts that were provided with students' responses (see Figure 10.4). Here, teachers can observe and make sense of their own prompts that were used in their classroom and how those prompts were taken up by students. This can provide teachers with the ability to make the kinds of instructional decisions they would need to make about standards, disciplinary learning, and collaborative problem-solving processes. This allows teachers to closely monitor students' interactions, especially with regard to scientific inquiry, that are otherwise challenging to observe. This space can serve to empower teachers to reflect on their own teaching practices and help them make the kinds of decisions they need to make to foster deep learning and orchestrate the class. When they engage in this reflective space for their teaching practice, they can provide their input through the text box, such as whether it was an effective prompt or not. Then, the system-generated guidance will be improved each time with the other collected classroom data in a tight feedback loop through the orchestration assistant. Together, these features may enable teachers to feel supported in both classroom management and more ambitious learning practices.

10.7 Discussion and Future Research

We offered several concrete decisions about the orchestration assistant design that we believe can balance the tensions we just discussed. Our hope is that as the orchestration assistant is increasingly deployed and thus more data is available about diverse teacher preferences, increased AI and machine learning should allow for greater and smoother automaticity, personalization, and effectiveness. For example, in the long term, the orchestration assistant should be able to combine a briefer teacher survey with teachers' decisions to make a profile for the teacher and automatically adjust its level of support accordingly.

Here, we discussed several design conjectures for the orchestration assistant. At the time of this writing, the orchestration assistant is being developed for

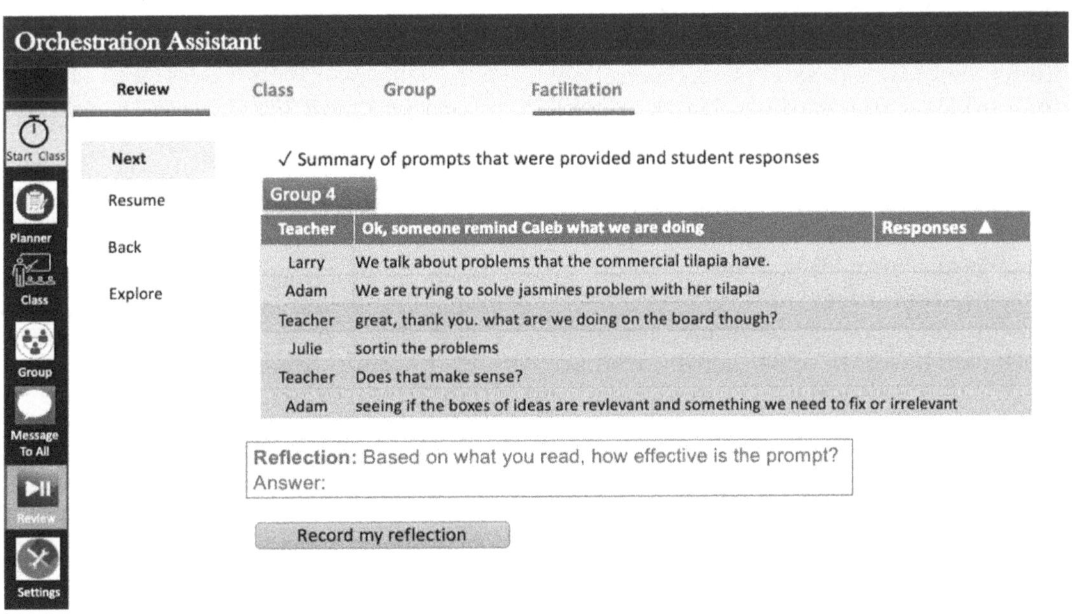

FIGURE 10.4
Teachers can reflect on their facilitation effectiveness.

piloting in a variety of classrooms. Through the process of constructing the orchestration assistant within technological limitations, as well as design-based research to improve its effectiveness, we hope to better understand the practical implications of AI technology for supporting classroom orchestration. For example, as AI inferential capabilities grow increasingly powerful, we anticipate that these will produce a concomitant increase in orchestration assistance capabilities. First, orchestration assistants will be able to more effectively interpret classroom situations. They will be able to understand the dynamics of students' problem-solving activities, and they will be better able to understand how collaborative learning is playing out in the classroom. Second, they will be able to reason about classroom situations to provide teachers with diagnostic information that can support their orchestration. In particular, they will be able to provide teachers with increasingly accurate assessments of students, learning processes, and interactions. For example, they will be able to use natural language processing to understand student dialogue during collaborative learning. Third, they will be able to provide guidance to teachers that is tailored not only to classroom conditions and student competencies but also to teachers' preferences and needs for support. Thus, rather than relying solely on 'student models', they will also be able to leverage 'teacher models' to most effectively support orchestration

Considerations of privacy will play an increasingly important role in orchestration assistants. Designers of orchestration assistants will address privacy concerns through a variety of technologies. For example, it seems likely that 'edge machine learning' will be widely employed. Rather than transmitting personally identifying data to the cloud and running inference algorithms centrally on servers, orchestration assistants that utilize edge machine learning will perform computations locally to ensure that personally identifying data will be kept local. Orchestration assistants that use edge machine learning and other related techniques will preserve students' and teachers' privacy while also providing robust AI performance to support orchestration.

Moving forward, orchestration assistance research will investigate methods that apply across a broad range of subject matters. While the orchestration assistant presented here is being designed specifically for use with the EcoJourneys game-based learning environment, in future work it will be important to devise innovative orchestration assistant technologies that can support domain-general orchestration assistance.

Acknowledgments. This work is supported by the National Science Foundation through grants IIS-1839966 and SES-1840120. Any opinions, findings, and conclusions or recommendations expressed in this chapter are those of the authors and do not necessarily reflect the views of the National Science Foundation. We thank Seung Lee for his insights on AI as applied to an orchestration assistant.

References

Alkhatib, A., & Bernstein, M. (2019). Street-level algorithms. In *Proceedings of the 2019 CHI conference on human factors in computing systems - CHI '19* (pp. 1–13).

An, P., Holstein, K., d'Anjou, B., Eggen, B., & Bakker, S. (2020, April). The TA Framework: Designing real-time teaching augmentation for K-12 classrooms. In *Proceedings of the 2020 CHI conference on human factors in computing systems* (pp. 1–17).

Bae, H., Glazewski, K. D., Hmelo-Silver, C. E., Lester, J., Mott, B. W., & Rowe, J. (2019). Intelligent cognitive assistants to support orchestration in CSCL. In K. Lund, G. P. Niccolai, E. Lavoué, C. Hmelo-Silver, G. Gweon, & M. Baker (Eds.), *International conference on computer supported collaborative learning (CSCL) 2019* (Vol. 2, pp. 947–948). International Society of the Learning Sciences.

Bae, H., Saleh, A., Feng, C., Glazewski, K., Hmelo-Silver, C. E., Chen, Y., Scribner, A., Brush, T., Lee, S. Y., Mott, B. W., & Lester, J. (2020). Designing intelligent cognitive assistants with teachers to support classroom orchestration of collaborative inquiry. In M. Gresalfi & I. S. Horn (Eds.), *International conference of the learning sciences (ICLS) 2020* (Vol. 4, pp. 2101–2108). International Society of the Learning Sciences.

Bae, H., Glazewski, K., Brush, T., & Kwon, K. (2021). Fostering transfer of responsibility in the middle school PBL classroom: An investigation of soft scaffolding. *Instructional Science*, 1–27.

Bell, T., Urhahne, D., Schanze, S., & Ploetzner, R. (2010). Collaborative Inquiry Learning: Models, tools, and challenges. *International Journal of Science Education, 32*(3), 349–377. https://doi.org/10.1080/09500690802582241.

Braun, V., & Clarke, V. (2006). Using thematic analysis in psychology. *Qualitative Research in Psychology, 3*(2), 77–101.

Dillenbourg, P., Järvelä, S., & Fischer, F. (2009). The evolution of research on computer-supported collaborative learning. In N. Balacheff, S. Ludvigsen, T. de Jong, A. Lazonder, & S. Barnes (Eds.), *Handbook of technology-enhanced learning* (pp. 3–19). Springer.

Dillenbourg, P., Prieto, L. P., & Olsen, J. K. (2018). Classroom orchestration. In F. Fischer, C. E. Hmelo-Silver, S. R. Goldman, & P. Reimann (Eds.), *International handbook of the learning sciences* (pp. 180–190). Routledge.

Dobber, M., Zwart, R., Tanis, M., & van Oers, B. (2017). Literature review: The role of the teacher in inquiry-based education. *Educational Research Review, 22*, 194–214.

Engle, R. A., & Conant, F. R. (2002). Guiding principles for fostering productive disciplinary engagement: Explaining an emergent argument in a community of learners classroom. *Cognition and Instruction, 20*(4), 399–483.

Glazewski, K., Shuster, M., Brush, T., & Ellis, A. (2014). Conexiones: Fostering socioscientific inquiry in graduate teacher preparation. *Interdisciplinary Journal of Problem-Based Learning, 8*(1), 2.

Glazewski, K. D., & Hmelo-Silver, C. E. (2019). Scaffolding and supporting use of information for ambitious learning practices. *Information and Learning Sciences, 120*(1/2), 39–58.

Hall, R., & Jurow, A. S. (2015). Changing concepts in activity: Descriptive and design studies of consequential learning in conceptual practices. *Educational Psychologist, 50*(3), 173–189.

Heer, J. (2019). Agency plus automation: Designing artificial intelligence into interactive systems. *Proceedings of the National Academy of Sciences of the United States of America, 116*(6), 1844–1850.

Hmelo-Silver, C. E., Kapur, M., & Hamstra, M. (2018). Learning through problem solving. In F. Fischer, C. E. Hmelo-Silver, S. R. Goldman, & P. Reimann (Eds.), *International handbook of the learning sciences* (pp. 210–220). Routledge.

Hmelo-Silver, C. E., Bridges, S. M., & McKeown, J. M. (2019). Facilitating Problem-Based Learning. In *The Wiley handbook of problem-based learning* (pp. 297–319).

Hmelo-Silver, C. E., & Barrows, H. S. (2006). Goals and strategies of a problem-based learning facilitator. *Interdisciplinary Journal of Problem-Based Learning, 1*(1), 21–39. https://doi.org/10.7771/1541-5015.1004.

Holstein, K., McLaren, B. M., & Aleven, V. (2019). Co-designing a real-time classroom orchestration tool to support teacher AI complementarity. *Journal of Learning Analytics, 6*(2), 27–52.

Ng, H. W., Nguyen, V. D., Vonikakis, V., & Winkler, S. (2015, November). Deep learning for emotion recognition on small datasets using transfer learning. In *Proceedings of the 2015 ACM on international conference on multimodal interaction* (pp. 443–449).

Olsen, J., Rummel, N., & Aleven, V. (2018). *Co-designing orchestration support for social plane transitions with teachers: Balancing automation and teacher autonomy.* International Society of the Learning Sciences, Inc [ISLS].

Prieto, L. P., Sharma, K., & Dillenbourg, P. (2015, September). Studying teacher orchestration load in technology-enhanced classrooms. In G. Conole, T. Klobučar, C. Rensing, J. Konert, & E. Lavoué (Eds.), *European conference on technology enhanced learning* (pp. 268–281). Springer.

Prieto, L. P., Sharma, K., Kidzinski, Ł., & Dillenbourg, P. (2018). Orchestration load indicators and patterns: In-the-wild studies using mobile eye-tracking. *IEEE Transactions on Learning Technologies, 11*(2), 216–229.

Roschelle, J., Lester, J., & Fusco, J. (Eds.). (2020). AI and the future of learning: Expert panel report [Report]. *Digital Promise.* https://circls.org/reports/ai-report.

Rubel, A., & Jones, K. M. (2016). Student privacy in learning analytics: An information ethics perspective. *The Information Society, 32*(2), 143–159.

Saleh, A., Hmelo-Silver, C. E., Glazewski, K. D., Mott, B., Chen, Y., Rowe, J. P., & Lester, J. C. (2019). Collaborative inquiry play: A design case to frame integration of collaborative problem solving with story-centric games. *Information and Learning Sciences, 120*(9/10), 547–566.

Savery, J. R. (2015). Overview of problem-based learning: Definitions and distinctions. In A. Walker, H. Leary, C. E. Hmelo-Silver, & P. A. Ertmer (Eds.), *Essential readings in problem-based learning: exploring and extending the legacy of Howard S. Barrows* (pp. 5–15). Purdue University Press.

Serrano-Laguna, Á., Martínez-Ortiz, I., Haag, J., Regan, D., Johnson, A., & Fernández-Manjón, B. (2017). Applying standards to systematize learning analytics in serious games. *Computer Standards and Interfaces, 50*, 116–123.

Slotta, J. D., Tissenbaum, M., & Lui, M. (2013, April). Orchestrating of complex inquiry: Three roles for learning analytics in a smart classroom infrastructure. In *Proceedings of the third international conference on learning analytics and knowledge* (pp. 270–274).

Shu, M. (2019) *Deep learning for image classification on very small datasets using transfer learning* (Master's thesis). https://lib.dr.iastate.edu/creativecomponents/345/.

Uttamchandani, S., Bhimdiwala, A., & Hmelo-Silver, C. E. (2020). Finding a place for equity in CSCL: Ambitious learning practices as a lever for sustained educational change. *International Journal of Computer-Supported Collaborative Learning*, 373–382.

van Leeuwen, A., & Rummel, N. (2019). Orchestration tools to support the teacher during student collaboration: A review. *Unterrichtswissenschaft, 47*(2), 143–158.

van Leeuwen, A., Rummel, N., & Van Gog, T. (2019). What information should CSCL teacher dashboards provide to help teachers interpret CSCL situations? *International Journal of Computer-Supported Collaborative Learning, 14*(3), 261–289.

11

The Role of AI to Support Teacher Learning and Practice: A Review and Future Directions

Jennifer L. Chiu, James P. Bywater, and Sarah Lilly

CONTENTS

11.1 Background ... 164
11.1.1 Teacher Learning Perspectives ... 164
11.1.2 Technologies for Teacher Learning ... 165
11.1.2.1 AI-Based Teacher Tools ... 165
11.1.3 Design Guidance for AIEd Tools ... 165
11.1.4 The ICAP Framework ... 166
11.2 ICAP and AI-Based Technologies for Teacher Learning ... 167
11.2.1 Passive AIEd Technologies for Teacher Learning ... 167
11.2.2 Active AIEd Technologies for Teacher Learning ... 167
11.2.3 Constructive AIEd Technologies for Teacher Learning ... 168
11.2.4 Interactive AIEd Technologies for Teacher Learning ... 169
11.3 Discussion ... 169
11.4 Limitations ... 171
11.5 Conclusion ... 171
Acknowledgment ... 171
References ... 171

Artificial intelligence (AI) has been increasingly used in science, technology, engineering, and mathematics (STEM) classrooms over the past 30 years to augment learning experiences (Chen et al., 2020; Zawaki-Richter et al., 2019). We define AI as the use of computers to approximate tasks that can be done by humans or exceed the capabilities of humans. AI typically involves a computational system that reasons about data or the world and then makes some judgment or action based on the analysis of the data. Two subfields of AI commonly used in educational settings are machine learning (ML) and Natural Language Processing (NLP). Machine learning involves computational systems learning and improving from experience or detecting patterns in data without being explicitly programmed. ML algorithms examine datasets and use what they learn to make predictions. Some ML algorithms are used to identify key features (e.g., principal component analysis) or patterns (e.g., generalized sequential pattern algorithm) within data, or to cluster data into similar groups (e.g., k-means). Other ML algorithms are used to classify data into the learned categories (e.g., random forest, support vector machine) to make predictions. These techniques allow a variety of just-in-time analysis to occur about students such as identifying student groups based on performance, level of mastery, or patterns of activity within a learning platform.

Natural Language Processing algorithms use ML algorithms to learn from or make predictions about linguistic data. By learning from large quantities of text, NLP algorithms can build models that can identify the structure of language (e.g., part-of-speech tagging, syntactic parsing), the semantic meaning of words (e.g., word embeddings, semantic role labeling, word-sense disambiguation), and the emotional role of words (e.g., sentiment analysis). Building on ML algorithms, NLP algorithms can classify statements that students make or cluster students' responses into similar groups based on a variety of criteria. Many educational technologies use one or both ML and NLP approaches in learning settings. We characterize both as AI-based education (AIEd) technologies.

Many AIEd efforts in STEM have focused on trying to approximate human tutors, giving individualized feedback and responses to learners (e.g., VanLehn, 2011). AIEd has been used to assess what students know or do, to predict future learning, and to support

DOI: 10.1201/9781003181187-14

learners across a variety of cognitive, metacognitive, and behavioral dimensions. However, relatively fewer studies using AI have focused on providing feedback or learning experiences for teachers or instructors (e.g., Jensen et al., 2020).

AI-based technologies for teachers often focus on visualizations or dashboards of digested student data to aid with teachers' in-the-moment instructional decision-making, with demonstrated effects on teacher practice. However, many of these teacher-facing technologies do not take an explicit teacher learning frame to the development of their systems. For example, presenting data in a dashboard or visualization to a user without any subsequent interaction or action could be categorized as a somewhat passive learning experience, yet these are common experiences for teachers with AI-based tools. In addition, very few existing technologies place instructors 'in the loop' of the automated feedback to become aware of and act upon students' ideas, processes, or behaviors before feedback is given to students. These kinds of teacher-in-the-loop technologies can help teachers learn from the automated feedback to students, augmenting the noticing of and responding to students' ideas (Bywater et al., 2019a).

In this chapter, we argue for the utility of taking a teacher learning perspective for the design of teacher-facing, AI-based technologies. We posit that centering teacher learning can inspire new directions for AI-based technologies in STEM by offering opportunities for teachers to develop skills of noticing, interpreting, and responding to students' ideas and by offering teachers feedback on their own practice. This chapter leverages the Interactive-Constructive-Active-Passive (ICAP) framework (Chi & Wylie, 2014; Chiu & Chi, 2014) to outline a spectrum of AI-based technologies to support teacher learning. The chapter presents a review of existing work and work-in-progress to help the field move toward creating teacher learning technologies featuring AI that help augment learning experiences for teachers as well as students.

11.1 Background

11.1.1 Teacher Learning Perspectives

Teacher learning can be viewed from a situational perspective where learning involves participation in the community of teaching where an individual can use and develop knowledge within specific sociocultural contexts (e.g., Borko, 2004). Teachers learn in their classrooms, school communities, workshops, in formal, informal, and independent settings that are located within specific schools, school districts, and other educational systems (e.g., Jones & Dexter, 2014). Professional learning communities within and across these spaces offer ways for teachers to develop their own practice (e.g., Webster-Wright, 2009). Within these spaces, teachers hold knowledge about their content, pedagogical knowledge, pedagogical content knowledge, knowledge about their learners, the instructional context, as well as their own beliefs and interests that shape the kinds of instructional decisions and choices that teachers make in classrooms (e.g., Gess-Newsome, 2015).

Crucial pedagogical skills for teachers involve effectively noticing, interpreting, and responding to students' ideas and behaviors in their classrooms (Sherin, 2002). Noticing involves what teachers deem important to pay attention to during classroom interactions, how teachers connect classroom events to pedagogical knowledge, and use context to reason about events in classrooms (van Es & Sherin, 2002). Noticing also involves domain-specific understanding such as understanding the nuance within students' mathematical ideas and the variety of non-normative approaches that students may take (Ball et al., 2008). Research demonstrates that many teachers struggle with noticing in their classrooms, especially attending to the diversity and subtlety of students' ideas (Sherin et al., 2011).

Even if teachers can effectively notice students' ideas and behaviors in their classrooms, teachers may still need support to respond effectively to students' ideas (Jacobs et al., 2010). Effective responding or feedback to students greatly depends on the content of the response or feedback (e.g., Black & Wiliam, 1998; Hattie & Timperley, 2007; Shute, 2008). Effective responding also involves instructional 'moves' or strategies to help students express more of their ideas, such as revoicing (O'Conner & Michaels, 1993), or asking questions that press for students' reasoning and explanation (Chapin et al., 2009; Herbel-Eisenmann et al., 2013). However, teachers need learning opportunities to rehearse and practice their developing noticing and responding skills.

Research reveals certain key elements of effective learning opportunities for teachers (Kennedy, 2016). Studies demonstrate the importance of time and sustained interactions for teacher learning (e.g., Garet et al., 2001; Guskey, 2000). Effective teacher learning opportunities include active, not passive, activities and mirror the desired pedagogical skills or strategies. Additionally, learning experiences that provide records of classroom practice are important tools for teacher learning. Artifacts from classroom practice can be especially beneficial when used in communities of practice where teachers can share, critique, and generate ideas about the externalized students' ideas that they can then use to take back to their individual classrooms (e.g., Kazemi & Franke, 2004).

11.1.2 Technologies for Teacher Learning

Given that noticing and responding to students' ideas can be difficult for teachers, prior research has investigated how technologies, both AI-based and non-AI-based, can help teachers develop these kinds of pedagogical skills. Non-AI-based technologies that support teachers to engage in noticing and interpreting students' ideas include reflections on classroom video (e.g., van Es & Sherin, 2010; Walkoe, 2015) and multimedia storyboarding tools (Walkoe & Levin, 2018) as well as simulations of classrooms presented as text scenarios (e.g., Hillaire et al., 2020). Interactive avatar-based environments where humans control virtual agents, such as TeachLivE/Mursion and Second Life, can support the development of evidence-based instructional practices (e.g., Brown et al., 2011; Dieker et al., 2017; Ersozlu et al., 2021). Additionally, technologies that have provided live coaching to teachers during instruction using video and wireless earpieces ('bug-in-the-ear') have been found to be beneficial to supporting a variety of pedagogical practices (e.g., Rock et al., 2014; Wake et al., 2017).

11.1.2.1 AI-Based Teacher Tools

With recent advances in AI, several AI-based teacher-facing technologies have emerged to augment teachers' noticing of students' ideas. For example, a range of studies have provided teachers with information about students in intelligent tutoring or other computer-based learning environments (e.g., Verbert et al., 2014). Many of these tools use machine learning techniques to notice patterns in student data and relay the patterns to the instructor in a visualization or dashboard (Casamayor et al., 2009). Other studies have investigated how to augment teaching in-the-moment using environmental cues such as sound and light (d'Anjou et al., 2019). Others have explored how technologies that automatically process and display student's or teachers' actions can help teachers more effectively respond to students during instruction (e.g., Ahuja et al., 2019; Ramakrishnan et al., 2021).

Several studies provide evidence of AIEd systems that give automated feedback on teacher practice on the cusp of development. For example, Jensen et al. (2021) describe the foundation of a system that automatically captures and gives feedback on teachers' discourse and questioning practices during instruction. Similarly, Suresh et al. (2019) report on the use of different models to classify teachers' talk moves as well as ways to provide automated feedback to teachers based on those classifications (Suresh et al. 2021). Advances in NLP and deep learning tools have opened doors to providing near real-time feedback to teachers.

11.1.3 Design Guidance for AIEd Tools

Several studies provide design guidance for AI-based teacher tools. For example, An et al. (2020) describe a Teaching Augmentation Framework for technologies that complement teachers' practice during classroom implementation. The framework consists of five dimensions: target, attention, social visibility, presence over time, and interpretation, as well as design tensions for each dimension. For example, for the target of attention, the tension presented is the need for autonomy versus automation, or what tasks should be performed by the teacher and what should be automated for them. However, the focus of An et al.'s (2020) framework is specifically meant to complement teacher's practice during instruction, not offer learning opportunities for teachers that may occur before, during, or after instructional episodes.

Gerritsen et al. (2018) provide a framework that centered the instructor as the learner, focusing on planning, action, and reflection upon the teaching practice. They approximated AI systems by using people to log in-class behaviors and actions, which they then visualized for graduate and undergraduate teaching assistants (TAs). Case study results found that three of the TAs engaged in meaningful planning and reflection and tried to improve their practice. Two of the TAs did not interact beneficially with planning or reflection. Although they did not test the system with AI-based tools, the paper emphasizes the need for teacher-centered learning tools and highlights the challenges inherent in AI-based tools, including the motivation and engagement of the instructor as precursors to teacher learning.

Holstein et al. (2020) present a conceptual framework that emphasizes the ways in which humans and AIEd systems can provide synergistic benefits to each other to enhance educational experiences. Instead of the AIEd systems enacting the goals of the designers, AIEd systems could inform the goals of the instructors, or instructors could inform the goals of the AIEd system. For example, teachers may hold critical contextual or other on-the-ground information that can provide a large benefit to the AIEd system. Similarly, Holstein et al. (2020) argue that AIEd–human systems can synergistically enhance decision-making during instruction, helping instructors make effective pedagogical choices as well as humans enhancing AIEd decision-making. For instance, AIEd systems can help nudge instructors to make instructional decisions that align with pedagogical goals, or instructors can provide feedback to AIEd systems to customize or adapt to specific classroom contexts. This chapter complements the general framework put forth by Holstein et al. (2020) in that we specifically focus on AIEd–human synergies through the lens of teacher learning.

Additionally, many AIEd studies have called for teachers to co-design AIEd systems to create effective learning technologies (e.g., Prieto et al., 2020). By inviting teachers as co-creators and co-developers of the technologies, teachers provide valuable insight into their classroom practice and ground developers in the kinds of design decisions that would work in authentic settings (e.g., Holstein et al., 2019). But even co-design experiences from an HCI perspective are not enough. Although teachers bring expertise about their classroom contexts and their own pedagogical skills and strategies, they themselves are not experts on teacher learning.

Many of these studies also come from an HCI perspective that focuses on metrics of usability or user experience for success (e.g., Mavrikis et al., 2016), and not necessarily teacher learning as an explicit outcome. This chapter argues that taking a *teacher learning* perspective can provide additional value for designing AIEd technologies. We argue that the design of AIEd systems for use by teachers need to draw upon what we know about teacher learning to understand how to create technologies that provide learning opportunities for both students and teachers. Specifically, we leverage the ICAP framework to frame the development of AIEd systems from a teacher learning perspective.

11.1.4 The ICAP Framework

The interactive-constructive-active-passive framework (ICAP) is based on decades of research and provides guidance on the kinds of experiences that have been shown to be effective for students' learning. *Passive* learning experiences are those that require minimal action from the learner. Listening to a lecture, watching a video, or reading a text is an example of passive learning experiences. *Active* experiences are when learners take part in physical action or manipulation of the information. Active experiences include opportunities for learners to engage with the material by underlining, highlighting, or pausing and replaying a video. *Constructive* experiences involve the generation of learning artifacts that introduce new ideas or connections. Examples of constructive activities include generating self-explanations, concept maps that represent new ideas or connections, or making inferences or connections across different resources. *Interactive* activities involve dialogic exchange where the learner responds to a prompt or question, receives a response based on learner's response, and the exchange continues in this fashion. Learners can interact with peers, teachers, facilitators, or computer agents. Interactive activities require two participants to engage constructively. That is, instead of two people just clicking through a video together, they need to elaborate, critique, or build on each other's ideas or ideas from the video. Interactive activities also require both participants to contribute, instead of only one person generating ideas. Other interactive activities include opportunities to dialogue with peers, co-construct knowledge, or engage with peer critique or comparison. In this way, ICAP resonates with effective practices for teacher learning, with active, generative, and interactive activities helping teachers develop individually as well as encouraging dialoguing within their community of practice.

ICAP predicts active experiences to have more benefit for learners than passive experiences, constructive than active, and interactive than constructive (e.g., interactive > constructive > active > passive) based on underlying knowledge processes (Chi & Wylie, 2014). Several studies in STEM have demonstrated the utility of the framework (Chi et al., 2018; Wiggins et al., 2017). For example, metanalysis of undergraduate STEM courses reveals that on average, students in active learning classrooms outperform students in passive classrooms with lower failure rates (Freeman et al., 2014). Studies have also shown active learning approaches benefit students from backgrounds underutilized in STEM (Theobald et al., 2020). Studies also provide evidence of the general benefit of interactive processes on students' learning (Meneske et al., 2013).

ICAP has also shown value for learning with technology-enhanced environments. For example, Wang et al. (2016) examined learning with online discussion forums with undergraduate students. Results demonstrated that constructive and interactive contributions have more benefit to learning outcomes than active contributions. Similarly, Henderson (2019) used a clicker system to explore relationships among passive, constructive, and interactive learning experiences and found that constructive and interactive processes were better than passive activities for students' learning. Likewise, Wekerle et al. (2020) found that technology-supported classes were more likely to have active, constructive, or interactive activities than non-technology supported classes. Applied to teachers, ICAP has been used as a framework for professional development with demonstrated effect on students' learning (Chi et al., 2018) and to increase the quality of teachers' questions.

A few recent studies have leveraged ICAP with AI-based tools. Farrow (2019) used ICAP as a rubric to automatically assess students' engagement in online discussions. Fahid et al. (2021) used ICAP as a frame for adaptive scaffolding for cognitive engagement within learning environments for operational command skills (Fahid et al., 2021). These studies highlight the promise of combining ICAP with AIEd systems.

11.2 ICAP and AI-Based Technologies for Teacher Learning

The ICAP framework can also provide guidance to the design of AI-based technologies for *teacher learning*. When thinking about teachers as learners, ICAP presents a complementary view of the affordances and constraints of current and future AI-based systems. For example, presenting data to teachers about their students without teachers taking any action on that data would fall into passive engagement with the information. Understandably, these passive kinds of systems are designed for ease of use by teachers in the moment in the classroom, to ease any kind of orchestration load or burden (Prieto et al., 2015). We do not argue with the demonstrated effect on students' learning from these kinds of passive systems, nor do we argue that some teachers are able to learn from viewing data from their classroom, similar to some students being able to learn from simply watching a video or attending a lecture. However, we do posit that the design of these AIEd environments can also provide rich learning experiences for teachers by designing for ICAP engagement processes. For example, teacher learning experiences can be enhanced by enabling the manipulation of dashboard data (e.g., active), by supporting teachers to reflect upon and generate self-explanations of the classroom data (constructive), or by helping teachers to dialogue with a peer where both parties are generating new pedagogical ideas based upon visualized data (interactive). The next section reframes existing and works-in-progress AIEd technologies from a student's perspective into the ICAP categories for teacher learning and discusses affordances and constraints for teacher learning. Table 11.1 provides an overview of selected studies sorted into ICAP categories.

11.2.1 Passive AIEd Technologies for Teacher Learning

Passive AIEd technologies for teachers encompass systems that enable instructors to pay attention to various aspects of their classroom or students but do not enable teachers to interact with student's data, ideas, or the system itself. With passive systems, teachers observe or notice students or classroom data that has been digested or analyzed in some way. In most cases, these technologies are designed for use in-the-moment during classroom instruction. Typical examples of passive AIEd technologies are dashboards of students' progress without the ability for teachers to interact with the system (e.g., Leony et al., 2012).

For example, FireFlies-VLE (d'Anjou et al., 2019) is a system that provides light displays ranging from green to red for each student in a classroom. FireFlies-VLE uses a computer-based learning environment to track students' progress and then uses a light display to output student's progress. Given the only output of the system is the light that teachers can notice across the room, the system enables teachers to observe a representation of student's understanding by walking around the room.

Given the importance of teacher noticing of students' ideas (Sherin, 2002), passive AIEd systems enable teachers to be able to notice various aspects of their student or classroom that were not previously available without AI-based technologies. In this way, passive systems can help teachers provide more differentiated and equitable instruction for all students. However, the onus is on the teacher to be able to use that information to inform a change in teaching practice, as passive systems do not provide any guidance or support for teachers in what to notice or how to respond to the data presented. In addition, since the material is presented to the instructor, the teacher has little input or understanding of how the data is analyzed within the system.

11.2.2 Active AIEd Technologies for Teacher Learning

Active AIEd technologies enable teachers to interact or manipulate the system in simple ways, without teachers providing or creating any new information. In active systems, teachers can click on different student icons to see additional data from the system or choose a different view or representation of the analyzed data. Basic classroom management functionality, including freezing student's screens, is also part of active functionality in AIEd systems. Most importantly, in active systems, the teacher is not generating or creating anything themselves or for the learners, but instead manipulating or interacting with the system in ways that create different ways to display data.

For example, with the MiGen TA system (Mavrikis et al., 2016), teachers can see students' progress in a computer-based mathematics learning environment. Teachers can click on students to see individual

TABLE 11.1

The Definition of ICAP Categories and Example AI-Based Technologies for Teachers

ICAP Category	Definition	Authors	AIEd Technology and Description
Interactive	Teachers dialogue with a responsive AIEd system in which both parties generate new ideas	Bywater et al., 2021a; Datta et al., 2021	AI-based Classroom Teaching Simulator (ACTS) where in-service and preservice teachers engage with a virtual student to practice mathematical questioning strategies
Constructive	Teachers generate learning artifacts or new ideas with an AIEd system	Bywater et al., 2019a, 2019b	Teacher Responding Tool (TRT) that uses NLP to provide recommendations for teacher responses to students' mathematical ideas
		Gerard et al., 2020	Teacher Action Planner (TAP) that recommends ways for teachers to customize instruction based on patterns in student explanations
		Martinez-Maldonado et al., 2013	MTDashboard that enables teachers to control students' interactions with tabletop activities as well as monitor progress with activity visualizations
Active	Teachers able to interact with an AIEd system that provides classroom data	Mavrikis et al., 2016	MiGenTA system where teachers can select what they want to display on their dashboard
Passive	Teachers view an AIEd system that provides classroom data	Ahuja et al., 2019	EduSense system that provides instructors with sensor data about their classroom, including body position and student versus instructor speech
		Casamayor et al., 2009	An intelligent agent for teachers to help detect situations during online collaborative distance learning activities
		d'Anjou et al., 2019	FireFlies-VLE system uses lights on top of student computers to provide a snapshot of students' progress during an activity
		Leony et al., 2012	Gradient's Learning Analytics System (GLASS) that provides a visualization of students' progress in learning environments

Note: These classifications were based on the functionalities described in the papers and may not represent the totality of interactions afforded by the specific tool.

data of what steps that student is currently working on as well as move student icons to put students in similar positions to the physical classroom layout. Instructors are also able to click on other views to see the overall progress by specific goals of specific students. Thus, the information is provided to the instructor on-demand, based on the interests or needs of the instructor without overwhelming visual displays. Despite the interactivity of the system, however, instructors are not able to generate or create anything new for themselves or for the learners; thus, this is considered an active AIEd learning system for teacher learning.

By providing ways for teachers to interact with and manipulate the data and displays in AIEd systems, teachers can customize the tools to enhance and augment what they choose to notice or attend to in the class. Instead of having a predetermined set of what ideas, data, or information is important to show, instructors can tailor the system to investigate what they deem as important based on their pedagogical knowledge, classroom context, or adjustments based on how the class is enacted. In this way, active AIEd technologies can facilitate more robust teacher noticing and may help teachers provide more nuanced and targeted feedback or instruction to students.

11.2.3 Constructive AIEd Technologies for Teacher Learning

Constructive AIEd systems enable teachers to generate externalized ideas or create artifacts that go beyond information given to them by the system. For example, a constructive AIEd teacher learning system could provide instructors the ability to generate feedback based on the types of students' responses or ideas, or prompt teachers to generate explanations or reflections about instructional decisions.

A simple example of a constructive AIEd technology could enable teachers to send messages to students based upon what they see in the visualization or dashboard. For example, in MTDashboard, teachers can send simple text messages to all students based on what they see in their teacher dashboard during class time (Martinez-Maldonado et al., 2013). Through this kind of functionality, the technology also facilitates the teacher responding to what they notice with the system.

Another example of a constructive system involves using AI tools to help teachers both notice and respond to students' ideas in the classroom. The Teacher Responding Tool (TRT; Bywater et al., 2019a) uses Natural Language Processing to provide recommendations to teachers for how to respond to students' written mathematical ideas in a computer-based learning environment. The TRT provided three recommended choices that teachers could click to use to respond to the student, create their own feedback to the student entirely, or edit the suggested choices to create their own tailored feedback. Results found that about half of the time teachers selected responses and edited them, a quarter of the time selected responses and used them as-is, and the other quarter created their own response entirely but used similar wording to one of the recommendations. Findings also revealed that the degree of teachers' responses alignment to students' mathematical ideas was greater with the TRT than in their responses without the TRT (Bywater et al., 2019b). Although merely clicking on a suggested response may be classified as active, the opportunity for teachers to create their own response and the classroom data that suggested teacher created their own responses based on suggestions places the TRT in the constructive category.

11.2.4 Interactive AIEd Technologies for Teacher Learning

Interactive AIEd technologies involve collaborative interactions in which two partners engage in mutually generative dialogue. Two partners could be an instructor and another instructor, an instructor and a student, or an instructor and a virtual agent. The mutually generative dialogue must be constructive, that is, going beyond procedures, revoicing, or repeating, so that each partner's contributions responds to, addresses, or engages the other partner's contributions. According to ICAP, interactive exchanges do not discriminate in terms of types of dialogue, reasoning, or output, but need to go beyond what is presented in the instructional material. In this way, interactive AIEd systems can follow constructive systems in that the system enables multiple, follow-up responses to information or feedback in some fashion.

Although many non-AI-based technologies provide interactive systems for teacher learning (e.g., Bondie et al., 2021), a very limited but growing amount of research describes the field approaching interactive AIEd technologies. For example, the AI-based Classroom Teaching Simulation (ACTS) uses deep learning models and advances in machine learning techniques to create an open-source virtual conversational agent that focuses on helping teachers promote meaningful mathematical discourse (Bywater et al., 2021a; Datta et al., 2021). With ACTS, instructors interact with a virtual student on a specific mathematical task such as understanding how the scale factor changes parameters of a three-dimensional object. ACTS enables instructors to talk to a virtual student who is having difficulty with a mathematical task. The instructor can also manipulate the mathematical visualization on the screen and use the values within the visualization to anchor conversations. Based upon the content of the input, the virtual student adapts and responds in real time. This chain of instructor–virtual student dialogue continues until the instructor chooses to stop or feels that the virtual student understands the underlying concepts and can solve the problem. Through this interaction, instructors can rehearse instructional moves or mathematical questioning that supports ambitious mathematics instruction.

Thus, ACTS is an interactive system where teachers can co-generate dialogue with a virtual student that goes beyond the task presented. With such a system, pedagogical moves and practices such as teacher questioning are automatically categorized and can be given back to the user to support reflection upon performance as well as to provide teacher educators with important data to guide instruction. Pilot results demonstrate that the system was robust, resulted in minimal errors, and was perceived as realistic by teachers interacting with the virtual student.

Taken together, reinterpreting the affordances of AIEd tools through the ICAP lens provides insight into the different kinds of learning opportunities offered to teachers.

11.3 Discussion

This chapter leverages the ICAP framework to reframe the discussion of AIEd technologies around teacher learning. Given the growing power and relative ease of use of AI-based technologies in educational settings, we argue that the ICAP framework helps to situate current contributions and highlight opportunities for future work with teacher learning. Although there is a growing number of studies that demonstrate the value of ICAP for student learning, few studies have used ICAP with AI-based technology or with teachers, and very few to investigate teacher learning. Future studies can look to investigate the benefit of the ICAP framework with technology and teacher learning and the extent to which offering these kinds of affordances to teachers results in change in instructional practices.

One interesting trend emerging from this chapter is that the large majority of published AIEd tools as described seem to be passive or active tools for teachers. Understandably, researchers made these specific design choices to assist in the orchestration of classroom interactions in-the-moment. However, recent studies have pressed for more constructive and interactive forms of tools for teachers (e.g., Prieto et al., 2020). This chapter aimed to illustrate the kinds of design decisions that can lead to more generative and collaborative tools for teachers. For instance, enabling teachers to respond to what they see in dashboards and provide additional support for students outside of what systems provide can help to move a tool from an active to a constructive system for teacher learning. This chapter also helps to envision moving beyond the metaphor of providing an intelligent tutor or coach for teachers. For example, interactive AIEd systems could analyze classroom information from different teachers with similar lessons to highlight patterns or differences across classrooms and invite collaborating teachers to discuss, reflect upon, or generate explanations of these classroom comparisons.

This chapter builds upon and complements existing frameworks such as the Real-Time Teaching Augmenting framework (An & Holstein et al., 2020) in that it centers teacher learning and presents options outside of in-the-moment classroom activities (e.g., Gerritson et al., 2018). This chapter also builds upon HCI-centered approaches where usability is a primary metric for success (e.g., Martinez-Maldonado et al., 2015). We posit that specifying the kinds of learning opportunities provided to teachers can lead to teacher learning outcomes and more exploration of how AI based tools can help teachers rehearse evidence-based practices such as noticing and responding (Jacobs et al., 2010).

The ICAP framework offers opportunities to enhance AIEd with non-technical or human solutions to push toward constructive or interactive engagement with material. Similar to research on learning with simulations, games, and visualizations (e.g., McElhaney et al., 2015), the cognitive load during instruction or enactment may be too high to facilitate teacher learning in-the-moment. Constructive or interactive activities do not necessarily need to be entirely AI-based but could also be a blend of human and computer interaction. For example, systems could encourage teachers to plan, enact, and reflect upon the AI-based feedback of classroom data. Taking advantage of AIEd systems to provide opportunities to reflect after instructional sessions may be a promising avenue for future AIEd systems (e.g., Prieto et al., 2020). Future research can explore these kinds of nuanced approaches to AIEd systems for teacher learning.

Similarly, this chapter builds upon research demonstrating the benefit of co-designing AIEd systems with teachers. Co-designing AIEd systems with teachers incorporates their understanding of the classroom context as well as pedagogical strategies that would be of interest into the design of AIEd systems. However, most of the co-design literature focuses on the development of systems to be used during classroom instruction, and few studies consider the ways that tools could be of benefit outside of class time (e.g., Wylie et al., 2020) or part of ongoing professional development efforts. By envisioning more nuanced interactions among teachers and AIEd tools, we may discover and create new functionality for teachers, perhaps some previously thought to be either too complex or cumbersome for teachers.

The use of the ICAP framework may also be able to provide guidance to the field about ways to introduce and empower instructors to understand and use AI in their own classrooms. For instance, active systems in which teachers select different assumptions or different underlying models to analyze their classroom data may help teachers understand how AI tools introduce their own biases and assumptions that may or may not resonate with their own goals (e.g., Dietvorst et al., 2018). Constructive systems could scaffold the use of AI-based tools such that teachers can create their own recommendation tools or algorithms for their own classrooms. Future research can explore the relationship of how enabling various levels of teachers' engagement with AI-based tools can support teacher use and learning of AI concepts and tools (e.g., Kaufmann, 2021).

Providing learning opportunities for teachers with AIEd tools also has the potential to promote more equitable classroom practices. For example, providing opportunities for teachers to reflect or explain 'unbiased' classroom data that shows inequities in teacher attention or quality of teacher responses or questions has the potential to not only uncover but also help teachers learn how to address different biases in their classrooms.

This chapter also highlights how a spectrum of tools may be needed to support teacher learning with teachers of various experience or expertise. For example, teachers with rich pedagogical knowledge may benefit from passive technologies because they have the experience to take the information presented to them and make effective instructional decisions. On the other hand, constructive or interactive technologies may be more effective for teachers who understand what to notice from their students but may need support to respond in productive ways. Interactive systems may be of greatest utility in teacher preparation programs with preservice teachers with minimal experience. Future research to explore how AI-based tools with different levels of engagement may relate to teachers with different levels of pedagogical experience and expertise.

11.4 Limitations

The ICAP framework was originally envisioned to provide a theoretical hierarchy of cognitive processes that are of increasing value to learners. ICAP may not have the same kind of utility when applied to AIEd systems for teacher learning. Although this chapter highlights the potential and possibilities of AIEd systems, we acknowledge the biases, limitations, and potential harm that AIEd systems may incur in educational settings. For example, one crucial tension lies in the judgment of what is important to capture, notice, or bring to the attention of the instructor (e.g., An et al., 2020). We hope that through the kinds of opportunities presented in this chapter, we can invite teachers to take a more central role in this conversation.

11.5 Conclusion

This chapter argues for the design of more AIEd tools that provide rich teacher learning opportunities. Given the recent advances in Natural Language Processing and deep learning methods, novel AIEd tools can support teachers not only to notice and respond to students' ideas more effectively but also engage in reflection, discussion, and interaction around the content of students' ideas. The chapter uses ICAP to frame examples of existing work and work-in-progress with the aim to make progress in creating AIEd systems that help teachers and students engage in learning.

Acknowledgment

This work has been supported in part by the Robertson Foundation, grant no. 9909875.

References

Ahuja, K., Kim, D., Xhakaj, F., Varga, V., Xie, A., Zhang, S., Eric Townsend, J., Harrison, C., Ogan, A., & Agarwal, Y. (2019). EduSense: Practical classroom sensing at scale. *Proceedings of the ACM on interactive, mobile, wearable and ubiquitous technologies*, *3*(3), 1–26.

An, P., Holstein, K., d'Anjou, B., Eggen, B., & Bakker, S. (2020, April). The TA framework: Designing real-time teaching augmentation for K-12 classrooms. In R. Bernhaupt, F. Mueller, D. Verweij, & J. Andres (Chairs.), *Proceedings of the 2020 CHI conference on human factors in computing systems* (pp. 1–17).

Ball, D. L., Thames, M. H., & Phelps, G. (2008). Content knowledge for teaching: What makes it special. *Journal of Teacher Education*, *59*(5), 389–407.

Black, P., & Wiliam, D. (1998). Assessment and classroom learning. *Assessment in Education: Principles, Policy and Practice*, *5*(1), 7–74.

Bondie, R., Mancenido, Z., & Dede, C. (2021). Interaction principles for digital puppeteering to promote teacher learning. *Journal of Research on Technology in Education*, *53*(1), 107–123.

Borko, H. (2004). Professional development and teacher learning: Mapping the terrain. *Educational Researcher*, *33*(8), 3–15.

Brown, I. A., Davis, T. J., & Kulm, G. (2011). Pre-service teachers' knowledge for teaching algebra for equity in the middle grades: A preliminary report. *Journal of Negro Education*, 266–283.

Bywater, J., Chiu, J. L., Hong, J., & Sankaranarayanan, V. (2019a). The teacher responding tool: Scaffolding the teacher practice of responding to student ideas in mathematics classrooms. *Computers and Education*, *139*, 16–30.

Bywater, J. P., Chiu, J. L., & Watson, G. S. (2019b). Assessing the effectiveness of an intelligent tool that supports targeted teacher responses to student ideas. In *Proceedings of the international convention of the association for educational communications and technology* (Vol. 1, pp. 45–53). October 21–25.

Bywater, J., Datta, D., Phillips, M., Watson, G., Lilly, S., Brown, D., & Chiu, J. L. (2021, April). *Deep learning approaches to classifying teacher questions within the AI-based classroom teaching simulation (ACTS)*. Annual meeting of the American Educational Research Association. Virtual Conference.

Casamayor, A., Amandi, A., & Campo, M. (2009). Intelligent assistance for teachers in collaborative e-learning environments. *Computers and Education*, *53*(4), 1147–1154.

Chapin, S. H., O'Connor, C., & Anderson, N. C. (2009). *Classroom discussions: Using math talk to help students learn*. Math Solutions Grades K-6.

Chen, L., Chen, P., & Lin, Z. (2020). Artificial intelligence in education: A review. *IEEE Access*, *8*, 75264–75278.

Chi, M. T., Adams, J., Bogusch, E. B., Bruchok, C., Kang, S., Lancaster, M., Levy, R., Li, N., McEldoon, K., Stump, G., Wiley, R., Xu, D., & Yaghmourian, D. L. (2018). Translating the ICAP theory of cognitive engagement into practice. *Cognitive Science*, *42*(6), 1777–1832.

Chi, M. T., & Wylie, R. (2014). The ICAP framework: Linking cognitive engagement to active learning outcomes. *Educational Psychologist*, *49*(4), 219–243.

Chiu, J. L., & Chi, M. T. H. (2014). Supporting self-explanation in the classroom. In V. A. Benassi, C. E. Overson, & C. M. Hakala (Eds.), *Applying science of learning in education: Infusing psychological science into the curriculum*. Retrieved from the Society for the Teaching of Psychology web site: http://teachpsych.org/ebooks/asle2014/index.php

d'Anjou, B., Bakker, S., An, P., & Bekker, T. (2019, June). How peripheral data visualisation systems support secondary school teachers during VLE-supported lessons. In *Proceedings of the 2019 on designing interactive systems conference* (pp. 859–870).

Datta, D., Phillips, M., Bywater, J. P., Chiu, J., Watson, G. S., Barnes, L., & Brown, D. (2021). Virtual pre-service teacher assessment and feedback via conversational agents. In *Proceedings of the 16th workshop on innovative use of NLP for building educational applications* (pp. 185–198).

Dieker, L. A., Hughes, C. E., Hynes, M. C., & Straub, C. (2017). Using simulated virtual environments to improve teacher performance. *School–University Partnerships, 10*(3), 62–81.

Dietvorst, B. J., Simmons, J. P., & Massey, C. (2018). Overcoming algorithm aversion: People will use imperfect algorithms if they can (even slightly) modify them. *Management Science, 64*(3), 1155–1170.

Ersozlu, Z., Ledger, S., Ersozlu, A., Mayne, F., & Wildy, H. (2021). Mixed-reality learning environments in teacher education: An analysis of TeachLivE™ research. *Sage Open, 11*(3), 21582440211032155.

Fahid, F. M., Rowe, J. P., Spain, R. D., Goldberg, B. S., Pokorny, R., & Lester, J. (2021, June). Adaptively scaffolding cognitive engagement with batch constrained deep Q-networks. In *International conference on artificial intelligence in education* (pp. 113–124). Springer.

Farrow, E. (2019, July). Modelling student participation using discussion forum data. In *14th European conference on technology enhanced learning: Transforming learning with meaningful technologies*. CEUR Workshop Proceedings (CEUR-WS.org).

Freeman, S., Eddy, S. L., McDonough, M., Smith, M. K., Okoroafor, N., Jordt, H., & Wenderoth, M. P. (2014). Active learning increases student performance in science, engineering, and mathematics. *Proceedings of the National Academy of Sciences, 111*(23), 8410–8415.

Garet, M. S., Porter, A. C., Desimone, L., Birman, B. F., & Yoon, K. S. (2001). What makes professional development effective? Results from a national sample of teachers. *American Educational Research Journal, 38*(4), 915–945.

Gerard, L., Wiley, K., Bradford, A., Chen, J. K., Lim-Breitbart, J., & Linn, M. (2020). Impact of a teacher action planner that captures student ideas on teacher customization decisions. In *Proceedings of the 14th international society for learning sciences conference* (pp. 2077–2084).

Gerritsen, D., Zimmerman, J., & Ogan, A. (2018, July). Towards a framework for smart classrooms that teach instructors to teach. In *International conference of the learning sciences* (Vol. 3).

Gess-Newsome, J. (2015). A model of teacher professional knowledge and skill including PCK: Results of the thinking from the PCK Summit. In *Re-examining pedagogical content knowledge in science education* (pp. 38–52). Routledge.

Guskey, T. R. (2000). *Evaluating professional development*. Corwin Press.

Hattie, J., & Timperley, H. (2007). The power of feedback. *Review of Educational Research, 77*(1), 81–112.

Henderson, J. B. (2019). Beyond "active learning": How the ICAP framework permits more acute examination of the popular peer instruction pedagogy. *Harvard Educational Review, 89*(4), 611–634.

Herbel-Eisenmann, B. A., Steele, M. D., & Cirillo, M. (2013). (Developing) teacher discourse moves: A framework for professional development. *Mathematics Teacher Educator, 1*(2), 181–196.

Hillaire, G., Larke, L., & Reich, J. (2020, April). Digital storytelling through authoring simulations with teacher moments. In *Society for information technology & teacher education international conference* (pp. 1736–1745). Association for the Advancement of Computing in Education (AACE).

Holstein, K., Aleven, V., & Rummel, N. (2020, July). A conceptual framework for human–AI hybrid adaptivity in education. In *International conference on artificial intelligence in education* (pp. 240–254). Springer.

Holstein, K., McLaren, B. M., & Aleven, V. (2019). Co-designing a real-time classroom orchestration tool to support teacher–AI complementarity. *Journal of Learning Analytics, 6*(2).

Jacobs, V. R., Lamb, L. L., & Philipp, R. A. (2010). Professional noticing of children's mathematical thinking. *Journal for Research in Mathematics Education, 41*(2), 169–202.

Jensen, E., Dale, M., Donnelly, P. J., Stone, C., Kelly, S., Godley, A., & D'Mello, S. K. (2020, April). Toward automated feedback on teacher discourse to enhance teacher learning. In *Proceedings of the 2020 CHI conference on human factors in computing systems* (pp. 1–13).

Jensen, E., Pugh, S. L., & D'Mello, S. K. (2021, April). A deep transfer learning approach to modeling teacher discourse in the classroom. In *LAK21: 11th international learning analytics and knowledge conference* (pp. 302–312).

Jones, W. M., & Dexter, S. (2014). How teachers learn: The roles of formal, informal, and independent learning. *Educational Technology Research and Development, 62*(3), 367–384.

Kaufmann, E. (2021). Algorithm appreciation or aversion? Comparing in-service and pre-service teachers' acceptance of computerized expert models. *Computers and Education: Artificial Intelligence*. https://doi.org/10.1016/j.caeai.2021.100028

Kazemi, E., & Franke, M. L. (2004). Teacher learning in mathematics: Using student work to promote collective inquiry. *Journal of Mathematics Teacher Education, 7*(3), 203–235.

Kennedy, M. M. (2016). How does professional development improve teaching? *Review of Educational Research, 86*(4), 945–980.

Leony, D., Pardo, A., de la Fuente Valentín, L., de Castro, D. S., & Kloos, C. D. (2012, April). GLASS: A learning analytics visualization tool. In *Proceedings of the 2nd international conference on learning analytics and knowledge* (pp. 162–163).

Martinez-Maldonado, R., Kay, J., Yacef, K., Edbauer, M. T., & Dimitriadis, Y. (2013). MTClassroom and MTDashboard: Supporting analysis of teacher attention in an orchestrated multi-tabletop classroom. In *Proceedings of the computer supported collaborative learning* (pp. 119–128).

Mavrikis, M., Gutierrez-Santos, S., & Poulovassilis, A. (2016, April). Design and evaluation of teacher assistance tools for exploratory learning environments. In *Proceedings of the sixth international conference on learning analytics & knowledge* (pp. 168–172).

McElhaney, K., Chang, H. Y., Chiu, J. L., & Linn, M. C. (2015). Evidence for effective uses of dynamic visualizations in science curriculum materials. *Studies in Science Education, 51*(1), 49–85.

Menekse, M., Stump, G. S., Krause, S., & Chi, M. T. (2013). Differentiated overt learning activities for effective instruction in engineering classrooms. *Journal of Engineering Education, 102*(3), 346–374.

O'Conner, M., & Michaels, S. (1993). Aligning academic task and participation status through revoicing: Analysis of a classroom discourse strategy. *Anthropology and Education Quarterly, 24*(4), 318–335.

Prieto, L. P., Magnuson, P., Dillenbourg, P., & Saar, M. (2020). Reflection for action: Designing tools to support teacher reflection on everyday evidence. *Technology, Pedagogy and Education, 29*(3), 279–295.

Prieto, L. P., Sharma, K., & Dillenbourg, P. (2015, September). Studying teacher orchestration load in technology-enhanced classrooms. In *European conference on technology enhanced learning* (pp. 268–281). Springer.

Ramakrishnan, A., Zylich, B., Ottmar, E., LoCasale-Crouch, J., & Whitehill, J. (2021). Toward automated classroom observation: Multimodal machine learning to estimate class positive climate and negative climate. In *IEEE transactions on affective computing.*

Rock, M. L., Schumacker, R. E., Gregg, M., Howard, P. W., Gable, R. A., & Zigmond, N. (2014). How are they now? Longer term effects of e coaching through online bug-in-ear technology. *Teacher Education and Special Education, 37*(2), 161–181.

Sherin, M., Jacobs, V., & Philipp, R. (Eds.). (2011). *Mathematics teacher noticing: Seeing through teachers' eyes.* Routledge.

Sherin, M. G. (2002). A balancing act: Developing a discourse community in a mathematics classroom. *Journal of Mathematics Teacher Education, 5*(3), 205–233.

Shute, V. J. (2008). Focus on formative feedback. *Review of Educational Research, 78*(1), 153–189.

Suresh, A., Jacobs, J., Clevenger, C., Lai, V., Tan, C., Martin, J. H., & Sumner, T. (2021, June). Using AI to promote equitable classroom discussions: The TalkMoves application. In *International conference on artificial intelligence in education* (pp. 344–348). Springer.

Suresh, A., Sumner, T., Jacobs, J., Foland, B., & Ward, W. (2019, July). Automating analysis and feedback to improve mathematics teachers' classroom discourse. In *Proceedings of the AAAI conference on artificial intelligence* (Vol. 33, No. 1, pp. 9721–9728).

Theobald, E. J., Hill, M. J., Tran, E., Agrawal, S., Arroyo, E. N., Behling, S., Chambwe, N., Cintrón, D. L., Cooper, J. D., Dunster, G., Grummer, J. A., Hennessey, K., Hsiao, J., Iranon, N., Jones, L. 2nd, Jordt, H., Keller, M., Lacey, M. E., Littlefield, C. E., … Freeman, S. (2020). Active learning narrows achievement gaps for underrepresented students in undergraduate science, technology, engineering, and math. *Proceedings of the National Academy of Sciences, 117*(12), 6476–6483.

Van Es, E. A., & Sherin, M. G. (2002). Learning to notice: Scaffolding new teachers' interpretations of classroom interactions. *Journal of Technology and Teacher Education, 10*(4), 571–596.

Van Es, E. A., & Sherin, M. G. (2010). The influence of video clubs on teachers' thinking and practice. *Journal of Mathematics Teacher Education, 13*(2), 155–176.

VanLehn, K. (2011). The relative effectiveness of human tutoring, intelligent tutoring systems, and other tutoring systems. *Educational Psychologist, 46*(4), 197–221.

Verbert, K., Govaerts, S., Duval, E., Santos, J. L., Van Assche, F., Parra, G., & Klerkx, J. (2014). Learning dashboards: An overview and future research opportunities. *Personal and Ubiquitous Computing, 18*(6), 1499–1514.

Wake, D., Dailey, D., Cotabish, A., & Benson, T. (2017). The effects of virtual coaching on teacher candidates' perceptions and concerns regarding on-demand corrective feedback. *Journal of Technology and Teacher Education, 25*(3), 327–357.

Walkoe, J. (2015). Exploring teacher noticing of student algebraic thinking in a video club. *Journal of Mathematics Teacher Education, 18*(6), 523–550.

Walkoe, J., & Levin, D. M. (2018). Using technology in representing practice to support preservice teachers' quality questioning: The roles of noticing in improving practice. *Journal of Technology and Teacher Education, 26*(1), 127–147.

Wang, X., Wen, M., & Rosé, C. P. (2016, April). Towards triggering higher-order thinking behaviors in MOOCs. In *Proceedings of the sixth international conference on learning analytics & knowledge* (pp. 398–407).

Webster-Wright, A. (2009). Reframing professional development through understanding authentic professional learning. *Review of Educational Research, 79*(2), 702–739.

Wekerle, C., Daumiller, M., & Kollar, I. (2020). Using digital technology to promote higher education learning: The importance of different learning activities and their relations to learning outcomes. *Journal of Research on Technology in Education, 54*(1), 1–17.

Wiggins, B. L., Eddy, S. L., Grunspan, D. Z., & Crowe, A. J. (2017). The ICAP active learning framework predicts the learning gains observed in intensely active classroom experiences. *AERA Open, 3*(2), 2332858417708567.

Wiley, K. J., Dimitriadis, Y., Bradford, A., & Linn, M. C. (2020, March). From theory to action: Developing and evaluating learning analytics for learning design. In *Proceedings of the tenth international conference on learning analytics & knowledge* (pp. 569–578).

Zawacki-Richter, O., Marín, V. I., Bond, M., & Gouverneur, F. (2019). Systematic review of research on artificial intelligence applications in higher education–where are the educators? *International Journal of Educational Technology in Higher Education, 16*(1), 1–27.

12

Learning Outcome Modeling in Computer-Based Assessments for Learning

Fu Chen and Chang Lu

CONTENTS

12.1 Psychometric Measurement for Learning Outcome Modeling 176
12.1.1 Classical Test Theory 176
12.1.2 Item Response Theory 177
12.1.3 Cognitive Diagnosis 178
12.2 Bayesian Networks 179
12.3 Bayesian Knowledge Tracing 181
12.4 Additive Factors Model 181
12.5 Deep Learning for Learning Outcome Modeling 182
12.5.1 Deep Knowledge Tracing 182
12.5.2 Other Deep Learning Approaches for Learning Outcome Modeling 183
12.6 Collaborative Filtering for Learning Outcome Modeling 183
12.6.1 Matrix Factorization 184
12.6.2 Collaborative Filtering-Based Approaches for Learning Outcome Modeling 185
12.6.3 Deep Learning–Based Collaborative Filtering 186
12.7 An Overview of Approaches for Learning Outcome Modeling 186
12.8 Conclusion 187
References 189

Assessment is a fundamental element of teaching and learning. It is an important way to collect learner information and measure learners' performance and understanding with respect to learning goals. Generally, we distinguish two types of assessments: summative assessments (i.e., assessments *of* learning) and formative assessments (i.e., assessments *for* learning; Black & Wiliam, 2009). The former refers to evaluating individuals' learning outcomes at the end of a teaching and learning unit in contrast to an established standard or benchmark. The latter refers to ongoing evaluations of learners' progress with prescriptive feedback for improving teaching and learning. Summative and formative assessments are developed to address different educational purposes. For example, traditional school learning is often evaluated with standards-based summative assessments for accountability purposes (e.g., final or midterm exams, standardized tests for admissions, and standardized assessments for informing educational policy). The results of summative assessments can be used to inform learners if they have achieved educational goals, to make comparison between different populations, to promote the accountability at different educational levels, and to inform educational policy (Shute & Rahimi, 2017). However, in learner-centered scenarios, formative assessments are used frequently as a supportive approach. With formative assessments, instructors can evaluate individuals' learning progress in a timely manner while learners receive individualized instruction and feedback to improve their learning outcomes. Compared with summative assessments, formative assessments demonstrate greater potential in supporting learning and they are successfully used for different audiences across various content domains and educational sectors (e.g., Davies & Ecclestone, 2008; Gikandi et al., 2011; Meek et al., 2017; Shute et al., 2008; Tsai et al., 2015). Moreover, it was found that learners, especially struggling learners, instructed with formative assessments are more likely to increase their academic performance than those instructed with standard pedagogical approaches (e.g., Kleitman & Costa, 2014; López-Pastor et al., 2013; Pastor, 2011). In summary, formative assessment, or assessment *for* learning, is playing an increasingly important

DOI: 10.1201/9781003181187-15

role in education given its advantages in supporting learning.

Among different applications of formative assessments, computer-based assessment (CBA) for learning is extensively used nowadays. The explosive ICT developments in recent years are changing learning from instructor-centered to learner-centered and have given rise to the emerging use of new pedagogical approaches, such as project-based learning (Bell, 2010), game-based learning (Kiili, 2005), and more recently, personalized learning (Shute et al., 2016). Inevitably, benefiting from ICT advances, CBA for learning is also transforming a variety of educational processes (e.g., Chatzopoulou & Economides, 2010; Joosten-ten Brinke et al., 2007; Peat & Franklin, 2002; Terzis & Economides, 2011). Incorporated with new technologies, CBAs can be designed to approximate real-life problem-solving environments with more integrative and interactive tasks (e.g., Azevedo et al., 2010; Blanchard et al., 2012), a desired feature of assessment for the twenty-first century (Shute & Becker, 2010). Consequently, more complex and multidimensional learner competencies such as creativity, critical thinking, and problem-solving skills can be evaluated with CBAs in the digital age (e.g., Greiff et al., 2014; Pásztor et al., 2015; Rosen & Tager, 2014).

Irrespective of the content areas and the design features of CBAs, making accurate inferences about learners' cognitive states, or learner modeling, should be of high priority. Generally, for CBAs with multiple independent assessment questions measuring a single latent skill, learners' cognitive states are inferred by modeling all question–answers or item responses simultaneously (e.g., IRT models). However, for ITSs or other similar CBAs, it is required to trace learners' cognitive states based on the performance data. That being said, learners' history of problem-solving attempts or item responses would affect their current or future problem-solving success because their cognitive states continuously change across multiple learning opportunities. In this regard, a sequential modeling technique (e.g., BKT) is required to monitor learners' cognitive states at each time point. For both sequential modeling and non-sequential modeling, from the methodological perspective, because inferences are made based on elicited learning outcomes by the system, a model is deemed effective for estimating learners' cognitive states if it is capable of accurately predicting or recovering item responses. For example, an effective sequential modeling technique should accurately predict a learner's present and future item responses given his or her previous item responses; an effective non-sequential modeling technique should result in non-significant differences between observed item responses and model-predicted item responses.

In addition to learner modeling, learning outcome modeling is also used to make inferences on item–skill associations (i.e., domain modeling). Contrast to the mature methodological developments for learner modeling, approaches for domain modeling in the context of CBAs for learning are relatively underdeveloped. For most CBAs, domain experts play a role in specifying the skills, knowledge components, and production rules for the assessment. As such, item–skill associations are prespecified in most CBAs. However, as mentioned earlier, human judgments are not always guaranteed to result in precise item–skill associations, and it is costly and less feasible to rely purely on human efforts in the case of a great number of assessment items. This is a call for more data-driving approaches developed to account for item–skill associations in the context of CBAs for learning.

The following sections review a wide range of approaches for learner and domain modeling. Generally, these approaches can be categorized as psychometric measurement models, Bayesian networks, BKT, additive factors models, deep learning–based approaches, and collaborative filtering (CF) approaches.

12.1 Psychometric Measurement for Learning Outcome Modeling

Educational measurement is a discipline focusing on the use of methodologies for assigning scores obtained from educational assessments to students, based on which inferences about the abilities, knowledge, and skills of students can be made. In terms of analytic approaches, educational measurement overlaps psychometrics, a discipline focusing on the theory and methodologies of psychological measurement. Essentially, an analytical approach in psychometric measurement is a type of learner modeling techniques estimating learners' latent ability levels or presence/absence of latent skills based on test scores. However, the majority of psychometric measurement approaches are theory-driven, which is a distinctive feature in comparison with data-driven approaches in computing science. In the following, the chapter briefly introduces the most basic modeling techniques in psychometric measurement. These techniques were developed based on three psychometric measurement theories: classical test theory (CTT), item response theory (IRT), and cognitive diagnosis models (CDMs).

12.1.1 Classical Test Theory

CTT was the dominant approach prior to IRT and yet it is still used widely in practice due to its simplicity and interpretability. A key assumption of CTT is that a learner's observed test score X is equal to the sum of the learner's innate true score T and the measurement error E (Spearman, 1904):

$$X = T + E. \tag{12.1}$$

For example, if a learner has actually mastered 50% of the knowledge required by a test and 50% is the learner's true score, the learner might have an observed test score between 45% and 55% because there is 5% discrepancy from the true score due to errors of measurement. The errors of measurement E are assumed to follow a normal distribution with a mean of zero, which indicates that the average score of the distribution of observed test scores for a learner who takes a test an infinite number of times would be equal to that test-taker's true score. In CTT, a learner's cognitive state is typically calculated as the total score of the test. That said, multiple test items are assumed to measure a single latent skill, and the sum of learners' scores on each item indicates their proficiency levels on the skill. Regarding item–skill associations, CTT uses item discrimination indices to indicate the associations between items and the latent skill. Item discrimination refers to the degree to which an item is capable of differentiating learners with high proficiency levels on the targeted skill from learners with low proficiency levels on the targeted skill, which is used as the hallmark of a good test item in practice. Item discrimination is within a range of 0–1 and expected to be as large as possible given that higher discrimination levels indicate stronger affinity of an item to the latent skill.

Despite its simplicity, CTT has several disadvantages. The item parameters and learner ability estimates approximated by CTT are greatly dependent on the test items and the examinee group. That said, the difficulty and discrimination of items are likely to be different given different groups of learners tested by the items, and the learners' estimated ability levels are likely to vary if they are tested with different sets of items measuring the same latent skill. Moreover, the measurement error in CTT is the same for all learners given that they are estimated at the test level.

12.1.2 Item Response Theory

Compared with CTT, IRT demonstrates several advantages. In IRT, item parameters are invariant to the examinee groups and learner ability levels are invariant to the test items. In addition, the measurement error in IRT is estimated for different learner ability levels, which implies that the extent to which each test item precisely measures each learner's latent ability can be informed by IRT. Also, IRT assumes that only one dominant skill is allowed to be measured in a test and the probability of a learner answering an item correctly is independent of his or her odds of success on other items (Reise et al., 2005).

IRT models are a type of latent variable models which estimate learners' probabilities of answering an item correctly through a set of item and learner parameters. Specifically, item parameters in IRT are item difficulty, item discrimination, and item guessing, and the learner parameter indicates a learner's proficiency level on the targeted skill. Different IRT models assume different degrees of item parameterization. For example, the three-parameter logistic (3PL) model parameterizes item difficulty, item discrimination, and item guessing. Given the 3PL model, learner i's probability of correctly answering item j can be formulated as follows:

$$P(R_{ij} = 1 \mid \theta_i) = c_j + \frac{1 - c_j}{1 + e^{-a_j(\theta_i - b_j)}}, \tag{12.2}$$

where c_j denotes the guessing parameter of item j and the other parameters are the same as the 2PL model. For IRT models, the item discrimination parameter can be used to indicate the item–skill associations. Higher item discrimination level implies stronger affinity of the item to the latent skill, and as a result, the item is more capable of differentiating high-performing learners from low-performing learners. Given the formulation of IRT models, it can be seen that the item–learner interaction is modeled by the linear combination of learner ability, item difficulty, item discrimination, and item guessing, which is then non-linearly converted to a predicted probability of correct item response ranging from 0 to 1 through a sigmoid transformation.

The above IRT models are all assumed to be unidimensional. However, in reality, the majority of educational assessments are designed to evaluate multiple skills or knowledge components, and traditional IRT models fail to deal with multidimensional data. More recently, the approach of multidimensional item response modeling (MIRT) was proposed to address multiple latent skills of multidimensional data (Reckase, 1997; Yao & Boughton, 2007). Compared with conventional IRT models, the MIRT approach allows items to measure different skills which enables a finer-grained analysis of learner data. In MIRT, different item difficulties and learner abilities are estimated for multiple latent skills. For example, given the multidimensional 3PL model, learner i's probability of correctly answering item j can be formulated as follows:

$$P(R_{ij}=1\mid\vec{\theta}_i)=c_j+\frac{1-c_j}{1+e^{-\vec{a}_j\odot\vec{\theta}_i^T+b_j}},\qquad(12.3)$$

where b_j and c_j denote item difficulty and item guessing of item j, respectively, which are scalar parameters. $\vec{\theta}_i$ is a vector parameter indicating learner i's proficiency levels on multiple latent skills, and $\vec{a}_j$ is a vector parameter indicating item j's discrimination levels on multiple latent skills. As such, given a vector of item discrimination, the MIRT approach is capable of providing each item with item–skill associations for multiple skills, and given a vector of learner ability, each learner can be estimated with multiple proficiency levels on multiple latent skills. However, given that MIRT models bear greater complexity, they have not been widely used in the digital learning environments (Desmarais & Baker, 2012).

12.1.3 Cognitive Diagnosis

Cognitive diagnosis is an approach for profiling learners with information on mastery or non-mastery of multiple skills (Rupp et al., 2010). CDMs calculate the probability of a correct response based on learners' mastery profile of the skills that are measured by an item (e.g., Henson et al., 2009; Tatsuoka, 1983). Given the mastery profile of the required skills, learners can be evaluated with fine-grained diagnostic information, which in turn supports targeted interventions for learning. Similar to MIRT, CDMs also address multiple latent skills. However, unlike MIRT, CDMs requires a prespecified mapping of items to skills, which is called the Q-matrix, for item parameterization and model estimation. In a Q-matrix, the columns and rows represent the required skills and the test items respectively with matrix entries of 0's or 1's indicating the mapping of one item to one skill. Table 12.1 presents a sample Q-matrix which involves five items and three skills. It can be seen that item 1 requires only skill 2, whereas item 5 requires both skills 1 and 3. Moreover, the entries of Q-matrix can be polytomous (e.g., 0, 1, and 2), indicating the degree to which an item measures a skill (von Davier, 2005). In this sense, the Q-matrix naturally represents the item–skill associations for learning outcome modeling.

TABLE 12.1

A Sample Q-Matrix with Five Items and Three Skills

	Skill 1	Skill 2	Skill 3
Item 1	0	1	0
Item 2	0	0	1
Item 3	1	0	0
Item 4	1	1	1
Item 5	1	0	1

Over the past several decades, a great number of CDMs have been proposed in the psychometric measurement literature. Despite a variety of modeling techniques designed for various purposes, the majority of CDMs can be characterized by several general modeling frameworks. With the general modeling frameworks, other specific CDMs can be derived through statistical constraints on model parameters. Therefore, given the limit of the space, the chapter describes CDMs by a brief introduction to a general modeling framework, the log-linear cognitive diagnosis model (LCDM; Henson et al., 2009). However, the general modeling frameworks are saturated models developed at the sacrifice of model simplicity, which might not be optimal for practice. Given LCDM, learner i's probability of correctly answering item j can be formulated as follows:

$$P(R_{ij}=1\mid\vec{\alpha}_i)=\frac{1}{1+e^{-\left(\lambda_{j,0}+\vec{\lambda}_j'\mathbf{h}(\vec{\alpha}_i,\vec{q}_j)\right)}},\qquad(12.4)$$

where $\vec{\alpha}_i=(\alpha_{i1},\ldots,\alpha_{iK})$ denotes the skill mastery profile of learner i on K latent skills, $\lambda_{j,0}$ represents the intercept parameter of item j, $\vec{q}_j=(q_{j1},\ldots,q_{jK})$ indicates the Q-matrix entries for item j, and $\mathbf{h}$ is a mapping function that linearly combines $\vec{\alpha}_i$ and $\vec{q}_j$:

$$\vec{\lambda}_j'\mathbf{h}(\vec{\alpha}_i,\vec{q}_j)=\sum_{k=1}^{K}\lambda_{j,1,(k)}\alpha_{ik}q_{jk}+\sum_{k=1}^{K-1}\sum_{k'>k}\lambda_{j,2,(k,k')}\alpha_{ik}\alpha_{ik'}q_{jk}q_{jk'}+\cdots.\qquad(12.5)$$

For item j and a total K latent skills, the probability of a correct response is affected by the main effect of each skill and the interaction effects between skills. As such, in the equation, $\lambda_{j,1,(k)}$ indicates the main effect of skill k, and $\lambda_{j,2,(k,k')}$ refers to the two-way interaction effect between skills k and k'. Moreover, because item j might not require and learner i might not master all the K latent skills, the terms $\alpha_{ik}q_{jk}$ and $\alpha_{ik}\alpha_{ik'}q_{jk}q_{jk'}$, which are the product of the skill mastery profile of learner i and the Q-matrix entries for item j, are used to control which main and interaction effects would be present for the learner–item interaction between learner i and item j. The LCDM accounts for all possible effects of the presence or absence of skills on item responses, which involves a great number of model parameters to be estimated. In addition, for a real-world educational assessment, its required skills might not pose all main and interaction effects on learners' item response. Therefore, more parsimonious models are typically desired in practice. Given LCDM, removing all main and lower order interaction effects and only retaining the highest-order interaction effect result in the deterministic inputs noisy and gate model (DINA; Haertel, 1989; Junker & Sijtsma, 2001), and removing all interaction effects and retaining the main effects lead to the compensatory re-parameterized unified model (C-RUM; Hartz, 2002), which are parsimonious models with much fewer model parameters. Contrast to LCDM and its variants, another type of CDMs uses a statistical pattern recognition approach to diagnose learners' skill mastery profiles. The most well-known models are the rule space model (Tatsuoka, 1983) and the attribute hierarchy model (Leighton et al., 2004).

Regarding challenges of psychometric measurement models, in general, they require strong theoretical assumptions regarding how skills are measured by items. Particularly, CTT and IRT models assume a single skill to be measured, which is largely infeasible for fine-grained cognitive diagnosis and multiple skill modeling. However, for cognitive diagnosis models, they typically require an accurate Q-matrix prespecified by domain experts, which inevitably brings flaws and limits the scalability for modeling. Furthermore, given the strong theoretical assumptions, domain modeling is typically infeasible given the standard forms of psychometric models.

12.2 Bayesian Networks

Bayesian networks were widely used to make probabilistic inferences based on the learners' performance data of CBAs (de Klerk et al., 2015). Bayesian networks are a type of probabilistic graphic models, which graphically represent a joint distribution of a set of random variables (Koller & Friedman, 2009). Essentially, building a Bayesian network requires the specification of a directed acyclic graph and a table of probability distributions for each variable, or called node, in the graph. Figure 12.1 presents an example Bayesian network with three nodes. In the network, each node represents a random variable and directional edge represents the dependency or the causal relationship between two random variables. The two bottom nodes of squares represent two assessment items indicating learners' item responses (i.e., correct/incorrect) and the top node of oval represents the latent skill measured by the two items (i.e., mastery/non-mastery). As such, the example network depicts that learners' probabilities of giving correct or incorrect responses to item 1 and item 2 are dependent on their probabilities of having the latent skill mastered or not. To know the joint distribution of the network, it is required to define the distribution of the latent skill and the conditional distributions for item 1 and item 2. Mathematically, the joint distribution of the three nodes shown in the example network is given by

$$P(\text{skill}, \text{item 1}, \text{item 2}) = P(\text{item 1} | \text{skill})\, P(\text{item 2} | \text{skill})\, P(\text{skill}), \tag{12.6}$$

where $P(\text{item 1}|\text{skill})P(\text{item 2}|\text{skill}) = P(\text{item 1}, \text{item 2} \,|\, \text{skill})$ given that item 1 and item 2 are conditionally independent of each other. Practically, making inferences about learners' mastery levels on the latent skill given the information of their item responses to item 1 and item 2 can be characterized as the process of finding $P(\text{skill} \,|\, \text{item 1}, \text{item 2})$; predicting learners' item responses to item 1 or item 2 can be represented as the process of finding $P(\text{item 1}|\text{skill})$ or $P(\text{item 2}|\text{skill})$. Estimating the conditional probabilities mentioned above can be formularized as a maximum likelihood estimation problem as other statistical learning models (Heckerman et al., 1995). More concretely, for the network shown in Figure 12.2, the parameters to be estimated include two conditional probabilities for the two possible values of item 1 (i.e., correct and incorrect), two conditional probabilities for the two possible values of item 2 (i.e., correct and incorrect), and two probabilities for the latent skills (i.e., mastery and non-mastery). The problem of estimating all the above parameters Θ given a dataset D can be formulated as follows:

$$\Theta_{ML} = \arg\max\{L(\Theta : D)\}. \tag{12.7}$$

Heckerman et al. (1995) provide more details regarding the parameter estimation for Bayesian networks.

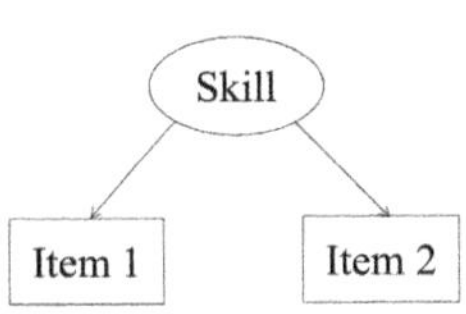

FIGURE 12.1
An example Bayesian network with two items and one latent skill.

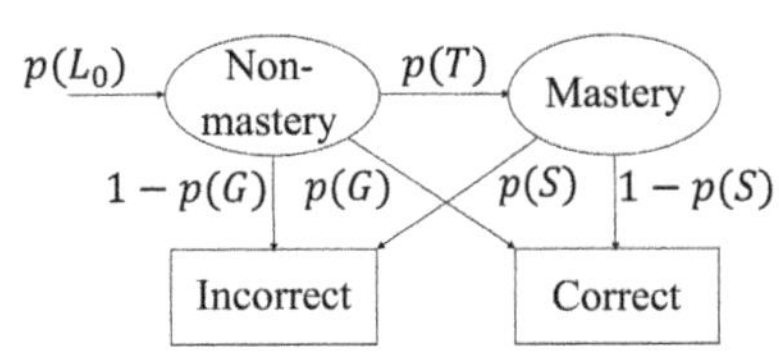

FIGURE 12.2
Graphical representation of Bayesian knowledge tracing.

For the majority of CBAs, multiple learning opportunities are designed for learners to practice the latent skills. Bayesian networks, however, cannot address the temporal dependencies between the multiple learning opportunities presented through a CBA. For events occurring over a period, dynamic Bayesian networks (DBNs) can be used to account for the temporal dependencies between multiple timesteps in making inferences on the conditional probabilities of random variables. Essentially, DBN is an extended version of Bayesian networks with time information. Figure 12.3 presents an example DBN with one latent skill and one observable item across a total of T learning opportunities. Each learning opportunity represents one timestep. The T timesteps are connected by the temporal relationships of the latent skill between one timestep and its subsequent timestep. It should be noted that, without the temporal connection, the nodes and edges for each timestep constitute a simple Bayesian network, and the DBN can be considered as T connected copies of the simple Bayesian network. By modeling the temporal dependencies between timesteps, the state of the latent skill changes over time can be inferred. This feature of DBN is especially useful for CBAs because individuals' learning progress can be tracked by analyzing their item responses with DBN. Given the DBN shown in Figure 12.4, the state of the latent skill at a certain timestep (i.e., mastery/non-mastery) is dependent on both the state of the latent skill at the previous timestep and the current state of the item (i.e., correct/incorrect). Therefore, the joint distribution of the latent

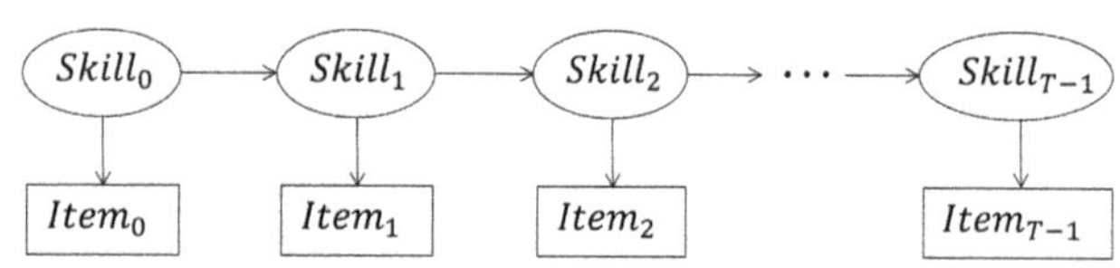

FIGURE 12.3
An example dynamic Bayesian network with one item and one latent skill.

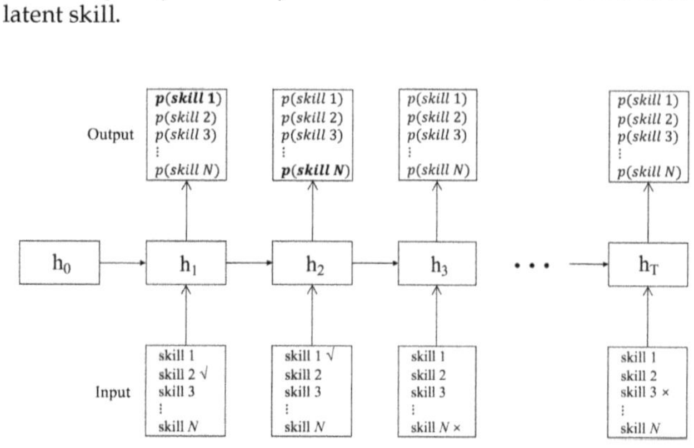

FIGURE 12.4
Graphical representation of deep knowledge tracing.

skill, $\text{Skill} = \{\text{Skill}_0, \text{Skill}_1, \ldots, \text{Skill}_{T-1}\}$, and the item, $\text{tem} = \{\text{Item}_0, \text{Item}_1, \ldots, \text{Item}_{T-1}\}$, over T timesteps is given by

$$P(\text{Skill}, \text{Item}) = \prod_{t=1}^{T-1} P(\text{Skill}_t \mid \text{Skill}_{t-1}) \prod_{t=0}^{T-1} P(\text{Item}_t \mid \text{Skill}_t) P(\text{Skill}_0), \tag{12.8}$$

where $P(Skill_0)$ is the prior distribution of the skill, $P(\text{Item}_t \mid \text{Skill}_t)$ indicates the observation distribution of the item dependent on the skill at timestep t, and $P(\text{Skill}_t \mid \text{Skill}_{t-1})$ refers to the state transition distribution presenting how the state of the latent skill at timestep t is affected by its state at the previous timestep t–1. Given the above formulation, the problem of estimating how the state of the latent skill changes over time can be solved by finding the conditional probabilities $P(\text{Skill} \mid \text{Item})$, where $\text{Skill} = \{\text{Skill}_0, \text{Skill}_1, \ldots, \text{Skill}_{T-1}\}$ and $\text{Item} = \{\text{Item}_0, \text{Item}_1, \ldots, \text{Item}_{T-1}\}$. It should be noted the DBN is established based on the Markov assumption, which states that the conditional probability of the latent skill at timestep t is only dependent on the state of the latent skill at timestep $t-1$; the states of the latent skill at timesteps prior to $t-1$ are of no influence (Koller & Friedman, 2009).

Regarding domain modeling, standard Bayesian networks and DBNs fail to automatically estimate the item–skill associations because a correspondence between each item and the skill it measures is required to be prespecified to construct the graphical model. Therefore, similar to psychometric measurement models, domain expertise is required for Bayesian networks and DBNs.

Bayesian networks or DBNs showed great potential for learning outcome modeling in the literature because of their strong flexibility, high expressiveness, and sound computations (Desmarais & Baker, 2012). For example, Bayesian networks have been used to model the performance data of CBAs on computer networking skills (Levy & Mislevy, 2004; Levy, 2013; Rupp et al., 2012; West et al., 2012), dental practice (Mislevy et al., 2002), systems thinking (Mislevy et al., 2014; Shute et al., 2010), creative problem-solving (Shute et al., 2009), causal reasoning (Shute, 2011), and twenty-first century skills (Shute & Ventura, 2013). For the CBAs with multiple learning opportunities, DBNs have been used to address the temporal dependencies and model the performance data for CBAs on air combat (Poropudas & Virtanen, 2007), weather phenomenon (Cui et al., 2019), mathematics (Levy, 2014), and Navy damage control operations

(Iseli et al., 2010; Koenig et al., 2010). Bayesian networks and DBNs were successfully applied in various CBAs in a wide range of application domains. However, in pursuit of strong flexibility, Bayesian networks are used subject to the curse of dimensionality. That is, a Bayesian network might involve a great number of latent variables, which results in complex computation of the conditional probabilities. To address this, Bayesian networks can be constructed with simplifying assumptions or in a data-driven way that handles the observable variables only and reduces the network complexity (Desmarais & Baker, 2012).

12.3 Bayesian Knowledge Tracing

BKT (Corbett & Anderson, 1994) is a learning outcome modeling approach extensively used in the community of educational data mining, especially for tracking learners' changes of cognitive states over time in intelligent tutoring systems. Essentially, BKT is a constrained and simplified version of DBN, where the number of conditional probabilities is reduced for modeling.

Concretely, given the DBN case shown in Figure 12.3, the item has either correct or incorrect state, whereas the skill has either a mastery or a non-mastery state. As such, the horizontal directional edges represent the transition probability from non-mastery to mastery of the skill, denoted as $p(T)$, and the transition probability from mastery to non-mastery of the skill, denoted as $p(F)$. The transition probability, $p(T)$, indicates the learning of the skill and the transition probability, $p(F)$, indicates the forgetting of the skill. The vertical directional edges represent the emission probability of incorrectly answering the item given a mastery of the skill, denoted as $p(S)$, and the emission probability of correctly answering the item given a non-mastery of the skill, denoted as $p(G)$. The emission probability $p(S)$ is the *slip* probability and the emission probability $p(G)$ is the *guess* probability. Moreover, the DBN requires a definition of the prior probability of mastering the skill, which is denoted by $p(L_0)$. These conditional probabilities can be used to calculate the conditional probabilities of mastery or non-mastery of the skill given correct or incorrect responses, which are in turn used to infer the probabilities of mastery or non-mastery of the skill at each timestep.

BKT is a special case of the DBN described above, where the transition probability $p(F)$ is fixed as 0, indicating an assumption that learners will never forget the learned skill. The model parameters of BKT are graphically represented in Figure 12.2. Despite two ovals and two squares shown in Figure 12.2, they represent the two states of the skill and the item, rather than two skills and two items. Given the above definitions of model parameters, BKT estimates the conditional probabilities of mastering the skill given either correct or incorrect responses at the timestep t as follows:

$$p(L_t \mid \text{response} = \text{correct}) = \frac{p(L_{t-1})[1-p(S)]}{p(L_{t-1})[1-p(S)]+p(G)[1-p(L_{t-1})]}$$

$$p(L_t \mid \text{response} = \text{incorrect}) = \frac{p(L_{t-1})p(S)}{p(L_{t-1})p(S)+[1-p(G)][1-p(L_{t-1})]}. \quad (12.9)$$

Having the two conditional probabilities, BKT proceeds to estimate the probability of mastering the skill at timestep t given a learner's correct or incorrect response as follows:

$$p(L_t) = p(L_t \mid \text{response}) + [1-p(L_t \mid \text{response})]p(T). \quad (12.10)$$

It should be noted that each latent skill should be estimated and updated by a different BKT model, indicating that different skills must work independently in influencing learners' item response in BKT.

Consistent with DBN, standard BKT cannot be used to directly model item–skill associations because the items for each skill are required to be known for the estimation of conditional probabilities. Nevertheless, Lindsey et al. (2014) showed the potential of BKT for discovering item–skill associations by developing a BKT-based generative probabilistic model with experts' knowledge as a prior.

12.4 Additive Factors Model

Additive Factors Model (AFM; Cen et al., 2005, 2006) is a statistical model proposed in the community of educational data mining for modeling learners' probabilities of correctly answering items. Similar to IRT models, AFM estimates learners' cognitive states for a given skill, which are converted to probabilistic predictions of item responses by a logistic function (i.e., the sigmoid transformation). However, unlike IRT models, AFM can be considered as an alternative to BKT for sequential modeling of learners' changes of cognitive states over time. Concretely, the AFM models learner i's probability of correctly answering item j as follows:

$$P(R_{ij}=1\mid\theta_i)=\frac{1}{1+e^{-(\theta_i+\sum_{k=1}^{K}\beta_k q_{jk}+\sum_{k=1}^{K}\gamma_k q_{jk} t_{ik})}}, \quad (12.11)$$

where θ_i indicates learner i's latent ability level, β_k indicates the easiness of the skill $k \in \{1,\ldots,K\}$, γ_k denotes the learning rate of the skill k, q_{jk} indicates whether skill k is measured by item j, t_{jk} denotes the total number of learning opportunities learner i has previously accessed for practicing skill k, and K is the number of latent skills measured by the assessment.

Given the above formulation, it can be seen that, because AFM accounts for multiple learning opportunities (i.e., t_{ik}), learners' progress of learning the latent skills can be tracked with AFM, which is a major difference from other logistic function-based models. Moreover, a prespecified mapping of items to skills (i.e., q_{jk}) is required for constructing AFM, indicating that AFM cannot be used for domain modeling which learns item–skill associations from scratch.

12.5 Deep Learning for Learning Outcome Modeling

In recent years, deep learning has received a great deal of attention for its predictive capacity in a wide range of applications domains. Deep learning can be defined as 'a class of machine learning techniques that exploit many layers of non-linear information processing for supervised or unsupervised feature extraction and transformation, and for pattern analysis and classification' (Deng & Yu, 2014). As a subfield of machine learning, deep learning automatically makes predictions or decisions based on learning labeled or unlabeled sample data. In terms of learning outcome modeling, a deep learning–based model is often developed to predict learners' item responses for each item. Therefore, the output of the deep learning model should be item responses, such as correct or incorrect scores. However, the input for the deep learning model can be various. For example, given a simple item response matrix without any other information, a deep learning model can simply learn the identifications of each item and each learner as input. Given the availability of more information regarding the items and the learners (e.g., item text, learner background information), a deep learning model can learn the additional information as input. In deep learning, how the input is integrated, analyzed, and learned for outputting the final predictions is determined by deep learning architectures.

In a nutshell, deep learning is essentially an extension of artificial neural networks with multiple hidden layers. There are a variety of deep learning architectures developed for different application domains. Two most fundamental architectures of deep learning, which are closely pertinent to the topic of learning outcome modeling, are deep neural networks and recurrent neural networks (RNNs). The former is typically used to capture the complexity of the input data and the latter is typically used to model the temporal dependencies between multiple timesteps of the input data.

Both deep neural networks and RNNs can be considered as a neural network with more than one hidden layer. The predictive capacity and the model complexity are controlled by the number of hidden layers and the number of hidden nodes in each layer, both of which are hyperparameters to be tuned for a deep learning architecture. For a simple classification problem, according to the recommendations by Goodfellow et al. (2016, p. 192) and Lippmann (1987), one or two hidden layers is sufficient for a neural network. However, to capture the complexity of training data, a deep learning architecture typically uses more than one hidden layer. Regarding how to determine the optimal numbers of hidden layers and the hidden nodes for each hidden layer, for a simple model structure and a small training sample size, a k-fold cross-validation technique can be possibly used to optimize the model. However, in practice, given the complexity of model architecture and the large sample size, the k-fold cross-validation is infeasible and there are no rules of thumb for determining the two hyperparameters. Alternatively, it is suggested to configure the numbers of hidden layers and the hidden nodes by trial and error until satisfactory prediction accuracy is met (Goodfellow et al., 2016).

12.5.1 Deep Knowledge Tracing

In the context of educational data mining, a representative deep learning–based approach for learning outcome modeling is deep knowledge tracing (DKT; Piech et al., 2015). Given a sequence of learners' item responses for multiple skills, in essence, DKT predicts a specific item response through learning the temporal dependencies between item responses prior to the current one based on RNNs. Figure 12.4 demonstrates a graphical representation of DKT. In the diagram, the multiple opportunities of applying each skill are connected with an RNN for learning their temporal dependencies. The input data for the DKT framework is the information regarding which skills a learner accessed and what the outcomes of answering their associated items were (i.e., correct or incorrect). It should be noted that DKT also works at the item level. That said, the inputs for DKT can also be the items and associated responses. If modeling at the skill level, DKT does not recognize the differences between items measuring the same skill. In DKT, each item can be considered as a learning opportunity for practicing its associated skill. For example, for the first item response in the diagram, it shows that the learner practiced skill 2 and had a correct answer on an item measuring skill 2. Subsequently, the learner proceeded to practice skill 1 and answered an item measuring skill 1 correctly. Then the learner practiced skill N but incorrectly answered its associated item. The outputs of DKT are the predicted probabilities of correctly answering the items for each skill. For example, at the first timestep, the learner's item response for skill 2 is fed into DKT to produce his or her probabilities of getting items correct for each skill. Because at the second timestep the learner practiced skill 1, the first element of the vector of predicted probabilities can be used to infer his or her success likelihood of getting an item for skill 1 correct. In addition to dropout rate, batch size, and learning rate, the key hyperparameter to be tuned for DKT is the hidden dimensionality of the RNN. In addition, for domain modeling, standard DKT can be used to model the relationships among skills or item–skill associations depending on skill- or item-level modeling.

After DKT was proposed, in recent years, several studies have examined the performance of DKT and extended the DKT framework for more complex learning outcome modeling problems. For example, Xiong et al. (2016) reexamined the performance of DKT with performance factors analysis and BKT as baselines on multiple datasets and found that DKT outperformed the baselines. Moreover, DKT was adapted to model open-ended item responses such as programming exercises (Wang et al., 2017), revised by introducing regularized loss function to enhance the prediction consistency (Yeung & Yeung, 2018), and integrated with psychometric models such as IRT to improve its interpretability (Yeung, 2019).

12.5.2 Other Deep Learning Approaches for Learning Outcome Modeling

In addition to DKT, in recent years, deep learning architectures are often incorporated in other approaches for learning outcome modeling. For example, as mentioned in the section 'Additive Factors Model', the LSTM network can be incorporated with AFM to dynamically model learners' item responses for refining or learning from scratch the expert-specified item–skill associations (Pardos & Dadu, 2018). Moreover, based on deep learning architectures, the item–skill associations can be derived without learning learner product data. For example, Chaplot et al. (2018) proposed a framework named Cognitive Representation Learner to automatically extract the skills required by each item through learning the representations of item text or item content based on convolutional neural networks or RNNs. As stated by the authors, their framework is capable of discovering item–skill associations without any learner product data, which is especially beneficial for items in ill-structured domains where data and human knowledge both are not available.

In contrast to learning outcome modeling without product data, another stream of research focused on how to exploit a variety of auxiliary information along with learner data for enhanced learning outcome modeling. For example, based on the IRT framework, Cheng et al. (2019) proposed that item content and item-associated latent skills can be learned by deep neural networks and LSTM networks to automatically generate item difficulties, item discriminations, and learners' latent ability levels. These learned item and learner parameters are then used to produce the predicted probabilities of correct item responses. Their framework was found to outperform conventional IRT models because more information is exploited for estimating the model parameters. Moreover, Su et al. (2018) developed a sequential modeling framework based on the LSTM network for predicting learners' item responses based on their history item responses. Particularly, their framework integrates a representation learning architecture for exploiting the item content associated with each item response, which contributes to higher predictive capacity in comparison with conventional approaches such as IRT, BKT, and DKT. Furthermore, deep learning techniques showed potential of detecting learners' affective states for learner modeling. In contrast to traditional affective detection approaches leveraging physical and physiological sensors, Botelho et al. (2017) developed a novel sensor-free affect detector based on multiple RNN variants to automatically recognize learners' affective states from their interactions with the system for learner modeling, which demonstrated higher prediction performance than conventional machine learning–based approaches.

12.6 Collaborative Filtering for Learning Outcome Modeling

Collaborative filtering is a promising approach for learning outcome modeling examined by an increasing number of studies in recent years (e.g., Almutairi et al., 2017; Desmarais & Naceur, 2013; Durand et al., 2015; Lan et al., 2014; Matsuda et al., 2015). CF is originally used for recommender systems, but its idea has been extended to address issues in other domains such as disease diagnosis (Shen et al., 2017), and online learning (Wang & Yang, 2012). CF makes recommendations for a user on new items based on the fundamental assumption that if two users have similar behaviors on items (e.g., similar item responses, buying or watching decisions), their behaviors on other items are also similar (Goldberg et al., 2001). The CF approaches deal with a dataset to make recommendations in the following form: there are a set of items and a list of users, and each user has a value on partly or all of the items. Represented as a user by item matrix, the dataset looks like a sparse matrix (i.e., a matrix with many missing entries). Those missing entries are the values to be predicted by the CF approaches. It should be noted that the values can be both explicit and implicit. Explicit values refer to quantified item responses such as ratings ranging from 1 to 5 or scores of 0 and 1. Implicit values refer to unquantified item responses such as actions of buying an item, watching a movie, or clicking an item.

According to the review by Su and Khoshgoftaar (2009), there are three categories of CF approaches: memory-based CF, model-based CF, and hybrid recommenders. The memory-based CF approach makes recommendations through computing the similarity between users or items. For example, given the user-based top-N recommendation algorithm, a user's predicted rating on an item is simply the aggregated ratings on the item provided by some other users who are most similar to the user in the dataset. Likewise, given the item-based top-N recommendation algorithm, an item's predicted rating by a user is simply the aggregated ratings by the user on some other items which are most similar to the item in the dataset. The key idea of memory-based CF approaches is to quantify the similarity between users and items, which can be calculated by a variety of measures (see Su & Khoshgoftaar, 2009). Memory-based CF approaches are easy to implement, but suffers disadvantages such as depending on human ratings, not working well for sparse data, cold-start problems, and limited scalability (Wang et al., 2014; Zhang et al., 2020). The model-based CF approaches can be used to address these disadvantages as they are developed based on a variety of data mining and machine learning models such as Bayesian networks, matrix factorization, clustering algorithms, and regression models (Aggarwal, 2016; Mehta & Rana, 2017). Compared with the memory-based CF approaches, the model-based CF approaches are more complex for computation, but they are more scalable, more capable of dealing with sparse data, and more accurate in prediction. The hybrid recommenders are combinations of CF approaches and content-based recommenders (Dong et al., 2017; Kumar & Fan, 2015; Zhang et al., 2017). The content-based recommenders make recommendations based on the analysis of a variety of contextual information and item content but are not scalable and suffer the cold-start problem. As such, the hybrid recommenders integrate the advantages of CF approaches and the use of contextual information for elevated prediction accuracy.

In the next section, a widely used model-based CF approach, matrix factorization, is introduced, given its great popularity in recommenders systems. In the following, 'users' will be replaced by 'learners' when describing the technical details because of the context of the current research topic.

12.6.1 Matrix Factorization

Matrix factorization is exceptionally effective for building recommender systems (Koren et al., 2009). In essence, matrix factorization deals with a sparse high dimensionality learner–item matrix with missing responses by introducing a set of latent factors for dimensionality reduction. The association between latent factors and learners and the association between latent factors and items are two lower dimensionality matrices factorized from a complete or incomplete learner–item matrix. Mathematically, through matrix factorization, an item response matrix $R \subset \mathbb{R}^{m \times n}$ of m learners and n items can be decomposed into two low-rank matrices $U \in \mathbb{R}^{m \times k}$ and $V \in \mathbb{R}^{n \times k}$:

$$R \approx UV^T. \tag{12.12}$$

The former is a learner-by-factor matrix representing the learner–skill associations and the latter is an item by factor matrix representing the item–skill associations. The dimension k indicates that there is a total of k latent factors modeled by matrix factorization, which is a key hyperparameter to be tuned in training.

In the matrix U, the ith row $\overrightarrow{u_i}$ indicates the associations between a learner and the k latent factors; in the matrix V, the jth row $\overrightarrow{v_j}$ indicates the associations between an item and the k latent factors. Therefore, an entry of the item response matrix, r_{ij}, can be approximately recovered by the dot product of the learner factor $\overrightarrow{u_i}$ and the item factor $\overrightarrow{v_j}$:

$$r_{ij} \approx \overrightarrow{\mathrm{u_i}} \cdot \overrightarrow{v_j}. \tag{12.13}$$

Estimating the lower dimensionality matrices in matrix factorization can also be formulated as a maximum likelihood estimation problem like most machine learning models. That is, we seek two lower dimensionality matrices U and V that minimize the differences between the original item response matrix entries and the predicted item response matrix entries given by Equation 23. The gradient descent algorithms can be used for optimization to solve the problem. Similar to other machine learning or deep learning models, as mentioned in the section of deep neural networks, the regularization technique can be used to decay large latent factor values to prevent or reduce overfitting in matrix factorization. For example, the L_2 regularization technique can be applied to the matrices U and V, which changes the problem as minimizing the following objective function:

$$\arg\min_{U,V} J = \frac{1}{2}R - UV^{T2} + \frac{\lambda}{2}U_F^2 + \frac{\lambda}{2}V_F^2, \qquad (12.14)$$

where λ denotes the regularization weight controlling the degree to which the latent factor values are decayed, and $\bullet_F^2$ represents the Frobenius norm.

The above formulation of matrix factorization stands for its most basic form, which is not typically used in practice. This is due to that given no model constraints, the solution of lower dimensionality matrices U and V are hard to be fixed, which leads to an ill-posed problem. Therefore, various model constraints per application domain are typically imposed on matrix factorization to enhance prediction performance and interpretability. Some representative model constraints for matrix factorization are non-negativity, orthogonality, and sparseness of model weights (e.g., Ding et al., 2006; Hoyer, 2004; Lee & Seung, 2001). For example, the L_1 regularization technique is typically used to encourage the matrix sparseness. However, regardless of model constraints, the matrix factorization based approaches are generally developed to stably estimate meaningful lower dimensionality representations of users/learners and items in terms of how they are connected with a limited number of latent factors.

Moreover, the matrix factorization approach, in addition to being popular, is mainly described for the purpose of demonstrating how the CF approaches make inferences about item responses based on learning the lower dimensionality representations of learners and items. How learner and item latent representations contribute to the predicted probabilities of item responses is not necessarily modeled by the matrix multiplication. For example, the interaction between the two representations can also be learned through deep neural networks. More details about other relevant work will be given in the next section.

12.6.2 Collaborative Filtering-Based Approaches for Learning Outcome Modeling

Overall, the application of CF in education is still in its infancy. In the past ten years, a growing number of studies are advancing the field by proposing a variety of CF-based approaches for analyzing learner data. Particularly, since matrix factorization can represent items with latent factors, the majority of these approaches were developed with a focus on learning from scratch or refining item–skill associations based on learner data. For example, several studies (e.g., Desmarais, 2012; Desmarais & Naceur, 2013; Durand et al., 2015; Matsuda et al., 2015; Sun et al., 2014) used the CF framework to evaluate expert-specified Q-matrices or automatically generate data-driven Q-matrices. In their work on evaluating the predictive capacity of expert-specified Q-matrices, Durand et al. (2015) proposed an evaluation method based on cross-validation and found that their approach was capable of efficiently and quickly evaluating expert-specified Q-matrices without complex computation of multiple model parameters as in sophisticated CDMs. Desmarais (2012) examined the potential of non-negative matrix factorization for recovering the Q-matrix and found that it is a highly effective approach for deriving the Q-matrix given the assumption of skill independence, but it is less effective if the values of learner, item, and skill parameters vary a lot in the data. Desmarais and Naceur (2013) compared the performance between the expert-specified Q-matrix and the data-driven Q-matrix by matrix factorization and found that they shared similar patterns between item–skill mappings, but the matrix factorization approach lightly outperformed the expert-specified Q-matrix. Similarly, Sun et al. (2014) found that their proposed approach based on the Boolean matrix factorization could successfully recover the original Q-matrix from learner data. In the context of large-scale online courses, compared with the expert-specified Q-matrix, the approach developed by Matsuda et al. (2015) based on the matrix factorization framework was found to be faster, more predictive, and more scalable for discovering item–skill associations.

To sum up, the previous work on learning the Q-matrix from learner data generally show that CF approaches, especially matrix factorization, could be successfully used for domain modeling and they had the potential of outperforming the expert knowledge in some contexts (e.g., Desmarais & Naceur, 2013; Matsuda et al., 2015). However, researchers also indicated that matrix factorization–based approaches such as alternating least square, non-negative matrix factorization and Boolean matrix factorization still showed limited capacities to learn the expert-specified

Q-matrix from scratch (Desmarais, 2011; Desmarais & Naceur, 2013), and they could be more effectively used to refine expert-specified Q-matrices.

The above studies mainly focused on learning the Q-matrix from the learner data in comparison with the original expert-specified Q-matrix. Studies also strived to learn item–skill associations from scratch leveraging the idea of matrix factorization. A representative work is the sparse factor analysis algorithm proposed by Lan et al. (2014), which is capable of learning item–skill and learner–skill associations and item difficulties from binary-valued item responses without any auxiliary information. Their approach showed strong predictive capacity and interpretability. Moreover, another stream of research emphasized the usefulness of contextual information in learning outcome modeling. For example, based on matrix factorization and tensor factorization under the CF framework, Almutairi et al. (2017) proposed three methods to model students' grade data and found that the time when a learner was graded was helpful for improving the prediction performance. Similarly, Sahebi et al. (2016) used learners' interactions with the learning resources to model their learning progress based on the tensor factorization approach and found that their approach was significantly more predictive of learner performance than BKT and another tensor factorization approach. In addition, the sequential modeling approach based on tensor factorization proposed by Thai-Nghe et al. (2012) was successfully used to predict learners' future item responses based on learning history item responses.

12.6.3 Deep Learning–Based Collaborative Filtering

In recent years, informed by the deep learning advances, more and more studies have focused on incorporating the CF framework with deep learning architectures to improve model predictive capacity. Despite not being proposed specific to learning outcome modeling, these deep learning–based CF approaches are very promising in learner modeling and domain modeling. Most of the deep learning–based CF approaches utilize a deep neural network architecture to learn more complex or non-linear interactions between learner and item representations for prediction. For example, in the 'two-stream neural network architecture for matrix completion' proposed by Nguyen et al. (2018), rows and columns of a user–item matrix, which represent user and item vectors, are separately fed into multiple neural network layers to learn more effective item and learner representations, which can be extended to new users and new items. In the deep CF framework proposed by Li et al. (2015), the item and user representations are learned through marginalized denoising stacked autoencoders based on additional sources of information on users and items, which are in turn incorporated into the matrix factorization framework for prediction. In the neural CF framework developed by He et al. (2017), solely based on the user–item rating matrix, the concatenation of user and item representations (user and item embeddings) is fed into multiple neural network layers to learn the non-linear interactions between users and items, which are incorporated into a generalizable matrix factorization framework for prediction. These CF approaches based on deep learning architectures generally outperformed other state-or-art methods in terms of prediction performance.

In summary, compared with other types of CF approaches, deep learning–based CF approaches are more capable of capturing the complexity of interactions between learners and items in affecting item responses. This means that the model predictive capacity benefits from the finer-grained representations learned by deep learning as more information can be extracted to know about learners and items for prediction. Moreover, deep learning architectures are exceptionally effective for learning additional information about learners and items, such as learners' background information, item content, and potentially, the process data associated with item responses. Leveraging the representation of these additional information, the CF approaches can be improved in terms of two aspects. First, they can predict missing responses with higher accuracy because the system knows items and learners better. Second, with respect to learner modeling and domain modeling, the learner and item representations can be refined as well through learning additional information. For example, item–skill associations can be more accurately estimated through learning learners' actions and time durations for answering each item. Methodologically, adding more information in model learning can be considered as a regularization technique which makes the weights of item–skill associations more stable, interpretable, and generalizable. This is a desirable feature for learner modeling and domain modeling. Unfortunately, very few, if any at all, established deep learning–based CF approaches were developed specifically for learning outcome modeling and process data learning. Investigations of deep learning–based CF approaches for learning outcome modeling are acutely needed.

12.7 An Overview of Approaches for Learning Outcome Modeling

In previous sections, we introduced the mainstream approaches for learning outcome modeling in the literature (see an overview in Table 12.2). In general, these approaches can be categorized as either psychometric or educational data mining approaches. Both types of approaches aim to infer the degree to which learners understand or master the constructs or skills required by an assessment based on their interactions with the assessment questions. However, how they indicate learner differences in cognitive states may differ. For example, psychometric approaches typically assume that an assessment requires one or several latent constructs and learners' cognitive states are quantified by a continuous scale or indicated by either presence or absence of the required constructs. Bayesian approaches, however, estimate the probabilities of mastery or non-mastery for each skill based on probabilistic inferences. Regarding machine learning and deep learning approaches such as DKT- and CF-based approaches, learners' cognitive states are typically indicated by their predicted probabilities of correctly solving assessment questions. This feature of machine learning and deep learning approaches is particularly useful for large-scale learning and assessment settings where learners' future learning outcomes need to be inferred based on their current and history learning activities, which are used to facilitate recommendations of tailored learning materials and assessment questions. However, compared with psychometric approaches, these approaches are less interpretable in learner modeling as practitioners are not able to directly quantify and interpret learners' cognitive states.

Another major distinction between the two types of approaches for learning outcome modeling is how they model the associations between assessment questions and latent skills. In general, psychometric models and Bayesian approaches require a prespecified mapping of assessment questions to latent constructs or skills. For example, standard cognitive diagnosis models require an accurate Q-matrix to be defined by domain experts, otherwise the estimation of model parameters and learner skill mastery profiles would be biased. However, for most machine learning and deep learning approaches, a predefined mapping of items to skills, or a predefined set of latent skills, is not necessarily required. This feature of machine learning and deep learning approaches is especially beneficial for large-scale learning and assessment settings because very few human experts are needed to investigate and label a huge number of assessment questions, which is cost-effective for assessment development.

In terms of data variety, in general, psychometric and Bayesian approaches are less capable than machine learning and deep learning approaches for addressing unstructured learner data. Typically, psychometric models require dichotomous or polytomous item responses and some can deal with unstructured learner data such as response time (Lee & Chen, 2011). However, in the context of computer-based assessments for learning, learner data might include a variety of information on learner–assessment interaction, such as learner background information, item content, and process data associated with final item responses. This type of data cannot be directly analyzed by psychometric and Bayesian models but can be handled by machine learning and deep learning approaches in a flexible way. Therefore, psychometric and Bayesian approaches are limited in generalizability and scalability in terms of unstructured learner data modeling.

The computational demands required by each approach are dependent on the model complexity, the availability of computing power, and the data volume and variety. In general, compared with psychometric approaches, Bayesian approaches are much more computationally intensive when complex models are needed to handle multiple items and skills. This is because they require complex computation of great amounts of conditional probabilities, which is a challenge for computational resources. However, given a simple model with model constraints (e.g., BKT), computational demands of Bayesian approaches might decrease. Compared with other approaches for learning outcome modeling, machine learning approaches, especially deep learning approaches, are intrinsically more computationally demanding. This is because deep learning models typically have a great number of parameters, which require a big amount of data for training and testing. This problem becomes especially severe when a deep learning model has more model parameters than available data points, which is very common in real-life deep learning applications. For example, given a deep learning–based CF model specific to learning outcome modeling, a complex model architecture involving multiple layers of deep neural networks and RNNs might be needed to address both structured and unstructured learner data such as item responses, item content, and in-process actions and time for solving each item. In this case, data of a great number of learners from a variety of assessment settings are needed to address the model complexity to learn an accurate, reliable, and generalizable model.

12.8 Conclusion

Over the past decades, computer-based assessments for learning are increasingly popular in the education

TABLE 12.2

A Summary Table of Key Approaches for Learning Outcome Modeling

Approach	Temporal Modeling	Learner Modeling	Domain Modeling	Multiple Skills	Multiple Items	Major Challenges	Citations
CTT	No	Yes	No	No	Yes	Item and learner dependent; untestable assumptions; high measurement error; limited use in the context of CBAs	Spearman (1904)
IRT	No	Yes	No	No	Yes	Unidimensionality; strong assumptions; requires structured and complete data; no temporal modeling	Reise et al. (2005); Reckase (1997); Yao and Boughton (2007); Desmarais and Baker (2012)
Cognitive diagnosis	No	Yes	No	Yes	Yes	Hard to specify accurate Q-matrices; strong assumptions; requires structured and complete data; no temporal modeling	Rupp et al. (2010); Henson et al. (2009); Tatsuoka (1983); von Davier (2005); Haertel (1989); Junker and Sijtsma (2001); Hartz (2002); Leighton et al. (2004)
Bayesian network	No	Yes	No	Yes	Yes	High demands on computational resources; requires expert knowledge on model construction; no temporal modeling	de Klerk et al. (2015); Koller and Friedman (2009); Heckerman et al. (1995); Desmarais and Baker (2012); Levy and Mislevy (2004); Levy (2013); Rupp et al. (2012); West et al. (2012); Mislevy et al. (2002); Mislevy et al. (2014); Shute et al. (2010); Shute et al. (2009); Shute (2011); Shute and Ventura (2013)
DBN	Yes	Yes	No	Yes	Yes	High demands on computational resources; requires expert knowledge on model construction	Koller and Friedman (2009); Poropudas and Virtanen (2007); Cui et al. (2019); Levy (2014); Iseli et al. (2010); Koenig et al. (2010)
BKT	Yes	Yes	No	No	Yes	Only models one skill; skill-level modeling without item parameters	Corbett and Anderson (1994); Lindsey et al. (2014)
AFM	Yes	Yes	No	Yes	Yes	Hard to specify accurate Q-matrices; skill-level modeling without item parameters	Cen et al. (2005); Cen et al. (2006)
DKT	Yes	Yes	Yes	Yes	Yes	Deals with structured item responses only	Piech et al. (2015); Xiong et al. (2016); Wang et al. (2017); Yeung and Yeung (2018); Yeung (2019)
Deep learning	Yes	Yes	Yes	Yes	Yes	Requires sophisticated design of model architecture; requires large amounts of data; requires large amounts of computational resources	Deng and Yu (2014); Pardos and Dadu (2018); Chaplot et al. (2018); Cheng et al. (2019); Su et al. (2018); Botelho et al. (2017)
Matrix factorization	Yes	Yes	Yes	Yes	Yes	Limited in recovering item–skill associations; limited interpretability of model weights	Koren et al. (2009); Desmarais (2012); Desmarais and Naceur (2013); Durand et al. (2015); Matsuda et al. (2015); Sun et al. (2014); Desmarais (2011); Lan et al. (2014); Almutairi et al. (2017); Sahebi et al. (2016); Thai-Nghe et al. (2012)
Deep learning–based CF	Yes	Yes	Yes	Yes	Yes	Requires sophisticated design of model architecture; requires large amounts of data; requires large amounts of computational resources	Nguyen et al. (2018); Li et al. (2015); He et al. (2017)

Note: The approaches summarized in the table mostly refer to their standard forms. Their variants might have different features with respect to the indicators in the table. Domain modeling refers to estimating or refining item–skill associations.

sector. As a technical underpinning, learning outcome modeling techniques play a vital role in successful learner evaluation. In this chapter, we reviewed the mainstream learning outcome modeling techniques from both communities of psychometric measurement and educational data mining. We pointed out that psychometric models such as IRT and cognitive diagnosis require strong theoretical assumptions about how latent skills take effects on learners' item responses, and they are limited in dealing with unstructured learner data which is most accessible in the context of digital learning. This chapter also highlighted the computational demands and prior human knowledge about item and skill relations required by the Bayesian family approaches. Moreover, we emphasized the effectiveness of deep learning, especially deep learning–based CF approaches, for learning outcome modeling given their higher prediction accuracy, flexibility, generalizability, and scalability than conventional approaches. Furthermore, an increasing body of research has focused on modeling learner process data for enhanced learning outcome modeling in recent years (e.g., Chen & Cui, 2020). We encourage future studies to further investigate how deep learning approaches can be better used to model unstructured learner data.

References

Aggarwal, C. C. (2016). Model-based collaborative filtering. In *Recommender systems* (pp. 71–138). Springer. https://doi.org/10.1007/978-3-319-29659-3_3

Almutairi, F. M., Sidiropoulos, N. D., & Karypis, G. (2017). Context-aware recommendation-based learning analytics using tensor and coupled matrix factorization. *IEEE Journal of Selected Topics in Signal Processing*, *11*(5), 729–741. https://doi.org/10.1109/JSTSP.2017.2705581

Azevedo, R., Johnson, A., Chauncey, A., & Burkett, C. (2010). Self-regulated learning with MetaTutor: Advancing the science of learning with metacognitive tools. In M. Khine & I. Saleh (Eds.), *New science of learning* (pp. 225–247). Springer.

Bell, S. (2010). Project-based learning for the 21st century: Skills for the future. *Clearing House*, *83*(2), 39–43. https://doi.org/10.1080/00098650903505415

Black, P., & Wiliam, D. (2009). Developing the theory of formative assessment. *Educational Assessment, Evaluation and Accountability*, *21*(1), 5–31. https://doi.org/10.1007/s11092-008-9068-5

Blanchard, E. G., Wiseman, J., Naismith, L., & Lajoie, S. P. (2012). A realistic digital deteriorating patient to foster emergency decision-making skills in medical students. In *Society 12th IEEE international conference on advanced learning technologies* (pp. 74–76). IEEE Communications Society. https://doi.org/10.1109/ICALT.2012.44

Botelho, A. F., Baker, R. S., & Heffernan, N. T. (2017). Improving sensor-free affect detection using deep learning. In *International conference on artificial intelligence in education* (pp. 40–51). Springer. https://doi.org/10.1007/978-3-319-61425-0_4

Cen, H., Koedinger, K., & Junker, B. (2005). Automating cognitive model improvement by A*search and logistic regression. In *Proceedings of the AAAI 2005 educational data mining workshop*. https://www.aaai.org/Papers/Workshops/2005/WS-05-02/WS05-02-007.pdf

Cen, H., Koedinger, K., & Junker, B. (2006). Learning factors analysis – A general method for cognitive model evaluation and improvement. In *Intelligent tutoring systems. ITS 2006. Lecture notes in computer science* (Vol. 4053, pp. 164–175). Springer. https://doi.org/10.1007/11774303_17

Chaplot, D. S., MacLellan, C., Salakhutdinov, R., & Koedinger, K. (2018). Learning cognitive models using neural networks. In *International conference on artificial intelligence in education* (pp. 43–56). Springer. https://doi.org/10.1007/978-3-319-93843-1_4

Chatzopoulou, D. I., & Economides, A. A. (2010). Adaptive assessment of student's knowledge in programming courses. *Journal of Computer Assisted Learning*, *26*(4), 258–269. https://doi.org/10.1111/j.1365-2729.2010.00363.x

Chen, F., & Cui, Y. (2020). LogCF: Deep collaborative filtering with process data for enhanced learning outcome modeling. *Journal of Educational Data Mining*, *12*(4), 66–99. https://doi.org/10.5281/zenodo.4399685

Cheng, S., Liu, Q., Chen, E., Huang, Z., Huang, Z., Chen, Y., Ma, H. & Hu, G. (2019). Dirt: Deep learning enhanced item response theory for cognitive diagnosis. In *Proceedings of the 28th ACM international conference on information and knowledge management* (pp. 2397–2400). Association for Computing Machinery. https://doi.org/10.1145/3357384.3358070

Corbett, A. T., & Anderson, J. R. (1994). Knowledge tracing: Modeling the acquisition of procedural knowledge. *User Modeling and User-Adapted Interaction*, *4*(4), 253–278. https://doi.org/10.1007/BF01099821

Cui, Y., Chu, M. W., & Chen, F. (2019). Analyzing student process data in game-based assessments with Bayesian knowledge tracing and dynamic Bayesian networks. *Journal of Educational Data Mining*, *11*(1), 80–100. https://doi.org/10.5281/zenodo.3554751

Davies, J., & Ecclestone, K. (2008). 'Straitjacket' or 'springboard for sustainable learning'? The implications of formative assessment practices in vocational learning cultures. *Curriculum Journal*, *19*(2), 71–86. https://doi.org/10.1080/09585170802079447

de Klerk, S., Veldkamp, B. P., & Eggen, T. J. (2015). Psychometric analysis of the performance data of simulation-based assessment: A systematic review and a Bayesian network example. *Computers and Education*, *85*, 23–34. https://doi.org/10.1016/j.compedu.2014.12.020

Deng, L., & Yu, D. (2014). Deep learning: Methods and applications. *Foundations and Trends in Signal Processing*, *7*(3–4), 197–387. http://doi.org/10.1561/2000000039

Desmarais, M. (2011). Conditions for effectively deriving a q-matrix from data with non-negative matrix factorization. In *Proceedings of the 4th international conference on educational data mining* (pp. 41–50). International Educational Data Mining Society.

Desmarais, M. C. (2012). Mapping question items to skills with non-negative matrix factorization. *ACM SIGKDD Explorations Newsletter, 13*(2), 30–36. https://doi.org/10.1145/2207243.2207248

Desmarais, M. C., & Baker, R. S. D. (2012). A review of recent advances in learner and skill modeling in intelligent learning environments. *User Modeling and User-Adapted Interaction, 22*(1–2), 9–38. https://doi.org/10.1007/s11257-011-9106-8

Desmarais, M. C., & Naceur, R. (2013). A matrix factorization method for mapping items to skills and for enhancing expert-based q-matrices. In *International conference on artificial intelligence in education* (pp. 441–450). Springer. https://doi.org/10.1007/978-3-642-39112-5_45

Ding, C., Li, T., Peng, W., & Park, H. (2006). Orthogonal nonnegative matrix t-factorizations for clustering. In *Proceedings of the 12th ACM SIGKDD international conference on knowledge discovery and data mining* (pp. 126–135). Association for Computing Machinery. https://doi.org/10.1145/1150402.1150420

Dong, X., Yu, L., Wu, Z., Sun, Y., Yuan, L., & Zhang, F. (2017). A hybrid collaborative filtering model with deep structure for recommender systems. In *Proceedings of the thirty-first AAAI conference on artificial intelligence* (pp. 1309–1315).

Durand, G., Belacel, N., & Goutte, C. (2015). Evaluation of expert-based Q-matrices predictive quality in matrix factorization models. In *Design for teaching and learning in a networked world* (pp. 56–69). Springer. https://doi.org/10.1007/978-3-319-24258-3_5

Gikandi, J. W., Morrow, D., & Davis, N. E. (2011). Online formative assessment in higher education: A review of the literature. *Computers and Education, 57*(4), 2333–2351. https://doi.org/10.1016/j.compedu.2011.06.004

Goldberg, K., Roeder, T., Gupta, D., & Perkins, C. (2001). Eigentaste: A constant time collaborative filtering algorithm. *Information Retrieval, 4*(2), 133–151.

Goodfellow, I., Bengio, Y., & Courville, A. (2016). *Deep learning*. MIT Press.

Greiff, S., Kretzschmar, A., Müller, J. C., Spinath, B., & Martin, R. (2014). The computer-based assessment of complex problem solving and how it is influenced by students' information and communication technology literacy. *Journal of Educational Psychology, 106*(3), 666–680.

Haertel, E. H. (1989). Using restricted latent class models to map the skill structure of achievement items. *Journal of Educational Measurement, 26*(4), 301–323. https://doi.org/10.1111/j.1745-3984.1989.tb00336.x

Hartz, S. (2002). *A Bayesian framework for the unified model for assessing cognitive abilities: Blending theory with practicality* [Unpublished doctoral dissertation]. University of Illinois at Urbana-Champaign.

He, X., Liao, L., Zhang, H., Nie, L., Hu, X., & Chua, T. S. (2017). Neural collaborative filtering. In *Proceedings of the 26th international conference on world wide web* (pp. 173–182). International World Wide Web Conferences Steering Committee. https://doi.org/10.1145/3038912.3052569

Heckerman, D., Geiger, D., & Chickering, D. M. (1995). Learning Bayesian networks: The combination of knowledge and statistical data. *Machine Learning, 20*(3), 197–243. https://doi.org/10.1023/A:1022623210503

Henson, R. A., Templin, J. L., & Willse, J. T. (2009). Defining a family of cognitive diagnosis models using log-linear models with latent variables. *Psychometrika, 74*(2), 191–210. https://doi.org/10.1007/s11336-008-9089-5

Hoyer, P. O. (2004). Non-negative matrix factorization with sparseness constraints. *Journal of Machine Learning Research, 5*, 1457–1469.

Iseli, M. R., Koenig, A. D., Lee, J. J., & Wainess, R. (2010). *Automatic assessment of complex task performance in games and simulations*. CRESST Report 775. National Center for Research on Evaluation, Standards, and Student Testing (CRESST).

Joosten-ten Brinke, D., Van Bruggen, J., Hermans, H., Burgers, J., Giesbers, B., Koper, R., & Latour, I. (2007). Modeling assessment for re-use of traditional and new types of assessment. *Computers in Human Behavior, 23*(6), 2721–2741. https://doi.org/10.1016/j.chb.2006.08.009

Junker, B. W., & Sijtsma, K. (2001). Cognitive assessment models with few assumptions, and connections with nonparametric item response theory. *Applied Psychological Measurement, 25*(3), 258–272. https://doi.org/10.1177/01466210122032064

Kiili, K. (2005). Digital game-based learning: Towards an experiential gaming model. *Internet and Higher Education, 8*(1), 13–24. https://doi.org/10.1016/j.iheduc.2004.12.001

Kleitman, S., & Costa, D. S. (2014). The role of a novel formative assessment tool (Stats-mIQ) and individual differences in real-life academic performance. *Learning and Individual Differences, 29*, 150–161. https://doi.org/10.1016/j.lindif.2012.12.001

Koenig, A. D., Lee, J. J., Iseli, M., & Wainess, R. (2010). *A conceptual framework for assessing performance in games and simulations*. CRESST Report 771. National Center for Research on Evaluation, Standards, and Student Testing (CRESST).

Koller, D., & Friedman, N. (2009). *Probabilistic graphical models: Principles and techniques*. The MIT Press.

Koren, Y., Bell, R., & Volinsky, C. (2009). Matrix factorization techniques for recommender systems. *Computer, 42*(8), 30–37. https://doi.org/10.1109/MC.2009.263

Kumar, N. P., & Fan, Z. (2015). Hybrid user-item based collaborative filtering. *Procedia Computer Science, 60*, 1453–1461. https://doi.org/10.1016/j.procs.2015.08.222

Lan, A. S., Waters, A. E., Studer, C., & Baraniuk, R. G. (2014). Sparse factor analysis for learning and content analytics. *Journal of Machine Learning Research, 15*(1), 1959–2008.

Lee, D. D., & Seung, H. S. (2001). Algorithms for non-negative matrix factorization. In *Advances in neural information processing systems* (pp. 556–562). MIT Press. http://papers.nips.cc/paper/1861-algorithms-for-non-negative-matrix-factorization.pdf

Lee, Y. H., & Chen, H. (2011). A review of recent response-time analyses in educational testing. *Psychological Test and Assessment Modeling, 53*(3), 359–379.

Leighton, J. P., Gierl, M. J., & Hunka, S. M. (2004). The attribute hierarchy method for cognitive assessment: A variation on Tatsuoka's rule-space approach. *Journal of Educational Measurement, 41*(3), 205–237. https://doi.org/10.1111/j.1745-3984.2004.tb01163.x

Levy, R. (2013). Psychometric and evidentiary advances, opportunities, and challenges for simulation-based assessment. *Educational Assessment, 18*(3), 182–207. https://doi.org/10.1080/10627197.2013.814517

Levy, R. (2014). *Dynamic Bayesian network modeling of game based diagnostic assessments.* CRESST Report 837. National Center for Research on Evaluation, Standards, and Student Testing (CRESST).

Levy, R., & Mislevy, R. J. (2004). Specifying and refining a measurement model for a computer-based interactive assessment. *International Journal of Testing, 4*(4), 333–369.

Li, S., Kawale, J., & Fu, Y. (2015). Deep collaborative filtering via marginalized denoising auto-encoder. In *Proceedings of the 24th ACM international on conference on information and knowledge management* (pp. 811–820). https://doi.org/10.1145/2806416.2806527

Lindsey, R. V., Khajah, M., & Mozer, M. C. (2014). Automatic discovery of cognitive skills to improve the prediction of student learning. In Z. Ghahramani, M. Welling, C. Cortes, N. D. Lawrence, & K. Q. Weinberger (Eds.), *Advances in neural information processing systems* (pp. 1386–1394). Curran Associates, Inc. http://papers.nips.cc/paper/5554-automatic-discovery-of-cognitive-skills-to-improve-the-prediction-of-student-learning.pdf

López-Pastor, V. M., Pintor, P., Muros, B., & Webb, G. (2013). Formative assessment strategies and their effect on student performance and on student and tutor workload: The results of research projects undertaken in preparation for greater convergence of universities in Spain within the European Higher Education Area (EHEA). *Journal of Further and Higher Education, 37*(2), 163–180. https://doi.org/10.1080/0309877X.2011.644780

Matsuda, N., Furukawa, T., Bier, N., & Faloutsos, C. (2015). Machine beats experts: Automatic discovery of skill models for data-driven online course refinement. In *Proceedings of the 8th international conference on educational data mining* (pp. 101–108). International Educational Data Mining Society.

Meek, S. E., Blakemore, L., & Marks, L. (2017). Is peer review an appropriate form of assessment in a MOOC? Student participation and performance in formative peer review. *Assessment and Evaluation in Higher Education, 42*(6), 1000–1013. https://doi.org/10.1080/02602938.2016.1221052

Mehta, R., & Rana, K. (2017). A review on matrix factorization techniques in recommender systems. In *2017 2nd international conference on communication systems, computing and IT applications (CSCITA)* (pp. 269–274). IEEE. https://doi.org/10.1109/CSCITA.2017.8066567

Mislevy, R. J., Oranje, A., Bauer, M. I., von Davier, A. A., & Hao, J. (2014). *Psychometric considerations in game-based assessment.* GlassLab Games. http://www.instituteofplay.org/work/projects/glasslab-research/

Mislevy, R. J., Steinberg, L. S., Breyer, F. J., Almond, R. G., & Johnson, L. (2002). Making sense of data from complex assessments. *Applied Measurement in Education, 15*(4), 363–389. https://doi.org/10.1207/S15324818AME1504_03

Nguyen, D. M., Tsiligianni, E., & Deligiannis, N. (2018). Extendable neural matrix completion. In *2018 IEEE international conference on acoustics, speech and signal processing (ICASSP)* (pp. 6328–6332). IEEE. https://doi.org/10.1109/ICASSP.2018.8462164

Pardos, Z. A., & Dadu, A. (2018). dAFM: Fusing psychometric and connectionist modeling for Q-matrix refinement. *Journal of Educational Data Mining, 10*(2), 1–27. https://doi.org/10.5281/zenodo.3554689

Pastor, V. M. L. (2011). Best practices in academic assessment in higher education: A case in formative and shared assessment. *Journal of Technology and Science Education, 1*(2), 25–39. http://doi.org/10.3926/jotse.20

Pásztor, A., Molnár, G., & Csapó, B. (2015). Technology-based assessment of creativity in educational context: The case of divergent thinking and its relation to mathematical achievement. *Thinking Skills and Creativity, 18*, 32–42. https://doi.org/10.1016/j.tsc.2015.05.004

Peat, M., & Franklin, S. (2002). Supporting student learning: The use of computer–based formative assessment modules. *British Journal of Educational Technology, 33*(5), 515–523. https://doi.org/10.1111/1467-8535.00288

Piech, C., Bassen, J., Huang, J., Ganguli, S., Sahami, M., Guibas, L. J., & Sohl-Dickstein, J. (2015). Deep knowledge tracing. In *Advances in neural information processing systems 28* (pp. 505–513). Curran Associates, Inc. http://papers.nips.cc/paper/5654-deep-knowledge-tracing.pdf

Poropudas, J., & Virtanen, K. (2007). Analyzing air combat simulation results with dynamic Bayesian networks. In *Proceedings of the 2007 winter simulation conference* (pp. 1370–1377). IEEE. https://doi.org/10.1109/WSC.2007.4419745

Reckase, M. D. (1997). The past and future of multidimensional item response theory. *Applied Psychological Measurement, 21*(1), 25–36. https://doi.org/10.1177/0146621697211002

Reise, S. P., Ainsworth, A. T., & Haviland, M. G. (2005). Item response theory: Fundamentals, applications, and promise in psychological research. *Current Directions in Psychological Science, 14*(2), 95–101. https://doi.org/10.1111/j.0963-7214.2005.00342.x

Rosen, Y., & Tager, M. (2014). Making student thinking visible through a concept map in computer-based assessment of critical thinking. *Journal of Educational Computing Research, 50*(2), 249–270. https://doi.org/10.2190/EC.50.2.f

Rupp, A. A., Nugent, R., & Nelson, B. (2012). Evidence-centered design for diagnostic assessment within digital learning environments: Integrating modern psychometrics and educational data mining. *Journal of Educational Data Mining, 4*(1), 1–10. https://doi.org/10.5281/zenodo.3554639

Rupp, A. A., Templin, J., & Henson, R. (2010). *Diagnostic measurement: Theory, methods, and applications*. Guilford Press.

Sahebi, S., Lin, Y. R., & Brusilovsky, P. (2016). Tensor factorization for student modeling and performance prediction in unstructured domain. In *Proceedings of the 9th international conference on educational data mining* (pp. 502–506). International Educational Data Mining Society.

Shen, F., Liu, S., Wang, Y., Wang, L., Afzal, N., & Liu, H. (2017). Leveraging collaborative filtering to accelerate rare disease diagnosis. In *AMIA annual symposium proceedings* (Vol. 2017, p. 1554). American Medical Informatics Association.

Shute, V. J. (2011). Stealth assessment in computer-based games to support learning. In S. Tobias & J. D. Fletcher (Eds.), *Computer games and instruction* (Vol. 55, pp. 503–524). Information Age Publishers.

Shute, V. J., & Becker, B. J. (2010). Prelude: Issues and assessment for the 21st century. In V. J. Shute & B. J. Becker (Eds.), *Innovative assessment for the 21st century: Supporting educational needs* (p. 1e11). Springer-Verlag.

Shute, V. J., Hansen, E. G., & Almond, R. G. (2008). You can't fatten a hog by weighing it–or can you? Evaluating an assessment for learning system called ACED. *International Journal of Artificial Intelligence in Education, 18*(4), 289–316.

Shute, V. J., Leighton, J. P., Jang, E. E., & Chu, M. W. (2016). Advances in the science of assessment. *Educational Assessment, 21*(1), 34–59. https://doi.org/10.1080/10627197.2015.1127752

Shute, V. J., Masduki, I., Donmez, O., Dennen, V. P., Kim, Y. J., Jeong, A. C., & Wang, C. Y. (2010). Modeling, assessing, and supporting key competencies within game environments. In D. Ifenthaler, P. Pirnay-Dummer, & N. M. Seel (Eds.), *Computer-based diagnostics and systematic analysis of knowledge* (pp. 281–309). Springer.

Shute, V. J., & Rahimi, S. (2017). Review of computer-based assessment for learning in elementary and secondary education. *Journal of Computer Assisted Learning, 33*(1), 1–19. https://doi.org/10.1111/jcal.12172

Shute, V. J., & Ventura, M. (2013). *Measuring and supporting learning in games: Stealth assessment*. The MIT Press.

Shute, V. J., Ventura, M., Bauer, M., & Zapata-Rivera, D. (2009). Melding the power of serious games and embedded assessment to monitor and foster learning. In U. Ritterfeld, M. Cody, & P. Vorderer (Eds.), *Serious games: Mechanisms and effects* (pp. 295–321). Routledge, Taylor and Francis.

Spearman, C. (1904). General intelligence: Objectively determined and measured. *American Journal of Psychology, 15*(2), 202–259.

Su, X., & Khoshgoftaar, T. M. (2009). A survey of collaborative filtering techniques. *Advances in Artificial Intelligence, 2009*, 421425. https://doi.org/10.1155/2009/421425

Su, Y., Liu, Q., Liu, Q., Huang, Z., Yin, Y., Chen, E., Ding, C., Wei, S., & Hu, G. (2018). Exercise-enhanced sequential modeling for student performance prediction. In *Thirty-second AAAI conference on artificial intelligence* (pp. 2435–2443).

Sun, Y., Ye, S., Inoue, S., & Sun, Y. (2014). Alternating recursive method for Q-matrix learning. In *Proceedings of the 7th international conference on educational data mining* (pp. 14–20). International Educational Data Mining Society.

Tatsuoka, K. (1983). Rule space: An approach for dealing with misconceptions based on item response theory. *Journal of Educational Measurement, 20*(4), 345–354. https://doi.org/10.1111/j.1745-3984.1983.tb00212.x

Terzis, V., & Economides, A. A. (2011). The acceptance and use of computer based assessment. *Computers and Education, 56*(4), 1032–1044. https://doi.org/10.1016/j.compedu.2010.11.017

Thai-Nghe, N., Drumond, L., Horváth, T., Krohn-Grimberghe, A., Nanopoulos, A., & Schmidt-Thieme, L. (2012). Factorization techniques for predicting student performance. In *Educational recommender systems and technologies: Practices and challenges* (pp. 129–153). IGI Global.

Tsai, F. H., Tsai, C. C., & Lin, K. Y. (2015). The evaluation of different gaming modes and feedback types on game-based formative assessment in an online learning environment. *Computers and Education, 81*, 259–269. https://doi.org/10.1016/j.compedu.2014.10.013

von Davier, M. (2005). *A general diagnostic model applied to language testing data* (RR-05-16). Educational Testing Service. https://doi.org/10.1002/j.2333-8504.2005.tb01993.x

Wang, L., Sy, A., Liu, L., & Piech, C. (2017). Deep knowledge tracing on programming exercises. In *Proceedings of the fourth annual ACM conference on learning at scale* (pp. 201–204). https://doi.org/10.1145/3051457.3053985

Wang, P. Y., & Yang, H. C. (2012). Using collaborative filtering to support college students' use of online forum for English learning. *Computers and Education, 59*(2), 628–637. https://doi.org/10.1016/j.compedu.2012.02.007

Wang, Z., Yu, X., Feng, N., & Wang, Z. (2014). An improved collaborative movie recommendation system using computational intelligence. *Journal of Visual Languages and Computing, 25*(6), 667–675. https://doi.org/10.1016/j.jvlc.2014.09.011

West, P., Rutstein, D. W., Mislevy, R. J., Liu, J., Levy, R., Dicerbo, K. E., Crawford, A., Choi, Y., Chapple, K., & Behrens, J. T. (2012). A Bayesian network approach to modeling learning progressions. In *Learning progressions in science* (pp. 255–292). Brill Sense.

Xiong, X., Zhao, S., Van Inwegen, E. G., & Beck, J. E. (2016). Going deeper with deep knowledge tracing. In *Proceedings of the 9th international conference on educational data mining* (pp. 545–550). International Educational Data Mining Society.

Yao, L., & Boughton, K. A. (2007). A multidimensional item response modeling approach for improving subscale proficiency estimation and classification. *Applied Psychological Measurement, 31*(2), 83–105. https://doi.org/10.1177/0146621606291559

Yeung, C. K. (2019). Deep-IRT: Make deep learning based knowledge tracing explainable using item response theory. arXiv preprint arXiv:1904.11738. https://arxiv.org/abs/1904.11738

Yeung, C. K., & Yeung, D. Y. (2018). Addressing two problems in deep knowledge tracing via prediction-consistent regularization. In *Proceedings of the fifth annual ACM conference on learning at scale* (pp. 1–10). https://doi.org/10.1145/3231644.3231647

Zhang, S., Yao, L., & Xu, X. (2017). AutoSVD++ an efficient hybrid collaborative filtering model via contractive auto-encoders. In *Proceedings of the 40th international ACM SIGIR conference on research and development in information retrieval* (pp. 957–960).

Zhang, Z., Zhang, Y., & Ren, Y. (2020). Employing neighborhood reduction for alleviating sparsity and cold start problems in user-based collaborative filtering. *Information Retrieval Journal*, *23*(4), 449–472. https://doi.org/10.1007/s10791-020-09378-w

13

Designing Automated Writing Evaluation Systems for Ambitious Instruction and Classroom Integration

Lindsay Clare Matsumura, Elaine L. Wang, Richard Correnti, and Diane Litman

CONTENTS

13.1 Introduction 195
13.1.1 Overview of eRevise System 196
13.1.2 Chapter Overview 197
13.2 Considerations for Automated Writing Evaluation (AWE) System Design 197
13.2.1 Authentic Tasks That Aid Ambitious Teaching 198
13.2.2 Features of a Construct Are Identified and Named 199
13.2.3 Information Is at a 'Grain Size' to Support Learning 199
13.2.4 Underlying Algorithms Are Fair 199
13.2.5 System Supports Instructional Interactions and Discourse 200
13.3 Designing to Facilitate Widespread Adoption of AWE Systems and Integration in Classroom Routines 201
13.3.1 Education Policies 202
13.3.2 Values and Goals Held by School Leaders 202
13.3.3 Values and Goals Held by Teachers 203
13.4 Concluding Thoughts 204
References 205

13.1 Introduction

Artificial intelligence (AI) generally is growing in sophistication. It has the potential to increase students' learning opportunities 'at scale'. Uptake of tools leveraging AI capabilities in classrooms, however, has been limited. In the United States at least, schools have traditionally been slow to adopt new technologies, or teachers implement new technologies in ways that reify rather than transform educational practice (Cuban, 2009). As a result, technological advancements have generally resulted in limited widespread effects on students' learning.

One reason for the limited uptake of AI-driven technologies is that these have seldom been designed to be integrated into teachers' daily work or instructional routines. As we argue in this chapter, this is, in part, because technology development and implementation have not always been responsive to the larger education context in which schools are situated. Instead, technologies often are 'one more thing' added to a roster of district reform efforts and teachers' already packed classroom schedules. As such, despite the intention of such tools to unburden teachers or provide a unique learning opportunity for students, school leaders and teachers can regard the tools as disruptive to, rather than facilitative of, teaching and learning goals.

Teachers have also been tentative to take up AI-powered technologies because many of the extant systems (e.g., cognitive tutors) are designed exclusively for students. Thus, they do not take into account, or they underspecify, the role of the teacher or peers in students' sense-making of new content and development of new skills. The default position of AI-supported systems being student-facing is problematic as it can undermine the essential identity of teachers as active participants in students' learning. This too, makes technologies something 'extra' that can seem disconnected from the essential tasks of teaching and less likely to be implemented consistently.

More importantly, systems that ignore teaching and/or the role of teacher–student interactions can have limited utility for student learning because the development of complex thinking processes – such as those included in the Next Generation Science Standards

DOI: 10.1201/9781003181187-16

(NGSS, 2013) and Common Core State Standards (CCSSO, 2010) – typically occur in the context of rich social interactions (Cohen et al., 2003; Goldman & Pellegrino, 2015; Tharp & Gallimore, 1991). The extent to which AI tools can support people to talk and create together is at the heart of their potential 'value-added' for education. Understanding these issues, recent imaginings of how AI can advance learning (see, for example, Baker & Boser, 2021; Roschelle et al., 2020) call for the design of AI-supported learning systems to be better integrated in instructional routines – the methods, procedures, or patterns of moves that teachers carry out in regular configurations and sequences for specific instructional activities (e.g., Yinger, 1979; Lampert & Graziani, 2009). Such routines also should be based in robust learning theory to support the forms of instruction that prepare students for academic success.

In this chapter, we consider the role of AI tools to support ambitious teaching practices on a sustained basis. We anchor our discussion in the use of automated writing evaluation (AWE) systems that provide formative feedback, a common way that AI is used in education settings (Deeva et al., 2021). We focus specifically on the use of automated formative feedback to improve students' writing skills, drawing on our research developing an AWE system (*eRevise*) (Wang et al., 2020; Zhang et al., 2019). (See Table 13.1 for a description of how the *eRevise* system is situated in the larger body of work on automated feedback systems.)

13.1.1 Overview of eRevise System

Our AWE system, *eRevise*, supports students' argument writing in the upper elementary and middle school grades by providing automated formative feedback on drafts of students' essays targeted toward improving their use of text evidence. The idea is that students use this feedback to guide their revisions, thus improving the quality of their essays and, in an ideal world, strengthening their conceptual understanding of the features of 'good' evidence use that they can apply in future essays. We note that writing is a 'gateway' skill for entry into college and completion of college-level coursework. Writing about disciplinary content also increases students' learning in social studies, science, and mathematics (Graham et al., 2020). Writing is a difficult skill to teach and learn, however. In the United States, for example, many teachers report feeling underprepared to teach writing well, and infrequently implement research-based practices for writing instruction (Brindle et al., 2016; Kiuhara et al., 2009). Unsurprisingly, a very large majority of students do not have proficient writing skills, and this is especially the case for minoritized students and students from low-income families

TABLE 13.1

Contextualizing Properties of *eRevise* Relative to Systematic Review of Studies on Automated Feedback Systems (see Deeva et al., 2021)

Classification Properties	Properties of *eRevise*	% of Studies with Similar Properties
Architecture		
Domain (construct within discipline)	Argument writing	<13
Expert knowledge (knowledge of important elements for expert use of evidence)	Amount and specificity of text evidence to support a claim	N/A
Student data (predefined answer; open answer; in-system behavior; real-life behavior; student characteristics)	Open answer	55
Feedback generation model (expert-driven; data-driven; mixed)	Data-driven/mixed	50
Implementation (stand-alone; plug-in)	Stand-alone	67
Feedback		
Adaptiveness (non-adaptive; task-adaptive; student-adaptive)	Student-adaptive	34
Timing (immediate; on request; end-of-task)	End-of-task	34
Learner control over feedback (no control; mild control; high control)	No control	72
Purpose of feedback (corrective; suggestive; informative; motivational)	Suggestive; informative	6
Educational context		
Domain (STEM; languages: medicine; art; special education; other skills; non-specific)	English language arts	<13
Level (primary/elementary; secondary; tertiary; adult; non-specific)	Grades 4–6 (elementary)	5
Setting (online; blended; in-class; not-specified)	In-class	54

(McFarland et al., 2017). Classroom supports that increase the quality of writing instruction and learning are clearly needed to support equitable access to salaried positions in all fields, including STEM.

eRevise was designed around the Response-to-Text Assessment (RTA) (Correnti et al., 2013). Teachers read an article aloud to students focused either on a United Nation's initiative to eradicate poverty in rural Africa (the Millennium Villages project) or government investment in space travel. The prompts respectively ask students, 'Based on the article, did the author provide a convincing argument that "winning the fight against poverty is achievable in our lifetime"?' and 'Did the author convince you that "space exploration is desirable when there is so much that needs to be done on earth"?' For both tasks, students are asked to support their answer with multiple examples from the text. The RTA rubric for human raters focuses on five dimensions – evidence use, analysis, organization, academic style, and mechanics. Each is scored on a scale from '1 = low' to '4 = high'. *eRevise* focuses specifically on the dimension of text evidence use. As illustrated in Figure 13.1, *eRevise* uses NLP features generated during automatic scoring of students' initial essays to identify strengths and weaknesses of evidence use in order to provide students with formative feedback to guide their revisions (Correnti et al., 2019; Rahimi et al., 2014, 2017; Wang et al., 2020; Zhang & Litman, 2017; Zhang et al., 2019). As discussed later in this chapter, we also have used the feature scores as outcomes in our validity arguments (Correnti, Matsumura, Wang, & Litman, 2022).

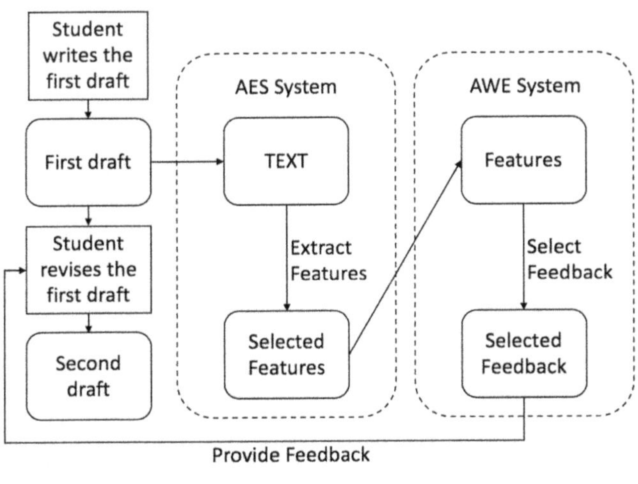

FIGURE 13.1
Architecture of *eRevise* system (Wang et al., 2020).

13.1.2 Chapter Overview

To inform our specific suggestions for system design, in the first section of our chapter we lean on sociocultural theory (Vygotsky, 1980, 2012). From this perspective, we argue that, to serve as learning tools, AWE systems should communicate the features of authentic tasks, provide information that is transparent, actionable, and fair, and open up avenues for student-centered classroom discussions and collaboration. In the second section, we draw on education policy implementation research to take up the issue of classroom integration and widespread adoption of AWE systems (Coburn, 2005; Spillane et al., 2002). We argue that it is necessary to expand traditional conceptions of usability and feasibility as drivers of design choices to include the contextual and subjective factors that influence classroom routines. These factors include the policy context in which classrooms are situated, as well as the values and goals held by individuals within that educational system (National Academies of Science, Engineering and Medicines, 2018). We summarize the key considerations in Table 13.2.

13.2 Considerations for Automated Writing Evaluation (AWE) System Design

Broadly speaking, sociocultural theory emphasizes the distributed and social nature of cognition (Vygotsky, 2012; Tharp & Gallimore, 1991; Wertsch et al., 1984). From this perspective, learners acquire new strategies and knowledge by engaging in activities with others. These activities are shaped by cultural norms, traditions, and institutions, and mediated by tools and objects in the environment (Greeno et al., 1996). More specifically, learners develop through watching others engage in an activity (modeling), as well as engaging in a task with others (joint activity) with only as much assistance as needed to close the gap between what they can do independently and what is needed for task completion. Importantly, this assistance (often referred to as 'scaffolding') fades over time, leading to the learner's ability to complete a task independently (Collins et al., 1991).

In classrooms, the productive social processes that support learning then are characterized by activities that draw students into using the cognitive and communicative functions of a discipline. These functions are nurtured and developed through intentionally constructed interactions between a knowledgeable other and a novice (i.e., teacher and student) or among peers. Specifically, sociocultural theory emphasizes the importance of 'student-centered' classroom discussions

TABLE 13.2
Overview of Key Considerations for Design of AI-Based Formative Feedback Tools

Aspect	Consideration for Design
Features of effective formative feedback tools	
Task authenticity and complexity	Learning tools targeting complex thinking processes should be based in authentic and complex tasks, tasks that resemble the kinds of thinking and norms for communication that characterize a discipline
Construct features	Effective formative assessments assist learning by making salient the gap between the performance task and next steps for improvement, and provide scaffolding to support skills development. Formative assessments and systems should provide information that teachers can use to improve their teaching. Both of these aims require that features of the targeted construct be transparent and communicable
Grain size of information	Teachers can better adjust their instruction and students can better improve their work products if the information provided by the formative assessment tool is at a grain size that is actionable
Underlying algorithms	Formative assessment systems that are fair and unbiased, or at least anticipate and mitigate bias to the extent possible, more equitably serve all student populations and their learning needs
Instructional interactions and discourse	Learners acquire new skills and knowledge by engaging in activities with others. Teacher–student or peer interactions and discussions lead to deeper understanding of disciplinary content and development of skills
Factors influencing implementation and integration	
Education policies	Policies to which teachers are held accountable, including disciplinary standards, curricula, and state assessments, altogether shape the content that teachers teach and how instructional time is used
Values and goals held by school leaders	School leaders' content knowledge and assumptions (values) for what good teaching should look like underlie their interpretation of education reforms and innovations and is consequential for subsequent implementation of reform by teachers
Values and goals held by teachers	Teachers hold different conceptions of disciplinary practices, understanding of their role in student learning, trust in AI-driven systems, and beliefs about students' capabilities that affect how they shape instructional routines

that empower students to assert their thinking and reasoning, and to work collaboratively with peers to make sense of ideas and concepts (Kim & Wilkinson, 2019; Tharp & Gallimore, 1991). This is in contrast to 'teacher-centered' classroom discussions that are characterized by students recalling facts and procedures rather than interpreting and analyzing information, and students providing short, unelaborated answers to known-answer questions posed by a teacher (Cazden, 1988; Mehan, 1979). Sociocultural theory also emphasizes the importance of students engaging in authentic academic tasks that are aligned with the intellectual work in which skilled adults engage within their field (Newmann et al., 2001). The learning tasks typically assigned to students, however, limit thinking to recapitulation of facts and procedures, require unelaborated response, and overall, devalue students' experiences and ideas (Ibid., Matsumura et al., 2015).

Applying the principles of sociocultural theory to the design of AWE systems in specific, we present the following five considerations.

13.2.1 Authentic Tasks That Aid Ambitious Teaching

As we noted earlier, to be useful as a learning tool for complex thinking processes, educational technologies (specifically AWE systems) need to be based in authentic and complex tasks, that is, tasks that resemble the kinds of thinking and norms for communication that characterize a discipline and empower students to (1) apply knowledge and skills in original ways; (2) interrogate the artifacts and ideas produced by others (including output generated by AWE systems); and (3) produce elaborated responses (Newmann et al., 2001). Furthermore, to be useful as a tool for learning, the educational technology should be 'high leverage', that is, support teachers to teach content or support students to perform a task that would be difficult to do without such support.

In developing *eRevise*, for example, we focused on argument writing in response to text. The ability to marshal evidence in support of claims in writing is especially critical to college readiness and especially hard to teach and learn. Younger writers often lack familiarity with the discursive features associated with argumentation, such as identifying convincing evidence and explaining how it connects to a claim (Ferretti & Graham, 2019; O'Hallaron, 2014; Schleppegrell, 2004; Snow & Ucellli, 2009). Because argument writing is a relatively new addition to the elementary grade curriculum, teachers can lack content knowledge of pertinent concepts related to effective argumentation, such as the importance of providing reasons linking evidence to claims (Langer & Applebee, 1987; Shanahan & Shanahan, 2008).

Moreover, as we noted earlier, teachers rarely assign tasks that require analysis and use of text evidence (Kiuhara et al., 2009; Matsumura et al., 2015). A goal then, in creating *eRevise*, was to model a task (text-based argument essay) that could increase students' exposure to this form of writing.

Revising in response to substantive feedback is critical to writing development, but faced with stacks of student assignments, teachers often are reluctant to assign many tasks that require multiple drafts, or else they resort to providing feedback that focuses on mechanics (grammar, spelling, and punctuation) (Matsumura et al., 2002; Patthey-Chavez et al., 2004). Our second goal, then, was to reduce teachers' burden and increase students' opportunities to receive substantive feedback to develop their conceptual understanding of text evidence use.

13.2.2 Features of a Construct Are Identified and Named

Like all formative assessments, AWE systems are intended to assist learning by making salient the gap between the performance task and next steps for improvement, and provide scaffolding (e.g., hints or suggestions) to support students' revision activity (Shepard, 2005; Shute, 2008). To serve a learning purpose, AWE systems also would ideally provide information that teachers can use to improve their teaching. Toward these ends, systems should be explicit about the features of a targeted construct that is being assessed, and these should be instantiated in scoring algorithms that are transparent and communicable to actors across the school system including school leaders, teachers, and students.

In our AWE system, we use the acronym CASE (complete, accurate, specific, and explained) to communicate the features of effective evidence use to users (Wang et al., 2017). These are reflected in the automated feedback messages provided to students. Keyed to the quality of students' responses, the individual feedback messages guide students to provide evidence that is *complete* (i.e., to use appropriate amount of evidence from the text), *accurate* (i.e., to identify evidence that could support the claim), and *specific* (i.e., to provide sufficient details about the evidence referenced), and to *explain* the evidence they provided and connect the evidence to their overall argument. The features are likewise aligned with this conception of effective evidence use so that what is being assessed can be 'mapped back' to students' writing. To determine the feedback to provide to each essay, *eRevise* draws on a feedback selection algorithm that includes a focus on the number of pieces of evidence (NPE) (i.e., unique topics) and specificity (SPC) – a count of the number of non-duplicate, unique evidence of words or phrases from four primary topics (see Rahimi et al., 2017; Zhang & Litman, 2017; Zhang et al., 2019).

13.2.3 Information Is at a 'Grain Size' to Support Learning

To serve as learning tools, AWE systems should provide specific information about the strengths and weaknesses of students' abilities. Much of the past research on the quality of AWE systems has focused on comparing human and machine-generated rubric scores (Attali & Burstein, 2006; Elliot, 2003; Page, 2003). We argue, however, that a smaller 'grain size' for assessing change in students' writing is likely more useful for formative assessment purposes. Our research on *eRevise*, for example, showed that changes in the features of evidence use provided a more accurate picture of students' growth in evidence use. We found that human raters had moderate to good inter-rater reliability with the AWE system when rating the NPE and SPC features, as well as good inter-rater reliability for measures of change in those features. Furthermore, in comparison to the rubric score, which was only capable of indicating improvement for 41% of the students, we established that feature scores also improved for students whose scores showed no change in rubric scores. These improvements would be overlooked if *eRevise* focused on rubric scores alone.

Feature scores, then, were more sensitive to incremental differences in student writing than rubric scores, and thus, may better serve a formative assessment or learning purpose because they provide teachers with more information about students' revision efforts and progress. This may be especially important in the context of writing produced by younger students, many of whom struggle to revise effectively (MacArthur, 2018; Wang et al., 2020). A student who struggles with writing might provide one more piece of evidence in their revision, a change that likely would not meet the 'bar' for a change in rubric score. This small improvement would be information a teacher would want to have because it suggests that the student was attempting to implement the feedback. In all, scores on specific features are an important avenue to explore in developing AWE systems intended as formative assessments.

13.2.4 Underlying Algorithms Are Fair

The fact that algorithms often exhibit bias has become an increasing concern to users of educational (and other) AI-based technologies. For most systems, the level of bias is not studied or known, which is another potential obstacle to their usefulness in providing formative information. It is thus important that system designers anticipate, analyze, and attempt to mitigate any sources of bias throughout the technology development life cycle (Baker & Hawn, 2021). Explorations should seek to expand both the types of biases examined (e.g., beyond highly studied categories such as gender and race/ethnicity to less studied categories such as disability) (Ibid.) and the measures used to examine them (Loukina et al., 2019). Determining ways to productively communicate fairness properties of a technology to users also is important. Since most current AI technologies use machine learning to train predictive models from data, this analysis begins with data collection. Example data properties to examine include mismatches between development and user populations (e.g., with respect to properties of students, teachers, schools, and districts) as well as unequal demographic representations within a well-matched development population. Potential mitigation methods for the latter include resampling the training data to balance the demographics, or training demographic-specific models.

Turning to data representation, a collected training example is often converted into a formal representation consisting of a set of predictive features (e.g., the NPE and SPC features used to represent a student essay in *eRevise*), which are often unexamined for bias or correlations with student demographics. One method for mitigating this type of bias is to manipulate the data representation so that it uses 'fairer' features. We note, however, that there is a potential trade-off if the less fair features have high construct validity. In other words, we are most concerned about bias when it contains construct irrelevant bias. In a fairness evaluation of *eRevise*, we also found that while changing the feature representation could indeed increase fairness, there was a trade-off in that the new representation reduced predictive accuracy (Litman et al., 2021). As a last example, many fairness definitions and associated measures for evaluating predictive AI-driven models have been proposed in the literature. Researchers in the area of automated essay scoring, including Loukina et al. (2019), have advocated for evaluating fairness using multiple measures, arguing that total algorithmic fairness may not be achievable and that different measures can identify different issues, some of which might or might not have implications in particular use cases. Different insights were in fact obtained across the multiple measures used in our fairness evaluation of *eRevise* with respect to race, gender, and socio-economic status (Litman et al., 2021).

13.2.5 System Supports Instructional Interactions and Discourse

In an AWE system designed in accordance with sociocultural theory, the work that students perform in the system, along with system messages and outputs (e.g., automated scores and feedback on student writing), all become artifacts that cross the boundary of the screen and enter the classroom space. These become objects with the potential to support teacher–student or student–student discourse, leading to deeper understanding of disciplinary content and acquisition of skills. Findings from our recent research supports the use of AI systems aligned with the sociocultural perspective. In a pilot study of *eRevise*, we provided no explicit instructions to teachers (n = 16) about how to engage with their students around the automated feedback. We had teachers log the questions students asked and their responses to those questions. Results showed that teachers varied in their use of the system. About one-third of the teachers treated the essay revision task as an assessment to be completed by students on their own, without assistance; these teachers either did not interact with students at all during the revision process or only fielded procedural questions. A smaller group of teachers generally responded to student questions by referring them back to the task description. Finally, a third group of teachers engaged in substantive ways with the students, often offering support in interpreting feedback messages. The students of this group of teachers had the highest improvement scores, defined as the difference between the feature scores of the first and the second drafts (Correnti et al., 2022). In all, then, interacting with students around the task and system outputs appeared to be critical for learning.

To achieve the vision of an AWE-integrated sociocultural classroom, it is not enough for systems to provide outputs and teacher-accessible dashboards. These are essential boundary objects, but as passive supports, they may not lead to productive discourse. Rather than leave the possibility of quality instructional interactions to happenstance, we propose that AWE systems provide discussion protocols that orient teachers toward and support them to use system outputs (e.g., feedback messages, error corrections, scores, or annotations) as objects of discourse and learning. For example, such protocols could help teachers and students co-examine the feedback messages provided by the AWE system. Teachers could elicit students' understanding of the feedback itself (e.g., What do you think the feedback means? What does [key feedback term] mean in argument writing?). Furthermore, teachers could also prompt discussion about the feedback's application to the student's draft essay (e.g., Where does the feedback apply? How can that part of the essay be revised? Why would that revision improve the overall argument?). Beyond the initial questions, the discussion protocols should help anticipate possible student misconceptions or misinterpretations of system outputs, provide probing or follow-up questions, identify points of connections with the curriculum or disciplinary concepts, and offer guidance for instructional next steps. In light of the importance of peer assistance, particularly for writing development (Graham et al., 2012; Graham & Perin, 2007; Harris et al., 2006), similar discussion scripts could be developed to scaffold peer interactions. Overall, such interaction protocols would help situate the use of the technology within the greater classroom ecosystem. The resulting discussions would build students' conceptual understanding of key disciplinary constructs that are the focus of the feedback (e.g., evidence use in argument writing), which is necessary for transfer to future writing situations.

Student agency is a critical component of instruction aligned with sociocultural theory (Clarke et al., 2016). In the context of an AWE system designed to support discourse and interactions, student agency could be supported in discussion protocols that guide students to interrogate – not just accept – the authority of the feedback and system. The questions could prompt students to reflect upon and critique or problematize the system outputs (e.g., automated rubric scores or the feedback they received) in the context of their learning outside of the system (i.e., in the curriculum and classroom instruction). This would serve two key purposes. First, encouraging students to discuss the feedback in light of their emerging disciplinary knowledge helps to surface for the teacher potential misunderstandings the student holds. It also helps the students to solidify their own conceptual understanding. It provides an opportunity for students to work through any perceived disconnect between system feedback and other sources. Second, research indicates that actively engaging with others to make sense of feedback is important for successful application and revision (Adie et al., 2018; Carless & Boud, 2018). Thus, such interactions would mitigate against students either ignoring system feedback or merely engaging in procedural implementation of feedback. Indeed, research has found that large proportions of students do not revise their writing in response to AWE feedback or do so in nonsubstantive ways (Wilson et al., 2014). In essence then, providing scripts or protocols that support discourse around system outputs will help to position students as sense-makers in their own learning process and support their use of the system toward deeper learning and construction of knowledge.

Scripts or protocols that position the system outputs not as 'truths', but as objects of inquiry can also help address a common issue – lack of trust (or over-trust) in AWE systems. Research shows that teachers and students are often skeptical of the feedback provided by AWE technologies (Chen & Cheng, 2008; Roscoe et al., 2017). This is often due to perceived inaccuracies in the scoring or feedback provided. Less trust results in less inclination to use the tool. Guided by a well-designed script, teachers and students can confront and work through perceived inaccuracies. This may lead to a change in the user's perception that is more favorable toward the AWE (e.g., the student recognizes the validity of the feedback after working through a discussion protocol that helps interpret the feedback). Or, it may lead users to understand and accept a reasonable degree of fallibility in the automated system. This alignment of user expectations and system capability may be important to establish and may help instill greater trust in AWE systems.

Beyond offering protocols and supporting materials then, system developers should also design and provide a program of professional learning. Decades of research on education reform indicate that provision of curricula and standards in and of themselves can have minimal impact on teachers' instructional behaviors and routines absent significant opportunities for teacher learning (e.g., Cohen, 1990; Spillane & Zeuli, 1999; Stein & Kaufman, 2010; Van Driel et al., 2001). We emphasize that such learning should extend far beyond how to navigate and operationally use the system or platform. It should focus substantively on supporting teachers to use the technology to enact dialogic instruction. To this end, the professional learning program should introduce teachers to the scripts and protocols that support conversations with students around the task and outputs of the system, train them on the goal of the conversation, and coach them on talk moves that build students' conceptual understanding and agency.

Yet even a well-designed AWE system reflective of sociocultural principles for learning and instruction is not guaranteed to result in widespread adoption and integration into classroom routines. Multiple factors in the implementing context play a role in supporting or thwarting the adoption and use of new technologies. We turn to these considerations in the next section.

13.3 Designing to Facilitate Widespread Adoption of AWE Systems and Integration in Classroom Routines

A wealth of research shows that education policies at the federal, state, and local levels exert considerable influence on classroom practice, including the tasks and texts assigned to students, teaching interactions, and the amount of time teachers spend on particular content (Coburn, 2004, 2005; Cohen, 1990; studies reviewed in Neumerski, 2013; Russell & Bray, 2013; Spillane et al., 2002; Spillane & Zeuli, 1999). Within the bounds of policy, however, school leaders and teachers exhibit significant variation in their uptake and implementation of new reforms (Hamilton et al., 2003; Kisa & Correnti, 2015; Matsumura et al., 2010). This is because pedagogical leadership and decision-making are shaped as well by subjective factors – the knowledge, values, and goals held by school leaders and teachers, which in turn, shape their scripts for interacting with teachers and students, respectively. We elaborate on these factors that are likely to impact the adoption and use of AWE systems (and all reforms) below.

13.3.1 Education Policies

In the United States at least, test-based accountability is a driving force in education reform. Codified in the No Child Left Behind legislation (2001), and updated in the Every Child Succeeds Act (2015), federal policy requires annual assessments of students in Grades 3–8 and one year of high school in reading and math. The idea is to hold districts and schools accountable for ensuring that all populations of students (e.g., children from all racial groups, English learners, children with special needs) attain or show 'adequate yearly progress' toward proficiency on state standardized assessments.

While on one hand, pressure to improve students' scores has contributed to rising interest in education reforms, these policies also have contributed to a narrowing of the curricula to focus on the content and format of state standardized accountability tests that – for the most part – run counter to forms of instruction that develop complex thinking and reasoning (e.g., Howe & Correnti, 2020; Mathison & Freeman, 2003; Wang & Matsumura, 2018). Pressure to meet accountability targets for test scores and attendant negative effects on teaching are especially apparent in schools that serve minoritized students from low-income families (Au, 2010). All of this can reduce the likelihood that AWE systems will be adopted or implemented as intended in schools that serve low-income students especially.

The curriculum also exerts a significant influence on the implementation of education innovations and reforms including AWE systems and other AI-based learning tools. Teachers may have little wiggle room to add a new activity or content to their instructional routine, and this is especially the case when teachers are required to keep pace with highly structured curriculum guides (Bauml, 2015). In our research, we asked teachers what the *eRevise* system would need to 'look like' to be incorporated in their regular classroom routine. Teachers were overall positive about *eRevise*, reporting that the system was consonant with their district's standards and assessments that emphasized students using text evidence in support of claims (Matsumura et al., 2020). Teachers also reported, however, that they would more likely use *eRevise* on a regular basis if the source texts in the system that the writing tasks are based on were from their curricula. Our findings suggest that even when aligned with some district accountability policies, learning tools/AWE systems that are 'one more thing' on top of a curriculum are less likely to be accommodated in teachers' classroom routine.

13.3.2 Values and Goals Held by School Leaders

Some research shows that how school leaders interpret and subsequently communicate to teachers about the forms of instruction needed to meet accountability targets is consequential for the implementation of education reforms (Coburn, 2005; Louis & Robinson, 2012). These interpretations are driven by goals for instruction that are in large part driven by content knowledge and assumptions about what good teaching should look like. Innovations and reforms, including standards-based reforms targeting complex thinking and reasoning, are unlikely to be taken up in a school, or are only partially taken up, if school leaders do not see an innovation as directly relevant to meeting their targets or aligned with their vision for instruction (Matsumura & Wang, 2014).

School leaders exert influence on teachers' instruction in their managerial functions. For example, through their management of their school's professional development budget, they decide the amount, type, and content of professional development available to teachers (Goldenberg, 2004; McLaughlin & Talbert, 2006). Teachers are unlikely to seriously incorporate a new technology in their teaching or acquire new skills necessary for implementing a new technology without sustained and intensive supports.

School leaders also have a heavy influence in their role as evaluators of teachers' performance. Asking teachers to invest in new technologies is a tall order if this time and effort is not rewarded in their performance evaluations. This can be especially challenging in the context of reforms to support students' complex reasoning because implementation often depends on the principal's sense-making within contexts that are primarily focused on accountability to (low-level) assessments (Matsumura & Wang, 2014). Finally, districts, especially large urban districts, are often festooned with multiple education reforms simultaneously, sometimes with competing visions or goals. School leaders are often left to decide which initiatives to emphasize or de-emphasize to their teachers.

In all, then, policy mandates and their interpretation by school leaders play a significant role in how teachers construct their instructional routines. Designers of AWE systems need to understand that even sophisticated learning tools can expect little purchase in a policy context that communicates a competing vision of what instruction should look like. Designers need to be aware of existing accountability goals and/or curricula that might compete with the time and effort needed to engage with new technologies. Considering how technology will or will not facilitate accountability test goals and 'fit' within curricular aims will be important for widespread and sustained implementation.

13.3.3 Values and Goals Held by Teachers

Beyond school leaders, teachers, at least in the United States, ultimately have agency in how they shape instructional routines. Studies show that teachers' beliefs, values, and goals play a significant role in how they interpret and enact policies and reform practices (Coburn, 2004; Spillane et al., 2002; Wang & Matsumura, 2018; Yurekli et al., 2020). These influential factors include (1) teachers' conceptions of key skills or practices students are to learn in a content area, (2) assumption about their role in student learning, (3) trust in AI-driven systems, and (4) beliefs about students' capabilities. We address each of these below.

Research we have conducted on text-based writing shows that teachers recognized it as a skill emphasized in standards, but had *different conceptions* of what it meant, even though they taught in the same district policy context. Whereas some teachers regarded text-based writing as an application of reading comprehension skills and strategies (e.g., text-based writing as summarizing), others approached it as an opportunity to deepen students' inquiry into text ideas (e.g., analyzing themes) (Wang & Matsumura, 2018). The tasks teachers assigned to students reflected their conceptions. Even when teachers assigned the same or similar tasks, there was considerable variation in what they accepted as high-quality responses (Wang et al., 2018). These variations have consequences for students' learning opportunities.

To a great extent, once implemented, AWE systems help to mitigate against individual teacher-level variation because the tasks are predetermined and the algorithms are applied systematically across contexts. However, our research suggests that teachers may be disinclined to integrate these systems into their instructional routines if they do not perceive the tasks and automated feedback messages as aligned with their conception of what students should experience and focus on. We are not suggesting designing AWE systems to accommodate various possible teacher conceptions of a disciplinary practice; rather, aligned with our earlier call for transparency, we suggest that developers make explicit – through system documentation, outputs, and accompanying professional learning – the conception of disciplinary practice underlying the system. Transparency about the ultimate learning goals will help anchor specific learning objectives within a larger purpose, providing motivation for building students' complex reasoning.

In addition, the system can be built to allow for some teacher agency and interactivity. For example, the system could allow teacher-initiated input (e.g., task guidance, feedback messages) on top of what is provided by the AWE system to better align the feedback with learning goals the teacher has for individual students. Another possibility is to design a dynamic system with options that can be emphasized or muted. For example, in the case of an AWE tool for argument writing, perhaps teachers can turn off the grading or feedback on grammar and mechanics, but retain feedback related to use of text evidence. These ideas reflect our interview findings that teachers believed the teacher and AWE system should work together; specifically, they sought the option to layer comments on top of *eRevise*'s automated feedback or identify for students the points to focus on in their revision.

Teachers' *assumption about their role in student learning* is a related factor to consider in building AWE systems that teachers will likely use and integrate into instructional routines (e.g., Kim et al., 2013). Developers should incorporate ways for educators to interact with students *around the system*, in other words, position it not as a student-facing tool that supplants teachers, but as a technology whose output provides the subject matter (i.e., object) for in-person instructional interactions. In the case of *eRevise*, the object for interactions is the automated feedback; thus, a primary role of *eRevise* is to guide or steer a teacher–student interaction toward a concept of evidence use that students could work on during revision. This, then, facilitates teachers to construct a student learning opportunity in interaction with the teacher around that concept.

In all, the emphasis should be on the power of AWE systems to assist and support teachers in orchestrating classroom experiences (Roschelle et al., 2020). To these ends, in addition to dashboards that provide data for conversations with individual students about progress, systems can offer suggestions for grouping students – for example, based on the feedback they received – to address their individual needs; teachers may reteach certain concepts to groups needing that support. Supports in the form of teacher trainings and use-case scenarios can also help teachers integrate the technology into their routines in a way that reduces burden and enhances their interactions with students.

Teachers' use of AWE systems also hinge on their *attitudes toward and trust in technologies.* Studies have shown that teachers who did not regard the technology as relevant to their pedagogical aims or supportive of student learning tend to use it sparingly or in peripheral ways (see Williams & Beam, 2019; Ertmer, 2005; Ertmer et al., 2015). One way developers can address this concern is by including educators in the design process, especially those who may be tentative about integrating technology in classroom instruction. Doing so could help surface and target beliefs that hinder system implementation and thus could provide useful fodder for professional development. Second, developers should make transparent the underlying algorithm and other typically 'hidden' aspects of the system. In the case of *eRevise*, for example, this would include the key features the system used to select feedback messages (e.g., NPE, SPC). All this can help teachers assess the alignment between their aims and the extent to which an AWE system is based on relevant disciplinary constructs and practices. Third, documentation and professional learning that accompanies the system should report the reliability and validity of the automated feedback. This can help gain teachers' trust in the system while also conveying that the system is not infallible, thus positioning the system as a partner with, not replacement for, teachers.

A final set of beliefs we consider here that could influence whether and how teachers use AWE systems are *beliefs about student capabilities.* A general review of research on teachers' beliefs suggests that teachers' theories about students and students' abilities to learn affect their instructional behaviors (see Fang, 1996). Indeed, teachers often provide differential learning opportunities to students that they perceive as more or less capable (Prime & Miranda, 2006). This includes, for example, providing students perceived as less capable with tasks of lower cognitive demand or over-scaffolding tasks for students that they believe are struggling while assuming that higher-achieving students can manage largely on their own. One implication of teachers' beliefs about students' abilities for integrating AWE systems in classroom routines is that teachers may determine that the automated system is only appropriate for certain students, or that they will use the system differently for different learners. This tendency has been documented with respect to English learners; some teachers felt that certain technology-aided research and writing activities could not be attempted with non-proficient English speakers (see e.g., Warschauer et al., 2004). On the other hand, teachers may leave 'higher-achieving students' to interact with these systems without guidance, which may also compromise effective use of the system.

To address these concerns, system developers can work with students of special populations (e.g., English learners, students with special needs) and teachers of those students to ensure that the system is accessible and includes supports, such as a built-in dictionary or how-to tutorials for students in various languages. Developers can also ensure that the NLP tools within a system assessing substantive aspects of student writing (e.g., text evidence use) can account for misspellings, grammar, and other common linguistic errors that students produce (Zhang & Litman, 2017). These aspects of an AWE system should be explained or made apparent to teachers. While this may not change teachers' perceptions of students, it could help them understand that all students could potentially use and benefit from the AWE system.

13.4 Concluding Thoughts

In this chapter, we focused our discussion on AWE systems, a common application of AI for classroom use. We note, however, that AI can be used to provide automated feedback to increase the quality of students' learning opportunities in other ways as well, including for example, to support students' development of concept maps (Ryoo & Linn, 2016), engagement in simulations (Gobert et al., 2013), and complex problem-solving in engineering mechanics (Orthaber et al., 2020). We believe that the design considerations that we have highlighted here – presenting authentic tasks, ensuring algorithmic fairness, and privileging social interactions – are likely applicable across multiple types of AI-driven feedback systems that aim to develop students' complex thinking skills across disciplines.

Moreover, the contextual issues we have highlighted with respect to the interaction of policies and school leaders' and teachers' values and goals also are likely to play a role in the integration of all new AI-driven technologies in teachers' practice. As noted by others, the potential of learning technologies, regardless of their level of sophistication, is likely to be thwarted absent attention to these larger issues (Zhai et al., 2021). While designers may not be able to change district policies and educators' values and goals, they can work to make the system responsive to varied contexts and educator needs. To the extent that designers are transparent about the tasks, scoring features, bias mitigation strategies, and other aspects of the system, they are helping district administrators, who typically make resource purchase decisions, school leaders, and teachers envision how the system can be leveraged and integrated into instructional routines.

While system design typically brings to mind the development of tasks, scoring algorithms, dashboard, outputs, and the like, throughout this chapter, we have emphasized the need for system design to also include professional learning opportunities for teachers and leaders that support integration of technology in ambitious forms of instruction. Again, drawing on lessons from decades of reform, we argue that sustained professional learning is needed, not only to support users of the AI system to learn the technical skills necessary for implementation, but also to orient teachers toward its use as a tool that activates meaningful instructional interactions and discourse. Such professional learning opportunities can also surface and address potential tensions between the goals for technology use and the goals for achieving other ends such as accountability targets. It should also be noted that because NLP performance will never be perfect, professional development for educators should also include information about the limitations of AI technologies. This is important for supporting teachers to be productive partners with AWE systems; it helps them know when to trust, as well as when to correct, the system.

In sum, we believe that AI-supported formative feedback systems that attend to the considerations we have presented could have the potential for strong uptake by educators and, in turn, support teaching and learning of complex skills and cognitive processes necessary for academic success.

References

Adie, L., van der Kleij, F., & Cumming, J. (2018). The development and application of coding frameworks to explore dialogic feedback interactions and self-regulated learning. *British Educational Research Journal, 44*(4), 704–723.

Advancement Project. (2010). *Test, punish and pushout: How 'zero tolerance' and high-stakes testing funnel youth into the school-to-prison pipeline*. Authors.

Attali, Y., & Burstein, J. (2006). Automated essay scoring with e-rater v2. *Journal of Technology, Learning and Assessment*, 4(3). http://www.jtla.org

Au, W. (2010). *Unequal by design: High-stakes testing and the standardization of inequality*. Routledge.

Baker, R. S., & Boser, U. (2021). *High-leverage opportunities for learning engineering*. Penn Center for Learning Analytics, University of Pennsylvania.

Baker, R. S., & Hawn, A. (2021). Algorithmic bias in education. edArxiv.org

Bauml, M. (2015). Beginning primary teachers' experiences with curriculum guides and pacing calendars for math and science instruction. *Journal of Research in Childhood Education, 29*(3), 390–409.

Brindle, M., Graham, S., Harris, K. R., & Hebert, M. (2016). Third and fourth grade teacher's classroom practices in writing: A national survey. *Reading and Writing, 29*(5), 929–954.

Carless, D., & Boud, D. (2018). The development of student feedback literacy: Enabling uptake of feedback. *Assessment and Evaluation in Higher Education, 43*(8), 1315–1325.

Cazden, C. B. (1988). *Classroom discourse: The language of teaching and learning*. Heinemann.

Chen, C. F. E., & Cheng, W. Y. E. C. (2008). Beyond the design of automated writing evaluation: Pedagogical practices and perceived learning effectiveness in EFL writing classes. *Language Learning and Technology, 12*(2), 94–112.

Clarke, S. N., Howley, I., Resnick, L., & Rosé, C. P. (2016). Student agency to participate in dialogic science discussions. *Learning, Culture and Social Interaction, 10*, 27–39.

Coburn, C. E. (2004). Beyond decoupling: Rethinking the relationship between the institutional environment and the classroom. *Sociology of Education, 77*(3), 211–244.

Coburn, C. E. (2005). Shaping teacher sensemaking: School leaders and the enactment of reading policy. *Educational Policy, 19*(3), 476–509.

Cohen, D. K. (1990). A revolution in one classroom: The case of Mrs. Oublier. *Educational Evaluation and Policy Analysis, 12*(3), 311–329.

Cohen, D. K., Raudenbush, S. W., & Ball, D. L. (2003). Resources, instruction, and research. *Educational Evaluation and Policy Analysis, 25*(2), 119–142.

Collins, A., Brown, J. S., & Holum, A. (1991). Cognitive apprenticeship: Making thinking visible. *American Educator, 15*(3), 6–11.

Correnti, R., Matsumura, L. C., Hamilton, L. S., & Wang, E. (2013). Assessing students' skills at writing in response to texts. *Elementary School Journal, 114*(2), 142–177.

Correnti, R., Matsumura, L. C., Wang, E. L., Litman, D., & Zhang, H. (2022). Building a validity argument for an automated writing evaluation system (eRevise) as a formative assessment. *Computers and Education Open, 3*, 100084.

Correnti, R., Matsumura, L. C., Wang, E. L., Litman, D., Rahimi, Z., & Kisa, Z. (2019). Automated scoring of students' use of text evidence in writing. *Reading Research Quarterly, 55*(3), 493–520.

Council of Chief State School Officers (CCSSO). (2010). *Common core state standards: Standards for mathematical practice*. Authors.

Cuban, L. (2009). *Oversold and underused: Computers in the classroom.* Harvard University Press.
Deeva, G., Bogdanova, D., Serral, E., Snoeck, M., & De Weerdt, J. (2021). A review of automated feedback systems for learners: Classification framework, challenges, and opportunities. *Computers & Education, 162.*
Elliot, S. (2003). IntelliMetric: From here to validity. In M. D. Shermis & J. Burstein (Eds.), *Automated essay scoring: A cross-disciplinary perspective* (pp. 43–54). Routledge.
Ertmer, P. A. (2005). Teacher pedagogical beliefs: The final frontier in our quest for technology integration? *Educational Technology Research and Development, 53*(4), 25–39.
Ertmer, P. A., Ottenbreit-Leftwich, A. T., & Tondeur, J. (2015). Teachers' beliefs and uses of technology to support 21st-century teaching and learning. In H. Fives & M. G. Gills (Eds.), *International handbook of research on teacher beliefs* (pp. 403–418). Routledge.
Fang, Z. (1996). A review of research on teacher beliefs and practices. *Educational Research, 38*(1), 47–65.
Ferretti, R. P., & Graham, S. (2019). Argumentative writing: Theory, assessment, and instruction. *Reading and Writing, 32*(6), 1345–1357.
Gobert, J. D., Sao Pedro, M., Raziuddin, J., & Baker, R. S. (2013). From log files to assessment metrics: Measuring students' science inquiry skills using educational data mining. *Journal of the Learning Sciences, 22*(4), 521–563.
Goldenberg, C. N. (2004). *Successful school change: Creating settings to improve teaching and learning.* Teachers College Press.
Goldman, S. R., & Pellegrino, J. W. (2015). Research on learning and instruction: Implications for curriculum, instruction, and assessment. *Policy Insights from the Behavioral and Brain Sciences, 2*(1), 33–41.
Graham, S., Kiuhara, S. A., & MacKay, M. (2020). The effects of writing on learning in science, social studies, and mathematics: A meta-analysis. *Review of Educational Research, 90*(2), 179–226.
Graham, S., McKeown, D., Kiuhara, S., & Harris, K. R. (2012). A meta-analysis of writing instruction for students in the elementary grades. *Journal of Educational Psychology, 104*(4), 879–896.
Graham, S., & Perin, D. (2007). A meta-analysis of writing instruction for adolescent students. *Journal of Educational Psychology, 99*(3), 445–476.
Greeno, J. G., Collins, A. M., & Resnick, L. B. (1996). Cognition and learning. *Handbook of Educational Psychology, 77,* 15–46.
Hamilton, L. S., McCaffrey, D., Klein, S. P., Stecher, B. M., Robyn, A., & Bugliari, D. (2003). Studying large-scale reforms of instructional practice: An example from mathematics and science. *Educational Evaluation and Policy Analysis, 25*(1), 1–29.
Harris, K. R., Graham, S., & Mason, L. H. (2006). Improving the writing, knowledge, and motivation of struggling young writers: Effects of self-regulated strategy development with and without peer support. *American Educational Research Journal, 43*(2), 295–340.
Howe, E., & Correnti, R. (2020). Negotiating the political and pedagogical tensions of writing rubrics: Using conceptualization to work toward sociocultural writing instruction. *English Education, 52*(4), 335–360.
Kim, C., Kim, M. K., Lee, C., Spector, J. M., & DeMeester, K. (2013). Teacher beliefs and technology integration. *Teaching and Teacher Education, 29,* 76–85.
Kim, M. Y., & Wilkinson, I. A. (2019). What is dialogic teaching? Constructing, deconstructing, and reconstructing a pedagogy of classroom talk. *Learning, Culture and Social Interaction, 21,* 70–86.
Kisa, Z., & Correnti, R. (2015). Examining implementation fidelity in America's choice schools: A longitudinal analysis of changes in professional development associated with changes in teacher practice. *Educational Evaluation and Policy Analysis, 37*(4), 437–457.
Kiuhara, S. A., Graham, S., & Hawken, L. S. (2009). Teaching writing to high school students: A national survey. *Journal of Educational Psychology, 101*(1), 136–160.
Lampert, M., & Graziani, F. (2009). Instructional activities as a tool for teachers' and teacher educators' learning. *Elementary School Journal, 109*(5), 491–509.
Langer, J. A., & Applebee, A. N. (1987). *How writing shapes thinking: A study of teaching and learning.* NCTE Research Report No. 22. National Council of Teachers of English.
Litman, D., Zhang, H., Correnti, R., Matsumura, L. C., & Wang, E. (2021). A fairness evaluation of automated methods for scoring text evidence usage in writing. In *Proceedings of the 22nd artificial intelligence in education (AIE)* (pp. 255–267), June.
Louis, K. S., & Robinson, V. M. (2012). External mandates and instructional leadership: School leaders as mediating agents. *Journal of Educational Administration, 50*(5), 629–665.
Loukina, A., Madnani, N., & Zechner, K. (2019). The many dimensions of algorithmic fairness in educational applications. In *Proceedings of the fourteenth workshop on innovative use of NLP for building educational applications* (pp. 1–10). Association for Computational Linguistics.
MacArthur, C. A. (2018). Evaluation and revision. In S. Graham, C. A. MacArthur, & M. Hebert (Eds.), *Best practices in writing instruction* (pp. 287–308). Guilford.
Mathison, S., & Freeman, M. (2003). Constraining elementary teachers' work: Dilemmas and paradoxes created by state mandated testing. *Education Policy Analysis Archives, 11*(34). Retrieved October 16, 2021, from http://epaa.asu.edu/epaa/v11n34/
Matsumura, L. C., Correnti, R., & Wang, E. (2015). Classroom writing tasks and students' analytic text-based writing skills. *Reading Research Quarterly, 50*(4), 417–438.
Matsumura, L. C., Garnier, H., & Resnick, L. B. (2010). Implementing literacy coaching: The role of school social resources. *Educational Evaluation and Policy Analysis, 32*(2), 249–272.
Matsumura, L. C., Patthey-Chavez, G. G., Valdés, R., & Garnier, H. (2002). Teacher feedback, writing assignment quality, and third-grade students' revision in lower-and higher-achieving urban schools. *Elementary School Journal, 103*(1), 3–25.
Matsumura, L. C., & Wang, E. (2014). Principals' sensemaking of coaching for ambitious reading instruction in a high-stakes accountability policy environment. *Education Policy Analysis Archives, 22*(51). Retrieved October 16, 2021, from http://doi.org/10.14507/epaa.v22n51.2014

Matsumura, L. C., Wang, E., Correnti, R., & Litman, D. (2020, July 24). *What do teachers want to see in automated writing evaluation systems?* e-school News.

McFarland, J., Hussar, B., De Brey, C., Snyder, T., Wang, X., Wilkinson-Flicker, S., Semhar, G., Jijun, Z., Amy, R., Amy, B., Farrah, B. M., & Hinz, S. (2017). *The condition of education 2017. NCES 2017–144.* National Center for Education Statistics.

McLaughlin, M. W., & Talbert, J. E. (2006). *Building school-based teacher learning communities: Professional strategies to improve student achievement* (Vol. 45). Teachers College Press.

National Academies of Sciences, Engineering, and Medicine. (2018). *How people learn II: Learners, contexts, and cultures.* The National Academies Press.

Neumerski, C. M. (2013). Rethinking instructional leadership, a review: What do we know about principal, teacher, and coach instructional leadership, and where should we go from here? *Educational Administration Quarterly, 49*(2), 310–347.

Newmann, F. M., Bryk, A. S., & Nagaoka, J. K. (2001). *Authentic intellectual work and standardized tests: Conflict or coexistence? Improving Chicago's schools.* Consortium on Chicago School Research.

Next Generation Science Standards (NGSS) Lead States. (2013). *Next generation science standards: For states, by states.* The National Academies Press.

O'Hallaron, C. L. (2014). Supporting fifth-grade ELLs' argumentative writing development. *Written Communication, 31*(3), 304–331.

Orthaber, M., Stütz, D., Antretter, T., & Ebner, M. (2020). Concepts for e-assessments in STEM on the example of engineering mechanics. *International Journal of Emerging Technologies in Learning (iJET), 15*(12), 136–152.

Page, E. B. (2003). Project essay grade: PEG. In M. D. Shermis & J. Burstein (Eds.), *Automated essay scoring: A cross-disciplinary perspective* (pp. 43–54). Routledge.

Patthey-Chavez, G. G., Matsumura, L. C., & Valdes, R. (2004). Investigating the process approach to writing instruction in urban middle schools. *Journal of Adolescent and Adult Literacy, 47*(6), 462–476.

Prime, G. M., & Miranda, R. J. (2006). Urban public high school teachers' beliefs about science learner characteristics. *Urban Education, 41*(5), 506–532.

Rahimi, Z., Litman, D., Correnti, R., Matsumura, L. C., Wang, E., & Kisa, Z. (2014). Automatic scoring of an analytical response-to-text assessment. In S. Trausan-Matu, K. Boyer, M. Crosby, & K. Panourgia (Eds.), *Intelligent tutoring systems.* Paper presented at the 12th International Conference on Intelligent Tutoring Systems (ITS), (pp. 601–610). Springer.

Rahimi, Z., Litman, D., Correnti, R., Wang, E., & Matsumura, L. C. (2017). Assessing student's use of evidence and organization in response-to-text writing: Using natural language processing for rubric-based automated scoring. *International Journal of Artificial Intelligence in Education, 27*(4), 694–728.

Roschelle, J., Lester, J., & Fusco, J. (Eds.). (2020). *AI and the future of learning: Expert panel report* [Report]. Digital Promise. https://circls.org/reports/ai-report

Roscoe, R. D., Wilson, J., Johnson, A. C., & Mayra, C. R. (2017). Presentation, expectations, and experience: Sources of student perceptions of automated writing evaluation. *Computers in Human Behavior, 70,* 207–221.

Russell, J. L., & Bray, L. E. (2013). Crafting coherence from complex policy messages: Educators' perceptions of special education and standards-based accountability policies. *Education Policy Analysis Archives, 21*(12). Retrieved October 16, 2021, from http://epaa.asu.edu/ojs/article/view/1044

Ryoo, K., & Linn, M. C. (2016). Designing automated guidance for concept diagrams in inquiry instruction. *Journal of Research in Science Teaching, 53*(7), 1003–1035.

Schleppegrell, M. J. (2004). *The language of schooling: A functional linguistics perspective.* Routledge.

Shanahan, T., & Shanahan, C. (2008). Teaching disciplinary literacy to adolescents: Rethinking content-area literacy. *Harvard Educational Review, 78*(1), 40–59.

Shepard, L. A. (2005). Linking formative assessment to scaffolding. *Educational Leadership, 63*(3), 66–70.

Shute, V. J. (2008). Focus on formative feedback. *Review of Educational Research, 78*(1), 153–189.

Snow, C. E., & Uccelli, P. (2009). The challenge of academic language. In D. R. Olson & N. Torrance (Eds.), *The Cambridge handbook of literacy* (pp. 112–133). Cambridge University Press.

Spillane, J. P., Reiser, B. J., & Reimer, T. (2002). Policy implementation and cognition: Reframing and refocusing implementation research. *Review of Educational Research, 72*(3), 387–431.

Spillane, J. P., & Zeuli, J. S. (1999). Reform and teaching: Exploring patterns of practice in the context of national and state mathematics reforms. *Educational Evaluation and Policy Analysis, 21*(1), 1–27.

Stein, M. K., & Kaufman, J. H. (2010). Selecting and supporting the use of mathematics curricula at scale. *American Educational Research Journal, 47*(3), 663–693.

Tharp, R. G., & Gallimore, R. (1991). *Rousing minds to life: Teaching, learning, and schooling in social context.* Cambridge University Press.

Van Driel, J. H., Beijaard, D., & Verloop, N. (2001). Professional development and reform in science education: The role of teachers' practical knowledge. *Journal of Research in Science Teaching, 38*(2), 137–158.

Vygotsky, L. S. (1980). *Mind in society: The development of higher psychological processes.* Harvard University Press.

Vygotsky, L. S. (2012). *Thought and language.* MIT Press.

Wang, E. L., & Matsumura, L. C. (2018). Text-based writing in elementary classrooms: Teachers' conceptions and practice. *Reading and Writing, 32*(2), 405–438.

Wang, E. L., Matsumura, L. C., Correnti, R., Litman, D., Zhang, H., Howe, E., Magooda, A., & Quintana, R. (2020). *eRevis(ing)*: Students' revision of text evidence use in an automated writing evaluation system. *Assessing Writing, 44.* Retrieved October 16, 2021 from https://doi.org/10.1016/j.asw.2020.100449

Wang, E., Matsumura, L. C., & Correnti, R. (2017). Making a CASE: Improving use of text evidence in students' writing. *Reading Teacher, 70*(4), 479–484.

Wang, E., Matsumura, L. C., & Correnti, R. (2018). Student writing accepted as high-quality responses to analytic text-based writing tasks. *Elementary School Journal, 118*(3), 357–383.

Warschauer, M., Knobel, M., & Stone, L. (2004). Technology and equity in schooling: Deconstructing the digital divide. *Educational Policy, 18*(4), 562–588.

Wertsch, J. V., Minick, N., & Arns, F. A. (1984). The creation of context in joint problem solving. In B. Rogoff & J. Lave (Eds.), *Everyday cognition: Its development in social contexts* (pp. 151–171). Harvard University Press.

Williams, C., & Beam, S. (2019). Technology and writing: Review of research. *Computers and Education, 128*, 227–242.

Wilson, J., Olinghouse, N. G., & Andrada, G. N. (2014). Does automated feedback improve writing quality? *Learning Disabilities: A Contemporary Journal, 12*, 93–118.

Yinger, R. (1979). Routines in teacher planning. *Theory into Practice, 18*(3), 163–169.

Yurekli, B., Stein, M. K., Correnti, R., & Kisa, Z. (2020). Teaching mathematics for conceptual understanding: Teachers' beliefs and practices and the role of constraints. *Journal for Research in Mathematics Education, 51*(2), 234–247.

Zhai, X., Krajcik, J., & Pellegrino, J. W. (2021). On the validity of machine learning-based next generation science assessments: A validity inferential network. *Journal of Science Education and Technology, 30*(2), 298–312.

Zhang, H., & Litman, D. (2017). Word embedding for response-to-text assessment of evidence. In *Proceedings of the student research workshop of the annual meeting of the association for computational linguistics* (pp. 75–81).

Zhang, H., Magooda, A., Litman, D., Correnti, R., Wang, E. L., Matsumura, L. C., & Quintana, R. (2019). *eRevise*: Using natural language processing to provide formative feedback on text evidence usage in student writing. In *Proceedings of the AAAI conference on artificial intelligence* (Vol. 33, pp. 9619–9625).

Section IV

Learning Analytics and Educational Data Mining in AI and STEM Education

14

Promoting STEM Education through the Use of Learning Analytics: A Paradigm Shift

Shan Li and Susanne P. Lajoie

CONTENTS

14.1 Introduction ... 211
14.2 Learning Analytics as an Enabler for a Paradigm Shift in STEM Education ... 212
14.2.1 The Advancement of Data Collection Techniques ... 212
14.2.2 The Expansion of Data Analysis Methods ... 214
14.2.3 The Emergence of Innovative Visualization Strategies ... 215
14.3 Facilitating Change in STEM Education with Theory-Driven Learning Analytics ... 216
14.4 Case Example: Examining Students' STEM Learning with SRL Theory ... 218
14.4.1 Theory-Driven Data Collection ... 219
14.4.2 Theory-Driven Data Analysis and Visualization ... 219
14.4.3 Theory-Driven Interpretation of Results ... 220
14.5 The Future of STEM Education with Learning Analytics ... 220
References ... 222

14.1 Introduction

The origin of the STEM (Science, Technology, Engineering, and Math) movement can be traced back to the 1990s when the US National Science Foundation (NSF) highlighted the impact of information technology on undergraduate education in the domains of science, mathematics, engineering, and technology (Martín-Páez et al., 2019). The STEM movement was initiated as a response to the need for improving the innovation capacities of US students and maintaining American economic competitiveness. However, the idea of STEM quickly drew attention from many countries, policymakers, educators, and researchers worldwide. The research on STEM education has grown rapidly and it continues to be an area of research of high interest. It is noteworthy that STEM education has evolved since its inception, due to the development of learning theories and the prevalence of technologies in learning environments.

Compared to lecture-based instructions that are prevalent in traditional STEM education, learning in STEM disciplines has begun to provide more active and engaged forms, such as inquiry learning, problem-based learning, project-based learning, learning by self-explanation, and peer-tutoring (de Jong, 2019). Consequently, growing attention has shifted to students' learning experiences within STEM contexts rather than performance outcomes or learning products. In particular, there is an increased interest in studying the behavioral, cognitive, metacognitive, affective, and social processes of learning in STEM contexts, whether it be self-regulated learning or collaborative learning. As an illustration, Lamb et al. (2015) examined the interplay of cognition, affect, and content within a STEM integrated curriculum to develop a deep understanding of students' STEM learning processes. Moreover, the design and implementation of an integrative STEM curriculum is becoming more common since students need to use knowledge and skills from multiple disciplines to solve real-world STEM problems (Thibaut et al., 2018). In addition, the integration of advanced technologies has become an increasingly prominent feature in the landscape of STEM education. Educational institutions and companies can provide accessible and affordable STEM education at scale, due to the widespread adoption of learning management systems (LMSs), massive open online courses (MOOCs), intelligent tutoring systems (ITSs), and educational video games. Meanwhile, the development of computer simulations and immersive learning technologies present new opportunities for students to engage in STEM learning. For instance, online labs or simulations are

DOI: 10.1201/9781003181187-18

widely used in science education (de Jong, 2019). The Go-Lab is a good example that offers educators and students a federation of virtual and remote laboratories for inquiry-based STEM learning (Jong et al., 2014). As noticed by Sırakaya and Alsancak Sırakaya (2020) in a systematic review, recent years have witnessed an explosion of research on augmented reality (AR) in STEM education.

As learning occurs in various forms with technology-rich environments in the STEM field, researchers have come to realize the need to embrace the practices of learning analytics, such as multimodal data collection and the use of machine learning and data mining techniques, to gain insights into what works and what does not work in STEM teaching and learning. Regarding learning analytics, researchers generally adopt the definition proposed in the First International Conference on Learning Analytics held in 2011 (Gašević et al., 2015; Siemens, 2013). In particular, learning analytics is defined as 'the measurement, collection, analysis, and reporting of data about learners and their contexts, for the purposes of understanding and optimizing learning and the environments in which it occurs'. From a practical perspective, learning analytics provides an integrated solution that can address many challenges that have arisen in the recent development of STEM education concerning data collection, analysis, and visualization.

In this chapter, we first synthesize the use of learning analytics in STEM contexts with the primary aim of understanding how and in what ways learning analytics has been enabling a paradigm shift in STEM education. Particular emphasis will be placed on the advancement of data collection techniques, the expansion of data analysis methods, and the emergence of innovative visualization strategies. Then, we introduce a theory-driven learning analytics model, which has the potential to further promote the evolution of STEM education and research. We present an example study to illustrate how theory-driven learning analytics can be applied into practice in a STEM learning context. We close the chapter by discussing the future trends of learning analytics in STEM education.

14.2 Learning Analytics as an Enabler for a Paradigm Shift in STEM Education

Learning is essentially a complex process, which consists of different phases (e.g., goal setting, strategic planning, execution, and self-reflection) that are continuously affected by external and internal conditions such as one's cognition, metacognition, motivation, and emotion. Learning in STEM contexts is no exception. To promote a deep understanding of STEM learning, it is vital to design research projects using a systematic approach that can capture the richness of students' responses, reveal the underlying mechanisms of learning, and generate visual presentations of complex datasets. In fact, the use of advanced data collection techniques, sophisticated data analysis methods, and innovative visualization strategies has been reflected in a growing number of STEM-related studies. These new data collection, analysis, and visualization methodologies are essential research topics in the field of learning analytics, and meanwhile, are important drivers that could dramatically shift the paradigm of STEM education.

14.2.1 The Advancement of Data Collection Techniques

With the increasing accessibility to advanced data collection techniques such as log files, eye-tracking, and physiological sensors, researchers have been exploring different data modalities to gain fine-grained insights into the learning process. Di Mitri et al. (2018) proposed a taxonomy of multimodal data, in which two main branches of data modality were analyzed: behavioral motoric and behavioral physiological modalities. The behavioral motoric modalities are concerned with the movements of a specific part of the body, such as the head, torso, legs, arms, and hands. In particular, the movements of the head include analysis of facial expressions, eye-tracking, and speech analysis. Regarding the physiological modalities, Di Mitri et al. (2018) mainly referred to the heart, brain, and skin as the main organs, which produce electrical signals known as electrocardiogram (ECG), electroencephalograph (EEG), and electrodermal activity (EDA), respectively. It is noteworthy that Di Mitri et al. (2018) mentioned a third branch of data modality, and they framed it as the contextual modality but provided no further descriptions. We contend that the contextual modalities include, but are not limited to, digital trace (log files), text analysis, and discourse analysis. Relying on the taxonomy of multimodal data (Di Mitri et al., 2018), we briefly introduce each type of data modality and provide examples of use to demonstrate how the advancement in data collection techniques expands research opportunities in STEM contexts (see Table 14.1).

The behavioral motoric category of data modalities has received considerable attention within the STEM community since the data modalities involved in this category are directly observable and are informative

TABLE 14.1

The Taxonomy of Multimodal Data for Studying STEM Learning

Category	Modality	Examples of Use in STEM Learning
Behavioral motoric	Body language (posture, gestures, and motion)	It is ubiquitous in STEM fields that students reason about complex relational systems by creating spatial analogies with gestures (Cooperrider et al., 2016)
	Facial expression	The output of facial expression analysis includes facial action units and potentially emotions, which could be used for building affect-enhanced student models in science education (Sawyer et al., 2017)
	Eye movements	Eye movements are important indicators of students' attentional readiness and engagement in STEM learning (Pande & Chandrasekharan, 2014)
	Speech analysis	The elements of speech analysis, including linguistic, textual, and prosodic features, predicted students' level of expertise on engineering design tasks (Worsley & Blikstein, 2011)
Behavioral physiological	Electroencephalogram (EEG) for detecting the electrical signals of the brain	Students' engagement, memory retention, and social dynamics are reflected by brain-to-brain synchrony (i.e., the similarity in brain responses), which can be captured by EEG technology (Davidesco, 2020)
	Electrodermal activity (EDA) for measuring skin conductance	EDA can be used to identify attentional processing among engineering students. EDA is related to idea generation and creativity (Günay et al., 2020)
	Electrocardiogram (ECG) for heart rate tracking	The synchronization of heart rate signals among group members was associated with adaptation in socially shared regulation of learning (SSRL) in a physical course (Sobocinski et al., 2021)
	Electromyogram (EMG) for identifying muscles activities	EMG sensors make the levels of muscle contraction visible to students when designing human-interactive robots (Knop et al., 2017)
Contextual	Digital trace (log files)	Log files can be used to reveal the temporal dynamics of self-regulated learning behaviors in engineering design (Li et al., 2020)
	Text analysis	It is possible to identify different types of scientific explanations through text analysis and topic modeling in particular, as students described the existence of the four seasons (Sherin, 2013)
	Discourse analysis	Discourse analysis was used to explore power and positioning in small group interactions as students work together on an engineering design challenge (Wieselmann et al., 2021)

of students' learning experiences. First of all, students' body language, such as posture, gestures, and motion, are useful for detecting and assessing expertise (Worsley & Blikstein, 2013). For example, experts were found to have more two-handed actions than novices as they accomplished an engineering design task (Worsley & Blikstein, 2013). As pointed out by Cooperrider et al. (2016), it is ubiquitous in STEM fields that students perform abundant spatial gestures when reasoning about complex phenomena; therefore, it is plausible for researchers to infer students' reasoning activities by analyzing the quantity and quality of their gestures. In terms of facial expressions, they are extensively studied for emotion recognition in STEM learning. Researchers have also explored the use of facial expressions for building affect-enhanced student models in science education (Sawyer et al., 2017). Eye movement is another type of data modality that falls in the motoric category of multimodal data. In particular, eye movement provides a unique opportunity to study students' cognitive and metacognitive processes that no other data source can afford (van Gog & Scheiter, 2010). For example, the eye-tracking technique allows researchers to identify the stimuli a student is paying attention to, the order of visual processing, and the distribution of attention as they engage in a STEM task. The last type of data modality in the motoric category is speech signal, whereby a range of features can be extracted by speech analysis. For instance, Worsley and Blikstein (2011) predicted students' level of expertise on engineering design tasks, using the linguistic, textual, and prosodic features of speech signals. They found that uncertainty words were more common among novices than experts.

Moreover, the collection and analysis of physiological signals, including but not limited to electroencephalogram, electrodermal activity, electrocardiogram, and electromyogram (EMG), is an area of growing study in STEM research. EEG is an important technology in neuroscience that allows the detecting of the electrical signals in the brain, which can be further analyzed for recognizing brain activities and patterns. Davidesco (2020) provided an excellent review of research on the use of EEG in STEM classrooms. In particular, Davidesco (2020) introduced recent developments in portable brain technologies, and the potentials of EEG technology for understanding learning phenomena in STEM contexts. For instance, the extent of brain-to-brain synchrony (i.e., the similarity in brain responses) between students and teachers, which can be captured by EEG technology, reflects students' engagement, memory retention, and social closeness in classrooms. According to Davidesco (2020), EEG enables 'a deeper, more mechanistic understanding of STEM learning by illuminating what is happening in students' brains during the learning process and what factors mediate learning' (p. 5). With respect to EDA, it refers to the variation of the electrical conductance of the skin, which is usually quantified as skin conductance response (SCR) and skin conductance level (SCL). EDA provides a sensitive psychophysiological index, whose changes are widely considered as being related to students' affective and cognitive processes in learning (Ahonen et al., 2018). As an example, Günay et al. (2020) used EDA to investigate the neurocognitive aspects of ideation and divergent thinking among engineering students. ECG is another type of physiological signal, which contains a wealth of information of cardiac features such as heart rate (Ahonen et al., 2018). Researchers have explored the use of ECG to study collaborative learning in STEM contexts. For example, Sobocinski et al. (2021) explored adaptation in socially shared regulation of learning (SSRL) as students collaborated in the same groups throughout a physics course. Sobocinski et al. (2021) collected students' heart rate data with the E4 bracelet, and the heart rate data were later synchronized between group members. They contended that the synchronization of heart rate signals among group members was associated with adaptation in the shared regulatory process in collaboration. Heart rate synchrony could be an important indicator of 'sharedness' regarding students' monitoring and adaptation processes in SSRL (Sobocinski et al., 2021). Lastly, educators have begun to integrate EMG sensors in STEM-related programs, although the research is rare on how EMG signals associate with learning experiences and outcomes. For example, in the study conducted by Knop et al. (2017), students first developed an understanding of the changes in their muscle contraction levels measured by EMG sensors. They were then instructed to develop a sense of how EMG signals could control the motion of the Neu-pulator, an educational robotics platform. With the help of EMG sensors and the Neu-pulator platform, students were asked to design, program, manufacture, and test a human-interactive robot (i.e., robotic arm) that could be controlled by their arms.

The contextual modalities refer to data channels or techniques that capture students' interactions with a specific learning context. As the use of technology-rich learning environments (e.g., intelligent tutoring systems and computer simulations) has become pervasive in STEM education, researchers are able to automatically capture students' learning behaviors in log files at a precise level of detail. The analyses of log files allow for a nuanced understanding of students' strategical behaviors and decision-making processes. As an example, Li et al. (2020) used log files to examine the temporal dynamics of self-regulated learning behaviors in an engineering design task. They revealed how learning behaviors inhibit or promote each other over time and how the behavioral interactions account for performance differences. Furthermore, we contend that text is another type of contextual data modality, which can be easily gathered from online discussions, think-aloud protocols, self-explanations, interviews, and tests. Researchers can identify students' cognitive, metacognitive, and affective activities from texts using various text analysis techniques. For instance, Sherin (2013) used topic modeling and clustering technique to explore the progression of students' ideas, as they made scientific explanations for the existence of the four seasons (Blikstein & Worsley, 2016). Discourse analysis is similar to text analysis in that it takes discourse context (e.g., social context) to study written or spoken language. Discourse analysis is particularly useful for examining students' small group STEM interactions (Wieselmann et al., 2021). For instance, Wieselmann et al. (2021) used discourse analysis to explore power and positioning in small group interactions as students work together on an engineering design challenge.

The aforementioned data modalities are by no means exhaustive, but they do represent strong examples that leverage the state-of-the-art techniques to assess students at a fine-grained temporal level. The advancement of data collection techniques helps researchers develop a deep understanding of students' STEM learning since different data sources offer unique perspectives on learning. Furthermore, there is a general call on researchers to use multimodal data for analyzing the STEM learning process, considering that multimodal data are deemed as being more accurate than a single data source. For example, Ahonen et al. (2018) investigated students' collaboration during a self-directed pair-programming exercise using EDA and ECG simultaneously.

14.2.2 The Expansion of Data Analysis Methods

In parallel to the advancement of data collection practices are the extensive use of sophisticated data mining and machine learning techniques, which have been catalytic in shifting the paradigm of STEM education and research. As an illustration, a wide range of advanced analytical methods are available to analyze students' learning trajectories and their underlying patterns and trends. Example methods include sequential mining, process mining, state-transition analysis, recurrence quantification analysis, co-occurrence analysis, and hidden Markov models. According to Alexander et al. (2009), the third generation of expertise research focuses on the trajectory from novice learners to experts. When learning trajectories are dynamically analyzed and made visible to learners, the transition from novices to experts can be accelerated (Lajoie, 2003). Moreover, the advanced analytical techniques allow students to gain immediate feedback about where they have gone, how much time they have spent on a task, and what they might try next, by comparing their problem-solving trajectories with that of experts in real time. Providing scaffoldings for the less proficient through dynamic assessment, along with the overlay model of novice–expert learning trajectories, can promote efficient learning (Lajoie, 2020).

More recent analytics techniques attempt to uncover the complex interplays among students' characteristics, learning components, and contextual factors using structural analyses such as network analysis, social network analysis, epistemic network analysis, and latent semantic analysis (Siemens, 2013). For instance, Li et al. (2020) conceptualized the process of self-regulated learning (SRL) as a network of mutually interacting learning behaviors. As a result, they applied network analysis to examine the interplays of students' SRL behaviors in a STEM task. Epistemic network analysis (ENA) is another emerging technique that identifies connections among elements in coded data and represents them in dynamic networks (Shaffer et al., 2016). In particular, ENA provides a promising solution to quantify qualitative data of students' knowledge, skills, habits of mind, and other cognitive elements over the course of a learning task (Shaffer et al., 2016). As pointed out by Linn et al. (2004), a knowledge web, which contains a repertoire of ideas and their connections, is crucial for students to develop STEM expertise. ENA enables researchers to assess students' knowledge webs, which ultimately promotes a deep understanding of the complex nature of STEM learning.

In addition, researchers have used machine learning algorithms, such as neural networks, natural language processing, decision tree, random forest, and support vector machine, to model learners, knowledge domains, and the development of expertise. Due to space restrictions, we do not provide a complete list of learning analytics techniques that are prevalent in researching STEM education. However, the presented examples have clearly demonstrated that advanced analytics methods and techniques have been opening up new possibilities for data analyses and generating new insights about STEM learning.

14.2.3 The Emergence of Innovative Visualization Strategies

According to Monroy et al. (2013), one of the biggest challenges facing STEM educators and stakeholders is turning a massive volume of data into meaningful information, and visualization is a crucial tool to address this challenge. The most commonly used visualization strategies include but are not limited to graph, statistical visualizations (e.g., bar plot, line plot, pie chart, scatter plot), spiral timeline, word cloud, interaction matrix, circular graph, heatmap, bubble plot, concept map, radar plot, glyph, and GeoMap (Vieira et al., 2018). As an example, Monroy et al. (2013) used the visualization strategy of timelines to understand the sequence and pacing of teaching activities in an online K-12 science curriculum. Moreover, heat maps were used to illustrate the level of students' involvement in the seven types of scientific inquiry activities, which include the 5E (engagement, exploration, explanation, elaboration, and evaluation) and intervention/acceleration activities. Visualization is an important research topic in learning analytics and is an emerging research field. To be specific, Vieira et al. (2018) proposed a new field of research, namely, visual learning analytics, based on a systematic literature view at the intersection of learning analytics and visual analytics. In particular, Vieira et al. (2018) defined visual learning analytics as 'the use of computational tools and methods for understanding educational phenomena through interactive visualization techniques' (p. 120). The emergence of visual learning analytics, as a powerful integrative field, affords students the opportunity to monitor their learning behaviors and progress in real time. Meanwhile, educators can gain direct insights about students' learning trajectories and performance from visualizations and use them as placeholders to provide personalized scaffoldings and feedback in a timely manner. For example, Minović et al. (2015) designed a new form of diagram to visualize student-overlapping models during educational game sessions. Based on the diagram, educators can provide hints or interact with students to improve their learning outcomes. As STEM researchers and educators continuously take advantage of visual learning analytics, a paradigm shift becomes a possibility. In addition to single visualization strategies, learning analytics dashboards (LAD) are being designed as an integrated approach that presents multiple views of various aspects of students' learning (Matcha et al., 2019). According to Schwendimann et al. (2016), LAD refers to 'a single display that aggregate different indicators about learner(s), learning process(es) and/or learning context(s) into one or multiple visualization' (p. 8). By visualizing abstract data into meaningful representations, LADs provide stakeholders the opportunities for awareness, reflection, sense-making, and behavioral change (Verbert et al., 2013).

In sum, the increasing adoption of learning analytics in STEM education is leading to a paradigm shift towards a new era of STEM research that is transdisciplinary, data-driven, and technology-intensive (see Figure 14.1). While data-driven approaches of learning analytics enable researchers and educators to discover patterns in learner-generated data in a holistic manner, some researchers argue that the results generated in such approaches are often recognized as not sufficiently informative (Gašević et al., 2015; Mangaroska & Giannakos, 2018). In fact, the field of learning analytics suffers from a critical criticism that it lacks theoretical foundations in terms of data collection, analysis, interpretation, and visualization (Dawson et al., 2015; Matcha et al., 2019; Rogers et al., 2016; Viberg et al., 2018). As an illustration, Viberg et al. (2018) conducted a systematic literature review of studies on learning analytics in higher education, yielding a number of 252 papers. They found that most papers (57%) were descriptive studies, which described a phenomenon without any use of theory. In terms of data visualization and learning analytics dashboards in particular, Matcha et al. (2019) found that most of the learning analytics dashboards (69%) were not grounded in any established educational theories, based on a systematic review of empirical studies on learning analytics dashboards. As pointed out by Mangaroska and Giannakos (2018), learning analytics has been making a gradual shift from technological toward educational orientation over the past ten years. For this reason, a theory-driven approach to learning analytics will become increasingly important to ensure its continuing impact on STEM education across various learning contexts.

14.3 Facilitating Change in STEM Education with Theory-Driven Learning Analytics

In this section, we describe a theory-driven learning analytics model, which is adapted from Siemens's (2013) model, to study students' learning in STEM contexts. Siemens (2013) proposed a learning analytics model (LAM), which includes seven components: collection, storage, data cleaning, integration, analysis, representation and visualization, and action. In particular, Siemens's (2013) model takes advantage of systemwide approaches to analytics, which provides researchers and practitioners with guidance on how to initiate analytics projects at an organizational level. For instance, Siemens (2013) claimed that learning management systems (LMS) and student information systems (SIS) are the two main sources of data collection, although it is imperative to increase the scope and quality of data capture for future research. The purposes of learning analytics vary, including marketing, advising, administration, learning, and institutional research. However, as discussed earlier, one limitation of Siemens's (2013) model is that it did not illustrate the role of learning theories in learning analytics practices. Moreover, Siemens's (2013) model offers limited insights on the optimal practice of learning analytics at the individual or task levels. We attempt to fulfill this gap by proposing the theory-driven learning analytic model (see Figure 14.2). The proposed model highlights the ultimate goal of enhancing students' learning experiences and promoting their learning performance with learning analytics technologies that are guided by theories.

As shown in Figure 14.2, in the hub of the theory-driven learning analytic model are theories, including

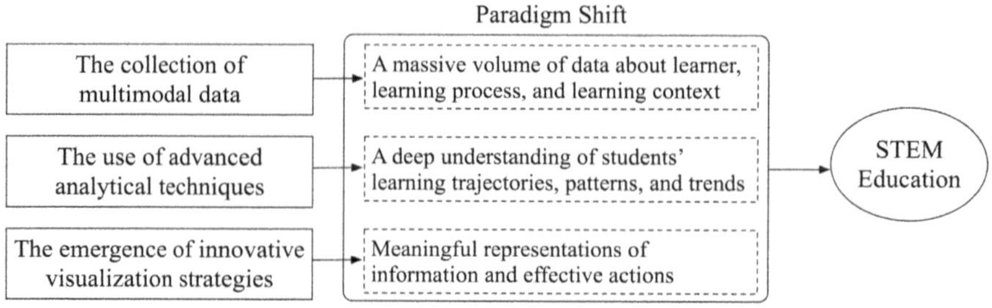

FIGURE 14.1
A paradigm shift in STEM education with learning analytics.

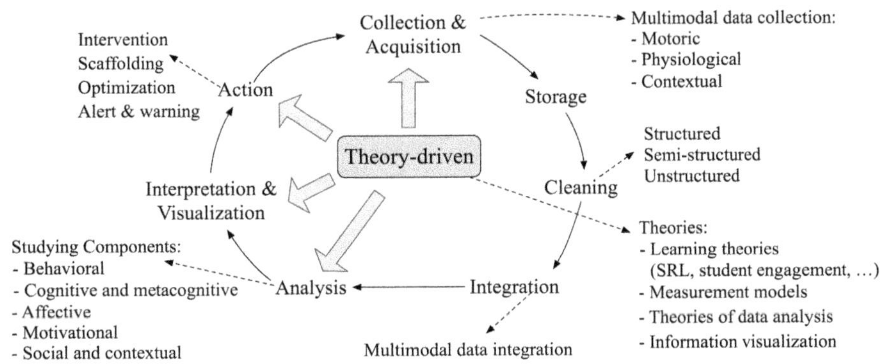

FIGURE 14.2
Theory-driven learning analytics model for examining students' learning. *Note:* This model was adapted from Siemens's (2013) learning analytics model (LAM).

learning theories, established measurement models, theories of data analysis, and information visualization theories. The selection of learning theories depends on the phenomenon under study and the nature of research questions. Some of the primary learning theories include self-regulated learning (SRL), socially shared regulation of learning (SSRL), social and emotional learning (SEL), student engagement, community of inquiry, knowledge building, and cognitive apprenticeship. The role of measurement models in learning analytics is to guarantee the reliability and validity of measurement instruments and to ensure the generalizability, comparability, and fairness of research findings. When the measurement instruments are tests or psychological questionnaires, a variety of measurement models exist, such as factor analysis, latent class and latent mixture models, item response theory, growth models, and cognitive diagnostic models (Bergner, 2017). Future research should focus on the development of measurement models that illustrate the connections between advanced data sources (e.g., eye movements, facial expression, EDA, and EEG) and psychological constructs. Moreover, the theory-driven learning analytic model highlights the importance of theories of data analysis, suggesting that data analysis methods and techniques need to align with the ontological assumptions of research questions. For instance, researchers are expected to use nonlinear techniques for data analysis (e.g., nonlinear time series analysis, dynamic modeling, and network analysis) when investigating phenomena conceptualized as complex dynamical systems (Hilpert & Marchand, 2018). Finally, as visualization in learning analytics becomes increasingly important and sophisticated, it is crucial to refer to information visualization theories when designing visual representations and interactive visual interfaces.

As we look into the theory-driven learning analytics model, it contains the same seven components as Siemens's (2013) model, but they do not necessarily share the same meanings. For instance, the data collection and acquisition processes of Siemens's (2013) model refer to the capture of data in scale or scope, whereas the data collection in the theory-driven learning analytics model deals with different modalities of data about students' learning at an individual level. In compliance with the taxonomy of multimodal data, researchers can collect motoric, physiological, and contextual types of data, as well as their combinations, to gain insights into students' learning. The collected data need to be deliberately stored, cleaned, and integrated before they can be used for analyses. The purposes of data analysis are typically to explore learning components in the dimensions of behavior, cognition, metacognition, emotion, motivation, and social and contextual factors. Afterward, researchers make interpretations and visualizations based on the results of data analyses, so that meaningful and effective action plans can be generated regarding learning intervention, scaffolding, optimization, alert, and warning. The seven components form a cyclical process, with each stage laying the foundation for the next. In order to fully realize the potential of learning analytics, Gašević et al. (2015) argued that 'the field of learning analytics needs to ground data collection, measurement, analysis, reporting and interpretation processes within the existing research on learning' (p. 66). Therefore, the proposed model suggests that the processes of data collection and acquisition, analysis, interpretation and visualization, and action should be guided by theories in particular.

In the following section, we introduce an example study conducted in the STEM context to illustrate how theory-driven learning analytics can be applied to practice. As highlighted by Joksimović et al. (2019), 'research in learning analytics has rapidly progressed from studies developing predictive models of student retention to more acute challenges linked to affect, self-regulation and feedback processes' (p. 52). In fact, self-regulated learning (SRL) is a commonly referenced theoretical framework in learning analytics research (Lim et al., 2021; Viberg et al., 2020). SRL skills have also proved to be crucial in determining students' performance in STEM learning (Li et al., 2020; Zheng et al., 2020). Therefore, the case example below demonstrates a practice of theory-driven learning analytics that relies on SRL theory to gain insights into students' problem-solving processes in engineering design (Li et al., 2020). We chose engineering design since it 'provides the opportunities to build connections among the STEM disciplines' (Kelley & Knowles, 2016, p. 5). The process of engineering design often involves scientific inquiry, technology considerations, and mathematical reasoning.

14.4 Case Example: Examining Students' STEM Learning with SRL Theory

Self-regulated learning (SRL) is a widely adopted theoretical framework that describes how students adaptively monitor and control the behavioral, cognitive, metacognitive, and affective aspects of learning process to accomplish personal goals (Greene & Azevedo, 2007; Pintrich, 2004; Schunk & Greene, 2017; Winne, 2017; Zimmerman, 2000). According to Zimmerman (2000), SRL is an active, cyclical process that consists of three phases: forethought, performance, and self-reflection. In the forethought phase, students develop an understanding of task conditions (i.e., task analysis), set goals for learning, and make strategic plans for achieving those goals. In the performance phase, students use a wide variety of techniques, such as self-instruction, imagery, attention focusing, and task strategies, to optimize their effort in learning or problem-solving. Meanwhile, students track their learning progresses, and monitor whether their effort will yield expected performance. The last phase of SRL, self-reflection, involves students self-evaluating their performance and making adaptations accordingly. Ample empirical evidence has demonstrated that the most effective SRL processes are 'structurally interrelated and cyclically sustained' (Zimmerman, 2000, p. 15).

In STEM education, there is a growing effort to promote students' performance by fostering their SRL skills. For example, Barak (2012) noticed that experienced engineering teachers tended to focus extensively on the technical side of project work, but they began to give attention to the development of SRL skills for enhancing learning. For researchers, SRL theory is becoming the main choice for investigating students' STEM learning since STEM tasks are usually ill-structured problems, which underscore the autonomous role of learners in making plans, decisions, and adaptations throughout the problem-solving process. In the example study, we relied on a situated model of SRL to explain students' learning in an engineering design task (see Figure 14.3) (Li et al., 2020; Zheng et al., 2020).

We first introduce the task and the learning environment to help readers better understand the adapted SRL model, given that the SRL behaviors involved in this model are task-dependent. In this study (Li et al., 2020), a total of 111 ninth-grade students were recruited to design a green building in a computer-simulated environment, Energy3D (Xie et al., 2018). Specifically, Energy3D is a simulation-based computer-aided design (CAD) environment, which allows users to sketch a realistic-looking house structure in a short time with its 3D modeling tools. Students can also obtain information about the cost and the energy performance of a house with the analysis tools embedded in the Energy3D platform. To successfully accomplish the task, students were asked to design an energy-efficient house, which consumed only as much energy as could be produced by solar panels for heating and cooling in a calendar year. Moreover, students were expected to design a Colonial-style house, which should meet the following requirements: the house needs to demonstrate curb appeal; the ratio of the total area of windows to the floor area must be

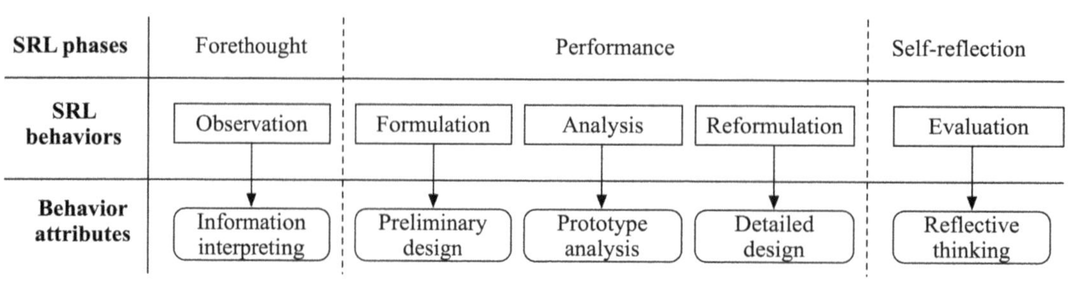

FIGURE 14.3
The situated model of self-regulated learning in engineering design.

between 0.05 and 0.15; the roof overhang must be less than 50 cm wide; the area of the house needs to be between 120 and 160 m^2, and the height needs to be between 8 and 10 m. In addition, the budget for the house should be within $200,000.

Students began the task by familiarizing themselves with the task environments and conditions, a process that involves extensive observation behaviors in the context of engineering design. For example, students explore available functions of the Energy3D platform, such as rotating the building, zooming in and out for different views, and showing shadow, axes, or heliodon. Meanwhile, students develop an understanding of the design specs. They may propose several ideas to accomplish the task and ruminate over a specific plan. All of these processes and beliefs occur before efforts to design, which fall in the forethought phase according to Zimmerman's (2000) three-phase model of SRL. Students then take actions to construct a house (i.e., formulation behaviors) by adding walls, windows, doors, roofs, solar panels, etc. To meet the design specs, students conduct a series of analyses, for instance, calculating the annual energy usage for heating and cooling. Based on the analysis results, students decide how to revise their design to improve the energy efficiency of the house, i.e., the reformulation process. They may resize the house, adjust the height for the selected wall, or change the color for the whole building. In the adapted SRL model for engineering design, the formulation, analysis, and reformulation behaviors comprise the performance phase of SRL. The last phase, self-reflection, involves students reflecting on their design performance, which is triggered by the embedded prompts throughout the design process.

In this example study, we examined the temporal dynamics of SRL behaviors (i.e., observation, formulation, analysis, reformulation, and evaluation) as students solved the engineering design task. We were particularly interested in how SRL behaviors interact with each other over time and how these interactions may lead to performance differences. For the purpose of this chapter, we demonstrate how the data were collected, analyzed, and interpreted in theory-driven approaches to address our research questions.

14.4.1 Theory-Driven Data Collection

In line with the situated model of SRL in engineering design, we collected the information on the sequential occurrences of the five types of SRL behaviors. In particular, we extracted the information from log files, which stored students' operational behaviors that were automatically recorded in real time in their designing processes. By leveraging the affordances of an SRL framework, we were able to understand the purposes and meanings of students' behaviors, even if some behaviors seemed to be isolated events or noisy data. It is also noteworthy that we used log files to study SRL, which resonated with the recent development to view SRL as events instead of aptitudes (Greene & Azevedo, 2007; Schunk & Greene, 2017). Log files provide fine-grained details of how students' behaviors in engineering design unfold, allowing researchers to explore the temporal dependence of SRL behaviors.

14.4.2 Theory-Driven Data Analysis and Visualization

We analyzed the interactions among SRL behaviors in a network approach, which is rooted in the complex systems (CS) theory (Hilpert & Marchand, 2018). According to the CS theory, the relationship among components of a CS is considered interdependent. The components of a CS augment or inhibit each other over time, making up the internal dynamics of the system and the very nature of the phenomenon studied (Bringmann et al., 2016). The internal dynamics of system components, rather than the components per se, produce an emergent outcome. As such, the outcome of a CS cannot be fully predicted with linear methods. We contended that SRL is essentially a complex system, consisting of complex interplays among behavioral, cognitive, metacognitive, affective, and environmental factors. At a microlevel, SRL behaviors can also be viewed as a CS of interacting components, meaning the role of SRL behaviors and the strength and direction of their relationships change over time. Based on this understanding, we did not examine the causal relationships between SRL behaviors and students' learning outcomes. Instead, we focused on the interactions among SRL behaviors and how the interactions influenced students' performance.

As aforementioned, we used a network method to study the system of SRL behaviors since 'the underlying structure of a CS is a network' (Hilpert & Marchand, 2018, p. 196). In fact, as pointed out by Bringmann et al. (2016), network analysis is one of the most popular approaches to study, analyze, and visualize the dynamic processes of a CS. Specifically, we used the multilevel vector autoregression (VAR) model (Bringmann et al., 2016) to build the network of SRL behaviors. The multilevel VAR model allows for the estimations of network edges at both individual and group levels, so that we can compare differences in the network of SRL behaviors within individuals, as well as between different performance groups. In this study, we classified students into three groups based on their performance profiles: unsuccessful,

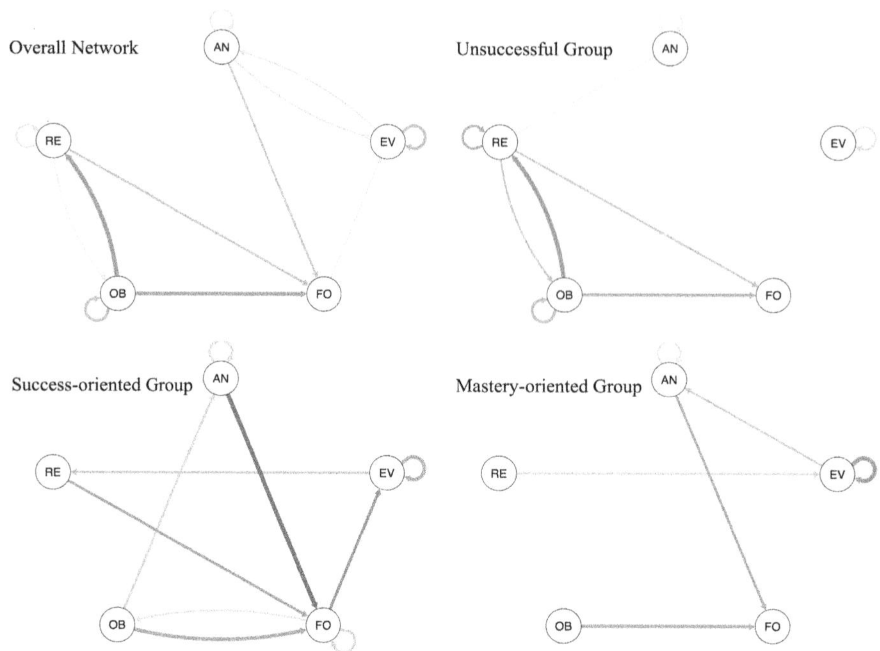

FIGURE 14.4
The SRL behavioral networks of different performance groups. *Note:* The figure in the top left shows the overall network of SRL behaviors of all participants. OB = observation, FO = formulation, AN = analysis, RE = reformulation, EV = evaluation. Only edges that surpass the significance threshold (i.e., $p = .05$) are shown.

success-oriented, and mastery-oriented groups. We built an SRL behavioral network for each group and compared their differences. We presented our results in the visualization of networks (see Figure 14.4), which enabled an immediate intuitive understanding of the complex interactions of SRL behaviors. Specifically, the directed weighted networks showed the strength and direction of the relationships between any of the two connected nodes. As shown in Figure 14.4, the thickness of the arrows indicated the weight of the edges. For this reason, researchers and practitioners could visually pinpoint the differences in SRL behavioral networks among the three performance groups.

14.4.3 Theory-Driven Interpretation of Results

When it came to the interpretation of our research findings, we referred back to SRL theory to understand how the features of students' behavioral networks might explain their performance differences. Due to space constraints, we will not discuss all of our findings but rather focus on one key finding and how we perceive this finding from an SRL perspective. Specifically, we found that the SRL behavioral network of the mastery-oriented group demonstrated a clear interaction pattern among SRL behaviors, i.e., reformulation behaviors stimulated evaluation behaviors, evaluation behaviors inhibited analysis behaviors, and analysis behaviors inhibited formulation behaviors. According to SRL theorists, SRL behaviors occur in a loose sequence, with behaviors at an earlier stage promoting the occurrences of later behaviors in an SRL cycle (Pintrich, 2004; Schunk & Greene, 2017; Winne, 2017; Zimmerman, 2000). While our finding is in line with this theoretical standpoint, our study suggested further that SRL behaviors restrained the occurrences of behaviors at an earlier stage. An explanation is that students performed quality SRL behaviors for each step of problem-solving so that they did not need to repeat previous behaviors.

14.5 The Future of STEM Education with Learning Analytics

As we envision the next generation of STEM education empowered by learning analytics, we should keep in mind that learning analytics per se is a young field that is fast-growing. While we talk about the future development of STEM education, we sometimes refer to the trends in learning analytics and see how they might be represented in the STEM contexts. In general, there is a trend toward integrating learning analytics into learning designs (Gašević et al., 2017; Joksimović et al., 2019). Whereas current STEM education practice introduces learning analytics because of its availability, future research will be more impactful when it carefully considers the role of learning analytics in each step of design (e.g., pedagogical strategies, use of technologies, and sequence of learning activities) and how learning analytics connects with specific instructional objectives (Gašević et al., 2017). When learning analytics becomes an integral part of STEM learning design, it can realize the promise of delivering optimal STEM learning experiences for each student.

A crucial aspect of learning analytics concerns the collection of multimodal data about learners and learning contexts. The collection of multimodal data enables researchers to get a whole picture of students' learning in STEM contexts since researchers can gather different pieces of evidence about the learning process in the dimensions of behavior, cognition, metacognition, motivation, and emotion. However, multimodal data collection faces several challenges such as the affordability of relevant hardware and software, the accessibility to expert guidance, the participant effects caused by the invasive nature of various physical and psychological sensors, and the constrained research potentials for large-scale studies (Hassan et al., 2021). Therefore, a promising direction for future research on learning analytics in STEM education is the collection of multimodal data through a web-based approach that leverages the affordances of computer vision and machine learning techniques (Hassan et al., 2021). For instance, Wu et al. (2012) developed a computer vision algorithm to extract and reveal the pulse signals (e.g., heart rate) of participants from video frames in real time based on the subtle temporal changes in face colors. This approach could be an important substitute for an electrocardiogram, which traditionally requires participants to wear sensors for measuring heart rate. As another example, Li et al. (2021) extracted three categories of facial behavioral cues (i.e., eye gaze, head pose, and facial action units) from students' facial expressions, which were recorded via a computer webcam as students solved a clinical reasoning problem in a technology-rich learning environment. They built a machine learning model to infer students' cognitive engagement states from their facial behavioral cues, which provided a novel and non-intrusive approach to capture cognitive engagement. Furthermore, Hassan et al. (2021) developed a web-based multimodal data collection system of the EZ-MMAL Toolkit, which 'runs entirely within the web browser' (p. 581). The system supports the extraction of multimodal data (i.e., body posture, face orientation, gaze location, emotion, and heart rate) from video and audio input that can either be uploaded or captured in real time. In sum, we contend that the use of computer vision and machine learning models, especially the web-based applications designed to collect multimodal data, will reshape how we teach STEM disciplines and how we research STEM education.

Moreover, learning analytics is a practical field, which is heavily influenced by the progress in data analysis methods and techniques. There is no doubt that new techniques for data analysis will continue to emerge and consequently help researchers generate new insights about learning in STEM contexts. For instance, Wieselmann et al. (2021) introduced what they called 'a microethnographic approach to critical multimodal discourse analysis' in a study of examining complex small group interactions within STEM activities. The approach consists of three phases of data analysis: microethnographic methods to develop individual and group storylines (Phase 1), critical multimodal discourse analysis of key events (Phase 2), and the integration of storylines and discourse analysis (Phase 3). In particular, Phase 2 involves the analyses of students' spoken discourse, prosodic features of speech (i.e., pauses, intonation and stress, tone, and volume), kinetic (e.g., gestures, posture, gaze, facial expressions), and proxemic features. As claimed by Wieselmann et al. (2021), the combination of ethnographic and multimodal methods allows for a nuanced analysis of students' power and positioning in collaboration, making the approach particularly useful in analyzing small group interactions within integrated STEM contexts. In sum, research on STEM education is being methodologically influenced by learning analytics, especially when it comes to data analysis. Meanwhile, it is reasonable to expect that advanced learning analytics tools that are unique for analyzing STEM phenomena will be gradually proposed by more and more researchers.

Additionally, as learning analytics dashboards are growing in popularity in STEM education, more research is needed regarding (1) the visualization of multimodal data (e.g., log files, emotions, bodily and verbal language) captured in STEM learning contexts; (2) the selection of effective indicators for visualization across different STEM domains and learning environments; (3) the impact of learning analytics dashboards on STEM education from the perspectives of diverse stakeholders such as administrators, educators, and students; (4) the applications of learning analytics dashboards in K-12 and informal settings instead of the university settings that are currently drawing the most attention; (5) the design of learning analytics dashboards that are capable of displaying data with different levels of complexity in a meaningful, discernible, and interpretable approach; (6) the innovations in visual displays of not only raw data but also the underlying patterns and trends about teaching and learning activities; (7) the criteria or standards for evaluating a learning analytics dashboard with respect to its effectiveness, efficiency, usability, and usefulness; and (8) the alignment between features of learning analytics dashboards and STEM-focused educational interventions (Klerkx et al., 2017; Matcha et al., 2019; Schwendimann et al., 2016). These promising directions in the field of visual learning analytics have the potential to facilitate the decision-making process, reflection, and adaptive performance in STEM education.

Most importantly, the field of STEM education can be substantially enhanced by learning analytics practices that are grounded in learning sciences and contemporary educational theories. The value proposition of learning analytics lies in the meaningful insights and actionable recommendations drawn from data rather than the data per se. As learning analytics is shifting its focus from the prediction of students' retention and performance to a more refined understanding of learning processes (Joksimović et al., 2019), it should rely upon what we know about students' learning and how to improve learning performance. Only in this way can learning analytics play a significant role in educational research. In sum, we are hopeful that learning analytics research in STEM contexts will continue to progress. However, to maximize the benefits of learning analytics, researchers must collectively realize the importance of the theory-driven approach of learning analytics and embrace this approach when addressing various challenges in STEM teaching and learning.

References

Ahonen, L., Cowley, B. U., Hellas, A., & Puolamäki, K. (2018). Biosignals reflect pair-dynamics in collaborative work: EDA and ECG study of pair-programming in a classroom environment. *Scientific Reports, 8*(1), 1–16.

Alexander, P. A., Murphy, P. K., & Kulikowich, J. M. (2009). Expertise and the adult learner: A historical, psychological, and methodological exploration. In M. C. Smith & N. DeFrates-Densch (Eds.), *Handbook of research on adult learning and development* (pp. 484–523). Routledge.

Barak, M. (2012). From 'doing'to 'doing with learning': Reflection on an effort to promote self-regulated learning in technological projects in high school. *European Journal of Engineering Education, 37*(1), 105–116.

Bergner, Y. (2017). Measurement and its uses in learning analytics. In C. Lang, G. Siemens, A. F. Wise, & D. Gaševic (Eds.), *The handbook of learning analytics* (1st ed., pp. 34–48). Society for Learning Analytics Research (SoLAR). http://solaresearch.org/hla-17/hla17-chapter1

Blikstein, P., & Worsley, M. (2016). Multimodal learning analytics and education data mining: Using computational technologies to measure complex learning tasks. *Journal of Learning Analytics, 3*(2), 220–238.

Bringmann, L. F., Pe, M. L., Vissers, N., Ceulemans, E., Borsboom, D., Vanpaemel, W., Tuerlinckx, F., & Kuppens, P. (2016). Assessing temporal emotion dynamics using networks. *Assessment, 23*(4), 425–435.

Cooperrider, K., Gentner, D., & Goldin-Meadow, S. (2016). Spatial analogies pervade complex relational reasoning: Evidence from spontaneous gestures. *Cognitive Research: Principles and Implications, 1*(1), 1–17.

Davidesco, I. (2020). Brain-to-brain synchrony in the STEM classroom. *CBE – Life Sciences Education, 19*(3), es8.

Dawson, S., Mirriahi, N., & Gasevic, D. (2015). Importance of theory in learning analytics in formal and workplace settings. *Journal of Learning Analytics, 2*(2), 1–4.

de Jong, T. (2019). Moving towards engaged learning in STEM domains; there is no simple answer, but clearly a road ahead. *Journal of Computer Assisted Learning, 35*(2), 153–167. https://doi.org/10.1111/jcal.12337

Di Mitri, D., Schneider, J., Specht, M., & Drachsler, H. (2018). From signals to knowledge: A conceptual model for multimodal learning analytics. *Journal of Computer Assisted Learning, 34*(4), 338–349.

Gašević, D., Dawson, S., & Siemens, G. (2015). Let's not forget: Learning analytics are about learning. *TechTrends, 59*(1), 64–71.

Gašević, D., Kovanović, V., & Joksimović, S. (2017). Piecing the learning analytics puzzle: A consolidated model of a field of research and practice. *Learning: Research and Practice, 3*(1), 63–78. https://doi.org/10.1080/23735082.2017.1286142

Greene, J. A., & Azevedo, R. (2007). A theoretical review of Winne and Hadwin's model of self-regulated learning: New perspectives and directions. *Review of Educational Research, 77*(3), 334–372.

Günay, E. E., Tapia, L. D., Gould, T., Raje, S., Scallon, J., Teberg, J., & Kremer, G. E. O. (2020). Electrodermal activity experiment to enhance divergent thinking in engineering students. In *IIE annual conference. Proceedings* (pp. 1383–1388).

Hassan, J., Leong, J., & Schneider, B. (2021). Multimodal data collection made easy: The EZ-MMLA toolkit. In *LAK21: 11th international learning analytics and knowledge conference* (pp. 579–585). Association for Computing Machinery. https://doi.org/10.1145/3448139.3448201

Hilpert, J. C., & Marchand, G. C. (2018). Complex systems research in educational psychology: Aligning theory and method. *Educational Psychologist, 53*(3), 185–202. https://doi.org/10.1080/00461520.2018.1469411

Joksimović, S., Kovanović, V., & Dawson, S. (2019). The journey of learning analytics. *HERDSA Review of Higher Education, 6*, 27–63.

Jong, T. De, Sotiriou, S., & Gillet, D. (2014). Innovations in STEM education: The Go-Lab federation of online labs. *Smart Learning Environments, 1*(3), 1–16.

Kelley, T. R., & Knowles, J. G. (2016). A conceptual framework for integrated STEM education. *International Journal of STEM Education, 3*(1), 1–11.

Klerkx, J., Verbert, K., & Duval, E. (2017). Learning analytics dashboards. In C. Lang, G. Siemens, A. F. Wise, & D. Gaševic (Eds.), *The handbook of learning analytics* (1st ed., pp. 143–150). Society for Learning Analytics Research (SoLAR). http://solaresearch.org/hla-17/hla17-chapter1

Knop, L., Ziaeefard, S., Ribeiro, G. A., Page, B. R., Ficanha, E., Miller, M. H., Rastgaar, M., & Mahmoudian, N. (2017). A human-interactive robotic program for middle school stem education. *2017 IEEE Frontiers in Education Conference (FIE)*, 1–7.

Lajoie, S. P. (2003). Transitions and trajectories for studies of expertise. *Educational Researcher, 32*(8), 21–25.

Lajoie, S. P. (2020). Student modeling for individuals and groups: The BioWorld and HOWARD platforms. *International Journal of Artificial Intelligence in Education*, 1–16. https://doi.org/10.1007/s40593-020-00219-x

Lamb, R., Akmal, T., & Petrie, K. (2015). Development of a cognition-priming model describing learning in a STEM classroom. *Journal of Research in Science Teaching, 52*(3), 410–437.

Li, S., Du, H., Xing, W., Zheng, J., Chen, G., & Xie, C. (2020). Examining temporal dynamics of self-regulated learning behaviors in STEM learning: A network approach. *Computers and Education, 158*, 103987. https://doi.org/10.1016/j.compedu.2020.103987

Li, S., Lajoie, S. P., Zheng, J., Wu, H., & Cheng, H. (2021). Automated detection of cognitive engagement to inform the art of staying engaged in problem-solving. *Computers and Education, 163*, 104114. https://doi.org/10.1016/j.compedu.2020.104114

Lim, L.-A., Gasevic, D., Matcha, W., Ahmad Uzir, N., & Dawson, S. (2021). Impact of learning analytics feedback on self-regulated learning: Triangulating behavioural logs with students' recall. In *LAK21: 11th international learning analytics and knowledge conference* (pp. 364–374). Association for Computing Machinery. https://doi.org/10.1145/3448139.3448174

Linn, M. C., Eylon, B.-S., & Davis, E. A. (2004). The knowledge integration perspective on learning. In M. C. Linn, E. A. Davis, & P. Bell (Eds.), *Internet environments for science education* (pp. 29–46). Routledge.

Mangaroska, K., & Giannakos, M. (2018). Learning analytics for learning design: A systematic literature review of analytics-driven design to enhance learning. *IEEE Transactions on Learning Technologies, 12*(4), 516–534.

Martín-Páez, T., Aguilera, D., Perales-Palacios, F. J., & Vílchez-González, J. M. (2019). What are we talking about when we talk about STEM education? A review of literature. *Science Education, 103*(4), 799–822.

Matcha, W., Gašević, D., & Pardo, A. (2019). A systematic review of empirical studies on learning analytics dashboards: A self-regulated learning perspective. *IEEE Transactions on Learning Technologies, 13*(2), 226–245.

Minović, M., Milovanović, M., Šošević, U., & Conde González, M. Á. (2015). Visualisation of student learning model in serious games. *Computers in Human Behavior, 47*, 98–107. https://doi.org/10.1016/j.chb.2014.09.005

Monroy, C., Rangel, V. S., & Whitaker, R. (2013). STEMscopes: Contextualizing learning analytics in a K-12 science curriculum. *Proceedings of the third international conference on learning analytics and knowledge* (pp. 210–219). https://doi.org/10.1145/2460296.2460339

Pande, P., & Chandrasekharan, S. (2014). Eye-tracking in STEM education research: Limitations, experiences and possible extensions. *2014 IEEE sixth international conference on technology for education* (pp. 116–119).

Pintrich, P. R. (2004). A conceptual framework for assessing motivation and self-regulated learning in college students. *Educational Psychology Review, 16*(4), 385–407. https://doi.org/10.1007/s10648-004-0006-x

Rogers, T., Dawson, S., & Gašević, D. (2016). Learning analytics and the imperative for theory-driven research. In C. Haythornthwaite, R. Andrews, J. Fransman, & E. M. Meyers (Eds.), *The SAGE handbook of e-learning research* (pp. 232–250). SAGE Publications Ltd. https://doi.org/10.4135/9781473955011

Sawyer, R., Smith, A., Rowe, J., Azevedo, R., & Lester, J. (2017). Enhancing student models in game-based learning with facial expression recognition. *Proceedings of the 25th conference on user modeling, adaptation and personalization* (pp. 192–201). https://doi.org/10.1145/3079628.3079686

Schunk, D. H., & Greene, J. A. (2017). Historical, contemporary, and future perspectives on self-regulated learning and performance. In P. A. Alexander, D. H. Schunk, & J. A. Greene (Eds.), *Handbook of self-regulation of learning and performance* (2nd ed., pp. 1–15). Routledge. https://doi.org/10.4324/9781315697048-1

Schwendimann, B. A., Rodriguez-Triana, M. J., Vozniuk, A., Prieto, L. P., Boroujeni, M. S., Holzer, A., Gillet, D., & Dillenbourg, P. (2016). Perceiving learning at a glance: A systematic literature review of learning dashboard research. *IEEE Transactions on Learning Technologies, 10*(1), 30–41.

Shaffer, D. W., Collier, W., & Ruis, A. R. (2016). A tutorial on epistemic network analysis: Analyzing the structure of connections in cognitive, social, and interaction data. *Journal of Learning Analytics, 3*(3), 9–45.

Sherin, B. (2013). A computational study of commonsense science: An exploration in the automated analysis of clinical interview data. *Journal of the Learning Sciences, 22*(4), 600–638. https://doi.org/10.1080/10508406.2013.836654

Siemens, G. (2013). Learning analytics: The emergence of a discipline. *American Behavioral Scientist, 57*(10), 1380–1400.

Sırakaya, M., & Alsancak Sırakaya, D. (2020). Augmented reality in STEM education: A systematic review. *Interactive Learning Environments*, 1–14.

Sobocinski, M., Malmberg, J., & Järvelä, S. (2021). Exploring adaptation in socially shared regulation of learning using video and heart rate data. *Technology, Knowledge and Learning*, 1–20. https://doi.org/10.1007/s10758-021-09526-1

Thibaut, L., Ceuppens, S., De Loof, H., De Meester, J., Goovaerts, L., Struyf, A., Boeve-de Pauw, J., Dehaene, W., Deprez, J., & De Cock, M. (2018). Integrated STEM education: A systematic review of instructional practices in secondary education. *European Journal of STEM Education, 3*(1), 2.

van Gog, T., & Scheiter, K. (2010). Eye tracking as a tool to study and enhance multimedia learning. *Learning and Instruction, 20*(2), 95–99. https://doi.org/10.1016/j.learninstruc.2009.02.009

Verbert, K., Duval, E., Klerkx, J., Govaerts, S., & Santos, J. L. (2013). Learning analytics dashboard applications. *American Behavioral Scientist, 57*(10), 1500–1509.

Viberg, O., Hatakka, M., Bälter, O., & Mavroudi, A. (2018). The current landscape of learning analytics in higher education. *Computers in Human Behavior, 89*, 98–110.

Viberg, O., Khalil, M., & Baars, M. (2020). Self-regulated learning and learning analytics in online learning environments: A review of empirical research. In *Proceedings of the tenth international conference on learning analytics & knowledge* (pp. 524–533).

Vieira, C., Parsons, P., & Byrd, V. (2018). Visual learning analytics of educational data: A systematic literature review and research agenda. *Computers and Education, 122*, 119–135. https://doi.org/10.1016/j.compedu.2018.03.018

Wieselmann, J. R., Keratithamkul, K., Dare, E. A., Ring-Whalen, E. A., & Roehrig, G. H. (2021). Discourse analysis in integrated STEM activities: Methods for examining power and positioning in small group interactions. *Research in Science Education, 51*(1), 113–133.

Winne, P. H. (2017). Cognition and metacognition within self-regulated learning. In P. A. Alexander, D. H. Schunk, & J. A. Greene (Eds.), *Handbook of self-regulation of learning and performance* (2nd ed., pp. 36–48). Routledge. https://doi.org/10.4324/9781315697048-3

Worsley, M., & Blikstein, P. (2011). What's an expert? Using learning analytics to identify emergent markers of expertise through automated speech, sentiment and sketch analysis. *EDM 2011 - Proceedings of the 4th international conference on educational data mining* (pp. 234–239).

Worsley, M., & Blikstein, P. (2013). Towards the development of multimodal action based assessment. In *Proceedings of the third international conference on learning analytics and knowledge* (pp. 94–101). https://doi.org/10.1145/2460296.2460315

Wu, H.-Y., Rubinstein, M., Shih, E., Guttag, J., Durand, F., & Freeman, W. (2012). Eulerian video magnification for revealing subtle changes in the world. *ACM Transactions on Graphics (TOG), 31*(4), 1–8.

Xie, C., Schimpf, C., Chao, J., Nourian, S., & Massicotte, J. (2018). Learning and teaching engineering design through modeling and simulation on a CAD platform. *Computer Applications in Engineering Education, 26*(4), 824–840.

Zheng, J., Xing, W., Zhu, G., Chen, G., Zhao, H., & Xie, C. (2020). Profiling self-regulation behaviors in STEM learning of engineering design. *Computers and Education, 143*, 103669. https://doi.org/10.1016/j.compedu.2019.103669

Zimmerman, B. J. (2000). Attaining self-regulation: A social cognitive perspective. In M. Boekaerts, P. R. Paul, & M. Zeidner (Eds.), *Handbook of self-regulation* (1st ed., pp. 13–39). Academic Press. https://doi.org/10.1016/B978-012109890-2/50031-7

15

Using Learning Analytics to Understand Students' Discourse and Behaviors in STEM Education

Gaoxia Zhu, Wanli Xing, Vitaliy Popov, Yaoran Li, Charles Xie, and Paul Horwitz

CONTENTS

15.1 Introduction 225
15.2 STEM Education 226
15.3 Technology-Enhanced Environments to Support STEM Education 226
15.3.1 Teaching Teamwork for Collaborative Problem-Solving 227
15.3.2 Energy3D for Engineering Design 228
15.4 Learning Analytics to Analyze Generated Data 229
15.4.1 Analytical Methods 230
15.5 Case Studies 232
15.5.1 Science 232
15.5.1.1 Transformative and Non-Transformative Discourse 232
15.5.1.2 Self-Regulation 232
15.5.1.3 Multi-Faceted Engagement 233
15.5.2 Engineering 233
15.5.3 Mathematics 234
15.6 Research Gaps and Future Trends 234
15.7 Conclusion 236
References 236

15.1 Introduction

STEM education aims to cultivate students' capacity to understand and resolve complex problems using STEM content knowledge or approaches (Washington STEM Study Group, 2011). It usually involves more active pedagogical approaches such as inquiry-based learning and problem-solving learning in which students take active roles in navigating multiple disciplines to form hypotheses, conduct experiments, collect data, interpret data, draw conclusions and, eventually, solve problems (de Jong, 2006). Active learning often takes place in technology-rich learning environments (e.g., National Academies of Sciences Engineering Medicine, 2018) that can generate rich learning process data such as log files, chat messages, and clickstreams. The great quantity and rich forms of data trails make it possible to explore learning (e.g., students' engineering design process, interactions between group members) from new perspectives. Learning analytics, an emerging field, provides new premises in studying learning process data to better understand learning in real time and in a more comprehensive manner.

This chapter reviewed and synthesized our studies on employing learning analytics to understand students' transformative discourse, learning engagement, self-regulation, and design process in the STEM education context. The outline of this chapter is as follows: first, we briefly review the history and development of STEM education and its associated pedagogical approaches and learning environments. Second, we describe the technology-enhanced learning environments used to support STEM education with an emphasis on the Teaching Teamwork simulation platform (Horwitz et al., 2017) and an Energy3D modeling tool (Xie et al., 2018) because they were more frequently used in our studies. Next, we introduce what is learning analytics and why it is a promising field for understanding and supporting STEM education. We also synthesize the analytical methods employed in our studies. Later, by integrating the technology-enhanced learning environments, data collected, and analytical methods adopted, we

DOI: 10.1201/9781003181187-19

elaborate on several cases concerning how students' transformative discourse, self-regulation, multiple-faceted engagement, and other constructs were examined in STEM education contexts. Finally, we discuss future directions in the field of learning analytics to respond to current research gaps or challenges.

15.2 STEM Education

The STEM movement started in the 1990s when there was a recognized need to better prepare US students to develop STEM identities, skills, and competencies to maintain economic competitiveness (Bøe et al., 2011; US National Academy of Sciences, National Academy of Engineering, Institute of Medicine, 2007). STEM education is about cultivating students' capacity to identify and apply STEM content knowledge to understand and resolve complex situations that cannot be addressed using a mono-disciplinary knowledge or approach (Washington STEM Study Group, 2011). Similarly, Bybee (2010a) suggested, 'A true STEM education should increase students' understanding of how things work' (p. 996). Previous studies suggest STEM education benefits the cognitive, procedural, and attitudinal aspects of students, such as boosting academic performance (Chang et al., 2015) and enriching STEM knowledge (Toma & Greca, 2018); developing creativity and technology skills (Duran et al., 2014; Lamb et al., 2015); and promoting positive attitude and commitment towards STEM disciplines (Toma & Greca, 2018). A review of the positive impacts of STEM education and relevant studies can be found in Martín-Páez et al. (2019).

STEM education, focusing on solving real problems, has an 'interdisciplinary nature' in its nucleus (Bybee, 2013). There is an increasing trend to call for STEM education to facilitate students to solve complex issues in integrated learning environments. Bybee (2010a, 2010b) suggested that STEM education should improve students' use of technologies and enhance the engineering component during precollege education because, currently, STEM is usually interpreted as Math or Science. Bybee (2010b) advocated to recognize the importance of Engineering and Technology as they are closely related to our daily life but people know less about them; furthermore, engineering is directly involved in problem-solving and innovation, which are critical to the economic priorities of every nation (Lichtenberg et al., 2008).

STEM education usually requires more student-centered pedagogical approaches such as inquiry-based learning, project-based learning, or problem-solving learning. In active or inquiry learning, students usually play more active roles in which they may form hypotheses, conduct experiments, collect data, interpret data, and draw conclusions (de Jong, 2006). This learning falls under the umbrella of constructivism and socio-constructivism that emphasizes students to construct their own knowledge through interacting with ideas, materials, and technologies in a particular learning and socio-cultural context (Woolfolk et al., 2009). Active learning has several benefits over teachers' direct instruction. A teacher-led directive educational approach may not always be effective to help students to develop deep conceptual knowledge in STEM learning (de Jong, 2019). In a meta-analysis of 225 studies, Freeman et al. (2014) found that students in active/engaged learning performed better than students in more direct forms of learning.

In the context of the STEM movement, many initiatives have taken place in K-12 and higher education, in- and outside of schools, at the individual or collaborative learning levels. For instance, at the K-12 level, Moreno et al. (2016) implemented after-school STEM activities (e.g., Think Like an Astronaut) to help elementary students, especially the ones who might otherwise lose interest in STEM, to transit to middle school and introduce them to STEM careers. At the higher education level, Van den Beemt et al. (2020) researched how undergraduate students in Industrial & Engineering Management, Civil Engineering, and Applied Mathematics programs collaborated to help hospitals to solve planning and routing problems. STEM education can happen at the level of individual learning or in collaborative learning contexts. Moore et al. (2014) advocated that 'teamwork' and 'communication-related to engineering' are key indicators of quality K-12 engineering education because of their importance to practicing engineers.

15.3 Technology-Enhanced Environments to Support STEM Education

Active/engaged learning or in particular inquiry learning usually takes place in technology-enhanced learning environments (e.g., National Academies of Sciences Engineering Medicine, 2018). Nowadays, experiential and inquiry learning is increasingly enabled by simulations (e.g., online laboratories), games, and modeling tools (de Jong et al., 2018). These technologies have several benefits. First, web-based or mobile applications (e.g., online labs or simulations) are more accessible to learners and make it possible for them to engage in self-directed learning anywhere, anytime, and with any device (de Jong, 2019). Second, technologies are often interactive and allow learners to engage with a domain and extend their knowledge through testing their ideas, exploring relationships between variables, and investigating the mechanisms of how things work (de Jong et al., 2018). Third, in some cases, technologies are more practical (de Jong et al., 2018). For instance, simulations make invisible things visible so people can see, for example, molecules and evolution. Augmented reality can integrate virtual content in the real world and allow students to interact with virtual objects in the real environment which may enhance their sense of presence (Sırakaya & Alsancak Sırakaya, 2020). Doing inquiry with technologies may also be less dangerous and can save chemical substances (Alessi & Trollip, 2001). Fourth, learning games make learning more interesting, and rewarding and can provide real-time challenges and feedback (Lyons, 2015).

Some widely used online labs and simulations include PhET, Go-Lab, WISE, etc. (de Jong et al., 2018). In addition to the simulation components, games add features such as competition, rules, goal setting, and rewards to make them more attractive and engaging (Leemkuil & de Jong, 2011). Modelings such as NetLogo and Scratch enable students to create models underlying simulations (de Jong & van Joolingen, 2008). This chapter does not aim to provide a comprehensive review of technologies that support STEM inquiry learning. Relevant details can be found in de Jong et al. (2018, 2019). Instead, we elaborate on a Teaching Teamwork simulation platform and an Energy3D modeling tool that was more frequently used in our studies.

15.3.1 Teaching Teamwork for Collaborative Problem-Solving

Web-based Teaching Teamwork platform features a series of interactive STEM simulations that support learners to collaboratively explore scientific phenomena and examine the relationships between different variables through interaction with the system. For instance, Figure 15.1 shows the interface of Ohm's Law activities. This interface mainly includes three parts: (1) the first part consists of group members' information, the goal of the activity, a calculator, and buttons for students to submit their results. In this shown task, the goal for each group member is to make their voltage 1.9 volts when the voltage E and resistor R0 are given. (2) The second part is a simulation of a series circuit including supply voltage E and resistors R0, R1, R2, and R3. In a group of three, each member can

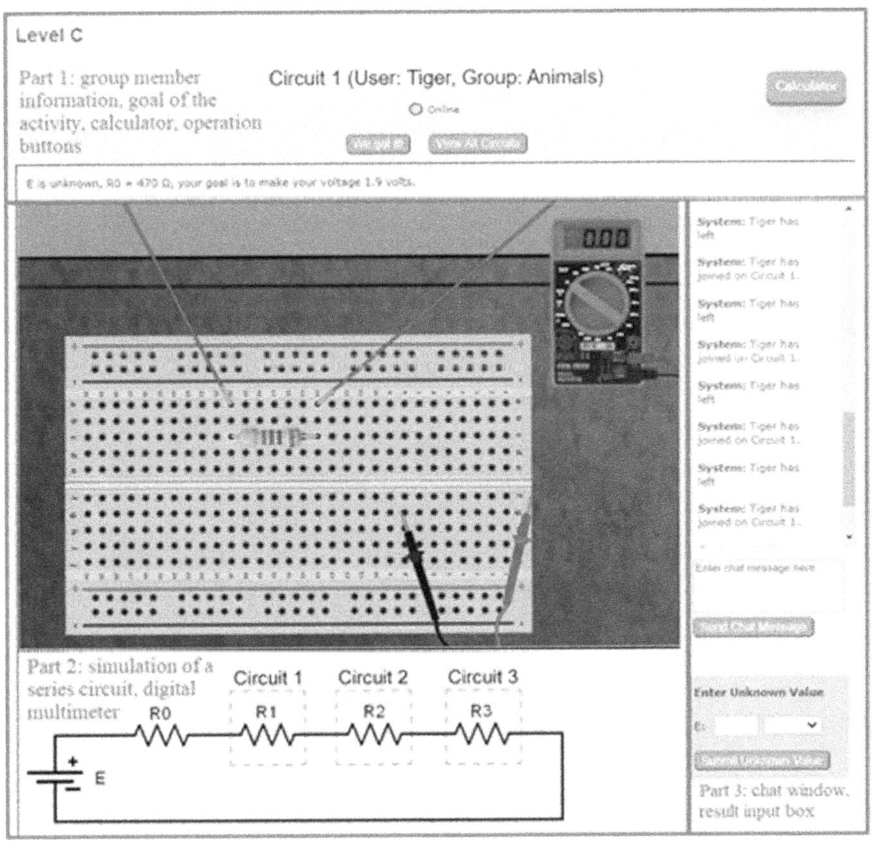

FIGURE 15.1
The interface of an Ohm's Law activity in Teaching Teamwork.

control a resistor – R1, R2, or R3 – and regulate the resistance to achieve their goal. A simulation of a digital multimeter with black and red probes is provided for students to measure the resistance and voltage of the resistor that they control and the current of the circuit. Several tasks with increasing complexity are provided. For example in the final task, the values of E and R0 are unknown, and the goal voltages across R1, R2, and R3 are not equal. The reason for such a design is to ensure students work at the complexity level appropriate for their knowledge and skills rather than overwhelming them. (3) The third part is a chat window in which group members can discuss their goal, use their relevant knowledge to make plans for solving the task, report and monitor progress, and adjust plans. This part also includes an area for students to submit their results. The series circuit requires the group members to communicate and collaborate through the chat window as the change of the resistance they control would influence the current across the circuit and, therefore, influence the voltage across other group members' resistors.

15.3.2 Energy3D for Engineering Design

Energy3D is an engineering design simulation and modeling tool for users to design buildings or power stations to harness renewable energy (https://energy.concord.org/energy3d/; please see its successor Aladdin at https://intofuture.org/aladdin.html). Here we take designing a green building as an example to illustrate the affordances of Energy3D. Figure 15.2 shows the interface of the tool. It includes four major parts. First, Part 1 includes toolbars such as 3D modeling tools (e.g., floor, wall, window, solar panel) for students to model their designed buildings; simulations that can show shadow or heliodon, and animate sun to support students' design by facilitating their understanding of the solar environment. Second, Part 2 visualizes students' design artifacts. The 'what you see is what you get' user interface design is likely to support students to timely adjust their design solutions and specific parameters. Third, Part 3 shows the parameters of the design project such as date, location, and temperature, which can be set by students. Furthermore, when students choose to run analysis on their design environments (e.g., showing shadow, showing heatmap, see Figure 15.3) or artifacts (e.g., computing energy and solar energy, and energy annual analysis), the analysis results are shown to help students understand their design environments or evaluate the energy performance of their design. Finally, Part 4 hides a reflection section with embedded prompts such as 'describe your design ideas and explain why you think they are good ideas'. Once students click the 'note' label, this area will be shown,

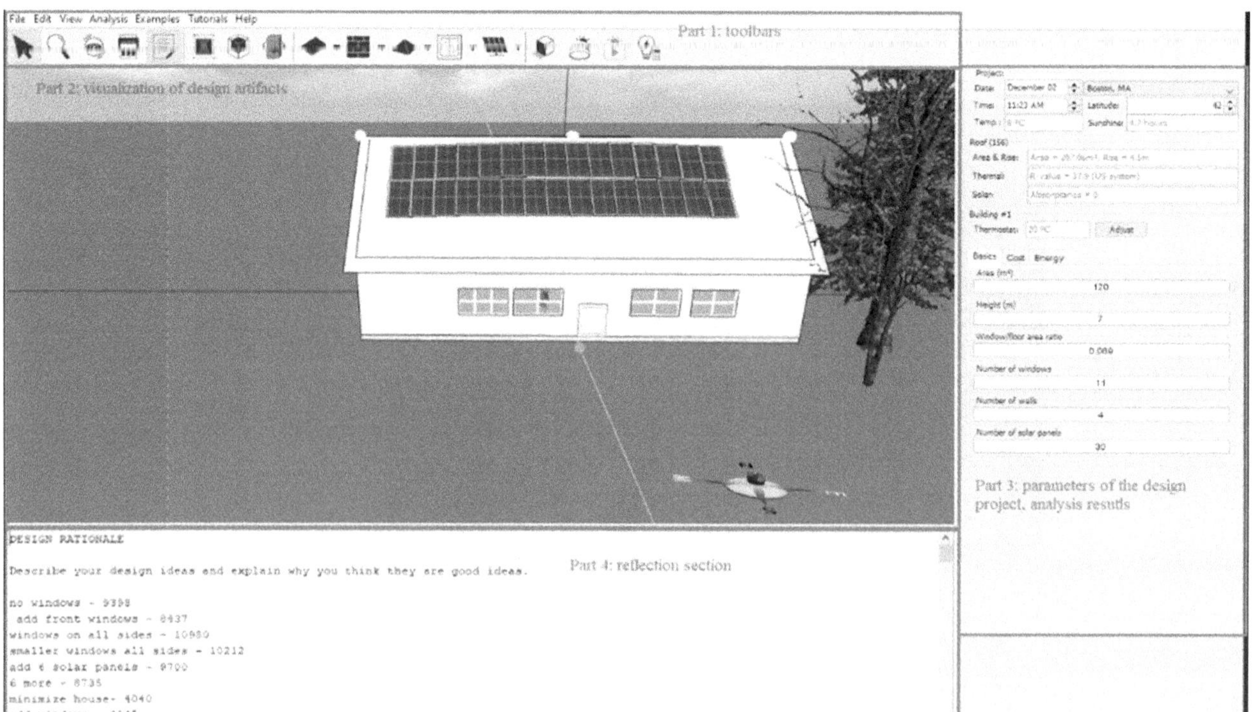

FIGURE 15.2
The interface of Enery3D.

FIGURE 15.3
Heatmap analysis and visualization (from Enery3D successor Aladdin).

and the prompts can guide students' reflections on their design.

15.4 Learning Analytics to Analyze Generated Data

Rich learning process data can be generated and logged for analysis in these environments. Learning analytics, an emerging field, provides new premises in studying learning process data, makes it possible to understand learning from new perspectives, and may provide comprehensive and real-time feedback to students and teachers to better understand students' learning. Learning analytics is about the 'measurement, collection, analysis and reporting of data about learners and their contexts, for purposes of understanding and optimizing learning and environments in which it occurs' (https://tekri.athabascau.ca/analytics/). The basic research objectives of learning analytics and educational data mining are student/student behavior modeling, prediction of performance, increase (self-)reflection and (self-)awareness, prediction of dropout and retention, improve feedback and assessment services, and recommendation of resources (Papamitsiou & Economid, 2014).

Learning analytics is mainly used in STEM education. Papamitsiou and Economid (2014) conducted a systematic review of the literature on learning analytics and educational data mining from 2008 to 2013 and found that STEM was the most dominant domain area. Li and Wong (2020) reviewed case studies of the use of learning analytics in STEM Education from 2013 to 2018 and suggested that learning analytics has been increasingly employed in STEM education in recent years. They summarized that enhancing understanding in STEM practices; providing tailored feedback and support; tracking learning progress; and increasing retention rate are the four most prominent purposes of adopting learning analytics in STEM practices (all greater than 10% of all the reviewed studies). Detailed analysis of the current research status of learning analytics in STEM education can be found in these reviews.

The types of data collected and analyzed by learning analytics are evolving. At an earlier period, static information such as demographic factors, participation in activities, and skills are usually collected to predict students' academic performance (e.g., assessment quizzes, final scores [Abdous et al., 2012; Huang & Fang, 2013]). At a later stage, researchers are more likely to collect learning processes-related data (Li & Wong, 2020). Discourse and behaviors may be the two most common learning process data sources recorded in learning logs. Discourse is usually generated in the collaborative learning context in which group members need to communicate to collaboratively solve problems. Cognitive, interactive, and metacognitive information can be manifested in group discourse. In technology-enhanced learning environments, problem-solving solutions need to be transformed into actions that can be performed in the environments. Accordingly, a behavioral stream is generated as learners navigate through different functions to solve problems. Currently, there is an increasing trend to collect bodily data such as face and hand tracking (e.g. Hossain et al., 2018) to better understand students' embodiment movement and engagement.

Recent educational studies on learning analytics have shown a growing interest in examining fair AI towards building accountable and trustworthy educational systems to intelligently support teaching and learning. A majority of educational studies have focused on evaluating the fairness of learning analytics models with fair AI metrics. These metrics can be used to examine predictive models' accuracy among demographic groups (e.g., females vs. males), offering insights on whether models will favor students with a specific demographic background. For example, Riazy and Simbeck (2019) built models to predict whether students pass or fail an online course. They found that their models could be biased against students with disabilities, overestimating their probability of failing the course. In another study, Hutt et al. (2019) examined if learning analytics models is biased in terms of race and socioeconomic background when used to predict students' graduation time. Their findings indicated that the predictive models they built showed minimal bias. Similarly, Sha et al. (2021) utilized natural language processing techniques to automatically classify students' forum posts and evaluated these classifiers' fairness on various demographic factors such as gender and native language. The results showed that their models tended to misunderstand students whose mother tongue was not English.

In our studies, the Teaching Teamwork platform saves both behavioral logs (e.g., changing circuits, using the digital multimeter, adjusting resistance, and performing calculations) of each group member and chat messages between group members. In Zhu et al. (2021), we categorized all the relevant engineering design behaviors into six dimensions: observation of surroundings (including five raw behaviors), building (41 raw behaviors), changing an environment parameter (five raw behaviors), reflection (one raw behavior only), evaluation (nine raw design behaviors), and reformulation (34 raw design behaviors). Details of the raw behaviors of each category can be found in Table 2 in Zhu et al. (2021).

15.4.1 Analytical Methods

Various methods have been adopted to examine students' learning process and performance. Papamitsiou and Economid (2014) classified the analysis method into classification, clustering, regression, text mining, association rule mining, social network analysis, discovery with models, visualization, and statistics. In our studies, we have mainly adopted text classification, clustering analysis, transition rate analysis and sequential pattern mining, multilevel modeling, and Multilevel Model Vector Autoregression to analyze the logged discourse and behaviors in Teaching Teamwork and Energy3D (see Table 15.1 for a summary of these methods). Next, we elaborate on these analytical methods.

We supervised text classification models to categorize the discourse generated during students' group problem-solving process into transformative and non-transformative dimensions (Popov et al., 2018; Xing et al., 2019; Zhu et al., 2019); self-regulation and socially shared regulation discourse (Zheng et al., 2019). The general process is that we first manually coded the discourse based on the coding schemes that we developed based on the relevant literature and our data. Then the manual coding results were used to supervise machine learning models using algorithms such as Decision Tree, Logistic Regression, Naïve Bayes, and Support Vector Machines in order to automatically classify the discourse. The performance of the models would be compared to select the best-performed one which could be applied to the remaining discourse. Such text classification reduces tedious manual coding work.

K-means clustering analysis can identify subgroups having similar properties and characteristics concerning input elements within groups and distinct features between groups (Na et al., 2010). K-means clustering is an unsupervised machine learning algorithm that requires a predetermined optimal number of clusters. There are several ways to help identify the optimal number of clusters. For instance, Silhouette is a visual and classic method that calculates the silhouette coefficients of observation for different k values and selects the k associated with the maximum average silhouette (Kaufman & Rousseeuw, 2009). Ball statistic is a classic measure for determining the optimal number (Milligan & Gooper, 1985). It gauges the dispersion of data points within a cluster and between clusters.

Both transition rate analysis and sequential pattern mining are analyses for uncovering the temporal characteristics of a set of sequences of events. Transition rate analysis measures the possibility for a state to transit to other states. Sequential pattern mining can identify the frequently occurring subsequences among a set of sequences based on a determined threshold. The sequences can be discourse states of groups' members, individuals' design behaviors, or any other states. Both methods are increasingly used

TABLE 15.1

A Brief Summary of the Analytical Methods Employed in Our Studies on STEM Education

Methods	Descriptions	Considerations (i.e., pros and cons)	Relevant examples
Text classification	Machine learning algorithms (e.g., Decision Tree, Logistic Regression, Naïve Bayes, and Support Vector Machines) can be employed to supervise models that can automatically classify students' chat messages or other text information into various categories.	Coding schemes need to be developed first based on relevant literature and data; manual coding is needed to serve as ground truth.	Popov et al., (2018), Zhu et al. (2019), and Zheng et al. (2019)
K-means cluster analysis	K-means cluster analysis is an unsupervised machine learning algorithm. It identifies subgroups that have similar features within groups but distinct characteristics between groups within a dataset based on input features.	This analysis is very sensitive to high-dimensional data, which can significantly compromise its performance (McCallum et al., 2000). Therefore, dimensionality usually needs to be reduced first using techniques such as principal component analysis or t-distributed stochastic neighbor embedding (t-SNE). Furthermore, the analysis requires a predetermined range to identify the optimal number of clusters.	Zheng et al., 2020
Transition rate analysis, sequential pattern mining	Both analyses uncover the temporal characteristics of a set of sequences of events. Transition rate analysis measures the transition probability from one state to any other state. Sequential pattern mining identifies all frequently occurring subsequences in a sequence database or the subsequence whose frequency is above a pre-input threshold.	Transition rate analysis computes the probability for other states to happen under the condition that the current state occurs. However, in reality, the probability for the current state to happen varies from very low to very high. All sequential pattern mining algorithms can return the same results if they analyze the same database using the same parameters; however, the computation speed can vary (see Fournier-Viger et al., 2017 for a review). Sequential pattern mining has limitations such as they assume the data sequences are static, and a huge number of subsequences can be identified and may need to be further analyzed by researchers.	Zhu et al., 2019
Epistemic network analysis (ENA)	ENA is a network and quantitative ethnographic technique that models the connections between codes based on their co-occurrence within conversations. It produces a weighted network of co-occurrences of codes (Shaffer, 2018).	ENA simultaneously analyzes all of the networks and allows researchers to compare a set of networks visually and statistically.	See the mathematics case study within this chapter.
Multilevel modeling (MLM)	MLM can address the variability associated with nesting effects (e.g., individuals nested in groups, time nested in individuals) and account for the non-independence of observations (Du et al., 2013) so that researchers can pursue multilevel questions (e.g., within-person vs between-person, individual vs group).	MLM and other multivariate approaches do not require researchers to assume that all participants in a group are the same. They make it possible to model heterogeneity by accounting for random effects(Gonzalez, 2009).	Xing et al., 2019

in social sciences and learning sciences research (e.g., Yang et al., 2018; Zhu et al., 2019).

Epistemic Network Analysis (ENA) is a network analysis technique that enables researchers to provide thick descriptions of large corpuses of data (Shaffer, 2018). It is a quantitative ethnographic technique for modeling the connections between codes by quantifying the co-occurrence of codes within conversations and producing a weighted network of co-occurrences (Shaffer, 2018). ENA is based on the theoretical idea that learners are embedded in cultures, discourse, interactions, and time (Shaffer, 2018). Therefore, it is possible and important to understand how codes (i.e., 'socially organized ways of seeing'; Goodwin, 1994, p. 606) are related to one another as learners interact with each other through discourse in a certain cultural context. Importantly, ENA analyzes all of the networks simultaneously, resulting in a set of networks that can be visually and statistically compared.

It is common to see datasets with multilevel structures in the field of social sciences, such as students in groups or classes, clients in therapists, and repeated measurements within persons. In the Teaching Teamwork project, students were nested in groups, and there were student-level variables and group-level variables that influence their learning performance. In this case, we cannot simply predict individuals' learning performance using traditional regression approaches but need to consider aggregation bias and heterogeneity of regression. Multilevel modeling is a statistical approach that can account for the non-independence of observations and address the variability associated with nesting effects (Du et al., 2013) by modeling the individual-level variances and group-level variances (Raudenbush & Bryk, 2002).

Multilevel Model Vector Autoregression (mlVAR) can be seen as a model that combines multilevel modeling and the Vector Autoregression (VAR) model. A VAR model can be used to model the evolution of multiple variables over time, or in more detail, how these variables at time point t can be regressed to the same variables at a previous time point t–1 (Bringmann et al., 2013). mlVAR enables researchers to consider the temporal dynamics of variables and their nested structures at different levels (e.g., individual and group levels) at the same time.

15.5 Case Studies

As described above, STEM education usually takes place in technology-enhanced learning environments. Such environments may enhance students' technology literacy. In this section, in different cases, we describe how technologies serve as environments that support students' STEM learning with a focus on science, math, and engineering, respectively; what data have been analyzed and what analytical methods have been adopted; and what we have found.

15.5.1 Science

15.5.1.1 Transformative and Non-Transformative Discourse

In Zhu et al. (2019), to analyze what characterizes students' successful collaborative problem-solving, we analyzed the chat messages (i.e., discourse) of small groups. The participants, namely 144 students of two-year colleges, were randomly assigned into groups of three. They worked on Ohm's Law-relevant tasks using the Teaching Teamwork simulation platform described above. The chat messages of each group for each task and whether each group has successfully solved a task were collected. We first manually coded the chat messages into transformative (i.e., orientation, proposition generation, experimentation, interpretation, and conclusion) and non-transformative (i.e., regulation, sustaining mutual understanding) discourse states. Then we supervised machine learning models to automatically classify the chat messages into the six transformative and non-transformative categories. Finally, we employed transition rate analysis and sequential pattern mining to analyze the temporal dimension of transformative and non-transformative discourse in tasks that were successfully solved and those not solved. We found that in tasks that were successfully solved, students tended to ensure that everyone in their group had a shared understanding of the relationship between the variables. However, in tasks that were not solved, the participants were more likely to regulate the process without reaching a shared understanding.

Similarly, Xing et al. (2019) examined the relationship between students' transformative and non-transformative learning processes and their learning outcomes at the individual and group levels. The participants were 106 college students who worked on Ohm's Law-related tasks and completed a post-survey evaluating their cognitive understanding of Ohm's Law. Multilevel modeling was adopted because individuals were nested in groups, and therefore, individuals' cognitive understanding after the tasks was influenced by their group transformative and non-transformative discourse. We conducted a two-level analysis with individual cognitive understanding after working on the tasks at the first level and a random intercept model of transformative and non-transformative group discussions at the second level. The results suggest that individual students tended to achieve greater cognitive understanding if their groups engaged in additional interpretation and sustained mutual understanding; in contrast, individual group members had statistically lower cognitive understanding when their groups engaged in additional orientation and proposition generation.

15.5.1.2 Self-Regulation

In collaborative learning, students not only need to intentionally adjust their individual cognitions, behaviors, and motivations to achieve learning goals (i.e., self-regulation learning; Zimmerman, 2000) but also need to regulate their group joint work (i.e., socially shared regulation; Volet et al., 2009). Incorporating students' behavioral and discourse log data collected using the Teaching Teamwork platform, Zheng et al. (2019) studied how individual self-regulation and socially shared regulation characterize their group performance. The participants were 156 high school and college students. We developed a coding framework to classify students' discourse and behavior logs into task analysis, planning, elaborating, and monitoring at the self-regulation and socially shared regulation levels and execution. Using sequential pattern mining, we found that in tasks that were successfully solved, students tended to start with self-executing and end with socially shared monitoring. Differently, in tasks that were not solved, students tended to start and end with self-executing.

15.5.1.3 Multi-Faceted Engagement

Recent literature considers engagement as a multi-faceted, dynamic, contextualized, and collective construct (e.g., Jung & Lee, 2018; Sinha et al., 2015). Behavioral engagement refers to students' sustained on-task behaviors when participating in academic activities (Fredricks et al., 2004). In the Ohm's Law problem-solving tasks, on-task behaviors including the number of chat messages, the number of performed calculations, and the number of times that a group member adjusted the resistance were used as indicators of the behavioral engagement of different groups. Social engagement is about the quality of group interactions to complete learning tasks (Sinha et al., 2015). We defined the time delay between sending messages, the difference between the number of pieces of messages sent by different group members, and the number of reciprocal interactions among group members as indicators of social engagement. Cognitive engagement refers to students applying domain-specific knowledge and disciplinary practices as well as investing cognitive efforts to solve a group task (Gresalfi & Barab, 2010). Accordingly, we first extracted students' elaboration on the relationships between variables in the chat messages and converted the information into formulas. Similarly, we extracted the values that students put in their calculators and compared the values with the specific numbers in potential columns to understand the meaning of the values (i.e., representing which variables) to extract the formulas embedded in their calculations. Finally, we compared the formulas we constructed based on students' chat messages and input information in calculators with various forms of Ohm's Law to examine students' cognitive engagement. Metacognitive engagement is about the planning, monitoring, and evaluating processes of accomplishing tasks (Zimmerman, 1990). Students' chat messages regarding discussing goals, planning, and monitoring problem-solving processes were used as metacognitive engagement indicators.

We supervised machine learning models to extract the indicators of each aspect of engagement and classified students' engagement concerning each aspect in each Ohm's Law task into low, medium, and high engagement levels. Using path modeling and mediation analysis, we examined how multi-faceted forms of engagement influence students' individual cognitive understanding and group problem-solving performance. We found that groups' behavioral and cognitive engagement positively affected groups' problem-solving performance, while social engagement negatively affected groups' problem-solving performance. Group's problem-solving performance partially mediates the impact of group behavioral engagement and fully mediates the impact of group social engagement on individual cognitive understanding.

15.5.2 Engineering

Our understanding of student engineering design behaviors is limited. A better understanding of how students engage in design iterations is fundamental to the research and practice of supporting students' engineering learning. Engineering design is an iterative process of analyzing the design problems and constraints and formulating, evaluating, and reformulating designs (Dym et al., 2005). Using the Energy3D simulation tool, we collected the logs of designing an energy-saving house from 111 high school students. As described above, we categorized students' design behaviors into observation, changing an environment parameter, formulation, evaluation, reflection, and reformulation. Evaluation behaviors are about using the analytical tools provided by Energy3D to assess the design artifacts. Reformulation behaviors involve editing the design elements to refine the design artifacts. In Zhu et al. (2021), we particularly examined the reciprocal relationships between the evaluation and reformulation behaviors of 108 high school students, and how the behaviors influence their design performance at the early, middle, and final design stages. The results suggested that there is a positive prediction relationship between students' evaluation and reformulation process over time; reformulation positively predicts design performance and mediates the impact of evaluation.

Students' self-regulation is critical to their problem-solving, especially for STEM tasks which usually require interdisciplinary knowledge and skills. Yet, what needs to be improved is our understanding of how self-regulation influences students' design performance and conceptual understanding in engineering design. In Zheng et al. (2020), using a K-means clustering analysis, we classified students based on their frequency into five self-regulation categories: observation, formulation, reformulation, analysis, and evaluation. The clustering analysis generated four distinct profiles. Based on the extreme low or high variables and other most salient features in each profile, we labeled the four groups as minimally self-regulated, reflective-oriented, cognitive-oriented, and competent learners. We compared the design performance and learning gains concerning scientific knowledge of the four profiles and found that the reflective-oriented group performed best regarding the energy efficiency of the green building, while the minimally self-regulated group performed worst, suggesting the importance of reflection and evaluation processes in design.

15.5.3 Mathematics

To understand the learning and collaboration processes of students co-constructing 3D objects, we conducted an exploratory study using multimodal learning analytics tools. Specifically, we used modular origami (i.e., Sonobe modular) as the design and problem-solving task and collected the learning process data through multiple sensors, such as Zoom cameras, wireless mics, and Neo smartpens. Fifteen undergraduate student dyads were asked to sketch a design of a cube made of two blue and four yellow Sonobe units. Their task was to sketch what the cube might look like. There were two phases to complete this task: the student first worked on this task individually for 10 min, and then they came together to produce a final sketch for another 10 min. We recorded students' discourse and actions for video analyses. We also scored their sketches and summarized the sketch details that contribute to a correct solution as well as the common mistakes they made. Figure 15.4 shows the research materials and setup.

Students' sketches produced at the end of each phase were scored based on a coding scheme informed by Jaeger et al. (2020). Students' discourse and action data in the collaborative problem-solving phase were coded based on a coding scheme informed by Kim and Maher's (2008) spatial cognition coding scheme and previous work on the collaborative problem-solving process (Stadler et al., 2020). Connections between codes were analyzed using Epistemic Network Analysis (ENA; Shaffer, 2018). ENA analysis helped to identify several differentiated relations between collaborative problem-solving behaviors and students' embodied engagement. Dyads with higher scores in their final sketches engaged more frequently in sketching and gesturing in their collaborative problem-solving process, while dyads with lower scores engaged more frequently in the manipulation of the origami units/cube (Figure 15.5).

15.6 Research Gaps and Future Trends

There are some research gaps or challenges concerning using learning analytics to analyze and support STEM education. First, as shown in the literature and our cases, although the importance of integrative STEM has been well recognized and efforts have been made to move the STEM education spectrum towards the interdisciplinary end, STEM education is usually conducted not in an interdisciplinary manner. Some possible reasons are that the current organization of curricula may not support students to navigate across disciplines (Baloche et al., 1996); students may not have developed the competencies to navigate complex problems across disciplinary boundaries (Ivanitskaya et al., 2002); and teachers need to be better prepared to support students to navigate complex conceptual

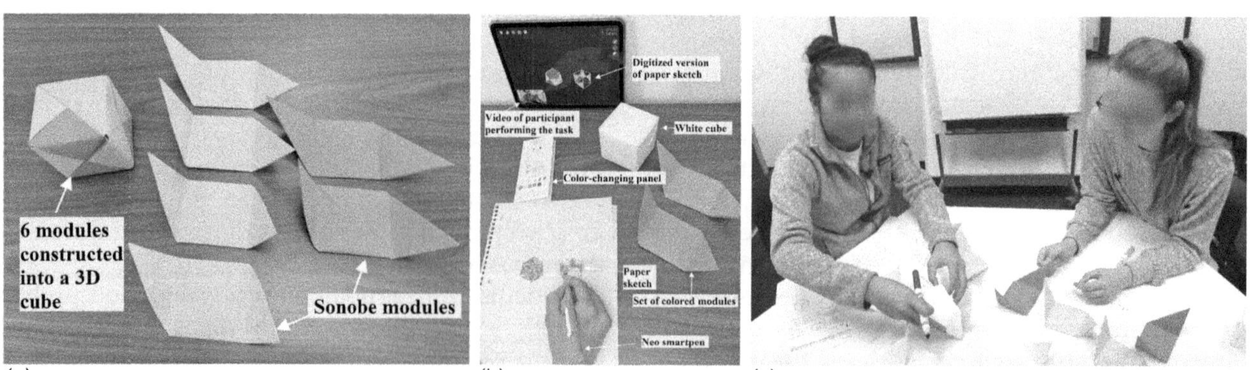

(a) (b) (c)

FIGURE 15.4
Left: Sonobe Units/Cube; mid: Phase I setup; and right: Phase II students collaborate.

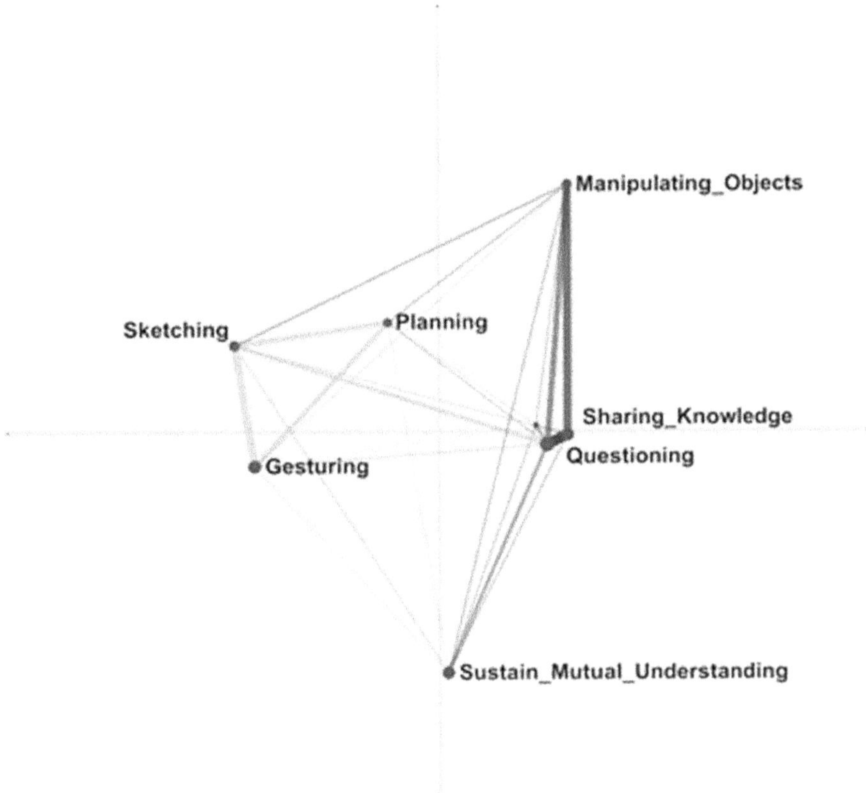

FIGURE 15.5
ENA by final sketch scores.

spaces to address real issues. Future research can work on these aspects to support integrative STEM education.

Second, currently, mainly discourse or behavior data has been collected and analyzed. It is relatively challenging to record other data, such as students' learning artifacts, bodily data, and psychological data. Some attempts have been made in this direction. For instance, Spikol et al. (2017) collected computer vision, user-generated content, and physical computing component data and employed multimodal learning analytics to examine what are good predictors of students' group success in open-ended STEM learning activities. In the math case described above, we collected the learning process data through multiple devices, such as Zoom cameras, wireless mics, and Neo smartpens to understand the processes of students co-designing and co-constructing 3D objects. Robles-Granda et al. (2021) leveraged various tools (e.g., PhoneAgent, wearable sensors, social media, beacons) to collect data from personal, social, and contextual sources to jointly predict individuals' psychological, physical and physiological, and job-performance constructs. Future research can study how to collect other sources of data and employ multimodal learning analytics to better understand learners' cognitive understanding, behaviors, and psychological characteristics.

A lot of the learning analytics described above used the data generated in the process after students have completed their learning tasks. Such a lag analysis fails to provide real-time feedback to stimulate students' reflection or to facilitate their learning. However, one of the future directions regarding guiding students' inquiry learning in technology-enhanced environments is providing dynamically adaptive guidance (de Jong, 2019). Future research should consider how to build on current research and transform the design into analytical tools that can be employed by students and teachers in real time to provide timely learning and teaching support.

Finally, as introduced above, increasing attention is paid to evaluate the fairness of learning analytics models for learners with different demographic backgrounds. Studies have conducted philosophical discussions on the fairness issues of learning analytics to conceptualize the future direction of educational research. For example, there was a detailed literature study on the concepts of model fairness led by Kizilcec and Lee (2020). Apart from the formulation of mathematical concepts to evaluate model fairness, the researchers also addressed the possible scenarios of fairness being breached in educational research during the data collection, model learning, and human-model collaboration phases. Baker and Hawn (2021) extended the work of Kizilcec and Lee to further discuss the origin of bias in learning analytics models. In the study, the researchers addressed bias coming before or during the data collection stage. For instance, the underrepresentation of certain demographic groups in the data or failure to ensure construct validity could lead to biased models. The researchers also went through the types of bias that could happen during model training and evaluation. Historical bias, for example, could happen when models are trained with both new data and historically biased data. There are also researchers framing the fairness issue from a qualitative perspective. Marcinkowski et al. (2020), for example, have proposed a protocol to qualitatively evaluate students' perceived fairness towards a college admission system powered by learning analytics through surveys and interviews.

15.7 Conclusion

STEM education usually requires students' active learning in technology-rich environments in order to support them to work on real and complex issues. Such a combination of learning approach, environment, and content makes it possible to collect digital trails for analyzing and understanding students' learning process. Learning analytics is an emerging field that aims to make meaning of this big data to understand and optimize learning and its occurring environments. This chapter discussed how learning analytics were employed to analyze students' transformative discourse, multi-faceted engagement, self-regulation, and evaluation and reformulation behaviors in STEM education. Several case studies were elaborated to illustrate the involved learning theories, technology environments, data collected, analytical methods, and results. Finally, we discussed several research gaps and challenges which imply future research directions such as supporting integrative STEM education, collecting other sources of data and developing multimodal learning analytics, creating analytical tools to provide timely learning and teaching feedback and support, and paying attention to the fairness of learning analytics models.

References

Abdous, M., He, W., & Yen, C.-J. (2012). Using data mining for predicting relationships between online question theme and final grade. *Educational Technology and Society, 15*(3), 77–88.

Alessi, S. M., & Trollip, S. R. (2001). *Multimedia for learning: Methods and development* (3rd ed.). Allyn and Bacon.

Baker, R. S., & Hawn, A. (2021). Algorithmic bias in education. https://doi.org/10.35542/osf.io/pbmvz

Baloche, L., Hynes, J. L., & Berger, H. A. (1996). Moving toward the integration of professional and general education. *Action in Teacher Education, 18*(1), 1–9. https://doi.org/10.1080/01626620.1996.10462817

Bøe, M. V., Henriksen, E. K., Lyons, T., & Schreiner, C. (2011). Participation in science and technology: Young people's achievement-related choices in late-modern societies. *Studies in Science Education, 47*(1), 37–72. https://doi.org/10.1080/03057267.2011.549621

Bringmann, L. F., Vissers, N., Wichers, M., Geschwind, N., Kuppens, P., Peeters, F., Borsboom, D., & Tuerlinckx, F. (2013). A network approach to psychopathology: New insights into clinical longitudinal data. *PLOS ONE, 8*(4), Article e60188. https://doi.org/10.1371/journal.pone.0096588

Bybee, R. W. (2010a). What is STEM education? *Science, 329*(5995), 996. https://doi.org/10.1126/science.1194998

Bybee, R. W. (2010b). Advancing STEM education: A 2020 vision. *Technology and Engineering Teacher, 70*(1), 30.

Bybee, R. W. (2013). *The case for STEM education challenges and opportunities.* National STEM Teachers Association.

Chang, S.-H., Ku, A.-C., Yu, L.-C., Wu, T.-C., & Kuo, B.-C. (2015). A science, technology, engineering and mathematics course with computer-assisted remedial learning system support for vocational high school students. *Journal of Baltic Science Education, 14*(5), 641–654. https://doi.org/10.33225/jbse/15.14.641

de Jong, T. (2006). Technological advances in inquiry learning. *Science, 312*(5773), 532–533. https://doi.org/10.1126/science.1127750

de Jong, T. (2019). Moving towards engaged learning in STEM domains; there is no simple answer, but clearly a road ahead. *Journal of Computer Assisted Learning, 35*(2), 153–167. https://doi.org/10.1111/jcal.12337

de Jong, T., Lazonder, A., Pedaste, M., & Zacharia, Z. (2018). Simulations, games, and modeling tools for learning. In *International handbook of the learning sciences* (pp. 256–266).

de Jong, T., & van Joolingen, W. R. (2008). Model-facilitated learning. In J. M. Spector, M. D. Merill, J. V. Merriënboer, & M. P. Driscoll (Eds.), *Handbook of research on educational communication and technology* (3rd ed., pp. 457–468). Lawrence Erlbaum.

Du, J., Xu, J., & Fan, X. (2013). Factors affecting online groupwork interest: A multilevel analysis. *Journal of Educational Computing Research, 49*(4), 481–499. https://doi.org/10.2190/EC.49.4.d

Duran, M., Höft, M., Lawson, D. B., Medjahed, B., & Orady, E. A. (2014). Urban high school students' IT/STEM learning: Findings from a collaborative inquiry- and design-based afterschool program. *Journal of Science Education and Technology, 23*(1), 116–137. https://doi.org/10.1007/s10956-013-9457-5

Dym, C. L., Agogino, A. M., Eris, O., Frey, D. D., & Leifer, L. J. (2005). Engineering design thinking, teaching, and learning. *Journal of Engineering Education, 94*(1), 103–120. https://doi.org/10.1002/j.2168-9830.2005.tb00832.x

Fournier-Viger, P., Lin, J. C. W., Kiran, R. U., Koh, Y. S., & Thomas, R. (2017). A survey of sequential pattern mining. *Data Science and Pattern Recognition, 1*(1), 54–77.

Fredricks, J., Blumenfeld, P., & Paris, P. (2004). School engagement: Potential of the concept and state of the evidence. *Review of Educational Research, 74*(1), 59–109. https://doi.org/10.3102/00346543074001059

Freeman, S., Eddy, S. L., McDonough, M., Smith, M. K., Okoroafor, N., Jordt, H., & Wenderoth, M. P. (2014). Active learning increases student performance in science, engineering, and mathematics. *Proceedings of the National Academy of Sciences, 111*(23), 8410–8415. https://doi.org/10.1073/pnas.1319030111

Goodwin, C. (1994). Professional vision. *American Anthropologist, 96*(3), 606–633. https://doi.org/10.1525/aa.1994.96.3.02a00100

Gresalfi, M., & Barab, S. (2010). Learning for a reason: Supporting forms of engagement by designing tasks and orchestrating environments. *Theory into Practice, 50*(4), 300–310. https://doi.org/10.1080/00405841.2011.607391

Horwitz, P., von Davier, A., Chamberlain, J., Koon, A., Andrews, J., & McIntyre, C. (2017). Teaching teamwork: Electronics instruction in a collaborative environment. *Community College Journal of Research and Practice, 41*(6), 341–343. https://doi.org/10.1080/10668926.2016.1275886

Hossain, Z., Bumbacher, E., Brauneis, A., Diaz, M., Saltarelli, A., Blikstein, P., & Riedel-Kruse, I. H. (2018). Design guidelines and empirical case study for scaling authentic inquiry-based science learning via open online courses and interactive biology cloud labs. *International Journal of Artificial Intelligence in Education, 28*(4), 478–507. https://doi.org/10.1007/s40593-017-0150-3

Huang, S., & Fang, N. (2013). Predicting student academic performance in an engineering dynamics course: A comparison of four types of predictive mathematical models. *Computers and Education, 61*, 133–145. https://doi.org/10.1016/j.compedu.2012.08.015

Hutt, S., Gardner, M., Duckworth, A. L., & D'Mello, S. K. (2019). Evaluating fairness and generalizability in models predicting on-time graduation from college applications. In C. Lynch, A. Merceron, M. Desmarais, & R. Nkambou (Eds.), *Proceedings of the 12th international conference on educational data mining (EDM 2019)* (pp. 79–88).

Ivanitskaya, L., Clark, D., Montgomery, G., & Primeau, R. (2002). Interdisciplinary learning: Process and outcomes. *Innovative Higher Education, 27*(2), 95–111. https://doi.org/10.1023/A:1021105309984

Jaeger, A. J., Marzano, J. A., & Shipley, T. F. (2020). When seeing what's wrong makes you right: The effect of erroneous examples on 3D diagram learning. *Applied Cognitive Psychology, 34*(4), 844–861. https://doi.org/10.1002/acp.3671

Jung, Y., & Lee, J. (2018). Learning engagement and persistence in massive open online courses (MOOCS). *Computers and Education, 122*, 9–22. https://doi.org/10.1016/j.compedu.2018.02.013

Kaufman, L., & Rousseeuw, P. J. (2009). *Finding groups in data: An introduction to cluster analysis* (Vol. 344). John Wiley & Sons. https://doi.org/10.1002/9780470316801

Kim, M. J., & Maher, M. L. (2008). The impact of tangible user interfaces on designers' spatial cognition. *Human-Computer Interaction, 23*(2), 101–137. https://doi.org/10.1016/j.destud.2007.12.006

Kizilcec, R. F., & Lee, H. (2020). Algorithmic fairness in education. ArXiv Preprint ArXiv:2007.05443.

Lamb, R., Akmal, T., & Petrie, K. (2015). Development of a cognition-priming model describing learning in a STEM classroom. *Journal of Research in Science Teaching, 52*(3), 410–437. https://doi.org/10.1002/tea.21200

Leemkuil, H., & de Jong, T. (2011). Instructional support in games. In S. Tobias & D. Fletcher (Eds.), *Can computer games be used for instruction?* (pp. 353–369). Information Age Publishers.

Li, K. C., & Wong, B. T. M. (2020). Trends of learning analytics in STE(A)M education: A review of case studies. *Interactive Technology and Smart Education, 17*(3), 323–335. https://doi.org/10.1108/ITSE-11-2019-0073

Lichtenberg, J., Woock, C., & Wright, M. (2008). *Ready to innovate: Are educators and executives aligned on the creative readiness of the US workforce?* Conference Board.

Lyons, E. J. (2015). Cultivating engagement and enjoyment in exergames using feedback, challenge, and rewards. *Games for Health Journal, 4*(1), 12–18. https://doi.org/10.1089/g4h.2014.0072

Marcinkowski, F., Kieslich, K., Starke, C., & Lünich, M. (2020). Implications of AI (un-) fairness in higher education admissions: The effects of perceived AI (un-) fairness on exit, voice and organizational reputation. In *Proceedings of the 2020 conference on fairness, accountability, and transparency* (pp. 122–130). https://doi.org/10.1145/3351095.3372867

Martín-Páez, T., Aguilera, D., Perales-Palacios, F. J., & Vílchez-González, J. M. (2019). What are we talking about when we talk about STEM education? A review of literature. *Science Education, 103*(4), 799–822. https://doi.org/10.1002/sce.21522

McCallum, A., Nigam, K., & Ungar, L. H. (2000, August). Efficient clustering of high-dimensional data sets with application to reference matching. In *Proceedings of the sixth ACM SIGKDD international conference on knowledge discovery and data mining* (pp. 169–178). ACM. https://doi.org/10.1145/347090.347123

Milligan, G. W., & Cooper, M. C. (1985). An examination of procedures for determining the number of clusters in a data set. *Psychometrika, 50*(2), 159–179. https://doi.org/10.1007/BF02294245

Moore, T. J., Glancy, A. W., Tank, K. M., Kersten, J. A., Smith, K. A., & Stohlmann, M. S. (2014). A framework for quality K-12 engineering education: Research and development. *Journal of Precollege Engineering Education Research (J-PEER), 4*(1), 2. https://doi.org/10.7771/2157-9288.1069

Moreno, N. P., Tharp, B. Z., Vogt, G., Newell, A. D., & Burnett, C. A. (2016). Preparing students for middle school through after-school STEM activities. *Journal of Science Education and Technology, 25*(6), 889–897. https://doi.org/10.1007/s10956-016-9643-3

Na, S., Xumin, L., & Yong, G. (2010, April). Research on k-means clustering algorithm: An improved k-means clustering algorithm. In *2010 third international symposium on intelligent information technology and security informatics* (pp. 63–67). IEEE.

Papamitsiou, Z. K., & Economides, A. A. (2014). Learning analytics and educational data mining in practice: A systematic literature review of empirical evidence. *Educational Technology and Society, 17*(4), 49–64.

Popov, V., Xing, W., Zhu, G., Horwitz, P., & McIntyre, C. (2018, June). The influence of students' transformative and non-transformative contributions on their problem solving in collaborative inquiry learning. In *The proceedings of the 13th international conference of the learning sciences* (pp. 855–862). International Society of the Learning Sciences. https://doi.org/10.22318/cscl2018.855

Raudenbush, S. W., & Bryk, A. S. (2002). *Hierarchical linear models: Applications and data analysis methods* (Vol. 1). Sage.

Riazy, S., & Simbeck, K. (2019). Predictive algorithms in learning analytics and their fairness. In N. Pinkwart & J. Konert (Eds.), *Proceedings of the DELFI 2019* (pp. 223–228). https://doi.org/10.18420/delfi2019_305

Robles-Granda, P., Lin, S., Wu, X., Martinez, G. J., Mattingly, S. M., Moskal, E., Striegel, A., Chawla, N. V., D'Mello, S., Gregg, J., Nies, K., Mark, G., Grover, T., Campbell, A. T., Mirjafari, S., Saha, K., De Choudhury, M., & Dey, A. K. (2021). Jointly predicting job performance, personality, cognitive ability, affect, and well-being. *IEEE Computational Intelligence Magazine, 16*(2), 46–61. https://doi.org/10.1109/MCI.2021.3061877

Sha, L., Rakovic, M., Whitelock-Wainwright, A., Carroll, D., Yew, V. M., Gasevic, D., & Chen, G. (2021). Assessing algorithmic fairness in automatic classifiers of educational forum posts. In I. Roll, D. McNamara, S. Sosnovsky, R. Luckin, & V. Dimitrova (Eds.), *AIED 2021: Artificial intelligence in education* (pp. 381–394). Springer. https://doi.org/10.1007/978-3-030-78292-4_31

Shaffer, D. W. (2018). Epistemic network analysis: Understanding learning by using big data for thick description. In *International handbook of the learning sciences* (pp. 520–531). Routledge.

Sinha, S., Rogat, T. K., Adams-Wiggins, K. R., & Hmelo-Silver, C. E. (2015). Collaborative group engagement in a computer-supported learning environment. *International Journal of Computer-Supported Collaborative Learning, 10*(3), 273–307. https://doi.org/10.1007/s11412-015-9218-y

Sırakaya, M., & Alsancak Sırakaya, D. (2020). Augmented reality in STEM education: A systematic review. *Interactive Learning Environments*, 1–14. https://doi.org/10.1016/j.compedu.2018.05.002

Spikol, D., Ruffaldi, E., Landolfi, L., & Cukurova, M. (2017, July). Estimation of success in collaborative learning based on multimodal learning analytics features. In *2017 IEEE 17th international conference on advanced learning technologies (ICALT)* (pp. 269–273). IEEE. https://doi.org/10.1109/ICALT.2017.122

Stadler, M., Herborn, K., Mustafić, M., & Greiff, S. (2020). The assessment of collaborative problem solving in PISA 2015: An investigation of the validity of the PISA 2015 CPS tasks. *Computers and Education, 157*, 103964. https://doi.org/10.1016/j.compedu.2020.103964

Toma, R. B., & Greca, I. M. (2018). The effect of integrative STEM instruction on elementary students' attitudes toward science. *Eurasia Journal of Mathematics, Science and Technology Education, 14*(4), 1383–1395. https://doi.org/10.29333/ejmste/83676

US National Academy of Sciences, National Academy of Engineering, Institute of Medicine. (2007). Rising above the gathering storm: Energizing and employing America for a brighter economic future. http://nap.edu/11463

Van den Beemt, A., MacLeod, M., Van der Veen, J., Van de Ven, A., van Baalen, S., Klaassen, R., & Boon, M. (2020). Interdisciplinary engineering education: A review of vision, teaching, and support. *Journal of Engineering Education, 109*(3), 508–555. https://doi.org/10.1002/jee.20347

Volet, S., Vauras, M., & Salonen, P. (2009). Self- and social regulation in learning contexts: An integrative perspective. *Educational Psychologist, 44*(4), 215–226. https://doi.org/10.1080/00461520903213584

Washington STEM Study Group. (2011). [Online]. What is STEM literacy? Retrieved July 14, 2021, from https://www.k12.wa.us/student-success/career-technical-education-cte/program-study-career-clusters-and-career-pathways/science-technology-engineering-mathematics-stem

Woolfolk, A., Winne, P. H., & Perry, N. E. (Eds.). (2009). *Educational psychology, fourth Canadian edition*. Allyn and Bacon.

Xie, C., Schimpf, C., Chao, J., Nourian, S., & Massicotte, J. (2018). Learning engineering design through modeling and simulation on a CAD platform. *Computer Applications in Engineering Education, 26*(4), 824–840. https://doi.org/10.1002/cae.21920

Xing, W., Popov, V., Zhu, G., Horwitz, P., & McIntyre, C. (2019). The effects of transformative and non-transformative discourse on individual performance in collaborative-inquiry learning. *Computers in Human Behavior, 98*, 267–276. https://doi.org/10.1016/j.chb.2019.04.022

Yang, X., Li, J., & Xing, B. (2018). Behavioral patterns of knowledge construction in online cooperative translation activities. *Internet and Higher Education, 36*, 13–21. https://doi.org/10.1016/j.iheduc.2017.08.003

Zheng, J., Xing, W., & Zhu, G. (2019). Examining sequential patterns of self-and socially shared regulation of STEM learning in a CSCL environment. *Computers and Education, 136*, 24–48. https://doi.org/10.1016/j.compedu.2019.03.005

Zheng, J., Xing, W., Zhu, G., Chen, G., Zhao, H., & Xie, C. (2020). Profiling self-regulation behaviors in STEM learning of engineering design. *Computers and Education, 143*, 103669. https://doi.org/10.1016/j.compedu.2019.103669

Zhu, G., Xing, W., & Popov, V. (2019). Uncovering the sequential patterns in transformative and non-transformative discourse during collaborative inquiry learning. *Internet and Higher Education, 41*, 51–61. https://doi.org/10.1016/j.iheduc.2019.02.001

Zhu, G., Zeng, Y., Xing, W., Du, H., & Xie, C. (2021). Reciprocal relations between students' evaluation, reformulation behaviors and engineering design performance over time. *Journal of Science Education and Technology*. https://doi.org/10.1007/s10956-021-09906-3

Zimmerman, B. J. (1990). Self-regulated learning and academic achievement: An overview. *Educational Psychologist, 25*(1), 3–17. https://doi.org/10.1207/s15326985ep2501_2

Zimmerman, B. J. (2000). Attaining self-regulation: A social cognitive perspective. In *Handbook of self-regulation* (pp. 13–39). https://doi.org/10.1016/B978-012109890-2/50031-7

16

Understanding the Role of AI and Learning Analytics Techniques in Addressing Task Difficulties in STEM Education

Sadia Nawaz, Emad A. Alghamdi, Namrata Srivastava, Jason Lodge, and Linda Corrin

CONTENTS

16.1 Introduction 241
16.2 The Multifaceted Aspects of Task Difficulty 242
16.3 LA and AI in Digital Learning Environments 244
16.3.1 Virtual Labs 244
16.3.2 Serious Games and Simulations 245
16.4 Opportunities 246
16.4.1 Real-Time Feedback and Intervention 246
16.4.2 Personalized Adaptive Learning 246
16.4.3 Leveraging the Advances in Sensor Technology 246
16.5 Challenges 247
16.5.1 Task Difficulty and Its Operationalization 247
16.5.2 Data Ethics and Student Privacy 249
16.5.3 What Data to Collect and How 249
16.5.4 Confounding Factors 250
16.6 Conclusion 251
Note 252
References 252

16.1 Introduction

Despite global recognition of the growing importance of science, technology, engineering, and mathematics (STEM) education, current STEM education is strewn with many problems, including lack of diversity in graduates (Ochoa & Henriques, 2021), low student enrolment and attrition rate (Winberg et al., 2019), teachers' professional development, inappropriate curriculum (English, 2016), and misalignment between STEM curricula and rapidly changing market requirements (Sithole et al., 2017). Of these challenges, students' disengagement and low retention in STEM education has been a subject of much debate and discussion among educators and policymakers (Watkins & Mazur, 2013). For many STEM students, task difficulty is a key barrier to their success in mastering the STEM subjects (Killpack & Melón, 2016). In a learning context, task difficulty can be defined as the extent or degree to which an instructional activity presents a personally demanding and meaningful situation – an activity which may require cognitive or physical efforts from learners to develop their knowledge and skills (Nawaz, 2021). For example, the perceived difficulty of classroom assignments can influence student motivation, their expectancy of success, and also the level of effort that students put forth (Hom & Maxwell, 1983). Perceptions of the difficulty of a learning activity can also influence students' self-efficacy beliefs, i.e., the probability of success in part can be determined by the level of difficulty of a task (Li et al., 2007). Researchers have also investigated the effects of difficulty judgments on students' off-task behaviors and have attributed task difficulty to as "one of the primary curricular variables that can set the occasion for problem behaviours in the classroom" (Umbreit et al., 2004, p. 1). In another study, Gickling and Armstrong (1978) reported that when the tasks were perceived to be "too difficult," task completion, task comprehension, and on-task behaviors all occurred at low levels (below 50%). By contrast, when the tasks were deemed "too easy," task completion and task comprehension rates were high (nearly 100%), but on-task behaviors were still quite low (around 50%). Therefore, it can be

DOI: 10.1201/9781003181187-20

expected that when the tasks are judged to be too difficult, they can almost immediately lead to students' disengagement and when the tasks are too easy, they may not push the students' knowledge in the area forward. Therefore, to promote students' learning and engagement, tasks should be designed at proper instructional levels that match with students' abilities. That is, for instruction to be effective, it must be aimed at students' proximal level of development where students may succeed with some assistance, i.e., a difficulty which is slightly harder than an exact match to students' abilities but not so hard that students fail to succeed (Vygotsky, 1980).

Ost (2010) analyzed longitudinal data from a large university in the United States and found that when STEM students encountered difficulty early on in their studies, they tended to drop out or change their majors to non-STEM subjects. As the vast majority of learning – including STEM learning – happens outside the classroom (Allen & Peter-Man, 2019), identifying what learning difficulties STEM students encounter in digital learning environments (DLEs) is crucial for providing effective learning support. One way to address this issue is by leveraging recent advances in learning analytics (LA) and artificial intelligence (AI) (Poquet & de Laat, 2021). Over the last decade, numerous studies have explored how LA and AI can be used in STEM education, yet their full potential is not fully realized.

While it is generally believed that STEM disciplines are difficult to learn and challenging to teach, little research has been conducted to address the specific issue of students' task difficulty (Winberg et al., 2019). In this chapter, we provide a discussion of the notion of task difficulty, and outline the key challenges inherent in defining and evaluating task difficulty in DLEs. We then highlight how the recent advances in learning analytics and artificial intelligence may help in assessing task difficulty in STEM as well as directions for future research.

16.2 The Multifaceted Aspects of Task Difficulty

Despite its prominent role in learning sciences and education, difficulty is one of those notions that is presumed to be known by everyone, and thus many researchers feel no obligation to define it. Notwithstanding decades of research, the notion of task difficulty is elusive and hard to pin down. It is often used interchangeably with terms such as task complexity, challengeability, and confusion (Campbell et al., 2007; D'Mello & Graesser, 2014; Lodge et al., 2018; Taylor, 1981). Researchers also used tautological terminologies to describe difficult tasks as those that are hard, problematic, or not easy. In this chapter, we explore some of the previous research on this concept and suggest the need to further explore and analyze various task difficulty measures so that students may be better supported in their learning journey.

In an early study, Feather (1961) associated the difficulty of the tasks with participants' subjective probability of success. That is, learning tasks were considered easy when the participants held a higher initial probability of success. However, when their subjective probability of success was lower, they found the tasks to be difficult. Willett and Eysenck (1962) attributed the difficulty of the tasks to learners' perception of familiarity with the concepts. Weeks and Gaylord-Ross (1981) defined the difficulty of tasks in terms of the errors that are produced by learners. In a study where college students were asked to solve anagrams, Eisenberger and Leonard (1980) defined task difficulty in terms of the time needed to solve the task. Tasks were considered simple when the median time to solve the anagrams ranged between 3-5 seconds. Tasks were considered complex when the median time to solve the anagrams ranged from 20 to 30 seconds and tasks were considered unsolvable when the solutions were too difficult to be attained by the students. Within the context of skill acquisition, Ahissar and Hochstein (1997) explored the differences in students' perceptions of task difficulty when stimuli were presented to the students for different durations. Tasks were considered easy when the stimuli were presented for longer durations, whereas the tasks were considered hard when the duration of the stimuli were brief.

Perceived task difficulty is another commonly used term that refers to one's beliefs regarding how much effort is needed to succeed at a task and whether success is attainable (Horvath et al., 2006). Studies have shown that perceived task difficulty can significantly affect students' behaviors, such as engagement, persistence, and achievement (Eccles & Wigfield, 2002; Pintrich & Schunk, 2002). For instance, the more complex the task, the more the students perceive the task as difficult. This can result in less active participation and less accurate performance. Similarly, Maynard and Hakel (1997) have suggested that task performance depends not only on the objective task complexity but also on subjective perceptions of task difficulty.

The role of students' perceptions in task difficulty has also been explored within medical education in the context of simulation-based practice and training.

Using the challenge point framework, task difficulty was defined as the cognitive challenge posed by a problem that often stems from an individual's perceptions of the task (Guadagnoli et al., 2012). According to this framework, an appropriate degree of challenge will result in some level of failure. However, such failure during practice can lead to eventual success during the test conditions, as depicted in Figure 16.1.

Importantly, therefore, task difficulty encompasses two key elements: (1) the inherent difficulty of the task itself due to, for example, the level of complexity or nature of the concepts, and (2) the students' judgment on the level of difficulty they experience while attempting to successfully complete the task and achieve the required outcomes. Tasks can be inherently difficult or not and can be perceived in alignment with this inherent difficulty, or not. For example, a learning activity focused on an inherently complex and counterintuitive concept (e.g. gravity) could be perceived incorrectly as not difficult. Perceptions do not always align with the level of difficulty inherent in a task or coming to understand a concept. This misalignment can have negative consequences for the decisions students make as they progress in a task or in their studies generally.

Task difficulty is not entirely detrimental to learning; some degree of difficulty is beneficial and desirable (Bjork & Bjork, 2011). Taylor (1981) observed that when subjects worked on a more challenging assignment, their learning performance improved.

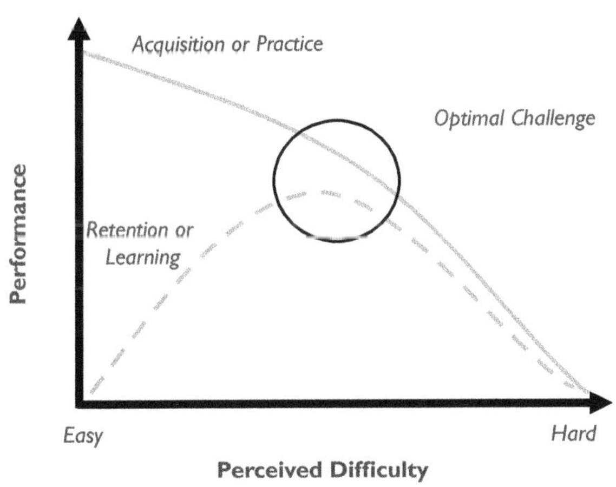

FIGURE 16.1
Challenge Point Framework – adapted from Guadagnoli et al. (2012). The dotted line demonstrates that with an increase in task difficulty, the learning also increases (at least up to the first half of the curve) until it reaches the optimal challenge condition. Once optimally challenged, learning can occur efficiently. This correspondingly occurs with decreased practice performance (as shown by solid line). Up to the optimal challenge point, practice performance may decrease but learning may improve. However, after the challenge point, practice performance as well as the learning performance can begin to suffer.

Within the STEM educational field, problem-based learning (PBL) is a popular learning design that, in the short term, can result in a slight underperformance, but can be effective in promoting long-term retention and performance improvement (Strobel & Van Barneveld, 2009). Moreover, works on impasse-driven theories of learning suggest that impasses, difficulties, and failures can motivate students to take an active role in constructing their knowledge and developing a better understanding of the concepts (VanLehn et al., 2003). Similarly, research on productive failure suggests that when students generate solutions to complex and novel problems without prior instruction, it can help students to learn from subsequent instruction and can be beneficial for their long-term memory and retention (Kapur, 2016; Kapur & Bielaczyc, 2012). Tawfik and Jonassen (2013) noted that "failure" is important in a learning process because it requires students to update their erroneous assumptions about previous actions and decisions. Yet failure does not lead students to gain new knowledge automatically; rather, students need to deliberately evaluate their choice of actions, practice for mastery, and understand why failure occurred (Schank, 1999). These conditions may partly explain why the evidence for the effectiveness of approaches like problem-based learning is mixed, at best (Kirschner et al., 2006).

It remains unclear what constitutes a beneficial difficulty in the learning process (Taylor et al., 2020; Yue et al., 2013). Among other things, it largely depends on the subject's perceptions or their estimates of task difficulties (DeLoache et al., 1985). Students are less likely to elicit strategic behaviors when presented with problems that are too easy or too difficult than when they are presented with moderately difficult problems. Similarly, Belmont and Mitchell (1987) suggested that if the task is perceived too easy or too hard, it can result in lower engagement levels than the task with medium or moderate level of difficulty – often referred to as the "law of optimum perceived difficulty." This also aligns with the concept of "cognitive fluency" (Taylor et al., 2020), which suggests that when knowledge or information feels easy to process, it is considered to be fluent. By contrast, when the information feels difficult to process, it is said to be disfluent. When learners experience a feeling of fluency, they typically take it positively, i.e., a sign of understanding. As a result, they go faster and engage in shallower learning processes. In comparison, when learners experience a feeling of disfluency, they take it as an alarming signal and, resultingly, switch to slower and deeper learning processes (Alter et al., 2007; Mitchell & Johnson, 2009; Oppenheimer, 2008).

16.3 LA and AI in Digital Learning Environments

Over the past decade, there has been a growing interest in modeling and analysis of students' behaviors and their performance. In particular, several algorithms and techniques have been proposed that rely on students' log data (as well as additional sensors, trackers, or physiological data) to trace moment by moment information on students' activities (Romero & Ventura, 2020). With ongoing technological advancements, the field of education has witnessed a rise in the usage of digital learning environments. For example, within the STEM field, simulation-based learning environments are becoming pervasive as they offer the students an opportunity to engage with learning activities in a safe, realistic, and authentic manner (Kearney, 2003). As the learning tasks become more interactive and complex, there is a need to create new methodologies and measures for task analyses. Frameworks or models to support these analyses need to be able to capture students' performance while considering the varying levels of skills that the students may possess.

In this regard, there is an opportunity to leverage the digital traces of learners' interaction data that they leave behind while interacting in digital learning environments. This has given rise to learning analytics which, according to Siemens and Long (2011), is defined as "the measurement, collection, analysis and reporting of data about learners and their contexts, for purposes of understanding and optimizing learning and the environments in which it occurs" (p. 32). Big data is considered a stimulus for increasing interest in learning analytics (Kennedy et al., 2017). Artificial intelligence and learning analytics offer many opportunities in educational environments, including the provision of automated tutoring and personalized learning.

Digital learning environments offer flexibility and convenience to students in terms of how and when they can access the course contents. These environments are typically student-centered, where the students have more control and autonomy in how they progress in the course, what learning material they access, and in what sequence. At the same time, these environments can pose an inherent challenge in that the students typically do not have face-to-face interaction with their instructors and peers. So, when students face a difficulty, a challenge, or an impasse, they may not be able to receive direct feedback as easily as they would if their peers and instructors are in the room with them. To encourage students to play a more active role in their learning, instructions should be tailored and personalized to meet the individual needs, goals, and skills of the students (Sampson & Karagiannidis, 2002). To achieve this goal, several technological streams are being pursued which we briefly discuss below.

16.3.1 Virtual Labs

Within the higher education sector, more broadly, distance learning and open universities are widely being used for teaching and learning. However, due to the nature of the subject domain, the STEM field is still relatively behind when using new technological approaches (especially for distance or online learning) (Potkonjak et al., 2016). One reason for this discrepancy may be due to the fact that the STEM fields often require laboratory exercises to provide effective skill acquisition and hands-on practice. It can be challenging to make these laboratories accessible and authentic for remote learning. Either the students should be enabled to access the real lab remotely which can be quite challenging or the virtual labs need to be fully replicated through software design.

Moreover, while the value of physical laboratories is generally recognized, the value of virtual labs as an alternative is often contested (National Science Teachers Association, 2007; Singer et al., 2005). Despite this, virtual labs can offer several potential benefits such as being a cost-effective way of enabling high-quality laboratory work, flexibility of creating different virtual experiments, scaling in terms of letting multiple students use the same virtual environment, and convenience with which the system or experimental configuration can be changed. Lastly, and perhaps most importantly, if designed correctly and carefully, these environments can help develop team work skills, promote conceptual understanding, and cultivate interest in science where young learners build enthusiasm about science by taking initiatives and engaging in interesting, challenging, and motivating experiences (Kollöffel & de Jong, 2013; McElhaney & Linn, 2011; Tsovaltzi et al., 2010; van der Meij & de Jong, 2006). Given their affordances, virtual labs can promote STEM education by offering hands-on learning experiences and feedback to facilitate learners in performing their required tasks (de Jong et al., 2014; Sergis et al., 2019; Zervas et al., 2015).

Previous research has compared physical and virtual labs in terms of students' conceptual understanding (Klahr et al., 2007; Zacharia & Constantinou, 2008) and inquiry skills and have reported no differences (Triona & Klahr, 2003). While reality can be adapted in virtual labs, one of the factors that designers of virtual labs should be mindful of is students' motivation and excitement about conducting the hands-on experiments. Further research is needed to understand the impact of virtual labs on students' interest in science as a career. Moreover, investigation of the design of virtual labs with automated guidance and support requires interdisciplinary teams of researchers, including domain experts, technologists, and learning scientists so that design is adaptive and considers students' task difficulties.

16.3.2 Serious Games and Simulations

Computer games are a great tool for learning in STEM. Extant research has investigated whether skills, attitudes, or cognitive capabilities acquired during game-play transfer to other non-game contexts such as schools, work, or daily life. This has resulted in the concept of "serious games," the games which are developed for purposes beyond just entertainment. Researchers report that the transfer of knowledge from computer games is expected when the games are designed around similar cognitive processes as the external tasks (Tobias & Fletcher, 2011a).

There are studies that report that learners who play games may be more flexible in allocating cognitive and perceptual resources (Anderson & Bavelier, 2011; Bailey et al., 2010). For example, in the medical context, surgeons who played games made fewer errors and worked faster than non-player surgeons (Rosser et al., 2007). Similarly, there is evidence that, compared to traditional lectures, game-based learning can promote higher-order thinking and deep learning (Crocco et al., 2016). Yet others have reported that the outcomes of the game-based learning studies are inconclusive (Kebritchi et al., 2010; Wouters et al., 2009), because the enrichment of the game-based environments have their own trade-offs and pitfalls (Greipl et al., 2018). Therefore, the evidence for knowledge or skills transfer from game-based learning to other contexts is weak and requires substantial further research (Tobias & Fletcher, 2011b). For a comparison between game-based studies and to better understand their learning effects, there is a need to develop a shared framework which can be used for classifying educational games (Greipl et al., 2020). In this regard, researchers should be encouraged to report what skills were learners taught using the games, how the computers were used, and what differences existed in terms of the instructional design (Vandercruysse et al., 2012; Wouters et al., 2009).

Simulation-based learning environments offer another avenue for students' learning. Simulations differ from computer games in that they focus more on reliability and authenticity, whereas computer games focus more on entertainment and challenge. Game-based environments generally have quests or story lines, whereas simulation-based environments focus on tasks and scenarios. Additionally, when game-based environments promote competition, simulation environments focus more on task completion. In this regard, simulations can promote stronger critical thinking and may result in developing a better understanding of the concepts. Learners can gradually infer the features of the concept under investigation while they proceed through the simulation. Kearney (2003) mentioned that in a classroom environment, it may not always be possible to provide students with realistic, authentic, and safe scientific phenomena. Virtual simulations can overcome this challenge and have been shown to be effective in promoting a conceptual change in students while offering them an opportunity to observe and learn abstract scientific concepts (de Jong & Van Joolingen, 1998).

Computer simulations can promote a constructivist viewpoint where students may learn through autonomous exposition and contribute to their own knowledge construction. Alternatively, simulations can offer opportunities for social constructivism where students can co-create knowledge through social interaction (Kent et al., 2016). Research on learning with interactive simulations has demonstrated that pure-inquiry learning (without guidance or support) does not benefit learners, whereas inquiry learning with occasional scaffolding, assistance, or support can result in learning gains (Alfieri et al., 2011).

The use of learning analytics to monitor and provide feedback on students' actions within virtual simulations is also gaining popularity. For example, learning analytics techniques have been used to investigate student behavior and learning outcomes in virtual laboratories such as LabLife3D (Qvist et al., 2015), students' interactions in business simulations (Hernández-Lara et al., 2019), and Java-based simulations in computer science (Esquembre, 2004). Similarly, the Habitable Worlds simulation environment has been developed to help students understand concepts related to physics, chemistry, and biology in the exploration of other habitable planets (Horodyskyj et al., 2018). Within this simulation, environment learning analytics methods and techniques have been used to analyze students' behavioral patterns while developing learner profiles (Nawaz et al., 2018), and analyzing their affective engagement such as when they show signs of confusion (Nawaz et al., 2020).

It has been suggested that for virtual simulations, analytic systems and data collection procedures should be designed to capture maximum information on student behaviors, actions, and interactivity (Dodero et al., 2017). However, it is not clear that the policy of "collecting as much information as you can" is well-suited to gathering the most relevant features.

Rigorous studies through empirical research and strong theoretical framework are required to establish valid and reliable measures for students' learning behaviors, processes, and outcomes (Gašević et al., 2015; Wise & Shaffer, 2015). Only then can feedback and support be developed which not only can assist students in their learning journey, but may also help teachers in updating their teaching to tailor or adjust to common student behaviors. For example, if several students are making similar mistakes, it could be hint or cue for the teachers to either slow down their teaching pace or to add additional learning material and support.

16.4 Opportunities

16.4.1 Real-Time Feedback and Intervention

While DLEs offer more autonomy and agency to students, at the same time, these environments require students to have a higher degree of self-regulation and metacognition to change their learning strategies when they face an impasse or a failure (Sankey et al., 2010). In this regard, Learning Analytics dashboards (Park & Jo, 2015), Student Activity Meters (Govaerts et al., 2011; Govaerts et al., 2012), and Open Learning Initiative dashboards (Bodily et al., 2018) have been designed to deliver feedback directly to students. These dashboards allow the learners to view their progress in the course. By capturing and visualizing learners' traces from multiple data sources, it is expected that learners' awareness and reflection can be promoted where they can see how they are doing in the course and how their progress compares to their peers. Eventually, learners can be enabled to engage in goal-setting and goal-achievement behaviors (Verbert et al., 2013). In a student-centered learning environment, however, the provision of real-time feedback and intervention is inherently challenging as the system should be sensitive to the nuances and changes in students' perceptions (Greer & McCalla, 2013). For the automated feedback to be effective, the underlying system should be flexible and responsive and should capture students' behavioral data at the finest granularity.

16.4.2 Personalized Adaptive Learning

Recently, personalized adaptive learning has been promoted to address the shortage of people trained in STEM fields. Though intelligent tutoring systems (ITS) are by no means a new innovation, recent advances in AI provide exciting opportunities to transform the manner in which students learn and how they are assessed. For example, through analyzing students' past performance, an adaptive system (e.g., AutoTutor) (Nye et al., 2014) can adjust the learning content to meet the students' needs and the desired learning outcomes. Similarly, ASSISTments have been widely explored to promote mathematical thinking in school students and to predict students' eventual success in school and college (Ocumpaugh et al., 2016). Cognitive tutors have been used to provide automated instruction to students and have been shown to be effective in developing students' conceptual understanding (Blessing et al., 2009).

In order to individualize the interaction, ITSs must keep track of several aspects of students' learning such as what concepts have been learned and to what extent, what learning approaches seem effective for a student, and what affective and motivational aspects impact that student. Therefore, developing ITS that can model and track all aspects of student learning is cumbersome, time-consuming, and expensive (Greer & McCalla, 2013). To promote the usage of ITSs, a group of researchers developed the Cognitive Tutor Authoring Tool (CTAT), which allows non-programmers to develop intelligent tutoring systems (see Aleven et al., 2016, for further details). A possible future direction could be designing systems that can predict the task difficulty of the learners in real time and provide personalized feedback and recommendations, as shown in Figure 16.2.

16.4.3 Leveraging the Advances in Sensor Technology

For DLEs to be more effective and responsive to task difficulties, they need to be able to detect the emotions that learners encounter during their learning (Arguel et al., 2017). DLEs should be able to react to learners' responses once detected, ideally through system-generated, personalized assistance (Lodge et al., 2018). While it remains a challenge to automate this, one approach that can assist in detecting and responding to learners' emotions is the use of "multimodal data collection and analysis techniques" (Multimodal Learning Analytics or MMLA) (Blikstein & Worsley, 2016). Unlike traditional learning analytics techniques such as learning system–generated clickstream and log-based data, MMLA methods capture massive amounts of heterogeneous data by recording students' behavioral patterns such as their body language (e.g., posture, gesture, and motion), facial expressions, gaze, speech, writing, and physiological patterns (e.g., heart rate, skin conductance, brain signals, facial temperature) (Ochoa & Worsley, 2016; Oviatt et al., 2018).

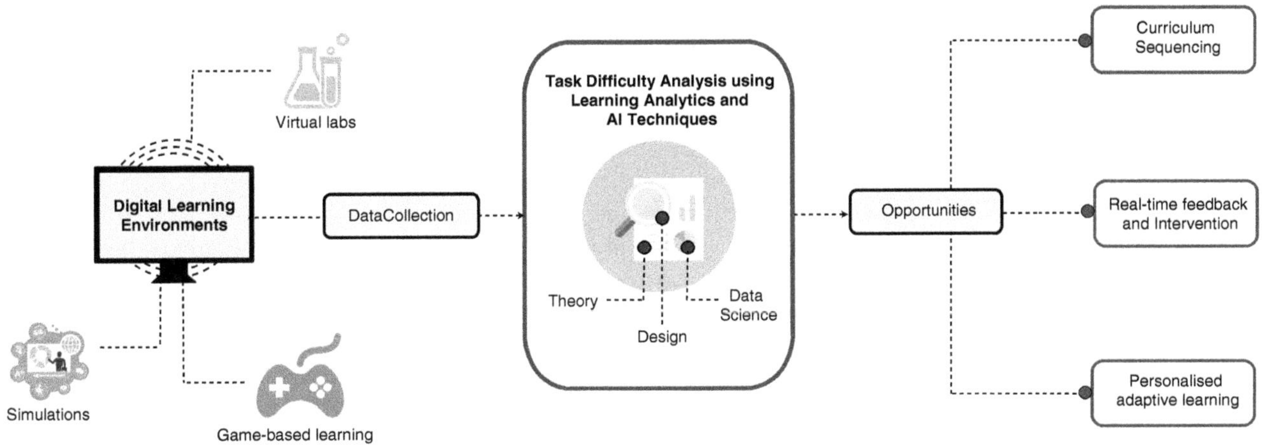

FIGURE 16.2
A possible AI-based system that can collect data across a range of digital learning environments. The system should be designed in line with existing educational theories and the data collection should be informed by the underlying educational design. The possible applications of such a system would be in terms of curriculum sequencing, where, e.g., based on the difficulty of the tasks and each student's zone of proximal development, the contents of the course can be personalized and made adaptive. Similarly, if the difficulty of the tasks gets "too" high for the students, such knowledge can be used for the provision of real-time feedback and interventions.

Due to the sedentary nature of the digital learning tasks, expensive devices such as functional magnetic resonance imaging (fMRI) scanners, electroencephalography (EEG) devices were traditionally used by the researchers to measure the cognitive workload of the learners (Anderson et al., 2010, 2011; Mazher et al., 2017; Spüler et al., 2016). As these devices are difficult to set up and fund, research using small, non-invasive, and low-cost sensors are gaining popularity in educational research. For instance, Romine et al. (2020) proposed a concept of a wearable educational fitness tracker (EduFit) that can measure students' cognitive load during problem-solving activities. Similarly, Larmuseau et al. (2019) collected physiological signals of the participants such as electrodermal activity (EDA) and skin temperature (ST) using a wearable device, and combined the data with subjective measurements of task difficulty to measure the cognitive load of 15 future primary school teachers in a complex learning task.

Recently, in an attempt to measure the cognitive workload of the learners using non-invasive "contactless" sensors, Srivastava et al. (2020, 2021) used an eye-tracker and thermal camera to measure how learners' attention direction and facial temperature change when they watch a video lecture. Using Sensor-based Learning Analytics (SbLA) techniques, Srivastava (2020) found that learners' facial temperature may change when the cognitive load associated with video lectures varies. Similarly, using low-cost eye-trackers, they measured learners' video-watching attention and found that most of the time learners' gaze followed the words on the slide as the instructors read them, however, there was a tendency of fixating more on the new content or concept that didn't fit with their prior knowledge.

To summarize, enrichment of digital learning environments with psycho-physiological sensor data and combining learning activities with indicators for cognitive states is a promising area of research in STEM education. However, this is a complex challenge. The history of research in psycho-physiology and cognitive neuroscience suggests that these kinds of measures generate a lot of noise relative to data that reliably allow for inferences to be made about cognitive and, particularly, metacognitive states.

16.5 Challenges

While the field of learning analytics has made great progress in the last few years due to the advancement of data collection techniques and the use of more sophisticated AI algorithms, there are still major challenges that remain to be addressed. In the following sections, we outline some of these challenges.

16.5.1 Task Difficulty and Its Operationalization

As we alluded to earlier, a perennial issue in assessing task difficulty and mitigating its impact in STEM education is the lack of an agreed-upon conceptualization. The existing definition and current measures of task difficulty are often too limited. Therefore, synthesizing the evidence about the role of task difficulty in STEM education becomes challenging.

In recent attempts to explore and analyze the effects of task difficulties, students' perceptions regarding tasks in simulation-based learning environments have been used (Nawaz, Srivastava et al., 2020; Nawaz et al., 2021). The tasks in this context consisted of one or more of the following activities: creating and examining virtual stars with varying mass, responding to multiple-choice and short-answer questions, creating hypotheses and testing them, providing textual reasoning for hypothesis generation, and watching short lecture-style videos. Students' task difficulties were inferred from their self-reported confidence and the perceived challenge associated with the tasks. As depicted in Figure 16.3, reporting higher confidence than challenge was considered to be indicative of *easy* task difficulties, reporting higher challenge than confidence was considered to indicative of *hard* task difficulties, and when students reported matching scores for confidence and challenge, the task difficulties were perceived as *medium* or moderately difficult. Note that this labeling of task difficulties matches with Csikszentmihalyi's flow theory (Csikszentmihalyi, 1990). When the flow theory is used to examine students' affective analysis in terms of their task challenge and skills, this research utilized students' confidence and task challenge to infer their task difficulties.

Interestingly, Nawaz, Srivastava et al. (2020) found that, in line with existing theories – such as Vygotsky's zone of proximal development (Vygotsky, 1980), Csikszentmihalyi's flow theory (Csikszentmihalyi, 1990), and the law of optimum perceived difficulty (Belmont & Mitchell, 1987) –*medium* task difficulties could lead students to achieve better learning outcomes than *easy* or *hard* task difficulties. This seems to suggest that when students perceive the current task to be moderately difficult, they perceive the learning materials, the learning environment, or perhaps both

FIGURE 16.3
A possible labeling of students' task difficulties in terms of their perceived confidence and challenge associated with the tasks.

are at an optimal level of difficulty. Such optimal difficulties can then lead these students to achieve better learning outcomes than their peers.

While students' perceptions of difficulties or their task difficulties have implications on students' learning outcomes, as discussed above. Can these perceptions influence students' task difficulties at the subsequent tasks? Various researchers have investigated this effect. For example, in their work, Schneider and Anderson (2010) reported that when individuals face a hard or challenging task, they need to allocate a greater amount of cognitive resources to it. When these individuals proceed to the next task, they may experience a depletion in the available resources, and as a result their performance in the next task may be affected. This idea is similar to Sweller's cognitive load theory. Sweller and Chandler (1991) suggested that when learners are presented with overly challenging tasks or environment, their working memory can be overloaded. As a result, the development and processing of conceptual knowledge may be obstructed. Contrary to these findings, when 76Nawaz et al. (2021) analyzed students' task difficulty sequences, they found that the students who perceived the current task to be *hard* were more likely than chance to perceive the subsequent task to be *medium*. Therefore, it can expected that *hard* and challenging tasks have the potential to lead students to optimally difficult learning conditions – conditions where students may find the difficulty of the task at match with their skills. But what does it mean for learning design? Should the tasks be made intentionally difficult or harder as they often lead to moderate levels of difficulties or should the following tasks be made easier by comparison? We believe that this question could benefit from further research.

Nawaz, Srivastava et al. (2020) have also investigated the effects of task difficulty transitions in relation to students' performance or learning outcomes. It has been found that the students who perceive two or more consecutive tasks as *medium* or moderately difficult could achieve significantly better performance than their peers. By contrast, the students who perceived two or more consecutive tasks as *hard* achieved significantly poorer scores. This finding combined with the previous results may suggest that while *hard* and challenging tasks have the potential to engage students and lead them to arrive at optimally difficult conditions, when the difficulty associated with tasks continues too long (i.e., it sustains over two or more tasks) then it can adversely affect students' learning and their eventual success.

Lastly, models have been developed to associate or establish learning analytics–based behavioral markers for different levels of task difficulties (Nawaz et al., 2021). It was found that, in terms of behavior, students who perceived the tasks to be *easy* generally adopted a "trial and error" approach or gaming behavior. These students seemed to make several attempts while spending less time on tasks. These students also wrote less while explaining their reasoning associated with the hypothesis explanation. In comparison, students who perceived the task difficulties to be *medium* spent more time on tasks and required fewer attempts to complete those tasks. Moreover, these students wrote longer texts to explain the reasoning associated with their hypotheses. Generally, these students seemed to reflect more on the learning tasks than their peers.

Therefore, we believe that task difficulty is a multifaceted and multidimensional construct rather than a unitary construct. Recently, a theoretical framework of task difficulty has been proposed that suggests that the interactions among learning and task parameters can result in productive or non-productive difficulties (Lodge et al., 2018). Another simple, yet comprehensive, typology of task complexity (Campbell, 1988) postulated that task difficulty can be seen as primarily (a) psychological experience, (b) an interaction between task and person characteristics, and (c) a function of objective task characteristics. The development of such frameworks can provide a strong foundation for the development of LA and AI tools that can enable constructive feedback for students to help them tackle and overcome their difficulties to meet their learning goals.

16.5.2 Data Ethics and Student Privacy

In order to provide more intelligent feedback and interventions, learning analytic solutions often rely on the use of *big data*. But as "these data commit to record details about human behavior, they have been perceived as a threat to fundamental values, including everything from autonomy, to fairness, justice, due process, property, soldiery, perhaps most of all privacy" (Barocas & Nissenbaum, 2014, p. 44). The challenges around the use of students' personal data are serious and need to be given due consideration by all the stakeholders. The data subjects (the students in this case) should have the right to access their data, be able to demand for rectification and data removal, as well as to decide upon what the data can be used for. According to the "General Data Protection Regulation"[1] (GDPR), these rights should exist as long as the data subject is identifiable. Despite these ethical issues, data is needed to develop and train computational models that can represent and support learners. Therefore, ethical AIED and LA implementation remains an important goal.

16.5.3 What Data to Collect and How

When developing data-informed learning tools, key questions arise: (1) What data should be collected for the purposes of analysis and interventions? (2) What implications do the different data formats have? Typically, learning analytics data consist of nominal features such as students' demographics, including country, ethnicity, or gender. For such data, there is no inherent ordering. Therefore, relational operations such as X > Y may not be reasonable to perform, nor would transitive statements apply like if X > Y and Y > Z, then X > Z. However, in such a case counting frequencies for each category would be a reasonable operation to perform such as counting the number of students who belong to country X. When the data is ordinal, there is an inherent ordering and statements such as X > Y can make sense, as in the context of course letter grades. However, operations such as addition, subtraction, or multiplication would still not be reasonable to perform. Some examples of ordinal data may include students' socioeconomic status such as yearly income, their highest degree completed (such as B.Sc., M.Sc., Ph.D.), or the time when an assessment was submitted such as 11 a.m., 12 p.m., 1 p.m., and so on.

While dealing with various data, it is imperative that the underlying properties and structure of the data are understood as they can often affect the validity of learning analytics–based interventions (Winne, 2017). For example, a model that predicts an outcome solely based on the gender, ethnicity, or demographic location of a learner may not offer strong grounds for interventions. In such a case, correlation between model features and outcome may not offer valid grounds to infer causation. Moreover, such demographics data cannot be manipulated which further limits the generalizability of the findings. At the same time, there is an increasing desire to analyze underlying student demographics as it can offer teachers an opportunity to understand their students and their needs (Nawaz et al., 2021).

Another fundamental challenge of learning analytics is in terms of sense-making. While the "big" data in education has been the driving force for learning analytics–based systems (Kennedy et al., 2017), the size of this data could stem from various factors such as the number of learners enrolled in a massive open online course (MOOC), or it could be a smaller number of students but at a more granular level such as moment-by-moment variation in learners' facial expressions, thermal temperature, eye-gaze, or body gestures. While such data may provide an excellent opportunity to investigate and understand learners' behaviors and learning processes, an inherent difficulty lies in terms of data interpretation "what counts as a meaningful finding when the number of data points is so large that something will always be significant" (Wise & Shaffer, 2015, p. 6). An underlying theory or learning design should guide the process of variable or feature selection. Analyzing learners' digital traces or access patterns without a theory-guided approach can be challenging because an action or a series of digital traces can be indicative of different intentions, processes, and outcomes. This means learners' digital traces can be "devoid of an intrinsically cognitive component" (Kennedy & Judd, 2004). Overall, this approach assumes that the psychological principles that underpin one area of human endeavor can translate to other areas (Weber & Johnson, 2009), and that the usage of existing theories can provide coherent set of statements, concepts, and ideas that can lead toward prediction and explanation of behaviors, events, and phenomena (Bem & De Jong, 2013; Eccles et al., 2005).

Another challenge stems from computing technologies. While AI can potentially transform the field of education, the mere use of advanced AI technologies may not warrant "good educational outcomes" (Ouyang & Jiao, 2021). Researchers and practitioners should be made aware of how different educational technologies imply different philosophical and pedagogical approaches, which in turn may influence the design and quality of learning and instruction (Hwang et al., 2020).

Similarly, from a data-analytics and algorithmic perspective, single data points may not capture students' learning processes at a fine enough level of detail. Therefore, analysis of learners' digital traces should be done over time, while considering other behaviors such as the details of the task, the errors made, the attempts required to rectify those errors, the number of clicks, the choice of tools selected by the learners, and the timing of such selection. Winne (2006) suggests that while there can be randomness around learners' digital traces, if they are the agents of their learning and the constructors of their own knowledge, then their actions (as stated above) can be cognitive. Typically, the teachers and learning designers provide specific tools to the students to be used in a particular context. It is up to the students how they use those tools. If students use the provided support and introduction to the tools in accordance with the learning design, they are likely to develop the expected skills. By contrast, if learners miss on the opportunity to use the tool in accordance with the learning design, then their knowledge construction may be less effective. Furthermore, while learners' digital traces (such as the frequency of an event or the time spent on a task) can be indicative of the degree to which they use a tool, "the high volumes of these measures cannot be directly interpreted as a high quality of learning" (Gašević et al., 2015, p. 68). To better understand the learning behaviors and to better support the learners, it is important to consider the particular strategies that students adopt. Only by developing rigorous and robust measures can the meaningful learner profiles and behavioral markers be established which can then advance the field of STEM education via advanced analytics techniques.

16.5.4 Confounding Factors

Learning does not occur in a vacuum. The learning environments in which the task is performed or practiced can impact how students perceive and interact with the learning tasks. Complex learning environments with a large number of alternatives presented in an inherently complex manner can be distracting and less useful for many students, especially for students who are unable to manage and self-regulate their learning. However, some students may benefit from a more flexible learning environment in terms of the different options that they can explore. Furthermore, task difficulties have implications for learners' affects and emotions. While it has been shown that under hard or difficult task conditions learners may undergo negative emotional experiences (Nawaz, Kennedy et al., 2020; Nawaz et al., 2018), the converse of this relation has also been shown empirically. Under different emotional states, tasks with varying difficulty levels may be approached differently by individuals. When motivated and in a flow state, learners tend to prefer hard tasks over the easy ones as the completion of hard tasks can result in more emotions than the completion of the easier tasks (Passyn & Sujan, 2012). Therefore, it can be suggested that the relation between task difficulties and emotions is a two-way relation. Not only can the emotions be a consequence of specific types of difficult tasks, but also the emotions may serve as antecedents. Against this background, untangling what makes a task difficult for students, and whether the difficulties that students encounter are due to learning task, the learning environments, or a combination of the two is not a trivial endeavor.

Among other factors that can influence task difficulties is task training. One might expect that when practice is done on tasks that are too hard, they can lead to better performance under the test conditions which are comparatively easier. However, empirical research has proven it otherwise (Power, 2019). Learners who practiced under highly difficult task conditions consistently performed poorer than their peers. By comparison, learners who practiced under easy task conditions consistently outperformed those who practiced under medium or high task difficulty conditions. So while, it may be expected that task practice or training can result in an improved performance (Kozlowski et al., 2001), the possible relation between task difficulty and learning specificity may mediate this effect. Ahissar and Hochstein (1997) also reported on the effects of learning specificity and task difficulties. They found that practicing *easy* tasks can result in the generalization of learning. However, as tasks increase in difficulty, they become more specific and less likely to generalize. This is possibly because different learning processes may get activated under different conditions and they could vary with different levels of task difficulties. Therefore, to make learning effective under difficult conditions, it is required that the onset of learning should have previously been enabled under easy conditions – a hypothesis often referred to as the "cognition hypothesis" (Jackson & Suethanapornkul, 2013; Robinson, 2001). In this regard, Robinson (2015) suggested that "complex cognition" is inevitably grounded in attempts to perform simpler tasks. To enable learners to excel in tasks with increasing complexity, they should first be supported and scaffolded on easier tasks, only then can learners manage task difficulties and complexities of more advanced material.

In technology-enabled courses, another issue that is becoming prevalent is learners' multitasking behavior, which can lead to interruption (Dindar & Akbulut, 2016). Multitasking means engaging in more than one tasks at any given time. In a digital context, it can happen when learners are exposed to different sources of information or when they switch between different media (e.g., read an online article, check news, or log onto social media accounts such as Facebook and Twitter). Such switching may happen by self-choice where individuals decide to take a break from their current task or it can happen due to an external interruption. The effects of multitasking in digital contexts have been analyzed with varying levels of task difficulties (Adler & Benbunan-Fich, 2015). Multitasking under difficult task conditions can be detrimental to students' performance possibly because it can result in too much arousal and an overload in students' working memory (Altmann & Trafton, 2002). By contrast, under easy task conditions, multitasking can lead to significantly better performance than no multitasking or voluntary multitasking. A possible reason for this effect is that when students get interrupted under easier task conditions, the amount of stimulation is increased that can result in students' performance improvement (Speier et al., 2003).

Based on the discussion in this chapter, there are several ways in which task difficulty within a STEM learning context can be conceptualized and addressed. The different definitions of task difficulty are interrelated, yet they do not overlap perfectly. Some approaches reflect task difficulty in a way that can be directly measured (e.g., response time, number of errors, performance or probability of success), while the other approaches reflect cognitive, psychological, or affective processes that are latent and cannot be measured directly. For example, low-demand tasks are considered easier than high-demand tasks where the demand of a task can be related to the complexity of that task which can then be associated with cognitive effort to solve or resolve that task. This definition can also be related to the developmental perspective in a learning context where an easier task would be the one that students can perform or attend to at an earlier developmental stage (Gilbert et al., 2012). For a clearer understanding and conceptualization, we have captured these different aspects of task difficulty in Figure 16.4.

From the literature and the elements shown in Figure 16.4, it is apparent that task difficulty has been used in several related but non-similar contexts. It is not possible to have a single definition of this construct that can precisely capture the various associated aspects. However, such conceptualization can pave way for researchers, learning designers, and teachers to understand how they can manipulate the difficulties, how they can sequence the learning tasks, how they can design assignments that match with students' skill levels, and how they can help students to use appropriate strategies to complete those assignments.

16.6 Conclusion

In this chapter, we reviewed the notion of task difficulty and how it has been defined and operationalized in the literature. Given that much of the onus of learning in STEM falls on learners themselves, there is a need for developing assignments that *slightly* exceed students' skill levels. So there is some difficulty which the students need to overcome by engaging in the learning tasks, but not so difficult that the students fail or disengage.

The design of assessments and tasks at the correct level of difficulty can promote students' on-task behaviors, increase task completions, and improve student comprehension. AI enabled, learner-centered environments offer a unique opportunity within the STEM domain where the students' performance and motivation may be optimized by adaptive

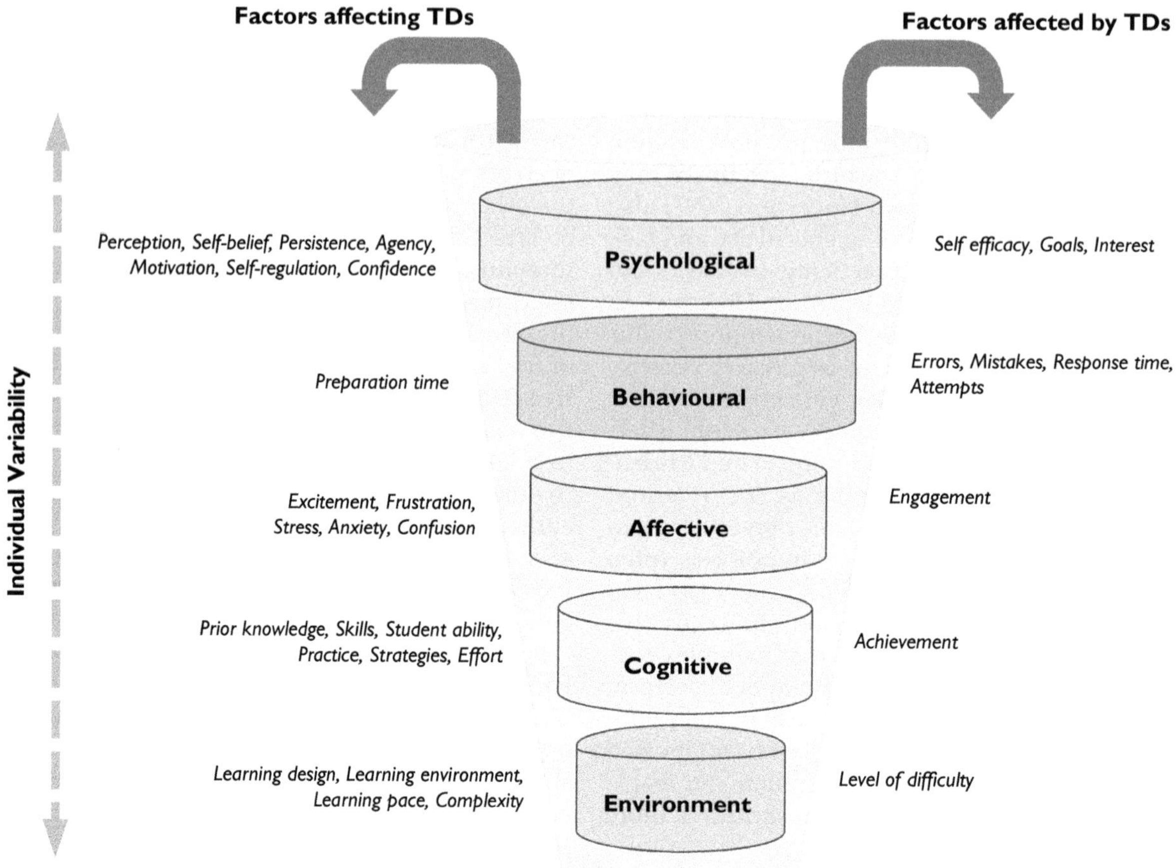

FIGURE 16.4
Task difficulties are multifaceted: not only can they depend on several factors, their effects can also be expected in more than one dimension. This is further complicated by the effects of individual variability which can be hard to capture in STEM, especially with growing class numbers.

manipulation of task difficulty conditions. In comparison to forced difficulty conditions, these environments have the potential to match students' skills and abilities.

In terms of educational implications, the instructors and learning designers should not design curricula to prevent student failure, as failure is not necessarily detrimental (Kapur, 2016). It is only when students struggle at the learning tasks, engage in effortful learning behaviors, and make a deliberate effort that they can overcome those challenges and develop a better understanding of the concepts. Such mastery-based approaches can produce STEM students who possess 21st-century skills and are ready to meet the challenges of the ever-changing job market.

Note

1. https://gdpr-info.eu – Legal obligation in Europe since May, 2018.

References

Adler, R. F., & Benbunan-Fich, R. (2015). The effects of task difficulty and multitasking on performance. *Interacting with Computers, 27*(4), 430–439.

Ahissar, M., & Hochstein, S. (1997). Task difficulty and the specificity of perceptual learning. *Nature, 387*(6631), 401–406. https://doi.org/10.1038/387401a0

Aleven, V., McLaren, B. M., Sewall, J., Van Velsen, M., Popescu, O., Demi, S., Ringenberg, M., & Koedinger, K. R. (2016). Example-tracing tutors: Intelligent tutor development for non-programmers. *International Journal of Artificial Intelligence in Education, 26*(1), 224–269.

Alfieri, L., Brooks, P. J., Aldrich, N. J., & Tenenbaum, H. R. (2011). Does discovery-based instruction enhance learning? *Journal of Educational Psychology, 103*(1), 1–18. https://doi.org/10.1037/a0021017

Allen, S., & Peterman, K. (2019). Evaluating informal STEM education: Issues and challenges in context. *New Directions for Evaluation, 161*, 17–33. https://doi.org/10.1002/ev.20354

Alter, A. L., Oppenheimer, D. M., Epley, N., & Eyre, R. N. (2007). Overcoming intuition: Metacognitive difficulty activates analytic reasoning. *Journal of Experimental Psychology: General, 136*(4), 569.

Altmann, E. M., & Trafton, J. G. (2002). Memory for goals: An activation-based model. *Cognitive Science, 26*(1), 39–83.

Anderson, A. F., & Bavelier, D. (2011). Action game play as a tool to enhance perception, attention and cognition. In S. Tobias & J. D. Fletcher (Eds.), *Computer games and instruction* (pp. 307–330). Information Age.

Anderson, J. R., Betts, S., Ferris, J. L., & Fincham, J. M. (2010). Neural imaging to track mental states while using an intelligent tutoring system. *Proceedings of the National Academy of Sciences, 107*(15), 7018–7023. https://www.pnas.org/content/107/15/7018

Anderson, J. R., Betts, S., Ferris, J. L., & Fincham, J. M. (2011). Cognitive and metacognitive activity in mathematical problem solving: Prefrontal and parietal patterns. *Cognitive, Affective, and Behavioral Neuroscience, 11*(1), 52–67. https://doi.org/10.3758/s13415-010-0011-0

Arguel, A., Lockyer, L., Lipp, O. V., Lodge, J. M., & Kennedy, G. (2017). Inside out: Detecting learners' confusion to improve interactive digital learning environments. *Journal of Educational Computing Research, 55*(4), 526–551. https://doi.org/10.1177/0735633116674732

Bailey, K., West, R., & Anderson, C. A. (2010). A negative association between video game experience and proactive cognitive control. *Psychophysiology, 47*(1), 34–42. https://doi.org/10.1111/j.1469-8986.2009.00925.x

Barocas, S., & Nissenbaum, H. (2014). Big Data's end run around anonymity and consent. In J. Lane, V. Stodder, S. Bender, & H. Nissenbaum (Eds.), *Privacy, big data and the public good* (pp. 44–75). Cambridge University Press.

Belmont, J. M., & Mitchell, D. W. (1987). The general strategies hypothesis as applied to cognitive theory in mental retardation. *Intelligence, 11*(1), 91–105. https://doi.org/10.1016/0160-2896(87)90029-8

Bem, S., & De Jong, H. L. (2013). *Theoretical issues in psychology: An introduction*. Sage.

Bjork, E. L., & Bjork, R. A. (2011). Making things hard on yourself, but in a good way: Creating desirable difficulties to enhance learning. In *Psychology and the real world: Essays illustrating fundamental contributions to society* (pp. 56–64). https://psycnet.apa.org/record/2011-19926-008

Blessing, S. B., Gilbert, S. B., Ourada, S., & Ritter, S. (2009). Authoring model-tracing cognitive tutors. *International Journal of Artificial Intelligence in Education, 19*(2), 189–210.

Blikstein, P., & Worsley, M. (2016). Multimodal learning analytics and education data mining: Using computational technologies to measure complex learning tasks. *Journal of Learning Analytics, 3*(2), 220–238. https://learning-analytics.info/index.php/JLA/article/view/4383

Bodily, R., Kay, J., Aleven, V., Jivet, I., Davis, D., Xhakaj, F., & Verbert, K. (2018). Open learner models and learning analytics dashboards: A systematic review. In *Proceedings of the 8th international conference on learning analytics and knowledge* (pp. 41–50). Association for Computing Machinery. https://doi.org/10.1145/3170358.3170409

Campbell, D. J. (1988). Task complexity: A review and analysis. *Academy of Management Review, 13*(1), 40–52. https://doi.org/10.5465/amr.1988.4306775

Campbell, J. P., DeBlois, P. B., & Oblinger, D. G. (2007). Academic analytics: A new tool for a new era. *Educause Review, 42*(4), 40. https://eric.ed.gov/?id=EJ769402

Crocco, F., Offenholley, K., & Hernandez, C. (2016). A proof-of-concept study of game-based learning in higher education. *Simulation and Gaming, 47*(4), 403–422.

Csikszentmihalyi, M. (1990). *Flow: The psychology of optimal experience* (Vol. 1990). Harper & Row.

D'Mello, S., & Graesser, A. (2014). Confusion and its dynamics during device comprehension with breakdown scenarios. *Acta Psychologica, 151*, 106–116. https://www.sciencedirect.com/science/article/pii/S0001691814001504

de Jong, T., Sotiriou, S., & Gillet, D. (2014). Innovations in stem education: The go-lab federation of online labs. *Smart Learning Environments, 1*(1), 3. https://doi.org/10.1186/s40561-014-0003-6

de Jong, T., & Van Joolingen, W. R. (1998). Scientific discovery learning with computer simulations of conceptual domains. *Review of Educational Research, 68*(2), 179–201. https://doi.org/10.3102/00346543068002179

DeLoache, J. S., Cassidy, D. J., & Brown, A. L. (1985). Precursors of mnemonic strategies in very young children's memory. *Child Development, 56*(1), 125–137. http://www.jstor.org/stable/1130180

Dindar, M., & Akbulut, Y. (2016). Effects of multitasking on retention and topic interest. *Learning and Instruction, 41*, 94–105.

Dodero, J. M., González-Conejero, E. J., Gutiérrez-Herrera, G., Peinado, S., Tocino, J. T., & Ruiz-Rube, I. (2017). Trade-off between interoperability and data collection performance when designing an architecture for learning analytics. *Future Generation Computer Systems, 68*, 31–37.

Eccles, J. S., & Wigfield, A. (2002). Motivational beliefs, values, and goals. *Annual Review of Psychology, 53*(1), 109–132. https://doi.org/10.1146/annurev.psych.53.100901.135153

Eccles, M., Grimshaw, J., Walker, A., Johnston, M., & Pitts, N. (2005). Changing the behavior of healthcare professionals: The use of theory in promoting the uptake of research findings. *Journal of Clinical Epidemiology, 58*(2), 107–112.

Eisenberger, R., & Leonard, J. M. (1980). Effects of conceptual task difficulty on generalized persistence. *American Journal of Psychology, 93*(2), 285–298. http://www.jstor.org/stable/1422233

English, L. D. (2016). Stem education k-12: Perspectives on integration. *International Journal of STEM Education, 3*(1), 1–8.

Esquembre, F. (2004). Easy Java simulations: A software tool to create scientific simulations in Java. *Computer Physics Communications, 156*(2), 199–204. https://doi.org/10.1016/S0010-4655(03)00440-5

Feather, N. T. (1961). The relationship of persistence at a task to expectation of success and achievement related motives. *Journal of Abnormal and Social Psychology, 63*(3), 552–561. https://doi.org/10.1037/h0045671

Gašević, D., Dawson, S., & Siemens, G. (2015). Let's not forget: Learning analytics are about learning. *TechTrends, 59*(1), 64–71. https://doi.org/10.1007/s11528-014-0822-x

Gickling, E. E., & Armstrong, D. L. (1978). Levels of instructional difficulty as related to on-task behavior, task completion, and comprehension. *Journal of Learning Disabilities, 11*(9), 559–566.

Gilbert, S., Bird, G., Frith, C., & Burgess, P. (2012). Does "task difficulty" explain "task-induced deactivation?" *Frontiers in Psychology, 3*, 125.

Govaerts, S., Verbert, K., & Duval, E. (2011). Evaluating the student activity meter: Two case studies. In H. Leung, E. Popescu, Y. Cao, R. W. H. Lau, & W. Nejdl (Eds.), *Advances in web-based learning – ICWL 2011* (pp. 188–197). Springer Berlin Heidelberg.

Govaerts, S., Verbert, K., Duval, E., & Pardo, A. (2012). The student activity meter for awareness and self-reflection. In *Chi '12 extended abstracts on human factors in computing systems* (pp. 869–884). Association for Computing Machinery. https://doi.org/10.1145/2212776.2212860

Greer, J. E., & McCalla, G. I. (2013). *Student modelling: The key to individualized knowledge-based instruction* (Vol. 125). Springer Science & Business Media.

Greipl, S., Moeller, K., & Ninaus, M. (2020). Potential and limits of game-based learning. *International Journal of Technology Enhanced Learning, 12*(4), 363–389.

Greipl, S., Ninaus, M., Bauer, D., Kiili, K., & Moeller, K. (2018). A fun-accuracy trade-off in game-based learning. In M. Gentile, M. Allegra, & H. Söbke (Eds.), *International conference on games and learning alliance* (pp. 167–177). Springer.

Guadagnoli, M., Morin, M.-P., & Dubrowski, A. (2012). The application of the challenge point framework in medical education. *Medical Education, 46*(5), 447–453. https://doi.org/10.1111/j.1365-2923.2011.04210.x

Hernández-Lara, A. B., Perera-Lluna, A., & Serradell-López, E. (2019). Applying learning analytics to students' interaction in business simulation games: The usefulness of learning analytics to know what students really learn. *Computers in Human Behavior, 92*, 600–612. https://doi.org/10.1016/j.chb.2018.03.001

Hom, H. L., & Maxwell, F. R. (1983). The impact of task difficulty expectations on intrinsic motivation. *Motivation and Emotion, 7*(1), 19–24.

Horodyskyj, L. B., Mead, C., Belinson, Z., Buxner, S., Semken, S., & Anbar, A. D. (2018). Habitable worlds: Delivering on the promises of online education. *Astrobiology, 18*(1), 86–99.

Horvath, M., Herleman, H. A., & Lee McKie, R. (2006). Goal orientation, task difficulty, and task interest: A multilevel analysis. *Motivation and Emotion, 30*(2), 169–176. https://doi.org/10.1007/s11031-006-9029-6

Hwang, G.-J., Xie, H., Wah, B. W., & Gašević, D. (2020). *Vision, challenges, roles and research issues of artificial intelligence in education*. Elsevier.

Jackson, D. O., & Suethanapornkul, S. (2013). The cognition hypothesis: A synthesis and meta-analysis of research on second language task complexity. *Language Learning, 63*(2), 330–367.

Kapur, M. (2016). Examining productive failure, productive success, unproductive failure, and unproductive success in learning. *Educational Psychologist, 51*(2), 289–299. https://doi.org/10.1080/00461520.2016.1155457

Kapur, M., & Bielaczyc, K. (2012). Designing for productive failure. *Journal of the Learning Sciences, 21*(1), 45–83. https://doi.org/10.1080/10508406.2011.591717

Kearney, M. (2003). A new tool for creating predict-observe-explain tasks supported by multimedia. *Science Education News, 52*(1), 13–17. https://search.informit.org/doi/10.3316/aeipt.129849

Kebritchi, M., Hirumi, A., & Bai, H. (2010). The effects of modern mathematics computer games on mathematics achievement and class motivation. *Computers and Education, 55*(2), 427–443. https://doi.org/10.1016/j.compedu.2010.02.007

Kennedy, G., Corrin, L., & de Barba, P. (2017). Analytics of what? Negotiating the seduction of big data and learning analytics. *Visions for Australian Tertiary Education*, 67–76.

Kennedy, G. E., & Judd, T. S. (2004). Making sense of audit trail data. *Australasian Journal of Educational Technology, 20*(1).

Kent, C., Laslo, E., & Rafaeli, S. (2016). Interactivity in online discussions and learning outcomes. *Computers and Education, 97*, 116–128. https://doi.org/10.1016/j.compedu.2016.03.002

Killpack, T. L., & Melón, L. C. (2016). Toward inclusive stem classrooms: What personal role do faculty play? *CBE – Life Sciences Education, 15*(3), es3.

Kirschner, P. A., Sweller, J., & Clark, R. E. (2006). Why minimal guidance during instruction does not work: An analysis of the failure of constructivist, discovery, problem-based, experiential, and inquiry-based teaching. *Educational Psychologist, 41*(2), 75–86. https://doi.org/10.1207/s15326985ep4102_1

Klahr, D., Triona, L. M., & Williams, C. (2007). Hands on what? The relative effectiveness of physical versus virtual materials in an engineering design project by middle school children. *Journal of Research in Science Teaching, 44*(1), 183–203. https://onlinelibrary.wiley.com/doi/abs/10.1002/tea.20152

Kollöffel, B., & de Jong, T. (2013). Conceptual understanding of electrical circuits in secondary vocational engineering education: Combining traditional instruction with inquiry learning in a virtual lab. *Journal of Engineering Education, 102*(3), 375–393. https://onlinelibrary.wiley.com/doi/abs/10.1002/jee.20022

Kozlowski, S. W., Gully, S. M., Brown, K. G., Salas, E., Smith, E. M., & Nason, E. R. (2001). Effects of training goals and goal orientation traits on multidimensional training outcomes and performance adaptability. *Organizational Behavior and Human Decision Processes, 85*(1), 1–31.

Larmuseau, C., Vanneste, P., Cornelis, J., Desmet, P., & Depaepe, F. (2019). Combining physiological data and subjective measurements to investigate cognitive load during complex learning. *Frontline Learning Research, 7*(2), 57–74.

Li, W., Lee, A., & Solmon, M. (2007). The role of perceptions of task difficulty in relation to self-perceptions of ability, intrinsic value, attainment value, and performance. *European Physical Education Review, 13*(3), 301–318.

Lodge, J. M., Kennedy, G., Lockyer, L., Arguel, A., & Pachman, M. (2018). Understanding difficulties and resulting confusion in learning: An integrative review. *Frontiers in Education, 3*, 49. https://doi.org/10.3389/feduc.2018.00049

Maynard, D. C., & Hakel, M. D. (1997). Effects of objective and subjective task complexity on performance. *Human Performance, 10*(4), 303–330.

Mazher, M., Abd Aziz, A., Malik, A. S., & Ullah Amin, H. (2017). An EEG-based cognitive load assessment in multimedia learning using feature extraction and partial directed coherence. *IEEE Access, 5*, 14819–14829.

McElhaney, K. W., & Linn, M. C. (2011). Investigations of a complex, realistic task: Intentional, unsystematic, and exhaustive experimenters. *Journal of Research in Science Teaching, 48*(7), 745–770. https://onlinelibrary.wiley.com/doi/abs/10.1002/tea.20423

Mitchell, K. J., & Johnson, M. K. (2009). Source monitoring 15 years later: What have we learned from FMRI about the neural mechanisms of source memory? *Psychological Bulletin, 135*(4), 638.

National Science Teachers Association. (2007). The integral role of laboratory investigations in science instruction. www.nsta.org/about/positions/laboratory.aspx

Nawaz, S. (2021). *Learning analytics: Towards analysing learner confusion and task difficulties* (Doctoral dissertation). The University of Melbourne.

Nawaz, S., Kennedy, G., Bailey, J., & Mead, C. (2020). Moments of confusion in simulation-based learning environments. *Journal of Learning Analytics, 7*(3), 118–137.

Nawaz, S., Kennedy, G., Bailey, J., Mead, C., & Horodyskyj, L. (2018). Struggle town? Developing profiles of student confusion in simulation-based learning environments. *Open Oceans: Learning Without Borders, 224*.

Nawaz, S., Srivastava, N., Yu, J. H., Baker, R. S., Kennedy, G., & Bailey, J. (2020). Analysis of task difficulty sequences in a simulation-based POE environment. In I. Bittencourt, M. Cukurova, K. Muldner, R. Luckin, & E. Millán (Eds.), *International conference on artificial intelligence in education* (pp. 423–436). Springer.

Nawaz, S., Srivastava, N., Yu, J. H., Khan, A. A., Kennedy, G., Bailey, J., & Baker, R. S. (2021). How difficult is the task for you? Modelling and analysis of students' task difficulty sequences in a simulation-based POE environment. *International Journal of Artificial Intelligence in Education*. https://doi.org/10.1007/s40593-021-00242-6

Nye, B. D., Graesser, A. C., & Hu, X. (2014). Autotutor and family: A review of 17 years of natural language tutoring. *International Journal of Artificial Intelligence in Education, 24*(4), 427–469.

Ochoa, M. N., & Henriques, B. (2021, March). Supporting students in career and technical education STEM programs using community of practice and open textbooks. In E. Langran & L. Archambault (Eds.), *Proceedings of society for information technology & teacher education international conference* (pp. 895–896). Association for the Advancement of Computing in Education (AACE).

Ochoa, X., & Worsley, M. (2016). Editorial: Augmenting learning analytics with multimodal sensory data. *Journal of Learning Analytics, 3*(2), 213–219. https://learning-analytics.info/index.php/JLA/article/view/5081

Ocumpaugh, J., San Pedro, M. O., Lai, H.-Y., Baker, R. S., & Borgen, F. (2016). Middle school engagement with mathematics software and later interest and self-efficacy for STEM careers. *Journal of Science Education and Technology, 25*(6), 877–887.

Oppenheimer, D. M. (2008). The secret life of fluency. *Trends in Cognitive Sciences, 12*(6), 237–241.

Ost, B. (2010). The role of peers and grades in determining major persistence in the sciences. *Economics of Education Review, 29*(6), 923–934. https://doi.org/10.1016/j.econedurev.2010.06.011

Ouyang, F., & Jiao, P. (2021). Artificial intelligence in education: The three paradigms. *Computers and Education: Artificial Intelligence, 2*, 100020.

Oviatt, S., Grafsgaard, J., Chen, L., & Ochoa, X. (2018). Multimodal learning analytics: Assessing learners' mental state during the process of learning. In S. Oviatt, B. Schuller, P. R. Cohen, D. Sonntag, G. Potamianos, & A. Krüger (Eds.), *The handbook of multimodal-multisensor interfaces: Signal processing, architectures, and detection of emotion and cognition* (Vol. 2, pp. 331–374). Association for Computing Machinery and Morgan & Claypool.

Park, Y., & Jo, I.-H. (2015). Development of the learning analytics dashboard to support students' learning performance. *Journal of Universal Computer Science, 21*(1), 110–133. http://www.jucs.org/jucs_21_1/development_of_the_learning

Passyn, K., & Sujan, M. (2012). Skill-based versus effort-based task difficulty: A task-analysis approach to the role of specific emotions in motivating difficult actions. *Journal of Consumer Psychology, 22*(3), 461–468.

Pintrich, P. R., & Schunk, D. H. (2002). *Motivation in education: Theory, research, and applications*. Prentice Hall.

Poquet, O., & de Laat, M. (2021). Developing capabilities: Lifelong learning in the age of AI. *British Journal of Educational Technology, 52*(4), 1695–1708. https://doi.org/10.1111/bjet.13123

Potkonjak, V., Gardner, M., Callaghan, V., Mattila, P., Guetl, C., Petrović, V. M., & Jovanović, K. (2016). Virtual laboratories for education in science, technology, and engineering: A review. *Computers and Education, 95*, 309–327. https://www.sciencedirect.com/science/article/pii/S0360131516300227

Power, J. (2019). The influence of task difficulty on engagement, performance and self-efficacy. In P. Williams & D. Barlex (Eds.), *Explorations in technology education research* (pp. 157–169). Springer.

Qvist, P., Kangasniemi, T., Palomäki, S., Seppänen, J., Joensuu, P., Natri, O., Närhi, M., Palomäki, E., Tiitu, H., & Nordström, K. (2015). Design of virtual learning environments: Learning analytics and identification of affordances and barriers. *International Journal of Engineering Pedagogy*. https://doi.org/10.3991/ijep.v5i4.4962

Robinson, P. (2001). Task complexity, task difficulty, and task production: Exploring inter-actions in a componential framework. *Applied Linguistics, 22*(1), 27–57. https://doi.org/10.1093/applin/22.1.27

Robinson, P. (2015). The cognition hypothesis, second language task demands, and the SSARC model of pedagogic task sequencing. *Domains and Directions in the Development of TBLT, 8*, 87–121.

Romero, C., & Ventura, S. (2020). Educational data mining and learning analytics: An updated survey. *WIREs Data Mining and Knowledge Discovery, 10*(3), e1355. https://doi.org/10.1002/widm.1355

Romine, W. L., Schroeder, N. L., Graft, J., Yang, F., Sadeghi, R., Zabihimayvan, M., Kadariya, D., & Banerjee, T. (2020). Using machine learning to train a wearable device for measuring students' cognitive load during problem-solving activities based on electrodermal activity, body temperature, and heart rate: Development of a cognitive load tracker for both personal and classroom use. *Sensors, 20*(17), 4833.

Rosser, J., James, C., Lynch, P. J., Cuddihy, L., Gentile, D. A., Klonsky, J., & Merrell, R. (2007). The impact of video games on training surgeons in the 21st century. *Archives of Surgery, 142*(2), 181–186. https://doi.org/10.1001/archsurg.142.2.181

Sampson, D., & Karagiannidis, C. (2002). Personalised learning: Educational, technological and standardization perspective. *Digital Education Review, 4*, 24–39.

Sankey, M., Birch, D., & Gardiner, M. (2010). Engaging students through multimodal learning environments: The journey continues. In *Proceedings of the ASCILITE 2010: 27th annual conference of the Australasian Society for Computers in learning in tertiary education: Curriculum, technology and transformation for an unknown future* (pp. 852–863). http://eprints.usq.edu.au/id/eprint/9100

Schank, R. (1999). *Dynamic memory revisited*. Cambridge: Cambridge University Press.

Schneider, D. W., & Anderson, J. R. (2010). Asymmetric switch costs as sequential difficulty effects. *Quarterly Journal of Experimental Psychology, 63*(10), 1873–1894.

Sergis, S., Sampson, D. G., Rodríguez-Triana, M. J., Gillet, D., Pelliccione, L., & de Jong, T. (2019). Using educational data from teaching and learning to inform teachers' reflective educational design in inquiry-based STEM education. *Computers in Human Behavior, 92*, 724–738. https://doi.org/10.1016/j.chb.2017.12.014

Siemens, G., & Long, P. (2011). Penetrating the fog: Analytics in learning and education. *Educause Review, 46*(5), 30.

Singer, S. R., Hilton, M. L., Schweingruber, H. A., & US, N. R. C. (2005). *America's lab report: Investigations in high school science* (Vol. 3). National Academies Press.

Sithole, A., Chiyaka, E. T., McCarthy, P., Mupinga, D. M., Bucklein, B. K., & Kibirige, J. (2017). Student attraction, persistence and retention in STEM programs: Successes and continuing challenges. *Higher Education Studies, 7*(1), 46–59. http://doi.org/10.5539/hes.v7n1p46

Speier, C., Vessey, I., & Valacich, J. S. (2003). The effects of interruptions, task complexity, and information presentation on computer-supported decision-making performance. *Decision Sciences, 34*(4), 771–797.

Spüler, M., Walter, C., Rosenstiel, W., Gerjets, P., Moeller, K., & Klein, E. (2016). Eeg-based prediction of cognitive workload induced by arithmetic: A step towards online adaptation in numerical learning. *ZDM, 48*(3), 267–278. https://doi.org/10.1007/s11858-015-0754-8

Srivastava, N. (2020). *Towards sensor-based learning analytics: A contactless approach* (Doctoral dissertation). The University of Melbourne.

Srivastava, N., Nawaz, S., Lodge, J. M., Velloso, E., Erfani, S., & Bailey, J. (2020). Exploring the usage of thermal imaging for understanding video lecture designs and students' experiences. In *Proceedings of the tenth international conference on learning analytics & knowledge* (pp. 250–259). Association for Computing Machinery. https://doi.org/10.1145/3375462.3375514

Srivastava, N., Nawaz, S., Newn, J., Lodge, J., Velloso, E., Erfani, M., Gasevic, D., & Bailey, J. (2021). Are you with me? Measurement of learners' video-watching attention with eye tracking. In *LAK21: 11th international learning analytics and knowledge conference* (pp. 88–98). Association for Computing Machinery. https://doi.org/10.1145/3448139.3448148

Strobel, J., & Van Barneveld, A. (2009). When is PBL more effective? A meta-synthesis of meta-analyses comparing PBL to conventional classrooms. *Interdisciplinary Journal of Problem-Based Learning, 3*(1), 4. https://doi.org/10.7771/1541-5015.1046

Sweller, J., & Chandler, P. (1991). Evidence for cognitive load theory. *Cognition and Instruction, 8*(4), 351–362.

Tawfik, A., & Jonassen, D. (2013). The effects of successful versus failure-based cases on argumentation while solving decision-making problems. *Educational Technology Research and Development, 61*(3), 385–406. http://www.jstor.org/stable/24546530

Taylor, A., Sanson, M., Burnell, R., Wade, K. A., & Garry, M. (2020). Disfluent difficulties are not desirable difficulties: The (lack of) effect of sans forgetica on memory. *Memory, 28*(7), 850–857.

Taylor, M. (1981). The motivational effects of task challenge: A laboratory investigation. *Organizational Behavior and Human Performance, 27*(2), 255–278. https://doi.org/10.1016/0030-5073(81)90049-0

Tobias, S. E., & Fletcher, J. D. (2011a). *Computer games and instruction*. IAP Information Age Publishing.

Tobias, S. E., & Fletcher, J. D. (2011b). Computer games, present and future. *Computer Games and Instruction*, 525–545.

Triona, L. M., & Klahr, D. (2003). Point and click or grab and heft: Comparing the influence of physical and virtual instructional materials on elementary school students' ability to design experiments. *Cognition and Instruction*, *21*(2), 149–173. https://doi.org/10.1207/S1532690XCI2102_02

Tsovaltzi, D., Rummel, N., McLaren, B. M., Pinkwart, N., Scheuer, O., Harrer, A., & Braun, I. (2010). Extending a virtual chemistry laboratory with a collaboration script to promote conceptual learning. *International Journal of Technology Enhanced Learning*, 2(1–2), 91–110.

Umbreit, J., Lane, K. L., & Dejud, C. (2004). Improving classroom behavior by modifying task difficulty: Effects of increasing the difficulty of too-easy tasks. *Journal of Positive Behavior Interventions*, *6*(1), 13–20.

van der Meij, J., & de Jong, T. (2006). Supporting students' learning with multiple representations in a dynamic simulation-based learning environment. *Learning and Instruction*, *16*(3), 199–212. Retrieved from https://www.sciencedirect.com/science/article/pii/S0959475206000260

Vandercruysse, S., Vandewaetere, M., & Clarebout, G. (2012). Game-based learning: A review on the effectiveness of educational games. In *Handbook of research on serious games as educational, business and research tools* (pp. 628–647). http://doi:10.4018/978-1-4666-0149-9.ch032

VanLehn, K., Siler, S., Murray, C., Yamauchi, T., & Baggett, W. B. (2003). Why do only some events cause learning during human tutoring? *Cognition and Instruction*, *21*(3), 209–249. https://doi.org/10.1207/S1532690XCI2103_01

Verbert, K., Duval, E., Klerkx, J., Govaerts, S., & Santos, J. L. (2013). Learning analytics dashboard applications. *American Behavioral Scientist*, *57*(10), 1500–1509.

Vygotsky, L. (1980). *Mind in society: The development of higher psychological processes*. Harvard University Press.

Watkins, J., & Mazur, E. (2013). Retaining students in science, technology, engineering, and mathematics (stem) majors. *Journal of College Science Teaching*, *42*(5), 36–41. http://www.jstor.org/stable/43631580

Weber, E. U., & Johnson, E. J. (2009). Mindful judgment and decision making. *Annual Review of Psychology*, *60*, 53–85.

Weeks, M., & Gaylord-Ross, R. (1981). Task difficulty and aberrant behavior in severely handicapped students. *Journal of Applied Behavior Analysis*, *14*(4), 449–463. https://doi.org/10.1901/jaba.1981.14-449

Willett, R. A., & Eysenck, H. J. (1962). Experimentally induced drive and difficulty level in serial rote learning. *British Journal of Psychology*, *53*(1), 35–39. https://doi.org/10.1111/j.2044-8295.1962.tb00812.x

Winberg, C., Adendorff, H., Bozalek, V., Conana, H., Pallitt, N., Wolff, K., Olsson, T., & Roxå, T. (2019). Learning to teach STEM disciplines in higher education: A critical review of the literature. *Teaching in Higher Education*, *24*(8), 930–947. https://doi.org/10.1080/13562517.2018.1517735

Winne, P. H. (2006). How software technologies can improve research on learning and bolster school reform. *Educational Psychologist*, *41*(1), 5–17.

Winne, P. H. (2017). Learning analytics for self-regulated learning. In C. Lang, G. Siemens, A. Wise, & D. Gaševć (Eds.), *Handbook of learning analytics* (pp. 241–249). Society for Learning Analytics Research.

Wise, A. F., & Shaffer, D. W. (2015). Why theory matters more than ever in the age of big data. *Journal of Learning Analytics*, 2(2), 5–13. http://doi.org/10.18608/jla.2015.22.2

Wouters, P., van der Spek, E. D., & van Oostendorp, H. (2009). Current practices in serious game research: A review from a learning outcomes perspective. In *Games-based learning advancements for multi-sensory human computer interfaces: Techniques and effective practices* (pp. 232–250). http://doi:10.4018/978-1-60566-360-9.ch014

Yue, C. L., Castel, A. D., & Bjork, R. A. (2013). When disfluency is – And is not – A desirable difficulty: The influence of typeface clarity on metacognitive judgments and memory. *Memory and Cognition*, *41*(2), 229–241. https://doi.org/10.3758/s13421-012-0255-8

Zacharia, Z. C., & Constantinou, C. P. (2008). Comparing the influence of physical and virtual manipulatives in the context of the Physics by Inquiry curriculum: The case of undergraduate students' conceptual understanding of heat and temperature. *American Journal of Physics*, *76*(4), 425–430. https://doi.org/10.1119/1.2885059

Zervas, P., Sergis, S., Sampson, D. G., & Fyskilis, S. (2015). Towards competence-based learning design driven remote and virtual labs recommendations for science teachers. *Technology, Knowledge and Learning*, *20*(2), 185–199. https://doi.org/10.1007/s10758-015-9256-6

17

Learning Analytics in a Web3D Based Inquiry Learning Environment

Guangtao Xu, Yingqian Li, Zhouyang Zhu, Yihui Hu, and Wenting Zhou

CONTENTS

17.1 Introduction ... 260
17.1.1 Learning Analytics ... 260
17.1.2 Web3D Technology ... 260
17.1.2.1 What Is Web3D ... 260
17.1.2.2 Implementation Technologies of Web3D ... 261
17.1.3 Web3D-Based Inquiry Learning Environment ... 262
17.1.4 Web3D and Learning Analytics ... 264
17.2 Theoretical-Basis of Learning Analytics ... 264
17.2.1 Knowledge Graph ... 264
17.2.1.1 The Origin of the Knowledge Graph ... 265
17.2.1.2 Educational Knowledge Graph ... 265
17.2.1.3 Theoretical Basis of Educational Knowledge Graph ... 266
17.2.1.4 Principles of Educational Knowledge Graph Construction ... 267
17.2.2 Learner Profile ... 267
17.2.2.1 What Is the Learner Profile ... 267
17.2.2.2 Research Status of Learner Profile ... 268
17.2.3 Theory and Record of Learning Behavior ... 268
17.2.3.1 Theoretical Basis ... 268
17.2.3.2 Standards for Recording Behavior Data ... 269
17.3 Technical Path of Learning Analytics ... 269
17.3.1 Realization Methods of Knowledge Graph ... 269
17.3.1.1 Construction of Educational Knowledge Graph ... 269
17.3.1.2 Storage and Presentation of Educational Knowledge Graph ... 270
17.3.2 How to Create a Learner Profile ... 270
17.3.2.1 Knowledge Level Assessment ... 271
17.3.2.2 Learning Behavior Assessment ... 271
17.3.2.3 Emotion and Attitude Assessment ... 271
17.3.3 Definition and Analytics Method of Learning Behavior ... 271
17.3.3.1 Definition of Learning Behavior Analytics ... 271
17.3.3.2 Analytics Method of Learning Behavior ... 272
17.4 Concrete Realization of Learning Analysis in Web3D Based Inquiry Learning Environment ... 273
17.4.1 Construction of Knowledge Graph of Chemistry Experiments in Middle School ... 273
17.4.2 Generation and Presentation of Learner Profile in Web-Based Inquiry Learning Environment ... 274
17.4.3 Analysis and Evaluation of Inquiry Learning Behavior ... 275
17.5 Summary and Outlook ... 275
References ... 276

DOI: 10.1201/9781003181187-21

17.1 Introduction

The use of inquiry-based learning methods is the first step to introduce STEM education, which is based on problem-solving in actual situations and promotes cooperation between learners. This way of learning fosters students' motivation and interest in science, that they learn to perform steps of inquiry similar to scientists, and that they gain knowledge of scientific processes. Many educators have carried out practical explorations based on inquiry-based learning. However, under the background of today's artificial intelligence era, new technologies are emerging, and the education field is facing unprecedented opportunities and challenges.

At the same time, to promote the development of personalized education, learning analytics is gradually implemented in an inquiry-based learning environment. As an important branch of learning analytics and as an important tool to promote educational informatization, knowledge graph, learner profile, and learning behavior analytics have attracted extensive attention from educational researchers in recent years. However, at present, there is still a lack of systematic review of case-based technology implementation of learning analytics, and how to integrate knowledge graph, learner profile, and learning behavior analytics is an unanswered question. Therefore, based on the theoretical basis of this field, this study aims to elaborate on the basic characteristics, research status, and future prospects of the field of learning analytics. Based on the three specific application branches which are knowledge graph, learner portrait, and learning behavior analytics, and based on the framework of theoretical basis, technical path, and concrete implementation in Web3D based inquiry learning environment, this chapter systematically introduces learning analytics in Web3D based inquiry learning environment based on a specific case.

17.1.1 Learning Analytics

Since entering the 21st century, with the development of mobile Internet technology, cloud computing, and Internet of Things technology, the appearance of big data has drawn public attention. The research and application of big data in the field of education are remarkably promoting the transformation of education informatization. In October 2012, the U.S. Department of Education issued 'Enhancing Teaching and Learning through Educational Data Mining and Learning Analytics'. The report pointed out that the current application of big data in the education field mainly includes Educational Data Mining (EDM) and Learning Analytics and Knowledge (LAK) (Bienkowski et al., 2012). The International Educational Data Mining Society defines EDM as follows: 'Educational Data Mining is an emerging discipline, concerned with developing methods for exploring the unique types of data that come from educational environments, and using those methods to better understand students, and the settings which they learn in'. Learning Analytics (LA) is a typical application of big data in education in recent years. As early as 2011, Professor George Siemens of Athabasca University in Canada defined LA as "measurement, collection, analytics and reporting of data about learners and their contexts, for purposes of understanding and optimizing learning and the environments in which it occurs" at First International Conference on Learning Analytics & Knowledge, LAK11 (Gašević et al., 2015, p. 59). As more and more educational data are collected and recorded, LA has been used to process this type of educational data to help advance education. As the study conducted by Wong et al. indicated, LA has been applied to many teaching institutions to provide individualized assistance to learners (Wong et al., 2018).

17.1.2 Web3D Technology

17.1.2.1 What Is Web3D

The use of virtual reality (VR) as an educational tool has been proposed and discussed by many scholars. VR provides the possibility to reproduce the real world or create a new world. In the field of education, VR provides an experience to help people understand concepts and learn specific experiences. They can learn and practice repeatedly in a safe environment according to the needs of educational objectives. However, the use of virtual reality technology still faces great challenges. It is not easy to develop a virtual environment with VR technology (because the technology is hard to be used and it costs much time to develop a VR app). The wearable devices providing a virtual reality environment are very expensive and hard to carry, which is difficult to be used by massive learners. Web3D provides a new solution, which is also to create an educational virtual environment. Different from VR, the virtual environment created by Web3D only needs a smart device (e.g., PC, smartphone, or tablet computer). The device makes it possible to surf the Internet and bring a Web3D supporting browser to access the 3D environment at any time, so as to use the Internet to enable more learners to participate in learning activities. Moreover, the virtual environment built by Web3D technology can make up for the problem of unspecified information by traditional

FIGURE 17.1
A virtual chemical laboratory by Web3D.

2D technology and the limited resources in the real world (Figure 17.1).

Web3D is the product of the combination of web technology and 3D model technology. It is the general name of Internet page rendering 3D graphics technology. Web3D technology has a similar principle to other web technologies. 3D virtual environments or media are saved on the server in an appropriate storage format, and the client can send HTTP requests to the server to get the resources. After the server makes the corresponding requests, the corresponding content is presented from the user's browser. Compared with the VR environment with expensive equipment and high development difficulty, Web3D technology can experience the 3D virtual environment by using a standard smart device with mature technology and low cost. Although the sense of immersion will be reduced, it can enable more learners to experience it. Moreover, Web3D Virtual environment can also be strongly coupled with other web resources, such as linking to other websites in the virtual environment provided by Web3D or using web hypertext resources. The specific comparison is shown in Table 17.1.

It can be predicted that when VR technology is not fully mature and popularized, it is more feasible to use Web3D technology for the construction of a virtual environment.

17.1.2.2 Implementation Technologies of Web3D

Network multimedia engine technology is developing rapidly with the innovation of information technology. The earliest Web3D technology can be traced back to an open 3D graphics standard, Virtual Reality Modeling Language (VRML), proposed in 1995. The standard enables developers to build interactive 3D models and scenes locally or on the Internet through modeling technology, and users can enter the virtual environment by downloading the corresponding browser plug-ins. It is also a file format for the general exchange of 3D graphics and multimedia technology so that it can directly import model materials for interaction. Thus, compared with video files, VRML can greatly reduce the size of transmission files when providing a vivid model image display. In the spring of 2000, the Web3D alliance completed the transformation from VRML to the new interface Extensible 3D (X3D). X3D standard integrates XML, Java, streaming media, and other technologies and has more powerful rendering capabilities, but it still does not solve the problem that plug-ins are required to operate. In addition to VRML and X3D, there are many similar types of Web3D technologies, such as Cult3D, Viewpoint, Shout3D, Flash, etc. Although these technologies have their own advantages, they still do not solve the problem of requiring browser plug-ins. Moreover, due to the existence of too many standards, low compatibility, and inconvenient actual use, they are gradually eliminated by the market (Figure 17.2).

Now the mainstream solution is WebGL technology, which is a 3D drawing standard. It combines the scripting language JavaScript (an interpretive language developed by the web) and OpenGL technology (an interface for drawing vector graphics, which can be accelerated by hardware graphics). WebGL can provide hardware acceleration function for browser with the help of graphics card, which makes the loading of model faster and the rendering efficiency

TABLE 17.1
The Differences between VR and Web3D Technology

	VR	Web3D
Required devices	Specialized wearable devices (e.g., VR glasses)	Smart devices (e.g., PC or tablet computer)
Development difficulty	Harder	Easier
Technology maturity	Not mature	Mature
Immersion	Very strong	Strong but weaker than VR
Interactivity	Strong	Normal
Scale of learner	Niche	Massive
Resource expansion	Few of resources	Lots of web resources

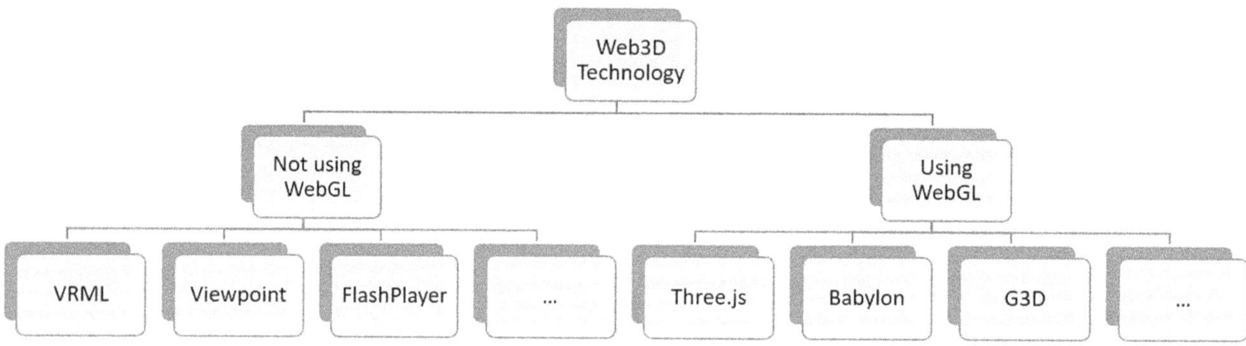

FIGURE 17.2
Technologies of Web3D.

improved. It is a great advantage of WebGL technology that there is no need to install any plug-in in the browser for 3D model display. Also it can use an OpenGL technology interface to call the underlying graphics hardware acceleration function under the same hardware conditions, so as to improve the display performance and rendering effect of 3D data as much as possible. The emergence of WebGL perfectly solves the two problems of low rendering efficiency and poor compatibility of Web3D interactive function. The two commonly used engine technologies in WebGL technology are three.js and babylon.js. Their specific differences are shown in Table 17.2.

Virtual 3D page display technology has been more and more recognized in the fields of commercial product display, urban construction, experimental simulation, and so on. Similarly, the teaching scene with an urgent demand for the concrete expression of abstract problems is an important scene covered by Web3D technology. Web3D has a significant application value for the construction of a virtual teaching environment. It can break through the limitations of time and space, overcome the shortcomings of various real teaching conditions, and help students better understand and study the teaching content. Using Web3D technology for virtual laboratory work, improving students' experimental skills, and saving experimental supplies are convincing examples.

17.1.3 Web3D-Based Inquiry Learning Environment

Generally, the learning environment is considered to be the surrounding environment for learners to carry out learning activities, and more scholars define the learning environment from the perspective of learning activities. Hannafin (1992, p. 40) believes that the learning environments are comprehensive, integrated systems that promote engagement through student-centered activities, including guided presentations, manipulations, and explorations among interrelated learning themes. Peppler and Solomou (2011) consider a learning environment as a community and the values of the community influence the behaviors of the members – teachers and students influence each other towards the support or constraint of creativity. Therefore, the learning environment is a learning space that supports the effective development of learning activities and promotes the development of students' knowledge and ability. It contains elements such as interpersonal environment, material environment, learning resources, methods, and strategies and is an organic and unified system. With the development of modern information technology and the transformation of education and teaching concepts, the focus of research in the field of education has gradually shifted from "teaching" to "learning". More and more learning environments have begun to

TABLE 17.2

The Differences between Three.js and Babylon.js Technology

	Three.js	Babylon.js
Technology objective	3D animation	3D games
Engine features	Focus on animation	Focus on physical collisions
Use difficulty	Easier	Harder
Developing document	Low integrity	High integrity
Development community	More active	Less active

support inquiry-based learning and advocate discovery learning and collaborative learning.

Inquiry learning is a guided learning method that allows students to actively ask questions, explore and make discoveries with the help of digital or physical exploration tools, and conduct rigorous arguments in the process of exploration and discovery, centering on a topic in the fields of nature, society, and humanities. Web3D is a new technology to realize the virtual reality in web pages. Its origin can be traced back to VRML, which is an open standard of 3D graphics on the Internet. VRML is a file format for the general exchange of 3D graphics and multimedia technology. Based on modeling technology, VRML describes interactive 3D objects and scenes. It is not only used on the Internet but also can be used in local customer systems, with a wide range of applications. It is the model file that is transmitted on the Internet, so the transmission quantity is much less than the video image, which greatly improves the loading rate. The inquiry learning environment based on Web3D is a new form of learning environment that emerged after the full development of information technology represented by web technology. This form of learning environment uses the network as an interactive medium and resource carrier to realize various teaching activities, including knowledge transfer, learning activity organization, and other learning support services.

Based on the theoretical research and application practice of inquiry learning environment based on Web3D, several outstanding characteristics can be summarized:

(1) Learning Autonomy

Learning autonomy means that the overall learning activities are mainly arranged and controlled by the learners themselves. The openness of the inquiry learning environment based on Web3D has brought great convenience to learners, who can learn anytime and anywhere. However, its learning effect also depends to a large extent on the learner's self-management ability.

(2) Personalization of Learning

Personalization of learning means that learners can determine the progress, content, and order of learning by themselves and choose which learning media and resources they need to use. The inquiry learning environment based on Web3D gives learners diversified choices. They can choose appropriate learning resources according to their own preferences or the degree of knowledge mastery.

(3) Combination with the Concept of Inquiry Learning

The new educational concept advocates that students should adopt independent learning, inquiry learning, and other learning methods to carry out learning activities. Paying attention to students' innovative ability and practical ability, and paying attention to the cultivation of students' initiative as well as the sustainable development of students' discipline ability have become the key points. The learning environment based on Web3D combined with the concept of inquiry-based learning can not only make up for the shortcomings of traditional learning methods but also focus on cultivating students' inquiry ability and even comprehensive ability.

(4) Providing Rich Cognitive Tools

In traditional school education, paper and pen are the main learning tools for students and listening, memorizing, retelling, practicing, and testing are the main ways of students' cognitive activities. Inquiry learning is easily limited by other conditions, and it is difficult to obtain better teaching results in such an environment. In the process of inquiry learning based on Web3D technology, with the help of powerful visualization technology and interactive technology. Students can conduct inquiry and get real-time feedback in the inquiry learning environment. In other words, manipulation of variables, collection of data, and observation of results were carried out through virtual experiments. It can be seen that the learning environment based on Web3D provides students with rich cognitive tools and truly realizes learner centered inquiry learning.

(5) The Web Page Presenting a Virtual Inquiry Learning Scene

Web3D technology uses 3D real-time distributed rendering technology to achieve real-time rendering of an infinite large-scale scene and has the function of compression and network streaming transmission. You can directly browse and download inside the web page without downloading the client or external devices. Therefore, in the practical application of computer equipment requirements are not high, conducive to the information resources in underdeveloped areas to promote the application. In addition, Web3D technology can restore the problem situation in real 3D and support learners to watch scenes and phenomena from multiple angles.

17.1.4 Web3D and Learning Analytics

Using Web3D technology can simulate a highly simulated virtual learning environment, but it goes far beyond that. Web3d has more advantages like low cost, high compatibility, low development difficulty, strong resource scalability, and online learning support. Such advantages not only ensure the possibility of massive online learning but also bring better immersion experience to learners, improve students' knowledge and skills more effectively. Similarly, the new learning environment built by Web3D also brings new opportunities to the learning analytics.

At present, the learning analytics technology in online learning environments (such as MOOC) is very mature, thanks to the development of Internet technology, making data collection easy. As we all know, a huge amount of data sets is necessary to the learning analytics. The learning environment constructed by Web3D, like other online learning environments, adopts online web technology, so it can support the needs of massive learning and ensure that the data volume of the data set to be analyzed is sufficient. Due to using web technology, it is easy to use scripts to track the event to mine behavior data of learners such as operation time, page dwell time, click times, and so on. Moreover, compared with the ordinary online learning environment, Web3D has stronger human-computer interaction. Learners' behavior is not limited to ordinary clicking buttons, but also more complex behavior operations (e.g., learners' meaningful operation sequence). There are more directions and deeper degrees that can be analyzed, and the analyzed results are more explanatory.

Web3D brings a more real and repeatable learning environment, so it can bring learners a more immersive experience and deepen their situational memory. Moreover, at present, most intelligent devices have integrated cameras. In addition to mining the surface learners' web-based learning behaviors such as clicking, touching, and dragging on the web, multimodal data analytics has also become possible, such as expression and action analytics. Therefore, when analyzing learners using the Web3D learning environment, scholars can not only excavate the learners' surface learning behavior mechanism but also pay attention to the lower cognitive mechanism or increase the emotional analytics.

Learning analytics focuses on the application of learning feedback and personalized resource recommendation to learners. At the level of learning feedback, that is, the presentation of analytics results, after modeling and analytics of learners, Web3D technology can be used to construct a virtual avatar with reference to the image of learners. This virtual avatar contains what learners should know about their own learning after being explained, which is the digital twin of learners. Compared with simple graphics and text presentation, students have more sense of empathy and identification with avatars. Also, they can improve their learning methods or strategies according to the information and prompts displayed by the avatars. At the same time, learners can also update the information of the number twin through multiple learning behaviors. So that the data results become more accurate so that learners can see their own changes, so as to strengthen the motivation of learning. In terms of personalized resource recommendation, Web3D technology can be strongly integrated with other existing web resources. So after analyzing the learning data of learners, it can be more convenient to recommend corresponding learning resources to meet the personalized needs of learners.

The learning environment constructed by Web3D is an online virtual learning environment combining online web technology and part of virtual reality technology. It also retains most of the characteristics of online learning environment and VR learning environment. So in terms of learning analytics technology, we can use the characteristics of both to conduct data mining and obtain more unique and in-depth learner information.

17.2 Theoretical Basis of Learning Analytics

17.2.1 Knowledge Graph

Inquiry learning is practical, process oriented, and open ended, which reflects the need to provide learners with personalized learning resources and learning assessments. Therefore, it is vital to build and apply knowledge graphs to achieve personalized learning recommendations and path planning. This is the basis for each learner to receive unique learning assistance before analyzing their learning behavior.

17.2.1.1 The Origin of the Knowledge Graph

When it comes to the origin of the knowledge graph, we have to pay attention to "concept map", which is firstly put forward by Joseph D. Novak in 1970. Since the publication of *Learning How to Learn* written by Novak and Gowin, 'concept map' is being increasingly reported in the literature. A concept map is a representation of meaning or ideational frameworks specific to a domain of knowledge, for a given context of meaning. We define concept as a perceived regularity in event or object, or a record of event or object, designated by a label. The meaning of any concept for a person would be represented by all of the propositional linkages the person could construct that include that concept. Novak announced that here is no domain of knowledge (or 'skills') for which concept maps cannot be used as a representational tool.

Knowledge graph is a combination of concept map and information technology. They can be distinguished from each other in mainly three aspects: the method of drawing, the form of presentation, and their functions. In short, it is the computer that caused the innovation. Instead of concentrating on human brain, knowledge map is aimed at the representation of knowledge in computer and finally trying to enable computers to store and even reason knowledge nonlinearly.

Knowledge graph is a method to describe the concept, entity, event, and relationship between the objective world. Concept refers to the conceptual expression of objective things formed in the process of understanding the world, such as people, animals, organizations, etc. Entity refers to the concrete things in the objective world. Events refer to activities in the objective world, such as employment, buying and selling, etc. Relationships describe the objective correlation between entities and events. Accordingly, triplet is a general representation of knowledge graph. Its basic form is mainly composed of entity 1, relation, entity 2 (or concept, attribute, attribute-value), etc. The entity is the most basic element among all of them. Different entities have different relationships. Each attribute-value pair can be used to represent an inherent attribute of an entity, while the relationship can be used to connect two entities and describe the association between them.

Therefore, knowledge graph is essentially a large-scale semantic network. From the perspective of artificial intelligence, knowledge graph is a kind of knowledge base to understand human language, so it can be regarded as a technology to develop cognitive intelligence. From the perspective of database, knowledge graph is a brand new kind of knowledge storage structure. From the perspective of knowledge representation, knowledge graph is a method for a computer to understand knowledge. And from the perspective of computer network, knowledge graph is a semantic interconnection between knowledge data (Mo, 2020).

In 2012, Google pioneered the concept of knowledge graph, which is designed to enhance the understanding ability of search engines, improve the quality and experience of user search, and finally achieve a successful application effect. Knowledge graph lays the foundation for knowledge interconnection on the World Wide Web and its powerful open interconnection and semantic processing power. On the one hand, the study of knowledge graph explores the methods and theories of obtaining knowledge from Internet language information. On the other hand, it promotes the research of knowledge-driven language understanding. At present, the research on knowledge graph mainly involves the type of knowledge graph, the construction method, and the construction process.

17.2.1.2 Educational Knowledge Graph

With regard to educational knowledge graph, the academic community has not yet formed a unified understanding of this concept. Based on the existing literature review and analysis, this study defines the educational knowledge graph as a method to personalize the learners' knowledge and skills by using subject knowledge graph (subject knowledge structure network) and individual knowledge graph (capacity development structure network) as tools. The specific content and function of the subject knowledge graph and the individual knowledge graph are shown in Table 17.3. Subject knowledge graph is an objective structure in a discipline. On the one hand, it is formulated by teachers, textbook writers, and subject experts based on their mastery of education, teaching, and knowledge organization of the subject. On the other hand, it can be constructed from bottom to top based on the Internet. Individual knowledge graph changes dynamically based on learners' learning progress and knowledge mastery. It is a dynamically changing structure that varies from subject to subject. Generally speaking, there is a certain connection between the two. Subject knowledge graph is the basis of the individual knowledge graph. Learners form a real-time updated individual knowledge graph in the process of learning based on subject knowledge graph. Then,

TABLE 17.3

Contents and Functions of Different Types of the Knowledge Graph

Tool	Content	Function
Subject knowledge graph	Visual structure with leading guidelines, learning objectives, learning methods, learning tasks, learning resources, etc.	To visualize the knowledge structure of subjects, to provide assistance for teachers' teaching and to provide guidance for students' learning
Individual knowledge graph	Learners' learning progress, mastery of knowledge, the next steps in the learning path	Reflect learner literacy and present learners' basic information, learning preferences, knowledge and capability status

based on the content and information of individual knowledge graph as feedback, learners can adjust learning plan and learning path planning to start a new round of learning. In the short term, the two sessions are carried out alternately, with the level of learners' knowledge and ability spiraling upward. In the long term, the two sessions are parallel and inseparable. Learners can look at the individual knowledge graph, refer to the subject knowledge graph, and ultimately harvest the ideal learning outcomes. The interaction process is shown in Figure 17.3.

17.2.1.3 Theoretical Basis of Educational Knowledge Graph

(1) **Theory of Basic Structure of Disciplines**

With regard to teaching content, Bruner's theory of the basic structure of disciplines points out that it is impossible to include all knowledge in a subject in the syllabus. It is also impossible for students to master all the scientific and cultural knowledge of mankind in a limited time. Therefore, in order to make the knowledge learned by students more valuable in the future, Bruner believes that learning and mastering the concepts, definitions, principles, and rule systems that are widely available in each discipline are the best approach for learners. He also stresses the importance of making students understand the basic structure of a subject. Bruner emphasizes the importance of the 'basic structure of discipline' – basic concepts, basic principles, and their interrelationships – as a whole showing the universal connections between them. This is important in an era of 'knowledge explosion'.

With regard to teaching methods, Brunner emphasizes the method of discovery. That is, by taking the basic structure of the discipline as the content and using the materials and information provided, students can move from a passive state of receiving knowledge to an active state of discovering it. It helps students develop creative thinking skills so that they can eventually master not only the basic structure of a subject but also the basic attitudes and methods of the subject.

In summary, we can draw the following two conclusions. Firstly, presenting the basic structure of a discipline – that is, the concept of the subject, definition, and the link between them – is a way of presenting quality content and a means of effectively promoting lifelong learning. Secondly, organizing the basic concepts and definitions in relation to relevant learning resources helps to guide students in developing their own higher-order thinking skills.

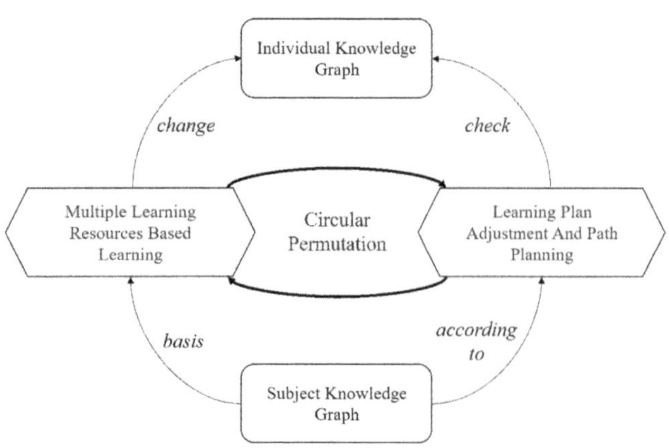

FIGURE 17.3
Diagram of the interaction between knowledge graphs.

(2) **Connectivism Learning Theory**

As a kind of learning theory in the digital age, the main theoretical basis of connectivism includes chaos theory, self-organization theory, and network theory. Connectivism provides theoretical guidance for learners to learn in the new era of online learning environment. The theory of connectivism learning holds that learning is a kind of network construction and generation of knowledge. This network is the interconnection of cognition within a person, and it builds connections between old and new knowledge.

Connectivism learning theory points out the characteristics of learning content, including variability, reality, continuity, and relevance. It proposes a variety of learning styles as opposed to a single learning channel and suggests the superiority of insight into the connections between concepts and definitions in different domains as opposed to simply understanding a single concept. Connectivism regards learning as a process of connecting knowledge nodes with information resources and believes that highly networked personal knowledge can greatly improve learning efficiency.

In summary, we can conclude that connectivism learning theory advocates the establishment of a network structure between knowledge and learning resources and points out that isolated knowledge is no longer competitive to meet the needs of individuals or society in the current era, while highly connected networked knowledge structure can improve learning efficiency and effectiveness.

17.2.1.4 Principles of Educational Knowledge Graph Construction

Based on the previous section, this section draws the following principles from the perspectives of subject knowledge graph and individual knowledge graph, based on Bruner's theory of the basic structure of disciplines and connectivism learning theory:

(1) **Subject Knowledge Graph**

- The principle of integrity. The basic concepts, basic principles, and their correlation with each other in the content of each discipline make it necessary for knowledge graph constructor to consider the structure, nature, change, and application of matter as a network of subject knowledge structures. They are integrated into a web of discipline knowledge structures that are in a constant state of unity, rather than a collection of isolated and fragmentary factual conclusions.
- The principle of diversity. In addition to constructing a scientific and complete knowledge framework, integrating and linking multi-source of heterogeneous teaching resources with existing entities can reduce the disorderly stacking of learning resources. Meanwhile, it helps learners plan their learning paths and match the appropriate learning resources.
- The principle of complexity. There should be a one-to-many or many-to-many relationship among knowledge nodes. Since the knowledge points are organized based on the principle of gradual differentiation and deep understanding, knowledge points can form a staggered and complex knowledge network, which has the inclusion relationship and the hyponymy relationship between the upper and lower levels.

(2) **Individual Knowledge Graph**

- The principle of uniqueness. Individual knowledge graph should present learners' basic information, learning preference, knowledge state, and ability state, and even reflect learners' literacy. The individual knowledge graph must be unique, as learners learn according to their own goals, interests, and pace and get the most out of their learning at a reasonable pace.
- The principle of accuracy. Learners should not only get a standard single score after studying the course but should also have an understanding of their current learning situation based on the content of their individual knowledge graph. They should have as accurate an understanding as possible of the structure of knowledge they have acquired.
- The principle of real time. Individual knowledge graph should dynamically reflect learners' learning progress and enable path planning for the next step of learning. This allows the system platform, which assumes the role of 'teacher', to guide every step of learning and give feedback on the basis of subject knowledge graph.

17.2.2 Learner Profile

17.2.2.1 What Is the Learner Profile

The concept of learner profile comes from user profile, which is a tool that records the data of consumers in a comprehensive and three-dimensional way and is widely used in the business field. Learner profile is the application of user profile in the field of education. It is a method of modeling and analyzing students' characteristic data. Learner profile takes the group characteristics as the core, classifies the group of students, and forms a label summary description. In the inquiry learning environment, learner profile can be used to classify learners according to their group feature. Based on this classification, we can optimize teaching design, provide targeted learning services, and facilitate learners' personalized learning.

Learner profile has three main characteristics:

In terms of construction process, with the help of quantitative analysis, the learner profile performs data mining and effective analysis of the big data of the learner's process and learning results and then presents them in a visual way;

In terms of data processing, learner profile is a process of labeling after data mining;

In terms of presentation, learner profile focuses on identifying learning groups with different attribute characteristics and presenting them in the form of labeling.

17.2.2.2 Research Status of Learner Profile

At present, most learner profile studies collect students' background, performance data, platform interaction behavior data, and questionnaire data mainly based on the open learning environment. Regression, classification, and clustering techniques are used to classify groups of students according to their motivation, engagement, and online motivation, in order to provide personalized guidance and improve learning effect. Some studies use association rules, collaborative filtering, and sequential pattern mining to construct the profile model based on learners' basic attributes, learning process, and learning results data. Next, an accurate and personalized learning path planning framework can be designed. Other studies have pointed out that the portrayal of learner profile should include three aspects: individual characteristics, personal performance, and personal development vision.

This shows that although learner profile has received a lot of attention from researchers, there are still limitations. On the one hand, most studies only focus on the characterization of students' learning behavior but lack attention to the internal cognitive characteristics such as ability and thinking. On the other hand, the data selected for profile construction is mainly based on the characteristics of online behavior data, without considering the knowledge level or ability performance reflected by students' behavior.

17.2.3 Theory and Record of Learning Behavior

17.2.3.1 Theoretical Basis

(1) Behavioral Science Theory

The American Encyclopedia of Management defines behavioral science as: 'Behavioral science refers to any science that studies the behavior of human beings (and lower animals) in natural and social environments by using experimental and observational methods similar to natural sciences'. What we usually call behavioral science, or behavioral science in the narrow sense, is a science that uses knowledge from various disciplines such as psychology, sociology, and anthropology to study the laws of human behavior and how to properly manage crowd relations and motivate people. Behavioral science considers behavior to be the purposeful activity of an individual in response to the environment and is the product and expression of the interaction between the person and the environment. Behavioral scientists are more concerned about explicit behavior. A behavior is observable if it can be seen, and measurable if it is possible to count the frequency or duration of its occurrence. In order for our directly observed behavior to be meaningful and reliable, both criteria are essential.

(2) Social Cognitive Theory

Social cognitive theory was proposed by a prominent American psychologist, Bandura, in the late 1970s and early 1980s. After a detailed investigation of various behavioral models and theories, including unconscious determinism in psychodynamics, psychological trait theory in cognitive psychology, and environmental determinism in behaviorism, Bandura proposed a new behavioral model theory – social cognitive theory. Firstly, social cognitive theory analyzes and reveals the dilemma of simple internal psychology and external environmental determinism, pointing out that both individual cognition and environment are of great significance to behavior. Secondly, social cognitive theory also denies the one-way determinism view that behavior is only affected by environment or psychological traits. It points out that behavior, environment, and individual cognition are mutually influenced and interact with each other (Figure 17.4).

This relationship is also common in web-based learning behavior systems. In this system, web-based learning platforms, learning partners, etc., can be considered 'environments' in the Bandura model. 'Individual' refers to a specific learner, whose learning style, learning motivation, belief, expectation, and so on can be considered his or her individual characteristics. The learning behaviors are obviously affected by learning style and motivation, as well as web-based learning environments. What's more, the selection and arrangement of learning resources by web-based

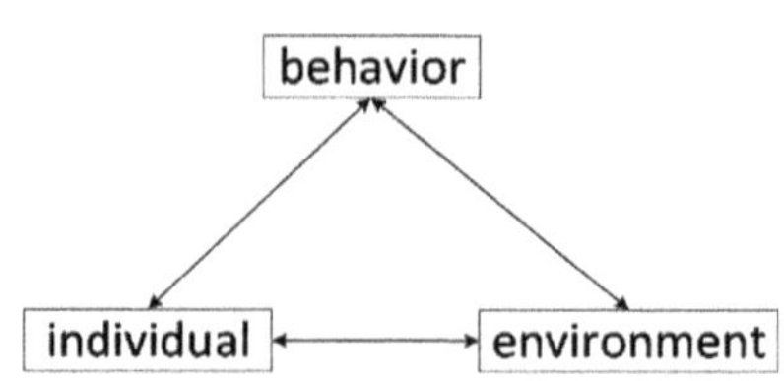

FIGURE 17.4
Bandura's diagram of individual, environment, and behavior.

learning platforms have a direct impact on learners' learning behavior. Conversely, learning behavior can also affect the learning environment. For example, the behavior of a learner can affect the evaluation of the learner by other learners or learning tutors in the web-based learning system and then affect the interaction between them, which may further affect other environmental factors. Of course, the change of learning environment can also affect the learning expectation, cognitive attitude, learning belief, learning emotion, and even learning preference of specific learners, which are the main parts of learners' internal characteristics. Bandura's basic principle of the tripartite influence of behavior, environment, and individual traits embodies in the web-based learning behavior system.

17.2.3.2 Standards for Recording Behavior Data

The data for online learning behavior analysis is mostly derived from the system log data of the Learning Management System (LMS). With the gradual improvement of learning methods such as mobile learning, learning behavior is no longer limited to the PC. Data on learners' learning behavior may come from different platforms and different devices. Therefore, only recording LMS data cannot fully represent the entire learning process of learners. Advanced Distributed Learning launched xAPI (Experience API) in 2013 to better describe learners' learning behaviors and enable data communication across platforms and devices. xAPI standardizes the description of learning experience, making education data services more flexible and easy to expand and providing a new digital service paradigm for educational big data.

xAPI learning behavior statement consists of three elements: actor, verb, and object. Among them, the actor is the participant in the learning activity, such as learners, teachers, and study groups, the verb describes the behavior of learning activities, and the object is object with which the actor interacts during the learning behavior. The data recorded by xAPI includes not only some traditional learning experience data but also the behavior of learners on a platform. Because of the interactive, autonomous, open, and process-oriented nature of learning in an inquiry environment, learners' learning behavior is also diverse and complex. In an inquiry environment, recording and analyzing student behavior provides evidence to support the improvement of the environment and the modification of teaching strategies. Therefore, using xAPI to document learning behavior in a web-based inquiry learning environment is also well suited.

17.3 Technical Path of Learning Analytics

17.3.1 Realization Methods of Knowledge Graph

17.3.1.1 Construction of Educational Knowledge Graph

(1) **Discipline Knowledge Graph**

Step 1: Identifying the domain and the data source.

The method of constructing domain knowledge graph is usually a semi-artificial construction method. The data source has two parts. One part is unstructured data such as textbooks and tutorials of selected subjects, and the other part is semi-structured data such as Wiki.

Step 2: Data acquisition and entity extraction

Firstly, for unstructured data, a top-down approach is adopted to define ontologies. After defining, entities should be added one by one to form a basically well-structured knowledge network.

Secondly, the semi-structured data contained in the web pages are acquired using the Python3-based crawler framework. The data, after cleaning, can be performed to compose the data set.

Finally, the sentences obtained by the crawler are pattern matched by the collected entities. Each sentence can correspond to one or more entities and assign relationship categories to each entity in the sentence.

Step 3: Identifying entity naming and extracting relationships

After defining entities, entity naming identification and entity relationship extraction are carried out. The entities and relationships are extracted from semi-structured and unstructured data and finally expressed in the form of 'entity-relationship-entity' triad, so as to initially realize the construction of the subject knowledge structure network. Then, it is handed over to first-line teachers or subject experts for guidance and revision.

Step 4: Linking knowledge points, knowledge blocks, knowledge units, and learning resources

Relate knowledge points to each other. Knowledge blocks that can solve practical problems are formed from single knowledge points. The knowledge points are further compounded into knowledge units to form a contextual knowledge map. After repeating these operations, we can get a complex, highly correlated mesh structure. Finally, we should link relevant texts, pictures, videos, and other learning resources in the learning platform to the knowledge points.

(2) **Individual Knowledge Graph**

Collect data of established facts such as learner's gender, age, and school number and generate fact labels to establish a learner's database management system.

Based on the online learning platform's data on learning behavior and the corresponding evaluation criteria, knowledge mastery in analyzing and predicting can be used. The knowledge points, experiments, and related resources and their connections are stored as nodes and links in the individual knowledge graph.

The updated method of personal knowledge graph is real time, and with path planning and resource learning, learners' personal knowledge graph will be closer to the subject knowledge graph.

17.3.1.2 Storage and Presentation of Educational Knowledge Graph

(1) **Knowledge Storage Based on the Graph Database Neo4j**

In general, we only need a query for a short-term relationship (an association within two layers) when we use a database to find a connection between things. However, if we need a larger range of relational queries, graph database functions are necessary. With the rapid development of social, e-commerce, finance, retail, Internet of Things, and other industries, a huge and complex network of relationships has been woven between things in the real world. In this situation, traditional databases are often helpless in the face of such complex relationships. As a result, graph database, a database that stores and queries data in the form of a graph data structure, has emerged.

Unlike the relational database that is used to store data and associations in the form of a two-dimensional table structure, Neo4j is a native graph database. It is also a kind of non-relational database which abstracts graphs into basic elements such as points and lines and stores them in a certain topological data structure. It breaks through the dilemma caused by complex relationships and dynamic queries of big data. The storage back end is customized and optimized for the storage and management of graph structure data. The nodes on the graph that are linked with each other have their physical addresses in the database. Therefore, the advantages of graph structure data can be brought into play.

In knowledge graph, the organization of knowledge is in the form of a graph structure. It's suitable for storage with Neo4j. Neo4j's data storage takes the form of nodes and edges to organize data. The nodes can represent entities in a knowledge graph, and the edges can be used to represent relationships between entities or properties of entities. Both nodes and edges can have their directions that correspond to the beginning and ending nodes. In addition, you can add one or more labels (Node Labels) to the node to represent the classification of an entity. Not only that, a collection of key value pairs to represent some additional properties of the entity in addition to relational properties. Relationships can also come with additional properties.

(2) **Knowledge Graph Visualization Based on D3.js**

D3 (Data-Driven Documents or D3.js) is a JavaScript framework that uses web standards to visualize data. The D3.js is able to load the following types of external data sources: XML HTTP Request, text file, JSON file, HTML file, XML file, CSV (comma separation value, comma-separated values files), and TSV (tab-separated values, tab-separated values) files. The data returned by these data sources can be processed by d3.js, and the only thing to watch out for is to make sure that you create an array from that data. D3 helps to transform the data to SVG, Canvas, and HTML. In short, D3 combines powerful visualization and interaction technologies with data-driven Document Object Model (DOM) operations. It can support the right visual interface for the data.

17.3.2 How to Create a Learner Profile

Learner profile needs to accurately reflect learners' learning situations and provide effective learning feedback and evaluation. It not only helps learners improve learning effectiveness but also helps teacher to improve their teaching methods. Generally, learner profiles are constructed by evaluating learners from three dimensions: knowledge level, behavior level, and emotion level.

17.3.2.1 Knowledge Level Assessment

Learning is a process in which learners construct knowledge based on their own prior knowledge in a given learning situation. Accurate learning feedback and personalized learning recommendations require real-time and accurate monitoring of students' knowledge and skill mastery levels. In order to match learning content of different difficulty levels to different learners' learning situations.

The assessment of learners' mastery level of knowledge is based on the labels of student groups with high or low knowledge or skill levels by clustering learners' achievement levels in most research. According to the level of knowledge or skill, learning resources with different difficulties are pushed. However, such simple clustering is based on the tag features obtained from the final summary evaluation of students. It may lead to a low interpretability classification result.

Knowledge tracking model is one of the important models to evaluate learners' knowledge level. Compared with clustering, knowledge tracking can dynamically adjust learners' mastery of knowledge points according to time development and learners' learning progress. According to different methods, it can be divided into three types of knowledge tracking models: ① Bayesian knowledge tracking (BKT) based on hidden Markov model, ② additive factor model (AFM) based on logistic regression model, and ③ deep knowledge tracking (DKT) based on cyclic neural network. BKT takes the historical achievement as the observation variable, takes the mastery of knowledge points as the state variable, considers the probability of guess and error, and uses the BKT probability formula to calculate the learner's knowledge mastery probability under the current observation variable. AFM takes learners' ability, the difficulty of knowledge points, and learning rate as parameters to establish a logistic regression model. DKT constructs a neural network model of long-term and short-term memory by using the resulting sequence of learners in historical tasks.

17.3.2.2 Learning Behavior Assessment

In the online learning environment, we can better monitor learners' learning behavior data by collecting students' online interactive data such as mouse click times, duration of watching demonstration video, the actions in online learning. After sorting and clearing the data sets, we can use the method of cluster analysis to classify students' learning style or learning participation. As well, regression and classification algorithms can be used to predict the next behavior of students. It's better to achieve more accurate personalized resource recommendation, so as to plan personalized learning paths for learners.

In the offline learning environment, the offline test can be used to collect students' data in combination with relevant questionnaires or scales. The regression or clustering method also can be used to analyze students' data and predict. In addition, multimodality data interaction equipment such as eye tracker, video analytics, and brain waves can be applied to analyze students' learning attention and learning preferences in offline learning environment.

17.3.2.3 Emotion and Attitude Assessment

In the online learning platform where communication activities can be conducted, learners' text data such as comments can be collected. These text data sets also can be analyzed and categorized based on Natural Language Processing (NLP). Thus, it can be easy to get the analysis results of learners' preferences, satisfaction, and attitude massively. Based on the data results of NLP, personalized online resources can be provided to meet students' learning interests and improve the construction of learning resources.

In offline inquiry learning activities, students' expressions and demeanor during learning can be analyzed with the help of multi-modal video expression analytics. Based on this analytics, it's possible to uncover what kind of learning contents and learning styles students like or are confused about, so as to improve the construction of learning resources and the perfection of teaching methods.

17.3.3 Definition and Analytics Method of Learning Behavior

17.3.3.1 Definition of Learning Behavior Analytics

Initially, learning behavior was a concept of educational psychology. In the era when behaviorism learning theory was prevalent, it originally referred to the actions or responses that learners exhibited during the learning process. However, with the development of cognitivism and continuous exploration of learning mechanism, the essence of learning behavior and its relationship with learning have gradually been clarified. Some researchers believe that learning behavior refers to the forms and methods of behavior adopted by learners in learning activities and is the concretization and realization of learners' learning activities under specific situations. They define learning behavior as the specific behavior expression of learners' thoughts, emotions, emotions, motivations, abilities, and operational procedures (Hu & Zhao, 2005). In fact, learning behavior is a complex concept, which includes both external performance that can be observed or measured, called explicit behavior, and internal mental activities such as thinking or introspecting by learners. During the learning process, the behavioral potential changes according to the situation, and this potential is usually transformed into the corresponding behavior after a period of time. Therefore, changes in behavior can be used as indicators to show and portray the learning state of learners. Data-driven learning behavior analytics refers to taking appropriate models or methods to analyze learning behavior data on the basis of reasonably quantifying and comprehensively collecting learner behavior data to gain insight into learners' learning psychology, discover learning rules, optimize learning process, and improve learning efficiency (Liu & Yang, 2016). As for network learning behavior, it is characterized by the ability to record and analyze behavior by making full use of the mechanism created by modern information technology. The results of the analytics are supported by data and are therefore more accurate. The recording and analytics of network learning behavior should be given a richer connotation while being more effectively and widely used as a product of the new learning mode.

17.3.3.2 Analytics Method of Learning Behavior

In the process of learning behavior analytics, the selection and research of analytics methods are important factors in data analytics. According to different analytics needs, appropriate analytics methods should be selected to achieve the effectiveness, accuracy, and comprehensiveness of analytics results. There are many methods of learning behavior analytics, and researchers often use statistical analytics methods, data mining, social network analytics, cluster analytics, and other methods to find the rules and links between behavior and learning effect. By quantifying and analyzing behavioral characteristics, the relationship between behavioral characteristics and learning effect can be better established, and the learning process and learning effect can be optimized.

The analytics of individual learning behavior mainly determines the individual learning situation of learners through the prediction of individual learning effect, analytics of cognitive ability, mining of learning interest, and identification of emotional state, so as to give personalized suggestions and recommendations. The group learning behavior analytics mainly obtains the general behavior pattern of the learner group by collecting and analyzing the behavioral data of the group, so as to understand the learning process of such learners. In other words, while individual learning behavior analytics emphasizes tracking individual learning processes, group learning behavior analytics stresses group process learning assessment by detecting group behaviors with the same or similar characteristics, forming a group behavior pattern with common characteristics, and thus deriving strategies for such groups that can compensate for the shortcomings of traditional summative assessment. Researchers can collect data on web-based exploratory platforms in such ways as the length of time spent on a particular page, the number of clicks, and the order in which different functions are used in the platform. The methods of learning behavior pattern analytics mainly include clustering, sequence analytics, sequence pattern mining, social network analytics, and so on.

(1) **Cluster Analytics**

Cluster analytics refers to the process of grouping a collection of physical or abstract objects into multiple classes of similar objects. The goal of cluster analytics is to collect data for classification based on similarity. The purpose of group learning behavior analytics is to identify learners with similar behavior patterns, or similar patterns of behavior. For example, Amershi and Conati (2006) proposed a data-based user modeling framework through a clustering approach to identify patterns of learners' behavior in inquiry learning environments, and by analyzing these behavioral patterns, researchers can determine the effectiveness of learning strategies.

(2) **Sequence Analytics**

Sequence analytics refers to recording and analyzing the activity types, functions, and corresponding relationships of the man-machine components in a production system in chronological order. Sequence analytics emphasizes the direct response between actions. The main data sources for traditional sequence analytics are observation, text, and video, but with the development of online education, the learning behavior of learners can be recorded for a long time, and thus some researchers also study the click flow in web-based learning platforms through sequence analytics.

(3) **Sequential Pattern Mining**

Sequential pattern mining is to mine the patterns formed by the combination of sequences that often appear in a data set composed of a group of ordered data columns. Compared with sequence analytics, sequence pattern mining emphasizes more on the sequence of behaviors in time. Sequential pattern mining can be used in conjunction with cluster analytics. For example, first use the clustering method to group learners with good learning effect into several clusters and then use sequential pattern mining technology to extract the most frequent learning patterns from the learning sequences, so as to achieve the purpose of personalized learning.

(4) **Social Network Analytics**

Social network analytics is a quantitative analytics method developed by sociologists based on mathematical methods and graph theory. Using relationships as the basic unit of analytics, it describes and measures the relationships between actors and the various tangible or intangible things that flow through these relationships, such as information, resources, etc. Social network analytics was first used in anthropological research, and in recent years, it has been widely used in sociology, psychology, education, and other fields. Some scholars are also beginning to use this approach to study social networks in online learning environments and analyze the relationships between activists. For example, Palonen and Hakkarainen (2013) used social network analytics tools to analyze the interactive process of learners' learning behavior on the CSILE network collaboration platform and studied the relationship between students' online collaborative learning and gender and grades.

In addition to the above analytics methods, learning behavior analytics is also constantly absorbing and integrating other techniques and methods. The introduction of these new analytical methods has greatly enriched the data processing methods and strategies of learning behavior analytics, as shown in Table 17.4.

17.4 Concrete Realization of Learning Analysis in Web3D Based Inquiry Learning Environment

17.4.1 Construction of Knowledge Graph of Chemistry Experiments in Middle School

Using chemical discipline knowledge graph in an existing Web3D based inquiry learning environment as an example, this section describes in detail the key steps and methodological descriptions in the construction process. This section mainly consists of three parts: determination of domain scope as well as data sources, ontology definition, and knowledge association.

First of all, by analyzing chemistry in the context of teaching objectives, it can be found that the learning of chemistry in middle school is mainly enlightening and basic in terms of teaching level, with the understanding of substances and their properties as the main focus; in terms of knowledge structure, it is characterized by a wide range of concepts and complex relationships. Secondly, chemistry originates from and depends on experiments. Chemistry is an experiment-based natural science. In order to clearly present the key knowledge points and their associations and thus facilitate teaching, the domain scope is determined as the experimental contents in middle school chemistry and the learning resources in the virtual experiment platform. The data sources mainly include unstructured data such as chemistry textbooks and tutorial books.

In this case, the knowledge points in unstructured data such as teaching materials and tutorial books are sorted out. Ontology is defined in a top-down way. After literature research and discussion with experts

TABLE 17.4

Analytics Methods of Learning Behavior

Teaching institutions	Analytics methods	Introduction of analytics methods	Research purpose
Turkey Middle East University	Statistical analytics	Study the scale, speed, scope, degree, and other quantitative relationships of the object to understand and reveal the relationship between things, the law of change and development trend	To examine the extent to which variables can affect the effectiveness of learners' online learning
Boise State University	Statistical machine learning	Based on the analytics of the data, select appropriate mathematical model and formulate the hyperparameters. According to certain strategies, train the model with appropriate learning algorithms and use the trained model to analyze and predict the data	To predict students' learning outcomes and their satisfaction with the curriculum and teachers
University of California, Berkeley	Time sequential analytics	Arrange the values of the same statistical indicators into a series in chronological order. Make predictions about the future based on existing historical data	To predict students' dropping out behavior in MOOC by using hidden Markov model

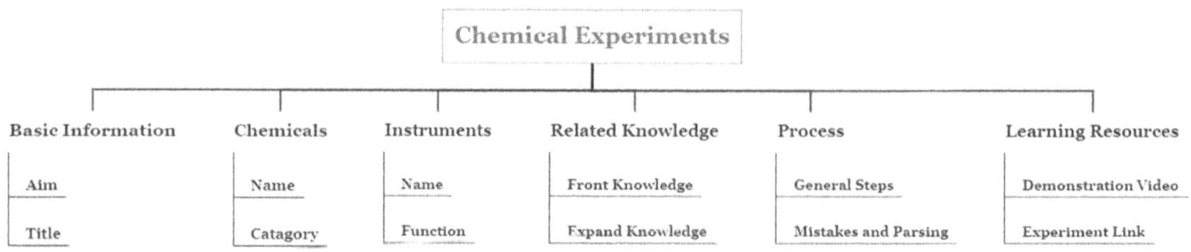

FIGURE 17.5
Elements of chemistry experiment knowledge graph.

and front-line teachers, the following entity categories were defined around the experiment: experiment purpose, experiment name, experiment instrument, instrument function, experimental drug, drug category, prior knowledge, extended knowledge, experiment steps, error prone point, analysis, demonstration video, virtual experiment, etc. The entity definition and main logical relationship are shown in Figure 17.5.

To realize the construction of subject knowledge graph, we also need to comb the relationship between entity categories after defining them. It also called entity links. This case demarcates the relationship between the precursor and the successor according to the teaching order, the relationship between the upper and lower positions according to the category, and the partial overall relationship according to the conceptual scope. In addition, there are special relationships in chemical experiments, such as the relationship between experimental steps, error prone points, and analysis.

Finally, the relevant texts, pictures, videos, and experimental resources in the online experiment platform are linked with knowledge points, thus forming a mesh structure with a certain degree of complexity and relevance, and completing the construction of knowledge based on experimental contents of chemistry subjects. In addition, functions and contents of the virtual experiment platform learning resources can be classified and managed.

In this case, the subject knowledge graph has showed the relationship between knowledge points. But the subject knowledge graph has not yet realized the formation of knowledge blocks. Knowledge blocks can solve practical problems from a single knowledge point. Knowledge blocks can further be compounded into knowledge units to finally form a contextual knowledge map. So, they are yet to be designed and developed in the future.

17.4.2 Generation and Presentation of Learner Profile in Web-Based Inquiry Learning Environment

The generation and presentation of learner profile requires the following four steps: data collection, data pre-processing, profile generation, and profile visualization.

The first step is data collection. In the web-based inquiry learning environment, learners need to actively explore and construct knowledge in a given learning situation. Because it is an online environment, a large amount of behavior data generated by users can be collected and stored in a database according to the constructed data format and rules.

The second step is data pre-processing. Data pre-processing is a very important part of data mining and knowledge discovery. Data pre-processing is also the basis for student behavior feature extraction. Data pre-processing refers to a series of data processing processes, such as data collection, data cleaning, feature extraction, and data conversion. The main purpose of data cleaning is to deal with missing and abnormal values in students' basic data and maintain the consistency of data. Feature extraction refers to selecting or extracting data features from the original data set that are relevant to the research objectives and forming new features, so as to retain the classification information of data objects and reduce the analysis dimension as much as possible. Data conversion is to normalize the data of different dimensions to make the data comparable and convenient for analysis.

The third step is to generate the profile. The generation of profile needs to analyze the pre-processed data sets, split them into three dimensions: knowledge level, learning behavior level, and learning emotion level, and turn them into labels. The assessment of knowledge level can be combined with knowledge graph. BKT model is used to analyze students' test scores, correct rates, and error rates on a knowledge point to get learners' accurate proficiency on knowledge points and then form label entries of 'master' or 'not masted'. In terms of learning behavior analysis, cluster analysis method is used to classify learners' learning style and learning participation and generate labels.

The fourth step is to visualize the profile. The labels depicted by the profile are generally expressed by word cloud. Word cloud shows text information and extracts keywords from a large number of clustering results. And word cloud visually highlights the 'keywords' with high frequency. Similarly, profile visualization can be achieved through the knowledge graph to show learners' mastery of each knowledge node.

17.4.3 Analysis and Evaluation of Inquiry Learning Behavior

Web-based inquiry learning environment is a new learning environment form emerging after the full development of information technology represented by web technology. This kind of learning environment uses the network as the interactive medium and resource carrier to realize all kinds of teaching activities. It includes knowledge transfer, learning activity organization, and other learning support services. Learners' behavioral data in this environment can be well recorded, which provides data for learning behavior analysis. Learning behavior analysis has obvious advantages in evaluating students' online learning performance and finding problems hidden in massive learning process data. This study can also make up for the deficiency of learning analysis in problem interpretation ability. From the perspective of analysis objectives, learning behavior analysis mainly meet two major needs of learners: first, learners actively adapt to various learning styles, make and implement learning plans, and independently choose learning strategies. The second is adaptive learning, i.e., the system predicts and judges learners' characteristics such as style, interest preference, knowledge level, and learning culture, implements corresponding teaching strategies, and adaptatively presents personalized and visual learning paths, learning resources, peers, tools, etc.

In traditional teaching activities, teachers mainly use observation and test to diagnose learners. The intervention of teachers to learners mainly adopts artificial means, such as individual tutoring, increasing practice, and so on. In the web-based inquiry learning environment, learners are mostly engaged in independent inquiry activities, so teachers cannot diagnose the correctness of students' operations in this environment in real time. Lack of teacher supervision may result in learners making too many mistakes or not knowing how to operate in this environment; therefore, learning diagnosis and learning intervention have significance in web-based inquiry learning environment. The goal of learning diagnosis is to identify the learner's mastery of knowledge or skills and weaknesses and provide the basis for subsequent learning remediation, intervention, or personalized learning. If the result of learning behavior analysis shows that the learner is in a poor learning state, the system will automatically send a reminder to urge the learner to learn; if the result of learning behavior analysis shows that the learner has seriously deviated from the learning track, the system will inform the teacher to conduct manual intervention on the learner. There has been research on what interventions are most valuable for students in web-based inquiry learning environments using learning behavior analytics results, but the results are not satisfactory (Brenner et al., 2017). Therefore, the research on how to conduct learning diagnosis and learning intervention through learning behavior analytics in web-based inquiry learning environment and what kind of intervention is most valuable to students using learning behavior analysis results need to be carried out.

17.5 Summary and Outlook

The theory and technology of learning analytics provide certain support for the realization of personalized education in the inquiry learning environment. That makes it possible to carry out different teaching based on learners' own learning style, cognitive style, and learning level. To achieve inquiry learning assessment and adaptive learning, this chapter describes the construction method and scheme of a personalized learning system from three aspects: (1) knowledge graph, (2) learning behavior analytics, and (3) learner profile.

In the case of Web3D virtual experiment platform, the organic combination of these three technologies based on different theories has basically realized the process of '(1) data collection and processing – (2) learning behavior analytics – (3) learner modeling – (4) learning recommendation'. Knowledge graph maps the relationship between subject content (concepts or knowledge points and their relationships) and learning resources (teaching videos, virtual experiments, etc.). It establishes complex and diversified meaningful associations to prepare for use in the subsequent stages. A large amount of heterogeneous data with different structural forms and semantic representations are stored in the subject knowledge graph, so it is necessary to overcome the difficulties of data definition and presentation while storing rich information. The core of learning behavior analytics is to obtain learning history, experiment records, and the large amount of information hidden behind them, including the login and browsing situation of learners, experiment operations and their action details, etc., to model this fragmented information and finally integrate it into a unique learning path for each learner. The results of learning behavior analytics and learner profile to evaluate the knowledge level, learning behavior, and emotional attitude of learner are collected. After that, the result and subject knowledge graph come into play in the process of labeling, visualizing, and finally generating individual knowledge graph. Finally, the recommendation engine provides learners with further learning suggestions and pushes corresponding learning resources according to the information on several aspects: (1) learning objectives, (2) learning progress, (3) learning commitment, (4) learning preferences, and (5) other indicators.

Learning analytics has become one of the important technologies in the field of technology-enhanced learning. It can not only explain learning situation by analyzing learning process, behavioral data, and learning trajectory by using a variety of analytics methods and data models but also provide personalized guidance and adaptive help. From the perspective of data sources, most of the learning and analytics data of existing research come from the Internet. Yet now, with the maturity of various sensors and identification technologies, offline data collection turns out to be strong support. The fusion of multi-modal and multi-scene data makes the acquisition of data suitable for learning analytics more complete and authentic to reflect the learning status of learners. In terms of analytics objectives, learning analytics has a wide range of application areas, including learning diagnosis, learning prediction, learning intervention, personalized learning recommendation, etc. How to extract valuable information from the heterogeneous 'big data' before analytics, and how to efficiently summarize meaningful strategies for teachers, students, and developers during the analytics process are also the key points need to be studied.

References

Amershi, S., & Conati, C. (2006, June). Automatic recognition of learner groups in exploratory learning environments. In *International conference on intelligent tutoring systems* (pp. 463–472).

Bienkowski, M., Feng, M., & Means, B. (2012). *Enhancing teaching and learning through educational data mining and learning analytics: An issue brief.* Office of Educational Technology, US Department of Education.

Brenner, D. G., Matlen, B. J., Timms, M. J., Gochyyev, P., Grillo-Hill, A., Luttgen, K., & Varfolomeeva, M. (2017). Modeling student learning behavior patterns in an online science inquiry environment. *Technology, Knowledge and Learning, 22*(3), 405–425.

Gašević, D., Dawson, S., & Siemens, G. (2015). Let's not forget: Learning analytics are about learning. *TechTrends, 59*(1), 64–71.

Hannafin, M. J. (1992). Emerging technologies, ISD, and learning environments: Critical perspectives. *Educational Technology Research and Development, 40*(1), 49–63.

Hu, W. X., & Zhao, M. (2005). An experimental study on students' Learning Behavior in multimedia Teaching process [J]. *Elementary and Middle School Audio-visual Education, 11*, 50–51.

Liu, S. Y., & Yang, Z. K. (2016). *Quantitative learning: A data-driven analytics of learning behavior[M].* Science Press.

Mo, H. (2020). *Introduction to artificial intelligence.* Post&Telecom Press.

Palonen, T., & Hakkarainen, K. (2013, April). Patterns of interaction in computer supported learning: A social network analytics. In *Fourth international conference of the learning sciences* (pp. 334–339). Erlbaum.

Peppler, K. A., & Solomou, M. (2011). Building creativity: Collaborative learning and creativity in social media environments. *On the Horizon, 19*(1), 13–23.

Wong, B. T. M., Li, K. C., & Choi, S. P. M. (2018). Trends in learning analytics practices: A review of higher education institutions. *Interactive Technology and Smart Education, 15*(2), 132–154.

18

On Machine Learning Methods for Propensity Score Matching and Weighting in Educational Data Mining Applications

Juanjuan Fan, Joshua Beemer, Xi Yan, and Richard A. Levine

CONTENTS

18.1 Introduction 277
18.2 Methods 278
18.2.1 Propensity Score 278
18.2.1.1 Assumptions for Propensity Score-Based Methods 278
18.2.1.2 Propensity Score Matching 279
18.2.1.3 Inverse Probability of Treatment Weighting 279
18.2.2 Random Forest 279
18.2.3 Ensemble Learning 281
18.3 Simulation Study 282
18.3.1 Data Generation 282
18.3.1.1 Generating Covariates (X) 282
18.3.1.2 Generating Treatment Assignment Indicator (Z) 282
18.3.1.3 Generating Outcome (Y) 283
18.3.2 Simulation Study Results 283
18.4 Students' Success Case Study 284
18.5 Discussion 286
Acknowledgment 287
References 287

18.1 Introduction

A well-designed randomized experiment can yield unbiased treatment effect because the randomized treatment allocation balances the baseline covariates between treated and control subjects. However, a randomized trial is not always feasible due to ethical or practical reasons. The observational study, as an alternative to the randomized experiment, can not only avoid the possible moral hazard (for example, when the treatment of interest is smoking) but also be less expensive. However, the observational study may produce biased estimates of the treatment effect due to treatment self-selection. Observational studies are at the heart of data analytics for institutional and student success research since students are often allowed to decide for themselves whether or not to take part in educational interventions.

Propensity score (Rosenbaum & Rubin, 1983), defined as the probability of being treated conditional on observed covariates, is a useful tool for deriving unbiased estimates of the treatment effect based on observational study data. Subjects having similar values of the propensity score share the same distribution of characteristics (covariates). Therefore, one can eliminate the treatment-selection bias in the observational study by controlling for the propensity score. In this chapter, we will evaluate the effectiveness of propensity score adjustment by matching and weighting based on propensity score estimates from a few different approaches, as detailed below.

Logistic regression (LR) is typically used to estimate propensity scores where treatment status is regressed on a set of observed covariates (Austin & Stuart, 2015). LR is a strong tool for statistical analysis; however, as McCaffrey et al. (2005) points out, large numbers of covariates tend to hurt its ability to accurately estimate propensity scores, as a result of multicollinearity. Non-linearities and interaction terms can also increase the number of covariates and can add to overfitting if iterative model-building and variable

DOI: 10.1201/9781003181187-22

selection are not performed, further affecting propensity score estimation. Due to the potentially abundant demographic and academic preparation variables in student success studies, these can be common hurdles in analysis. In order to account for large numbers of covariates, interactions, and non-linear terms, we look to the use of random forest (RF). We have found great success in applying random forest in educational research and for analyzing student success from pedagogical interventions (Spoon et al., 2016; He et al., 2018).

Random forest is an ensemble of many decision trees and, as a non-parametric method, overcomes all the issues of LR that are mentioned above. RF has been recognized as an excellent predictive tool compared to other machine learning methods, see, for example, Fernandez-Delgado et al. (2014), and He et al. (2018). In addition, the superior performance of random forest can be achieved with little model tuning and/or calibration by the user, making it an ideal tool for education researchers when estimating the propensity score. In this chapter, we also propose an additional ensemble learning method that combines predictions from eight popular machine learning methods (Hastie et al., 2016; James et al., 2013), including logistic regression, random forest, boosting, bagging, *k*-nearest neighbor, support vector machines (SVM), neural network, and naive Bayes. A combination of these base learner predictions should provide a more accurate propensity score estimation than any one base learner such as random forest (for discussion, see Beemer et al., 2017).

The goal of this research is threefold. The first goal is to compare accuracy of the propensity score estimates from three approaches: logistic regression, random forest, and the ensemble learner (EL). The second goal is to compare precision of the treatment effect estimates based on propensity score estimates from these three approaches, in tandem with propensity score adjustments from the literature, including matching, weighting, variance stabilization, and truncation. The first two goals are achieved by a large-scale simulation study. The third goal is achieved by applying select methods to evaluate the effectiveness of an educational intervention from San Diego State University.

The chapter is organized as follows. In Section 18.2, we provide relevant information about propensity score, random forest, and the proposed ensemble learner. In Section 18.3, we present the design and results of the simulation study. In Section 18.4, we provide a student success case study of observational data. We conclude the chapter in Section 18.5 with a summary of results and some discussions.

18.2 Methods

18.2.1 Propensity Score

Propensity score, e_i, is the probability of a subject's assignment to the treatment, while taking into account the subject's characteristics:

$$e_i = P(Z_i = 1 \mid X_i). \tag{18.1}$$

Here Z_i is a binary treatment indicator, $Z_i = 1$ if a subject is in the treatment group and $Z_i = 0$ if a subject is in the control group, and X_i is a vector of all observed variables other than treatment assignment.

18.2.1.1 Assumptions for Propensity Score-Based Methods

Propensity score based methods rely on four primary assumptions, as described below.

First, we assume that the treatment does not change in application across subjects. We assume no spillover effects, so a subject's treatment is not impacted by the treatment application on another subject. This assumption is standard in causal inference and named SUTVA – the stable unit treatment value assumption. In educational data mining applications, we must ensure interventions, instructional designs, intelligent tutoring systems, etc. follow a consistent template or rubric. We put particular attention on teacher training, if not common instructor, and intervention system version control to satisfy SUTVA. One also must be careful with student collaborations as that may lead to spillover effects.

Second, we assume that the outcome observed is the potential outcome that would be observed under the applied treatment. This assumption is labeled consistency and falls into counterfactual thinking for causal inference. A subject may receive any of the treatments under study. Each subject thus would realize a potential outcome for each of the treatment applications. In typical educational data mining practice of course, a subject is exposed to only one treatment. Consistency assumes that under the treatment actually received, a subject's observed outcome is the same as the subject's potential outcome. We find this assumption is always satisfied in studies of AI in education given careful definition of, protocols for, and application of the treatment regimes.

Third, we assume that every subject has a non-zero probability of receiving each treatment:

$$0 < P(Z = 1 \mid X) < 1. \tag{18.2}$$

This assumption is called positivity. In typical efficacy studies of AI in education, students have access or may receive any of the treatments under consideration, satisfying positivity.

Fourth, we assume that the treatment assignment is independent of the outcome conditional on the observed covariates:

$$\{Y(1), Y(0)\} \perp\!\!\!\perp Z \mid X, \tag{18.3}$$

where $Y(1)$ and $Y(0)$ are the possible subject outcomes for the treatment and control groups. This assumption is known as the "no unmeasured confounders" assumption, meaning all variables that affect the outcome and treatment assignment have been measured. Together, these latter two assumptions establish if the treatment assignment is strongly ignorable. Unfortunately, this fourth assumption cannot be verified in practice (Zhang et al., 2012). The usual tactic in causal inference is to collect and curate as many relevant inputs as possible, and run careful thought exercises on potential missing confounders.

For more details and mathematical buildup of these assumptions in observational studies, we refer the reader to Wilke et al. (2021) and the references therein. The bottom line for studies of AI in education is that with these assumptions, conditioning on the propensity score supports obtaining unbiased average treatment effect estimates (Rosenbaum & Rubin, 1983).

18.2.1.2 Propensity Score Matching

The propensity score matching method entails the following steps. First, the propensity score is estimated for each subject via a predictive model. Second, starting with a randomly selected treated subject, the subject is matched to a subject in the control group with the nearest propensity score; both the treated and control subjects are removed from the pool of future matches. This process is continued until all treated subjects are matched with a control subject. The final matched set will have an equal number of treated to control subjects, with the goal of having a balanced distribution for each covariate between the two treatment groups.

18.2.1.3 Inverse Probability of Treatment Weighting

Austin and Stuart (2015) review inverse probability of treatment weighting, variance stabilization, and truncation of weights as ways to better estimate treatment effects in observational studies. These methods give an educational researcher alternatives to propensity score matching, while still accounting for observed covariates in the study. This section will walk-through the three propensity score weighting methods.

Inverse probability of treatment weighting (IPTW) adjusts the underrepresented and overrepresented subjects within the control and treatment groups by assigning weights:

$$w_i = \frac{Z_i}{e_i} + \frac{1 - Z_i}{1 - e_i}, \tag{18.4}$$

where Z_i and e_i are the treatment indicator and propensity score ($i = 1, \ldots, n$, for n observations), respectively. IPTW gives higher weights to those in the treatment group with low propensity scores and those in the control group with high propensity scores, giving a more accurate estimation of the treatment effect (Rosenbaum, 1987). We propose to use IPTW in combination with regression models, as referenced in Austin and Stuart (2015), to improve estimation of causal treatment effects compared to propensity score matching.

When the propensity scores are close to zero in the treatment group and close to one in the control group, IPTW will assign a large weight to those subjects. Thus, a small group of subjects may carry a large proportion of the propensity score weight leading to potentially poor treatment effect estimation. To counter the increase in variability, we stabilize the weights by multiplying the treatment indicator, Z_i, by the marginal probability of treatment, $Pr(Z = 1)$, and multiplying the control indicator, $1 - Z_i$, by $Pr(Z = 0)$ (Robins et al., 2000). The adjusted weight is

$$w_i = \frac{Z_i Pr(Z = 1)}{e_i} + \frac{(1 - Z_i) Pr(Z = 0)}{1 - e_i}. \tag{18.5}$$

Lee et al. (2010) propose trimming or truncating the weights assigned by IPTW, again to prevent extreme weights from being assigned when the propensity scores are close to zero or one. The truncation is done by designating a minimum and maximum threshold, and if weights exceed the threshold, they are set to that threshold (Cole & Hernan, 2008; Lee et al., 2010).

18.2.2 Random Forest

Random forest uses bootstrap samples as training data to grow individual trees and then combines predictions from all trees by model averaging. Figure 18.1 presents an illustrative decision tree for presentation of the key terminology and concepts. This tree classifies students based on whether they went to a success program or not in an introductory statistics course. A tree consists of the root node (oval) and internal nodes (rectangles), each characterized by decision rules by which students are sent down the tree either to the left or to the right. For example, the root node splits students according to performance on a beginning

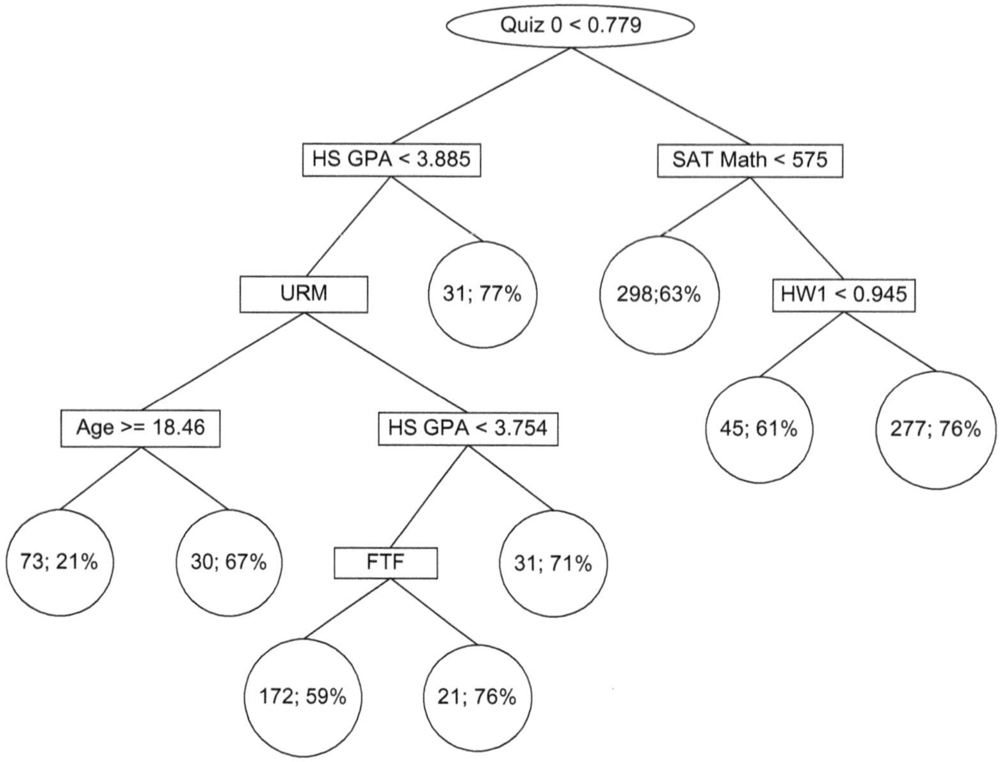

FIGURE 18.1
Illustrative decision tree to exposit the role of the root node (oval), internal nodes (rectangles), and terminal nodes (circles), and the decision rule determining splits and progression of observations down a tree. This graphic is of a classification tree with outcome (output) being treatment assignment in an introductory statistics course. The variables (inputs) on which the tree is grown are beginning of semester math readiness (Quiz 0), high school grade point average (GPA), SAT math score, underrepresented minority status (URM), student age, identification as a first-time freshman (FTF), and score on the first homework of the semester (HW1).

of semester math readiness assessment (Quiz 0). Students scoring below 0.779 are sent left down the tree, while students scoring at least 0.779 on the assessment are sent to the right down the tree. In this split, the root node is called the "parent," and the two internal nodes to which students may be sent are called the "children." The students continue to progress down the tree in this way until reaching a terminal node, the circles in Figure 18.1. For illustration purposes, in each terminal node, we show the number of students and the percentage of students attending the success program (treated).

At each split of a tree, a subset of covariates are chosen randomly. All the splitting rules based on these variables are assessed. A decision rule is chosen to split the data from this set of splitting rules according to a specified optimization criterion (see more on this below). Each split within a tree is binary and leads to another split or terminal node. The tree continues to grow until the terminal nodes are homogeneous, a preset tree depth is met, or a minimum number of observations is reached in a terminal node. For random forest, this process is repeated to grow many trees, thus creating a forest of decision trees. As an additional randomization element to construct a forest, each tree is grown on data randomly sampled with replacement from the original dataset. This bootstrap sample is the same size as the original dataset, mimicking data replication to allow for variety among the trees (decision rules) in the forest.

For each tree, the response variable (outcome/output) is typically predicted as a majority vote or average, for classification or regression respectively, based on the observations in each terminal node (Breiman, 2001). However, Malley et al. (2012) found that using the average outcome, as implemented in our study, outperformed the majority vote for classification problems. For the forest, the predicted values from each tree are averaged across all trees. Unlike logistic regression, random forest is unaffected by monotonic transformations of input variables, accommodates interactions among input variables through the recursive bisecting of the data, and reduces overfitting by taking bootstrap samples for growing each tree (Lee et al., 2010). A useful feature of random forest is the so-called out-of-bag (OOB) sample, which are the observations not included in the bootstrap sample and can be used as a built-in validation sample to assess prediction accuracy.

For propensity score estimation, the response variable we intend to predict is the treatment assignment. The best cut point for each node is determined by maximizing the reduction in the within-node impurity between the parent node and children nodes. In this project, we utilize the Gini index–based splitting criterion to evaluate each candidate split (Breiman et al., 1984). For binary response variable, the Gini index of node t can be written as follows:

$$i(t) = 1 - \{p(1 \mid t)\}^2 - \{p(0 \mid t)\}^2, \quad (18.6)$$

where $p(0 \mid t)$ and $p(1 \mid t)$ are the proportion of treated and untreated subjects in the node t. Smaller values of $i(t)$ indicate purer nodes with respect to the response. The goodness-of-split criterion $\Delta i(s,t)$ is defined as follows:

$$\Delta i(s,t) = i(t) - \{p_L i(t_L) + p_R i(t_R)\}, \quad (18.7)$$

where t_L and t_R are the descendants of node t split by s, and the weights, p_L and p_R, are the proportion of subjects in node t that are partitioned into the left child node t_L and right child node t_R, respectively. This splitting criterion shows the impurity difference between the parent node and two weighted child nodes, and the best split is the one having the largest $\Delta i(s,t)$.

After all trees are constructed, information provided by the terminal nodes is used to estimate the propensity score for each observation in the dataset. Observations ending up in the same terminal node have the same estimated propensity score, which is equal to the percentage of treated subjects in the node. The propensity score estimate for each subject based on the random forest is the average propensity score over the classification trees in the forest. In this chapter, we grow forests of 500 trees.

18.2.3 Ensemble Learning

Random forest in itself is an ensemble learner as we are making predictions across a set of trees in the forest. More generally, ensemble learning entails fitting multiple individual learners to the data, and combining predictions from these individual learners into an ensemble prediction. The idea of an ensemble learner is as follows. Using cross-validation and looping over the data, a predicted value is obtained for each observation based on each individual learner. Wolpert (1992), Breiman (1996), and LeBlanc and Tibshirani (1996) propose that the ensemble may combine the predictions from individual or base learners using a ridge regression, the so-called meta-learner. Ridge regression uses a penalty function to minimize coefficients of covariates that are weak predictors of the outcome (James et al., 2013, Chapter 6). In this case, the predictions from the individual learners are the covariates and the coefficients act as the weights, which become the ensemble learner via linear combination.

We propose a modified version of the ensemble learner presented in Beemer et al. (2017). The ensemble learning method starts with K-fold cross-validation: randomly split the data into K subsets of approximately equal sizes, removing one subset and training the base learners on the remaining subsets. The trained base learners are then used to make predictions for the subset that was removed. This "leave-one-out" method is repeated using the next subset until predictions are made for all observations. The predictions from base learners are then stacked, with predictions from each base learner forming a column. But instead of using a ridge regression to weigh the predictions as described above, a random forest is built to regress the true outcome against the predictions. Random forest has proven to be a reliable meta-learner in medical research (Wang et al., 2019), and a good alternative to regression methods in educational research (Spoon et al., 2016; He et al., 2018). We expect this modified ensemble learner to achieve strong predictive performance.

Algorithm 1: Ensemble Learner

1. Identify L base learners.
2. Randomly partition data into K subsets of approximately equal sizes.
3. **for** $k = 1,\ldots,K$ **do**
4. Leave the kth subset out as test set, remaining subsets are the training set.
5. **for** $j = 1,\ldots,L$ **do**
6. Train the jth base learner on training set.
7. Predict test set using trained base learner.
8. Stack predictions from each base learner as a column in a data frame.
9. Build a random forest using the stacked predictions from the base learners as inputs.

The algorithm for the ensemble learner is presented in Algorithm 1. The ensemble will have the ability to combine the predictions from L machine learning methods. In our software, we provide the user with eight base learners: logistic regression, random forest, boosting, bagging, k-nearest neighbor, support vector machines, neural network, and naive Bayes.

All codes performed in the project are developed in the statistical freeware R. Random forest is constructed through the R package PartyKit with modifications. The R package MatchIt is used to obtain the samples matched by propensity scores estimated using LR, RF, and our custom-made ensemble learner. The R software we wrote for the ensemble learner is publicly available at our GitHub depository (Beemer, 2021).

18.3 Simulation Study

In this section we present a simulation study to compare the propensity score estimates from logistic regression, random forest, and the ensemble learner. The ultimate goal is to evaluate the precision in treatment effect estimates when using different modeling approaches for PS estimation, combined with various propensity score adjustment methods. We start with data generation for the observational study.

18.3.1 Data Generation

We generate data in the following order.

18.3.1.1 Generating Covariates (X)

Eight covariates (X_1–X_8) were generated independently from specific probability distributions. Variable X_1 was generated as a binary variable from a Bernoulli distribution with probability of success at $p = 0.5$. Variable X_2 was generated as a nominal variable with five categories (A, B, C, D, E), with each category at different likelihoods to occur, (10%, 20%, 30%, 20%, 20%). Variables X_3 and X_4 were generated independently from a discrete uniform distribution from 0 to 1 with increments of 0.2, and treated as ordinal variables with five levels (0.2, 0.4, 0.6, 0.8, 1.0). The last four covariates (X_5–X_8) were designed to mimic continuous variables and were simulated by discrete uniform distributions from 0 to 1 with increments of 0.02.

18.3.1.2 Generating Treatment Assignment Indicator (Z)

Following Setoguchi et al. (2008), the true propensity score, or the probability of treatment assignment given covariates, is assumed to follow the logistic regression model:

$$P(Z=1\mid X)=\frac{1}{1+e^{-\beta f(X)}}. \tag{18.8}$$

Setoguchi et al. (2008) uses seven models in their study, in which the function $f(\cdot)$ has varying degrees of additivity and linearity, with non-additivity and non-linearity comprising two-way interactions and quadratic terms. In this chapter, we consider four models from Setoguchi et al. (2008) and add one additional model (model E) which contains three-way interactions and non-linearity terms other than the quadratic form. The five treatment-selection models are given as follows:

A. *Additivity and linearity (main effects only)*

$$P(Z=1\mid X)=(1+\exp[-(\beta_0+\beta_1X_1+\beta_2X_2+\beta_3X_3+\beta_4X_4+\beta_5X_5+\beta_6X_6+\beta_7X_7+\beta_8X_8)])^{-1}. \tag{18.9}$$

B. *Moderate non-linearity (three quadratic terms)*

$$P(Z=1\mid X)=(1+\exp[-(\beta_0+\beta_1X_1+\beta_2X_2+\beta_3X_3+\beta_4X_4+\beta_5X_5+\beta_6X_6+\beta_7X_7+\beta_8X_8+\beta_9X_3^2+\beta_{10}X_5^2+\beta_{11}X_7^2)])^{-1}. \tag{18.10}$$

C. *Mild non-additivity (four two-way interaction terms)*

$$P(Z=1\mid X)=(1+\exp[-(\beta_0+\beta_1X_1+\beta_2X_2+\beta_3X_3+\beta_4X_4+\beta_5X_5+\beta_6X_6+\beta_7X_7+\beta_8X_8+\beta_9X_3X_4+\beta_{10}X_4X_5+\beta_{11}X_5X_6+\beta_{12}X_6X_7)])^{-1}. \tag{18.11}$$

D. *Moderate non-additivity and non-linearity (ten two-way interaction terms and three quadratic terms)*

$$P(Z=1\mid X)=(1+\exp[-(\beta_0+\beta_1X_1+\beta_2X_2+\beta_3X_3+\beta_4X_4+\beta_5X_5+\beta_6X_6+\beta_7X_7+\beta_8X_8+\beta_9X_3^2+\beta_{10}X_5^2+\beta_{11}X_7^2+\beta_{12}X_3X_4+\beta_{13}X_4X_5+\beta_{14}X_5X_6+\beta_{15}X_6X_7+\beta_{16}X_7X_8+\beta_{17}X_3X_8+\beta_{18}X_5X_7+\beta_{19}X_4X_8+\beta_{20}X_3X_5+\beta_{21}X_6X_8)])^{-1}. \tag{18.12}$$

E. *Severe non-additivity and non-linearity (six two-way interaction terms and four three-way interaction terms; one quadratic term, one cubic polynomial, and one square root term)*

$$P(Z=1|X)=(1+\exp[-(\beta_0+\beta_1X_1+\beta_2X_2+\beta_3X_3 +\beta_4X_4+\beta_5X_5+\beta_6X_6+\beta_7X_7+\beta_8X_8 +\beta_9X_4^2+\beta_{10}X_6^3+\beta_{11}\sqrt{X_8}+\beta_{12}X_3X_4 +\beta_{13}X_4X_5+\beta_{14}X_5X_6+\beta_{15}X_6X_7 +\beta_{16}X_7X_8+\beta_{17}X_3X_8+\beta_{18}X_3X_5X_7 +\beta_{19}X_4X_6X_8+\beta_{20}X_3X_4X_5+\beta_{21}X_6X_7X_8)])^{-1}. \quad (18.13)$$

The true propensity score was used as the parameter, p, in a Bernoulli distribution to generate the treatment assignment for each subject in the dataset. In our experience with educational interventions, the proportion of students in the treatment group is typically within the 20–30% range. The coefficients (shown in Table 18.1) in the treatment-selection models were chosen so that the probability of being allocated to treatment was about 25%.

18.3.1.3 Generating Outcome (Y)

Based on covariates (X_1-X_8) and treatment assignment indicator (Z), the continuous outcome variable, Y, was generated as follows:

Model 1

$$Y=\alpha_{00}+\alpha_0Z+\alpha_1X_1+\alpha_2X_2+\alpha_3X_3+\alpha_4X_4+\alpha_5X_5 +\alpha_6X_6+\alpha_7X_7+\alpha_8X_8+\varepsilon, \quad (18.14)$$

Model 2

$$Y=\alpha_{00}+\alpha_0Z+\alpha_1X_1+\alpha_2X_2+\alpha_3X_3^2+\alpha_4X_4^2+\alpha_5\ln X_5 +\alpha_6\sqrt{X_6}+\alpha_7X_7X_8+\alpha_8X_3X_7+\varepsilon, \quad (18.15)$$

where $\varepsilon \sim N(0,1)$. In Model 1, a simple linear association is assumed between predictors and the outcome, while Model 2 involves several non-linear terms and two-way interactions, in order to examine the performance of different methods for treatment effect estimation in a more complex data structure. The true treatment effect (α_0) was fixed at 1.5 for both outcome models. The other coefficients in the models ($\alpha_1-\alpha_8, \alpha_{00}$) were set to 0.5, 0.3, 0.7, 0.6, 0.1, –1.2, –0.5, –1, and 0.5, respectively.

With five models (A–E) for treatment assignment, combined with two models (1–2) for the outcome, a total of ten models were used in our simulation study. These models are denoted as models A–E and A′–E′ hereafter. For example, model A assumes perfect additivity and linearity for both the treatment assignment model and the outcome model, while model E′ assumes most severe non-additivity and non-linearity for both the treatment assignment and outcome models. For each simulation scenario, 100 datasets of size n = 500 were simulated.

18.3.2 Simulation Study Results

Propensity scores were predicted from logistic regression, random forest, and the ensemble learner for the true propensity score models A–E, and from these predictions a mean squared error (MSE) was computed. Table 18.2 shows that the ensemble learner performs the best, and random forest performs the second best, for every treatment assignment model. Logistic regression underperforms even for Model A, which is a logistic regression model with perfect additivity and linearity. The excellent performance of the ensemble learner, especially for more complex models, supports the idea of using an ensemble learner over logistic regression for propensity score matching and weighting techniques.

Table 18.3 presents bias and mean squared error (MSE) for the estimated treatment effect using

TABLE 18.1

Coefficients Used in the Data Generation Models

Model	β_0	β_1	β_2	β_3	β_4	β_5	β_6	β_7	β_8	β_9	β_{10}
A	–0.5	–0.5	1.2	–1.0	–0.62	–0.7	–0.4	0.6	0.2	•	•
B	–0.5	–0.5	–1.2	–1.2	–0.72	0.7	0.4	0.6	0.2	0.3	–0.4
C	–0.5	–0.5	–1.2	–1.2	–0.72	0.7	0.4	0.6	0.2	0.3	–0.4
D	–0.5	–0.5	–1.2	–1.2	–0.62	0.7	0.4	0.6	0.2	0.3	–0.4
E	–0.5	–0.5	–1.2	–1.2	–0.72	0.7	0.4	0.6	0.2	0.3	–0.4
Model	β_{11}	β_{12}	β_{13}	β_{14}	β_{15}	β_{16}	β_{17}	β_{18}	β_{19}	β_{20}	β_{21}
A	•	•	•	•	•	•	•	•	•	•	•
B	1.1	•	•	•	•	•	•	•	•	•	•
C	1.1	0.46	•	•	•	•	•	•	•	•	•
D	1.1	–0.2	0.42	–0.8	0.9	–1	0.32	–0.45	0.36	–0.47	0.35
E	1.1	–0.2	0.42	–0.8	0.9	–1	–0.32	–0.45	–0.36	0.47	0.35

TABLE 18.2

Mean Squared Error (MSE) for the Propensity Score Estimates by Logistic Regression, Random Forest, and Ensemble Learner for Models A–E ($n = 500$)

	A	B	C	D	E
LR	0.041	0.052	0.054	0.048	0.047
RF	0.039	0.048	0.049	0.044	0.044
EL	0.018	0.017	0.017	0.017	0.018

propensity score matching and weighting, including four different weighting schemes (inverse probability of treatment weighting, variance stabilization of weights, weight truncation, variance stabilization with truncation), as detailed in Section 18.2.1. The methods for propensity score estimation include logistic regression, random forest, and ensemble learner. These results offer comparisons among models (logistic regression, random forest, and ensemble learning) and between matching and weighting.

For each model in Table 18.3, the smallest MSE across all methods is highlighted by an underscore and also bold-faced, while the smallest MSE that is not from the ensemble learner is highlighted by underscore but not bold-faced. It can be seen from Table 18.3 that the best performance, as signified by the smallest MSE, is almost always by the ensemble learner, especially when the ensemble learner is combined with propensity score weighting using variance stabilization and/or truncation of weights. Random forest, combined with propensity score weighting, also performs well. Comparing propensity score matching and weighting, weighting appears to be the clear winner, especially when using variance stabilization and/or truncation of weights.

18.4 Students' Success Case Study

In a "State of the CSU" address, Dr. Timothy White, the former Chancellor of the California State University system, states "The California State University is key to California's brightest and most hopeful future, opening the door to educational opportunities for all and transforming the lives of students and their families." With this mandate, San Diego State University offers many student success interventions such as supplemental instruction (see Guarcello et al., 2017, and the references therein), which is offered to students currently enrolled, or even pre-enrollment interventions that aim to help students to attend SDSU. We look at one such student success intervention that is offered to students who are from underprivileged communities and are given a path that optimally could afford them the opportunity for entrance into higher education.

Table 18.4 presents a comparison of students supported by the student success intervention and their peers not assisted by the intervention. This snapshot of a few student background characteristics shows that those in the intervention have a higher rate of being first-generation college students (i.e., first in their immediate family to go to college) and underrepresented minorities (an ethnicity categorization defined by the California State University System). They have lower mean SAT scores, slightly lower mean high school grade point average (GPA), and tend to earn, and transfer, fewer college-level course units than their peers at the university.

Table 18.5 investigates the balance in the demographic and background variables before and after matching, between treated and control groups. Before propensity score matching, the standardized mean difference (SMD), defined as the difference between the two sample means divided by the pooled standard deviation, is large for all but two variables in Table 18.5, using a value of SMD below 20 for balanced samples (Austin, 2009; Hillis et al., 2021). After propensity score matching, the standardized mean difference decreases for all covariates. Matching in this study does a very good job of balancing the background characteristics between the treated and control groups, with all SMD values below 20 after matching.

To evaluate the success of the intervention, we examine the effect of the intervention on the student GPA at the end of their second semester at the university, specifically the GPA for courses taken on campus. Table 18.6 presents the estimated treatment effect of the intervention, and the associated p-value and 95% confidence interval. Based on the results using propensity score weighting with variance stabilization and truncation, those students who participated in the student success intervention had on average an increase of 0.053 (with a 95% confidence interval of −0.024 to 0.130) in their end of second semester GPA compared to those students who did not participate in the intervention, accounting for all other possible factors. The results based on propensity score matching are similar to those from weighting, and the ensemble learner was used for propensity score estimation.

We note that the treatment effect confidence intervals cover zero, suggesting a failed student success intervention. However, since students provided with the intervention are perceived to be at a distinct academic disadvantage due to their socioeconomic background, an "on par" result is a success. These results show that by participating in the student success intervention, program students were able to match their peers in GPA at the end of their second semester.

TABLE 18.3

Bias and Mean Squared Error (MSE) for the Treatment Effect Estimates by Logistic Regression, Random Forest, and Ensemble Learner Using Various Propensity Score Matching and Weighting Techniques for All Models A–E and A′–E′ ($n = 500$)

		Matching			IPTW			Variance Stabilization			Weight Truncation			Variance Stabilization with Truncation		
		LR	RF	EL	LR	RF	EL	LR	RF	EL	LR	RF	EL	LR	RF	EL
A	BIAS	–0.016	–0.021	0.005	–0.028	–0.017	–0.004	–0.027	–0.015	0.003	–0.030	–0.025	–0.013	–0.029	–0.022	–0.007
	MSE	0.028	0.035	0.020	0.022	0.027	0.010	0.023	0.027	0.009	0.020	0.021	**0.008**	0.020	0.026	0.010
B	BIAS	0.012	0.029	0.044	0.020	0.025	–0.025	0.021	0.027	–0.004	0.009	0.015	0.007	0.013	0.026	0.019
	MSE	0.043	0.024	0.027	0.045	0.042	0.051	0.047	0.041	0.042	0.036	0.035	0.023	0.037	0.038	**0.020**
C	BIAS	–0.049	–0.037	–0.060	–0.041	–0.037	0.002	–0.040	–0.04	0.006	–0.055	–0.052	–0.039	–0.045	–0.048	–0.043
	MSE	0.033	0.057	0.023	0.026	0.026	0.011	0.026	0.028	**0.010**	0.029	0.030	**0.010**	0.029	0.028	0.012
D	BIAS	–0.005	0.001	0.028	–0.007	–0.005	–0.020	–0.004	0.001	0.001	–0.005	–0.008	–0.002	–0.009	–0.002	0.010
	MSE	**0.016**	0.029	**0.016**	0.045	0.034	0.040	0.047	0.031	0.032	0.028	0.028	0.022	0.028	0.027	0.017
E	BIAS	–0.029	–0.049	–0.020	–0.011	–0.011	–0.016	–0.009	–0.016	–0.033	–0.012	–0.016	–0.015	–0.011	–0.017	–0.031
	MSE	0.029	0.015	0.013	0.010	0.009	0.016	0.010	0.010	0.015	0.010	0.010	**0.007**	0.010	0.010	**0.007**
A′	BIAS	0.063	0.049	0.054	0.098	0.075	0.101	0.098	0.069	0.094	0.064	0.068	0.067	0.068	0.065	0.058
	MSE	0.029	0.043	0.019	0.039	0.035	0.032	0.040	0.034	0.030	0.028	0.030	0.017	0.031	0.033	**0.016**
B′	BIAS	–0.078	–0.093	–0.074	–0.102	–0.095	–0.033	–0.104	–0.092	–0.016	–0.101	–0.095	–0.089	–0.088	–0.085	–0.079
	MSE	0.025	0.019	0.010	0.026	0.018	0.007	0.026	0.015	**0.006**	0.021	0.018	0.014	0.016	0.014	0.011
C′	BIAS	–0.044	–0.043	–0.044	0.042	0.009	0.031	0.047	0.001	0.019	0.004	0.005	0.021	0.006	–0.003	0.012
	MSE	0.015	0.014	0.019	0.012	0.011	0.010	0.013	0.012	0.011	0.010	0.011	**0.006**	0.012	0.011	0.007
D′	BIAS	0.007	0.014	0.025	0.021	0.011	–0.033	0.022	0.013	–0.041	0.014	0.014	0.019	0.010	0.016	0.024
	MSE	0.012	0.016	0.008	0.019	0.014	0.040	0.020	0.013	0.042	0.013	0.012	0.008	0.012	0.011	**0.007**
E′	BIAS	0.055	0.048	0.043	0.034	0.035	0.080	0.033	0.038	0.082	0.034	0.039	0.036	0.038	0.037	0.041
	MSE	0.009	0.015	0.006	0.009	0.011	0.017	0.008	0.010	0.018	0.009	0.010	**0.005**	0.008	0.009	**0.005**

TABLE 18.4

Summary of Student Characteristics for Treatment and Control Groups

	Control	Treated
First generation	15.8%	21.0%
Underrepresented minority	32.4%	62.9%
SAT score	1209.1 (156.3)	1145.3 (85.1)
High school GPA	3.7 (0.3)	3.5 (0.3)
Transfer units	22.3 (26.3)	10.8 (11.2)

Mean and (standard deviation) reported for continuous variables, and percentage reported for categorical variables.

TABLE 18.5

Standardized Mean Difference before and after Matching

Covariate	Before	After
Age	62.8	6.7
Gender	10.4	3.2
SAT score	50.8	7.3
High school GPA	32.8	18.4
Incoming units	57.1	8.3
First Generation	13.3	2.6
Underrepresented minority	64.1	7.7
Hispanic	72.2	15.2

18.5 Discussion

Evaluations and assessments of student success interventions often require observational studies. Propensity score–based adjustments are powerful tools to derive unbiased estimates of the treatment effect from observational studies. In this chapter, we review two methods for propensity score estimation, including logistic regression and random forest, and propose our own custom-made ensemble learner combining prediction results from eight popular machine learning methods. In addition, we discuss propensity score matching and inverse probability of treatment weighting, including variance stabilization and truncation of weights, as means to improve performance of propensity score weighting.

A large-scale simulation study is conducted to compare the three modeling approaches for propensity score estimation: LR, RF, and EL; and to compare performance of propensity score matching and weighting in conjunction with the modeling approaches. The simulation results show that the ensemble learner provides the most accurate estimates of the propensity score under all model configurations, followed by random forest as the second best performer. In terms of accuracy of treatment effect estimation, the ensemble learner combined with propensity score weighting, incorporating variance stabilization and truncation, is an overall top performer. Random forest combined with propensity score weighting also performs reasonably well. Between propensity score matching and weighting, we recommend propensity score weighting using variable stabilization and truncation of weights.

The ensemble learner based propensity score matching and weighting methods are applied to a student success intervention at San Diego State University for underserved students before enrollment at SDSU. The ensemble learner–based propensity score matching is able to largely eliminate the imbalance in student background variables between students in the intervention and their peers not in the intervention. A study of the effect of intervention, using propensity score matching and weighting, shows that the intervention successfully removes academic disadvantage among participates, a consequence of their lower socioeconomic status and less prepared academic background, so that the participants are able to perform as well as the general student population at SDSU by the end of their second semester at the university.

In terms of performance comparisons of existing machine learning methods, Fernandez-Delgado et al. (2014) performed a broader study of 179 classifiers from 17 machine learning families and their results found that random forest performed the best overall, followed by boosted trees, neural network, and SVM. In comparison, naive Bayes, logistic regression, and decision tree do not perform as well. Considering propensity score weighting specifically, Lee et al. (2010) conducted a well-cited simulation study comparing the performance of logistic regression, decision tree, bagged and boosted trees, and random forest. Their recommendation was to use boosted trees and random forest for their consistent superior performance. More recently, Cannas and Arpino (2019) extended

TABLE 18.6

Student Success Treatment Effect, *p*-Value, and 95% Confidence Interval from Matching and Variance Stabilization with Truncation

Method	Treatment Effect	*p*-Value	95% Confidence Interval
Matching	0.063	0.196	−0.033, 0.158
Variance stabilization with truncation	0.053	0.179	−0.024, 0.130

the simulation study by Lee et al. to include both propensity score matching and weighting, while adding two new machine learning methods for comparison: neural network and naive Bayes. They also found random forest to have the best overall and most consistent performance, followed by neural network and logistic regression. In summary, random forest, boosted trees, and neural network all seem to perform well in general, with boosted trees and neural network requiring more user input and calibration. For the less experienced user of machine learning methods, without having to code their own ensemble or performing extensive calibration, we recommend random forest for propensity score estimation for its superior predictive power and relative ease of implementation.

We used a sample size of 500 with eight features in simulations presented in this chapter. For the simulation studies presented in Lee et al. (2010) and Cannas and Arpino (2019), sample sizes of 500, 1,000, and 2,000 were considered with ten features. These sample sizes and number of features were selected as they were similar to the observational study data under consideration and the recommended machine learning methods do pretty well under these configurations. Lee et al. point out that as the sample size increases, the comparative performance of the machine learning algorithms did not change, while the accuracy of treatment effect estimates improved for all methods. This generally agrees with our own experiences, see, for example, Autenrieth et al. (2021), in which an in-depth simulation study was provided. In addition, machine learning methods can generally perform well over a wide range of sample sizes and feature space dimensions.

It is important to point out that no method performs the best for all situations. Traditional methods such as logistic regression perform well when the models can be correctly specified, while machine learning methods such as random forest and ensemble learner have a distinct advantage with complex data structure since they are non-parametric in nature and hence more flexible. This trend can be seen from the simulation results presented in Table 18.3. Since one does not know the true model formats in the real world, we recommend that different methods should be evaluated for the specific dataset at hand with performance judged based on cross-validation or a test sample.

Quantification of uncertainty with machine learning methods is more complex and much less routinely performed compared to classical regression methods. However, there have been some recent advances on statistical inferences for random forests, interested readers are referred to Mentch and Hooker (2016), Wager and Athey (2018), Athey et al. (2019), Lu and Hardin (2021), and references therein. In the context of propensity score weighting and matching, machine learning methods are used only to obtain more accurate estimates of the propensity score. Since the ultimate goal is to reduce or eliminate the selection bias from the observational study, it is crucial that the predicted propensity scores can help achieve well-balanced covariates between treatment groups.

Acknowledgment

This research was supported in part by the National Science Foundation grant 1633130.

References

Athey, S., Tibshirani, J., & Wager, S. (2019). Generalized random forests. *Annals of Statistics, 47*(2), 1148–1178.

Austin, P. C. (2009). Balance diagnostics for comparing the distribution of baseline covariates between treatment groups in propensity score matched samples. *Statistics in Medicine, 28*(25), 3083–3107.

Austin, P. C., & Stuart, E. A. (2015). Moving towards best practice when using inverse probability of treatment weighting (IPTW) using the propensity score to estimate causal treatment effects in observational studies. *Statistics in Medicine, 34*(28), 3661–3679.

Autenrieth, M., Levine, R. A., Fan, J. J., & Guarcello, M. A. (2021). Stacked ensemble learning for propensity score methods in observational studies. *Journal of Educational Data Mining, 13*, 24–189.

Beemer, J. (2021). Ensemble learner codes: GitHub repository. https://github.com/jbeemer05/match_ED

Beemer, J., Spoon, K., He, L., Fan, J., & Levine, R. A. (2017). Ensemble learning for estimating individualized treatment effects in student success studies. *International Journal of Artificial Intelligence in Education, 28*(3), 315–335.

Breiman, L. (1996). Stacked regressions. *Machine Learning, 24*(1), 49–64.

Breiman, L. (2001). Random forests. *Machine Learning, 45*(1), 5–32.

Breiman, L., Friedman, J., Olsen, R., & Stone, C. (1984). *Classification and regression trees.* Wadsworth International Group.

Cannas, M., & Arpino, B. (2019). A comparison of machine learning algorithms and covariate balance measures for propensity score matching and weighting. *Biometrical Journal, 61*(4), 1049–1072.

Cole, S. R., & Hernan, M. A. (2008). Constructing inverse probability weights for marginal structural models. *American Journal of Epidemiology, 168*(6), 656–664.

Fernandez-Delgado, M., Cernadas, E., Barro, S., & Amorim, D. (2014). Do we need hundreds of classifiers to solve real world classification problems? *Journal of Machine Learning Research, 15*, 3133–3181.

Guarcello, M. A., Levine, R. A., Beemer, J., Frazee, J. P., Laumakis, M. A., & Schellenberg, S. A. (2017). Balancing student success: Assessing supplemental instruction through coarsened exact matching. *Technology, Knowledge and Learning, 22*(3), 335–352.

Hastie, T., Tibshirani, R., & Friedman, J. (2016). *The elements of statistical learning: Data mining, inference, and prediction* (2nd ed.). Springer.

He, L., Levine, R. A., Fan, J. J., Beemer, J., & Stronach, J. (2018). Random forest as a predictive analytics alternative to regression in institutional research. *Practical Assessment, Research and Evaluation, 23*(1), 1–16.

Hillis, T., Guarcello, M. A., Levine, R. A., & Fan, J. J. (2021). Causal inference in the presence of missing data using a random forest based matching algorithm. *Stat, 10*(1), e326.

James, G., Witten, D., Hastie, T., & Tibshirani, R. (2013). *An introduction to statistical learning.* Springer.

LeBlanc, M., & Tibshirani, R. (1996). Combining estimates in regression and classification. *Journal of the American Statistical Association, 91*(436), 1641–1650.

Lee, B. K., Lessler, J., & Stuart, E. A. (2010). Improving propensity score weighting using machine learning. *Statistics in Medicine, 29*(3), 337–346.

Lu, B., & Hardin, J. (2021). A unified framework for random forest prediction error estimation. *Journal of Machine Learning Research, 22*, 1–41.

Malley, J., Kruppa, J., Dasgupta, A., Malley, K., & Ziegler, A. (2012). Probability machines: Consistent probability estimation using nonparametric learning machines. *Methods of Information in Medicine, 51*(1), 74–81.

Mccaffrey, D., Ridgeway, G., & Morral, A. (2005). Propensity score estimation with boosted regression for evaluating causal effects in observational studies. *Psychological Methods, 9*(4), 403–425.

Mentch, L., & Hooker, G. (2016). Quantifying uncertainty in random forest via confidence intervals and hypothesis tests. *Journal of Machine Learning Research, 17*, 1–41.

Robins, J. M., Hernan, M., & Brumback, B. (2000). Marginal structural models and causal inference in epidemiology. *Epidemiology, 11*(5), 550–560.

Rosenbaum, P. R. (1987). Model-based direct adjustment. *Journal of the American Statistical Association, 82*(387), 394.

Rosenbaum, P. R., & Rubin, D. B. (1983). The central role of the propensity score in observational studies for causal effects. *Biometrika, 70*(1), 41–55.

Setoguchi, S., Schneeweiss, S., Brookhart, M. A., Glynn, R. J., & Cook, E. F. (2008). Evaluating uses of data mining techniques in propensity score estimation: A simulation study. *Pharmacoepidemiology and Drug Safety, 17*(6), 546–555.

Spoon, K., Beemer, J., Whitmer, J. C., Fan, J. J., Frazee, J. P., Stronach, J., Bohonak, A. J., & Levine, R. A. (2016). Random forests for evaluating pedagogy and informing personalized learning. *Journal of Educational Data Mining, 8*(2), 20–50.

Wager, S., & Athey, S. (2018). Estimation and inference of heterogeneous treatment effects using random forests. *Journal of the American Statistical Association, 113*(523), 1228–1242.

Wang, Y., Wang, D., Geng, N., Wang, Y., Yin, Y., & Jin, Y. (2019). Stacking-based ensemble learning of decision trees for interpretable prostate cancer detection. *Applied Soft Computing, 77*, 188–204.

Wilke, M. C., Levine, R. A., Guarcello, M. A., & Fan, J. (2021). Estimating the optimal treatment regime for student success programs. *Behaviormetrika, 48*(2), 309–343.

Wolpert, D. (1992). Stacked generalization. *Neural Networks, 2*(2), 241–259.

Zhang, B., Tsiatis, A. A., Laber, E. B., & Davidian, M. (2012). A robust method for estimating optimal treatment regimes. *Biometrics, 68*(4), 1010–1018.

19

Situating AI (and Big Data) in the Learning Sciences: Moving toward Large-Scale Learning Sciences

Danielle S. McNamara, Tracy Arner, Reese Butterfuss, Debshila Basu Mallick, Andrew S. Lan, Rod D. Roscoe, Henry L. Roediger III, and Richard G. Baraniuk

CONTENTS

19.1 Introduction....289
19.2 A Landscape of Learning Sciences and Some Challenges It Faces....290
19.2.1 Outcomes....290
19.2.2 Student Factors....291
19.2.3 Contextual Factors....292
19.2.4 Replication Crisis? (Or Maybe Context Matters)....293
19.3 AI and Its Affordances for the Learning Sciences....294
19.3.1 Deep Student Models....295
19.3.2 Causal Learning Outcome Models....295
19.3.3 Natural Language Processing....296
19.3.4 Sensor-Free Student Factor Measures....297
19.3.5 Instructional Policy Learning....298
19.4 Promoting Equity....299
19.5 Conclusion....300
Acknowledgments....300
References....300

19.1 Introduction

The learning sciences inherently involve interdisciplinary research with an overarching objective of advancing theories of learning and to inform the design and implementation of effective instructional methods and learning technologies. In these endeavors, learning sciences encompass diverse constructs, measures, processes, and outcomes pertaining to both learning, motivation, and social interactions. These complex goals are further influenced by a large array of factors stemming from the learning context, learning task, and the characteristics of the individual learners. Learning occurs within a multitude of interacting contextual factors spanning variations between schools, teachers, classrooms, peers, and available technologies. These contexts also differ widely in diverse factors such as the social support that students receive, instructor engagement, demographic and ideological diversity, as well as instructional design strategies and affordances offered by educational technologies (Anderson & Dron, 2011). The learners themselves vary across a host of fixed factors such as age, grade level, ethnicity, and cultural background, as well as malleable individual differences such as engagement, interests, learning strategies, reading skills, and prior knowledge (Cantor et al., 2019; Jonassen & Grabowski, 2012; Winne, 1996).

Figure 19.1 visually conceptualizes this deeply complex landscape of learning science research with regard to student factors, contextual factors, and their interactive effects, with the ultimate objective of understanding how these factors impact short-term and long-term outcomes (e.g., grades, motivation, persistence, and test performance). A multitude of invaluable studies has examined these factors individually. Increasingly, however, researchers are diving into the complex interactions that emerge among individual differences (Yukselturk & Bulut, 2007) and then, in turn, investigating how the impact of such individual differences depends on the learning context (e.g., Coiro, 2021; Snow, 2002).

A large body of extant research already informs our understanding of how to improve learning across a

DOI: 10.1201/9781003181187-23

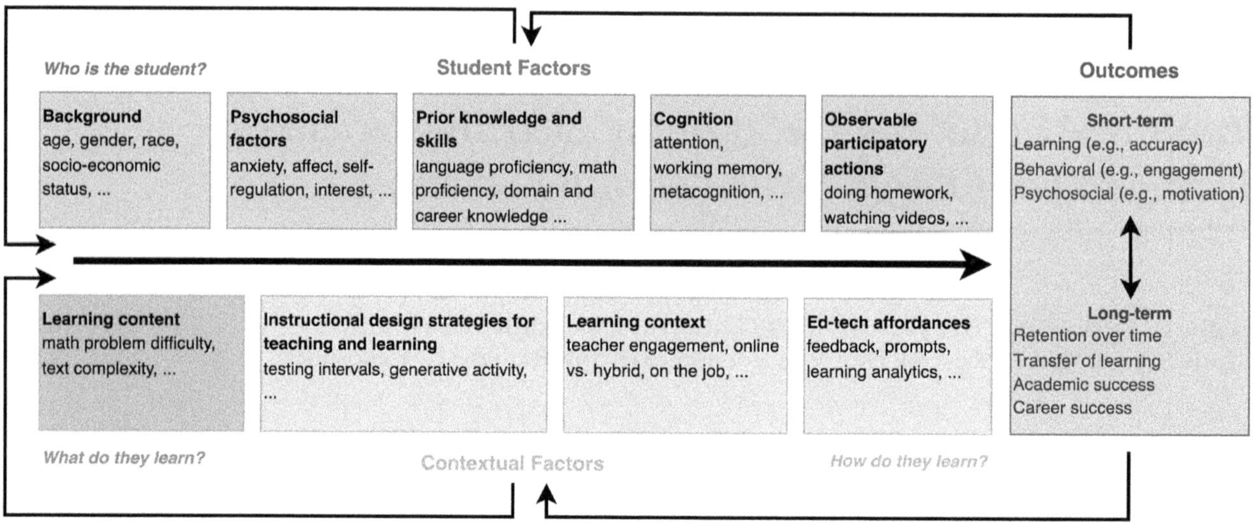

FIGURE 19.1
A landscape of learning science research. Numerous factors emerge from the characteristics of individual learners, learning contexts, and their interdependencies. Learners differ in fixed factors such as age, grade, ethnicity, and cultural background, as well as malleable individual differences such as engagement, interests, learning strategies, reading skills, and knowledge. Many studies that examine digital learning also assess observable participatory actions such as students' propensity to complete homework or attend class. In turn, the impact of individual differences depends on the learning context. Educational contexts differ widely in terms of instructor engagement, learner diversity, as well as instructional design strategies and affordances offered by educational technologies. The components included in each factor highlight only a subset of the variables that impact authentic student learning experiences.

variety of constrained contexts (e.g., laboratory experiments and targeted classroom studies). However, many of these assumptions, interventions, and findings have not yet been tested at scale or have failed to replicate beyond their initial, controlled environments. Moreover, existing work insufficiently draws from diverse theories and disciplines to examine the combined impact of multiple factors (Koedinger et al., 2013). Such narrow approaches are partially due to the combinatorial explosion that erupts when researchers study multiple complex factors and their relationships (Taber, 2019). In addition, many studies have been forced to assume or test linear relations between student and contextual factors, even though nonlinear relations are both theoretically and practically important. Finally, examining the impact of educational interventions in classrooms can be slow, as they inherently depend on academic timeframes spanning semesters or years.

Our supposition is that Artificial Intelligence (AI), naturally in combination with big data, provides multiple affordances with substantial potential to tackle the aforementioned challenges facing the learning sciences. AI – and artificial intelligence in education (AIEd) – is sometimes viewed as antithetical to the learning sciences. In education, AI is often viewed as a tool to build applications and educational technologies that are designed based on learning science theories and evidence. Indeed, AI-based educational technologies have been used as a means to modify or enhance contextual factors. However, in contrast to this utilitarian perspective, AI is rarely viewed as a means to directly augment or enhance our understanding of learning itself or to advance the learning sciences. The purpose of this chapter is to situate AI and AIED within the learning sciences and outline ways that AI can play a significant role in advancing the learning sciences.

In Section 19.2, we describe the landscape of learning science research (see Figure 19.1), including the broad range of individual and contextual factors and their relations to targeted educational outcomes as well as challenges that potentially inhibit significant advances in understanding and promoting learning. In response, Section 19.3 discusses affordances that arise from AI in this landscape and how AI can be leveraged to enhance, augment, and address challenges within the learning sciences. Finally, Section 19.4 proposes that leveraging AI in this manner has the potential to confront one of our biggest threats to education and society: inequitable systems that negatively impact the growth and success of learners and educators.

19.2 A Landscape of Learning Sciences and Some Challenges It Faces

19.2.1 Outcomes

One principal goal of education is to prepare students to become contributing members of a global knowledge economy (Dewey, 1934) and, in turn, empower students to be agentic participants in pursuing their own goals and contributing to the betterment of society. In the learning sciences, attainment of these goals has been measured in a variety of ways, including standardized assessments (e.g., NAEP), completion of educational levels (e.g., high school, undergraduate degrees, and graduate degrees), course completion, grades, and demonstrated skill or knowledge development. For instance, students' retention of content and skills often serve as critical outcome measures of the effectiveness of contextual manipulations (i.e., instruction or intervention), wherein retention is typically operationalized as performance on multiple-choice tests or free responses to content questions (Hunt, 2003). Researchers can also examine how individual factors interact with contextual factors to identify for whom, and under what conditions, learning activities may be successful and thus contribute to desired outcomes of skill acquisition or knowledge retention.

Students' motivation also plays an integral role in the learning process. Students who engage more deeply with learning activities or persist through challenges show greater improvement in both short-term outcomes (e.g., retention; Alarcon & Edwards, 2013) and long-term outcomes (e.g., transfer; Cormier & Hagman, 2014; Haskell, 2000; Murayama et al., 2013). To assess motivation, researchers often collect self-report measures (Liu et al., 2012) to evaluate how students are internally and externally driven to complete learning tasks in varying contexts (Howard et al., 2021). Additionally, researchers have measured task engagement by tracking students' and teachers' behavioral indicators (e.g., time on task, participation, and communication patterns) and exploring how such behaviors relate to outcomes (e.g., Ocumpaugh et al., 2015). Lastly, measuring the transfer of learning involves the design of learning sequences that allow researchers to track how earlier experiences and contexts influence later successes or struggles (Huang et al., 2009), as well as the learning strategies that students use and their long-term effectiveness. Overall, the ultimate goals of much of learning sciences research are to understand how student factors and contextual factors influence different learning outcomes at different timescales.

19.2.2 Student Factors

Learning inherently depends on what the learner brings to the table. There are multiple aspects of individual learners that are correlated with educational outcomes and potentially moderate or mediate the impact of instructional strategies or interventions, educational contexts, and affordances from technologies (see Figure 19.1; 40 Preacher & Sterba, 2019). Investigations of student success have often considered conveniently accessible individual differences such as gender, race, and socioeconomic background (Wang, 2013). Students' academic history, including enrollment in Advanced Placement courses and performance indicators (i.e., GPA), can also be important for evaluating or predicting future success (Ma & Johnson, 2008; NASEM, 2017). Student success and persistence can also be predicted by malleable factors such as general cognitive abilities, domain knowledge, and literacy skills, along with motivational and social factors (e.g., engagement, perceived self-efficacy; Ackerman et al., 2013). Indeed, learners' prior knowledge is widely recognized as the single most important individual difference factor in education (Mayer, 2011; McCarthy & McNamara, 2021).

Although substantial research has examined the importance of individual differences (e.g., gender, motivation, intelligence, knowledge), less research has investigated how the effects of contextual factors (e.g., interventions) depend on those individual differences. Moreover, there are as many theories regarding individual differences as there are measures, and even the term "individual differences" evokes a wide range of constructs. Some theories focus on purportedly inherent abilities (e.g., working memory or general intelligence) and assume that students who perform well have more resources to process information (Alloway & Alloway, 2010; Cowan, 2014; Just & Carpenter, 1992). Other theorists focus on malleable skills such as reading skills and domain knowledge (Alexander et al., 1995; Perfetti, 2007) and assume that training such skills will improve students' learning outcomes. Yet another focus is on differences in motivation, based on the theoretical assumption that intrinsically motivated students are more likely to perform well and succeed (Linnenbrink & Pintrich, 2002; Pintrich, 2003; Schunk & Zimmerman, 2012). In the context of instruction, one assumption is that *key* individual differences must be controlled for statistically. Alternatively, it is assumed that the effects of an instructional strategy or intervention might *depend* on a few key individual differences.

In the learning sciences, the question remains – how can the *right* individual differences be identified and measured to more comprehensively account for student learning? How can we match appropriate interventions to learners' needs? One significant challenge in answering this question is that the field has advanced primarily by testing hypotheses in small-scale studies across narrow cross-sections of learners (Clarke & Dede, 2009; Dede, 2006; Kenny & Judd, 2019). Within these studies, limited sets and combinations of student and contextual variables are manipulated in tightly controlled laboratory studies (Makel & Plucker, 2014) because it is not possible to administer all of the measures that would cover the spectrum of meaningful individual differences. Thus, crucial individual differences (e.g., prior knowledge, literacy skills) that potentially influence the benefits of learning activities (McNamara, 2004, 2017; O'Reilly et al., 2004) are often not accounted for. A particularly high hurdle is obtaining a sufficiently large, representative, and diverse sample: a statistical interaction in a 2 x 2 factorial design requires approximately four times as many students as a simple main effect of the same magnitude (Fleiss, 1986). Without access to large pools of diverse students, it becomes impractical or impossible to answer the kinds of scientific questions that will drive personalized learning forward.

19.2.3 Contextual Factors

The impact of individual differences on learning outcomes is inherently influenced by the learning context. Contextual components of educational environments include what students learn (e.g., STEM domains, literacy skills), how students learn (e.g., instructional strategies), the instructional providers (e.g., instructors, pedagogical agents, intelligent tutoring systems), as well as affordances offered by educational technologies (e.g., precision education; Yang, 2021). For decades, learning science has investigated these components across multiple domains in an effort to reveal the best alignment between strategies, learners, and contexts (Dunlosky et al., 2013; Koedinger et al., 2013; Mayer, 2011). A number of robust effects have emerged from this work. The first of these is the generation effect (Bertsch et al., 2007). While the majority of generation effect studies have been restricted to episodic memory for familiar words (Gardiner, 1988; Slamecka & Katsaiti, 1987), unfamiliar words or phrases (Lutz et al., 2003), or answers to simple equations (McNamara & Healy, 1995a), generating has also been shown to enhance learning and skill acquisition (McNamara, 1995; McNamara & Healy, 1995b; Rittle-Johnson & Kmicikewycz, 2008).

The generation effect is comparable to the testing effect, wherein students attempt to retrieve content in a test-like format (e.g., cued recall or multiple choice), usually multiple times (Pyc & Rawson, 2009; Dunlosky & Rawson, 2013; Roediger & Butler, 2011). Compared to less active strategies (e.g., rereading text or notes), students benefit from practice testing even when retrieval is not successful (i.e., the answer is incorrect; Kornell et al., 2009). Relatedly, students also benefit from effortful retrieval when introducing space between study episodes (Bjork, 2014; cf. Soderstrom et al., 2015). During study sessions, students retrieve to-be-learned content which is activated in memory. Subsequent retrieval after some time lag (i.e., minutes, days, weeks) reactivates content in memory but allows for irrelevant details to fall away, thus improving retention of target content (Cepeda et al., 2009, Vlach & Sandhofer, 2012). The benefit to students' subsequent recall varies with the type of content and the amount of time (e.g., minutes, days, weeks, or months) between study sessions such that longer gaps between retrievals are more beneficial to long-term retention (Rohrer, 2015). While the effect size varies, this effect is robust across grade levels and domains (Cepeda et al., 2009; Hintzman, 1974).

While various manipulations such as generating and repeated testing are helpful in enhancing memory, they are less helpful when the student is unable to understand the content. Often, students are challenged by the complexity of the content and the difficulty of the text. Self-explanation (explaining text while reading) combined with reading comprehension strategies (e.g., paraphrasing, generating inferences) helps students to better comprehend challenging, unfamiliar content (McNamara, 2004, 2017). Producing self-explanations prompts students to make inferences between sentences in the text (i.e., bridging inferences) or between the text and prior knowledge (i.e., elaborative inferences).

Learning scientists generally agree that learning should not be passive. Learning contexts should include active, constructive activities (Chi & Wylie, 2014; Ebert-May et al., 1997; Mayer, 2009; Prince, 2004). Active learning has been operationalized differently across several disciplines and contexts (e.g., engineering, mathematics, medical education, science education, and engineering; Chamberland & Mamede, 2015; Crouch & Mazur, 2001; Freeman et al., 2014; Prince, 2004). Nonetheless, across studies, learning is generally enhanced to the degree that students engage in effortful, generative, constructive, participatory, and social learning activities (Fiorella & Mayer, 2016; Trafton & Trickett, 2001; Wittrock, 1989). The benefits of active learning have been evaluated in multiple contexts including problem-based or inquiry-based learning (Hung et al., 2008), team-based learning (Sisk, 2011), collaborative learning (Menekse et al., 2013), and peer tutoring (Roscoe, 2014; Roscoe & Chi, 2008). Active learning, depending on its definition, is consistently found to be a key element for student success in developing essential 21st-century skills such as collaboration and inquiry (Buitrago-Flórez et al., 2021; Christensen & Knezek, 2015).

The manner or mode in which instruction and feedback are delivered adds additional layers of complexity beyond the interactions between individual differences, content, and strategies. For example, the type (e.g., correct/incorrect or elaborative) and timing of feedback (e.g., immediate or delayed) have different effects based on the content type and the skill level of the learner (Fyfe & Rittle-Johnson, 2016; Koedinger et al., 2013; Kulik & Kulik, 1988). One key challenge is to identify for whom and under what conditions various instructional techniques are most effective for developing students' skills. Learning science researchers currently lack efficient tools for measuring the effectiveness of active learning at sufficiently large scales, which is needed to identify methods and conditions that maximize effectiveness. Moreover, learning sciences research is often conducted with internal constraints (e.g., student factors, teacher training, intervention fidelity) and external constraints (e.g., administrative support, cost, materials) imposed by limited funding, limited personnel, and a lack of infrastructure to support collaboration (Sabelli & Dede, 2013). Addressing these challenges requires valid and reliable measures of active learning to collect large-scale data from students' engagement and learning in online learning environments (Bryan et al., 2021).

19.2.4 Replication Crisis? (Or Maybe Context Matters)

Unfortunately, studies demonstrating the impact of instructional practice often fail to demonstrate effectiveness beyond their initial learner populations, contexts, and scales (Bird et al., 2019; Dobronyi et al., 2019; Kizilcec et al., 2020; Lortie-Forgues & Inglis, 2019; Oreopoulos & Petronijevic, 2019; Yeager et al., 2019). This *replication crisis* has hit a number of fields, with many experiencing the pains of unreplicable findings far more sorely than the learning sciences.

Many of the challenges faced in the learning sciences are consequences of being restricted to small-scale studies, often in environments where behaviors and actions are untrackable. For example, randomized controlled trials (RCTs) have been a particularly popular assessment of efficacy; yet, Makel and Plucker (2014) found a paucity of published replications in the learning sciences. Given the complexity of bringing learning sciences innovations to scale, it is not surprising that replication attempts fail. Albeit not an exhaustive list, some reasons for such failures have included lack of fidelity (Nelson et al., 2012), questionable intervention design (Rodgers, 2016), and the assumption that students in authentic educational contexts will demonstrate benefits similar to those found in the context of laboratory settings (Clarke & Dede, 2009; Walker, 2004).

Another explanation for the absence of replication and generalization stems from the overarching assumption that *individual differences matter* and that the impact of context depends on the learner. Inherent to learning sciences is that learning contexts are complex and organic and that individuals vary in what they bring to the table and what they need. While the small-scale studies approach makes sense, the multivariate and multidimensional nature of learning creates a combinatorial explosion that traditional experimental approaches just cannot accommodate. Thus, it has constrained our understanding of the complex interdependencies of students' individual differences and contextual factors that influence learning (Toh et al., 2016). Traditional studies do not yield sufficient data to inform the impact of various learning activities and interventions across diverse learner profiles and contexts (Bryan et al., 2021). This lack of data reflects a serious shortcoming of traditional small-scale studies in the learning sciences.

Challenges of replication and scale-up of learning sciences research may be partially mitigated by reducing reliance on traditional RCTs. RCTs are extremely valuable experimental approaches in many situations. Depending on the design and choice of control/comparison conditions, an RCT provides the cleanest means of detecting the impact of an intervention if you expect an intervention to work the same way across most contexts and most individuals. It is the scientist's stethoscope. However, RCTs are also costly and time consuming and can lack generalizability due to variations across samples and contexts (Lortie-Forgues & Inglis, 2019). Perhaps the largest shortcoming of RCTs in the context of learning sciences is the focus on manipulated factors without consideration of natural (important) variations in contexts and individuals. RCTs follow the assumption that all factors other than those that are under examination must be controlled. As illustrated in Figure 19.2, the scientist considers the classroom to provide something like a sterile petri dish, wherein they can observe behaviors in nicely controlled classrooms, as a function of some experimental manipulation. But, classrooms are not petri dishes. And, students are not cells to be cultured. The notion of *controlling* classroom environments, as if we were randomly administering medication, is, well, just a bit ludicrous at best.

Alternatively, resources may be better invested in developing infrastructure to support rapid, large-scale testing of interventions with diverse learners in a variety of authentic contexts (i.e., *Large-Scale Learning Sciences* or LSLS). Within industrial research (e.g., Google), rapid A/B (or A/B/n) experimentation is the process of simultaneously deploying variants of the same interface or design to different individuals and comparing which variant drives more activity or conversions. In doing so, A/B experimentation provides a wealth of data and is relatively cost effective compared to implementing a series of traditional small-scale studies. Harnessing the power of big data has strong

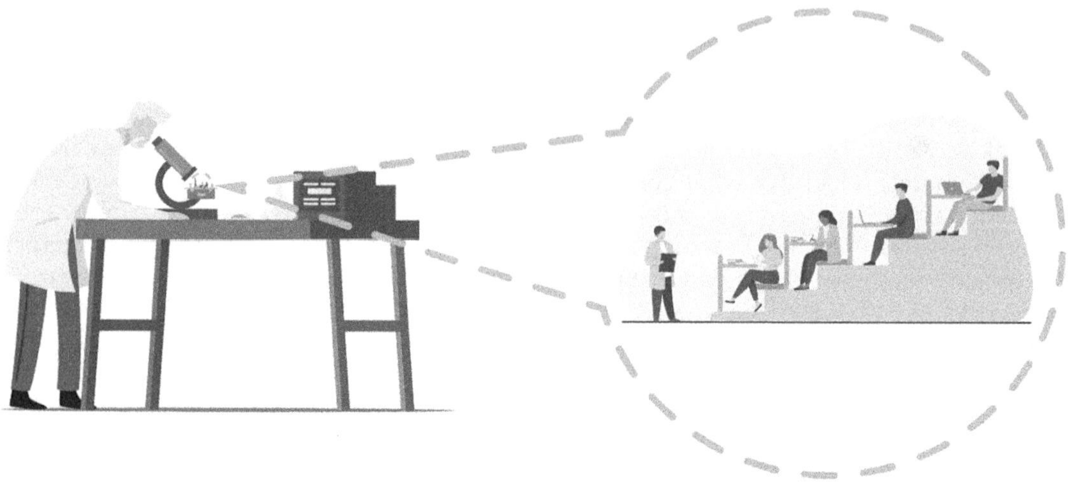

FIGURE 19.2
A scientist studies a *well-controlled* classroom. Following the medical analogy, classrooms can be well-controlled environments, like petri dishes, and students are identifiably pure, like cells to be cultured. However, classrooms must be organic and flexible, and students vary along a multitude of individual differences. This fundamental misconception regarding education reflects a potential shortcoming of relying so heavily on RCTs in the context of learning sciences (Illustration by Chris Kennedy).

promise for addressing the shortcomings of small-scale learning sciences research. Namely, big data methodologies afford multiple, rapid, iterative, and cost-effective replications. Rapid A/B experiments also show promise for evaluating the effectiveness of interventions or educational systems across diverse learners and contexts (Lortie-Forgues & Inglis, 2019). This approach is particularly useful when testing the impact of variations in educational technologies and AI algorithms.

The past four decades of learning sciences research have greatly advanced our understanding of crucial learning outcomes, learners, and learning contexts (Ben-Eliyahu & Bernacki, 2015; Lee & Shute, 2010). Nonetheless, we have also observed that traditional learning sciences approaches cannot account for the breadth of student factors that influence learning. *We do not fully understand learning in context*, principally because it is highly complex – perhaps too complex for existing approaches. Variations in context combined with the inherent complexities of learning and individual differences impose challenges in replicating, confirming, and applying the insights of these studies across different institutional contexts and broader demographic cross-sections of learners (Dede, 2006). Any combination of the variables discussed thus far can manifest within natural learning contexts, but they are difficult to detect, track, or measure.

19.3 AI and Its Affordances for the Learning Sciences

Artificial Intelligence techniques are often leveraged in the context of educational technologies to provide information about individual factors (e.g., students' knowledge and skills), assess students' input as they engage with educational technologies, iteratively adapt contexts in attempts to modify students' performance, and in turn inform learning sciences (Corbett et al., 1997; Mousavinasab et al., 2021). Applications of machine learning in educational contexts often leverage educational datasets and use machine learning algorithms to build models that describe or predict student performance and other constructs of interest (Urbina Nájera & Calleja Mora, 2017). Advances in machine learning have enriched the design of computer-based learning environments because appropriate learning strategies and pedagogies provided to the learner can be derived specifically from attributes such as behaviors and performance within learning environments (Gamboa & Fred, 2002; Schiaffino et al., 2008). In turn, automated (intelligent) systems leverage AI to adjust task difficulty and learning paths and provide feedback or support based on the information that the systems acquire about students (Peirce et al., 2008; Van Eck, 2007).

One type of widely used educational technology that adapts the learning context based on student information is intelligent tutoring systems (ITSs). ITSs are computer programs that provide customized instruction, and immediate feedback to learners (Psotka et al., 1988; VanLehn, 2006). Key features of ITSs include AI-driven feedback, real-time cognitive diagnosis, and adaptive remediation (Shute & Psotka, 1996). When students interact with the ITS, the user interface presents learning materials to students through various media (e.g., pedagogical agents) and uses AI to interpret student input (i.e., speech, keystrokes, mouse clicks). Specifically, the student input data is processed and computationally analyzed, and then the analyzed input is used to update the student model and adjust the learning context to better suit the student. The "intelligent" aspects of systems like these are driven by AI to collect data about students' performance, conduct analyses, make instructional decisions to meet students' needs, and communicate with students (i.e., provide instruction or feedback; Shute & Psotka, 1996).

Opportunities to leverage AI in education have increased due to changes in the ways that instruction is delivered over the past two decades. We have seen dramatic shifts from a solely classroom-based instructional models to approaches that include some form of technology enhancement (i.e., hybrid) or to completely online instruction (e.g., synchronous, self-paced asynchronous; Clark & Mayer, 2011; Hamdan et al., 2013). This shift to technology-enhanced instruction provides new learning contexts and rich, personalized learning experiences (e.g., ITSs, virtual science labs, simulations) that were not previously available to students (Linn et al., 2004; Peffer et al., 2015; Slotta, 2010). Data science, data engineering, and machine learning provide powerful ways to collect and computationally analyze massive datasets and to untangle the complex interdependencies among large numbers of variables. The advent and proliferation of online learning in combination with advances in AI open up a wealth of opportunities, many of which are discussed in this volume of chapters. Here, we list five areas where we believe that AI (and big data) has strong potential to advance the learning sciences.

19.3.1 Deep Student Models

Student modeling (VanLehn, 1988) is an important research area as it provides key information on individual student factors that drive personalization. While a wide array of student models exists, we highlight two types to illustrate their importance. First, static item response theory models (van der Linden & Hambleton, 2013) analyze student responses to questions to estimate their knowledge levels on target knowledge components/skills/concepts. These models assume that student knowledge is static and are best suited for assessment purposes such as computerized, adaptive testing (Wainer et al., 2000). Second, knowledge tracing models (Corbett & Anderson, 1994) trace the evolution of student knowledge over time as they learn and are most effective for data collected over a longer time period.

The classic versions of these models are highly interpretable but cannot fully leverage large-scale student data made available recently, often consisting of tens of millions of student responses (Choi et al., 2020; Wang et al., 2021). Therefore, recent approaches have focused on using deep neural networks (Goodfellow et al., 2016) to improve the modeling capacity of student models (Piech et al., 2015; Wang et al., 2020). These methods achieve state-of-the-art performance in predicting next answer's correctness but lose some interpretability, presenting challenges when deploying them in practice.

Therefore, future research on deep student models may benefit from considering the following two directions: first, integrating cognitive theory into large-scale, data-driven models to improve interpretability (Ghosh et al., 2020). Second, it is important to systematically study algorithmic biases and, in turn, devise more effective ways of detecting and mitigating biases in data (Gardner et al., 2019; Kizilcec & Lee, 2020). A paramount and growing concern regards the need to impose constraints during model training that promote fairness across different student groups and protect vulnerable participants (Yao & Huang, 2017).

19.3.2 Causal Learning Outcome Models

Most studies on student modeling focus on predictive models that are *correlational* in nature, based on observed student and contextual factors. However, in order to make discoveries regarding instructional strategies and content design that promote learning, we need *causal* models that directly relate student learning outcomes to changes in a contextual factor (Spirtes et al., 2000). Such models in turn will more effectively inform rigorous A/B experiments (de Carvalho et al., 2018). However, existing studies on causal modeling in education are limited by the scale as well as the structure and affordances of the datasets currently available (Sales et al., 2018).

As more and more platforms experiment with different instructional strategies and educational technology affordances, there is a possibility to obtain large-scale, quasi-experimental data that provide us with golden opportunities to develop these causal models (Gopalan et al., 2020). For example, one can use deep latent variable modeling approaches to learn both a representation of potential confounders and the strengths of causal effects between measured variables (Louizos et al., 2017). In the early stages of a study, when students experience a potential experimental condition, this type of analysis of quasi-experimental data will afford estimates of causal effect strengths between experimental variables and various individual factors. This can be achieved by including in the model both potential confounding variables and the estimated impact of the hypothesized causal factor. This requires that the wealth of existing data is mined to identify potential causal relations among various factors such that these factors can then be examined via subsequent, confirmatory rapid A/B experimentation.

19.3.3 Natural Language Processing

Natural language processing (NLP) is the combination of computational linguistics and machine learning/AI to predict various aspects of the meaning or quality of the language (McNamara et al., 2017). Most learning environments include language, and for the most part, much of the rich interactions between students and among students goes unmined. Nonetheless, NLP has been used in various contexts to support student learning and assessment (Litman, 2016) and automatically score speech (Wang et al., 2018). The language that students use while learning is often key to understanding learning processes and predicting learning outcomes.

The most common use of NLP in the education domain is its use in automated scoring and assessment of writing quality (Burstein et al., 2017; Crossley et al., 2015). Automated essay scoring (AES) and automated writing evaluation (AWE) automatically assess students' essays and provide formative and summative feedback to writers during or after essay drafts (see Figure 19.3; Burstein et al., 2003; Foltz et al., 2013; Warschauer & Ware, 2006).

To develop these algorithms, machine learning is used to predict the quality of a corpus of essays that has been graded by experts based on linguistic and semantic features automatically extracted from the essays (McNamara et al., 2017). AES algorithms provide estimates of the summative scores related to the quality of essays (e.g., mechanics, content, cohesion, evidence). AWE algorithms, in turn, provide formative feedback that scaffolds the writer in improving that essay, or future essays, with the objective of enhancing writing skills (Proske et al., 2014; Roscoe & McNamara, 2013; Roscoe et al., 2013; Wilson & Roscoe, 2020; Wilson et al., 2017).

NLP has also been leveraged to assess various aspects of learning materials, such as the difficulty of the readings and textbooks (McNamara et al., 2014). The most common approaches to assessing text difficulty (e.g., Flesch-Kincaid Grade Level) are based on features of the words and individual sentences, such as the length of word and sentence, or signals for word difficulty (e.g., word frequency) and sentence difficulty (e.g., complex syntax). Coh-Metrix was designed to go beyond word- and sentence-level features to consider text cohesion (McNamara et al., 2014). Cohesion refers to the overlap in words, concepts, and ideas that impact the coherence and flow of the document. Cohesion is an important component of text difficulty because readers who are less

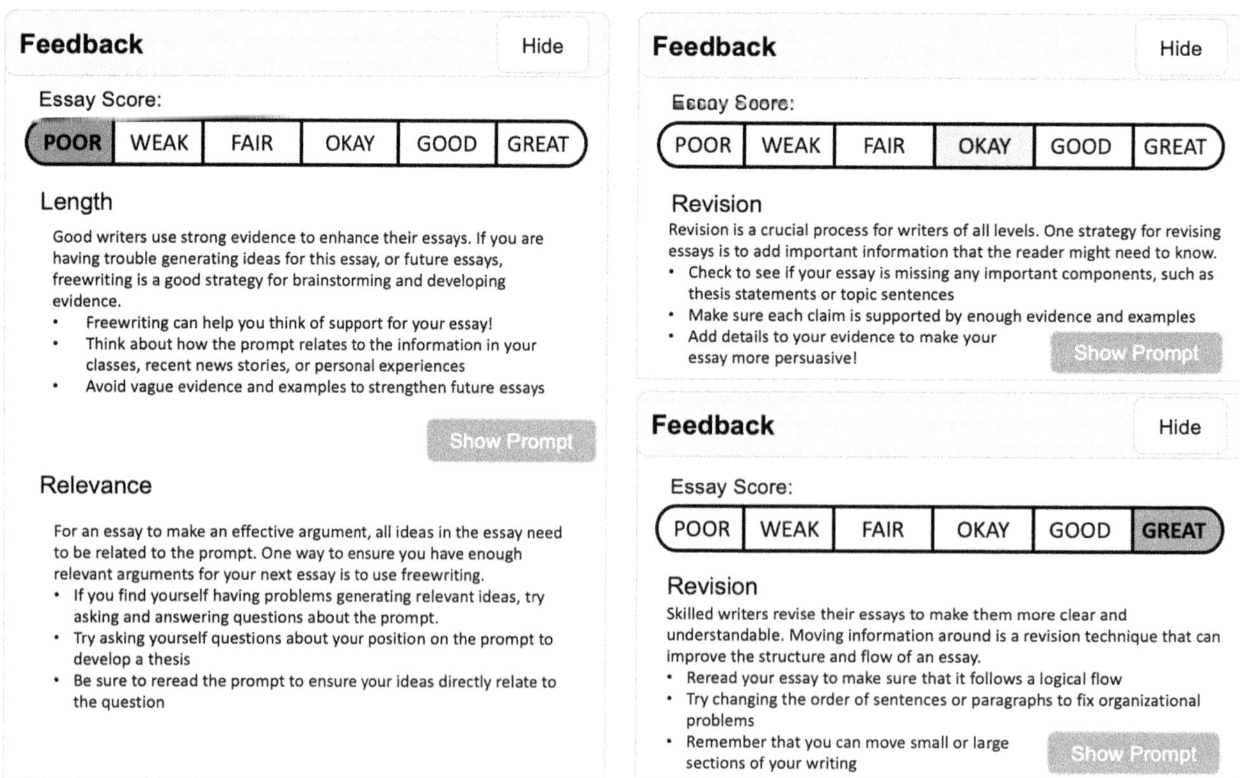

FIGURE 19.3
The Writing Pal intelligent tutoring system. The Writing Pal automatically assesses students' essays using NLP-based algorithms and provides feedback to guide students' revisions as well as improve writing skills.

knowledgeable about the topic have greater difficulties in filling in the cohesion gaps in text and, consequently, are less likely to understand or learn from the reading materials (McNamara, 2017). AI comes into play here because it has been essential in developing the algorithms that combine the multitude of features within texts and, in turn, predict whether they match the needs of the reader (Graesser & McNamara, 2011).

NLP can also be used examine specific aspects and attributes of the learner such as the relations of language sophistication, engagement, and collaboration on course performance (Crossley et al., 2015, 2016a, 2016b, 2017a, 2017b; Dascalu et al., 2018, 2020; Liu et al., 2016; San Pedro et al., 2015). Various NLP tools and methods can be used to extract dimensions and features of language within students' constructed responses (e.g., self-explanations, summaries, essays) and in turn infer various aspects of students' performance and individual differences (Allen & McNamara, 2015; McNamara et al., 2017). While approaches that leverage machine learning to combine multiple sources of information within educational contexts are promising, the bulk of this work has been limited to a narrow set of learning contexts. There is a need to develop infrastructure to help us examine learning beyond a few populations and learning contexts at a time.

19.3.4 Sensor-Free Student Factor Measures

To develop AI-based educational technologies, developers and educators need to make decisions about what information to collect about students in order to meet individual educational goals. AI offers the capability to measure student knowledge and skills covertly and unobtrusively using sensor-free measures. Many learning platforms have data on some student individual factors, such as fixed factors (e.g., grade level) and observables (e.g., past system interactions). However, we often do not have scalable measures of individual difference factors, such as motivation and social factors (e.g., engagement and affect) and ability factors (e.g., reading skill, prior knowledge). Traditional measures of these factors can be broadly categorized into two types. The first type resorts to student self-reports (Fredricks & McColskey, 2012), in-class observations (Ocumpaugh et al., 2015), or standardized assessments for individual factors such as reading skill. All of these require additional human effort and, thus, are not immediately scalable. The second type uses physiological sensors that require external devices such as sensors (Suárez-Pellicioni et al., 2014) or cameras (Dragon et al., 2008), which can be expensive and invasive and may pose privacy concerns. In contrast, "sensor-free" detectors (e.g., embedded or stealth assessments) of these student characteristics gathered from activity and behavior logs have the potential to scale up measurement of these factors. After a relatively small number of self-reports, expert observations, or standardized assessments are administered to students to collect "ground truth" values, we can use AI and machine learning to predict these constructs using students' activity and behavior logs in online learning environments.

Stealth assessments are sensor-free assessments that are embedded in digital games (Shute, 2011; Shute & Ventura, 2013). In stealth assessment, student evaluations are embedded within gaming tasks and activities such that learners are unaware of being evaluated as they are not interrupted with overt tests or quizzes. Games inherently include rich sequences of actions; students' actions and performance during the gameplay are used to predict various aspects of students' knowledge and skills, as well as attributes like persistence and creativity (Shute et al., 2013, 2016; Shute & Rahimi, 2021; Ventura & Shute, 2013).

NLP can also be leveraged to provide sensor-free assessments of knowledge and skills (Allen & McNamara, 2015; Yang et al., 2019). For example, reading skill as measured by Gates-MacGinitie (MacGinitie & MacGinitie, 2006) can be predicted using linguistic indices extracted from self-explanations (e.g., Allen et al., 2015; McCarthy et al., 2020) and essays (e.g., Allen et al., 2016). Skilled readers are more likely to generate self-explanations containing greater semantic overlap and more explicit connections between their self-explanations (Allen et al., 2015). Additionally, readers who are prompted to self-explain produce more diverse language and global connectives in their responses, providing evidence that the cohesion of these constructed responses can serve as a proxy for coherence-building processes during comprehension (Allen et al., 2016; Flynn et al., 2022). Fang et al. (2021) recently examined the number of student-generated self-explanations needed to predict reading comprehension skills during iSTART training (McCarthy et al., 2020). They found that the power of the linguistic features of self-explanations to predict reading comprehension skill increased as more self-explanations were included in the model, but only nine self-explanations were needed to explain 21% of the variance in reading comprehension skill. Ultimately, the objective is to combine *multiple* stealth measures of literacy skills (each separately accounting for different sources of variance) to provide a comprehensive profile of literacy, without the burden of administering standardized tests.

A principal benefit of AI-based assessments of individual factors is the ability to provide more precise individualized learning experiences for students. For

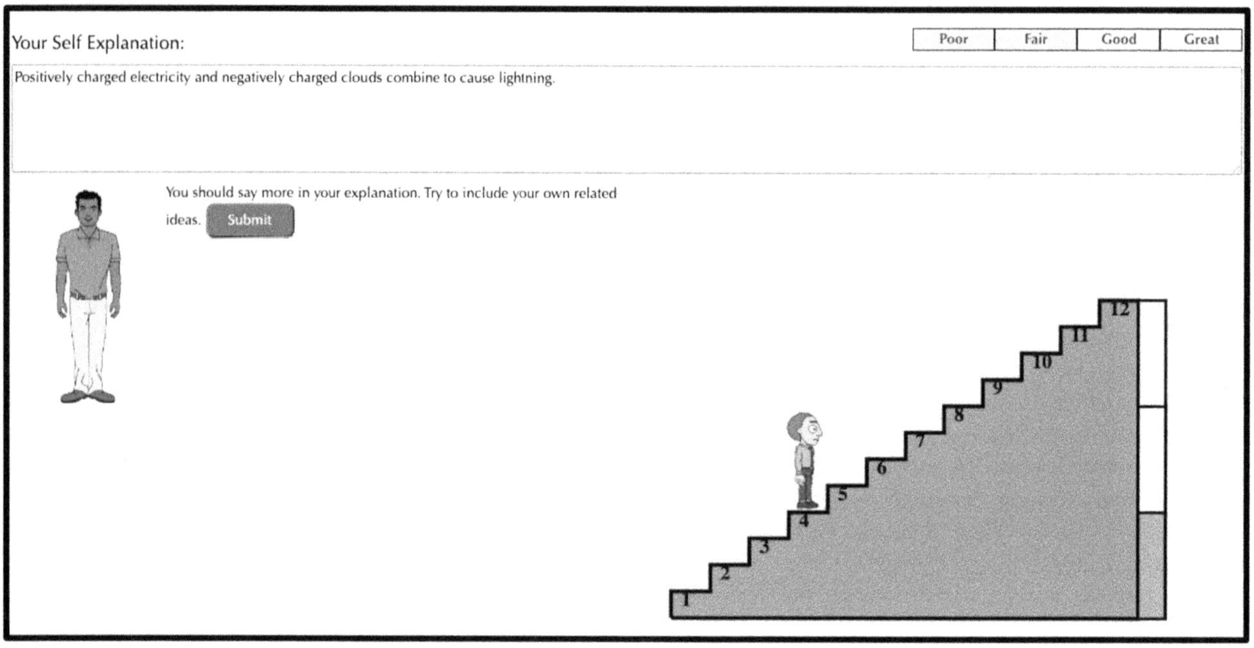

FIGURE 19.4
StairStepper game in iSTART intelligent tutoring system. The StairStepper game in the iSTART intelligent tutoring system is an example of an AI-based stealth assessment of students' literacy skill level. NLP-based algorithms assess students' self-explanations and either increase or decrease the difficulty of the next text based on the quality of their self-explanation.

example, AI-based stealth assessment of students' literacy skills in the context of an intelligent tutoring system can increase the precision of the system's adaptive features. Consequently, the system can provide automated literacy instructions and practice opportunities that are better suited to each student's skill profile (see Figure 19.4).

Additionally, AI-driven system analytics can provide early detection for students who are struggling before they are severely behind (Lu et al., 2018). In turn, teachers can use information from educational technologies to inform instructions they provide in the classroom, potentially promoting more efficient use of instructional time and teachers' unique expertise. The wealth of data collected in AI-driven educational technologies can, and should, be leveraged to generate more informative learning analytics for teachers, thereby providing them with readily available and more comprehensive windows into their students' current skill levels and progress.

19.3.5 Instructional Policy Learning

Instructional policies can contain typical low-level instructional planning (e.g., which learning a student should study at a point in time given their past learning record; Lan & Baraniuk, 2016) and high-level instructional strategies such as scaffolded support (Puntambekar & Kolodner, 2005), using spacing in retrieval practice (Roediger & Pyc, 2012), generative learning strategies (Mayer & Fiorella, 2015), monitoring and coping with student anxiety (Carver et al., 1989; Huang & Mayer, 2016), and utility-value interventions (Canning & Harackiewicz, 2015). Learning instructional policies from student data will reveal what particular instructional or curricular strategy should be used, in what sequence, and to what depth. Thus, constructing personalized instructional policies based on each student's individual student factors by changing their contextual factors can maximize students' learning outcomes.

Learning an instructional policy can typically be viewed as a decision-making problem; reinforcement learning (RL) has been widely adopted for such problems in many applications from robot path planning to learning to play games where an agent can learn by interacting with the environment (Sutton & Barto, 2018). However, its application in real-world educational settings faces practical challenges, especially the problem of insufficient training data (see Doroudi et al., 2019, for an overview). Researchers have developed several ways to work around this issue and have found some success. First, one can reduce the complexity of the problem by restricting the set of instructional actions from which to select, usually down to two or three distinct actions (Zhou et al., 2019). Second, one can use student models learned from real student data to synthetically generate unlimited amounts of training data for RL algorithms (Choffin et al., 2020). Additionally, one can use other less data-hungry decision-making algorithms such as bandits (Lan & Baraniuk, 2016) and cognitive model-based algorithms in certain domains involving mainly memorization tasks (Upadhyay et al., 2021).

19.4 Promoting Equity

To conduct valid research on understanding and improving learning requires scientists to confront the (in)equitable and (un)ethical realities of educational contexts. Participants in learning environments pursue diverse and personal goals, have access to different capabilities and resources, and represent a variety of backgrounds and identities. Consequently, providing equitable and effective learning contexts is a persistent challenge. Successful learning requires culturally responsive approaches that recognize student agency and draw upon assets in a local context. Within the realm of AI and education, challenges of responsible computing are front and center as learning applications need to take extra care to be unbiased, transparent, and equitable. Indeed, most learning scientists are concerned with potential inequalities and inequities that data-driven AI methods may introduce. Many well-meaning interventions specifically address students who are most likely to struggle or be left behind. Yet, learning technologies are often less effective for historically underrepresented learners (Mayfield et al., 2019), and algorithms to predict and guide performance are often biased – perpetuating and even magnifying inequities (Baker & Hawn, 2021; Kizilcec & Lee, 2020; Perry & Lee, 2019). Capturing and centering the "average" learning experience neglects the individuality of learners, which can be deleterious when the average ignores important variance.

Studies investigating aptitude-by-treatment interactions (and individual differences research, more generally) suggest that targeting an assumed average often fails to meet the needs of any one individual (e.g., Connor & Morrison, 2016). Small population (<1,000 students) research has limited ability to adjust for variation in individual factors. Hence, results reported based on population means can limit applicability to historically underrepresented populations. Indeed, such limited research may be a causal factor underlying such historical underrepresentation (i.e., if peoples' identities and needs cannot be examined within the data, then the data excludes those people). If important individual or contextual factors vary systematically with individual fixed factors (e.g., gender, ethnicity, learning English as a second language, etc.), then substantial explanatory elements for a particular student population are likely to go unnoticed under conventional research paradigms.

The bulk of the research focusing on educational inequalities or inequities has examined one or two social identities (e.g., race or gender) within narrowly defined contexts. For example, various studies have explored questions such as why and how gender gaps have evolved in STEM (Cheryan et al., 2017; Cimpian et al., 2020; Kanny et al., 2014; Soylu Yalcinkaya & Adams, 2020), why so few women major in Physics (Lewis et al., 2016; Walton et al., 2015), why Black and Hispanic students appear to achieve lower grades in middle school and beyond (Cohen et al., 2006), why STEM persistence rates differ by race/ethnicity (Riegle-Crumb et al., 2019), and the extent of such differences in online learning (Joosten & Cusatis, 2020; Mead et al., 2020; Wladis et al., 2016). These studies have yielded promising explanations and thus represent reasonable approaches when the study data are limited in size and scope.

By contrast, the use of AI and big data methods afford researchers the capacity to conduct research that examines multiple social identities simultaneously – including intersectional identities (e.g., Else-Quest & Hyde, 2016a, 2016b) – and explore the consequences to student outcomes and experiences longitudinally and comprehensively from their first day in college to graduation. We anticipate research that is able to combine multiple data sets and leverage AI to derive inferences across multiple populations and contexts will contribute to theories of human development, social support, self-directed and social learning, and computer-based learning in context, to name but a few.

To meet the needs of a broader array of students, researchers must conduct research with more diverse and inclusive samples of students and gather larger data sets that enable studying moderated and mediated relations across interventions and individual differences. In addition to diverse data sets, it is essential to collect *context-specific* data sets such that systems can be more responsive to a greater range of users who may have different needs and experiences (e.g., Bryan et al., 2021; Dolan, 2016). We must move beyond the petri dish analogy.

Issues of diversity, equity, and inclusion can no longer be an afterthought; instead, they must be a critical consideration at every step of exploration, design, development, and implementation. Further, equity and inclusion is not limited to the technologies themselves. There must also be greater diversity across those involved in the design, development, and evaluation of these systems. AI and big data have the potential to help learning scientists more thoroughly understand learning and improve outcomes for *all* learners when diversity, equity, and inclusion are consistently included throughout all phases of learning sciences research. Working toward making educational data readily available to more researchers and providing instruction and training on how to leverage and interpret those data is vital to equity and diversity, and to developing a more comprehensive understanding of learning.

19.5 Conclusion

In the context of education, data science methods have provided important insights, particularly through the emergence of educational data mining (Baker & Siemens, 2014; Romero & Ventura, 2010) and learning analytics (Lang et al., 2017) methods. Data science and data engineering provide powerful new ways to collect, curate, and computationally analyze massive datasets, and to disentangle the complex interdependencies among large numbers of variables. Modern educational technologies also facilitate the creation of engaging activities that may be further enhanced by AI and data science to adapt to students' specific needs to more effectively support learning, a practice that is not possible with traditional approaches to education (Nagar et al., 2019; Rau et al., 2017). However, as exciting as this possibility appears, adapting instruction to an individual in any sort of meaningful way is difficult because the existing science is not sufficiently advanced to provide guidance in all situations.

AI has a great deal to offer to the learning sciences. However, without deep domain knowledge and relevant disciplinary expertise, it is often difficult to explain and interpret the relationships revealed by data science. Indeed, data science is typically agnostic about causal interpretations, which depend on theoretical frameworks outside of data science. AI (and data science) can reveal nuanced patterns of student retention, persistence, and performance related to demographic variables, but expertise in learning theory and psychological sciences is needed to suggest mechanisms and explanations for these patterns. Merging these domains and the knowledge and techniques therein is crucial to understanding learning.

Acknowledgments

The research reported here was partially supported by the Institute of Education Sciences, U.S. Department of Education, through Digital Learning Network Grants R305N210041 to Arizona State University and R305N210064 to Rice University. Funding was also partially provided by IES grants R305A190050 and R305A190063 and Office of Naval Research, through Grants N00014-17-1-2300 and N00014-20-1-2623 to Arizona State University. The opinions expressed are those of the authors and do not represent views of the U.S. Department of Education or the Office of Naval Research.

References

Ackerman, P. L., Kanfer, R., & Beier, M. E. (2013). Trait complex, cognitive ability, and domain knowledge predictors of baccalaureate success, STEM persistence, and gender differences. *Journal of Educational Psychology, 105*(3), 911–927.

Alarcon, G. M., & Edwards, J. M. (2013). Ability and motivation: Assessing individual factors that contribute to university retention. *Journal of Educational Psychology, 105*(1), 129–137.

Alexander, P. A., Kulikowich, J. M., & Jetton, T. L. (1995). Interrelationship of knowledge, interest, and recall: Assessing a model of domain learning. *Journal of Educational Psychology, 87*(4), 559–575.

Allen, L. K., Dascalu, M., McNamara, D. S., Crossley, S., & Trausan-Matu, S. (2016). Modeling individual differences among writers using ReaderBench. In *Proceedings of the 8th international conference on education and new learning technologies (EDULearn16)* (pp. 5269–5279). IATED.

Allen, L. K., & McNamara, D. S. (2015). Promoting self-regulated learning in an intelligent tutoring system for writing. In A. Mitrovic, F. Verdejo, C. Conati, & N. Heffernan (Eds.), *Doctoral consortium within the proceedings of the 17th international conference on artificial intelligence in education (AIED 2015)* (pp. 827–830). Springer.

Allen, L. K., Snow, E. L., & McNamara, D. S. (2015). Are you reading my mind? Modeling students' reading comprehension skills with natural language processing techniques. In J. Baron, G. Lynch, N. Maziarz, P. Blikstein, A. Merceron, & G. Siemens (Eds.), *Proceedings of the 5th international learning analytics & knowledge conference (LAK'15)* (pp. 246–254). ACM.

Alloway, T. P., & Alloway, R. G. (2010). Investigating the predictive roles of working memory and IQ in academic attainment. *Journal of Experimental Child Psychology, 106*(1), 20–29.

Anderson, T., & Dron, J. (2011). Three generations of distance education pedagogy. *International Review of Research in Open and Distributed Learning, 12*(3), 80–97.

Baker, R. S., & Hawn, A. (2021). Algorithmic bias in education. edarxiv.org

Baker, R. S., & Siemens, G. (2014). Educational data mining and learning analytics. In K. Sawyer (Ed.), *The Cambridge handbook of the learning sciences* (2nd ed., pp. 253–274). Cambridge University Press.

Ben-Eliyahu, A., & Bernacki, M. L. (2015). Addressing complexities in self-regulated learning: A focus on contextual factors, contingencies, and dynamic relations. *Metacognition and Learning, 10*(1), 1–13.

Bertsch, S., Pesta, B. J., Wiscott, R., & McDaniel, M. A. (2007). The generation effect: A meta-analytic review. *Memory and Cognition, 35*(2), 201–210.

Bird, K. A., Castleman, B. L., Denning, J. T., Goodman, J., Lamberton, C., & Rosinger, K. O. (2019). *Nudging at scale: Experimental evidence from FAFSA completion campaigns*. NBER Working Paper No. 26158. National Bureau of Economic Research.

Bjork, R. A. (2014). Forgetting as a friend of learning. In D. S. Lindsay, C. M. Kelley, A. P. Yonelinas, & H. L. Roediger (Eds.), *Remembering: Attributions, processes, and control in human memory-papers in honour of Larry L. Jacoby* (pp. 15–28). Psychology Press.

Bryan, C. J., Tipton, E., & Yeager, D. S. (2021). Behavioural science is unlikely to change the world without a heterogeneity revolution. *Nature Human Behaviour, 5*(8), 980–989.

Buitrago-Flórez, F., Danies, G., Restrepo, S., & Hernández, C. (2021). Fostering 21st century competences through computational thinking and active learning: A mixed method study. *International Journal of Instruction, 14*(3), 737–754.

Burstein, J., Chodorow, M., & Leacock, C. (2003). Criterion online essay evaluation: An application for automated evaluation of student essays. In *Proceedings of the 15th annual conference on innovative applications of artificial intelligence* (pp. 3–10).

Burstein, J., McCaffrey, D., Beigman Klebanov, B., & Ling, G. (2017). Exploring relationships between writing and broader outcomes with automated writing evaluation. In J. Tetreault, J. Burstein, C. Leacock, & H. Yannakoudakis (Eds.), *Proceedings of the 12th workshop on innovative use of NLP for building educational applications (BEA)*. Association for Computational Linguistics.

Canning, E. A., & Harackiewicz, J. M. (2015). Teach it, don't preach it: The differential effects of directly-communicated and self-generated utility–value information. *Motivation Science, 1*(1), 47.

Cantor, P., Osher, D., Berg, J., Steyer, L., & Rose, T. (2019). Malleability, plasticity, and individuality: How children learn and develop in context. *Applied Developmental Science, 23*(4), 307–337.

Carver, C. S., Scheier, M. F., & Weintraub, J. K. (1989). Assessing coping strategies: A theoretically based approach. *Journal of Personality and Social Psychology, 56*(2), 267.

Cepeda, N. J., Coburn, N., Rohrer, D., Wixted, J. T., Mozer, M. C., & Pashler, H. (2009). Optimizing distributed practice: Theoretical analysis and practical implications. *Experimental Psychology, 56*(4), 236–246.

Chamberland, M., & Mamede, S. (2015). Self-explanation, an instructional strategy to foster clinical reasoning in medical students. *Health Professions Education, 1*(1), 24–33.

Cheryan, S., Ziegler, S. A., Montoya, A. K., & Jiang, L. (2017). Why are some STEM Fields more gender balanced than others? *Psychological Bulletin, 143*(1), 1–35.

Chi, M. T., & Wylie, R. (2014). The ICAP framework: Linking cognitive engagement to active learning outcomes. *Educational Psychologist, 49*(4), 219–243.

Choffin, B., Popineau, F., & Bourda, Y. (2020). Modelling student learning and forgetting for optimally scheduling skill review. *ERCIM News, 120*, 12–13.

Choi, Y., Lee, Y., Shin, D., Cho, J., Park, S., Lee, S., Baek, J., Bae, C., Kim, B., & Heo, J. (2020). Ednet: A large-scale hierarchical dataset in education. In *International conference on artificial intelligence in education* (pp. 69–73). Springer.

Christensen, R., & Knezek, G. (2015). Active learning approaches to integrating technology into a middle school science curriculum based on 21st century skills. In X. Ge, D. Ifenthaler, & J. Spector (Eds.), *Emerging technologies for STEAM education* (pp. 17–37). Springer.

Cimpian, J. R., Kim, T. H., & McDermott, Z. T. (2020). Understanding persistent gender gaps in STEM. *Science, 368*(6497), 1317–1319.

Clark, R. C., & Mayer, R. E. (2011). *E-learning and the science of instruction proven guidelines for consumers and designers of multimedia learning* (3rd ed.). Pfeiffer.

Clarke, J., & Dede, C. (2009). Robust designs for scalability. In L. Moller, J. B. Huett, & D. M. Harvey (Eds.), *Learning and instructional technologies for the 21st century*. Springer.

Cohen, G. L., Garcia, J., Apfel, N., & Master, A. (2006). Reducing the racial achievement gap: A social-psychological intervention. *Science, 313*(5791), 1307–1310.

Coiro, J. (2021). Toward a multifaceted heuristic of digital reading to inform assessment, research, practice, and policy. *Reading Research Quarterly, 56*(1), 9–31.

Connor, C. M., & Morrison, F. J. (2016). Individualizing student instruction in reading: Implications for policy and practice. *Policy Insights from the Behavioral and Brain Sciences, 3*(1), 54–61.

Corbett, A. T., & Anderson, J. R. (1994). Knowledge tracing: Modeling the acquisition of procedural knowledge. *User Modeling and User-Adapted Interaction, 4*(4), 253–278.

Corbett, A. T., Koedinger, K. R., & Anderson, J. R. (1997). Intelligent tutoring systems. In M. Helander, T. K. Landauer, & P. Prabhu (Eds.), *Handbook of human-computer interaction* (pp. 849–874). North-Holland.

Cormier, S. M., & Hagman, J. D. (Eds.). (2014). *Transfer of learning: Contemporary research and applications*. Academic Press.

Cowan, N. (2014). Working memory underpins cognitive development, learning, and education. *Educational Psychology Review, 26*(2), 197–223.

Crossley, S., Barnes, T., Lynch, C., & McNamara, D. S. (2017a). Linking language to math success in an online course. In X. Hu, T. Barnes, A. Hershkovitz, & L. Paquette (Eds.), *Proceedings of the 10th international conference on educational data mining (EDM)* (pp. 180–185). International Educational Data Mining Society.

Crossley, S. A., Dascalu, M., Baker, M., McNamara, D. S., & Trausan-Matu, S. (2017b). Predicting success in massive open online courses (MOOC) using cohesion network analysis. In B. K. Smith, M. Borge, K. Y. Lim, & E. Mercier (Eds.), *Proceedings of the 12th international conference on computer-supported collaborative learning (CSCL 2017)* (pp. 103–110). ISLS.

Crossley, S. A., Kyle, K., Davenport, J., & McNamara, D. S. (2016a). Automatic assessment of constructed response data in a chemistry tutor. In T. Barnes, M. Chi, & M. Feng (Eds.), *Proceedings of the 9th international conference on educational data mining (EDM 2016)* (pp. 336–340). International Educational Data Mining Society.

Crossley, S. A., Paquette, L., Dascalu, M., McNamara, D. S., & Baker, R. (2016b). Combining click-stream data with NLP tools to better understand MOOC completion. In D. Gašević, G. Lynch, S. Dawson, H. Drachsler, & C. P. Rosé (Eds.), *Proceedings of the 6th international learning analytics & knowledge conference (LAK'16)* (pp. 6–14). ACM.

Crossley, S., McNamara, D. S., Baker, R., Wang, Y., Paquette, L., Barnes, T., & Bergner, Y. (2015). Language to completion: Success in an educational data mining massive open online class. In O. C. Santos, J. G. Boticario, C. Romero, M. Pechenizkiy, A. Merceron, P. Mitros, J. M. Luna, C. Mihaescu, P. Moreno, A. Hershkovitz, S. Ventura, & M. Desmarais (Eds.), *Proceedings of the 8th international conference on educational data mining (EDM 2015)* (pp. 388–391). International Educational Data Mining Society.

Crouch, C. H., & Mazur, E. (2001). Peer instruction: Ten years of experience and results. *American Journal of Physics, 69*(9), 970–977.

Dascalu, M. D., Dascalu, M., Ruseti, S., Carabas, M., Trausan-Matu, S., & McNamara, D. S. (2020). Cohesion network analysis: Predicting course grades and generating sociograms for a Romanian Moodle course. *Proceedings of the 16th international conference on intelligent tutoring systems* (pp. 174–183). Springer.

Dascalu, M., Sirbu, M. D., Gutu-Robu, G., Ruseti, S., Crossley, S. A., & Trausan-Matu, S. (2018). Cohesion-centered analysis of sociograms for online communities and courses using ReaderBench. In *European conference on technology enhanced learning* (pp. 622–626). Springer.

de Carvalho, W. F., Couto, B. R. G. M., Ladeira, A. P., Gomes, O. V., & Zarate, L. E. (2018). Applying causal inference in educational data mining: A pilot study. In *Proceedings of the 10th international conference on computer supported education* (Vol. 1, pp. 454–460).

Dede, C. (2006). Scaling up: Evolving innovations beyond ideal settings to challenging contexts of practice. In R. K. Sawyer (Ed.), *Cambridge handbook of the learning sciences* (pp. 1–37). Cambridge University Press.

Dewey, J. (1934). Individual psychology and education. *Philosopher, 12*(1), 1–6.

Dobronyi, C. R., Oreopoulos, P., & Petronijevic, U. (2019). Goal setting, academic reminders, and college success: A large-scale field experiment. *Journal of Research on Educational Effectiveness, 12*(1), 38–66.

Dolan, J. E. (2016). Splicing the divide: A review of research on the evolving digital divide among K–12 students. *Journal of Research on Technology in Education, 48*(1), 16–37.

Doroudi, S., Aleven, V., & Brunskill, E. (2019). Where's the reward? *International Journal of Artificial Intelligence in Education, 29*(4), 568–620.

Dragon, T., Arroyo, I., Woolf, B. P., Burleson, W., El Kaliouby, R., & Eydgahi, H. (2008). Viewing student affect and learning through classroom observation and physical sensors. In *International conference on intelligent tutoring systems* (pp. 29–39). Springer.

Dunlosky, J., Rawson, K. A., Marsh, E. J., Nathan, M. J., & Willingham, D. T. (2013). Improving students' learning with effective learning techniques: Promising directions from cognitive and educational psychology. *Psychological Science in the Public Interest, 14*(1), 4–58.

Ebert-May, D., Brewer, C., & Allred, S. (1997). Innovation in large lectures: Teaching for active learning. *Bioscience, 47*(9), 601–607.

Else-Quest, N. M., & Hyde, J. S. (2016a). Intersectionality in quantitative psychological research: I. Theoretical and epistemological issues. *Psychology of Women Quarterly, 40*(2), 155–170.

Else-Quest, N. M., & Hyde, J. S. (2016b). Intersectionality in quantitative psychological research: II. Methods and techniques. *Psychology of Women Quarterly, 40*(3), 319–336.

Fang, Y., Li, T., Roscoe, R. D., & McNamara, D. S. (2021). Predicting literacy skills via stealth assessment in a simple vocabulary game. In *Proceedings of the international conference on human-computer interaction* (pp. 32–44). Springer.

Fiorella, L., & Mayer, R. E. (2016). Eight ways to promote generative learning. *Educational Psychology Review, 28*(4), 717–741.

Fleiss, J. L. (1986). Reliability of measurement. In J. L. Fleiss (Ed.), *The design and analysis of clinical experiments* (pp. 1–32). John Wiley and Sons.

Flynn, L. E., McNamara, D. S., McCarthy, K. S., Magliano, J. P., & Allen, L. K. (2022). The appearance of coherence: Using cohesive properties of readers' constructed responses to predict individual differences. *Revista Signos. Estudios de Lingüística, 54*(107), 1061–1088.

Foltz, P. W., Streeter, L. A., Lochbaum, K. E., & Landauer, T. K. (2013). Implementation and applications of the intelligent essay assessor. In M. D. Shermis & J. Burstein (Eds.), *Handbook of automated essay scoring: Current applications and future directions* (pp. 68–88). Routledge.

Fredricks, J. A., & McColskey, W. (2012). The measurement of student engagement: A comparative analysis of various methods and student self-report instruments. In S. Christenson, A. Reschly, & C. Wylie (Eds.), *Handbook of research on student engagement* (pp. 763–782). Springer.

Freeman, S., Eddy, S. L., McDonough, M., Smith, M. K., Okoroafor, N., Jordt, H., & Wenderoth, M. P. (2014). Active learning increases student performance in science, engineering, and mathematics. *Proceedings of the National Academy of Sciences, 111*(23), 8410–8415.

Fyfe, E. R., & Rittle-Johnson, B. (2016). Feedback both helps and hinders learning: The causal role of prior knowledge. *Journal of Educational Psychology, 108*(1), 82–97.

Gamboa, H., & Fred, A. (2002). Designing intelligent tutoring systems: A bayesian approach. In J. Filipe, B. Sharp, & P. Miranda (Eds.), *Enterprise information systems III* (pp. 146–152). Kluwer Academic Publishers.

Gardiner, J. M. (1988). Generation and priming effects in word-fragment completion. *Journal of Experimental Psychology: Learning, Memory, and Cognition, 14*(3), 495–501.

Gardner, J., Brooks, C., & Baker, R. (2019). Evaluating the fairness of predictive student models through slicing analysis. In *Proceedings of the 9th international conference on learning analytics & knowledge* (pp. 225–234).

Ghosh, A., Heffernan, N., & Lan, A. S. (2020). Context-aware attentive knowledge tracing. In *Proceedings of the 26th ACM SIGKDD international conference on knowledge discovery & data mining* (pp. 2330–2339).

Goodfellow, I., Bengio, Y., & Courville, A. (2016). *Deep learning*. MIT Press.

Gopalan, M., Rosinger, K., & Ahn, J. B. (2020). Use of quasi-experimental research designs in education research: Growth, promise, and challenges. *Review of Research in Education, 44*(1), 218–243.

Graesser, A. C., & McNamara, D. S. (2011). Computational analyses of multilevel discourse comprehension. *Topics in Cognitive Science, 3*(2), 371–398.

Hamdan, N., McKnight, P., McKnight, K., & Arfstrom, K. (2013). *A review of flipped classroom*. Flipped Learning Network.

Haskell, R. E. (2000). *Transfer of learning: Cognition and instruction*. Elsevier.

Hintzman, D. L. (1974). Theoretical implications of the spacing effect. In R. L. Solso (Ed.), *Theories in cognitive psychology: The Loyola symposium* (pp. 77–97). Lawrence Erlbaum Associates.

Howard, J. L., Bureau, J., Guay, F., Chong, J. X., & Ryan, R. M. (2021). Student motivation and associated outcomes: A meta-analysis from self-determination theory. *Perspectives on Psychological Science*. https://doi.org/10.1177/1745691620966789.

Huang, X., & Mayer, R. E. (2016). Benefits of adding anxiety-reducing features to a computer-based multimedia lesson on statistics. *Computers in Human Behavior, 63*, 293–303.

Huang, Y. M., Huang, T. C., Wang, K. T., & Hwang, W. Y. (2009). A Markov-based recommendation model for exploring the transfer of learning on the web. *Journal of Educational Technology and Society, 12*(2), 144–162.

Hung, W., Jonassen, D. H., & Liu, R. (2008). Problem-based learning. In M. Spector, D. Merrill, J. van Merrienböer, & M. Driscoll (Eds.), *Handbook of research on educational communications and technology* (pp. 485–506). Erlbaum.

Hunt, D. P. (2003). The concept of knowledge and how to measure it. *Journal of Intellectual Capital, 4*(1), 100–113.

Jonassen, D. H., & Grabowski, B. L. (2012). *Handbook of individual differences, learning, and instruction*. Routledge.

Joosten, T., & Cusatis, R. (2020). Online learning readiness. *American Journal of Distance Education, 34*(3), 180–193.

Just, M. A., & Carpenter, P. A. (1992). A capacity theory of comprehension: Individual differences in working memory. *Psychological Review, 99*(1), 122–149.

Kanny, M. A., Sax, L. J., & Riggers-Piehl, T. A. (2014). Investigating forty years of STEM research: How explanations for the gender gap have evolved over time. *Journal of Women and Minorities in Science and Engineering, 20*(2), 127–148.

Kenny, D. A., & Judd, C. M. (2019). The unappreciated heterogeneity of effect sizes: Implications for power, precision, planning of research, and replication. *Psychological Methods, 24*(5), 578.

Kizilcec, R. F., & Lee, H. (2020). Algorithmic fairness in education. arXiv preprint arXiv:2007.05443.

Kizilcec, R. F., Reich, J., Yeomans, M., Dann, C., Brunskill, E., Lopez, G., Turkey, S., Williams, J. J., & Tingley, D. (2020). Scaling up behavioral science interventions in online education. *Proceedings of the National Academy of Sciences, 117*(26), 14900–14905.

Koedinger, K. R., Booth, J. L., & Klahr, D. (2013). Instructional complexity and the science to constrain it. *Science, 342*(6161), 935–937.

Kornell, N., Hays, M. J., & Bjork, R. A. (2009). Unsuccessful retrieval attempts enhance subsequent learning. *Journal of Experimental Psychology: Learning, Memory, and Cognition, 35*(4), 989–998.

Kulik, J. A., & Kulik, C. L. C. (1988). Timing of feedback and verbal learning. *Review of Educational Research, 58*(1), 79–97.

Lan, A. S., & Baraniuk, R. G. (2016). A contextual bandits framework for personalized learning action selection. In *EDM* (pp. 424–429).

Lang, C., Siemens, G., Wise, A., & Gasevic, D. (Eds.). (2017). *Handbook of learning analytics*. SOLAR, Society for Learning Analytics and Research.

Lee, J., & Shute, V. J. (2010). Personal and social-contextual factors in K–12 academic performance: An integrative perspective on student learning. *Educational Psychologist, 45*(3), 185–202.

Lewis, K. L., Stout, J. G., Pollock, S. J., Finkelstein, N. D., & Ito, T. A. (2016). Fitting in or opting out: A review of key social-psychological factors influencing a sense of belonging for women in physics. *Physical Review Physics Education Research, 12*(2), 020110.

Linn, M. C., Davis, E. A., & Bell, P. (2004). Inquiry and technology. In M. C. Linn, E. Davis, & P. Bell (Eds.), *Internet environments for science education* (pp. 3–28). Erlbaum.

Linnenbrink, E. A., & Pintrich, P. R. (2002). Motivation as an enabler for academic success. *School Psychology Review, 31*(3), 313–327.

Litman, D. (2016). Natural language processing for enhancing teaching and learning. *Proceedings of the AAAI conference on artificial intelligence, 30*(1). https://ojs.aaai.org/index.php/AAAI/article/view/9879

Liu, O. L., Bridgeman, B., & Adler, R. M. (2012). Measuring learning outcomes in higher education: Motivation matters. *Educational Researcher, 41*(9), 352–362.

Liu, Z., Brown, R., Lynch, C. F., Barnes, T., Baker, R., Bergner, Y., & McNamara, D. S. (2016). MOOC learner behaviors by country and culture: An exploratory analysis. In T. Barnes, M. Chi, & M. Feng (Eds.), *Proceedings of the 9th international conference on educational data mining (EDM 2016)* (pp. 127–134). International Educational Data Mining Society.

Lortie-Forgues, H., & Inglis, M. (2019). Rigorous large-scale educational RCTs are often uninformative: Should we be concerned? *Educational Researcher, 48*(3), 158–166.

Louizos, C., Shalit, U., Mooij, J., Sontag, D., Zemel, R., & Welling, M. (2017). Causal effect inference with deep latent-variable models. In I. Guyon, U. V. Luxburg, S. Bengio, H. Wallach, R. Fergus, S. Vishwanathan, & R. Garnett (Eds.), *Proceedings of the 31st international conference on neural information processing systems* (pp. 6449–6459). Curran Associates Inc.

Lu, O., Huang, A., Huang, J., Lin, A., Ogata, H., & Yang, S. (2018). Applying learning analytics for the early prediction of students' academic performance in blended learning. *Journal of Educational Technology and Society, 21*(2), 220–232.

Lutz, J., Briggs, A., & Cain, K. (2003). An examination of the value of the generation effect for learning new material. *Journal of General Psychology, 130*(2), 171–188.

Ma, X., & Johnson, W. (2008). Mathematics as the critical filter: Curricular effects on gendered career choices. In H. M. G. Watt & J. S. Eccles (Eds.), *Gender and occupational outcomes: Longitudinal assessments of individual, social, and cultural influences* (pp. 55–83). American Psychological Association.

MacGinitie, W. H., & MacGinitie, R. K. (2006). *Gates-MacGinitie reading tests* (4th ed.). Houghton Mifflin.

Makel, M. C., & Plucker, J. A. (2014). Facts are more important than novelty: Replication in the education sciences. *Educational Researcher, 43*(6), 304–316.

Mayer, R. E. (2009). Constructivism as a theory of learning versus constructivism as a prescription for instruction. In S. Tobias & T. M. Duffy (Eds.), *Handbook of constructivist instruction* (pp. 196–212). Routledge.

Mayer, R. E. (2011). *Applying the science of learning.* Pearson.

Mayer, R. E., & Fiorella, L. (2015). *Learning as a generative activity.* Cambridge University Press.

Mayfield, E., Madaio, M., Prabhumoye, S., Gerritsen, D., McLaughlin, B., Dixon-Román, E., & Black, A. W. (2019, August). Equity beyond bias in language technologies for education. In *Proceedings of the fourteenth workshop on innovative use of NLP for building educational applications* (pp. 444–460).

McCarthy, K. S., Allen, L. K., & Hinze, S. R. (2020). Predicting reading comprehension from constructed responses: Explanatory retrievals as stealth assessment. In *Proceedings of the international conference on artificial intelligence in education* (pp. 197–202). Springer.

McCarthy, K. S., & McNamara, D. S. (2021). The multidimensional knowledge in text comprehension framework. *Educational Psychologist, 56*(3), 196–214.

McNamara, D. S. (1995). Effects of prior knowledge on the generation advantage: Calculators versus calculation to learn simple multiplication. *Journal of Educational Psychology, 87*(2), 307–318.

McNamara, D. S. (2004). SERT: Self-explanation reading training. *Discourse Processes, 38*(1), 1–30.

McNamara, D. S. (2017). Self-explanation and reading strategy training (SERT) improves low-knowledge students' science course performance. *Discourse Processes, 54*(7), 479–492.

McNamara, D. S., Allen, L. K., Crossley, S. A., Dascalu, M., & Perret, C. A. (2017). Natural language processing and learning analytics. In G. Siemens & C. Lang (Eds.), *Handbook of learning analytics and educational data mining* (pp. 93–104). Society for Learning Analytics Research.

McNamara, D. S., Graesser, A. C., McCarthy, P. M., & Cai, Z. (2014). *Automated evaluation of text and discourse with Coh-Metrix.* Cambridge University Press.

McNamara, D. S., & Healy, A. F. (1995a). A procedural explanation of the generation effect: The use of an operand retrieval strategy for multiplication and addition problems. *Journal of Memory and Language, 34*(3), 399–416.

McNamara, D. S., & Healy, A. F. (1995b). A generation advantage for multiplication skill and nonword vocabulary acquisition. In A. F. Healy & L. E. Bourne, Jr. (Eds.), *Learning and memory of knowledge and skills* (pp. 132–169). Sage.

Mead, C., Supriya, K., Zheng, Y., Anbar, A. D., Collins, J. P., LePore, P., & Brownell, S. E. (2020). Online biology degree program broadens access for women, first-generation to college, and low-income students, but grade disparities remain. *PLOS ONE, 15*(12), e0243916.

Menekse, M., Stump, G. S., Krause, S., & Chi, M. T. (2013). Differentiated overt learning activities for effective instruction in engineering classrooms. *Journal of Engineering Education, 102*(3), 346–374.

Mousavinasab, E., Zarifsanaiey, N., Niakan Kalhori, S. R., Rakhshan, M., Keikha, L., & Ghazi Saeedi, M. (2021). Intelligent tutoring systems: A systematic review of characteristics, applications, and evaluation methods. *Interactive Learning Environments, 29*(1), 142–163.

Murayama, K., Pekrun, R., Lichtenfeld, S., & Vom Hofe, R. (2013). Predicting long-term growth in students' mathematics achievement: The unique contributions of motivation and cognitive strategies. *Child Development, 84*(4), 1475–1490.

Nagar, N., Shachar, H., & Argaman, O. (2019). Changing the learning environment: Teachers and students' collaboration in creating digital games. *Journal of Information Technology Education: Innovations in Practice, 18*(1), 61–85. Retrieved August 13, 2021, from https://www.learntechlib.org/p/216644/

National Academies of Sciences, Engineering, and Medicine. (2017). *The economic and fiscal consequences of immigration.* National Academies Press. https://www.jstor.org/stable/pdf/44202635.pdf?casa_token=NNMEle4_gSkAAAAA:7ew66pDIAulpMdPVdG205XFaRBlwLwcPhBCKSUZwiwbjeIpy1RmGXZjgxRJNT0HwhI7_L3iOpUokQ9sq7yetVuanYAwFjX-QGDlc1cRyyP25wvoWb9nh

Nelson, M. C., Cordray, D. S., Hulleman, C. S., Darrow, C. L., & Sommer, E. C. (2012). A procedure for assessing intervention fidelity in experiments testing educational and behavioral interventions. *Journal of Behavioral Health Services and Research, 39*(4), 374–396.

Ocumpaugh, J., Baker, R. S., & Rodrigo, M. M. T. (2015). *Baker Rodrigo Ocumpaugh monitoring protocol (BROMP) 2.0 technical and training manual.* Columbia University, Manila University.

O'Reilly, T., Best, R., & McNamara, D. S. (2004). Self-explanation reading training: Effects for low-knowledge readers. In *Proceedings of the annual meeting of the cognitive science society* (Vol. 26, No. 26).

Oreopoulos, P., & Petronijevic, U. (2019). *The remarkable unresponsiveness of college students to nudging and what we can learn from it.* National Bureau of Economic Research. https://www.nber.org/system/files/working_papers/w26059/w26059.pdf

Peffer, M. E., Beckler, M. L., Schunn, C., Renken, M., & Revak, A. (2015). Science classroom inquiry (SCI) simulations: A novel method to scaffold science learning. *PLOS ONE, 10*(3), e0120638.

Peirce, N., Conlan, O., & Wade, V. (2008). Adaptive educational games: Providing non-invasive personalised learning experiences. In *2008 second IEEE international conference on digital game and intelligent toy enhanced learning* (pp. 28–35). IEEE.

Perfetti, C. A. (2007). Reading ability: Lexical quality to comprehension. *Scientific Studies of Reading, 11*(4), 357–383.

Perry, A. M., & Lee, N. T. (2019). *AI is coming to schools, and if we're not careful, so will its biases.* https://www.brookings.edu/blog/the-avenue/2019/09/26/ai-is-coming-to-schools-and-if-were-not-careful-so-will-its-biases/

Piech, C., Spencer, J., Huang, J., Ganguli, S., Sahami, M., Guibas, L., & Sohl-Dickstein, J. (2015). Deep knowledge tracing. arXiv preprint arXiv:1506.05908.

Pintrich, P. R. (2003). A motivational science perspective on the role of student motivation in learning and teaching contexts. *Journal of Educational Psychology, 95*(4), 667–686.

Preacher, K. J., & Sterba, S. K. (2019). Aptitude-by-treatment interactions in research on educational interventions. *Exceptional Children, 85*(2), 248–264.

Prince, M. (2004). Does active learning work? A review of the research. *Journal of Engineering Education, 93*(3), 223–231.

Proske, A., Roscoe, R. D., & McNamara, D. S. (2014). Game-based practice versus traditional practice in computer-based writing strategy training: Effects on motivation and achievement. *Educational Technology Research and Development, 62*(5), 481–505.

Psotka, J., Massey, L. D., & Mutter, S. A. (1988). *Intelligent tutoring systems: Lessons learned.* Lawrence Erlbaum Associates.

Puntambekar, S., & Kolodner, J. L. (2005). Toward implementing distributed scaffolding: Helping students learn science from design. *Journal of Research in Science Teaching: The Official Journal of the National Association for Research in Science Teaching, 42*(2), 185–217.

Pyc, M. A., & Rawson, K. A. (2009). Testing the retrieval effort hypothesis: Does greater difficulty correctly recalling information lead to higher levels of memory? *Journal of Memory and Language, 60*(4), 437–447.

Rau, M. A., Kennedy, K., Oxtoby, L., Bollom, M., & Moore, J. W. (2017). Unpacking "active learning": A combination of flipped classroom and collaboration support is more effective but collaboration support alone is not. *Journal of Chemical Education, 94*(10), 1406–1414.

Rawson, K. A., & Dunlosky, J. (2012). When is practice testing most effective for improving the durability and efficiency of student learning? *Educational Psychology Review, 24*(3), 419–435.

Riegle-Crumb, C., King, B., & Irizarry, Y. (2019). Does STEM stand out? Examining racial/ethnic gaps in persistence across postsecondary fields. *Educational Researcher, 48*(3), 133–144.

Rittle-Johnson, B., & Kmicikewycz, A. O. (2008). When generating answers benefits arithmetic skill: The importance of prior knowledge. *Journal of Experimental Child Psychology, 101*(1), 75–81.

Rodgers, E. (2016). Scaling and sustaining an intervention: The case of reading recovery. *Journal of Education for Students Placed at Risk (JESPAR), 21*(1), 10–28.

Roediger III, H. L., & Butler, A. C. (2011). The critical role of retrieval practice in long-term retention. *Trends in Cognitive Sciences, 15*(1), 20–27.

Roediger III, H. L., & Pyc, M. A. (2012). Inexpensive techniques to improve education: Applying cognitive psychology to enhance educational practice. *Journal of Applied Research in Memory and Cognition, 1*(4), 242–248.

Rohrer, D. (2015). Student instruction should be distributed over long time periods. *Educational Psychology Review, 27*(4), 635–643.

Romero, C., & Ventura, S. (2010). Educational data mining: A review of the state of the art. *IEEE Transactions on Systems, Man, and Cybernetics, Part C (Applications and Reviews), 40*(6), 601–618.

Roscoe, R. D. (2014). Self-monitoring and knowledge-building in learning by teaching. *Instructional Science, 42*(3), 327–351.

Roscoe, R. D., & Chi, M. T. (2008). Tutor learning: The role of explaining and responding to questions. *Instructional Science, 36*(4), 321–350.

Roscoe, R. D., & McNamara, D. S. (2013). Writing Pal: Feasibility of an intelligent writing strategy tutor in the high school classroom. *Journal of Educational Psychology, 105*(4), 1010.

Roscoe, R. D., Varner, L. K., Crossley, S. A., & McNamara, D. S. (2013). Developing pedagogically-guided algorithms for intelligent writing feedback. *International Journal of Learning Technology, 8*(4), 362–381.

Sabelli, N., & Dede, C. (2013). Empowering design based implementation research: The need for infrastructure. *National Society for the Study of Education, 112*(2), 464–480.

Sales, A., Botelho, A. F., Patikorn, T., & Heffernan, N. T. (2018). Using big data to sharpen design-based inference in A/B tests. In *Proceedings of the eleventh international conference on educational data mining* (pp. 479–485). National Science Foundation.

San Pedro, M. O. Z., Snow, E. L., Baker, R. S., McNamara, D. S., & Heffernan, N. T. (2015). Exploring dynamical assessments of affect, behavior, and cognition and math state test achievement. In O. C. Santos, J. G. Boticario, C. Romero, M. Pechenizkiy, A. Merceron, P. Mitros, J. M. Luna, C. Mihaescu, P. Moreno, A. Hershkovitz, S. Ventura, & M. Desmarais (Eds.), *Proceedings of the 8th international conference on educational data mining (EDM 2015)* (pp. 85–92). International Educational Data Mining Society.

Schiaffino, S., Garcia, P., & Amandi, A. (2008). eTeacher: Providing personalized assistance to e-learning students. *Computers and Education, 51*(4), 1744–1754.

Schunk, D., & Zimmerman, B. (2012). *Motivation and self-regulated learning: Theory, research, and applications*. Routledge.

Shute, V., & Ventura, M. (2013). *Stealth assessment: Measuring and supporting learning in video games* (p. 102). The MIT Press.

Shute, V., Ventura, M., Small, M., & Goldberg, B. (2013). Modeling student competencies in video games using stealth assessment. In R. Sottilare, A. Graesser, X. Hu, & H. Holden (Eds.), *Design recommendations for intelligent tutoring systems* (Vol. 1, pp. 141–152). Learner Modeling.

Shute, V. J. (2011). Stealth assessment in computer-based games to support learning. *Computer Games and Instruction, 55*(2), 503–524.

Shute, V. J., & Psotka, J. (1996). Intelligent tutoring systems: Past, present and future. In D. Jonassen (Ed.), *Handbook of research on educational communications and technology* (pp. 570–600). Scholastic Publications.

Shute, V. J., & Rahimi, S. (2021). Stealth assessment of creativity in a physics video game. *Computers in Human Behavior, 116*, 106647.

Shute, V. J., Wang, L., Greiff, S., Zhao, W., & Moore, G. (2016). Measuring problem solving skills via stealth assessment in an engaging video game. *Computers in Human Behavior, 63*, 106–117.

Sisk, R. J. (2011). Team-based learning: Systematic research review. *Journal of Nursing Education, 50*(12), 665–669.

Slamecka, N. J., & Katsaiti, L. T. (1987). The generation effect as an artifact of selective displaced rehearsal. *Journal of Memory and Language, 26*(6), 589–607.

Slotta, J. D. (2010). Evolving the classrooms of the future: The interplay of pedagogy, technology and community. In K. Mäkitalo-Siegl, F. Kaplan, J. Zottmann, & F. Fischer (Eds.), *Classroom of the future: Orchestrating collaborative spaces* (pp. 215–242). Sense Publishers.

Snow, C. (2002). *Reading for Understanding: Toward an R&D program in reading comprehension*. Rand Education. https://apps.dtic.mil/sti/pdfs/ADA402712.pdf

Soderstrom, N., Kerr, T., & Bjork, R. L. (2015). The critical importance of retrieval-and spacing-for learning. *Psychological Science, 27*(2), 223–230.

Soylu Yalcinkaya, N., & Adams, G. (2020). A cultural psychological model of cross-national variation in gender gaps in STEM participation. *Personality and Social Psychology Review, 24*(4), 345–370.

Spirtes, P., Glymour, C. N., Scheines, R., & Heckerman, D. (2000). *Causation, prediction, and search*. MIT Press.

Suárez-Pellicioni, M., Núñez-Peña, M. I., & Colomé, À. (2014). Reactive recruitment of attentional control in math anxiety: An ERP study of numeric conflict monitoring and adaptation. *PLOS ONE, 9*(6), e99579.

Sutton, R. S., & Barto, A. G. (2018). *Reinforcement learning: An introduction*. MIT Press.

Taber, K. S. (2019). Experimental research into teaching innovations: Responding to methodological and ethical challenges. *Studies in Science Education, 55*(1), 69–119.

Toh, Y., Lee, J. Y. L., & Ting, K. S. W. (2016). Building synergies: Taking school-based interventions to scale. In C. S. Chai, C. P. Lim, & C. M. Tan (Eds.), *Future learning in primary schools: A Singapore perspective* (pp. 177–197). Springer.

Trafton, J. G., & Trickett, S. B. (2001). Note-taking for self-explanation and problem solving. *Human-Computer Interaction, 16*(1), 1–38.

Upadhyay, U., Lancashire, G., Moser, C., & Gomez-Rodriguez, M. (2021). Large-scale randomized experiments reveals that machine learning-based instruction helps people memorize more effectively. *npj Science of Learning, 6*(1), 1–3.

Urbina Nájera, A. B., & Calleja Mora, J. D. L. (2017). Brief review of educational applications using data mining and machine learning. *Revista Electrónica de Investigación Educativa, 19*(4), 84–96.

van der Linden, W. J., & Hambleton, R. K. (Eds.). (2013). *Handbook of modern item response theory*. Springer Science & Business Media.

Van Eck, R. (2007). Building artificially intelligent learning games. In D. Gibson, C. Aldrich, & M. Prensky (Eds.), *Games and simulations in online learning: Research and development frameworks* (pp. 271–307). IGI Global.

VanLehn, K. (1988). Student modeling. *Foundations of Intelligent Tutoring Systems, 55*, 78.

VanLehn, K. (2006). The behavior of tutoring systems. *International Journal of Artificial Intelligence in Education, 16*, 227–265.

Ventura, M., & Shute, V. (2013). The validity of a game-based assessment of persistence. *Computers in Human Behavior, 29*(6), 2568–2572.

Vlach, H. A., & Sandhofer, C. M. (2012). Distributing learning over time: The spacing effect in children's acquisition and generalization of science concepts. *Child Development, 83*(4), 1137–1144.

Wainer, H., Dorans, N. J., Flaugher, R., Green, B. F., & Mislevy, R. J. (2000). *Computerized adaptive testing: A primer*. Routledge.

Walker, H. M. (2004). Commentary: Use of evidence-based interventions in schools: Where we've been, where we are, and where we need to go. *School Psychology Review, 33*(3), 398–407.

Walton, G. M., Logel, C., Peach, J. M., Spencer, S. J., & Zanna, M. P. (2015). Two brief interventions to mitigate a "chilly climate" transform women's experience, relationships, and achievement in engineering. *Journal of Educational Psychology, 107*(2), 468–485.

Wang, F., Liu, Q., Chen, E., Huang, Z., Chen, Y., Yin, Y., Huang, Z. & Wang, S. (2020). Neural cognitive diagnosis for intelligent education systems. In *Proceedings of the AAAI conference on artificial intelligence* (Vol. 34, No. 4, pp. 6153–6161).
Wang, X. (2013). Modeling entrance into STEM fields of study among students beginning at community colleges and four-year institutions. *Research in Higher Education, 54*(6), 664–692.
Wang, Z., Lamb, A., Saveliev, E., Cameron, P., Zaykov, Y., Hernandez-Lobato, J. M., Turner, R. E., Baraniuk, R. G., Barton, C., Peyton Jones, S., Woodhead, S., & Zhang, C. (2021). Results and insights from diagnostic questions: The NeurIPS 2020 education challenge. arXiv preprint arXiv:2104.04034.
Wang, Z., Zechner, K., & Sun, Y. (2018). Monitoring the performance of human and automated scores for spoken responses. *Language Testing, 35*(1), 101–120.
Warschauer, M., & Ware, P. (2006). Automated writing evaluation: Defining the classroom research agenda. *Language Teaching Research, 10*(1), 1–24.
Wilson, J., Roscoe, R., & Ahmed, Y. (2017). Automated formative writing assessment using a levels of language framework. *Assessing Writing, 34*, 16–36.
Wilson, J., & Roscoe, R. D. (2020). Automated writing evaluation and feedback: Multiple metrics of efficacy. *Journal of Educational Computing Research, 58*(1), 87–125.
Winne, P. H. (1996). A metacognitive view of individual differences in self-regulated learning. *Learning and Individual Differences, 8*(4), 327–353.
Wittrock, M. C. (1989). Generative processes of comprehension. *Educational Psychologist*, 24(4), 345–376.
Wladis, C., Conway, K. M., & Hachey, A. C. (2016). Assessing readiness for online education–Research models for identifying students at risk. *Online Learning, 20*(3), 97–109.
Yang, S. J. H. (2021). Guest editorial: Precision education - A new challenge for AI in education. *Educational Technology and Society*, 24(1), 105–108.
Yang, T. Y., Baker, R. S., Studer, C., Heffernan, N., & Lan, A. S. (2019). Active learning for student affect detection. In C. F. Lynch, A. Merceron, M. Desmarais, & R. Nkambou (Eds.), *Proceedings of the 12th international conference on educational data mining (EDM 2019)* (pp. 208–217). International Educational Data Mining Society.
Yao, S., & Huang, B. (2017). Beyond parity: Fairness objectives for collaborative filtering. arXiv preprint arXiv:1705.08804.
Yeager, D. S., Hanselman, P., Walton, G. M., Murray, J. S., Crosnoe, R., Muller, C., Tipton, E., Schneider, B., Hulleman, C. S., Hinojosa, C. P., Paunesku, D., Romero, C., Flint, K., Roberts, A., Trott, J., Iachan, R., Buontemp, J., Yang, S. M., Carvalho, C. M., … Dweck, C. S. (2019). A national experiment reveals where a growth mindset improves achievement. *Nature, 573*(7774), 364–369.
Yukselturk, E., & Bulut, S. (2007). Predictors for student success in an online course. *Educational Technology and Society, 10*(2), 71–83.
Zhou, G., Azizsoltani, H., Ausin, M. S., Barnes, T., & Chi, M. (2019). Hierarchical reinforcement learning for pedagogical policy induction. In *International conference on artificial intelligence in education* (pp. 544–556). Springer.

20

Linking Natural Language Use and Science Performance

Scott Crossley, Danielle S. McNamara, Jennifer Dalsen, Craig G. Anderson, and Constance Steinkuehler

CONTENTS

20.1 Introduction 309
20.1.1 Language Ability and Science Scores 310
20.1.2 Computer-Based Science Education 310
20.1.3 The Current Study 310
20.2 Method 311
20.2.1 Procedure 311
20.2.2 Participants 311
20.2.3 Data Collected 311
20.2.3.1 Pretest/Posttest Assessments 311
20.2.3.2 Individual Differences Data 312
20.2.3.3 Verbal Data 312
20.2.4 Transcriptions 312
20.2.5 Linguistic Variables 312
20.2.5.1 TAALES 312
20.2.5.2 TAACO 312
20.2.5.3 SEANCE 312
20.2.6 Statistical Analysis 313
20.3 Results 313
20.3.1 Non-Linguistic Model 313
20.3.2 Linguistic Model 313
20.3.3 Full Model 313
20.4 Discussion 314
20.5 Conclusion 315
Acknowledgments 316
References 316

20.1 Introduction

Success in science classrooms is often assumed to be related to students' language ability. Accordingly, science performance is argued to be (at least) partially based on the development of language skills that afford students the ability to understand scientific content (O'Reilly & McNamara, 2007). Language skills are assumed to affect or be related to students' knowledge of domain specific lexical terms, abstract reasoning skills, argument development, and the ability to report scientific findings (Lee et al., 2013). In addition, science success is partially related to affective features such as motivation (Liu et al., 2011).

While connections between language ability and science performance are generally accepted, very few studies have explicitly linked science success to language use within the classroom. Links between language skills and science success are most frequently demonstrated in studies that have examined associations between standardized science scores and instructional interventions or differences in scientific reasoning skills and language proficiency levels. A large portion of these studies has focused on second language (L2) learners whose language difficulties have been identified as a key obstacle to success in the science classroom (Tong et al., 2014; Torres & Zeidler, 2002). To our knowledge, however, no studies have examined the language produced by students within an instructional science setting. Specifically, no studies to date have examined relationships between

DOI: 10.1201/9781003181187-24

language complexity and language affect in student discourse to success on science assessments.

The purpose of this study is to fill that gap by examining the language used by students engaged in a week-long collaborative problem-solving task using natural language processing (NLP) techniques to automatically collect language data. The task was contextualized within an instructional science event (i.e., Game-a-Palooza [GaP]) that included a collaborative computer-based science education game (i.e., Virulent) providing instruction on the topic of virology (i.e., the branch of science that deals with the study of viruses). Students' conversations were recorded while they participated in the game, and a subset of this discourse was transcribed. Their language was analyzed using NLP tools to examine linguistic features related to text cohesion, lexical sophistication, and sentiment. In this study, we examine the extent to which the derived linguistic features were predictive of students' pretest and posttest science scores. We also examine a number of non-linguistic factors that are potentially predictive of science success including age, gender, event, and prior experience with technology. Our goal is to directly investigate links between linguistic production and science success using NLP techniques to better understand science education.

20.1.1 Language Ability and Science Scores

Success in the science classroom depends on students' language skills in various ways. Students with more knowledge of the language, including knowledge of the world and scientific concepts, better understand the material in science classrooms and better perform in classroom activities (O'Reilly & McNamara, 2007). For instance, Lee et al. (2013) identified a number of language skills cited within the Common Core State Standards that are necessary for success in the science classroom. Most importantly, students need to build their science knowledge by engaging with context rich texts as well as reading, writing, and speaking using scientific terms and approaches such as reasoning abstractly, engaging in argumentation, and reporting scientific findings.

A number of studies have also noted that students with limited English skills are also at a disadvantage in terms of science success. The basic premise behind these studies is that students with lower language skills have not reached the language threshold needed to be successful in content-based classes. For example, Torres and Zeidler (2002) demonstrated that L2 learners with low English language proficiency scored lower in terms of science content knowledge than their higher-proficiency peers. Likewise, improvements in language proficiency have been linked to improved course performance. For instance, Tong et al. (2014) examined the effects of providing instruction in science vocabulary, verbal and oral academic language skills, and language strategies to second language (L2) students in science and language arts classrooms. When compared to a control condition, L2 students who received science language instruction were reported to have higher academic science and reading scores than the control group.

20.1.2 Computer-Based Science Education

The context of this study is a collaborative computer-based science education game called Virulent. Computer games serve as powerful tools for learning new information (Wouters & Van Oostendorp, 2013) and gaining scientific literacy (Gee, 2007). Through games, students can build identities (Gee, 2003), participate in mentoring opportunities (Steinkuehler & Oh, 2012) and reciprocal apprenticeship (Steinkuehler, 2004), and tackle scientific challenges (Squire & Patterson, 2009). Computer games create authentic scenarios for situated learning (Lave & Wenger, 1991), allowing individuals to learn through practice and connect content to students in new and innovative ways.

A number of computer-based science games have shown success in teaching scientific topics and practices. For example, the game River City, which simulates an epidemic, can help students gain an understanding of scientific principles through active problem-solving in realistic environments (Spires et al., 2011). Research on computer-based science games such as Crystal Island, which allows players to investigate an unknown, infectious disease, reports positive relationships between in-game problem solving and engagement among students (Rowe et al., 2010). Similar findings for engagement have been reported for Quest Atlantis, a game where students seek to improve the lives of citizens, which has been found to increase engagement, inquiry, and scientific knowledge (Barab et al., 2009). In terms of learning, games such as Virtual Cell Experiment, which challenges students to complete activities relating to organic identification and viral respiration, have shown increases in scientific knowledge (McClean et al., 2001).

20.1.3 The Current Study

Collaborative science education games provide an effective and motivational approach for students to learn science. In addition, they provide a means to garner samples of students' natural language in socially interactive contexts. In this study, we capitalize on this context to examine the linguistic features of students' language production. We further capitalize on natural language processing tools that can be used to extract linguistic features that have strong potential to reveal cognitive, social, and affective processes engaged during learning. Here we focus on linguistic features related to text cohesion, lexical sophistication, and sentiment. Our criterion variables are students' performance on an assessment of virology knowledge before and after engaging with Virulent. In addition to examining relations between linguistic features of student language production and gains in science scores, we also control for a number of non-linguistic (demographic) factors including gender, age, race, and survey data including the number of hours spent using a tablet per week, and the number of hours spent gaming per week. Thus, in this study, we address the following research questions:

1. Are non-linguistic factors significant predictors of students' performance in the virology tests?
2. Are linguistic factors related to lexical sophistication and cohesion, and do they affect significant predictors of test performance?

20.2 Method

20.2.1 Procedure

Data for this study came from three week-long events. The first was a GaP event. The second took place at a private middle school, and the third was hosted by a local Boys and Girls Club. In all cases, participants roleplayed as scientists recruited by the Centers for Disease Control and Prevention (CDC) to stop the fictional 'Raven Virus' from spreading across the world. A key component of the event was a 'digiscope' device, which was an iPad loaded with the educational game Virulent whose game levels were called dynamic 'microscopic slides'. Throughout the week, small groups of students of around three or four children played through the 'microscopic slides' of the game to figure out how the Raven virus functioned and how the human immune system reacted. In their small groups, participants also created shared paper models of cell and viral systems, debated within and between small groups to decide which model and possible solution were most viable, wrote persuasive letters to the CDC, and recorded science presentation videos for the final group solutions to the threatening apocalypse.

20.2.2 Participants

A total of 87 elementary and middle school students participated in the study. Of these, 42 took part in the GaP event, 11 took part in the private school event, and the remaining 34 took part in the Boys and Girls Club event. The participants ranged from 9 to 14 years old with an average age of 11.4. Fifty-seven of the participants were male, and the remaining 27 were female. The majority of the students identified as Caucasian ($n = 37$). Other represented ethnicities included African-American ($n = 18$), Hispanic ($n = 14$), Asian ($n = 2$), and multiple ethnicities ($n = 3$). As a result of absences, only a subset of students ($N = 31$) completed the full study (pretest, posttest, and five days of the science event), completed the survey data, and consented to an audio recording of their dialogues. Thus, our analyses were conducted on this subset of 31 students. Of these 31 students, age was not reported for four and was imputed using multiple imputations. The multiple imputation model relied on all available demographic, survey, and linguistic data. The average age for the 31 students after imputation was 12 years. Of the 31 students, 17 took part in the GaP event, 9 took part in the private school event, and 5 took part in the Boys and Girls Club event. For the subset, there were 21 males and 10 females. With the subset, the majority of the students identified as white ($n = 21$) followed by Hispanic ($n = 15$), African-American ($n = 3$), and Asian ($n = 2$).

20.2.3 Data Collected

While a rich, diverse stream of data was collected across these contexts during the five-day event, for the purposes of this study, we are only interested in pretest/posttest assessments, demographic data, and verbal data. These are discussed in the following sections.

20.2.3.1 Pretest/Posttest Assessments

Each participant was administered a pre-/post-assessment on (1) general science content knowledge and attitudes and (2) content knowledge specific to virology. Both were framed as assessments of the materials used in the event and as feedback on the research agenda and game/material design. They were not framed as formal assessments of individual performance. The general science assessment was paper based, whereas the virology assessment was completed on a tablet. In this study, we only focus on the virology assessment, which was designed to assess the participants' content knowledge related to virology, specifically.

20.2.3.2 Individual Differences Data

A survey of demographic data was administered prior to data collection. Standard information related to gender, age, and ethnicity, as well as participants' prior experience with technology (e.g., estimated number of hours spent using a table or gaming per week), game genre preferences, motivations in gameplay (e.g., collecting items, winning the endgame), interest in science, favorite school subject, and average grades received were collected. All information was self-reported.

20.2.3.3 Verbal Data

Verbal data was captured via small recorders worn around a participant's neck (or placed nearby on their desk) on a lavalier nametag to avoid obstructing movement or conversation. Facilitators helped verify that recorders were functioning before group activities, but each participant was in control of their own recorder and its use. The conversation was recorded for all research team (small group) conversations throughout the curriculum and all cohort (large group) activities and debates. Small group research team conversations included a general discussion of the game and game strategies, identification of virus or immune system parts, questions raised, discussion around model making, and social banter. Cohort (large group) conversations included model presentations and cohort debates. Model presentations included small group explanations of virus and immune system functioning, with discussion and feedback about each model. Conversational chunks from the small group discussions were randomly sampled to be transcribed from a superset of discussions judged to be on-topic. All samples from the first session and the final session of the event (i.e., the first and last days of the event) were selected for analysis.

20.2.4 Transcriptions

A professional transcriber transcribed each of the speech samples collected from the participants. The transcriptions contained the speaker's words, some metalinguistic data (singing, laughing, sighing), and filler words (e.g., ummm, ahhhh). Disfluencies that were linguistic in nature (e.g., false starts, word repetition, repairs) were also retained. If any portion of the audio was not transcribable, the words were annotated either with an underscore or the flag 'INAUDIBLE'. The files were cleaned so that metalinguistic data, filler words, and indecipherable portions were removed prior to analysis. The transcripts were captioned to individual speakers within the groups.

20.2.5 Linguistic Variables

The transcripts were separated by the student and ran through a number of natural language processing tools including the Tool for the Automatic Analysis of Lexical Sophistication (TAALES; Kyle & Crossley, 2015; Kyle et al., 2018), the Tool for the Automatic Analysis of Lexical Sophistication (TAACO; Crossley et al., 2016; Crossley et al., 2019), and the SEntiment ANalysis and Cognition Engine (SEANCE, Crossley et al., 2017). The selected tools reported on language features related to lexical sophistication, text cohesion, and sentiment analysis, respectively.

20.2.5.1 TAALES

TAALES is a computational tool that is freely available and easy to use, works on most operating systems (Windows, Mac, Linux), allows for batch processing of text files, and incorporates over 150 classic and recently developed indices of lexical sophistication. These indices measure word frequency, bi-gram and tri-gram frequencies, and range counts taken from a number of corpora including the Corpus of Contemporary American English (COCA; Davies, 2009) as well as counts for academic words and phrases and word information measures like how imageable a word is use MRC vectors (Coltheart, 1981) and word association metrics.

20.2.5.2 TAACO

TAACO incorporates over 150 classic and recently developed indices related to text cohesion. For a number of indices, the tool incorporates a part of speech (POS) tagger and synonym sets from the WordNet lexical database. Specifically, TAACO calculates type token ratio (TTR) indices, sentence overlap indices that assess local cohesion, paragraph overlap indices, and a variety of connective indices including opposition connectives (e.g., but, however, nevertheless).

20.2.5.3 SEANCE

SEANCE is a sentiment analysis tool that relies on a number of pre-existing sentiment, social positioning, and cognition dictionaries. SEANCE contains a number of pre-developed word vectors developed to measure sentiment, cognition, and social order. These vectors are taken from freely available source databases. For many of these vectors, SEANCE also provides a negation feature (i.e., a contextual valence shifter) that ignores positive terms that are negated (e.g., not happy). SEANCE also includes a POS tagger.

20.2.6 Statistical Analysis

We conducted linear mixed effect (LME) models to answer our research questions. The purpose of the LME was to determine if linguistic features in the students' language output along with other fixed effects could be used to predict the students' pretest and posttest science scores. Thus, the LME model modeled the pretest and posttest results in terms of random factors (i.e., repeated variance explained by the students) and fixed factors (e.g., the linguistic features in their transcripts, gender, age, session).

Prior to the LME analysis, we first standardized all of the continuous data and checked that the variables were normally distributed. In addition, we controlled for multicollinearity between the variables (r > .700). We used R for our statistical analysis and the package lme4 to construct LME models. We also used the package lmerTest (Bates et al., 2015) to analyze the LME output and derive p values for individual fixed effects. Final model selection and interpretation were based on t and p values for fixed effects and visual inspection of residuals' distribution. To obtain a measure of effect sizes, we computed correlations between fitted and predicted residual values, resulting in an R^2 value for both the fixed factors and the fixed factors combined with the random factor (i.e., the repeated participant data from the pretest and the posttest). We first developed a model that included non-linguistic variables as fixed effects and participants as random effects. We next developed a model that only included linguistic features as fixed effects. We then developed a full model that included non-linguistic and linguistic fixed effects and participants as random effects. We examined differences in the predictive strength of the models using Analyses of Variance (ANOVA).

20.3 Results

20.3.1 Non-Linguistic Model

An LME model considering only the non-linguistic variables revealed a significant model for predicting the science scores. The model included three non-linguistic features: test (i.e., pretest, posttest), age, and the participant's estimated number of hours spent using a tablet per week (M = 7.45, SD = 4.41, Range = 0–14). The model indicated that virology test scores increased by 30% from pretest to posttest, by 6% with each increase of one year in age, and by 1% with an increase of one hour spent using a tablet. Table 20.1 displays the coefficients, standard error, t values, and p values for each of the fixed effects. Together, non-linguistic variables explained approximately 48% of the variance in the science scores (R^2 = .482), while the fixed and random variables combined to explain approximately 58% of the variance (R^2 = .579).

TABLE 20.1

Non-linguistic Model for Predicting Science Scores

Fixed Effect	Coefficient	Std. Error	t	p
(Intercept)	−0.802	0.264	−3.041	p < .001
Test	0.300	0.042	7.145	p < .001
Age	0.057	0.021	2.749	p < .050
Tablet hours	0.013	0.006	2.306	p < .050

20.3.2 Linguistic Model

An LME model considering only the linguistic variables revealed a significant model for predicting the virology test scores. The model included four linguistic indices: opposition connectives, word imageability, strength of associations in tri-grams (taken from the academic section of COCA), and a positive adjectives component score. The model indicated that science scores increased by 12% with an increase of 1 standard deviation above the mean number of opposition words (but, however), decreased by 8% with a decrease of 1 standard deviation below the mean average for imageable words, decreased by 5% with a decrease of 1 standard deviation below the mean number of tri-grams with greater association scores (i.e., participants who used less likely three-word combinations had higher science scores), and increased by 8% with an increase of 1 standard deviation above the mean number of positive adjectives. Table 20.2 displays the coefficients, standard error, t values, and p values for each of the fixed effects. Together, non-linguistic variables explained approximately 53% of the variance in the test scores (R^2 = .525), while the fixed and random variables combined to explain approximately 58% of the variance (R^2 = .576). A log-likelihood comparison yielded a significant difference between the non-linguistic and linguistic models ($\chi^2(2) = 4.802$, p < .050), indicating that linguistic features alone were better predictors of science success than non-linguistic features.

20.3.3 Full Model

A full model was developed to predict science scores that include both the non-linguistic and linguistic fixed effects. The model included three significant linguistic features (opposition connectives, word imageability, and positive adjectives) and two significant non-linguistic features (test and age). Table 20.3

TABLE 20.2

Linguistic Model for Predicting Science Scores of Linguistic Model

Fixed Effect	Coefficient	Std. Error	t	p
(Intercept)	0.439	0.024	18.592	p < .001
Opposition words	0.122	0.025	4.971	p < .001
MRC imageability all words	−0.080	0.024	−3.289	p < .001
Tri-gram association strength (COCA Academic)	−0.047	0.022	−2.098	p < .050
Positive adjective component score	0.076	0.023	3.293	p < .001

MRC = Medical Research Council, COCA = Contemporary Corpus of American English.

TABLE 20.3

Linguistic Model for Predicting Science Scores of Full Model

Fixed Effect	Coefficient	Std. Error	t	p
(Intercept)	−0.372	0.244	−1.521	p > .050
Session	0.152	0.051	2.958	p < .001
Age	0.043	0.018	2.376	p < .050
Tablet hours	0.009	0.005	1.795	p > .050
Opposition words	0.079	0.026	2.979	p < .001
MRC imageability all words	−0.052	0.023	−2.234	p < .050
Tri-gram association strength (COCA Academic)	−0.033	0.021	−1.580	p > .050
Positive adjective component score	0.055	0.022	2.510	p < .050

MRC = Medical Research Council, COCA = Contemporary Corpus of American English.

displays the coefficients, standard error, t values, and p values for each of the fixed effects included in the model. The percentage of increases and decreases patterned similarly to the two previous models. Together, the fixed factors including the linguistic and non-linguistic variables explained approximately 61% of the variance (R^2 = .609), while the fixed and random variables combined to explain approximately 65% of the variance (R^2 = .652). A log-likelihood comparison yielded a significant difference between the full model and the non-linguistic model ($\chi^2(2) = 20.204$, $p < .001$) and between the full model and the linguistic model ($\chi^2(2) = 15.401$, $p < .010$) such that the full model explained a greater amount of variance in the science scores.

20.4 Discussion

Previous studies that have examined relations between science success and language use have focused on examining associations between standardized science scores and instructional intervention or differences in scientific reasoning skills and language proficiency levels. This study takes a novel approach by examining the language produced by students within an instructional science setting that includes the use of computer-based learning tools and examining whether the language produced is predictive of scores in a science test focusing on virology. We accomplished this objective by using natural language processing tools which allow for automated linguistic analyses of language samples that can be easily replicated. We also move beyond language features alone and examine whether a number of non-linguistic factors including gender, age, and prior experience with technology are predictive of success in the science classroom.

Our findings indicate that students who participated in a week-long instructional event demonstrated significant gains in their virology pretest and posttest scores, indicating that learning did occur. Three non-linguistic factors were significant predictors of the virology test scores: test, age, and tablet hours. The findings indicate that as students progressed through the event, their virology scores increased. In addition, older students had higher virology scores, as did students with more experience using a tablet. This last finding is likely related to the use of the 'digiscope' device, which was an iPad loaded with the educational game Virulent. The use of this game was a key component of the event in that students used the game throughout the week to work out how the Raven virus functioned and how the human immune system reacted to it. Additionally, the tests were administered on tablets. The findings indicate that students who had more exposure to tablets performed better in the virology tests likely because they were more familiar with this technology. In total, the non-linguistic factors explained about 48% of the variance in the virology scores, while the random factor of students explained an additional 10% of the variance.

Our linguistic indices predicted about 53% of the variance in the virology test scores, and this increase was significant when compared to the non-linguistic model, indicating the linguistic indices were superior to non-linguistic indices in explaining the science scores. Four indices were significant predictors in the linguistic model. These included the use of opposition words, the use of imageable words, the use of tri-grams that had greater association strength, and the use of positive adjectives. The use of a greater number of opposition words likely indicates that students who have higher test scores use a greater number of words and phrases that connect ideas together by proposing competing ideas (i.e., through the use of words and phrases like in spite of, however, on the other hand). It is likely that such opposition terms are important components of reasoning abstractly and engaging in arguments, both of which are important elements of science language (Lee et al., 2013).

Two linguistic indices also indicated that increased science scores were related to the production of more complex language. For instance, students who produced less imageable words showed greater science scores as did students who used less common three-word combinations (i.e., lower tri-gram association strength scores which measure the strength of associations between words in a phrase). This finding shows that students with greater linguistic skills scored higher in the science test.

Lastly, an affective measure was predictive of science scores: positive adjectives. This finding demonstrated that students who produce more positive adjectives scored higher in the science test, potentially because students who viewed the events more positively were more successful.

A combined model using both the linguistic and non-linguistic features explained 61% of the variance in scores and was significantly better than the linguistic model. Within this model, only tablet hours and tri-gram association strength were not significant predictors of science scores. From an educational perspective, these results indicate that as students develop a better understanding of science words and ways to discuss science in a positive manner, their success in completing science problems increases. The findings indicate that science learning is likely not rote memorization of facts and the simple reiteration of these ideas, but the ability to discuss these ideas using appropriate language, especially language that is imageable and helps visualize ideas, and make connections between the ideas. Helping students to better understand how to discuss science using language would like to lead to greater science success. Using NLP tools such as those used in this study can help us better understand what science teachers should focus on in the classroom.

It should be noted that in all models, there were a number of linguistic and non-linguistic indices that were not predictive of science scores. These non-significant findings provide valuable evidence about elements of the study that may be unrelated to science success. For instance, gender and race were not predictive of science scores, indicating that females performed on par with males and all races performed proportionately. In addition, the number of hours spent using video games was not predictive indicating that familiarity with games specifically is not as important as familiarity with technology in general. From a linguistic standpoint, no features related to global cohesion (i.e., overlap of ideas across large segments of text) were retained in the final LME models, nor were a number of lexical features related to word frequency, range, and other lexical properties such as age of acquisition and academic word use.

20.5 Conclusion

The findings from this study have practical implications for understanding science performance and science instruction. Specifically, the findings provide support for the notion that language proficiency is a predictor of science performance such that a greater number of words that connect ideas together, the use of more complex language, and the production of more positive words predicts higher student performance across a week-long intervention study. In addition, the findings provide information about non-linguistic factors that predict science success, especially in environments where computer-based education is occurring. The findings reported in this study indicate that students learn over time, that older students perform better than younger students, and students with a greater background in the use of technology perform better in a computer-based learning environment. Future studies can build on the results presented here by sampling larger populations of students who come from more diverse backgrounds and more varied grade levels. Future replications of this work might also consider examining age differences in learning with older students, examining links between linguistic features and science success in environments that do not include a computer-based learning feature, and investigating learning in different science topics. While there is much left to do, this study is the first of its kind to link students' natural language to their performance in science.

Acknowledgments

This material is based upon work supported by the National Science Foundation under Grant Number #144 PRJ88JH. Any opinions, findings, conclusions, or recommendations expressed in this material are those of the author(s) and do not necessarily reflect the views of the National Science Foundation.

References

Barab, S. A., Scott, B., Siyahhan, S., Goldstone, R., Ingram-Goble, A., Zuiker, S. J., & Warren, S. (2009). Transformational play as a curricular scaffold: Using videogames to support science education. *Journal of Science Education and Technology, 18*(4), 305–320.

Bates, D., Mächler, M., Bolker, B., & Walker, S. (2015). Fitting linear mixed-effects models using lme4. *Journal of Statistical Software, 67*(1), 201–210.

Coltheart, M. (1981). The MRC psycholinguistic database. *Quarterly Journal of Experimental Psychology Section A, 33*(4), 497–505.

Crossley, S. A., Kyle, K., & Dascalu, M. (2019). The tool for the automatic analysis of cohesion 2.0: Integrating semantic similarity and text overlap. *Behavior Research Methods, 51*(1), 14–27.

Crossley, S. A., Kyle, K., & McNamara, D. S. (2016). The tool for the automatic analysis of text cohesion (TAACO): Automatic assessment of local, global, and text cohesion. *Behavior Research Methods, 48*(4), 1227–1237.

Crossley, S. A., Kyle, K., & McNamara, D. S. (2017). Sentiment analysis and social cognition engine (SEANCE): An automatic tool for sentiment, social cognition, and social order analysis. *Behavior Research Methods, 49*(3), 803–821.

Davies, M. (2009). The 385+ million word corpus of contemporary American English (1990–2008+): Design, architecture, and linguistic insights. *International Journal of Corpus Linguistics, 14*(2), 159–190.

Gee, J. P. (2003). What video games have to teach us about learning and literacy. *Computers in Entertainment, 1*(1), 20–20.

Gee, J. P. (2007). *Good video games plus good learning* (Vol. 27). Peter Lang.

Kyle, K., & Crossley, S. A. (2015). Automatically assessing lexical sophistication: Indices, tools, findings, and application. *TESOL Quarterly, 49*(4), 757–786.

Kyle, K., Crossley, S. A., & Berger, C. (2018). The tool for the automatic analysis of lexical sophistication version 2.0. *Behavior Research Methods, 50*(3), 1030–1046.

Lave, J., & Wenger, E. (1991). *Situated learning: Legitimate peripheral participation*. Cambridge University Press.

Lee, O., Quinn, H., & Valdés, G. (2013). Science and language for English language learners in relation to next generation science standards and with implications for common core state standards for English language arts and mathematics. *Educational Researcher*. https://doi.org/10.3102/0013189X13480524.

Liu, M., Horton, L., Olmanson, J., & Toprac, P. (2011). A study of learning and motivation in a new media enriched environment for middle school science. *Educational Technology Research and Development, 59*(2), 249–265.

McClean, P., Saini-Eidukat, B., Schwert, D., Slator, B., & White, A. (2001). Virtual worlds in large enrollment biology and geology classes significantly improve authentic learning. *12th international conference on college teaching and learning (ICCTL-01)* (pp. 111–118). JA Chambers, Center for the Advancement of Teaching and Learning.

O'Reilly, T., & McNamara, D. S. (2007). Reversing the reverse cohesion effect: Good texts can be better for strategic, high-knowledge readers. *Discourse Processes, 43*(2), 121–152.

Rowe, J. P., Shores, L. R., Mott, B. W., & Lester, J. C. (2010). Integrating learning and engagement in narrative-centered learning environments. In *International conference on intelligent tutoring systems* (pp. 166–177). Springer.

Spires, H. A., Rowe, J. P., Mott, B. W., & Lester, J. C. (2011). Problem solving and game-based learning: Effects of middle grade students' hypothesis testing strategies on learning outcomes. *Journal of Educational Computing Research, 44*(4), 453–472.

Squire, K., & Patterson, N. (2009). *Games and simulations in informal science education*. National Research Council Board on Science Education.

Steinkuehler, C., & Oh, Y. (2012). Apprenticeship in massively multiplayer online games. In C. Steinkuehler, K. Squire, & S. Barab (Eds.), *Games, learning and society: Learning and meaning in the digital age* (pp. 154–184). Cambridge University Press.

Steinkuehler, C. A. (2004). Learning in massively multiplayer online games. In Y. B. Kafai, W. A. Sandoval, N. Enyedy, A. S. Nixon, & F. Herrera (Eds.), *International conference of the learning sciences 2004: Embracing diversity in the learning sciences* (pp. 521–528). Lawrence Erlbaum Associates.

Tong, F., Irby, B., Lara-Alecio, R., & Koch, J. (2014). Integrating literacy and science for English language learners: From learning-to-read to reading-to-learn. *Journal of Educational Research, 107*(5), 410–426.

Torres, H. N., & Zeidler, D. L. (2002). The effects of English language proficiency and scientific reasoning skills on the acquisition of science content knowledge by Hispanic English language learners and native English language speaking students. *Electronic Journal of Science Education, 6*(3), 1–59.

Wouters, P., & Van Oostendorp, H. (2013). A meta-analytic review of the role of instructional support in game-based learning. *Computers and Education, 60*(1), 412–425.

Section V

Other Topics in AI and STEM Education

21

Quick Red Fox: An App Supporting a New Paradigm in Qualitative Research on AIED for STEM

Stephen Hutt, Ryan S. Baker, Jaclyn Ocumpaugh, Anabil Munshi, J.M.A.L. Andres, Shamya Karumbaiah, Stefan Slater, Gautam Biswas, Luc Paquette, Nigel Bosch, and Martin van Velsen

CONTENTS

21.1 Introduction 319
21.2 Design 321
21.2.1 Interview Triggers 322
21.2.2 Server-Side Platform 322
21.2.3 Client Side 323
21.2.3.1 Set-Up and Login 323
21.2.3.2 Presentation of Student and Trigger Information 323
21.2.3.3 Interview Recordings and Notes 323
21.2.3.4 Moving On (Next, Skip, End) 324
21.2.3.5 Data 324
21.3 Case Study 324
21.3.1 Betty's Brain 325
21.3.2 Developing Interview Triggers 325
21.3.3 Procedure 326
21.3.4 Data 326
21.3.5 Data Coding 326
21.3.6 Impact on Scholarly Work 327
21.4 General Discussion and Conclusions 328
21.4.1 Summary 328
21.4.2 Applications 328
21.4.3 Limitations 329
21.4.4 Future Development 329
Acknowledgments 330
Note 330
References 330

21.1 Introduction

Educational software and computer-based learning environments have become an increasingly prominent part of K-12 education. Even before the COVID-19 pandemic, there was a considerable increase in the use of these technologies (Marcus-Quinn & Hourigan, 2017), and this trend has amplified in the last year as teachers who previously did not have the technology (or the training to use it) were asked to convert to virtual teaching in a very short time. As students return to the classroom with these technological investments in place, interactive learning environments, such as intelligent tutoring systems, simulations, and problem-solving platforms are likely to be even more ubiquitous than they were two years ago.

As we seek to improve these technologies and create richer, more dynamic experiences, we must first gain a deeper understanding of how students learn with technology, and how these processes may be different than those encountered in traditional classroom learning. In order to truly understand, we must often tap into a student's internal cognitive and noncognitive processes, which may be hard for students (especially younger students) to articulate in traditional quantitative survey instruments. More qualitative

DOI: 10.1201/9781003181187-26

methods such as think-alouds, interviews, and open-ended self-reports, by contrast, can provide a clearer window into these processes.

Qualitative research can be critical in improving education as it gives information on 'how' and 'why' research questions that may be otherwise unanswerable (Cleland, 2017). Qualitative data can focus on thoughts, concepts, or experiences that may in turn be used to gain a deeper understanding of phenomena and context. By asking questions that cannot be boiled down to 'how many' or 'please rate', it becomes possible to collect a rich dataset that can complement the quantitative data that is already frequently collected (e.g., log data, student models, etc.) within artificial intelligence in education (AIED) research. However, conducting qualitative research on AIED technologies has proven challenging, and the logistics of collecting enough qualitative data (and the right qualitative data) have often proven highly resource-intensive (e.g. Schofield, 1995). For example, one option might be to use think-alouds or emote-alouds to get students to vocalize processes. However, this approach is resource-intensive, does not scale well, and is challenging to apply in real classrooms, often requiring researchers to pull students out of context and into a separate room (e.g., Kelley et al., 2015). Other researchers have conducted interviews or observational approaches in classrooms – however, doing so has also proven highly time consuming when attempting to exhaustively capture key events (e.g., Cobb et al., 2001).

Beyond this, in cases where a classroom observer is trying to study students' internal cognitive and affective processes as they use an AIED system, it may be difficult to capture events and processes of interest, without capturing huge amounts of data that is then difficult and time consuming to filter through. Both observation and interview methods are vulnerable to what Wessel (2015) describes as the 'one shot' problem, where events occur only once, and if missed, cannot be studied. Two sampling methods that are often used in classroom studies are momentary time sampling (as with the widely used BROMP quantitative observation protocol – Baker et al., 2020) and scan methods, where the observer monitors an entire classroom full of students at once. Momentary time sampling methods are known to bias towards events of longer duration and/or which occur more frequently (Meany-Daboul et al., 2007). Scan methods tend to bias towards more dramatic behaviors, while more subtle student actions may go unnoticed (Ostrov & Hart, 2013). This is of particular concern in the observation of classroom learning with technology, where specific uncommon events may be particularly essential to study, whether changes in student engagement (Andres et al., 2019) or critical but brief activities within a broader learning task (Bernacki, 2017; Jeong & Biswas, 2008).

Some researchers have instead focused on collecting qualitative data in the form of videos and then coding those videos in depth (Kane et al., 2014; Lehrer & Schauble, 2011). Videos can be re-watched an indefinite number of times and coded in terms of a variety of constructs, often using complex coding schemes. Video addresses the 'one shot' problem – at the cost of spending far more time coding data (Baker et al., 2006). In cases where video captures the rich dialogue between learners – or between learners and teachers – it is possible to make inferences about cognition as well as behavior. However, when learners work with a computer-based learning system (or individually, in general), this type of inference from video may be more challenging. A student may be working silently while complex activities are occurring internally. Even in video of classroom dialogue, we may not always have access to the complex reasoning (for instance, around self-presentation – Juvonen & Murdock, 1995) that impacts why students choose to say what they say.

The methods described so far provide detailed insights into *what* students are doing but may not always help understand *why* students make the choices they do. One method that gets at students' phenomenological understanding of why they do what they do is interviews. Interviews have a rich history in educational technology research. Perhaps the seminal work in this area was Schofield's book *Computers and Classroom Culture* (Schofield, 1995), which involved months of ethnographic embedding in schools, conducting interviews both during and outside of class. Shorter-term classroom interview studies of students using technology have investigated students' attitudes towards specific educational technologies and students' understanding of what they learned (Warren et al., 2005; Yoon et al., 2017). However, interview methods – like observational methods – suffer from the 'one shot' problem. While interviews can be conducted retrospectively, students may not always recall their exact reasoning around a decision made half an hour or a week earlier, or the emotions surrounding that decision. Even if interviews are conducted during class, in real time, it can be a challenge to identify relatively rare events of interest, within a class of 25 students working quietly on computers. Take, for instance, a researcher attempting to understand why a sequence of emotions occurs. Trying to spot a student going through a specific sequence of these emotions would be difficult for an observer to catch, especially while trying to monitor multiple students at the same time.

Finding the 'right time' to interview has often been a logistical barrier to efficient qualitative data collection; however, new technologies may provide a solution to this long-standing problem. Specifically, the last decade has seen major advances in recognizing complex student behaviors and states from technology. Detectors have now been developed that can recognize engagement indicators (e.g., boredom, confusion, engaged concentration, frustration, off-task behavior, on-task behavior, etc.) from students' interactions with learning software. These indicators have been validated to agree with human judgment for a wide variety of educational systems (Bosch & D'Mello, in press; Botelho et al., 2017; D'Mello, 2018; Wixon et al., 2014). These detectors of students' affect and behavior have been used in both fine-grained and coarse-grained analyses, from studying the characteristic shifts in engagement over a matter of seconds (D'Mello & Graesser, 2012) to studying how these measures correlate with long-term outcomes such as college attendance (San Pedro et al., 2014) or career choices (Makhlouf & Mine, 2020). As such, these detectors can be used to identify critical moments in a learning process (Lodge et al., 2018). Whereas previous work has considered using these detectors to drive in-the-moment automated intervention (e.g. Mills et al., 2021) or teacher reporting (Holstein et al., 2017), in this work we use these detectors to drive data collection for qualitative research. This approach can in turn inform future design work to adapt to students' behavior and affect.

In this chapter, we discuss an approach that attempts to address these limitations of existing qualitative methods – proposing a new way of conducting interviews, and a tool to facilitate this approach. In this approach, we leverage existing AIED technologies to target interviews so that the depth of understanding that interviews facilitate can be combined with the ability to capture key moments in the learning process.

In the remainder of the chapter, we will detail the design of Quick Red Fox (QRF), a new research tool developed to facilitate targeted in-the-moment interviews. QRF is an open-source[1] server-client Android app. Specifically, QRF optimizes researcher time by directing interviewers to users who have just displayed an interesting behavior (previously defined by the research team). QRF integrates with existing student modeling technologies (e.g., behavior-sensing, affect-sensing, detection of self-regulated learning) to alert researchers to key moments in a user's experience. QRF listens for events (e.g., interaction patterns) and identifies moments of interest, prioritizes them, and directs interviewers, accordingly, allowing the interviewer to record their interview directly in the app along with relevant metadata (e.g., participant ID). We demonstrate the efficacy of this approach through a case study involving classroom research on students' engagement and self-regulated learning. Finally, we discuss additional future applications of this tool in AIED technologies.

21.2 Design

QRF was designed based on principles for Minimal Attention User Interfaces (MAUIs). According to Pascoe et al. (2000), MAUIs for fieldwork should consider four characteristics that are important to observational research: (a) the *dynamic user configuration* (i.e., working conditions of the fieldwork researcher), which are unlikely to include a desk or even a chair, (b) the *limited attention capacity* of the fieldworker, who necessarily needs to observe the object of their research, (c) the need for *high speed interactions*, should the research subject suddenly have a spurt of relevant activities that need to be documented, and (d) the *context dependency* needs of the fieldwork, some of which (location, timestamps, etc.) can be automated by the system so that the researcher can focus on other things.

With the ubiquity of mobile phones, they present an attractive option for developing apps for field work, with existing mobile apps already being used in the classroom. Existing work has often linked classroom observations to student interaction data following the learning session in post hoc data processing. However, the QRF design requires knowledge of what students were doing before an observation or interview can be conducted, as data collection is targeted to events of interest, meaning that a stand-alone app would not be suitable. As such, QRF consists of two major components: (1) a server-side process that listens for events and assigns interviews, and (2) a client app (implemented in Java for Android mobile devices) that receives interview prompts and facilitates interview recordings. Interview triggers must also be defined but are integrated with the learning environment rather than QRF (see Section 21.2.1).

When designing the client side app for QRF, we align with the design considerations outlined by Pascoe et al. (2000) for MAUIs. In doing so we acknowledge that the classroom is a complex environment and observations and interviews need to be as simple as possible to record as researchers will likely have many other issues to negotiate. As such, QRF's design aligns carefully with an interview research protocol and facilitates context-aware coding (e.g., timestamps and the recording of triggering conditions) while allowing the researcher to focus their attention on the student. As described in more detail below, QRF displays details to direct them to a student, including the student's username and triggering conditions. The app also includes functionality for the researcher to take notes and/or record an audio file of an interview with the student. All data gathered is saved automatically, allowing the researcher to move on to the next observation with minimal effort devoted to the screen and reduced possibility of error.

In this section we outline the design for both major components of QRF as well as give detail on the process required for interview triggers.

21.2.1 Interview Triggers

In order for QRF to appropriately detect events, interview triggers must be defined. Detection services for QRF identify key moments in students' learning processes in real time. They do this by parsing students' log data or other available data streams. There are two components required to build a detection service: the individual detectors and (if necessary) the relevant patterns. First, automated detectors of constructs such as affect and behavior must have been trained beforehand, likely using a previously collected dataset often from the same learning system (i.e., Jiang et al., 2018). This process will also typically involve feature engineering, wherein a set of predictor variables was designed based on the student's activity in the system. An example of a detector is a simple logistic regression model inferring boredom, which is a weighted combination of the selected features. The model output is then thresholded to predict a binary outcome (e.g., bored or not bored).

Once the individual detectors are developed, they are embedded in the detection service code. If the researcher is interested in patterns of constructs/behaviors (e.g., bored then confused versus just bored), these patterns must be distilled from the data. These patterns are limited only by the detectors trained and are defined by the researcher for their specific study. Pattern detection then becomes an additional layer of detection that uses the output of individual detectors.

When a student starts interacting with the learning system, the detection service pulls in the student's activities from the system's interaction log data at a set interval of time (e.g., once every 20 seconds). Based on the student's activity in that interval, feature values are derived. Depending on the type of feature, the service may have to keep a history of the student's activity. For example, some features may only require student's activities in the current interval, while others may need data from the start of a session or the student's past sessions. These feature values are then fed into the individual detectors which in turn output predictions of relevant constructs (e.g., off-task behavior, frustration). According to the prediction in the current interval and past few intervals (depending on the length of the pattern), the detection service alerts the server of any detected pattern and its corresponding priority level (assigned beforehand). In addition, the service also keeps track of clearing the history and updating the past feature and prediction values as required. This is repeated regularly at the predefined time interval.

The detectors used to trigger interviews are (though necessary) separate processes from the main QRF infrastructure. Machine learning (or other model definition) is outside the control of the app. Instead, QRF applies these methods to drive data collection. That means that researchers have the flexibility to develop detectors in whatever language they choose, relevant to their broader platform. For example, if studying an application written in Python, researchers may wish to integrate detectors also written in Python, whereas that may not be appropriate for studying an iOS application. The only requirement is that the detectors be able to send a package over a network (a feature present in almost all contemporary programming languages). This flexibility is crucial to the QRF's design as it opens the door to a wider variety of future applications and research environments.

21.2.2 Server-Side Platform

QRF's server-side receives packages from the detectors (containing student ID, pattern/trigger, and priority) and assembles the interview queue. Each incoming pattern message is assigned a timestamp and inserted into a priority queue. The priority queue handles the selection and dispatch of pattern messages to the client side (described below). It sorts pattern messages (the detected triggers) based on their priorities so that interviewers will be notified of the highest priority interview first. The sorting algorithm also includes two parameters to ensure that the same student is not interviewed too frequently: (a) *'maxInterviews'* (default value: 4), i.e., the maximum number of interviews that can be conducted with a student in the current session, and (b) *'interviewsGap'* (default value: 10 minutes), i.e., the minimum time gap between two successive interviews with a student. A pattern message lives in the priority queue until (a) it is sent to one of the QRF apps, or (b) the message expires, in which case it is no longer relevant to the student's current activities and is hence expelled from the queue.

The framework also includes a QRF Ruby library which handles the registration, initialization, and communication between the server and the QRF app through a RabbitMQ message broker. Once the interviewer starts the QRF app and registers their handheld device on it, a direct communication line is established between the mobile app and the Betty's Brain server. Then pattern messages flow back and forth between the server and the app as requests for new patterns are made, accepted, or rejected by the interviewer.

21.2.3 Client Side

QRF's client side was implemented for Android devices using Java due to Android's strong support for app development and dissemination, and cost considerations (Android devices are often cheaper and more durable than other tablets). QRF synchronizes to internet time using a Network Time Protocol (NTP) server for logging purposes.

To make the process as smooth as possible, the app is heavily streamlined to require minimal interaction and thus allow the researcher to focus their resources elsewhere. The app was designed to avoid subpages that may be confusing to the researcher or result in erroneous recordings and aligns with several of the design principles from previous work discussed above. In addition to facilitating context awareness and limiting the number of times the researcher needs to enter the same information, QRF also allows the researcher to work with a small screen, thus reducing its obtrusiveness in the classroom. Finally, and perhaps most critically, QRF presents a user interface that aligns with the research protocol.

21.2.3.1 Set-Up and Login

QRF requires a login with username and password, during which time it registers with the server-side application. It then requests information about the research session (e.g., classroom name, etc.) to verify which data to receive and provide annotations for later research. Once the Android device is registered with the server, messages between the two flow back and forth (e.g., requests for new interviews are sent to the server, which responds as soon as a prioritized trigger event is identified). This process is completed once per research session (e.g., class period). Following set-up, the researcher is presented with the primary interface (see Figure 21.1).

21.2.3.2 Presentation of Student and Trigger Information

Figure 21.2 shows the primary QRF interface. From this screen, researchers receive information regarding which student to interview next and can record said interview from the same screen. When a prioritized trigger event is identified, QRF presents this to the researcher by displaying the User ID (i.e., what the student uses to log in with *or* a deidentified number, in cases where regulatory compliance requires it) at the top of the screen. Immediately below this information, QRF presents the trigger for the interviewer's use.

21.2.3.3 Interview Recordings and Notes

Should the researcher choose to interview the student, they can tap the 'start recording' button, and an

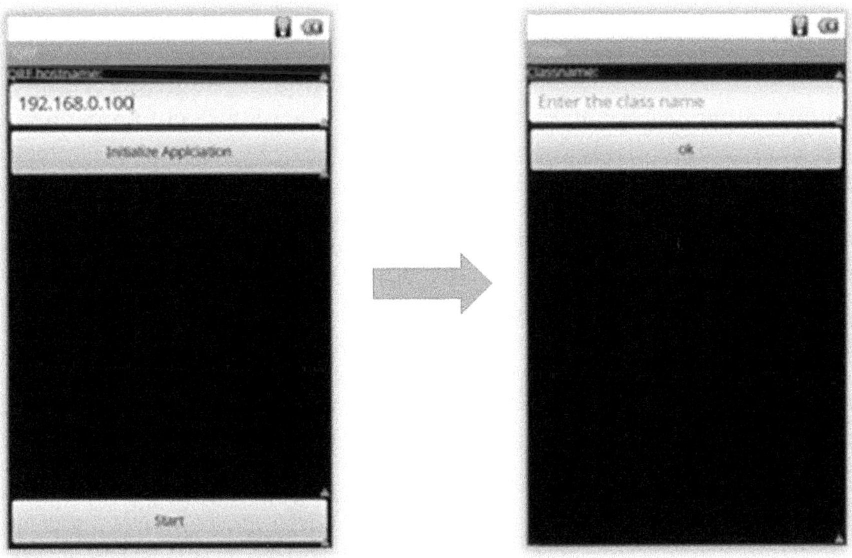

FIGURE 21.1
QRF Login screens where researchers login (left) and enter the class session ID (right).

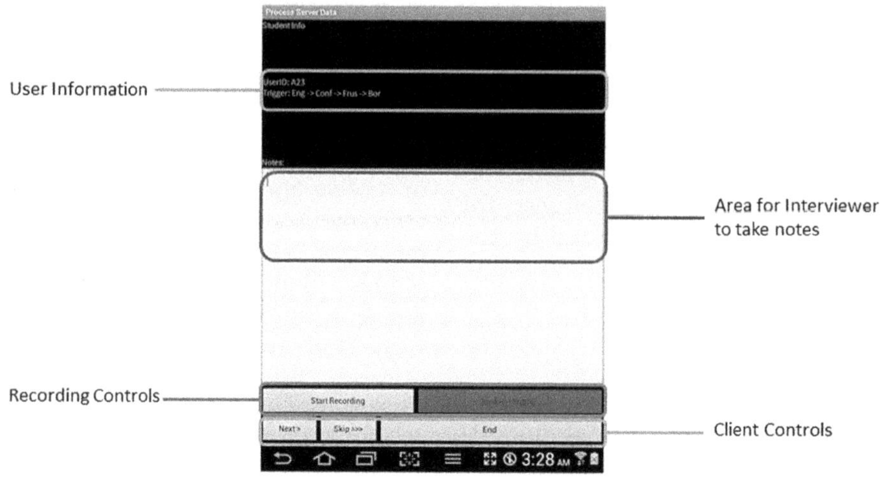

FIGURE 21.2
User interface.

integrated recording system records the material to the SD card on the Android device. The recording is then stopped when the interview taps the 'stop recording' button. A timestamp is recorded at both the start and end of the interview. Interviewers can generate more than one interview with a student from the same trigger (this functionality was rarely used in the case study). If the interviewer notices additional useful information, they can type notes into a textbox. This functionality can also be used to add contextual notes, including why a student was not interviewed (e.g., they were already talking to their teacher or if they declined to be interviewed). The text box was also occasionally used to note if a neighboring student also participated in an interview. These notes are automatically saved to the interview log file to avoid any potential data loss.

21.2.3.4 Moving On (Next, Skip, End)

Once the interview and/or observation concludes, the interviewer selects the 'next' button to advance. If the interviewer wishes to ignore a certain trigger (e.g., if a student should not be interrupted at that time), they can 'skip' that observation. Both 'next' and 'skip' send a new trigger request to the server. If the class (or observation) session has ended, the interviewer can deregister from the server-side process by pressing 'end'. Notably, there is not a back button to return to a previous interview, in keeping with design principles that suggest not allowing real-time corrections if they might introduce cascading errors.

21.2.3.5 Data

Data produced by the QRF app records all of the transactions made by the interviewer in CSV format. This includes timestamps of all button presses, the interviewer currently logged in, the student being observed or interviewed, the recorded affect and behavior, the trigger that prompted the interview recommendation, and any other notes that the interviewer provides to the system. Also stored are audio files for each of the recorded interviewer. Files' names are stored with the transaction data to allow recordings to be linked to the appropriate participant. These data can then be analyzed by researchers, or synchronized alongside system data using the timestamps from the app.

21.3 Case Study

We demonstrate the use of, and opportunities made possible by, QRF with a case study examining self-regulated learning (SRL) behaviors as students interact with AIED technologies. Understanding self-regulation lends itself to this data collection paradigm, as research has shown that multiple data sources are needed to evaluate SRL strategies (Azevedo et al., 2010; Winne, 2010) and that students may not be conscious enough of their approach to provide complete information in traditional self-report instruments.

This case study involved middle school students, interacting with a computer-based learning environment, Betty's Brain (Biswas et al., 2005), in an urban public school in Tennessee that serves approximately 700 fifth to eighth grade students. The school reports a student population that is 60% White, 25% Black, 9% Asian, and 5% Hispanic. Around 8% were enrolled in the free and reduced lunch program. Ninety-nine sixth graders used Betty's Brain during the 2018–2019 school year as part of their regular science classes. No demographic data was collected from individual students.

Students interacted with the Betty's Brain software in the classroom, as part of their regular science instruction. Students interacted with the software for 45–50 minutes per day for eight days total. As students interacted, two interviewers were directed to students at key moments in the learning process by QRF. QRF listened for two types of events: affective sequences and behaviors related to self-regulated learning strategies (more details in Section 21.3.2). Using previously integrated affect detectors, we set the server-side process to listen for affect sequences that are aligned with theoretical models of affect dynamics in educational contexts (D'Mello & Graesser, 2012), and predefined action sequences relevant to SRL (Jeong & Biswas, 2008).

21.3.1 Betty's Brain

Betty's Brain uses a learning-by-teaching model (Biswas et al., 2005), where students must teach a virtual agent named 'Betty' by creating a causal map of a scientific process (e.g., climate change or thermoregulation). Students then check their maps' validity by having Betty answer questions about scientific relationships, which she can only answer with the information they have entered into the map. Betty demonstrates her 'learning' by taking quizzes that are graded by a mentor agent, Mr Davis. As students construct Betty's map, they must navigate a variety of learning resources, including hypermedia resources about the subject matter and a teaching manual that explains how to represent causal reasoning. In this open-ended system, students choose how they build their maps, and how often they quiz Betty. They may also interact with Mr Davis, who can support their learning and teaching endeavors (Biswas et al., 2016).

Betty's Brain presents a suitable environment for examining SRL behaviors for two reasons. Firstly, students choose when and how to perform each step of the learning process (both their own and Betty's) (Kinnebrew et al., 2013; Roscoe et al., 2013). Indeed, the pedagogical agents in Betty's Brain are designed to facilitate the development of SRL behaviors by providing a framework for the gradual internalization of effective learning strategies. Secondly, students' interactions with Betty's Brain are logged to an online database with detailed timing information, enabling the microanalysis of students' actions (Siadaty et al., 2016) for the measurement of SRL strategies.

21.3.2 Developing Interview Triggers

The affect and behavior detectors used in Betty's Brain were developed using the data collected in 2017 from 93 sixth-grade students recruited from four urban public schools in the southeastern region of the United States (Jiang et al., 2018). The predictors were derived from interaction log data (146,141 actions). The outcome labels (five affective states and off-task behavior) were collected from real-time classroom observation by two coders using BROMP (Baker et al., 2020). Inter-rater reliability had Cohen's Kappa ≥ 0.60 for every construct, between the two coders. The 5,212 observations (~56 per student) of affect were distributed as follows: 78% engaged concentration, 6% confusion, 4.6% frustration, 4.2% boredom, and 2.9% delight. Off-task behavior comprised 10.2% of the total observations.

Using feature engineering, a set of meaningful features of student activity with Betty's Brain was generated as predictors for the automated detectors of affect and behavior. A total of 249 features were chosen from three categories. First, 41 basic features were designed, including time-based features (e.g., time spent reading a resource), count-based features (e.g., number of causal maps viewed), proportion features (e.g., percentage of effective actions), and descriptive features (e.g., average quiz score). Each feature was calculated in three different ways based on the time interval: since the student first started using the system (both total and normalized by time elapsed) and within the last 20 seconds. This led to a total of 123 basic features. Second, 30 sequence features were chosen based on the most frequent three action sequences (e.g., answer quiz -> read resource -> add concept). These were similarly conceptualized in three different ways (within a 20-second clip, thus far, and thus far divided by time elapsed), leading to a total of 90 sequence features. Third, 36 threshold features were developed using the optimized threshold values that led to the best correlation between the feature and student's post-test performance. The feature set, integrating across these three types of features, was then optimized to remove highly collinear features.

Finally, affect and behavior classifiers were built in RapidMiner 5.3 using selected features and binary outcome variables (e.g., off-task versus on-task, bored versus not bored). Due to the outcome labels being highly skewed, the data samples were resampled to balance the classes. In addition, a forward selection algorithm was used to only pick features that led to better model performance. The Logistic Regression, Step Regression, Naive Bayes, C4.5 (J48), and RIPPER (JRIP) algorithms were used to train the model (a selection based on the previous affect detectors). The models were evaluated using Cohen's kappa and AUC ROC on 10-fold student-level cross-validation. Experiments conducted by Jiang and colleagues (2018) showed that models with only basic features worked well for the detectors of engaged concentration, frustration, and delight. A combination of basic, sequence, and threshold features showed better results for confusion and boredom, and off-task behavior. In all cases model performance exceeded chance (average Kappa = 0.183, AUC ROC = 0.614). The final models were implemented in the server-side code to automatically detect students' affect and behavior based on the feature values that are continuously computed as the students interacted with Betty's Brain in real time.

Patterns selected to be interview triggers were a mixture of theoretically selected patterns (e.g., the affective pattern of engaged concentration -> confusion -> frustration -> boredom developed by D'Mello and Graesser, 2012) or patterns empirically identified as important (e.g., a high correlation between sustained boredom and poor post-test performance). SRL patterns, including both strategic behaviors and affect transitions, obtained from the detector algorithms in the Betty's Brain student-end, are packaged as *[pattern, priority, student_ID]* messages and communicated to a Betty's Brain data server via a router.

21.3.3 Procedure

Two sessions of data collection were conducted, over the course of seven school days each (not all days were spent interacting with Betty's Brain). The first data collection was conducted in December 2018, and the second occurred two months later. Students completed two different scenarios within the Betty's Brain system in these two sessions: climate change (session 1) and thermoregulation (session 2). Minor alterations were made to the feedback system within the platform between the first and second session, based on the findings from the interviews conducted in the first session (Ocumpaugh et al., 2021). The alterations consisted mainly of feedback providing scaffolds and prompts to users who followed previously identified sequences of actions or affect while using the system. More specifically, conversation trees between the user and the virtual agents (either Betty or Mr Davis) were adjusted to provide better guidance/hints and encouragement to students who ineffectively use within-platform resources or transition towards boredom. For example, students who incorrectly place or edit causal links on their concept map and take a quiz are prompted with one of three possible conversation scripts to help them recognize that the additions made to their concept maps were incongruent with the information they were given.

21.3.4 Data

As students interacted with Betty's Brain, automatic detectors of educationally relevant affective states (Jiang et al., 2018) and behaviors (Munshi et al., 2018), already embedded in the software, identified key moments in the students' learning processes, either from specific affective patterns or from theoretically aligned behavioral sequences. This detection was then used to prompt student interviews via QRF. Interviewers assumed a helpful but non-authoritative role when interacting with students. Interviews were open ended and occurred without a set script; however, students were often asked what their strategies were (if any) for getting through the system. As new information emerged in these open-ended interviews, questions that were designed to elicit information about intrinsic interest (e.g., 'What kinds of books do you like to read and why?') were added. Overall, however, students were encouraged to provide feedback about their experience with the software, their goals while using the software, and their choices.

A total of 594 interviews (358 from session 1 and 236 from session 2) were conducted during classroom sessions, and audio recordings were simultaneously collected during these interactions. These interviews lasted no longer than 260 seconds. Audio files were collected from the QRF app and stored on a secure file management system available only to members of the research team. Three members of the research team manually transcribed the interviews, having agreed upon formatting and style. Metadata, including associated timestamps and recording IDs, was preserved, but student-level information was deidentified (i.e., each student was assigned an alphanumeric identifier, used across data streams). Transcriptions of each interview were organized together along with their respective unique timestamps, filenames, interviewer, and student ID of the student being interviewed.

21.3.5 Data Coding

Interview transcripts were then coded for qualitative categories that correspond with SRL constructs. These constructs were based on several theoretical frameworks and perspectives (Bandura, 1986; Boekaerts, 1999; Efklides, 2011), primarily focusing on the COPES model (Winne & Hadwin, 1998). These previously published works were examined during the development process for the interview codes to identify relevant constructs that would support deeper analysis and understanding of the data in relation to experiences of self-regulated learning. Previously published findings and models guided our approach to the development of codes and their subsequent analysis. It is important, however, to note that individual contexts, implicit biases, and perspectives of the members of the research team inherently influence the entire process of the study and interpretations of the data collected (Constas, 1992; Howe & Eisenhart, 1990).

The process followed a recursive, iterative method used in Weston et al. (2001) that includes seven stages: conceptualization of codes, generation of codes, refinement of the first coding system, generation of the first codebook, continued revision and feedback, coding implementation, and continued revision of the codes (Weston et al., 2001). The conceptualization of codes included a review of related literature to capture meaningful experiences relevant to affect and SRL. Using grounded theory, we worked to identify categories that were (1) relevant to affective theory (i.e. D'Mello & Graesser, 2012) and self-regulated learning theory (e.g., Winne & Hadwin, 1998) and (2) likely to saliently emerge in the interviews. A draft lexicon and multiple criteria were generated for a coding system to help identify these constructs.

The draft lexicon was discussed with all members of the research team to build a common understanding of the constructs being examined and the features of the system. Feedback was provided by team members and the lexicon further refined. This process was repeated until the entire research team had reached a shared understanding of the criteria and constructs being examined by the codebook.

A total of 12 interview codes were developed and applied to the interview data (see Table 21.1).

Following the production of the codebook and accompanying manual, multiple coders simultaneously coded a subset of the data to reach inter-rater reliability between them before applying the coding system to all of the transcripts. The resulting kappa values for each of the interview codes are summarized in Table 21.2. Table 21.2 also summarizes the rates at which each interview code was observed across all students and all interviews. Throughout the coding process, the external coders met and clarified any concerns with the authors of the codebook to avoid misinterpretation or miscoding of the data. As these qualitative codes are not mutually exclusive, a single interview may be coded under multiple categories.

21.3.6 Impact on Scholarly Work

These interviews, the codes, and by extension, the QRF method have led to several scientific papers examining self-regulation, affect, and the interplay between the two. Bosch et al. (2021) used QRF interview transcripts to better understand metacognition and its affect in AIED technologies. This work leveraged the affect data collected as well as automatically analyzing interview transcripts for markers of metacognition. Baker et al. (2021) used the data collected for an in-depth analysis of frustration in AIED systems, considering both the causes and the effects of frustration in different students. Ocumpaugh et al. (2021)

TABLE 21.1

Interview Coding Categories

Code	Description
Difficult	Negative evaluations, confusion, or frustration while interacting with the platform
Helpfulness	Utility of within-game resources in learning, improvement, and positive evaluations of the resources
Interestingness	Interestingness of within-game resources in learning and a continued desire to use the platform
Strategic use	Indicates a plan for interacting with the platform and notes changes in strategy or interaction with the platform based on experiences
Perceived familiarity	The content has been previously learned or encountered and the student mentions ease in answering questions/ completing modules with familiar content
Positive Mr Davis attribution	Explicitly mentions interactions with Mr Davis as positive or negative experiences
Positive science attribution	Explicitly mentions science in relation to books read, future careers, subjects in school, expressed interest, and overall evaluations of science
Positive persistence	Expression of a desire for challenge and that the current task is a challenge, there is active pursuit of a goal, and repeated attempts to complete a step/problem
Procedural strategy	Step-by-step approach to the learning activity, active use of within-platform tools and interaction with the system, references a previous step or step following current actions
Motivational strategy	Explicit indication of an expected outcome from behaviors/actions, explicitly mentions a pursuit for mastery, contains a positive attribution/emotion towards completion of an activity, and mentions a desire to meet task demands
Task adaptation	Indicates a comparison between learning modules/activities and describes a change in activity in response to achievement or failure with a previous action
Self-confidence	Positive description of one's own progress or ability, implied monitoring of progress while learning, willingness to encounter challenges while learning, and recognition of helpful resources

TABLE 21.2

Inter-rater Reliability and Frequency of Each Interview Code across All Students and Interviews

		Student Level		Interview Level	
	κ	Study 1 (93)	Study 2 (89)	Study 1 (358)	Study 2 (236)
Difficult	.911	76.77%	73.74%	40.78%	59.32%
Helpfulness	.463	35.35%	63.64%	12.29%	50.85%
Interestingness	.726	8.08%	18.18%	2.23%	8.90%
Strategic use	.911	78.79%	77.78%	48.88%	73.73%
Perceived familiarity	.789	16.16%	10.10%	4.47%	4.24%
Positive Mr Davis attribution	.838	6.06%	48.48%	1.68%	30.08%
Positive science attribution	.837	21.21%	17.17%	6.70%	7.20%
Positive persistence	.911	48.48%	66.67%	22.35%	52.12%
Procedural strategy	.862	80.81%	79.80%	52.79%	75.85%
Motivational strategy	.870	65.66%	72.73%	37.99%	62.29%
Task adaptation	.808	75.76%	81.82%	45.81%	74.58%
Self-confidence	.877	71.72%	79.80%	41.62%	70.76%

demonstrated the potential of targeted interviews for identifying 'pain points' in the AIED software and subsequent iterative design process, refining the design of Mr Davis and Betty in ways that improved outcomes. Taken together, these articles demonstrate the wide potential of this data collection approach. By facilitating the collection of rich, time synchronized interviews with theoretically grounded triggers, we can pursue a wide variety of research questions.

21.4 General Discussion and Conclusions

21.4.1 Summary

This chapter introduces the Quick Red Fox (QRF) handheld app for targeted classroom observation and the associated backend software that enables its functionality. QRF informs a researcher when predefined events of interest occur in the classroom and provides support for collecting interviews and collecting qualitative observations. The key innovation in QRF is the idea of targeting qualitative data collection in real time, thus optimizing researcher's time.

We then presented a case study that used QRF to study self-regulated learning and affect in multiple classes as students interacted with the Betty's Brain learning system. This case study demonstrates the potential of this approach, yielding several findings around the manifestation of both affect and self-regulation that would be difficult to obtain using the previous methodology.

In general, QRF helps to address the 'one shot' problem by alerting researchers to infrequent or unseen behaviors. Though this approach does not fully solve the 'one-shot' issue – unexpected patterns may still be missed, and a single researcher still cannot be in two places at once – QRF comes closer to optimizing researcher time. In principle, any event that can be automatically detected (either through interaction analysis or more complex sensors) could be used as an interview trigger. Selecting triggers remains highly context-dependent and relies on researcher judgment, but the app and approach can support a wide range of use cases.

21.4.2 Applications

QRF can be used in a variety of education and training contexts. This is due to the combination of two factors. First, in classroom and training contexts, there are typically a number of people interacting with a given learning system at the same time. Second, research has consistently shown that internal cognitive and affective processes greatly influence how we learn but are often challenging to observe (Duckworth & Yeager, 2015; Linnenbrink, 2007). In K-12 educational research, there has been increased interest in understanding how complex internal processes such as emotion regulation, or engagement, impact learning, and how we might scaffold beneficial learning behaviors for students (Azevedo & Hadwin, 2005; Dumdumaya et al., 2017). QRF facilitates data collection that could allow interviewers to tap into a number of constructs that are crucial for effective learning but typically challenging to collect data on.

As such, QRF could be used for several potential applications, for both research and design. The case study above shows its potential usefulness for studying self-regulated learning and affect. In problem-solving domains such as mathematics and science, targeted interviews could be used to collect students' explanations of their problem-solving strategies, allowing researchers to better understand student misconceptions. QRF could be used to interview students who become stuck in a puzzle game, to figure out if the learner is not perceiving a key part of the interface or task. Furthermore, QRF could be used when a learner is wheel-spinning (Beck & Gong, 2013), to see what hints or scaffolding could get them back on track. QRF could also be used to evaluate new AIED technologies as they are developed, interviewing students about their experiences as software is refined.

Beyond education, QRF or apps like QRF could be used in usability research. QRF might be used to trigger interviews when users make actions not initially expected by developers or to better understand how users respond to error messages from the system. QRF could also be used to study usability outside the lab in real-world contexts, a crucial step for many projects, when real-world conditions may impact usability (Bevan & Macleod, 1994). For example, QRF may be useful in studying the usability of medical technologies (Acharya et al., 2010), a field where researchers would want to limit the number of interviews so as not to distract the users from their primary task of caregiving. Similar to educational environments, there are often multiple interactions happening at any given time (multiple caregivers each with multiple patients), thus optimizing interviewer's time would be critical.

Though we provide these sample applications, a key benefit of the QRF infrastructure is it can be used to address the 'one shot' problem in almost any environment, providing researchers can detect the event of interest. Detection must be timely, and somewhat accurate, and similarly, the environment must be suited to interviews (e.g., not a silent theatre). QRF leverages existing detection, likely machine learned models, but could also be triggered by other kinds of events such as rationally defined interaction patterns. Put simply, if you can define the event, and detect the event, you can interview after the event with QRF.

21.4.3 Limitations

The interviewing approach that QRF enables is not without its limitations. Many of those limitations center around the ways that QRF is targeted. QRF's targeting is based on predefined triggers. The approach is therefore limited by the triggers that are chosen. Interesting and useful opportunities may be missed if the research team was not aware in advance that a specific event would be important to study. This may occur either because of limited relevant theory or researchers' limited knowledge of the system. Therefore, it may be useful to conduct a round of more open observation or data analysis prior to commencing work with QRF. Similarly, if the detectors used are inaccurate and do not correctly identify the moments of interest, then the method is severely weakened. That said, it will still facilitate an interview approach that avoids students being interviewed too frequently. But in this situation the data may not provide the same level of insight on specific events of interest as it would with accurate detectors.

Furthermore, even if the research team knows what is relevant and important to study, the approach may be limited by the quality of detection available. QRF research in the context of Betty's Brain was largely enabled by the availability of high-quality detectors of self-regulated learning and affect. Learning systems for which sophisticated detection is unavailable may find that there are limits to what can be studied using QRF. It may still be possible to identify when a student spends substantially more time on a learning task than their peers or performs more poorly on a task relative to the average, but more complex constructs may be unavailable. In these cases, approaches such as clustering, sequence mining, or outlier detection may be used to provide more information for triggering interviews but may be unable to achieve the clarity of high-quality detectors of specific, well-understood constructs.

As such, an approach like QRF that focuses researcher time on key events is only as good as our ability to automatically detect that key event. Fortunately, the last decade has seen considerable progress within the educational data mining community on developing high-quality detectors of the types of constructs that might serve as triggers in QRF. A methodology like QRF's targeted interviews has only become feasible now due to that progress.

21.4.4 Future Development

The next key step for QRF is expansion: to a broader range of constructs, and a broader range of learning systems. Expanding the use of QRF in these fashions will naturally lead to enhancements to the app and infrastructure, to tailor their application for other uses. Potential extensions could include providing more information on the learner to the interviewer, and suggested questions for less experienced interviewers when the app is used at a greater scale. All a platform needs to be used with QRF is a high-quality interaction data stream and a server architecture where the communications architecture can be integrated. Our code for QRF is available online and fully open source, at https://github.com/pcla-code/QRF. We also invite researchers interested in using QRF to reach out to us to discuss potential collaborations.

Acknowledgments

This work was supported by NSF #DRL-1561567.

Note

1. https://github.com/pcla-code/QRF.

References

Acharya, C., Thimbleby, H., & Oladimeji, P. (2010). Human computer interaction and medical devices. *Proceedings of the HCI 2010, 24*, 168–176.

Andres, J. M. A. L., Ocumpaugh, J., Baker, R. S., Slater, S., Paquette, L., Jiang, Y., Karumbaiah, S., Bosch, N., Munshi, A., Moore, A., & Biswas, G. (2019). Affect sequences and learning in Betty's brain. In *Proceedings of the 9th international conference on learning analytics & knowledge* (pp. 383–390).

Azevedo, R., & Hadwin, A. F. (2005). *Scaffolding self-regulated learning and metacognition – Implications for the design of computer-based scaffolds*. Springer.

Azevedo, R., Johnson, A., Chauncey, A., & Burkett, C. (2010). Self-regulated learning with MetaTutor: Advancing the science of learning with metacognitive tools. In M. S. Khine & I. M. Saleh (Eds.), *New science of learning: Cognition, computers and collaboration in education* (pp. 225–247). Springer New York. https://doi.org/10.1007/978-1-4419-5716-0_11

Baker, R. S., Corbett, A. T., & Wagner, A. Z. (2006). Human classification of low-fidelity replays of student actions. *Proceedings of the Educational Data Mining Workshop at the 8th International Conference on Intelligent Tutoring Systems, 2002*, 29–36.

Baker, R. S., Nasiar, N., Ocumpaugh, J., Hutt, S., Andres, J. M. A. L., Slater, S., Schofield, M., Moore, A., Paquette, L., Munshi, A., & Biswas, G. (2021). Affect-targeted interviews for understanding student frustration. *Artificial Intelligence in Education, 2021.*

Baker, R. S., Ocumpaugh, J. L., & Andres, J. (2020). BROMP quantitative field observations: A review. In R. Feldman (Ed.), *Learning science: Theory, research, and practice*. McGraw-Hill.

Bandura, A. (1986). The explanatory and predictive scope of self-efficacy theory. *Journal of Social and Clinical Psychology, 4*(3), 359–373.

Beck, J. E., & Gong, Y. (2013). Wheel-spinning: Students who fail to master a skill. In *International conference on artificial intelligence in education, 7926 LNAI* (pp. 431–440). https://doi.org/10.1007/978-3-642-39112-5_44

Bernacki, M. L. (2017). Examining the cyclical, loosely sequenced, and contingent features of self-regulated learning: Trace data and their analysis. In D. H. Schunk & J. A. Greene (Eds.), *Handbook of self-regulation of learning and performance* (pp. 370–387). Routledge.

Bevan, N., & Macleod, M. (1994). Usability measurement in context. *Behaviour & Information Technology, 13*(1–2), 132–145.

Biswas, G., Leelawong, K., Schwartz, D., Vye, N., & The Teachable Agents Group at Vanderbilt. (2005). Learning by teaching: A new agent paradigm for educational software. *Applied Artificial Intelligence*. https://doi.org/10.1080/08839510590910200

Biswas, G., Segedy, J. R., & Bunchongchit, K. (2016). From design to implementation to practice a learning by teaching system: Betty's brain. *International Journal of Artificial Intelligence in Education*. https://doi.org/10.1007/s40593-015-0057-9

Boekaerts, M. (1999). Self-regulated learning: Where we are today. *International Journal of Educational Research, 31*(6), 445–457.

Bosch, N., Zhang, Y., Paquette, L., Baker, R. S., Ocumpaugh, J., & Biswas, G. (2021). Students' verbalized metacognition during computerized learning. *ACM SIGCHI: Computer-human interaction, 12*. Association for Computing Machinery. https://doi.org/10.1145/3411764.3445809

Botelho, A. F., Baker, R. S., & Heffernan, N. T. (2017). Improving sensor-free affect detection using deep learning. In *Artificial Intelligence in Education*. https://doi.org/10.1007/978-3-319-61425-0_4

Cleland, J. A. (2017). The qualitative orientation in medical education research. *Korean Journal of Medical Education, 29*(2), 61–71. https://doi.org/10.3946/kjme.2017.53

Cobb, P., Stephan, M., McClain, K., & Gravemeijer, K. (2001). Participating in classroom mathematical practices. *Journal of the Learning Sciences, 10*(1–2), 113–163.

Constas, M. A. (1992). Qualitative analysis as a public event: The documentation of category development procedures. *American Educational Research Journal, 29*(2), 253–266.

D'Mello, S. K. (2018). What do we think about when we learn? In K. K. Mills, D. Long, J. Magliano, & K. Wierner (Eds.), *Deep comprehension* (pp. 52–67). Routledge.

D'Mello, S. K., & Graesser, A. (2012). Dynamics of affective states during complex learning. *Learning and Instruction, 22*(2), 145–157. https://doi.org/10.1016/j.learninstruc.2011.10.001

Duckworth, A. L., & Yeager, D. S. (2015). Measurement matters: Assessing personal qualities other than cognitive ability for educational purposes. *Educational Researcher, 44*(4), 237–251. https://doi.org/10.3102/0013189X15584327

Dumdumaya, C. E., Banawan, M. P., Rodrigo, M., Mercedes, T., Ogan, A., Yarzebinski, E., & Matsuda, N. (2017). Investigating the effects of cognitive and metacognitive scaffolding on learners using a learning by teaching environment. In *International Conference on Computers in Education (ICCE).*

Efklides, A. (2011). Interactions of metacognition with motivation and affect in self-regulated learning: The MASRL model. *Educational Psychologist, 46*(1), 6–25.

Holstein, K., McLaren, B. M., & Aleven, V. (2017). Intelligent tutors as teachers' aides: Exploring teacher needs for real-time analytics in blended classrooms. In *Proceedings of the seventh international learning analytics & knowledge conference* (pp. 257–266).

Howe, K., & Eisenhart, M. (1990). Standards for qualitative (and quantitative) research: A prolegomenon. *Educational Researcher, 19*(4), 2–9.

Jeong, H., & Biswas, G. (2008). Mining student behavior models in learning-by-teaching environments. *EDM*, 127–136.

Jiang, Y., Bosch, N., Baker, R. S., Paquette, L., Ocumpaugh, J., Andres, A., Moore, A. L., & Biswas, G. (2018). Expert feature-engineering vs. deep neural networks: which is better for sensor-free affect detection? *Artificial Intelligence in Education*, 198–211. https://doi.org/10.1007/978-3-319-93843-1_15

Juvonen, J., & Murdock, T. B. (1995). Grade-level differences in the social value of effort: Implications for self-presentation tactics of early adolescents. *Child Development, 66*(6), 1694–1705.

Kane, T., Kerr, K., & Pianta, R. (2014). *Designing teacher evaluation systems: New guidance from the measures of effective teaching project*. John Wiley & Sons.

Kelley, T. R., Capobianco, B. M., & Kaluf, K. J. (2015). Concurrent think-aloud protocols to assess elementary design students. *International Journal of Technology and Design Education, 25*(4), 521–540.

Kinnebrew, J. S., Biswas, G., Sulcer, B., & Taylor, R. S. (2013). Investigating self-regulated learning in teachable agent environments. In *International handbook of metacognition and learning technologies* (pp. 451–470). Springer. https://doi.org/10.1007/978-1-4419-5546-3_29

Lehrer, R., & Schauble, L. (2011). Designing to support long-term growth and development. In T. Koschmann (Ed.), *Theories of learning and studies of instructional practice* (pp. 19–38). Springer.

Linnenbrink, E. A. (2007). The role of affect in student learning: A multi-dimensional approach to considering the interaction of affect, motivation, and engagement. In R. Pekrun (Ed.), *Emotion in education* (pp. 107–124). Elsevier. https://doi.org/10.1016/B978-012372545-5/50008-3

Lodge, J. M., Panadero, E., Broadbent, J., & de Barba, P. G. (2018). Supporting self-regulated learning with learning analytics. In J. M. Lodge, J. C. Horvath, & L. Corrin (Eds.), *Learning analytics in the classroom* (pp. 45–55). Routledge.

Makhlouf, J., & Mine, T. (2020). Analysis of click-stream data to predict STEM careers from student usage of an intelligent tutoring system. *Journal of Educational Data Mining, 12*(2), 1–18.

Marcus-Quinn, A., & Hourigan, T. (2017). *Handbook on digital learning for K-12 schools*. Springer.

Meany-Daboul, M. G., Roscoe, E. M., Bourret, J. C., & Ahearn, W. H. (2007). A comparison of momentary time sampling and partial-interval recording for evaluating functional relations. *Journal of Applied Behavior Analysis, 40*(3), 501–514.

Mills, C., Gregg, J. M., Bixler, R., & D'Mello, S. K. (2021). Eye-mind reader: An intelligent reading interface that promotes long-term comprehension by detecting and responding to mind wandering. *Human-Computer Interaction, 36*(4), 306–332. https://doi.org/10.1080/07370024.2020.1716762

Munshi, A., Rajendran, R., Ocumpaugh, J., Biswas, G., Baker, R. S., & Paquette, L. (2018). Modeling learners' cognitive and affective states to scaffold srl in open-ended learning environments. In *UMAP 2018 - Proceedings of the 26th conference on user modeling, adaptation and personalization* (pp. 131–138). https://doi.org/10.1145/3209219.3209241

Ocumpaugh, J., Hutt, S., Andres, J. M. A. L., Baker, R. S., Biswas, G., Bosch, N., Paquette, L., & Munshi, A. (2021). Using qualitative data from targeted interviews to inform rapid AIED development. In *Proceedings of the 29th international conference on computers in education*.

Ostrov, J. M., & Hart, E. J. (2013). Observational methods. In T. Little (Ed.), *The Oxford handbook of quantitative methods in psychology* (Vol. 1, pp. 286–319). https://doi.org/10.1093/oxfordhb/9780199934874.013.0015

Pascoe, J., Ryan, N., & Morse, D. (2000). Using while moving: HCI issues in fieldwork environments. *ACM Transactions on Computer-Human Interaction (TOCHI), 7*(3), 417–437.

Roscoe, R. D., Segedy, J. R., Sulcer, B., Jeong, H., & Biswas, G. (2013). Shallow strategy development in a teachable agent environment designed to support self-regulated learning. *Computers and Education*. https://doi.org/10.1016/j.compedu.2012.11.008

San Pedro, M. O., Ocumpaugh, J., Baker, R. S., & Heffernan, N. T. (2014). Predicting STEM and non-STEM college major enrollment from middle school interaction with mathematics educational software. *Educational Data Mining*, 276–279.

Schofield, J. W. (1995). *Computers and classroom culture*. Cambridge University Press.

Siadaty, M., Gasevic, D., & Hatala, M. (2016). Trace-based microanalytic measurement of self-regulated learning processes. *Journal of Learning Analytics, 3*(1), 183–214. https://doi.org/10.18608/jla.2016.31.11

Warren, D., Shen, E., Park, S., Baylor, A. L., & Perez, R. (2005). Adult learner perceptions of affective agents: Experimental data and phenomenological observations. In *Proceedings of the 2005 conference on artificial intelligence in education* (pp. 944–946).

Wessel, D. (2015). The potential of computer-assisted direct observation apps. *International Journal of Interactive Mobile Technologies, 9*(1).

Weston, C., Gandell, T., Beauchamp, J., McAlpine, L., Wiseman, C., & Beauchamp, C. (2001). Analyzing interview data: The development and evolution of a coding system. *Qualitative Sociology, 24*(3), 381–400. https://doi.org/10.1023/A:1010690908200

Winne, P. H. (2010). Improving measurements of self-regulated learning. *Educational Psychologist, 45*(4), 267–276. https://doi.org/10.1080/00461520.2010.517150

Winne, P. H., & Hadwin, A. F. (1998). Studying as self-regulated learning. In D. J. Hacker, J. Dunlosky, & A. C. Graesser (Eds.), *Metacognition in educational theory and practice* (pp. 277–304). Erlbaum.

Wixon, M., Arroyo, I., Muldner, K., Burleson, W., Rai, D., & Woolf, B. (2014). The opportunities and limitations of scaling up sensor-free affect detection. *Educational Data Mining, 2014.*

Yoon, S., Anderson, E., Lin, J., & Elinich, K. (2017). How augmented reality enables conceptual understanding of challenging science content. *Journal of Educational Technology & Society, 20*(1), 156–168.

22

A Systematic Review of AI Applications in Computer-Supported Collaborative Learning in STEM Education

Jingwan Tang, Xiaofei Zhou, Xiaoyu Wan, and Fan Ouyang

CONTENTS

22.1 Introduction and Review of Previous Work 333
22.2 Methodology 334
22.2.1 Research Purpose and Questions 334
22.2.2 Searching the Literature 335
22.2.3 Selecting Studies 335
22.2.4 Data Extraction 336
22.2.5 Data Synthesis and Analysis 336
22.3 Results 338
22.3.1 RQ1: The Overall Trend 338
22.3.1.1 Type of Research 338
22.3.1.2 Educational Contexts 338
22.3.1.3 Research Focus 339
22.3.1.4 Research Methods 339
22.3.2 RQ2: AI Techniques Used to Support Group Formation 340
22.3.2.1 Group Size and Grouping Mechanisms 341
22.3.2.2 AI Techniques and Evaluation 342
22.3.3 RQ3: AI-Enabled Systems to Support Student Interactions 343
22.3.3.1 Student Interaction Design 344
22.3.3.2 Student Interaction Evaluation 347
22.4 Discussions and Implications 349
22.4.1 Overall Trend of AI Applications in CSCL Research 349
22.4.1.1 Type of Research Paper 349
22.4.1.2 Research Focus 350
22.4.1.3 Educational Contexts 350
22.4.1.4 Research Design 350
22.4.2 AI Applications to Support Group Formation 350
22.4.2.1 Issues of Design 350
22.4.2.2 Issues of Evaluation 351
22.4.3 AI Applications to Support Student Interactions 351
22.4.3.1 Issues of Design 351
22.4.3.2 Issues of Evaluation 352
22.5 Conclusions 353
References 353

22.1 Introduction and Review of Previous Work

Computer-supported collaborative learning (CSCL) is a research field concerning joint meaning-making practices and how these practices are mediated by tools and languages (Stahl, 2015; Stahl et al., 2006; Wise & Schwarz, 2017). One of the primary goals of CSCL is to understand what and how technologies can help create a learning environment that supports the co-construction of knowledge-building activities in face-to-face or/and online learning spaces (Jeong

DOI: 10.1201/9781003181187-27

et al., 2019; Resta & Laferriere, 2007). The advancement of AI techniques (e.g., machine learning, deep learning, and natural language processing) has brought new opportunities to support CSCL (Zawacki-Richter et al., 2019), such as optimizing group formations, supporting student interactions, and facilitating domain understandings (Kumar & Kim, 2013; Magnisalis et al., 2011; Wise & Schwarz, 2017).

In CSCL, the formation of a group and student interaction within a group are two essential factors to guarantee productive collaboration (Suthers & Seel, 2012), which are also two primary targets of the design and implementation of AI techniques and tools as identified by previous review work (Magnisalis et al., 2011). On the one hand, group formation sets up a social learning environment for collaborative learning, which profoundly influences further group development, process, and performances (Putro et al., 2020). Previous studies have suggested that learner characteristics used for group formation critically influence collaborative quality, such as learning styles (Alfonseca et al., 2006) and interpersonal skills (Ren et al., 2012). Whether such characteristics should be grouped in a homogeneous or a heterogeneous way has been long debated (Noroozi et al., 2012; Cheng et al., 2008). Nevertheless, there is a consensus that a careful design of group composition can foster group cohesion, significantly influencing student interactions during collaborative learning (Magnisalis et al., 2011). On the other hand, students' interaction within a group is key to successful collaborative learning, through which learners share understandings, exchange ideas, and negotiate meanings (Dillenbourg & Hong, 2008; Hiltz, 1994). Previous studies have identified two factors of student interactions that help achieve desired learning outcomes: (a) engaging in knowledge construction through argumentations, such as reasoning, elaboration, and justification (Noroozi et al., 2012; Ohlsson, 1995); and (b) building a group atmosphere with positive interdependence and accountability (Johnson & Johnson, 2008).

A total of five literature reviews have been located, among which three of them systematically reviewed group formation (Maqtary et al., 2017; Odo et al., 2019; Purtro et al., 2020) and two reviewed student interactions (Harsely, 2015; Magnisalis et al., 2011). Regarding group formation, Odo et al. (2019) summarized AI techniques from 21 relevant studies between 2002 and 2017, including data mining, generic algorithm, approximation algorithms, semantic algorithm, ant colony optimization, and particle swarm optimization. They indicated, however, that no studies compared the effectiveness of AI algorithms. Maqtary et al. (2017) investigated 18 papers from 2005 to 2015 and reported that the commonly used AI technique was evolutionary algorithms, followed by clustering algorithms for homogeneous grouping. The most commonly used student attribute for the grouping was prior knowledge, followed by learning styles, personality traits, and relationships. Putro and colleagues (2020) analyzed 26 articles published between 2001 and 2019. They found that researchers tended to measure the performance of AI techniques on (a) formation performance (i.e., group formation time and optimization of member distribution in groups) and (b) students' performance (i.e., collaboration level, knowledge acquisition, and skill development). Furthermore, they suggested that research should consider both individual and group attributes to determine the most optimal approach for group formation. The other two review articles reviewed the applications of AI to support student interactions within groups during collaborative learning. Magnisalis et al. (2011) found that more than half of the research aimed to support peer interactions in CSCL settings. Furthermore, the findings suggested that the *implicit feedback* (i.e. actions taken by students that are provided implicitly by AI-enabled tools) was more optimal than the *explicit feedback* (i.e. actions taken by students that are explicitly described through AI supports) for developing conceptual understandings and improving collaborative skills (Walker et al., 2009). Harsley (2015) critically evaluated four intelligent tutoring systems (ITSs) (i.e., Cognitive Tutor Algebra, Rashi, Wayang Outpost, and Basilica) and concluded that the effects of ITSs on collaborative learning had not been proven yet. In summary, two gaps are identified in the existing literature review. First, there is a lack of a systematic review of applications of AI in collaborative learning under STEM education. Second, little information has been reported about the evaluation of the effect of associated AI techniques on collaborative learning in STEM education.

This literature review aims to gain an understanding of the applications and effects of AI in CSCL in STEM education from 2011 to 2021, including (a) the design of AI for supporting group formation and student interaction in CSCL in STEM education, and (b) the evaluation of proposed AI techniques for improving CSCL in STEM education.

22.2 Methodology

22.2.1 Research Purpose and Questions

The primary purpose of this literature review is to gain a comprehensive understanding of the AI applications in CSCL published between January 2011 and March 2021 in STEM education contexts. Three research questions (RQs) were posed:

RQ1: *What was the overall trend of research on applying AI to support CSCL in STEM education?*

RQ2: *How did AI apply to support group formation and what were the effects of AI on group formation?*

RQ3: *How did AI apply to support student interaction, and what were the effects of AI on supporting student interactions?*

22.2.2 Searching the Literature

This literature review followed the procedures described in Jesson et al. (2011). To identify relevant articles, we conducted a literature search in March 2021 across several databases, including Association for Computing Machinery (ACM), DBLP Computer Science Bibliography, IEEE, JSTOR, ScienceDirect, Taylor& Francis, and Wiley. Also, journals of educational technology recommended by *Google Scholar Metrics* were used for further search to ensure the comprehensiveness of the literature being reviewed.

Three sets of keywords were adopted: (a) keywords related to collaboration, including 'collaborative', 'cooperative', 'team', and 'group'; (b) keywords related to education, including 'learning' and 'CSCL'; and (c) keywords related to artificial intelligence, including 'artificial intelligence', 'AI', 'intelligent', 'adapt', and 'personalized'. These keywords were connected by the Boolean 'OR' operator within the set as well as the 'AND' operator between the sets.

22.2.3 Selecting Studies

We created inclusion and exclusion criteria (see Table 22.1) and selected articles following three steps:

Quick screening. We scanned the titles, keywords, and abstracts of each article resulting from the keywords searching. Papers that did not conform to the criteria of publication year, source, type, language, and topic were excluded. A total of 177 articles were included and downloaded for further examination.

Full paper screening. We read each article to ensure that its research topic and research domain were consistent with the review purposes. First, we excluded articles that were irrelevant to STEM education. This included non-educational purposes (e.g., collaboration in working places) or non-STEM contexts (e.g., language learning). We applied the NSF (National Science Foundation) definition of STEM, thus including research works from mathematics, natural sciences, engineering, computer, and information sciences, as well as the social and behavioral sciences (see Gonzalez & Kuenzi, 2012). Also, we kept articles that were subject independent and articles that did not specify subject information. Second, drawing upon the definition of CSCL environment (Dennen & Hoadley, 2013), we included the AI techniques aiming to support collaborative learning between learners and excluded research work that applied AI for educational assessments. A total of 112 articles were included during this step.

Quality appraisal. We evaluated whether the selected articles could answer our research questions to assure articles' quality. If several papers described the same AI techniques with few improvements in collaborative learning support, we selected the report with a more detailed description of the design information and the research process. Finally, a total

TABLE 22.1

Inclusion and Exclusion Criteria for Article Selection

	Inclusion criteria	Exclusion criteria
Year of publication	2011.01–2021.03	–
Publication source	Peer-reviewed conference proceedings or journals	Book chapter; Magazines; Newspapers
Type of paper	Research article	Review; Poster; Short paper; Expert opinion
Language	English	–
Topic	Collaborative learning and AI techniques	Self-study or individual learning; Collaborative learning supported without AI techniques
Educational domain	STEM education; Domain independent	–
Research focus	AI techniques applied for supporting group formation and student interactions	AI techniques used for analysis of social interactions from a research perspective or/and summative assessment of learning outcomes
Details	AI techniques and research process are described in details	–

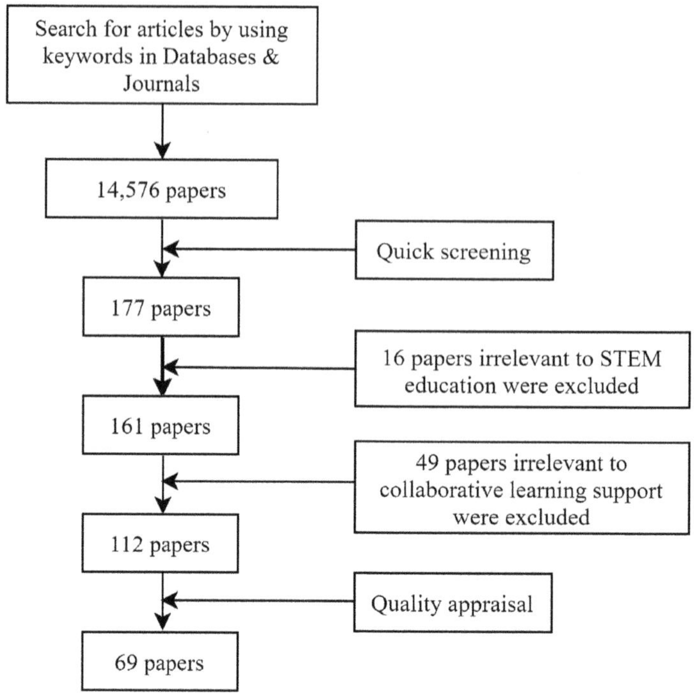

FIGURE 22.1
Search processes and results.

of 69 articles were included in this systematic review (see Figure 22.1).

22.2.4 Data Extraction

First, we extracted the bibliometric information of articles, including the year of publication, publication source, type of research, and keywords. The type of research was coded using Wieringa et al. (2005)'s classification themes. It includes (a) *proposal*, referring to articles that suggest a novel solution without a validation, (b) *validation*, referring to research conducted for investigating the effects of a proposal, and (c) *evaluation*, referring to papers about the implementation of an existing technique and its effects in practices. Second, we documented AI techniques for collaborative learning, including specific algorithms, systems, or frameworks informing that design process. Third, we recorded research results about the assessment of AI on collaborative learning, including methodologies and results.

22.2.5 Data Synthesis and Analysis

To answer RQ1, we developed a coding scheme (see Table 22.2) to synthesize information of each article, regarding (a) the type of research and publication source, (b) educational contexts, (c) research focus, and (d) research methods employed for validation/evaluation. We counted articles published from 2011 to the first quarter of 2021 (denoted as 2021 (Q1) in the rest of the article) for each code and then calculated the *growth rate* (i.e., a comparison of publications between 2011 and 2015 with publications between 2016 and 2020) (Zheng et al., 2019). Furthermore, we conducted a co-occurrence network analysis on research keywords using VOS viewer (van Eck & Waltman, 2009) to capture research interests. Two keywords that appeared together in at least three articles were contained in the keyword co-occurrence network.

Regarding RQ2, we open coded the data relevant to supporting group formation, including (a) group size, (b) learner characteristics used to group students, (c) grouping mechanisms, and (d) AI techniques proposed for generating groups. All generated codes were grounded in the selected articles. Moreover, we (e) evaluated the effects of those techniques on collaborative learning.

To answer RQ3, we open-coded data related to supporting student interactions, including (a) the end-users of AI-enabled tools, (b) interactive environments for collaborative learning, (c) the goal of design, and (d) proposed AI algorithms, tools, or systems. Furthermore, we open-coded (e) collaborative tasks, (f) durations for validation/evaluation research, (g) measures employed to assess the effects of AI technologies, and (h) evaluated outcomes. Similar to RQ2, all codes emerged from the data to represent a grounded picture of this research field.

TABLE 22.2

Coding Scheme for the Overall Trend of the Literature

Category		Codes	Definition
Information of articles	Type of papers	Proposal	Research paper that suggested a novel solution without validation
		Validation	Research paper that investigated the effects of a proposal
		Evaluation	Research paper that evaluated an existing proposal of AI techniques within different educational contexts and populations
	Publication source	Conference	–
		Journal	–
Educational contexts	Educational level	Elementary	Grade 1–6
		Middle	Grade 7–9
		High	Grade 10–12
		Higher education	College, university, graduate schools
	Educational domain	Natural science	Disciplines concerning natural phenomena such as biology, physics, and chemistry
		Mathematics	A discipline aiming for understanding patterns among numbers, quantities, and spaces
		Computer and information science	A blend of several STEM subjects, involving technology, engineering, and math
		Social and behavioral science	Disciplines of psychology, economics, sociology, and political science
Research focus	Target of research	Group formation	AI proposed for supporting group formation
		Student interactions	AI proposed for supporting student interactions
Research methods	Research design	RCT experiment	Employed a randomized controlled trial design for evaluation
		Quasi-experiment	Implemented an intervention without random sampling or/and control conditions
		Experiment (Simulation)	Simulated the real conditions of AI implementation with varied combinations of variables
		Descriptive	Reported the evaluated outcomes with descriptive data (such as frequency and percentage)
		Correlational	Assessed whether variables of interests (e.g. classification by AI vs classification by human experts) were correlated at a statistically significant level
		Design-based research	Developed AI tools or systems through iterative design and implementation
	Sample size	1~30	Small scale
		31~100	Small to medium
		101~1,000	Medium to large
		>1,000	Large scale
	Data collection	Survey	Evaluated learners' perceived experiences or attitudes towards AI and collaborative learning activities
		Pre- and post-tests	Assessed learners' gains of knowledge or skills before and after the intervention
		Log file data	Recorded learners' information while interacting with AI, including behaviors, chats, and posted messages
		Interview	Gained an in-depth understanding of learners' perspectives on AI and collaborative learning activities
		Observation	Captured user-user interaction or user-AI interaction through video and audio recording, or field notes
		Datasets	Involved multi-dimensional variables for characterizing the user models or the learning environment
	Measures	Performance of the learners (Cognitive)	Learners' performances related to knowledge acquisition and cognitive skill development

(Continued)

TABLE 22.2 (CONTINUED)

Coding Scheme for the Overall Trend of the Literature

Category	Codes	Definition
	Performance of the learners (Behavioral)	Learners' performances related to physical and verbal acts initiated during the collaborative learning process
	Performance of the learners (Affective)	Learners' performances related to feelings and emotions about the collaborative learning activities
	Experience of the learners	Learners' understanding of AI's functionalities for collaborative learning support
	Performance of algorithms/ systems	The ability of the proposed algorithms/systems for solving specific problems

TABLE 22.3

Numbers and the Growth Rate of Three Types of Research

Category	Sub-category	Total	2011–2015	2016–2020	2021(Q1)	Growth rate[a]
Type of research	Proposal and Validation	52	21	29	2	38.10%
	Evaluation	13	6	7	0	16.67%
	Proposal	4	1	2	1	100.00%

[a] The growth rate calculation excludes publications of 2021 (Q1).

TABLE 22.4

Numbers and the Growth Rate of Educational Contexts in the Recent Decade

Category	Sub-category	Total	2011–2015	2016–2020	2021(Q1)	Growth rate[a]
Educational level	Elementary	5	2	2	1	–
	Middle	6	3	3	0	–
	High	11	6	4	1	-33.33%
	Higher	35	13	21	1	61.54%
	Not specified	15	–	–		–
Educational domain	Computer science and technology and engineering	23	9	14	0	7.58%
	Natural science	11	4	5	2	1.52%
	Mathematics	9	5	4	0	−1.52%
	Social science	1	1	0	0	−1.52%
	Not specified	26	–	–		–

Note: Two articles reported studies in both middle and high school contexts; one article reported a study involving both high and higher education levels; one article reported a study conducted in both natural science and mathematics.

[a] The growth rate calculation excludes publications of 2021 (Q1).

22.3 Results

22.3.1 RQ1: The Overall Trend

22.3.1.1 Type of Research

Most of the reviewed articles reported both proposal and validation research (N = 52), followed by the evaluation research (N = 13), and the proposal research without validation (N = 4) (see Table 22.3). Of the four articles under the proposal type, one of them was recently published (Lee et al., 2021). One article (Anaya et al., 2013) has the evaluation research published later (Anaya et al., 2013), which is included in this review. Two other proposals, i.e., Gilbert et al. (2017) and Iantovics et al. (2016), lack empirical evidence to examine the effect of AI in CSCL.

22.3.1.2 Educational Contexts

Educational level. Most of the reviewed articles focused on the higher education level, which achieved the highest growth rate of 61.54% in the past five years (see Table 22.4). In contrast to higher education, elementary and middle school received the least attention. Although the high school had the second-largest publication, its growth rate decreased in the past five years.

Educational domain. Most of the reviewed articles targeted computer science, technology, and engineering, followed by natural science, and mathematics (see Table 22.4). Social science received the least attention, and only one paper on psychology was located (Hayashi, 2014). Also, the computer science domain had an evident increase in the growth rate.

22.3.1.3 Research Focus

The target of design. Among 69 reviewed articles, 19 articles targeted at the topic of group formation, 43 focused on student interactions, and two reported both group formation and student interaction. The remaining five articles belonged to neither the group formation category nor the student interaction category. For example, Challco et al. (2014) proposed a domain-independent authoring tool for collaborative scenario design. It did not directly facilitate the group development and process but helped the establishment of a CSCL environment. Lafifi et al. (2014) suggested a k-complementarity technique for grouping tutors and expected the collaboration among tutors would improve their tutoring skills for facilitating student groups. Iantovics et al. (2016) designed an intelligent system supporting problem-solving by matching students with a remote user capable of answering specific questions. This might not be considered as collaborative learning as collaboration rarely occurred between the inquirer and the replier. However, it could facilitate collaborative discussion with system improvements. Further, the number of research works related to student interaction is twice than that of group formation between January 2011 and March 2021 (see Table 22.5). Interestingly, the growth rate was significantly higher for group formation than student interaction, with 180.00% and 0.00%, respectively.

Keyword network analysis. From 2011 to 2015 (see Figure 22.2a), researchers mainly focused on (a) supporting collaborative learning by intelligent techniques in general, (b) online collaboration, and (c) student learning. There were two changes observed from 2016 to 2020 (see Figure 2b). First, a new network connected 'collaborative learning' with 'group formation' and relevant algorithms (the green network). Second, 'teaching' is closely related to 'education' and 'computer-aided instruction'.

22.3.1.4 Research Methods

Research design. Most of the reviewed articles evaluated the AI techniques with a quasi-experimental research design, which also achieved the highest growth rate of 83.33% in the past five years (see Table 22.6). This is followed by the randomized controlled trial (RCT) experimental design and the simulation design, which were reported by the equal number of selected articles (N = 18). RCT experiments, frequently applied in total though, decreased slightly in the second half compared to the first half of the decade. In contrast, the simulation methods experienced an increase with a growth rate of 57.14%. There are two types of simulation experiments. One is to simulate learners' behaviors and performance in collaboration (e.g., Saifa & Mala, 2012). Another is to evaluate if the grouping results meet specific requirements of the objective functions (Peng et al., 2020).

Sample size. Most studies selected participants with a small to medium sample size, followed by a sample of medium to large (see Table 22.6). The former had more considerable gains than the latter in the past five years regarding the growth rate. This indicates that the medium sample size became preferable to researchers. The large-scale sample was boosted in the second half of the decade, which all appeared in group formation research. These studies required large datasets to train learners' models for generating optimal collaborative learning groups (e.g., Anaya & Boticario, 2011; Procaci et al., 2015).

Data collection. The survey was the most frequently used instrument in selected articles and experienced a noticeable increase from the first half of the decade to the second half. The pre- and post-tests were the most-used methods between 2011 and 2015 but declined sharply in the past five years. Interviews and datasets became recognizable for data collection between 2016 and 2020. There are two types of

TABLE 22.5

Numbers and the Growth Rate of the Target of Research

Category	Sub-category	Total	2011–2015	2016–2020	2021(Q1)	Growth rate[a]
Target of research	Group formation	19	5	14	0	180.00%
	Student interactions	43	20	20	3	0.00%
	Both	2	0	2	0	–

Note: Two studies focusing on improving both student interactions and group formation counted within each sub-category separately.

[a] The growth rate calculation excludes publications of 2021 (Q1).

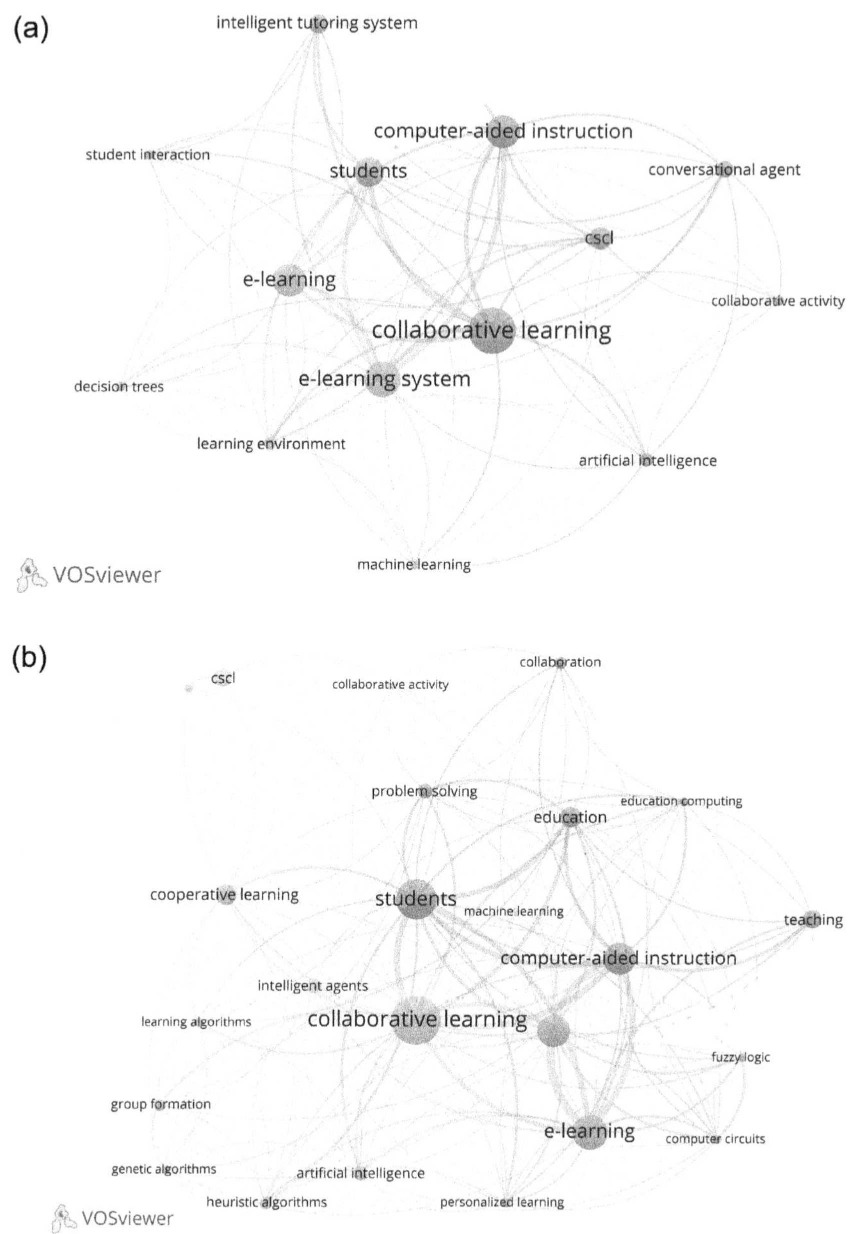

FIGURE 22.2
Keyword network analysis. (a) Keyword network analysis from 2011 to 2015. (b) Keyword network analysis from 2016 to 2020.

datasets, including the existing datasets of real users and the hypothetical profiles created by researchers. Some studies run both types of datasets to increase the accuracy of the established models (e.g., Liu et al., 2016; Procaci et al., 2015).

Measures. Most of the research works evaluated the effects of AI techniques by examining the performance of algorithms/systems (e.g., an accuracy rate of generating heterogeneous groups, the running time for producing optimal results). Also, it received significant gains in the past five years with a growth rate of 121.43%. The number of studies that measured the cognitive performance of learners remained nearly the same for the second half of the decade compared to the first half. Further, learners' affective performance experienced a growth spurt in the second half of the decade (N = 8) compared to the first half (N = 1).

22.3.2 RQ2: AI Techniques Used to Support Group Formation

All 21 articles reporting the AI techniques for group formation proposed new algorithms or systems followed by validation research.

TABLE 22.6

Numbers and the Growth Rate of Research Methods in the Recent Decade

Category	Sub-category	Total	2011–2015	2016–2020	2021(Q1)	Growth rate[a]
Research design	Quasi-experimental	19	6	11	2	83.33%
	RCT experimental	18	10	8	0	–20.00%
	Experimental (Simulation)	18	7	11	0	57.14%
	Descriptive	5	2	3	0	50.00%
	Correlational	3	1	2	0	100.00%
	Design-based research	2	1	1	0	-
Sample size	Small scale	15	8	7	0	–12.50%
	Small to medium	23	7	14	2	100.00%
	Medium to large	18	8	10	0	25.00%
	Large scale	5	1	4	0	300.00%
Data collection	Survey	27	8	18	1	125.00%
	Pre- and post-tests	15	11	3	1	–72.73%
	Log data	18	10	8	0	–20.00%
	Interview	11	3	7	1	133.33%
	Observation	6	2	3	1	50.00%
	Datasets	16	6	10	0	66.67%
Measures	Performance of algorithms/systems	45	14	31	0	121.43%
	Performance of the learners' (Cognitive)	28	14	13	1	–7.14%
	Performance of the learners' (Behavioral)	17	7	9	1	28.57%
	Performance of the learners' (Affective)	10	1	8	1	700.00%
	Experience of the learners	15	6	8	1	33.33%

[a] The growth rate calculation excludes publications of 2021 (Q1).

22.3.2.1 Group Size and Grouping Mechanisms

Group size. Seventeen studies proposed AI techniques for group formation of one size (i.e., the group size is fixed to a certain value) (see Table 22.7). Most of them favored small group sizes, namely dyad (N = 8) and four-member groups (N = 6). Four experimental studies created groups of more than two sizes for testing algorithms' capabilities in optimal group formation under different educational conditions (see Table 22.7). For example, Alhunitah and Menai (2016) analyzed and reported varying performances of algorithms for creating a group size ranging from 10 to 100.

Grouping mechanisms. Nine types of learner characteristics emerged from the selected articles to form groups (see Table 22.8). The most widely used trait was knowledge, followed by personality and cognitive style. In comparison, learners' demographic information and time availability for participation were less used by researchers to form groups. Most research works collected learner characteristics through pre-tests (e.g., personality traits) or existing data (e.g., GPA). Other studies extracted dynamic characteristics (e.g., preferred topics) from log data that recorded learners' learning and interaction behaviors to facilitate a real-time group formation.

TABLE 22.7

Numbers and Percentages of Group Formation Size

Group formation with one size		Group formation with multiple sizes	
Group size	N (%)	Group sizes	N (%)
2	8 (38.10)	2, 3, 4, 5, 6	1 (4.76)
4	6 (28.57)	4, 5	1 (4.76)
5	1 (4.76)	5, 6, 7, 8	1 (4.76)
6	1 (4.76)	10–100	1 (4.76)
10	1 (4.76)		
Total	17 (80.95)		4 (19.05)

Four types of grouping mechanisms were identified. Most studies (N = 14) aimed at enhancing intra-group heterogeneity (see Table 22.9). Six of them focused on merely forming heterogeneous groups, while another five made efforts to balance the inter-group homogeneity. For example, Peng et al. (2020) proposed a magic squares-based algorithm to enlarge the diversity of knowledge levels and personalities within groups while reducing the diversity between groups to ensure a balanced initial condition for all groups. Three studies intended to pair individual learners with more knowledgeable others (MKO) selected by (a) the similarity of problem-solving skills (Safia & Mala, 2012), or (b) the appropriateness of expertise and participation level in the domain topic (Procaci et al., 2015). The most frequently used learner characteristics for enhancing heterogeneity

TABLE 22.8

Codes of Learner Characteristic, Definition, and Descriptive Statistics

Learner Characteristic	Definition	N (%)
Knowledge	A learner's proficiency in certain knowledge domains.	13 (61.90)
Personality	A learner's personality traits, measured by personality assessments such as *Big Five, Belbin theory*, etc.	6 (28.57)
Cognitive styles	A learner's preferred ways of learning, thinking, and solving problems.	6 (28.57)
Communication skill	A learner's proficiency in communication regarding group coordination and idea exploration, such as reasoning and explaining.	5 (23.81)
Topic preference	A learner's interest in certain topics within the knowledge domain.	5 (23.81)
Participation level/Contribution	A learner's participation or contribution level of the group work.	4 (19.05)
Relationship	A learner's information that influences the building of interpersonal relationships within a group, such as friendship and collaboration history.	3 (14.29)
Demographic	A learner's demographic information such as gender and language.	2 (9.52)
Time availability	A learner's time availability for participating in collaborative learning.	1 (4.76)

TABLE 22.9

Codes of Grouping Mechanism, Definition, and Descriptive Statistics

Grouping mechanism	Definition	N (%)
Intra-group heterogeneity	Enhance intra-group heterogeneity only	6 (28.57)
	Enhance intra-group heterogeneity and balance inter-group homogeneity	5 (23.81)
	Pairing a learner with a more knowledgeable other (MKO)	3 (14.29)
Intra-group homogeneity	Enhance intra-group homogeneity	3 (14.29)
Hybrid	Enhance intra-group heterogeneity for a set of learner characteristics and enhance intra-group homogeneity for a different set of learner attributes	1 (4.76)
Other	Other grouping mechanisms which are not based on intra-group heterogeneity or homogeneity	3 (14.29)

within groups were knowledge, personality, and cognitive style.

Among three studies of homogeneous group formation, two grouped members together with similar knowledge levels and participation levels during the collaboration (Al-Tarabily et al., 2018; Huang & Huang, 2017). The other clustered students together with similar cognitive styles in perception, reception, processing, and understanding (Bernacki & Kozierkiewicz-Hetmańska, 2014). One research project created hybrid groups (Lambić et al., 2018). They promoted the diversity of prior knowledge of group members while encouraging the shared personal traits in prosocial behavior/openness, assessed by a questionnaire. Moreover, three studies employed different grouping mechanisms that did not belong to the aforementioned categories. It includes (a) grouping learners who had good relationships with each other (Chuang et al., 2012), (b) forming stable student-student matching based on knowledge level and personality traits (Yusri et al., 2020), and (c) sorting learners according to emergent topic preferences and collaborators (Tissenbaum & Slotta, 2019).

22.3.2.2 AI Techniques and Evaluation

Sample size and research design. Group formation research varied largely in sample size, ranging from 10 to more than 90,000 (see Table 22.10). The simulation research method tended to employ multiple datasets for evaluation, thus including the largest sample size among types of research design.

Measures. Most research evaluated the performance of algorithms or systems in achieving targeted grouping results, followed by the learners' performances of cognition (see Figure 22.3). Four studies evaluated the performance of both algorithms/systems and learners (Putro et al., 2020; Erkens et al., 2016; Lambić et al., 2018; Takači et al., 2017). For example, Putro et al.'s (2020) study not only assessed whether the proposed algorithm could successfully generate groups with optimal heterogeneity but also examined whether such grouping results encouraged students to perform productive collaborative behaviors (e.g., maintaining positive communications, sharing resources, or negotiating

TABLE 22.10

Research Design and Associated Sample Size

Research design	Average sample size (SD)
Experimental (simulation)	3650.13 (4430.00)
Experimental (RCT and Quasi)	135.29 (170.32)
Correlational	111.00 (103.24)
Design-based research	45 (-)

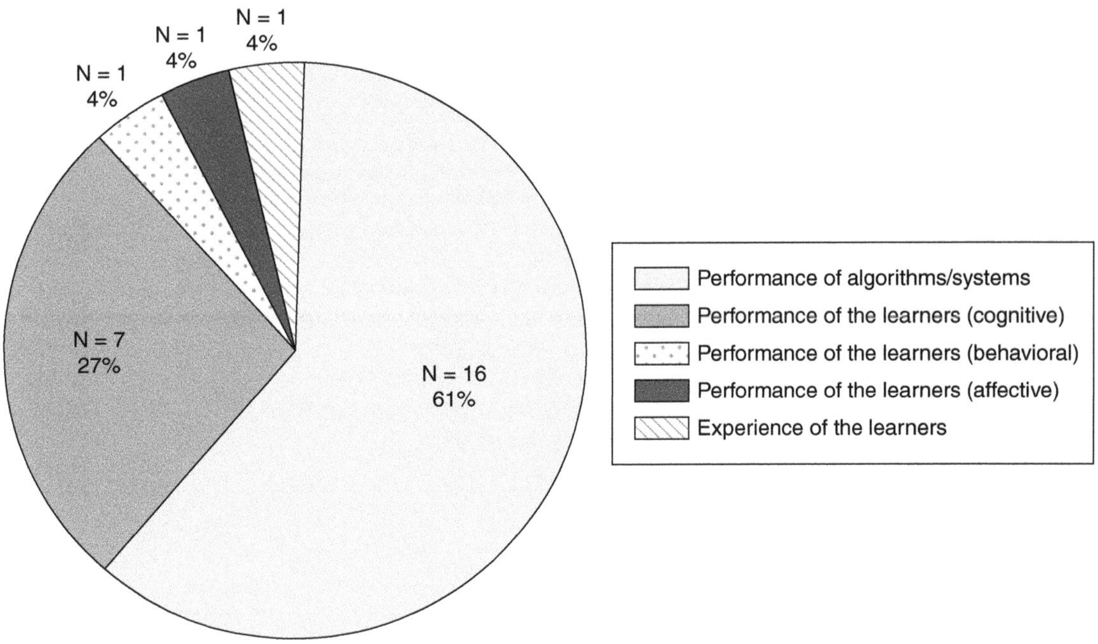

FIGURE 22.3
Distribution of group formation evaluation measures.

ideas). Researchers adopted various approaches to evaluating the algorithms' performances, including (a) goodness of heterogeneity (e.g., Putro et al., 2020), defined by an existing formula proposed by Graf and Bekele (2006); (b) quality ratio, calculated as the best solution generated by the proposed algorithms to the predefined optimal solution (Garshasbi et al., 2019); (c) self-defined optimization function such as the fit score of a particular student within a group (Yusri et al., 2020); and (d) accuracy rate, representing the number of correct identifications to the total number of student groups (e.g., Procaci et al., 2015).

AI techniques. The most regularly used AI techniques for group formation were (a) evolutionary algorithms, such as the multi-objective genetic algorithms (Garshasbi et al., 2019), and hybrid evolutionary algorithms (Yannibelli & Amandi, 2018), and (b) algorithms proposed by researchers themselves. Other major categories included (a) natural language processing (NLP)-based techniques, which were used generally for detecting meanings of learners' verbal contributions in group discussion whereby the knowledge level or/and the topic preferences of a learner could be modeled as learners' characteristics for group formation (Huang & Huang, 2017); (b) clustering, applied to identify and assign students with similar features to either the same group or different groups for enhancing intra-group homogeneity (Bernacki & Kozierkiewicz-Hetmańska, 2014) and intra-group heterogeneity (Liu et al., 2016), respectively; (c) metaheuristic, adopting especially the variable neighborhood search (VNS) to solve a set of combinatorial optimization and global optimization problems (Takači et al., 2017), (d) symbolic AI, utilized for acquiring specific learner characteristics from conversations between students; and (e) neural networks (Procaci et al., 2015) and fuzzy-based technique (Safia & Mala, 2012), used to identify appropriate matches for students in group work (see Table 22.11).

Effects. A total of 24 AI techniques from 21 research articles were evaluated (see Table 22.12). Six studies assessed (a) if the grouping results achieved the optimal outcomes and (b) if the group formation enhanced learners' performances. Nineteen AI techniques out of 24 were reported as effective to generate groups with optimality; nearly half of them were capable of improving learners' performances in collaborative learning (see Table 22.12). The latter, however, was not entirely based on empirical research conducted in authentic settings; three studies used simulation methods.

22.3.3 RQ3: AI-Enabled Systems to Support Student Interactions

A total of 45 articles depicted the design or implementation of AI for supporting student interactions, with 30 proposals followed by validation research, 13 evaluation studies, and 2 proposals only. This section provides a review of (a) student interaction design (proposal and proposal with validation studies, N = 32) and (b) student interaction evaluation (proposal with validation and evaluation studies, N = 43).

TABLE 22.11

Numbers and Percentages of AI Techniques for Group Formation

High-level category	Low-level category	N (%)
Proposed algorithm	STCP Solver + SynTeam, pairing strategy algorithm, magic square heterogeneous grouping algorithm (MASA), game-theory based partner selection algorithm, balanced gain based algorithm	5 (20.83)
Evolutionary algorithm	Particle swarm optimization (PSO), genetic algorithms, hybrid evolutionary algorithm	5 (20.83)
Natural language processing (NLP)-based technique	Latent Dirichlet Allocation (LDA), information retrieval, cosine similarity, the inverse-log-likelihood ratio for similarity	4 (16.67)
Clustering	K-means clustering related algorithm	4 (16.67)
Metaheuristic	Variable neighborhood search algorithm (VNS)	2 (8.33)
Symbolic AI	Conversation agent, Intelligent Software Agent (ISA)	2 (8.33)
Neural Networks	Artificial Neural Networks (ANN)	1 (4.17)
Fuzzy-based technique	The fuzzy associative matrix system	1 (4.17)

TABLE 22.12

Evaluation of AI Techniques for Group Formation

Higher-level category	N	Optimal grouping	Improve learning performance	Evaluated with learners/by simulation
Proposed algorithm	5	4	2	1/1
Evolutionary algorithm	5	5	1	1/0
NLP-based technique	4	3	2	2/0
Clustering	4	3	2	1/1
Metaheuristic	2	2	2	2/0
Symbolic AI	2	0	1	1/0
Neural Networks	1	1[a]	0	0/0
Fuzzy-based technique	1	1	1	0/1
Total	24	19	11	8/3

[a] Procaci et al. (2015) found good algorithm performance when using training data from larger communities with similar sizes.

22.3.3.1 Student Interaction Design

This section describes the AI design for student interaction support from (a) user type, (b) interaction setting, (c) design focus, and (d) AI techniques.

User type. The majority of AI-enabled systems targeted student users (N = 21) (see Figure 22.4), followed by tools serving both teachers and students (N = 6). For instance, Daradoumis et al. (2013) used an emotion labeling model to automatically detect and display affective behaviors within student interactions, which improved students' emotional awareness as well as enabled teachers to offer affective feedback. Two studies reported the AI-enabled systems for teacher users, which supported teachers as wandering facilitators in classrooms by reducing their workload through context-sensitive material assignment, appropriate group formation, and coordination of inquiry activities (Chounta & Avouris, 2014; Tissenbaum & Slotta, 2019).

Interaction setting. Online asynchronous (28.13%) and synchronous (21.88%) interactions achieved similar proportions in all studies (see Table 22.13). Asynchronous interactions include the interactions with the Learning Management Systems (LMS) (e.g., Chih-Ming & Ying-You, 2020) and forums (e.g., Anaya & Boticario, 2011), whereas synchronous student interactions happened with the chatbot or conversational agent (e.g., Adamson & Rosé, 2012; Chan & Fung, 2020), system-provided adaptive support (Walker et al., 2011), or real-time evaluation (Chounta & Avouris, 2014). Several research works (N = 3) mixed online interactions through both asynchronous and synchronous discussions (e.g., Arguedas et al., 2016; Caballé et al., 2013). For example, Caballé et al. (2013) proposed a virtualized collaborative session (VCS) system that supported students' live discussions and ongoing asynchronous discussions through forums. About 15.53% of papers supported blended learning, where both offline and online interactions occur during the application settings (e.g., Matazi et al., 2018; Putro et al., 2020). For example, Matazi et al. (2018) used a multi-intelligent agent to assess the level of learner collaboration in a hybrid web-technology

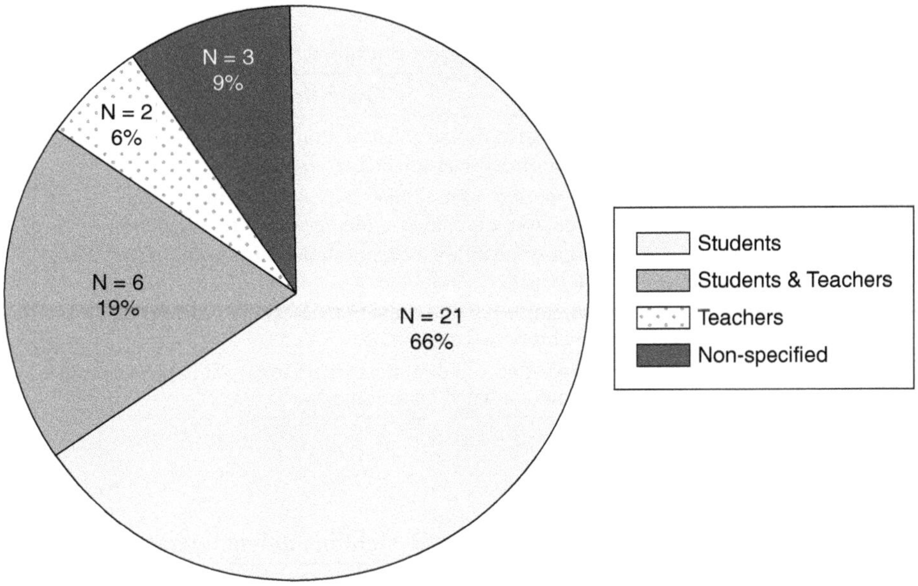

FIGURE 22.4
End-user of student interaction support systems.

TABLE 22.13
Codes of Interaction Settings, Definition, and Descriptive Statistics

Interaction Setting	Sub-category	Definition	N (%)
Online	Asynchronous	AI-supported online collaboration occurs in a delayed time. For example, forum, discussion board, email.	9 (28.13)
	Synchronous	AI-supported online collaboration occurs in real-time interaction. For example, video conference, chat room, phone.	7 (21.88)
	Non-specified	AI-supported online collaboration occurs without specifying an interaction mode.	5 (15.63)
	Mixed	AI-supported online collaboration occurs synchronously and asynchronously combined.	3 (9.38)
Blended		AI-supported collaboration occurs both online and offline.	5 (15.63)
Face-to-Face		AI-supported collaboration occurs in the same physical place without mediated communication.	3 (9.38)

course, including students' online discussions in an open forum and face-to-face interactions in the classroom. Several reported systems (N=3) enhanced face-to-face learning, such as social robot-supported facilitation of student group activities for problem-solving (Rosenberg-Kima et al., 2020).

Design focus. The largest percentage of studies (40.63%) provided multi-level support (see Table 22.14), related to the affective, behavioral, and cognitive aspects of student interactions, to enhance the quality of collaboration (e.g., Durán & Amandi, 2011; Tissenbaum & Slotta, 2019; Chounta & Avouris, 2014). Durán and Amandi (2011) constructed a mixed model based on the sociocultural perspective of collaboration, which combined characteristics of the individual, group, and collaborative situations. Rummel et al. (2012) adopted collaboration scripts on Cognitive Tutor Algebra (CTA) to improve student interactions by coordinating the activity flow, promoting student motivation for collaboration, and developing their problem-solving skills. This research indicated that researchers were inclined to situate students in a constructive learning environment to support their diverse needs.

The second largest proportion of research was on improving students' cognitive learning outcomes (see Table 22.14). Systems of this strand were intended to increase learning gains (e.g., Huang et al., 2015; Troussas et al., 2020), facilitate knowledge construction (Njenga et al., 2017; Putro et al., 2020), and correct

TABLE 22.14

Codes of the Design Focus, Definition, and Descriptive Statistics

Design Focus	Definition	N (%)
Multi-level support of interaction/quality of collaboration	AI-supported online collaboration occurs in a delayed time, e.g., forum, discussion board, email.	13 (40.63)
Cognitive learning outcome	AI-supported online collaboration occurs in real-time interaction, e.g., video conference, chat room, phone.	6 (18.75)
Personalized learning/Learning recommendation	AI-supported online collaboration occurs without specifying an interaction mode.	5 (15.63)
Affective/emotional awareness	AI-supported online collaboration occurs synchronously and asynchronously combined.	4 (12.50)
Quantity of interaction/engagement	AI-supported collaboration occurs in the same physical place without mediated communication.	4 (12.50)

TABLE 22.15

Numbers and Percentages of AI Techniques Used for Designing Student Interaction Support Systems

Category	Sub-category	N (%)
Machine learning	Classification	6 (13.33)
	Regression	3 (6.67)
	Clustering	3 (6.67)
	Neural Networks	3 (6.67)
	Evolutionary Algorithm	2 (4.44)
Symbolic AI	Intelligent Tutoring System (ITS)/Intelligent Pedagogical Agent (IPA)	7 (15.56)
	Rule-based intervention system/Intelligent Software Agent	3 (6.67)
	Robot	1 (2.22)
	Conversational Agent	3 (6.67)
NLP-based techniques		7 (15.56)
Fuzzy-based techniques		4 (8.89)
Graph-based techniques		2 (4.44)
Data mining		1 (2.22)

misconceptions in discussions (Troussas et al., 2020). The third one was personalized learning or learning recommendations (N = 5). For example, Chen and Demmans (2020) developed a recommendation system supported by the PageRank algorithm and NLP techniques in online educational forums to provide personalized advice for learners based on their discussion content with peers and viewed messages.

The rest of the categories were affective/emotional awareness and quantity of interaction/engagement (N = 4) (see Table 22.14). The former aimed to encourage students' positive feelings toward collaborative activities by constructing discourse analysis models for detecting emotional states (Daradoumis et al., 2013) or providing affective feedback to enhance emotion regulation (Caballé et al., 2013). The latter focused on improving the quantity of interaction, such as a message sent (Anaya et al., 2013; Anaya & Boticario, 2011), number of conversations (Matazi et al., 2018), word counts (Constapel et al., 2019), or temporal delay (Constapel et al., 2019).

Six out of nine research supported asynchronous collaboration falls into the category of personalized learning/learning recommendations (N = 3) or quantity of interaction/engagement (N = 4). They aimed to address (a) the lack of relevancy of posts read by students (e.g., Chen et al., 2020) and (b) retention issues in the participation of asynchronous communication in forums (Anaya & Boticario, 2011).

AI techniques of interaction systems. A total of 45 AI techniques were used in 32 studies, where 81.25% (N = 26) articles adopted a single AI technique and 18.75% (N = 6) used multiple AI techniques. *Machine learning (ML)* and *symbolic AI* are two main categories (see Table 22.15). Traditional ML methods (N = 14) such as classification, regression, and clustering were widely used. Researchers employed classification algorithms such as decision trees (e.g., Anaya et al., 2013; Anaya & Boticario, 2011) and support vector machine (SVM) (Huang et al., 2015) for two purposes. First, it classified students' interaction behaviors in the system to (a) identify collaborative or study delay indicators (e.g., Anaya

& Boticario, 2011; Constapel et al., 2019), (b) recommend appropriate collaboration scenarios (Anaya et al., 2013), or (c) assess collaboration qualities (Chounta & Avouris, 2014). Second, it classified students' emotional states (Tian et al., 2014) or learner arguments (Huang et al., 2015) to offer intervention (such as feedback and assistance) during the collaboration process.

Clustering methods were usually combined with classification methods to group students' behaviors (e.g., Anaya et al., 2013; Anaya & Boticario, 2011) or applied individually to detect communities of users sharing interests for learning resources recommendation (Khaled, Ouchani, & Chohra, 2019). Several studies used neural networks (N = 3). It is worthy to note that Chih-Ming and Ying-You (2020) tested five machine learning methods, which involved convolutional neural networks and artificial neural networks, to predict students' communication competence based on micro-learning behaviors. Evolutionary algorithms received the least attention (N = 2). Al-Tarabily et al. (2018) employed particle swarm optimization (PSO) to match students with appropriate collaboration tasks; Molina et al. (2011) adopted genetic classifier systems (complex systems including rule-based and genetic algorithm-based classifiers) to characterize learning situations for generating feedback to students.

Symbolic AI is an umbrella term referring to an intelligent agent that consists primarily of heuristics and rule-based systems adapted to enhance collaboration (Hildman & Hirsch, 2018). About 31.1% of techniques fall into this category, including intelligent agents, intelligent software, and robots. Some intelligent tutoring systems (ITS) adapted the existing system (e.g., CTA) for new purposes, such as peer-tutoring support (Walker et al., 2011) or collaboration support through collaborative scripts (Rummel et al., 2012). Conversational agents were mainly used to sustain students' dialogue or facilitate discussions (e.g., Adamson & Rosé, 2012; Adamson et al., 2014; Chan & Fung, 2020). For example, Tegos and colleagues (2014) designed MentorCat to trigger students' productive dialogue by eliciting more reasoning activities and transactive talks.

Some research adopted natural language processing (NLP)-based techniques to (a) capture students' emotional states via semantic analysis (Caballé et al., 2013), (b) detect students' discussion topics in an online forum using Latent Dirichlet Allocation (LDA) (Chen et al., 2020), and (c) track students' interests by a term frequency-inverse document frequency (TF-IDF) technique (Chen & Demmans, 2020).

22.3.3.2 Student Interaction Evaluation

This section provides a review of research that validates or evaluates AI's effects on supporting student interactions. First, it presents the collaborative contexts where researchers implemented AI techniques to enhance collaboration and learning (see Figure 22.5). Then, it reports the measures for evaluation and the associated outcomes in terms of the performance of the learners (cognitive, behavioral, and affective), the experience of the learners, and the performance of algorithms or systems (see Table 22.16). Since nearly half of the studies applied more than one method for assessments, findings were mixed, given different measures and conditions.

Collaborative contexts. Three types of collaborative tasks emerged (see Figure 22.5a): (a) problem-solving tasks, in which students complete a shared task by analyzing the problem and finding a solution together; (b) concept learning tasks, indicating the knowledge development of a specific domain through argumentation and discussion; and (c) designing tasks, aiming for composing an artifact via activities of planning, organizing, and developing. For example, Daradoumis et al. (2013) asked students to create a WiKi page explaining the installment of a content delivery platform through an Apache server. Most studies required problem-solving tasks (N = 18), while the designing tasks appeared the least (N = 3). Ten articles did not report the information on the tasks.

In terms of the duration (see Figure 22.5b), eleven papers did not specify the length of engagement in collaborative tasks. Many research projects lasted a short time, ranging from 20 minutes to 3 hours within one day (N = 14). One research extended more than six months, allowing students to thoroughly engage with a collaborative activity recommendation system in a college-level course and evaluate its effects on student learning at the end of the academic year (Troussas et al., 2020).

Performance of the learners (cognitive). Researchers mainly assessed four aspects of learners' cognitive performance (see Table 22.16): (a) learning gains in domain knowledge evaluated with the pre- and post-tests; (b) quality of in-task performance, measuring the level of knowledge construction (e.g., Njenga et al., 2017), completion (e.g., Tissenbaum & Slotta, 2019), the correctness of answers (e.g., Tegos & Demetriadis, 2017), and multiplicity of perspectives generated in discussion (e.g., Wang et al., 2011); (c) learning retention, determined by the knowledge transfer tests which was conducted several weeks after intervention (e.g., Rau et al., 2017; Rummel et al., 2012); and (d) development of cognitive skills, such as socio-scientific reasoning (Chen et al., 2020).

For articles reporting cognitive evaluation (N = 14), ten of them claimed the positivity of AI for (a) bringing significant learning gains resulting from a within-subject experimental design (e.g., Tissenbaum & Slotta, 2019), or (b) promoting learning for treatment

TABLE 22.16

Codes of Outcome Measures, Definition, and Descriptive Statistics

Measures	Sub-measures	Definition	N (%)
Performance of the learners (Cognitive)	Learning gains	Changes in domain knowledge	10 (23.26)
	In-task performance	Collaborative performance respecting completion, correctness, and comprehensiveness	5 (11.63)
	Learning retention	Transfer of new knowledge into the long-term memory	2 (4.65)
	Skill Development	Changes in cognitive skills	1 (2.33)
Performance of the learners (Behavioral)	Cognitive behaviors	Cognitive activities performed in collaborative tasks	11 (25.58)
	Engagement level	Quantity of participation in collaborative tasks	6 (13.95)
Performance of the learners (Affective)	Attitudes	Attitudes reported by learners toward collaborative learning supported by AI technologies	4 (9.30)
	Emotional states	Emotions experienced by learners while engaging in activities	2 (4.65)
	Perceived achievement	Self-report development respecting domain knowledge	2 (4.65)
Experience of the learners	Usability	Overall experiences of AI techniques	9 (20.93)
	Learnability	Functionalities of AI in supporting domain learning	7 (16.28)
	Sociability	Functionalities of AI in supporting social interactions between different actors	4 (9.30)
Performance of algorithms/systems	Accuracy	The precision of prediction/classification as well as relevancy of adaptive supports	11 (25.58)
	Efficiency	Time for producing correct results or relevant supports	3 (6.98)
	Stability	Few changes occur while adjusting the training data	1 (2.33)

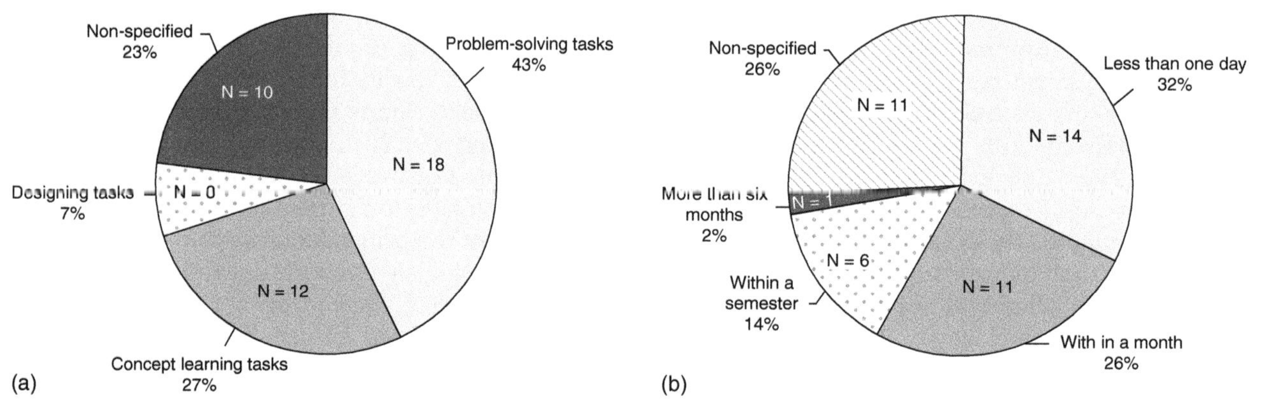

FIGURE 22.5
The summary of work related to student interaction. (a) Collaborative tasks used in research of student interaction support. (b) Duration of validation and evaluation research of student interaction support.

groups than the control groups (e.g., Tegos et al., 2015). Among these ten articles, three of them created different treatment conditions for comparison to explore what adaptive supports were more conducive to learning. They found that (a) no differences in group performance by interacting with multiple intelligent agents and with a single agent (Hayashi, 2014); (b) scripted roles were less helpful for knowledge construction than guided facilitation (Njenga et al., 2017); and (c) the quality of task completion was better for AI-supported a weaker student than it addressed all members of a group (Tegos et al., 2016).

Performance of the learners (behavioral). There are two measures for assessing learners' behavioral performances (see Table 22.16). One is the quality of cognitive behaviors, referring to students' participation in cognitive activities, such as requesting help from teachers or peers (e.g., Walker et al., 2011), correcting errors in discussion (e.g., Vanlehn et al., 2019), and idea elaborations (e.g., Rummel et al., 2012). Another measure is engagement level, representing the quantities of student participation in collaborative tasks, which included verbal behaviors (e.g., numbers of turn-taking) and physical actions (e.g., time spent on work) (e.g., Constapel et al., 2019; Hayashi, 2014). Ninety-three percent of studies on behavioral evaluation (N = 15) showed a positive effect of AI on learners' performances. One study presented mixed findings (VanLehn et al., 2019): students of treatment conditions and control groups did not differ in collaborative behaviors (such as engagement, explaining, and co-construction); but the percentage of error correction was significantly higher for students with adaptive supports.

Performance of the learners (affective). The measures of affective performance include (a) attitudes toward collaborative learning supported by AI technologies, referring to whether the activity was enjoyable (e.g., Han et al., 2021) or challenging (e.g., Rau et al., 2017); (b) emotional states, a dynamic feeling experienced by students when doing the activity, such as frustration, anxiety, and happiness (e.g., Daradoumis et al., 2013); (c) perceived achievement, meaning the development of domain knowledge claimed by students (e.g., Tegos et al., 2015) (see Table 22.16). Of eight articles reporting the affective information, five showed positive affectivity that students experienced when engaging in the collaborative activity, while the other three indicated no significant differences between the treatment and control conditions (Arguedas et al., 2016; Harsley et al., 2017; Rau et al., 2017).

Experience of the learners. Using Vatrapu, Suthers, and Medina's (2008) CSCL design evaluation framework, the measures of student perceptions of AI were grouped into three categories. This includes (see Table 22.16) (a) usability, assessing an overall experience with AI, such as whether it was exciting and easy to use (e.g., Caballé et al., 2013), or whether learners were satisfied with its functionality (e.g., Chan & Fung, 2020); (b) learnability, evaluating if AI helps domain learning and skill development; and (c) sociability, standing for whether AI supports social interactions of student-student, student-teacher, and student-computer. The least number of articles (N = 4) measured the sociability of AI and were all focusing on the perceived quality of student-computer interactions (e.g., Rosenberg-Kima et al., 2020). Regarding the outcome, 9 out of 14 papers agreed on the positive experience of using AI techniques, while 3 studies reported mixed findings. In one research, learners acknowledged that the resources recommended by the new system were interesting but not adaptable (Khaled et al., 2019). In two other studies, users criticized the AI's sociability as it was unnatural and robot-like (Tegos et al., 2014), impossible to develop interpersonal relationships, and lacked communication skills to behave in a socially acceptable manner (Rosenberg-Kima et al., 2020, p. 8).

Performance of algorithms/systems. Researchers examined the performance of algorithms and systems in terms of accuracy, efficiency, and stability (see Table 22.16). Accuracy stands for the correctness rate in prediction/classification for algorithms (e.g., Anaya & Boticario, 2011), and the relevancy of adaptive supports for systems. For example, Chen and Demmans (2020) evaluated whether their proposed system, aiming for offering tailored recommendations of forum posts for learners, successfully suggested messages with novelty and diversity based on students' needs. Efficiency stands for the execution time spent on computing solutions or establishing models. Stability, only assessed by Chih-Ming and Ying-You's (2020) study, refers to consistent results generated by algorithms regarding the slight changes in the training data. Of 13 papers reporting the performance of algorithms/systems, 10 presented positive findings. One study indicated the failure of AI: student ignorance of adaptive assistance provided by the intelligent system (Walker et al., 2011). Further, two research articles reported that the prediction of proposed algorithms achieved an accuracy rate of about 60%, which was less satisfactory (Molina et al., 2011; Tian et al., 2014).

22.4 Discussions and Implications

22.4.1 Overall Trend of AI Applications in CSCL Research

22.4.1.1 Type of Research Paper

AI applications have received particular attention in CSCL in recent years (Wise & Schwarz, 2017). By examining 69 studies on CSCL supported by AI technologies under STEM education, this literature review revealed (a) the overall trend of this research field, (b) the design and evaluation of AI techniques for supporting group formation, and (c) the design and evaluation of AI technologies for enhancing student interactions. Regarding the first research question, most selected articles justified proposals by providing validation results, but the number of follow-up evaluations (N = 13) was much less than the number of proposals that came forth in the recent decade (N = 56). This result indicates that more than half of AI techniques have not been evaluated regarding whether, how, and to what extent they could support collaborative learning. Since the evaluation research can result in novel knowledge of AI's affordances (Wieringa et al., 2005), future efforts in conducting this type of research are indeed necessary. Researchers might consider implementing the existing AI technologies in varied places and populations for gaining a broader understanding of AI application in practice.

22.4.1.2 Research Focus

Studies on student interaction support were more abundant than those on group formation support in the past decade. There are three theoretical perspectives on student interaction in collaborative learning. A cognitive view emphasizes developing students' epistemic fluency (Morrison & Collins, 1995) by enabling them to apply a set of epistemic tasks, including elaborating, questioning, and reasoning (Ohlsson, 1995). The social-cultural approach aims to promote equity in collaboration, caring about whether each member of a group can gain equal participation and equal access to the shared learning resources (Esmond, 2009). A social-emotional perspective concerns participants' feelings toward each other (e.g., trustiness), toward the group/community (e.g., belongingness), and the collaborative activity (e.g., recognize the collaboration as a valuable experience) (Kreijns et al., 2003). Therefore, researchers can design AI to address different aspects of student interaction according to different theoretical frameworks. This explanation is supported by the results showing the various design focuses of AI technologies supporting student interactions (see Table 22.14), including cognitive development, learning content recommendation, affective awareness, and quantity of engagement.

Nevertheless, studies on group formation had gained much attention in the past five years. This might be because, with the advancement of AI techniques, researchers gained a deeper understanding of generating optimal grouping results with high efficiency and accuracy. The close connection between group formation and algorithms shown in the keyword co-occurrence analysis of the 2016–2020 period might support this speculation (see Figure 22.2b). It should be noticed that only a few numbers of studies focused on both group formation and student interaction. Since group composition affects student interaction (Cohen, 1994; Webb, 1982), we suggest future researchers pursue this research direction as group formation and student interaction are inseparable for building an effective CSCL environment.

22.4.1.3 Educational Contexts

Most studies were conducted in higher education institutions and computer-related classes in the past decade. A potential reason is convenient sampling because most researchers are affiliated with universities/colleges and majored in computer science. Therefore, we suggest more research be conducted for K-12 students and subjects other than computer science to understand whether and how AI techniques support collaborative learning in broader contexts. In effect, the implementation of AI technologies in K-12 education can have far-reaching influences on student development. On the one hand, it provides K-12 students with valuable AI experiences to learn the *natural interaction* and *social impacts* of AI – the two fundamental ideas of AI education for K-12 audiences suggested by Touretzky et al. (2019). On the other hand, it serves to develop K-12 students' collaborative skills, which is critical for their success in academic study and future careers in the 21st century (Ananiadou & Claro, 2009).

22.4.1.4 Research Design

Researchers frequently selected experimental studies in the past decade, including RCT experiments, quasi-experiments, and simulation methods. Its popularity might be due to its capability of revealing whether an intervention is effective or not (Bryman, 2016). The results showed that the use of quasi-experimental design increased, while the adoption of RCT experiments decreased in the recent five years. This might be because most of the selected studies were conducted in natural classrooms, where randomly assigning students to different conditions was unrealistic and inefficient. Further, the utilization of the simulation approach increased for group formation research in particular. A potential reason is that the simulation method is more accurate and efficient to test whether an algorithm can generate optimal grouping results while minimizing the risk of errors and costs (Durán & Amandi, 2011). However, this approach fails to provide empirical evidence of the algorithms' performances in authentic settings. For example, does the proposed method for optimal group formation exactly enhance collaboration and learning? Therefore, we suggest combining simulation with methods applied to realistic contexts to gain an authentic understanding of AI applications in practice.

22.4.2 AI Applications to Support Group Formation

22.4.2.1 Issues of Design

Group size. Most studies aimed to form groups with fixed group sizes, while nearly 20% of research works supported the group formation of varied sizes. Because different collaborative contexts might demand different group sizes, we suggest developing more techniques for generating diverse group sizes to fulfill different teaching and learning purposes. Regarding the studies on group formation of fixed group sizes, about 38.10% of research targeted generating dyad groups, and 27.57% created four-member groups. This indicated that most researchers of selected studies preferred small group size. As indicated in the articles, researchers believed that the small group is better than the large group for engaging the participants in interactive and productive communication. Such an understanding resonated with previous studies, which showed that the smaller groups ensured better students' participation, more efficient communication, and more information exchange (Kooloos et al., 2011; Saqr et al., 2019). Further, it is interesting that few studies investigated how to produce three-member groups, which could be a research direction for future researchers.

Grouping mechanisms. The learner's knowledge level was the most regularly used characteristic for determining both homogeneous and heterogeneous groups. This indicated an inconsistency in the conceptual framework of optimal grouping advocated by researchers of the selected studies. Specifically, some researchers embraced Vygotsky's (1978) theory of Zone of Proximal Development, considering learners could benefit from interacting with more knowledgeable others (e.g., Safia & Mala, 2012). Others believed that learners with similar levels of knowledge were more likely to engage in meaningful talk and productive problem-solving (e.g., Huang & Huang, 2017). However, several studies in CSCL show that neither the similarities nor the differences in knowledge distribution between group members matter for group productivity (Fransen et al., 2013; Smith-Jentsch et al., 2008). Instead, they argue that the perspectives held by each member should be similar enough for mutual understanding, but not the exact sameness for allowing the exploration of multiple paths.

Only three studies considered the factors related to group relationship building, such as friendship and collaboration history when determining the relevant learner's characteristics for optimal group formation. Therefore, more research is needed to pursue this direction as the interpersonal relationship, by influencing group cohesion, can largely determine teamwork productivity (Peterson & Behfar, 2003). A key attribute of group relationship building is mutual trust, which forms positive interdependence between students and reinforces collaboration (Fransen et al., 2013). One strategy of building trustiness is to group students with shared personal history or experiences. Another strategy is to consider factors that might influence impression formation. For example, Kreijns et al. (2003) summarized several non-verbal cues determining an interpersonal attraction, including the visual sense (e.g., facial expression), olfaction (e.g., the use of perfume), and audition (e.g., voice volume). Since these cues are more accessible in face-to-face settings than in an online learning environment, future researchers might address the problem of positive impression formation in virtual learning spaces.

Most studies generated either heterogeneous or homogeneous groups in terms of the type of groups, while only one research created hybrid groups. Since it has been long debated whether homogeneous groups or heterogeneous groups are influential (Fransen et al., 2013), studies on the formation of hybrid groups and their effects on student learning are rare. Therefore, future researchers can pursue this research direction. Key questions that should be addressed are: what characteristics should be matched based on similarity for optimal groups, and what characteristics are better to be grouped based on heterogeneity?

22.4.2.2 Issues of Evaluation

Most AI techniques proposed in the selected studies were demonstrated as effective for producing optimal groups. Regarding their effects on students' learning, only about one-third of studies investigated the cognitive performance of learners. This indicated that researchers of the group formation were more concerned about the algorithm's performance in terms of its accuracy and efficiency while paying less attention to the algorithm's impacts on group work effectiveness. Therefore, we suggest that future evaluation of group formation techniques should examine learners' performances regarding the cognitive, behavioral, and affective aspects of authentic group tasks. Moreover, since the group dynamics change over time, we suggest a longitudinal design for evaluation research to understand whether an initial grouping model is conducive to group development during a long-term collaborative activity. Such an understanding can provide insights into the future design of grouping mechanisms.

22.4.3 AI Applications to Support Student Interactions

22.4.3.1 Issues of Design

Design focus. The majority of AI technologies proposed in the selected studies aimed at providing multi-level support for improving collaboration quality according to the theory of collaboration scripts. This is aligned with the CSCL research that aims to provide all-rounded support (i.e., cognitive, behavioral, and social-emotional supports) for improving student collaboration and learning performance (Kreijns et al., 2003) . There is an increase in studies on developing AI to support learning personalization/recommendation in the past five years, due to the advancement of adaptive learning technologies (see Table 22.17). Adaptive support for collaborative learning, embraced by many researchers of CSCL in this

TABLE 22.17

Research for Personalized Learning/Learning Recommendation

Authors	Year	Type of Research	Interaction Settings	Outcome Measure
Molina et al.	2011	Proposal and Validation	Online (Non-specified)	Performance of system/algorithm
Al-Tarabily et al.	2018	Proposal and Validation	Online (Non-specified)	Performance of system/algorithm
Khaled et al.	2019	Proposal and Validation	Online (asynchronous)	Experience of the learners
Chen and Demmans	2020	Proposal and Validation	Online (asynchronous)	Performance of system/algorithm
Troussas et al.	2020	Proposal and Validation	Online (asynchronous)	Experience of the learners

decade (Wise & Schwarz, 2017), is capable of accommodating individual learners' experiences based on their current statuses, such as cognition level and interests (Natriello, 2013; Plass & Pawar, 2020). The advancement of AI techniques enhances the computational power of adaptive technologies so that supporting a highly relevant adaptivity through automatic analysis becomes possible (Romero et al., 2008). Further, AI techniques for learning personalization/recommendation support were particularly favored in asynchronous online contexts. This is because it can enable learners to keep track of discussion topics of importance and interest, even facing massive messages in online forums (Chen & Demmans, 2020; Khaled et al., 2019).

Target users. The results illustrated that most AI-enabled systems supporting student interaction targeted student users, while only a few numbers of AI technologies were designed for teacher users. This is because the mainstream direction of this research field is to develop an AI that can mimic teachers' facilitation roles in collaborative learning through intelligent systems or pedagogical agents. However, this kind of AI failed to engage in social talks with students and promote their interpersonal skills, as shown in Rosenberg-Kima et al.'s study (2020). This means that, so far, the teacher's role in the social aspect of student interaction is not substitutable. The AI technologies designed for teachers mainly addressed teachers' instructional challenges in classroom orchestration. Recent research efforts have been made to enable the systems to control the activity flow, distribute learning materials, and monitor students' statuses (e.g., Tissenbaum & Slotta, 2019; VanLehn et al., 2019). These can free teachers from the workloads of classroom management, allowing them to merely focus on the facilitation of student interactions (Sharples, 2013). With the limited intelligence of the technology that could 'replace' teachers' roles, enhancing teachers' capabilities in group facilitation by AI applications is a promising research field.

22.4.3.2 Issues of Evaluation

Collaborative contexts. The results illustrated that most studies evaluated the effects of AI by engaging students to complete the collaborative problem-solving tasks, while only 6% of studies offered collaborative designing tasks. In effect, the designing tasks are more complex than problem-solving and concept learning tasks since it demands more abstract thinking skills (e.g., creation), contain more sub-goals, and follow much less predefined procedures (Kumar, 1996). In other words, the collaborative designing activity requires more interventions on group coordination and socio-cognitive regulation, which is a promising area for bringing AI technologies by future researchers. In addition, the duration of collaborative tasks shall be extended. Most research projects lasted less than one day, which is insufficient for learning to occur. As argued by sociocultural theorists, it is the long-term participation in knowledge-related practices that develop learners' understandings of a specific domain (Lave & Wenger, 1991; Wenger, 1998). Thus, we suggest a more extended time for evaluation research to understand whether and how AI techniques enhance learners' participation and the identity development of a community of practice. Besides, about 25% of studies did not specify the collaborative contexts in the report. Such information, however, is essential for understanding the specific situations in which AI technology takes effect.

Effects of AI techniques. Most studies showed that students' cognitive, behavioral, and affective performances were enhanced by interacting with AI technologies in CSCL, while seven research projects reported mixed findings. Two reasons might account for the mixed results. First, researchers employed different measurements to examine learners' performances. For example, Rau et al. (2017) evaluated participants' cognitive development by two types: a reproduction knowledge test and a transfer test. The results demonstrated that participants of AI conditions obtained higher gains in the transfer tests but not the reproduction knowledge tests. Second, researchers compared the students' performances in AI groups with others from more than one condition. For example, Walker et al. (2011) designed four conditions: students learned collaboratively under AI, students learned individually with AI, students learned collaboratively without

AI, and students learned individually without AI. In this case, the effects of AI in collaborative learning were positive compared to the individual AI condition and individual non-AI condition but not different compared to the collaborative non-AI condition. In effect, the varied measurements and comparison conditions utilized by the selected studies aid an understanding of to what extent, in which situation, and for whom the AI could be practical. Hence, more research employing diverse measurements and comparison conditions is needed to provide insights on the improvement and implementation of AI for specific learning contexts.

22.5 Conclusions

This chapter provides a systematic literature review of AI applications for CSCL in STEM education from 2011 to 2021. It (a) depicts the overall trend of this research field, particularly for group formation and student interaction, (b) identifies the critical issues for designing AI technologies for collaborative learning support, and (c) synthesizes the effects of AI techniques on learners' learning experiences and performances. Based on the review results, this chapter suggests that the future AI design should dedicate to (a) combining group formation techniques and the student interaction support within an AI-enabled CSCL environment, (b) utilizing individual's characteristics related to interpersonal relationship building for determining the optimality of groups, (c) developing techniques capable of forming hybrid group types and varied group sizes, and (c) promoting student interactions by enhancing teachers' capabilities of classroom orchestration. Further, more evaluation research is needed, and researchers shall (a) pay more attention to K-12 audiences and disciplinary domains of social science, natural science, and mathematics, (b) evaluate the AI's effects on natural learners in authentic settings, (c) employ varied measurements and comparison conditions in a study, and (d) provide the details of the collaborative context in the report.

References

*Review of Education Articles.

*Adamson, D., Dyke, G., Jang, H., & Rosé, C. P. (2014). Towards an agile approach to adapting dynamic collaboration support to student needs. *International Journal of Artificial Intelligence in Education*, *24*(1), 92–124. https://doi.org/10.1007/s40593-013-0012-6

Adamson, D., & Rosé, C. P. (2012, June). Coordinating multidimensional support in collaborative conversational agents. In *Proceedings of Intelligent Tutoring Systems* (pp. 346–351). Springer.

Alfonseca, E., Carro, R. M., Martín, E., Ortigosa, A., & Paredes, P. (2006). The impact of learning styles on student grouping for collaborative learning: A case study. *User Modeling and User-Adapted Interaction*, *16*(3–4), 377–401. https://doi.org/10.1007/s11257-006-9012-7

*Alhunitah, H., & Menai, M. E. B. (2016). Solving the student grouping problem in e-learning systems using swarm intelligence metaheuristics. *Computer Applications in Engineering Education*, *24*(6), 831–842. https://doi.org/10.1002/cae.21752

*Al-Tarabily, M. M., Abdel-Kader, R. F., Azeem, G. A., & Marie, M. I. (2018). Optimizing dynamic multi-agent performance in e-learning environment. *IEEE Access*, *6*, 35631–35645.

Ananiadou, K., & Claro, M. (2009). *21st century skills and competences for new millennium learners in OECD countries*. Organization for Economic Cooperation and Development. EDU Working paper no. 41. Available online at: http://repositorio.minedu.gob.pe/handle/20.500.12799/2529, accessed 10 September 2022.

*Anaya, A. R., & Boticario, J. G. (2011). Application of machine learning techniques to analyse student interactions and improve the collaboration process. *Expert Systems with Applications*, *38*(2), 1171–1181. https://doi.org/10.1016/j.eswa.2010.05.010

*Anaya, A. R., Luque, M., & García-Saiz, T. (2013). Recommender system in collaborative learning environment using an influence diagram. *Expert Systems with Applications*, *40*(18), 7193–7202. https://doi.org/10.1016/j.eswa.2013.07.030

*Arguedas, M., Daradoumis, T., & Xhafa, F. (2016). Analyzing the effects of emotion management on time and self-management in computer-based learning. *Computers in Human Behavior*, *63*, 517–529. https://doi.org/10.1016/j.chb.2016.05.068

*Bernacki, J., & Kozierkiewicz-Hetmańska, A. (2014). Creating collaborative learning groups in intelligent tutoring systems. In *Computational collective intelligence: Technologies and applications* (pp. 184–193). Springer International Publishing. http://doi.org/10.1007/978-3-319-11289-3_19

Bryman, A. (2016). *Social research methods*. Oxford University Press.

*Caballé, S., Mora, N., Feidakis, M., Gañán, D., Conesa, J., Daradoumis, T., & Prieto, J. (2013). CC-LR: Providing interactive, challenging and attractive collaborative complex learning resources. *Journal of Computer Assisted Learning*, *30*(1), 51–67. https://doi.org/10.1111/jcal.12021

*Challco, G. C., Gerosa, M. A., Bittencourt, I. I., & Isotani, S. (2014). Automated instructional design for CSCL: A hierarchical task network planning approach. *Expert Systems with Applications*, *41*(8), 3777–3798. https://doi.org/10.1016/j.eswa.2013.12.016

*Chan, H. C. B., & Fung, T. T. (2020, December 8). Enhancing student learning through mobile learning groups. In *2020 IEEE international conference on teaching, assessment, and learning for engineering (TALE)*. http://doi.org/10.1109/tale48869.2020.9368416

*Chen, X.-X., Li, M.-C., Chen, C.-M., & Chang, W. C. (2020, September). Developing a topic analysis instant feedback system to facilitate asynchronous online discussion performance. In *2020 9th international congress on advanced applied informatics (IIAI-AAI)*. http://doi.org/10.1109/iiai-aai50415.2020.00054

Chen, Z., & Demmans, C. (2020). CSCLRec: Personalized recommendation of forum posts to support socio-collaborative learning. In A. N. Rafferty, J. Whitehill, V. Cavalli-Sforza, & C. Romero (Eds.), *Thirteenth international conference on educational data mining (EDM 2020)* (pp. 364–373). Fully Virtual: *International Educational Data Mining Society*.

Cheng, R. W. Y., Lam, S. F., & Chan, J. C. Y. (2008). When high achievers and low achievers work in the same group: The roles of group heterogeneity and processes in project-based learning. *British Journal of Educational Psychology, 78*(2), 205–221. https://doi.org/10.1348/000709907x218160

*Chih-Ming, C., & Ying-You, L. (2020). Developing a computer-mediated communication competence forecasting model based on learning behavior features. *Computers and Education: Artificial Intelligence, 1*, 100004. https://doi.org/10.1016/j.caeai.2020.100004

*Chounta, I.-A., & Avouris, N. (2014). Towards the real-time evaluation of collaborative activities: Integration of an automatic rater of collaboration quality in the classroom from the teacher's perspective. *Education and Information Technologies, 21*(4), 815–835. https://doi.org/10.1007/s10639-014-9355-3

*Chuang, P. J., Chiang, M. C., Yang, C. S., & Tsai, C. W. (2012). Social networks-based adaptive pairing strategy for cooperative learning. *Journal of Educational Technology and Society, 15*(3), 226–239.

Cohen, E. G. (1994). Restructuring the classroom: Conditions for productive small groups. *Review of Educational Research, 64*(1), 1–35.

*Constapel, M., Doberstein, D., Hoppe, H. U., & Hellbruck, H. (2019, September). IKARion: Enhancing a learning platform with intelligent feedback to improve team collaboration and interaction in small groups. In *2019 18th international conference on information technology based higher education and training (ITHET)*. http://doi.org/10.1109/ithet46829.2019.8937348

*Daradoumis, T., Arguedas, M., & Xhafa, F. (2013, September). Building intelligent emotion awareness for improving collaborative e-learning. In *2013 5th international conference on intelligent networking and collaborative systems*. http://doi.org/10.1109/incos.2013.49

Dennen, V. P., & Hoadley, C. (2013). Designing collaborative learning through computer support. In C. E. Hmelo-Silver, C. A. Chinn, C. L. L. Chan, & A. O'Donnell (Eds.), *The international handbook of collaborative learning* (pp. 389–402). Routledge.

Dillenbourg, P. & Hong, F. (2008). The mechanics of CSCL macro scripts. *The International Journal of Computer-Supported Collaborative Learning, 3*(1), 5–23.

*Durán, E. B., & Amandi, A. (2011). Personalised collaborative skills for student models. *Interactive Learning Environments, 19*(2), 143–162. https://doi.org/10.1080/10494820802602667

*Erkens, M., Bodemer, D., & Hoppe, H. U. (2016). Improving collaborative learning in the classroom: Text mining based grouping and representing. *International Journal of Computer-Supported Collaborative Learning, 11*(4), 387–415. https://doi.org/10.1007/s11412-016-9243-5

Fransen, J., Weinberger, A., & Kirschner, P. A. (2013). Team effectiveness and team development in CSCL. *Educational Psychologist, 48*(1), 9–24. https://doi.org/10.1080/00461520.2012.747947

*Garshasbi, S., Mohammadi, Y., Graf, S., Garshasbi, S., & Shen, J. (2019). Optimal learning group formation: A multi-objective heuristic search strategy for enhancing inter-group homogeneity and intra-group heterogeneity. *Expert Systems with Applications, 118*, 506–521. https://doi.org/10.1016/j.eswa.2018.10.034

*Gilbert, S. B., Slavina, A., Dorneich, M. C., Sinatra, A. M., Bonner, D., Johnston, J., Holub, J., MacAllister, A., & Winer, E. (2017). Creating a team tutor using GIFT. *International Journal of Artificial Intelligence in Education, 28*(2), 286–313. https://doi.org/10.1007/s40593-017-0151-2

Gonzalez, H. B., & Kuenzi, J. J. (2012, August). *Science, technology, engineering, and mathematics (STEM) education: A primer*. Congressional Research Service, Library of Congress.

*Graf, S., & Bekele, R. (2006). Forming heterogeneous groups for intelligent collaborative learning systems with ant colony optimization. In *Intelligent tutoring systems* (pp. 217–226). Springer. http://doi.org/10.1007/11774303_22

*Han, J., Kim, K. H., Rhee, W., & Cho, Y. H. (2021). Learning analytics dashboards for adaptive support in face-to-face collaborative argumentation. *Computers and Education, 163*, 104041. https://doi.org/10.1016/j.compedu.2020.104041

Harsley, R. (2015). *When two heads are better than one: A critical review of four collaborative intelligent tutoring systems*. University of Illinois at Chicago. Recuperado de http://rachelharsley.com/docs/HARSLEY_Qual.Pdf.

*Harsley, R., Fossati, D., Di Eugenio, B., & Green, N. (2017, March 8). Interactions of individual and pair programmers with an intelligent tutoring system for computer science. In *Proceedings of the 2017 ACM SIGCSE technical symposium on computer science education*. http://doi.org/10.1145/3017680.3017786

*Hayashi, Y. (2014, June). Togetherness: Multiple pedagogical conversational agents as companions in collaborative learning. In *The international conference on intelligent tutoring systems* (pp. 114–123). Springer. https://doi.org/10.1007/978-3-319-07221-0_14

Hildmann, H., & Hirsch, B. (2018). Overview of artificial intelligence. In N. Lee (Ed.), *Encyclopedia of computer graphics and games*. Springer. https://doi.org/10.1007/978-3-319-08234-9_228-1

Hiltz, S. R. (1994). *The virtual classroom: Learning without limits via computer networks*. Ablex.

*Huang, C.-J., Chang, S.-C., Chen, H.-M., Tseng, J.-H., & Chien, S.-Y. (2015). A group intelligence-based asynchronous argumentation learning-assistance platform. *Interactive Learning Environments, 24*(7), 1408–1427. https://doi.org/10.1080/10494820.2015.1016533

*Huang, T.-C., & Huang, Y.-M. (2017). Where are my cooperative learning companions: Designing an intelligent recommendation mechanism. *Multimedia Tools and Applications, 76*(9), 11547–11565. https://doi.org/10.1007/s11042-015-2678-2

*Iantovics, L. B., Kovacs, L., & Fekete, G. L. (2016). Next generation university library information systems based on cooperative learning. *New Review of Information Networking, 21*(2), 101–116. https://doi.org/10.1080/13614576.2016.1247742

Jeong, H., Hmelo-Silver, C. E., & Jo, K. (2019). Ten years of computer-supported collaborative learning: A meta-analysis of CSCL in STEM education during 2005–2014. *Educational Research Review, 28*, 100284. https://doi.org/10.1016/j.edurev.2019.100284

Jesson, J., Matheson, L., & Lacey, F. M. (2011). *Doing your literature review: Traditional and systematic techniques*. Sage Publications.

Johnson, D. W., & Johnson, R. T. (2008). Social interdependence theory and cooperative learning: The teacher's role. In *The teacher's role in implementing cooperative learning in the classroom* (pp. 9–37). Springer.

*Khaled, A., Ouchani, S., & Chohra, C. (2019). Recommendations-based on semantic analysis of social networks in learning environments. *Computers in Human Behavior, 101*, 435–449.

Kreijns, K., Kirschner, P. A., & Jochems, W. (2003). Identifying the pitfalls for social interaction in computer-supported collaborative learning environments: A review of the research. *Computers in Human Behavior, 19*(3), 335–353. https://doi.org/10.1016/s0747-5632(02)00057-2

Kumar, R., & Kim, J. (2013). Special issue on intelligent support for learning in groups. *International Journal of Artificial Intelligence in Education, 24*(1), 1–7. https://doi.org/10.1007/s40593-013-0013-5

Kumar, V. S. (1996, April). Computer-supported collaborative learning: Issues for research. In *Eighth annual graduate symposium on computer science*. University of Saskatchewan.

Kooloos, J. G., Klaassen, T., Vereijken, M., Van Kuppeveld, S., Bolhuis, S., & Vorstenbosch, M. (2011). Collaborative group work: Effects of group size and assignment structure on learning gain, student satisfaction and perceived participation. *Medical Teacher, 33*(12), 983–988.

*Lafifi, Y., Bendjebar, S., & Zedadra, A. (2014). A k-complementarity technique for forming groups of tutors in intelligent learning environments. *Journal of Computing and Information Technology, 22*(2), 115. https://doi.org/10.2498/cit.1002377

*Lambić, D., Lazović, B., Djenić, A., & Marić, M. (2018). A novel metaheuristic approach for collaborative learning group formation. *Journal of Computer Assisted Learning, 34*(6), 907–916. https://doi.org/10.1111/jcal.12299

Lave, J., & Wenger, E. (1991). *Situated learning: Legitimate peripheral participation*. Cambridge University Press.

*Lee, S., Mott, B., Ottenbreit-Leftwich, A., Scribner, A., Taylor, S., Park, K., Rowe, J., Glazewski, K., Hmelo-Silver, C. E., & Lester, J. (2021, May). AI-infused collaborative inquiry in upper elementary school: A game-based learning approach. In *Proceedings of the AAAI conference on artificial intelligence* (Vol. 35, No. 17, pp. 15591–15599). https://www.intellimedia.ncsu.edu/wp-content/uploads/Lee-EAAI-2021.pdf

*Liu, Y., Liu, Q., Wu, R., Chen, E., Su, Y., Chen, Z., & Hu, G. (2016). Collaborative learning team formation: A Cognitive modeling perspective. In *Database systems for advanced applications* (pp. 383–400). Springer International Publishing. http://doi.org/10.1007/978-3-319-32049-6_24

Magnisalis, I., Demetriadis, S., & Karakostas, A. (2011). Adaptive and intelligent systems for collaborative learning support: A review of the field. *IEEE Transactions on Learning Technologies, 4*(1), 5–20. https://doi.org/10.1109/tlt.2011.2

*Maqtary, N., Mohsen, A., & Bechkoum, K. (2017). Group formation techniques in computer-supported collaborative learning: A systematic literature review. *Technology, Knowledge and Learning, 24*(2), 169–190. https://doi.org/10.1007/s10758-017-9332-1

*Matazi, I., Bennane, A., Messoussi, R., Touahni, R., Oumaira, I., & Korchiyne, R. (2018, November). Multi-agent system based on fuzzy logic for e-learning collaborative system. In *2018 international symposium on advanced electrical and communication technologies (ISAECT)*. http://doi.org/10.1109/isaect.2018.8618737

*Molina, A. I., Jurado, F., Duque, R., Redondo, M. A., Bravo, C., & Ortega, M. (2011). Applying genetic classifier systems for the analysis of activities in collaborative learning environments. *Computer Applications in Engineering Education, 21*(4), 704–716. https://doi.org/10.1002/cae.20517

Morrison, D., & Collins, A. (1995). Epistemic fluency and constructivist learning environments. *Educational Technology, 35*(5), 39–45.

Natriello, G. (2013). *Adaptive educational technologies: Tools for learning, and for learning about learning*. National Academy of Education.

*Njenga, S. T., Oboko, R. O., Omwenga, E. I., & Muuro, E. M. (2017, September). Regulating group cognitive conflicts using intelligent agents in collaborative M-learning. *2017 IEEE AFRICON*. http://doi.org/10.1109/afrcon.2017.8095452

Noroozi, O., Weinberger, A., Biemans, H. J. A., Mulder, M., & Chizari, M. (2012). Argumentation-based computer supported collaborative learning (ABCSCL): A synthesis of 15 years of research. *Educational Research Review, 7*(2), 79–106. https://doi.org/10.1016/j.edurev.2011.11.006

*Odo, C., Masthoff, J., & Beacham, N. (2019, June). Group formation for collaborative learning. In S. Isotani, E. Millán, A. Ogan, P. Hastings, B. McLaren, & R. Luckin

(Eds.), *Artificial Intelligence in Education* (pp. 206–212). Springer. https://doi.org/10.1007/978-3-030-23207-8_39

Ohlsson, S. (1995). Learning to do and learning to understand: A lesson and a challenge for cognitive modelling. In P. Reimann & H. Spada (Eds.), *Learning in humans and machines: Towards an interdisciplinary learning science* (pp. 37–62). Pergamon.

*Peng, C.-C., Tsai, C.-J., Chang, T.-Y., Yeh, J.-Y., & Lee, M.-C. (2020). Novel heterogeneous grouping method based on magic square. *Information Sciences, 517*, 340–360. https://doi.org/10.1016/j.ins.2019.12.088

Peterson, R. S., & Behfar, K. J. (2003). The dynamic relationship between performance feedback, trust, and conflict in groups: A longitudinal study. *Organizational Behavior and Human Decision Processes, 92*(1–2), 102–112. https://doi.org/10.1016/s0749-5978(03)00090-6

Plass, J. L., & Pawar, S. (2020). Adaptivity and personalization in games for learning. In J. L. Plass, R. E. Mayer, & B. D. Homer (Eds.), *Handbook of game-based learning* (pp. 263–282). MIT Press.

*Procaci, T. B., Siqueira, S. W. M., Braz, M. H. L. B., & Vasconcelos de Andrade, L. C. (2015). How to find people who can help to answer a question? – Analyses of metrics and machine learning in online communities. *Computers in Human Behavior, 51*, 664–673. https://doi.org/10.1016/j.chb.2014.12.026

Putro, B. L., Rosmansyah, Y., & Suhardi, S. (2020). An intelligent agent model for learning group development in the digital learning environment: A systematic literature review. *Bulletin of Electrical Engineering and Informatics, 9*(3), 1159–1166. https://doi.org/10.11591/eei.v9i3.2009

*Rau, M. A., Bowman, H. E., & Moore, J. W. (2017). An adaptive collaboration script for learning with multiple visual representations in chemistry. *Computers and Education, 109*, 38–55. https://doi.org/10.1016/j.compedu.2017.02.006

Ren, Y., Harper, F. M., Drenner, S., Terveen, L., Kiesler, S., Riedl, J., & Kraut, R. E. (2012). Building member attachment in online communities: Applying theories of group identity and interpersonal bonds. *Management Information Systems Quarterly, 36*(3), 841–864.

Resta, P., & Laferrière, T. (2007). Technology in support of collaborative learning. *Educational Psychology Review, 19*(1), 65–83.

Romero, C., Ventura, S., & García, E. (2008). Data mining in course management systems: Moodle case study and tutorial. *Computers and Education, 51*(1), 368–384.

*Rosenberg-Kima, R. B., Koren, Y., & Gordon, G. (2020). Robot-supported collaborative learning (RSCL): Social robots as teaching assistants for higher education small group facilitation. *Frontiers in Robotics and AI, 6*. https://doi.org/10.3389/frobt.2019.00148

*Rummel, N., Mullins, D., & Spada, H. (2012). Scripted collaborative learning with the cognitive tutor algebra. *International Journal of Computer-Supported Collaborative Learning, 7*(2), 307–339. https://doi.org/10.1007/s11412-012-9146-z

*Safia, A., & Mala, T. (2012, December). Ascertaining the More Knowledgeable Other among peers in collaborative e-learning environment. In *2012 fourth international conference on advanced computing (ICoAC)*. http://doi.org/10.1109/icoac.2012.6416852

Saqr, M., Nouri, J., & Jormanainen, I. (2019, September). A learning analytics study of the effect of group size on social dynamics and performance in online collaborative learning. In *European conference on technology enhanced learning* (pp. 466–479). Springer.

Sharples, M. (2013). Shared orchestration within and beyond the classroom. *Computers and Education, 69*, 504–506.

Smith-Jentsch, K. A., Cannon-Bowers, J. A., Tannenbaum, S. I., & Salas, E. (2008). Guided team self-correction. *Small Group Research, 39*(3), 303–327. https://doi.org/10.1177/1046496408317794

Stahl, G. (2015). A decade of CSCL. *International Journal of Computer-Supported Collaborative Learning, 10*(4), 337–344.

Stahl, G., Koschmann, T., & Suthers, D. D. (2006). Computer-supported collaborative learning: An historical perspective. In R. K. Sawyer (Ed.), *Cambridge handbook of the learning sciences* (pp. 409–426). Cambridge University Press.

Suthers, D. D., & Seel, N. M. (2012). Computer-supported collaborative learning. In N. M. Seel (Ed.), *Encyclopedia of the sciences of learning* (pp. 719–722). Springer.

*Takači, Đ., Marić, M., Stankov, G., & Djenić, A. (2017). Efficiency of using VNS algorithm for forming heterogeneous groups for CSCL learning. *Computers and Education, 109*, 98–108. https://doi.org/10.1016/j.compedu.2017.02.014

*Tegos, S., & Demetriadis, S. (2017). Conversational agents improve peer learning through building on prior knowledge. *Journal of Educational Technology and Society, 20*(1), 99–111. https://www.jstor.org/stable/10.2307/jeductechsoci.20.1.99

*Tegos, S., Demetriadis, S., & Karakostas, A. (2015). Promoting academically productive talk with conversational agent interventions in collaborative learning settings. *Computers and Education, 87*, 309–325. https://doi.org/10.1016/j.compedu.2015.07.014

*Tegos, S., Demetriadis, S., Papadopoulos, P. M., & Weinberger, A. (2016). Conversational agents for academically productive talk: A comparison of directed and undirected agent interventions. *International Journal of Computer-Supported Collaborative Learning, 11*(4), 417–440. https://doi.org/10.1007/s11412-016-9246-2

Tegos, S., Demetriadis, S., & Tsiatsos, T. (2014). A configurable conversational agent to trigger students' productive dialogue: A pilot study in the CALL domain. *International Journal of Artificial Intelligence in Education, 24*(1), 62–91.

*Tian, F., Gao, P., Li, L., Zhang, W., Liang, H., Qian, Y., & Zhao, R. (2014). Recognizing and regulating e-learners' emotions based on interactive Chinese texts in e-learning systems. *Knowledge-Based Systems, 55*, 148–164. https://doi.org/10.1016/j.knosys.2013.10.019

*Tissenbaum, M., & Slotta, J. (2019). Supporting classroom orchestration with real-time feedback: A role for teacher dashboards and real-time agents. *International Journal of Computer-Supported Collaborative Learning, 14*(3), 325–351. https://doi.org/10.1007/s11412-019-09306-1

Touretzky, D., Gardner-McCune, C., Martin, F., & Seehorn, D. (2019, July). Envisioning AI for K-12: What should every child know about AI? In *Proceedings of the AAAI conference on artificial intelligence* (Vol. 33, No. 1, pp. 9795–9799). AAAI Press.

*Troussas, C., Giannakas, F., Sgouropoulou, C., & Voyiatzis, I. (2020). Collaborative activities recommendation based on students' collaborative learning styles using ANN and WSM. *Interactive Learning Environments*, 1–14. https://doi.org/10.1080/10494820.2020.1761835

van Eck, N. J., & Waltman, L. (2009). Software survey: VOSviewer, a computer program for bibliometric mapping. *Scientometrics, 84*(2), 523–538. https://doi.org/10.1007/s11192-009-0146-3

*VanLehn, K., Burkhardt, H., Cheema, S., Kang, S., Pead, D., Schoenfeld, A., & Wetzel, J. (2019). Can an orchestration system increase collaborative, productive struggle in teaching-by-eliciting classrooms? *Interactive Learning Environments*, 1–19. https://doi.org/10.1080/10494820.2019.1616567

Vatrapu, R., Suthers, D., & Medina, R. (2008). Usability, sociability, and learnability: A CSCL design evaluation framework. In *Proceedings of the 16th international conference on computers in education (ICCE 2008)* (pp. 369–373). IOS Press.

Vygotsky, L. S. (1978). Socio-cultural theory. *Mind in Society, 6*, 52–58.

Walker, E., Rummel, N., & Koedinger, K. R. (2009). Beyond explicit feedback. In *Proceedings of the 9th international conference on computer supported collaborative learning - CSCL '09*. http://doi.org/10.3115/1600053.1600133

*Walker, E., Rummel, N., & Koedinger, K. R. (2011). Designing automated adaptive support to improve student helping behaviors in a peer tutoring activity. *International Journal of Computer-Supported Collaborative Learning, 6*(2), 279–306. https://doi.org/10.1007/s11412-011-9111-2

*Wang, H.-C., Rosé, C. P., & Chang, C.-Y. (2011). Agent-based dynamic support for learning from collaborative brainstorming in scientific inquiry. *International Journal of Computer-Supported Collaborative Learning, 6*(3), 371–395. https://doi.org/10.1007/s11412-011-9124-x

Webb, N. M. (1982). Student interaction and learning in small groups. *Review of Educational Research, 52*(3), 421–445.

Wenger, E. (1998). Communities of practice: Learning as a social system. *Systems Thinker, 9*(5), 2–3.

Wieringa, R., Maiden, N., Mead, N., & Rolland, C. (2005). Requirements engineering paper classification and evaluation criteria: A proposal and a discussion. *Requirements Engineering, 11*(1), 102–107. https://doi.org/10.1007/s00766-005-0021-6

Wise, A. F., & Schwarz, B. B. (2017). Visions of CSCL: Eight provocations for the future of the field. *International Journal of Computer-Supported Collaborative Learning, 12*(4), 423–467. https://doi.org/10.1007/s11412-017-9267-5

*Yannibelli, V., & Amandi, A. (2018). Collaborative learning team formation considering team roles: An evolutionary approach based on adaptive crossover, mutation and simulated annealing. *Research in Computing Science, 147*(4), 61–74. https://doi.org/10.13053/rcs-147-4-5

*Yusri, R., Abusitta, A., & Aïmeur, E. (2020). Teens-online: A game theory-based collaborative platform for privacy education. *International Journal of Artificial Intelligence in Education*. https://doi.org/10.1007/s40593-020-00224-0

Zawacki-Richter, O., Marín, V. I., Bond, M., & Gouverneur, F. (2019). Systematic review of research on artificial intelligence applications in higher education – Where are the educators? *International Journal of Educational Technology in Higher Education, 16*(1). https://doi.org/10.1186/s41239-019-0171-0

Zheng, L., Zhang, X., & Gyasi, J. F. (2019). A literature review of features and trends of technology-supported collaborative learning in informal learning settings from 2007 to 2018. *Journal of Computers in Education, 6*(4), 529–561. https://doi.org/10.1007/s40692-019-00148-2

23

Inclusion and Equity as a Paradigm Shift for Artificial Intelligence in Education

Rod D. Roscoe, Shima Salehi, Nia Nixon, Marcelo Worsley, Chris Piech, and Rose Luckin

CONTENTS

23.1 Inclusion and Equity as a Paradigm Shift for Artificial Intelligence in Education 359
23.2 AI and DEI: A Bidirectional Relationship 360
23.2.1 AI for DEI: How Can the Principles and Methods of Artificial Intelligence Support Diversity, Equity, and Inclusion? 360
23.2.2 Inclusivity in STEM Introductory Courses 361
23.2.2.1 Performance in Introductory STEM Courses 361
23.2.2.2 Active Learning in Introductory STEM Courses 361
23.2.3 Collaboration and Discourse 362
23.2.4 Learning Assessments 363
23.2.5 DEI for AI: How Can the Principles of Diversity, Equity, and Inclusion Transform Artificial Intelligence in Education? 364
23.2.6 Person-Centered Variables, Outcomes, and Ownership 364
23.2.7 Revealing, Mitigating, and Preventing Biases in Analysis and Interpretation 365
23.2.8 Transparency in Feedback, and Dissemination 365
23.2.9 Considering Who Will Use the AI 366
23.2.10 Educational Opportunities for AI 366
23.3 Ethics and Challenges 366
23.4 Conclusion 368
Acknowledgments 369
Note 369
References 369

23.1 Inclusion and Equity as a Paradigm Shift for Artificial Intelligence in Education

Artificial intelligence (AI) methods allow researchers and educators to assess complex patterns among diverse variables (e.g., learner backgrounds, behaviors, outcomes, learning context, and outcomes) to generate inferences and predictions for supporting learners and teachers (Baker & Inventado, 2014; Roll & Wylie, 2016). For example, various intelligent tutoring systems (ITSs, e.g., Kulik & Fletcher, 2016; Nye, 2015) have modeled learners based on factors such as task performance, behaviors, interaction patterns (Baker et al., 2010), natural language (Nye et al., 2014), and signals of affect (D'Mello et al., 2009; D'Mello & Graesser, 2012) and then used these AI-driven models to guide instruction and feedback. Similar approaches have contributed to automated scoring (Yan et al., 2020) and writing evaluation (AWE) technologies (McNamara et al., 2015; Shermis et al., 2016; Wilson & Roscoe, 2020), game-based learning and assessment (Shute, 2011; Shute et al., 2021), and social and collaborative learning (Schneider et al., 2021; Walker & Ogan, 2016). Collectively, the applications of AI in education (AIED) have enabled broad classes of adaptive and personalized educational technologies that facilitate students' learning.

As a paradigm shift, AI and AIED experts are increasingly attending to questions of diversity, inclusion, equity, ethics, belonging, and justice within their efforts (Blanchard, 2015; Holmes et al., 2021; Joyce et al., 2021). For brevity, we collectively and inclusively refer to these sweeping issues using the acronym 'DEI' (i.e., diversity, equity, and inclusion). Although AI applications for education are powerful and beneficial,

DOI: 10.1201/9781003181187-28

their development and implementation may have neglected social and societal factors related to bias, injustice, and how learners' identities and experiences affect their learning processes and environments. There is growing awareness of algorithmic bias, such that algorithms and automated systems can recreate or exacerbate discriminatory or oppressive outcomes. For instance, recent research has investigated biases in devices that use biometric sensors and measures (Drozdowski et al., 2020), algorithms used to inform decisions about criminal sentencing and recidivism (Miron et al., 2020; Wisser, 2019), and algorithms for guiding diagnosis and treatment in healthcare (Obermeyer et al., 2019; Panch et al., 2019; Walsh et al., 2020; Wien et al., 2020). Given that AI-driven technologies are increasingly ubiquitous in such everyday (e.g., phones and cars) and high-stakes (e.g., criminal justice and medicine) environments, there are substantial dangers associated with tools that do not work correctly, safely, or fairly for certain groups of people.

Educational contexts and technologies are not immune from bias. For example, in writing instruction and assessment, human ratings of writing can be biased based on presentation, dialect, content, perceived errors, and other aspects of linguistic diversity (Canz et al., 2020; Hammond, 2019; Johnson & VanBrackle, 2012; Johnson et al., 2017; Reaser et al., 2017). Although computer-based writing assessment may be perceived as 'objective' or 'fair' (i.e., the algorithms don't have feelings or personal agendas), this perspective ignores that algorithms are typically trained based on human annotations and ratings. Biases in training data are not automatically removed when developing computational algorithms. On the contrary, biases may become reified, reinforced, and even harder to inspect (Mayfield et al., 2019). A related challenge is training data that are sourced from exclusive or non-representative samples (see Roscoe, 2021) and thus fail to capture the true range or variation in student writers. Algorithms derived from limited samples can only be validly accurate or predictive within those limited samples. On a broader scale, there is increasing awareness that conclusions based on statistical averages can be misleading or exclusionary for learners who do not conform to 'average' or majority demographics (Rose, 2016).

To address these and related concerns, AIED scholars must carefully consider DEI challenges and alternative approaches to studying educational phenomena, analyzing data, and drawing meaningful educational conclusions without biases against a particular group(s). For instance, models may need to be disaggregated to include more nuanced variables and effects related to demographic factors and social identities (Kauh et al., 2021; Else-Quest & Hyde, 2016; Nichols & Stahl, 2019). Simultaneously, intersectional approaches (see Bauer et al., 2021; Bowleg, 2008; Cole, 2009; McKay et al., 2018; Rosenthal, 2016) are needed to represent learners' multiple identities (and associated power, privilege, and history) and to interpret these effects in findings and models.

Fortunately, AI methods have significant potential for investigating complex relationships among variables (Kizilcec et al., 2013; Piech et al., 2012) and characterizing learners at differing scales (e.g., from districts to individuals) with accuracy (Wang et al., 2021) if we are cautiously and mindfully inclusive throughout all stages. Therefore, AI methods can enhance DEI efforts in education through their power to carefully identify learners and their learning progression and needs. Hence, this paradigm shift in AIED is poised to empower personalized and effective educational outcomes for a much greater diversity of learners.

This chapter will discuss the *bidirectional relationship* between AI methods and DEI approaches. DEI approaches offer a valuable and necessary lens for conceptualizing, implementing, and interpreting AI while avoiding unintended but consequential biases. Synergistically, AI approaches and methods offer valuable ways for exploring complex data and nuanced relations – to enhance DEI in education. Together, this bidirectional relationship represents an important 'paradigm shift' for AIED as a field.

23.2 AI and DEI: A Bidirectional Relationship

The AI and DEI relationship is *bidirectional*: the analytical power of AI can enhance the DEI research through closer examination of learners, learning contexts, and learning outcomes (AI for DEI), and DEI lenses are necessary to improve AI approaches and avoid bias (DEI for AI).

23.2.1 AI for DEI: How Can the Principles and Methods of Artificial Intelligence Support Diversity, Equity, and Inclusion?

AI principles and methods can deepen our understanding of DEI and empower us to design and test interventions that address DEI issues and challenges. In the following, we present example research strands in which AI approaches have enhanced studying DEI challenges and generated new insights for interventions that improve DEI in learning environments.

23.2.2 Inclusivity in STEM Introductory Courses

23.2.2.1 Performance in Introductory STEM Courses

Women and underrepresented racial minority (URM) students remain underrepresented in STEM majors and careers (Dasgupta & Stout, 2014). The relative scarcity of women and URM students entering and persisting in STEM majors constrains opportunities to access high-demand STEM jobs and socioeconomic mobility. Performance in introductory STEM courses is also a key predictor and target for retention (Seymour & Hunter, 2019). However, it has been repeatedly observed that historically marginalized learners seem to be disadvantaged by such courses and underperform compared to peers from more privileged backgrounds (Chen, 2013). However, most of these analyses have been limited to descriptive statistics and comparisons of aggregate performance measures across different demographic groups. This approach arguably leads to further stigmatization of marginalized groups rather than insights about how to help. To address this challenge, Salehi and collaborators have employed large and longitudinal data sets across vastly different institutions to move beyond descriptive analysis to examine factors that impact the performance of students in large introductory physics courses (Salehi et al., 2019a, 2020).

The authors employed more nuanced quantitative approaches to discover that although marginalized groups (e.g. women, first-generation, and URM students) received lower grades in this physics course compared to their peers, almost all of these apparent performance gaps could be explained by variations in incoming STEM preparation (see Figure 23.1). Marginalized students received lower grades across all three institutions. However, when researchers controlled for incoming preparation, these performance gaps were no longer statistically significant – marginalized and non-marginalized students with the same level of STEM incoming preparation *performed the same* in their introductory physics courses.

Unfortunately, due to inequities in the United States' societal structure, along with systems of neglect and discrimination, marginalized students tend to receive reduced incoming preparation for STEM courses. Such students are more likely to attend under-resourced high schools and thus receive fewer opportunities for STEM exposure and preparation (Fahle et al., 2020). Moreover, typical STEM introductory courses often ignore variations in STEM incoming preparation and are tailored mostly to well-prepared students. Thus, underserved students who attend college and aspire to STEM careers may continue to be underserved, and perhaps underperform, compared to their peers.

23.2.2.2 Active Learning in Introductory STEM Courses

Previous analyses suggested that students from demographically marginalized groups in STEM (e.g., first generation, URMs, and women) were likely underserved with regard to STEM preparation (i.e., an indicator of inequity in social structures and educational infrastructure). Given that such preparation is an important predictor of performance in introductory STEM courses, and the importance of these courses

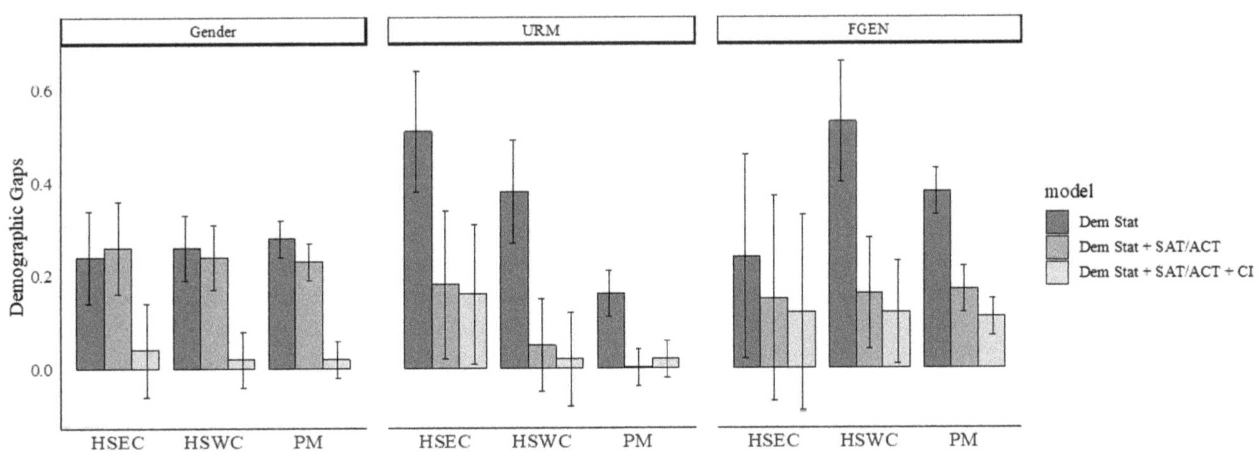

FIGURE 23.1
Size of coefficients for demographic variables in regression models that predict course performance at each institution. In each plot, the left-most bar (darkest) indicates the coefficient when *only the indicated demographic variable* (e.g., URM identity, first-generation status, or gender) is included. The central bar indicates the demographic coefficient when *math SAT or ACT scores are added predictors.* The right-most bar (lightest) indicates the coefficient when *concept inventory (CI) pretest scores are added as a predictor along with math SAT or ACT scores.* Error bars represent the standard error of the coefficients. Regression models that include only demographic status have *R*-squared values of 0.03 or less, but these increase to 0.2–0.3 when measures of incoming preparation are added to the model.

for retention, we need to identify equitable instructional practices that provide students with divergent levels of STEM preparation an equal opportunity to excel.

Introductory college STEM courses remain primarily lecture-based and grounded in foundational knowledge, and thus students' performance depends heavily on pre-college STEM preparation (Salehi et al., 2019a, 2020). This scenario further reinforces inequities in learning opportunities that result from disparities in STEM preparation quality and opportunities (Card & Rothstein, 2007; Fahle et al., 2020; Reardon & Owens, 2014). Furthermore, the effectiveness of lecture-based instruction has been challenged for *all* students. Previous meta-analyses have found that replacing lecture-based teaching with interactive, learner-centered instruction broadly improves the average performance of all students (Freeman et al., 2014; Haak et al., 2011). One caveat, however, is that many prior studies have relied on aggregate measures of students' performance that ignore demographic categories or variance. Thus, it remains unclear whether 'active learning' methods are effective in creating more equitable learning environments or addressing disparities. In a meta-analysis of more than 250 studies on the effects of active learning on academic performance, Theobald et al. (2020) found that only 15 studies reported results disaggregated by demographic group. Encouragingly, those few studies seemed to show that active learning instructional approaches particularly benefit marginalized students in STEM.

There are additional shortcomings in this literature (Theobald et al., 2020). First, although there is a disaggregation of performance outcomes across demographic groups, operationalization of instructional practices remains fairly coarse. Specifically, instructional approaches are broadly labeled as 'lecture-based' versus 'learner-centered' or 'active learning', which do not necessarily specify *how* or *which* instructional components of active learning methods actually benefit marginalized students. Consequently, it is not clear how to implement active learning (i.e., specific methods or activities) in ways that create an authentically inclusive environment.

To explore how active learning benefits marginalized students, Ballen et al. (2017) conducted structural equation modeling analyses using a large data set to explore potential mediating variables. The researchers found that active learning particularly benefits marginalized students by improving a sense of science self-efficacy, which in turn improves course performance. To further explore the specific components of active learning that benefit marginalized students, Ballen et al. (2017) also examined gender disparities across assessment methods in an introductory biology course. High-stakes exams were the most prone to gender disparities due to a disproportionately negative influence of test anxiety on performance of women. The researchers later replicated these results across a larger data set from 15 introductory STEM courses (Salehi et al., 2019b). These findings suggest that active learning improves equity in STEM courses through less reliance on inequitable high-stakes exam assessments.

Another important instructional component of active learning is frequent group activities. In the following sections we discuss how AI approaches help us better understand barriers and interventions for equitable collaborative learning.

23.2.3 Collaboration and Discourse

Collaboration is an essential aspect of learning, research, and modern work. This is particularly true in STEM where team science is responsible for highly impactful discoveries and multidisciplinary team research is the future of solving complex problems. Both educational and professional contexts require bringing together people with varying expertise to share knowledge, learn from each other, solve problems, and create products and ideas. Traditionally, teams have collaborated face-to-face, but contemporary collaboration increasingly occurs via virtual (i.e., online) platforms. Although digitally mediated collaborative problem-solving (CPS) environments hold the potential for creating more equitable and inclusive peer interactions, they are typically not characterized as such (Dasgupta et al., 2015; Du et al., 2015; Huang et al., 2014; Ke & Kwak, 2013). During technology-mediated STEM interactions, women and URMs face unique barriers such as feeling unwelcome to participate, having limited opportunities to contribute when conversations are dominated by a few members, and lacking perceived interpersonal power when attempts to engage are ignored. In each of these circumstances, complex and unique challenges are presented that can result in a detrimental impact on students' sense of belonging in the STEM milieu (Eddy et al., 2015).

To address such challenges, Dowell and colleagues have developed *Group Communication Analysis* (GCA), an innovative artificial intelligence-based methodology for quantifying and characterizing the discourse dynamics between learners in online multi-party interactions (Dowell et al., 2020; Dowell, Nixon, et al., 2019; Dowell & Poquet, 2021; Schneider et al., 2021). GCA applies automated computational linguistic analysis to the sequential interactions of participants in online group communication. GCA captures the structure of the group discussion and quantifies the complex semantic cohesion relationships between learners' contributions as they unfold over time, revealing intra- and interpersonal processes in group communication.

Dowell has used GCA and related NLP methods to study communication dynamics in online team interactions across gender and race to understand inclusivity in collaborative problem-solving (Dowell, 2019; Dowell, Lin, et al., 2019; Dowell et al., 2021; Lin et al., 2019; Lin & Dowell, 2019; Lin et al., 2020). Across several studies, Dowell discovered substantial intra- and interpersonal differences in women and URM's engagement that could influence their sense of belonging in online STEM environments. For example (see Figure 23.2), the difference between women and men in online STEM teams was not in how often they spoke. Instead, differences were evident in the extent to which they engage in productive discourse that responded to what other learners said previously (overall responsivity), provided meaningful contributions that warranted follow-up by peers (social impact), and monitored and built on their own previous contributions over the course of interaction (internal cohesion). Women's conversations showed greater overall responsivity, social impact, and internal cohesion than men's. In another study, Dowell, Lin, et al. (2019) examined how variations in team gender composition (female-minority, sex-parity, and female-majority) impacted socio-cognitive conversation patterns among team members using GCA. Results showed that the behavioral impact of men-dominated teams was more specific than simply gender differences in speaking up. Both men and women engaged in less productive collaborative problem-solving behaviors in men-dominated teams.

Across these illustrative examples and other studies, Dowell's team has revealed substantial intra- and interpersonal communication differences between women and men during CPS interactions. Moving forward, Dowell's team will be directing their efforts towards two important issues: (a) documenting the implications of observed differences for students' learning and psychological experience (e.g., sense of belonging, self-efficacy, propensity to remain in STEM majors) in teams, and (b) how to build sensitive, real-time feedback systems to best mitigate the detrimental impacts of certain team dynamics for marginalized populations.

23.2.4 Learning Assessments

Finally, an important consideration for research that centers on marginalized identities is how we assess and evaluate those learning experiences. This concern has been one focal aspect of the CrossMMLA (Multimodal Learning Analytics Across Spaces)

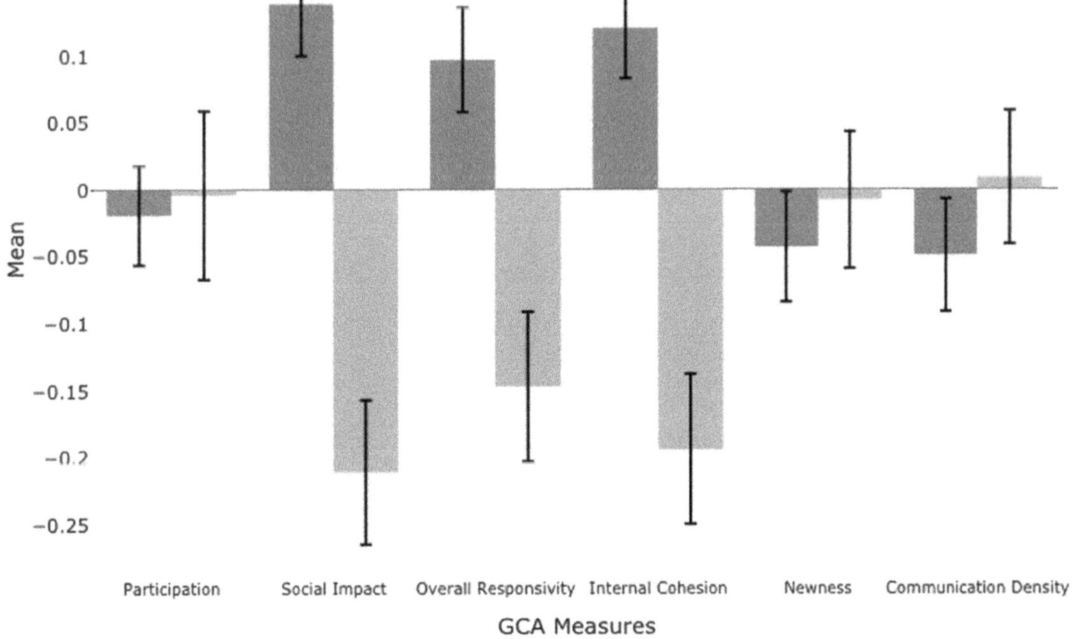

FIGURE 23.2
Collaborative group behaviors across participant's gender.

special interest group within SOLAR (Society of Learning Analytics Research). CrossMMLA highlights two important aspects of supporting and evaluating learning. First, learning takes place across innumerable locations. Although we often compartmentalize learning as primarily taking place within physical school buildings, we know that learning extends beyond school and other organized learning spaces. CrossMMLA thus emphasizes the need to develop systems and analytic approaches that can intelligibly chronicle learning as it unfolds across a variety of spaces (Blikstein & Worsley, 2016; Spikol et al., 2021). Doing so means that we open the door for learners to practice and engage in meaningful learning outside of the confines of schools, and potentially with a broader set of learning partners.

CrossMMLA also emphasizes that learners may demonstrate proficiency or knowledge growth in many ways. Multimodal sensors and analyses provide additional ways to surface student learning. This becomes increasingly imperative as we work to support learning across different spaces and use learning analytics (e.g., Sports Sense, Jones et al., 2020) that might not align with traditional instructional design paradigms. Hence, AI can provide crucial tools in not only opening the door for new patterns of engagement but also towards an appropriate set of metrics to honor the diverse ways that students demonstrate their learning. In turn, this approach can support inclusive learning environments for learners who may not resonate with traditional classroom activities or who carry a primary discourse that does not fully align with schooling practices.

23.2.5 DEI for AI: How Can the Principles of Diversity, Equity, and Inclusion Transform Artificial Intelligence in Education?

Principles of DEI can strengthen and expand AIED by informing research agendas and questions, operationalization and interpretation of variables, and revealing and mitigating biases in AI applications. These pursuits are challenging and require a deep understanding of inequitable outcomes along with underlying local and systemic causes and correlates. In turn, such efforts can inspire interventions to rectify harmful practices and environments, which must then be carefully evaluated for efficacy. Any of these goals – understanding, intervention, and evaluation – can become concrete research agendas for AIED scholarship. In other words, diversity, equity, inclusion, and related constructs can be a valid focus for research. Several of these commitments are specific to multimodal learning analytics, but others have applicability across research that bridges artificial intelligence and education (see Table 23.1).

23.2.6 Person-Centered Variables, Outcomes, and Ownership

DEI approaches enable more authentic consideration of rich, person-centered data. Such data extend beyond classic 'individual differences' like self-efficacy (Ballen et al., 2017) to encompass demographic, cultural, and contextual factors (e.g., race and gender, stereotypes and social norms, and power imbalances). Moreover, DEI conceptualizations emphasize the overlapping and contingent ways that such variables influence each other. For instance, the needs

TABLE 23.1

Twelve Commitments for Centering DEI in AIED Research

Data Collection	Analysis and Inference	Feedback and Dissemination
Multimodality: recognize that learning is a multimodal process.	*Multimodal data and human inference*: triangulate among different data sources and help inform interpretation of learner actions	*Transparency and benefit*: ensure that the research process is transparent to participants and that the experiences provide obvious benefits
Expansive learning experiences: advance opportunities to transcend traditional classroom activities.	*Limitations in prediction from multimodal data*: predictions should be about learner actions and not about assigning decontextualized and static labels to learners	*Multimodal feedback*: move beyond dashboards and consider ways to provide multimodal feedback to participants
Make learners' complexity visible: utilize sensors that can reveal hard to see interactions, actions, and states	*Participatory interpretation of multimodal data*: include participants within data analysis and inference processes	*Meaningful, usable feedback*: develop feedback that is both usable and understandable to people outside of the research community
Learning across spaces: people learn in a variety of contexts.	*Representation and multimodal data analysis*: the ways that data are represented and analyzed plays a major role in the inferences that we draw	
Multimodal data control: learners should have control of their data and how it is used		

and experiences of a first-generation, Autistic, White woman graduate student may be very different from those of a first-generation, Black woman graduate student. Even though they share 'first-generation' and 'woman' identities, both race and neurodiversity exert further mediating and moderating influences. Any analysis that focuses solely on one component of their identities would 'miss the mark'. A general algorithm can help to surface salient features, but the parameters or weights for those features may need to be updated to reflect individuals.

Critical and intersectional frameworks may provide a lens for describing and operationalizing DEI variables and their effects (Bauer et al., 2021; Bowleg, 2008; Cole, 2009; Else-Quest & Hyde, 2016). These approaches articulate that social identity categories (e.g., 'Black' and 'White', or 'man' and 'woman') entail substantial within-category variance stemming from the *intersection* of multiple categories. Every person embodies a multiplicative combination of identities. Intersectional frameworks also address how and why social identities are related to outcomes via differences in power and privilege. Thus, these perspectives offer the explanatory potential for predicting and testing relationships within our data. Instead of merely documenting disparities (e.g., a 'race gap'), we can study factors that might generate those gaps (e.g., disparities in academic preparation opportunities and resources) (Pierson et al., 2020).

It is important to note that the data alluded to above are not just 'person-centered' but also *personal*. This is especially true as we endeavor to make algorithms as individualized as possible to minimize bias. In some cases, AI algorithms might include log or clickstream data. In other instances, algorithms might include audio and video data, which might contain facial expressions, body poses, or information about shared joint attention. Given these challenges, it is essential that researchers carefully reflect on who should own the data within these systems and the implications that this has on participants. By and large, we advocate for student ownership of their data but recognize that this introduces additional challenges in terms of data analysis methodology and introduces an additional division between researchers and the data that can be used to support AIED research.

23.2.7 Revealing, Mitigating, and Preventing Biases in Analysis and Interpretation

Attention to DEI brings awareness of how people are historically ignored, neglected, or excluded, including the effects of systemic biases built into our technologies and human-technology interactions (Chen et al., 2020; Raji et al., 2020). Consequently, DEI frameworks provide a lens for inspecting the production and maintenance of inequities. Existing data might be (re)analyzed or (re)interpreted with respect to DEI principles regardless of whether DEI was the specific focus of the original research. If demographic data are collected, disaggregation allows researchers to explore differences, similarities, and disparities between groups along with crucial within-group variance. Likewise, equity-based approaches may help to account for observed findings such as explaining how and why individuals may perform differently as a result of experiences of exclusion.

A related concern is how we talk about the individuals who participate in research. This is particularly important for marginalized communities whose values, practices, and identities may often be treated as inferior (Williams & Gilbert, 2019). We need to avoid classifying or labeling individuals based on their experience or performance within a given environment. For instance, learners who receive lower grades (i.e., an event) should not be labeled as 'low achievers' (i.e., a trait). Doing so reiterates many of the approaches that educational institutions have used to exclude and oppress marginalized groups. This practice also fails to acknowledge the contextual nature of the data and is counter to the belief that people can learn and improve. A commitment to inclusive language also translates into the terms that we use to refer to different minoritized groups. Diverse groups have varying preferences, and researchers and practitioners should commit to and invest in genuinely learning how individuals prefer to be described (Dunn & Andrews, 2015).

23.2.8 Transparency in Feedback, and Dissemination

Research should provide meaningful benefits to the participants, and the processes and practices should be transparent. There is a long history of exploiting minoritized and marginalized groups to 'advance research' at significant risk to the participants. Although education research does not tend to produce noticeable physical or medical harm to participants, the AIED community must transcend merely avoiding harm and embrace practices that offer substantive benefits (i.e., beyond instruction or financial compensation). We must commit to sharing findings with participants in ways that they can reasonably interpret and proactively invite their feedback. Researchers also need to distill their findings into representations that can be understood by people outside of their discipline. This commitment increases the need for researchers to gather reflections and corroborate interpretations of data from participants, something that rarely happens within many AIED research projects.

In the following, we will present several examples in which DEI lenses can impact AI approaches. It should be mentioned that the implications of AIED-DEI extend beyond just research and how we do research, but also into how we diversify, educate, train, and prepare future scholars and practitioners in the field.

23.2.9 Considering Who Will Use the AI

One way to engage with DEI in AI is to intentionally and respectfully center marginalized identities within the design and implementation of the research. As a precursor to doing this work, researchers should reflect on their own positionality relative to the individuals that they will partner with and approach the setting from a place of bidirectional learning and value.

For example, Worsley and Bar-El (2020), and Bar-El and Worsley (2021), have described university courses that centered on disability both as the focus of the design space and as an important community for critiquing and improving practices within the AI community. One course engaged students in developing strategies for engaging with local organizations that serve people with disabilities and leveraging the capabilities of AI and digital fabrication in ways that are generative for their constituents. Example student-created designs included multimodal AI-based interfaces for navigating new physical spaces and for working with digital fabrication technologies. In other instances, students designed prototypes that used AI to support activities such as utilizing music apps within the Deaf and Hard of Hearing community. Others have explored ways to instrument loom technology with sensors and gesture detection algorithms to support blind and low-vision weavers.

All of the above examples were implemented in conjunction with marginalized communities, and the goals of the community were the driving focus of the work. Simultaneously, the class also challenged students to question underlying assumptions that may contribute to ongoing marginalization within the computer science communities and to recognize the various contributions that disabled people make to computer science. These contributions span a vast range of applications and prototypes, as well as important perspectives and critiques of the existing practices that are often used within the design community.

23.2.10 Educational Opportunities for AI

Another example of centering marginalized identities relates to working on putting the tools of artificial intelligence in the hands of younger people and youth of color as a means to equip and empower them as designers. Several researchers within the learning sciences and computer science community have been embarking on work within this space (Lee et al., 2015; Payne et al., 2021; Zimmermann-Niefield et al., 2019).

Jones et al. (2020) described one instantiation of this type of work that specifically sought to engage youth of color who participated in sports or other physical activities. Whereas many youth are socialized to see sports participation as distinct from (or even detrimental to) academic endeavors, the Sports Sense program (previously Data in Motion) positioned sports and sports participation assets for learning about artificial intelligence. Importantly, it did so from a bidirectional perspective, where sports could be a generative space to learn AI and AI could positively contribute to improving athletic performance. Within this program, youth were introduced to existing sports wearables and applications that utilized various types of artificial intelligence (e.g., machine learning, computer vision, gesture detection, and more). Students were supported as they explored these technologies and subsequently used low-cost tools to design their own AI-enabled sports wearables and applications.

Data from the first implementation of this program suggested that students possessed a number of compelling ideas for creating the next generation of sports wearables, and youth found this design space to be new, exciting, and something that they would be interested in pursuing long term. Arriving at this program design required the researchers to center the interests and motivations of youth from the start, and to consider ways that the learning experience could simultaneously teach them about artificial intelligence and contribute to meeting their goals.

23.3 Ethics and Challenges

Ensuring an effective bidirectional relationship between AI and DEI is complex, and there are many challenges to be addressed as well as significant benefits to be gained. Two examples of these challenges are (a) the ethical implications that are inherent in the analytical power that AI can provide and (b) the lack of understanding about AI among the vast majority of those whose data is and will be processed.

The work of the Institute for Ethical AI and Education[1] tackled the first of these challenges by developing a framework for practitioners and educational leaders to use when procuring AI for use in education. A key motivation for this work was to enable all learners to benefit optimally from AI and to be protected against known risks. In February 2020, an interim report, 'Towards a Shared Vision of Ethical AI in Education', was published by the IEAIED. This report outlined the risks and benefits posed by AI's use and suggested ways in which some of the tensions between the risks and the benefits might be ethically addressed. Suggestions from the report were used to drive a wide consultation with a cross-section of stakeholders through expert interviews and a series of roundtables – including three dedicated to participation by young people – and a Global Summit that brought together over 200 experts and authorities. The aim was to agree on a shared understanding of the ethical implications of using AI in education and to agree on a set of recommendations for how AI could be ethically designed and applied in education. The result from the consultation process was a four-page framework for educators that was organized around a set of nine objectives:

1. *Achieving Educational Goals.* AI should be used to achieve well-defined educational goals based on strong societal, educational, or scientific evidence that this is for the benefit of learners.
2. *Forms of Assessment.* AI should be used to assess and recognize a broader range of learners' talents.
3. *Administration and Workload.* AI should increase the capacity of organizations while respecting human relationships.
4. *Equity.* AI systems should be used in ways that promote equity between different groups of learners and not in ways that discriminate against any group of learners.
5. *Autonomy.* AI systems should be used to increase the level of control that learners have over their learning and development.
6. *Privacy.* A balance should be struck between privacy and the legitimate use of data for achieving well-defined and desirable educational goals.
7. *Transparency and Accountability.* Humans are ultimately responsible for educational outcomes and should therefore have an appropriate level of oversight of how AI systems operate.
8. *Informed Participation.* Learners, educators, and other relevant practitioners should have a reasonable understanding of artificial intelligence and its implications.
9. *Ethical Design.* AI resources should be designed by people who understand the impacts these resources will have.

Each objective is associated with criteria and questions that educators can pose to companies that are marketing an AI product or service. For example, the Equity objective has three criteria including, 'Develop and implement a strategy to reduce the digital divide among the cohort of learners for whom you have responsibility'. The associated question checklist item asked users to consider, 'Will the implementation of this strategy ensure that all learners for whom you are responsible are able to access and benefit from AI? (Pre-procurement)'. One important conclusion from the work completed by the Institute was the following theme:

> Only if well-intentioned people from diverse backgrounds continue to work together with the interests of learners in mind, especially the most disadvantaged, will we ensure that AI is truly going to find its optimal use, which maximize its potential and minimizes its downsides.

The clear need for input and engagement from a diverse population in the development of AI speaks to the second challenge identified at the start of this section: the lack of understanding about AI among the vast majority of those whose data is and will be processed.

The urgent need for people, and particularly educators, to better understand AI and the associated benefits of the participatory design were articulated by Luckin and Cukurova (2019). However, for educators and learners to confidently contribute to AI design, they need to understand basic AI concepts. For instance, why are data so important for machine learning AI? What data might be useful? How can data be accessed? How are new data collected? They need to understand that data has to be prepared by people before an AI can process it and need to understand the implications for the privacy of each individual and the security of their data at each step (i.e., from sourcing data to processing the data to outputting results). Most importantly, they need to understand the importance of the *imperative* for using AI: is the imperative one that will increase diversity, inclusion, and equity, or is it an imperative that will exacerbate existing inequalities?

In parallel, it is crucial for people to understand that the same data set or the same algorithms can produce dramatically different results due to the imperative of the AI application. For example, an algorithm might be deployed on data that has been harvested from public sources such as Twitter feeds, Facebook streams, LinkedIn profiles, or Instagram posts of a group of students taking an online course. The algorithm attempts to find patterns that are associated with cultural features in the students' data and produces a set of student profiles based on such cultural features. One imperative for the use of AI in this case might be to ensure that the course is sensitive to students' cultural context and that adaptations of the course material are guided by the AI-produced profiles. Alternatively, the imperative for the application of AI might be to ensure that students who are members of a particular profile are only allowed to interact with students from that same profile. Or, the aim might be to separate students who belong to a particular profile and provide them with a narrow course that will not broaden their horizons. These last two examples are obviously of concern, and yet the data and algorithm are remarkably similar to the first imperative case.

A possible tool for tackling this second challenge emerges from *AI Readiness*, which is a framework for providing educators with a contextualized practical experience of what AI is and what it can do. This framework is the basis for an AI Readiness course that helps educators and their leaders 'get inside' a machine learning algorithm – to explore what it can do with the type of data that the educators may have access to (Luckin & George, 2022).

23.4 Conclusion

In this chapter, we discussed how AI methods and DEI approaches mutually benefit one another and can advance work in each space. AI methods expand the tools that researchers can use to identify barriers to DEI and then design, implement, and evaluate interventions to address these challenges (i.e., AI for DEI). Similarly, DEI approaches are not a superficial addition to AI research but instead offer a core perspective that ensures AI asks meaningful questions and employs equitable methods across diverse populations (i.e., DEI for AI).

AI and DEI can synergize in numerous ways, and the research documented in the preceding sections highlights several relevant approaches. Existing techniques remain relevant and can generate important insights, and there is also ample opportunity for innovation in this nascent field. AI methods have exciting potential for investigating complex relationships among demographic, performance, and behavioral variables (Kizilcec et al., 2013; Piech et al., 2012). For instance, several studies have shown that classic multiple regression and structural equation modeling approaches when they incorporate demographic variables informed by DEI principles, can challenge our assumptions about 'performance gaps' and pathways for different populations (Salehi et al., 2020; Ballen et al., 2017). Other studies have pioneered new methods such as Group Communication Analysis (GCA) (Dowell & Poquet, 2021; Schneider et al., 2021) and Multimodal Learning Analytics Across Spaces (CrossMMLA) (Blikstein & Worsley, 2016). The former combines natural language data, sequence mining, dynamic models, and demographic variables to understand (in)equitable discourse patterns. The latter combines multimodal sensors, gestures, natural language, computer vision, and additional data streams over time via machine learning to develop rich models of learners and to improve accessibility.

An important consideration for bridging AI and DEI is considering how demographic data are appropriately and ethically included in analytic models. A growing literature is investigating how quantitative analyses can be authentically intersectional to respect multiple and overlapping demographic 'categories' (Bauer et al., 2021; Else-Quest & Hyde, 2016), and the ways such categories affect learners' lives and performance. Likewise, learning analytics models typically treat demographic factors as distinct categories (e.g., 'Black' or 'White' or 'Asian') but can also further explore more nuanced *intra*categorical and *inter*categorical variance (McKay et al., 2018) that better captures the range of human experience.

Finally, although this bidirectional relationship is very promising, there are many questions open to further exploration and optimization. For instance, data in equity-oriented research paradigms can be both a *treasure* and a *terror*. Specifically, rich data enable deeper and more contextualized characterization of learners, their needs, and their journeys. However, these data also entail privacy challenges (e.g., revealing personal and identifying information) and have the potential to further stigmatize learners if misused (e.g., uncritically interpreting a 'performance gap' as 'evidence' of inferiority). Similarly, we must contend with and challenge methodological assumptions about statistical power and sampling. In the era of 'big data', techniques have been developed to address thousands and millions of data points. However, increasingly personalized, contextualized, and intersectional analyses drive analytical methods in the opposite direction. Equitable analyses need to be feasible, valid, and reliable even for 'small' samples and populations (e.g., Hispanic, LGBTQ+, Autistic, first-generation college students). There is no clear guideline for how much data is necessary to conduct appropriate analyses.

As always, careful and creative power analyses can be conducted to estimate sufficient sample sizes for a given study, but the field would benefit from new and innovative methods that work at broader scales. Data collection will also need to pursue more representative sampling such that data sets include authentic diversity. Recruiting diverse participants and building inclusive corpora will require more proactive and strategic sampling strategies than mere convenience sampling (Roscoe, 2021).

Overall, as the synergy between AI and DEI continues to develop, we can be guided by heuristic questions such as 'What are the best practices for collecting data?'; 'Who should own the data or have access to it once collected?'; 'How can we ensure privacy and confidentiality of participants when triangulated data facilitates identification?'; and 'How can results and findings be disseminated in ways that are precise, insightful, and beneficial to participants while forestalling misrepresentation or harmful conclusions?' The authors hope this chapter can promote discussion about linking DEI and AI approaches and inspire researchers and practitioners to answer the above questions in their work.

Acknowledgments

This chapter summarizes and expands upon a discussion panel held at the online 2021 Artificial Intelligence in Education (AIED) conference. Drs Roscoe and Salehi served as the organizers and moderators, and Drs Dowell, Worsley, Piech, and Luckin contributed as expert panelists. The authors are grateful to Drs Ido Roll and Danielle S. McNamara, who made the AIED conference panel possible. The authors also appreciate feedback from the book editors and reviewers. Portions of this work were supported by grants to Dr Roscoe (Gates Foundation INV-006123), Dr Nixon (Gates Foundation INV-000752), and Dr Salehi (Stanford's IDEAL initiative). All opinions, findings, conclusions, or recommendations expressed in this work are those of the authors and do not necessarily reflect the views of any funding sources or agencies.

Note

1. https://www.buckingham.ac.uk/research-the-institute-for-ethical-ai-in-education/.

References

Baker, R. S., D'Mello, S. K., Rodrigo, M. M. T., & Graesser, A. C. (2010). Better to be frustrated than bored: The incidence, persistence, and impact of learners' cognitive–Affective states during interactions with three different computer-based learning environments. *International Journal of Human-Computer Studies, 68*(4), 223–241. https://doi.org/10.1016/j.ijhcs.2009.12.003

Baker, R. S., & Inventado, P. S. (2014). Educational data mining and learning analytics. In J. Larusson & B. White (Eds.), *Learning analytics* (pp. 61–75). Springer. https://doi.org/10.1007/978-1-4614-3305-7_4

Ballen, C. J., Wieman, C., Salehi, S., Searle, J. B., & Zamudio, K. R. (2017). Enhancing diversity in undergraduate science: Self-efficacy drives performance gains with active learning. *CBE—Life Sciences Education, 16*(4), ar56. https://doi.org/10.1187/cbe.16-12-0344

Bar-El, D., & Worsley, M. (2021). Making the maker movement more inclusive: Lessons learned from a course on accessibility in making. *International Journal of Child-Computer Interaction, 29,* 100285. https://doi.org/10.1016/j.ijcci.2021.100285

Bauer, G. R., Churchill, S. M., Mahendran, M., Walwyn, C., Lizotte, D., & Villa-Rueda, A. A. (2021). Intersectionality in quantitative research: A systematic review of its emergence and applications of theory and methods. *SSM – Population Health, 14,* 100798. https://doi.org/10.1016/j.ssmph.2021.100798

Blanchard, E. G. (2015). Socio-cultural imbalances in AIED research: Investigations, implications and opportunities. *International Journal of Artificial Intelligence in Education, 25*(2), 204–228. https://doi.org/10.1007/s40593-014-0027-7

Blikstein, P., & Worsley, M. (2016). Multimodal learning analytics and education data mining: Using computational technologies to measure complex tasks. *Journal of Learning Analytics, 3*(2), 220–238. https://doi.org/10.18608/jla.2016.32.11

Bowleg, L. (2008). When Black + lesbian + woman ≠ Black lesbian woman: The methodological challenges of qualitative and quantitative intersectionality research. *Sex Roles, 59*(5), 312–325. https://doi.org/10.1007/s11199-008-9400-z

Canz, T., Hoffmann, L., & Kania, R. (2020). Presentation-mode effects in large-scale writing assessments. *Assessing Writing, 45,* 100470. https://doi.org/10.1016/j.asw.2020.100470

Card, D., & Rothstein, J. (2007). Racial segregation and the black–white test score gap. *Journal of Public Economics, 91*(11–12), 2158–2184. https://doi.org/10.1016/j.jpubeco.2007.03.006

Chen, I. Y., Pierson, E., Rose, S., Joshi, S., Ferryman, K., & Ghassemi, M. (2020). Ethical machine learning in healthcare. *Annual Review of Biomedical Data Science, 4,* 123–144. https://doi.org/10.1146/annurev-biodatasci-092820-114757

Chen, X. (2013). STEM attrition: College students' paths into and out of STEM Fields. Statistical Analysis Report. NCES 2014–001. National Center for Education Statistics. https://eric.ed.gov/?id=ED544470

Cole, E. R. (2009). Intersectionality and research in psychology. *American Psychologist, 64*(3), 170. https://doi.org/10.1037/a0014564

D'Mello, S. K., Dowell, N. M., & Graesser, A. C. (2009). Cohesion relationships in tutorial dialogue as predictors of affective states. In V. Dimitrova, R. Mizoguchi, B. Du Boulay, & A. C. Graesser (Eds.), *Artificial intelligence in education: Building learning systems that care: From knowledge representation to affective modeling* (pp. 9–16). IOS Press.

D'Mello, S., & Graesser, A. C. (2012). Language and discourse are powerful signals of student emotions during tutoring. *IEEE Transactions on Learning Technologies, 5*(4), 304–317. https://doi.org/10.1109/TLT.2012.10

Dasgupta, N., Scircle, M. M., & Hunsinger, M. (2015). Female peers in small work groups enhance women's motivation, verbal participation, and career aspirations in engineering. *Proceedings of the National Academy of Sciences, 112*(16), 4988–4993. https://doi.org/10.1073/pnas.1422822112

Dasgupta, N., & Stout, J. G. (2014). Girls and women in science, technology, engineering, and mathematics: STEMing the tide and broadening participation in STEM careers. *Policy Insights from the Behavioral and Brain Sciences, 1*(1), 21–29. https://doi.org/10.1177/2372732214549471

Dowell, N. M. (2019). Preparing for the future: Group communication analysis as a tool to facilitate adaptive support during digitally-mediated team interactions. *International Conference on Artificial Intelligence + Adaptive Education*. Beijing, China.

Dowell, N. M. M., McKay, T. A., & Perrett, G. (2021). It's not *that* you said it, it's *how* you said it: Exploring the linguistic mechanisms underlying values affirmation interventions at scale. *AERA Open, 7*(1), 1–19. https://doi.org/10.1177/23328584211011611

Dowell, N. M. M., & Poquet, O. (2021). SCIP: Combining group communication and interpersonal positioning to identify emergent roles in scaled digital environments. *Computers in Human Behavior, 119*, 106709. https://doi.org/10.1016/j.chb.2021.106709

Dowell, N. M., Lin, Y., Godfrey, A., & Brooks, C. (2019). Promoting inclusivity through time-dynamic discourse analysis in digitally-mediated collaborative learning. In S. Isotani, E. Millán, A. Ogan, P. Hastings, B. M. McLaren, & R. Luckin (Eds.), *Proceedings of the 20th international conference on artificial intelligence in education* (pp. 207–219). ACM. https://doi.org/10.1007/978-3-030-23204-7_18

Dowell, N. M., Lin, Y., Godfrey, A., & Brooks, C. (2020). Exploring the relationship between emergent sociocognitive roles, collaborative problem-solving skills and outcomes: A group communication analysis. *Journal of Learning Analytics, 7*(1), 38–57. https://doi.org/10.18608/jla.2020.71.4

Dowell, N. M., Nixon, T., & Graesser, A. C. (2019). Group communication analysis: A computational linguistics approach for detecting sociocognitive roles in multi-party interactions. *Behavior Research Methods, 51*(3), 1007–1041. https://doi.org/10.3758/s13428-018-1102-z

Drozdowski, P., Rathgeb, C., Dantcheva, A., Damer, N., & Busch, C. (2020). Demographic bias in biometrics: A survey on an emerging challenge. *IEEE Transactions on Technology and Society, 1*(2), 89–103. https://doi.org/10.1109/TTS.2020.2992344

Du, J., Ge, X., & Xu, J. (2015). Online collaborative learning activities: The perspectives of African American female students. *Computers and Education, 82*, 152–161. https://doi.org/10.1016/j.compedu.2014.11.014

Dunn, D. S., & Andrews, E. E. (2015). Person-first and identity-first language: Developing psychologists' cultural competence using disability language. *American Psychologist, 70*(3), 255–264. https://doi.org/10.1037/a0038636

Eddy, S. L., Brownell, S. E., Thummaphan, P., Lan, M.-C., & Wenderoth, M. P. (2015). Caution, student experience may vary: Social identities impact a student's experience in peer discussions. *CBE Life Sciences Education, 14*(4), ar45. https://doi.org/10.1187/cbe.15-05-0108

Else-Quest, N. M., & Hyde, J. S. (2016). Intersectionality in quantitative psychological research: II. Methods and techniques. *Psychology of Women Quarterly, 40*(3), 319–336. https://doi.org/10.1177/0361684316647953

Fahle, E. M., Reardon, S. F., Kalogrides, D., Weathers, E. S., & Jang, H. W. (2020). Racial segregation and school poverty in the United States, 1999–2016. *Race and Social Problems, 12*(1), 42–56. https://doi.org/10.1007/s12552-019-09277-w

Freeman, S., Eddy, S. L., McDonough, M., Smith, M. K., Okoroafor, N., Jordt, H., & Wenderoth, M. P. (2014). Active learning increases student performance in science, engineering, and mathematics. *Proceedings of the National Academy of Sciences, 111*(23), 8410–8415. https://doi.org/10.1073/pnas.1319030111

Haak, D. C., HilleRisLambers, J., Pitre, E., & Freeman, S. (2011). Increased structure and active learning reduce the achievement gap in introductory biology. *Science, 332*(6034), 1213–1216. https://doi.org/10.1126/science.1204820

Hammond, J. W. (2019). Making our invisible racial agendas visible: Race talk in assessing writing, 1994–2018. *Assessing Writing, 42*, 100425. https://doi.org/10.1016/j.asw.2019.100425

Holmes, W., Porayska-Pomsta, K., Holstein, K., Sutherland, E., Baker, T., Shum, S. B., Santos, O. C., Rodrigo, M. T., Cukurova, M., Bittencourt, I. I., & Koedinger, K. R. (2021). Ethics of AI in education: Towards a community-wide framework. *International Journal of Artificial Intelligence in Education*. https://doi.org/10.1007/s40593-021-00239-1

Huang, J., Dasgupta, A., Ghosh, A., Manning, J., & Sanders, M. (2014, March). Superposter behavior in MOOC forums. In M. Sahami, A. Fox, M. A. Hearst, & M. T.

H. Chi (Eds.), *Proceedings of the first ACM conference on learning @ scale conference* (pp. 117–126). https://doi.org/10.1145/2556325.2566249

Johnson, A. C., Wilson, J., & Roscoe, R. D. (2017). College student perceptions of writing errors, text quality, and author characteristics. *Assessing Writing, 34*, 72–87. https://doi.org/10.1016/j.asw.2017.10.002

Johnson, D., & VanBrackle, L. (2012). Linguistic discrimination in writing assessment: How raters react to African American "errors," ESL errors, and standard English errors on a state-mandated writing exam. *Assessing Writing, 17*(1), 35–54. https://doi.org/10.1016/j.asw.2011.10.001

Jones, S. T., Thompson, J., & Worsley, M. (2020). Data in motion: Sports as a site for expansive learning. *Computer Science Education, 30*(3), 279–312. https://doi.org/10.1080/08993408.2020.1805287

Joyce, K., Smith-Doerr, L., Alegria, S., Bell, S., Cruz, T., Hoffman, S. G., Noble, S. U., & Shestakofsky, B. (2021). Toward a sociology of artificial intelligence: A call for research on inequalities and structural change. *Socius: Sociological Research for a Dynamic World, 7*. https://doi.org/10.1177/2378023121999581

Kauh, T. J., Read, J. G., & Scheitler, A. J. (2021). The critical role of racial/ethnic data disaggregation for health equity. *Population Research and Policy Review, 40*(1), 1–7. https://doi.org/10.1007/s11113-020-09631-6

Ke, F., & Kwak, D. (2013). Online learning across ethnicity and age: A study on learning interaction participation, perception, and learning satisfaction. *Computers and Education, 61*, 43–51. https://doi.org/10.1016/j.compedu.2012.09.003

Kizilcec, R. F., Piech, C., & Schneider, E. (2013, April). Deconstructing disengagement: Analyzing learner subpopulations in massive open online courses. In D. Suthers, K. Verbert, E. Duval, & X. Ochoa (Eds.), *Proceedings of the third international conference on learning analytics and knowledge* (pp. 170–179). ACM. https://doi.org/10.1145/2460296.2460330

Kulik, J. A., & Fletcher, J. D. (2016). Effectiveness of intelligent tutoring systems: A meta-analytic review. *Review of Educational Research, 86*(1), 42–78. https://doi.org/10.3102/0034654315581420

Lee, V. R., Drake, J., & Williamson, K. (2015). Let's get physical: K-12 students using wearable devices to obtain and learn about data from physical activities. *TechTrends, 59*(4), 46–53. https://doi.org/10.1007/s11528-015-0870-x

Lin, Y., & Dowell, N. M. (2019, July). Does gender really matter?: Exploring differences in emerging discourse styles during digitally-mediated collaborative interactions. Paper presented at the *29th Annual Meeting of the Society for Text and Discourse*.

Lin, Y., Dowell, N. M., Godfrey, A., Cho, H., & Brooks, C. (2019). Modeling gender differences in intra- and interpersonal dynamics during online learning collaborative interactions. In D. Azcona & R. Chung (Eds.), *Proceedings of the 9th international conference for learning analytics & knowledge* (pp. 431–435). ACM. https://doi.org/10.1145/3303772.3303837

Lin, Y., Yu, R., & Dowell, N. (2020, July). LIWCs the same, not the same: Gendered linguistic signals of performance and experience in online STEM courses. In I. I. Bittencourt, M. Cukurova, K. Muldner, R. Luckin, & E. Millán (Eds.), *Proceedings of the 21st artificial intelligence in education conference* (Vol. 12163, pp. 333–345). Springer. https://doi.org/10.1007/978-3-030-52237-7_27

Luckin, R., & Cukurova, M. (2019). Designing educational technologies in the age of AI: A learning sciences-driven approach. *British Journal of Educational Technology, 50*(6), 1–20. https://doi.org/10.1111/bjet.12861

Luckin, R., George, K., & Cukurova, M. (2022). *AI for school teachers*. Boca Raton, FL: CRC Press. https://doi.org/10.1201/9781003193173

Mayfield, E., Madaio, M., Prabhumoye, S., Gerritsen, D., McLaughlin, B., Dixon-Roman, E., & Black, A. W. (2019, August). Equity beyond bias in language technologies for education. In *Proceedings of the fourteenth workshop on innovative use of NLP for building educational applications* (pp. 444–460). Association for Computational Linguistics.

McKay, T., Grom, G., & Koester, B. (2018, March). Categorization, intersectionality, and learning analytics. In A. Pardo et al. (Eds.), *Proceedings of the 8th international conference on learning analytics and knowledge*. SOLAR. http://bit.ly/lak18-companion-proceedings

McNamara, D. S., Crossley, S. A., Roscoe, R. D., Allen, L. K., & Dai, J. (2015). A hierarchical classification approach to automated essay scoring. *Assessing Writing, 23*, 35–59. https://doi.org/10.1016/j.asw.2014.09.002

Miron, M., Tolan, S., Gómez, E., & Castillo, C. (2020). Evaluating causes of algorithmic bias in juvenile criminal recidivism. *Artificial Intelligence and Law*, 1–37. https://doi.org/10.1007/s10506-020-09268-y

Nichols, S., & Stahl, G. (2019). Intersectionality in higher education research: A systematic literature review. *Higher Education Research and Development, 38*(6), 1255–1268. https://doi.org/10.1080/07294360.2019.1638348

Nye, B. D. (2015). Intelligent tutoring systems by and for the developing world: A review of trends and approaches for educational technology in a global context. *International Journal of Artificial Intelligence in Education, 25*(2), 177–203. https://doi.org/10.1007/s40593-014-0028-6

Nye, B. D., Graesser, A. C., & Hu, X. (2014). AutoTutor and family: A review of 17 years of natural language tutoring. *International Journal of Artificial Intelligence in Education, 24*(4), 427–469. https://doi.org/10.1007/s40593-014-0029-5

Obermeyer, Z., Powers, B., Vogeli, C., & Mullainathan, S. (2019). Dissecting racial bias in an algorithm used to manage the health of populations. *Science, 366*(6464), 447–453. https://doi.org/10.1126/science.aax2342

Panch, T., Mattie, H., & Atun, R. (2019). Artificial intelligence and algorithmic bias: Implications for health systems. *Journal of Global Health, 9*(2), 020318, https://doi.org/10.7189%2Fjogh.09.020318

Payne, W. C., Bergner, Y., West, M. E., Charp, C., Shapiro, R. B. B., Szafir, D. A., Taylor, E. V., & DesPortes, K. (2021, May). danceON: Culturally responsive creative computing. In P. Bjørn & S. Drucker (Eds.), *Proceedings of the 2021 CHI conference on human factors in computing systems* (pp. 1–16). https://doi.org/10.1145/3411764.3445149

Piech, C., Sahami, M., Koller, D., Cooper, S., & Blikstein, P. (2012, February). Modeling how students learn to program. In L. S. King, D. R. Musicant, T. Camp, & P. Tymann (Eds.), *Proceedings of the 43rd ACM technical symposium on computer science education* (pp. 153–160). https://doi.org/10.1145/2157136.2157182

Pierson, E., Simoiu, C., Overgoor, J., Corbett-Davies, S., Jenson, D., Shoemaker, A., Ramachandran, V., Barghouty, P., Phillips, C., Shroff, R., & Goel, S. (2020). A large-scale analysis of racial disparities in police stops across the United States. *Nature Human Behaviour, 4*(7), 736–745. https://doi.org/10.1038/s41562-020-0858-1

Raji, I. D., Gebru, T., Mitchell, M., Buolamwini, J., Lee, J., & Denton, E. (2020, February). Saving face: Investigating the ethical concerns of facial recognition auditing. In A. Markham, J. Powles, T. Walsh, & A. L. Washington (Eds.), *Proceedings of the AAAI/ACM conference on AI, ethics, and society* (pp. 145–151). https://doi.org/10.1145/3375627.3375820

Reardon, S. F., & Owens, A. (2014). 60 years after Brown: Trends and consequences of school segregation. *Annual Review of Sociology, 40*(1), 199–218. https://doi.org/10.1146/annurev-soc-071913-043152

Reaser, J., Adger, C. T., Wolfram, W., & Christian, D. (2017). *Dialects at school: Educating linguistically diverse students.* Routledge. https://doi.org/10.4324/9781315772622

Roll, I., & Wylie, R. (2016). Evolution and revolution in artificial intelligence in education. *International Journal of Artificial Intelligence in Education, 26*(2), 582–599. https://doi.org/0.1007/s40593-016-0110-3

Roscoe, R. D. (2021). Designing for diversity: Inclusive sampling. *Ergodesign & HCI, 9*(1), 67–81. http://doi.org/10.22570/ergodesignhci.v9i1.1502

Rose, T. (2016). *The end of average: How to succeed in a world that values sameness.* Penguin UK.

Rosenthal, L. (2016). Incorporating intersectionality into psychology: An opportunity to promote social justice and equity. *American Psychologist, 71*(6), 474–485. https://psycnet.apa.org/doi/10.1037/a0040323

Salehi, S., Burkholder, E., Lepage, G. P., Pollock, S., & Wieman, C. (2019a). Demographic gaps or preparation gaps?: The large impact of incoming preparation on performance of students in introductory physics. *Physical Review Physics Education Research, 15*(2), 020114. https://doi.org/10.1103/PhysRevPhysEducRes.15.020114

Salehi, S., Cotner, S., Azarin, S. M., Carlson, E. E., Driessen, M., Ferry, V. E., Harcombe, W., McGaugh, S., Wassenberg, D., Yonas, A., & Ballen, C. J. (2019b). Gender performance gaps across different assessment methods and the underlying mechanisms: The case of incoming preparation and test anxiety. *Frontiers in Education, 4,* Article 107. https://doi.org/10.3389/feduc.2019.00107

Salehi, S., Cotner, S., & Ballen, C. J. (2020). Variation in incoming academic preparation: Consequences for minority and first-generation students. *Frontiers in Education, 5,* Article 552364. https://doi.org/10.3389/feduc.2020.552364

Schneider, B., Dowell, N., & Thompson, K. (2021). Collaboration analytics—Current state and potential futures. *Journal of Learning Analytics, 8*(1), 1–12. https://doi.org/10.18608/jla.2021.7447

Seymour, E., & Hunter, A. B. (2019). *Talking about leaving revisited: Persistence, relocation, and loss in undergraduate STEM education.* Springer Nature. https://doi.org/10.1007/978-3-030-25304-2

Shermis, M. D., Burstein, J., Elliot, N., Miel, S., & Foltz, P. W. (2016). Automated writing evaluation: An expanding body of knowledge. In C. A. MacArthur, S. Graham, & J. Fitzgerald (Eds.), *Handbook of writing research* (pp. 395–409). Guilford Press.

Shute, V. J. (2011). Stealth assessment in computer-based games to support learning. In S. Tobias & J. D. Fletcher (Eds.), *Computer games and instruction* (pp. 503–524). Information Age Publishing.

Shute, V., Rahimi, S., Smith, G., Ke, F., Almond, R., Dai, C. P., Kuba, R., Liu, Z., Yang, X., & Sun, C. (2021). Maximizing learning without sacrificing the fun: Stealth assessment, adaptivity and learning supports in educational games. *Journal of Computer Assisted Learning, 37*(1), 127–141. https://doi.org/10.1111/jcal.12473

Spikol, D., Ochoa, X., Worsley, M., Di Mitri, D., Cukurova, M., Martinez-Maldonado, R., & Schneider, J. (2021, April). CROSSMMLA futures: Collecting and analysing multimodal data across the physical and virtual. In M. Scheffel, N. Dowell, S. Joksimovic, & G. Siemens (Eds.), *Proceedings of the 11th international conference on learning analytics and knowledge* (pp. 381–385). ACM.

Theobald, E. J., Hill, M. J., Tran, E., Agrawal, S., Arroyo, E. N., Behling, S., Chambwe, N., Cintrón, D. L., Cooper, J. D., Dunster, G., Grummer, J. A., Hennessey, K., Hsiao, J., Iranon, N., Jones, L. 2nd, Jordt, H., Keller, M., Lacey, M. E., Littlefield, C. E., & Grummer, J. A. (2020). Active learning narrows achievement gaps for underrepresented students in undergraduate science, technology, engineering, and math. *Proceedings of the National Academy of Sciences, 117*(12), 6476–6483. https://doi.org/10.1073/pnas.1916903117

Walker, E., & Ogan, A. (2016). We're in this together: Intentional design of social relationships with AIED systems. *International Journal of Artificial Intelligence in Education, 26*(2), 713–729. https://doi.org/10.1007/s40593-016-0100-5

Walsh, C. G., Chaudhry, B., Dua, P., Goodman, K. W., Kaplan, B., Kavuluru, R., Solomonides, A., & Subbian, V. (2020). Stigma, biomarkers, and algorithmic bias: Recommendations for precision behavioral health with artificial intelligence. *JAMIA Open, 3*(1), 9–15. https://doi.org/10.1093/jamiaopen/ooz054

Wang, K. D., Salehi, S., Arseneault, M., Nair, K., & Wieman, C. (2021, June). Automating the assessment of problem-solving practices using log data and data mining techniques. In C. Meinel, M. Pérez-Sanagustín, M. Specht, & A. Ogan (Eds.), *Proceedings of the eighth ACM conference on learning@ scale* (pp. 69–76). ACM. https://doi.org/10.1145/3430895.3460127

Wiens, J., Price, W. N., & Sjoding, M. W. (2020). Diagnosing bias in data-driven algorithms for healthcare. *Nature Medicine, 26*(1), 25–26. https://doi.org/10.1038/s41591-019-0726-6

Williams, R. M., & Gilbert, J. C. (2019). "Nothing about us without us": Transforming participatory research and ethics in human systems engineering. In R. D. Roscoe, E. K. Chiou, & A. R. Wooldridge (Eds.), *Advancing diversity, inclusion, and social justice through human systems engineering* (pp. 113–134). CRC Press. https://doi.org/10.1201/9780429425905

Wilson, J., & Roscoe, R. D. (2020). Automated writing evaluation and feedback: Multiple metrics of efficacy. *Journal of Educational Computing Research, 58*(1), 87–125. https://doi.org/10.1177/0735633119830764

Wisser, L. (2019). Pandora's algorithmic black box: The challenges of using algorithmic risk assessments in sentencing. *American Criminal Law Review, 56*, 1811–1832.

Worsley, M., & Bar-El, D. (2020). Inclusive making: Designing tools and experiences to promote accessibility and redefine making. *Computer Science Education*. https://doi.org/10.1080/08993408.2020.1863705

Yan, D., Rupp, A. A., & Foltz, P. W. (Eds.). (2020). *Handbook of automated scoring: Theory into practice* (1st ed.). CRC Press. https://doi.org/10.1201/9781351264808

Zimmermann-Niefield, A., Turner, M., Murphy, B., Kane, S. K., & Shapiro, R. B. (2019, June). Youth learning machine learning through building models of athletic moves. In J. A. Fails (Ed.), *Proceedings of the 18th ACM international conference on interaction design and children* (pp. 121–132). ACM. https://doi.org/10.1145/3311927.3323139

Index

Note: Locators in *italics* represent figures and **bold** indicate tables in the text.

AA, *see* Archetypal analysis
Academic analytics, 128
  decision support in education, 128–129
  information system course, 131
    assessment, 132–133
    classroom, 132
    course delivery organization, 132
    course management, 133–134
    framework, 132
  platform, 134
  at Politecnico di Torino, 129–130
  research questions, 130–131
Actionable information, 158
Action units (AUs), 83
Active AIEd technologies, teacher learning, 167
ACT Programming Tutor system, 5
ACTS, *see* AI-based Classroom Teaching Simulation
Adaptive formative assessment (AFA), 78, 80–82
Adaptive learning profiles, in education domain, 127, *130*
  academic analytics, 128
    decision support in education, 128–129
    information system learning process, 131–134, *131*
    platform, 134
    at Politecnico di Torino, 129–130
    research questions, 130–131
  future work, 146–147
  information systems course data, 134
    association rules, 140–143
    clustering algorithm, 136–139
    dataset, 134–135
    dataset initial exploration, 135–136
    discussion on the results, 143–145
    performance-based clustering visualization, 139–140
    research questions and answers, 145–146
AddBlock, 30
Additive factor model (AFM), 181–182, 271
AES algorithms, *see* Automated essay scoring algorithms
AFA, *see* Adaptive formative assessment
AFM, *see* Additive factor model
AI-based Classroom Teaching Simulation (ACTS), 169
AI-directed STEM education, 5–6
AIEd, *see* Artificial intelligence-based education (AIEd) technologies
AI-empowered STEM education, 7–8
AI-supported STEM education, 6–7
AI-empowered education, 4
ALEAS (Adaptive LEArning system for Statistics) App, 62, 63, *64*, 65–67, *67*, 71, 73
Alex's learning trajectory, 29, *29*
Amazon Mechanical Turk (AMT), 102
Ambitious learning practices, 158
Ambitious teaching, 198–199
AMOEBA tool, 6
AMT, *see* Amazon Mechanical Turk
Analytic dashboards, learning, 84–86
ANCOVA analysis, 43–45
ANOVA testing, 44, 45, 313
Answer Prediction Model (APM), 98, 102, 103, **103**
APM, *see* Answer Prediction Model
AR, *see* Augmented reality
Archetypal analysis (AA), 66–67, 69, 70, 73
Artificial intelligence, for classroom orchestration, 152
Artificial intelligence-based education (AIEd) technologies
  active, 167–168
  constructive, 168–169
  design guidance for, 165–166
  interactive, 169
  passive, 167
Artificial Intelligence in Education (AIEd), 3, 4, *4*, *5*, 320, 359–360, 365–366
Artificial neural network, 7, 84, 182
ASSISTment Builder, 4, 5
ASSISTments, 246
Augmented reality (AR), 212
AUs, *see* Action units
Automated curriculum alignment, learning resources, 111–112
Automated essay scoring (AES) algorithms, 296
Automated scoring technology, 359
Automated systems *vs.* teacher agency, 157
Automated writing evaluation (AWE) systems, 195, 296, 359
  authentic tasks, ambitious teaching, 198–199
  construct, features of, 199
  design, 201–202
    education policies, 202
    school leaders, values and goals held by, 202–203
    teachers, values and goals held by, 203–204
  eRevise system, *196*, 196–197
  information, 199
  system supports instructional interactions and discourse, 200–201
  underlying algorithms, 199–200
Automatic detection, of students' engagement, 80, 82–84
Automatic generation, of formative assessments, 78
AWE systems, *see* Automated writing evaluation systems

BalanceScale, 47
Bandura model, 268, *268*
Basic local independence model (BLIM), 64
Bayesian approaches, 187
Bayesian causal networks, 16
Bayesian knowledge tracing (BKT), 78, 99, 181, 271
Bayesian networks, 179–181
Behavioral engagement, 79, 83
Behavioral science theory, 268
Behaviorism, 5
BERT model, 103
Beyond class assistance, 156
BIC index, 66
BKT, *see* Bayesian knowledge tracing
BLIM, *see* Basic local independence model
Blob detection, 41
Bloom's taxonomy, 81

Cardiopulmonary resuscitation (CPR) training, 6
CAT, *see* Content Assignment Tool
Causal learning outcome models, 295
CBA, *see* Computer-based assessment
CCLR, *see* Collaborative Complex Learning Resource

CDMs, *see* Cognitive diagnosis models
CERT, *see* Computer Expression Recognition Toolbox
CF, *see* Collaborative filtering
Chem Tutor, 57
Classical test theory (CTT), 176–177
Classroom context interviews, 154–155
Classroom management *vs.* ambitious learning practices, 158
Classroom orchestration, 151, 152
Cluster analytics, 272
Clustering algorithm, 136
  DBSCAN algorithm, 137–138
  k-means and k-medoids, 138–139
COCA, *see* Corpus of Contemporary American English
Cog Model Discovery Experiment Spring 2010, 105
Cognition hypothesis, 251
Cognitive diagnosis models (CDMs), 178–179, 185
Cognitive engagement, 79, 83, 167, 221, 233
Cognitive fluency, 243
Cognitive Tutor Algebra (CTA), 345
Cognitive Tutor Authoring Tool (CTAT), 246
Cognitive tutors, 5, 93
Cohen's Kappa, 116
Collaborative Complex Learning Resource (CCLR), 6
Collaborative filtering (CF), 183
  collaborative filtering-based approaches, 185–186
  deep learning–based collaborative filtering, 186
  matrix factorization, 184–185
Collaborative problem-solving (CPS) environments, 362
  teaching teamwork for, 227–228
Compensatory re-parameterized unified model (C-RUM), 179
Complex systems (CS) theory, 219
Computational block life cycle, *24*
Computational thinking, role of, 26
  computational modeling, effects on, 26–27
  engineering design, effects on, 27
  posttest scores, effects on, 27–28
Computer-based assessment (CBA), 176
Computer-based science education, 310
Computer Expression Recognition Toolbox (CERT), 83
Computer game, 40
Computer-supported collaborative learning (CSCL), 151, 333
  group formation, AI techniques used to support, 340
    AI techniques and evaluation, 342–343
    group size and grouping mechanisms, 341–342
    issues of design, 350–351
    issues of evaluation, 351
  methodology, 334
    data extraction, 336
    data synthesis and analysis, 336
    research purpose and questions, 334–335
    searching the literature, 335
    selecting studies, 335–336, **335**
  overall trend of AI applications in educational contexts, 338–339, 350
    research design, 350
    research focus, 339, 350
    research methods, 339–340
    type of research, 338, 349
  student interactions, AI-enabled systems to support, 343
    design, 344–347
    evaluation, 347–349
    issues of design, 351–352
    issues of evaluation, 352–353
Computer vision (CV) algorithms, 39, 40
ConnectBlock, 30–32
Construct, features of, 199
Constructionism, 40
Constructive AIEd technologies for teacher learning, 168
Constructivist classroom, 132
"Contactless" sensors, 247
Content Assignment Tool (CAT), 111
Context-based science education, 109
  automated curriculum alignment of learning resources, 111–112
  latent Dirichlet allocation (LDA), 112, *112*
  present study, 113
  science article preprocessing and vectorization, 114–115
  science article recommendation based on topic similarity, 116
    and item generation, 118–120
  science article topic analysis, 116–117
  science article topic modeling, 115
  science news articles and assessment in, 110–111
  topic modeling, 112–113
  topic prediction and topic evaluation units of study, 117–118
  topic structure prediction and evaluation of the units of study, 115–116
Corpus of Contemporary American English (COCA), 312
Correlation matrix, 136, *136*
Course management, 133–134
Covariates, generating, 282
CPR training, *see* Cardiopulmonary resuscitation training
CPS, *see* Collaborative problem-solving
CrossMMLA, 363–364, 368
C-RUM, *see* Compensatory re-parameterized unified model
CSCL, *see* Computer-supported collaborative learning
CSILE network collaboration platform, 273
CS theory, *see* Complex systems theory
CTA, *see* Cognitive Tutor Algebra
CTAT, *see* Cognitive Tutor Authoring Tool
CTT, *see* Classical test theory
CV algorithms, *see* Computer vision algorithms
CyberBook, 94–95, 106

Dashboard design, 85
Data analysis methods, expansion of, 214–215
Data collection, 274
  advancement of, 212–214
Data-driven learning behavior analytics, 272
Data ethics and student privacy, 249
Data generation, 282
  covariates, 282
  outcome, 283
  treatment assignment indicator, 282–283
Data hierarchy, 85
Data pre-processing, 274
DataShop, 101
Davies–Bouldin Index (DBI), 138, 139
DBI, *see* Davies–Bouldin Index
DBNs, *see* Dynamic Bayesian networks
DBSCAN algorithm, 136–138, **138**
Decision support, in education, 128–129
Decision trees, 84, *280*
Deep knowledge tracing (DKT), 73, 78, 182–183, 271
Deep learning–based collaborative filtering, 186
Deep learning models, 84
Deep student models, 295
DEI (diversity, equity, and inclusion), 360
  collaboration and discourse, 362–363
  commitments for centering, **364**

considering who will use the AI, 366
educational opportunities for AI, 366
introductory STEM courses
active learning in, 361–362
performance in, 361
learning assessments, 363–364
person-centered variables, outcomes, and ownership, 364–365
principles, 364
principles and methods of AI supporting, 360
revealing, mitigating, and preventing biases, 365
transparency in feedback and dissemination, 365–366
Dialogue-based tutoring systems (DTSs), 4
Differential Segmentation of Categorical Sequences (DiSCS), 18
Digital learning environments (DLEs), 242
Dirichlet distribution, 112–113
Discipline knowledge graph, 269–270
DisconnectBlock, 30, 33
Discrete-time Markov chain model, *31*
DiSCS, *see* Differential Segmentation of Categorical Sequences
DKT, *see* Deep knowledge tracing
DLEs, *see* Digital learning environments
Domain-specific modeling languages (DSML), 19–20, *21*
DSML, *see* Domain-specific modeling languages
D3 (Data-Driven Documents), 270
DTSs, *see* Dialogue-based tutoring systems
Dublin descriptors, 65
Dynamic Bayesian networks (DBNs), 180

EAP, *see* Expected a posteriori
EarthShake, 41, 43, 47, 48
ECD, *see* Evidence-centered design
ECG, *see* Electrocardiogram
EDA, *see* Electrodermal activity
EDM, *see* Educational data mining
EDM/LA, *see* Educational data mining and learning analytics
Educational data mining (EDM), 128, 260
Educational data mining and learning analytics (EDM/LA), 134
Educational knowledge graph, 265–266, 269
discipline knowledge graph, 269–270
individual knowledge graph, 270
principles of, 267
individual knowledge graph, 267
subject knowledge graph, 267
storage and presentation of
knowledge graph visualization based on D3.js, 270
knowledge storage based on graph database Neo4j, 270
theoretical basis of
connectivism learning theory, 266–267
theory of basic structure of disciplines, 266
Educational opportunities, 366
Educational video games, 211
Education policies, 202
EEG, *see* Electroencephalograph
Electrocardiogram (ECG), 212
Electrodermal activity (EDA), 212
Electroencephalograph (EEG), 212
ELEs, *see* Exploratory learning environments
EM, *see* Expectation-maximization
Emotional engagement, 79, 83
Emotion and attitude assessment, 271
ENA, *see* Epistemic network analysis
Energy3D modeling tool, 218, 225, 228–229, *228*
Engineering (case study), 233–234
Engineering design, effects on, 27
Engineering posttest scores, 28
Ensemble learning, 281–282
Epistemic network analysis (ENA), 215, 231, 234, *235*
eRevise system, *196*, 196–197
E-step, 66
Ethics and challenges, 366–368
Evidence-centered design (ECD), 19
Excessive training, 94
Expectation-maximization (EM), 66
Expected a posteriori (EAP), 66
Exploratory learning environments (ELEs), 4
Explore-construct mode of Intelligent Science Station, 42, *43*
Extensible 3D (X3D) standard, 261

FillField, 23, 30, 31, 33
Filters, 85
Flesch–Kincaid reading-ease test, 113
Formative assessments, automatic generation of, 78
Formative assessment scores, 26
Frequency heuristic, 99
Future considerations, AI in STEM education, 8–9

GCA, *see* Group Communication Analysis
GDPR, *see* General Data Protection Regulation
General Data Protection Regulation (GDPR), 249
Gensim, 115
GentleBoost, 83
Gini index–based splitting criterion, 281
Google, 265
GraphViz software, 31
Group Communication Analysis (GCA), 362–363, 368
Group formation, AI techniques supporting, 340
AI techniques and evaluation, 342–343
group size and grouping mechanisms, 341–342
issues of design, 350–351
issues of evaluation, 351
Guided-discovery activities, 46
Guided-discovery mode, Intelligent Science Station, 41, *41*

High-performance computing (HPC), 137
HPC, *see* High-performance computing
Human-in-the-loop technology, 154
Human–machine cooperation systems, 7

ICAP framework, *see* Interactive-constructive-active-passive framework
ICC, *see* Item Characteristic Curve
Impasse-driven adaptive remediation, 94
Inclusivity, in STEM introductory courses
active learning, 361–362
performance, 361
Individual knowledge graph, 267, 270
Information system learning process, 131
assessment, 132–133
classroom, 132
course delivery organization, 132
course management, 133–134
framework, 132
Information systems (IS) course, 132, 134
association rules, 140–143
clustering algorithm, 136
DBSCAN algorithm, 137–138
k-means and k-medoids, 138–139
dataset, 134–135

dataset initial exploration, 135–136
performance-based clustering visualization, 139–140
research questions and answers, 145–146
Innovative visualization strategies, emergence of, 215–216
Inquiry learning, 263
Inquiry learning behavior, analysis and evaluation of, 275
Instructional policy learning, 298–299
Intelligent learning dashboards, 77–80
Intelligent Science Stations, 39
AI guidance
making science exhibits yield better learning, 44–45
scaffolded science inquiry, 45–46
AI in 3D physical world *vs.* on flat screen, 42–44
physical setup and AI computer vision technology, 40–41
scenario, 41–42
Intelligent tutoring systems (ITSs), 4, 62, 211, 246, 334
Interactive AIEd technologies for teacher learning, 169
Interactive-constructive-active-passive (ICAP) framework, 164, 166, **168**
and AI-based technologies for teacher learning, 167
active AIEd technologies, 167–168
constructive AIEd technologies, 168–169
interactive AIEd technologies, 169
passive AIEd technologies, 167
Introductory STEM courses
active learning in, 361–362
performance in, 361
Inverse probability of treatment weighting (IPTW), 279
IPTW, *see* Inverse probability of treatment weighting
IRT model, *see* Item Response Theory model
IS course, *see* Information systems course
iSTART training, 297, *298*
Item Characteristic Curve (ICC), 65
Item Response Theory (IRT) model, 62, 65–66, 78, 177–178
ITSs, *see* Intelligent tutoring systems

Jensen–Shannon distance, 116

Kappa, 116
KC model, *see* Knowledge component model
Key performance indicator (KPI), 84, 85
Keyword network analysis, 339, *340*
KIKT, *see* Knowledge Interaction-enhanced Knowledge Tracing model
Kinect depth camera, 40
KL divergence, *see* Kullback–Leibler divergence
K-means, 138–139
K-means clustering model, 84, *85*, 101, 230
K-medoids, 138–139
Knowledge component (KC) model, 101
Knowledge graph, 73, 264, 266
connectivism learning theory, 266–267
contents and functions, **266**
discipline, 269–270
educational, 265–266
individual, 267, 270
interaction between, *266*
origin of, 265
subject, 267
theory of basic structure of disciplines, 266
Knowledge graph visualization, 270
Knowledge Interaction-enhanced Knowledge Tracing model (KIKT), 73
Knowledge level assessment, 271
Knowledge representation (KR) schemes, 19
Knowledge space theory (KST), 64, 67
Knowledge storage, graph database Neo4j, 270
Knowledge tracing, 99
KPI, *see* Key performance indicator
KR schemes, *see* Knowledge representation schemes
KST, *see* Knowledge space theory
K-12 students' CT and engineering design processes, 15
analysis methods, 23–26
background and motivation, 17
computational thinking, role of, 26
computational modeling, effects on, 26–27
engineering design, effects on, 27
posttest scores, effects on, 27–28
data sources, 22–23
future directions, 35–36
implementation, 22
multiple representations, learning with, 28–30
SPICE (Science Projects Integrating Computation and Engineering), 18, *23*
learning trajectory and curriculum, 20–22
system design perspectives, 18–20
STEM learning strategies of students, 17
computational modeling strategies, 18
engineering design strategies, 18
strategy use on learning, 30
computational modeling strategies, 30–33
engineering design strategies, 33–35
21st-century K-12 classrooms, 17
Kullback–Leibler (KL) divergence, 116

LAD, *see* Learning analytics dashboards
LAK, *see* Learning Analytics and Knowledge
LAM, *see* Learning analytics model
Language ability and science scores, 310
Latent Dirichlet Allocation (LDA), 110, 112, *112*, 347
Latent semantic analysis (LSA), 112
Law of optimum perceived difficulty, 243
LCDM, *see* Log-linear cognitive diagnosis model
LDA, *see* Latent Dirichlet Allocation
Learner profile, 267–268
creating, 270
emotion and attitude assessment, 271
knowledge level assessment, 271
learning behavior assessment, 271
research status of, 268
Learning analytics, 128, 211, 225, 260
and AI in digital learning environments, 244
serious games and simulations, 245–246
virtual labs, 244–245
to analyze generated data, 229
analytical methods, 230–232
data analysis methods, expansion of, 214–215
data collection techniques, advancement of, 212–214
as an enabler for paradigm shift, 212
engineering (case study), 233–234
future of STEM education with, 220–222
innovative visualization strategies, emergence of, 215–216

mathematics (case study), 234
research gaps and future trends, 234–236
science (case study), 232
multi-faceted engagement, 233
self-regulation, 232–233
transformative and non-transformative discourse, 232
self-regulated learning (SRL) theory (case), 218–219
STEM education, 226
technology-enhanced environments, 226
collaborative problem-solving, teaching teamwork for, 227–228
Energy3D, 228–229
theory-driven data analysis and visualization, 219–220
theory-driven data collection, 219
theory-driven interpretation of results, 220
theory-driven learning analytics, 216–218
Web3D and, 264
Learning analytics, technical path of
educational knowledge graph, construction of, 269
discipline knowledge graph, 269–270
individual knowledge graph, 270
knowledge graph visualization based on D3.js, 270
knowledge storage based on the graph database Neo4j, 270
Learning analytics, theoretical basis of
educational knowledge graph, 265–266
connectivism learning theory, 266–267
theory of basic structure of disciplines, 266
knowledge graph, 264
origin of, 265
learner profile, 267–268
research status of, 268
principles of, 267
individual knowledge graph, 267
subject knowledge graph, 267
Learning Analytics and Knowledge (LAK), 260
Learning analytics dashboards (LAD), 216
Learning analytics model (LAM), 216, *217*
Learning behavior
analytics method of, 272, **273**
cluster analytics, 272
sequence analytics, 272
sequential pattern mining, 272–273
social network analytics, 273
assessment, 271
definition of, 271–272
standards for recording behavior data, 269
theoretical basis
behavioral science theory, 268
social cognitive theory, 268–269
Learning environment, 274–275
Learning management systems (LMS), 83, 85–86, 211, 216, 269
Learning objectives (LOs), 80
Learning outcome modeling, 175
additive factors model (AFM), 181–182
approaches for, 186–187
Bayesian knowledge tracing (BKT), 181
Bayesian networks, 179–181
collaborative filtering for, 183
approaches, 185–186
deep learning–based collaborative filtering, 186
matrix factorization, 184–185
deep knowledge tracing (DKT), 182–183
deep learning approaches for, 183
psychometric measurement for, 176
classical test theory (CTT), 176–177
cognitive diagnosis models (CDMs), 178–179
item response theory (IRT), 177–178
Learning sciences, 289
AI and its affordances for, 294
causal learning outcome models, 295
deep student models, 295
instructional policy learning, 298–299
natural language processing (NLP), 296–297
sensor-free student factor measures, 297–298
challenges
contextual factors, 292–293
outcomes, 290–291
replication crisis, 293–294
student factors, 291–292
promoting equity, 299–300
Learning strategies, of STEM students, 17
computational modeling strategies, 18
engineering design strategies, 18
Learning trajectory graph (LTG), 99
Lemmatization, 114
Lewis structure, 52
LFA, *see* Logical framework analysis
Linear discriminant analysis, 84
Linear mixed effect (LME) models, 313
Linguistic variables, 312
SEANCE, 312
TAACO, 312
TAALES, 312
LME models, *see* Linear mixed effect models
LMS, *see* Learning management systems
Local Gray Code Patterns, 83
Logical framework analysis (LFA), 132
Logistic regression (LR), 277
Log-linear cognitive diagnosis model (LCDM), 178–179
LOs, *see* Learning objectives
LR, *see* Logistic regression
LSA, *see* Latent semantic analysis
LSTM networks, 183
LTG, *see* Learning trajectory graph

MAB algorithms, *see* Multi-armed bandit algorithms
Machine learning (ML), 163
Maker Movement, 45
Markov chain modeling, 25, *25*, 31, 35
Markov decision process (MDP), 99
Massive open online course (MOOC), 93, 211, 249
Mathematics (case study), 234
Matrix factorization, 184–185
MAUIs, *see* Minimal Attention User Interfaces
Maximum marginal likelihood (MML) approach, 66
MDP, *see* Markov decision process
MiGen TA system, 167
Minimal Attention User Interfaces (MAUIs), 323
Ministry of Education, University, and Research (MIUR), 147
MIRT, *see* Multidimensional item response modeling
MIUR, *see* Ministry of Education, University, and Research
ML, *see* Machine learning
mlVAR, *see* Multilevel Model Vector Autoregression
MML approach, *see* Maximum marginal likelihood approach
MMPASS, *see* Multimedia Probability and Statistics System
Moment of Inertia metric, 41

MOOC, *see* Massive open online course
M-step, 66
Multi-armed bandit (MAB) algorithms, 81, **83**
Multidimensional item response modeling (MIRT), 65–66, 177
Multi-faceted engagement, 233
Multilevel Model Vector Autoregression (mlVAR), 232
Multimedia Probability and Statistics System (MMPASS), 63
Multimodal computational model, 6
Multinominal logistic regression, 83
Multiple representations, learning with, 28–30

Naive Bayes classifier, 84
National Science Foundation (NSF), 211
Natural Language Processing (NLP), 9, 86, 163, 271, 296–297, 310, 347
Need-based adaptive assessment, 94
Neo4j, 270
Network Time Protocol (NTP) server, 323
NLP, *see* Natural Language Processing
NMF, *see* Non-negative matrix factorization
Non-negative matrix factorization (NMF), 112, 121
Non-STEM degrees, teaching STEM subjects in, 61
  archetypal analysis, 66–67
  implementation
    app interface and workflow, 67
    user model and data processing, 67–68
  limitations and future work, 71–73
  multidimensional Item Response Theory (IRT) model, 65–66
  simulation data results, 69
    design of the study, 69
    real-world data, preliminary application on, 71
    results, 69–71
  state of the art, 63
    Dublin descriptors, 65
    knowledge space theory (KST), 64
    statistical knowledge, hierarchical structure for, 63–64
NSF, *see* National Science Foundation
NTP server, *see* Network Time Protocol server
Number of pieces of evidence (NPE), 199

Ohm's Law activities, 227, *227*
OLI, *see* Open Learning Initiative
One-size-fits-all model, 79
OOB sample, *see* Out-of-bag sample
Open Learning Initiative (OLI), 100, 101
Orchestration assistant design, 155
  automated systems *vs.* teacher agency, 157
  classroom management *vs.* ambitious learning practices, 158–159
  during class *vs.* beyond class assistance, 156–157
  detailed information *vs.* actionable information, 158
  role of AI in, 153
Out-of-bag (OOB) sample, 280

Paper-and-pencil rule creation task, 20, *21*
Paradigm shift in STEM education, 4, 212, 216, *216*
  inclusion and equity as, 359–360
Passive AIEd technologies for teacher learning, 167
PASTEL techniques, 93
  CyberBook, 94–95, 106
  QUADL technology, **95**, 97–98
    evaluation, 101–103
  RADARS technology, **95**, 99–100, *100*
    evaluation, 105–106
  RAFINE technology, **95**, 98–99, *98*, 106
    evaluation, 103–105
  SMART technology, **95**, 96–97, *102*, 106
    evaluation, 101
  WATSON technology, 95–96, **95**, *96*
PBL, *see* Problem-based learning
PCH, *see* Principal convex hull
Perceptual fluency, 53–54
Performance-based clustering visualization, 139–140
Perplexity, 115
Personalized adaptive learning, 246
Person-centered variables, outcomes, and ownership, 364–365
pLSA, *see* Probabilistic latent semantic analysis
Policy action set, 99
POS document, *see* Programs of Study document
Posttest scores, effects on, 27
Predict–observe–explain cycle, 40
Principal convex hull (PCH), 67
Probabilistic latent semantic analysis (pLSA), 112
Problem-based learning (PBL), 152, 243
Profile, visualizing, 274–275
Profile generation, 274
Programs of Study (POS) document, 113
Project work syllabus (PWS), 130
Propensity score matching, 277, 279
  assumptions, 278–279
  data generation, 282
    covariates, 282
    outcome, 283
    treatment assignment indicator, 282–283
  ensemble learning, 281–282
  random forest, 279–281
  simulation study, 282–284
  students' success case study, 284
  treatment weighting, inverse probability of, 279
Psychological engagement, 79
Psychometric measurement, for learning outcome modeling, 176
  classical test theory (CTT), 176–177
  cognitive diagnosis models (CDMs), 178–179
  item response theory (IRT), 177–178
Psychometric models, 187
PWS, *see* Project work syllabus
Python NLTK library, 114

QCM, *see* Question Conversion Model
QFD, *see* Quality functional deployment
Q-matrix, 185–186
QRF, *see* Quick Red Fox
QUADL technology, **95**, 97–98
  evaluation, 101–103
Quality functional deployment (QFD), 132
Question Conversion Model (QCM), 98
Quick Red Fox (QRF), 319, 321
  applications, 328–329
  case study, 324
    Betty's Brain, 325
    data, 326
    data coding, 326–327
    impact on scholarly work, 327–328
    interview triggers, developing, 325–326
    procedure, 326
  client side, 323
    data, 324
    interview recordings and notes, 323–324
    moving on, 324
    presentation of student and trigger information, 323
    set-up and login, 323

design, 321
interview triggers, 322
server-side platform, 322–323
future development, 329
limitations, 329
Login screens, 323

RabbitMQ message broker, 323
RADARS technology, **95**, 99–100, *100*
evaluation, 105–106
RAFINE technology, **95**, 98–99, *98*, 106
evaluation, 103–105
Random forests (RFs), 278–281
RapidMiner operator, 136
Real-time feedback and intervention, 246
Recovery action plan, 131
Recurrent neural networks (RNNs), 182
Reinforcement learning (RL), 81, 298
RemoveBlock, 30
Representational competencies, 51
and adaptation to students' current knowledge level, 56–57
perceptual fluency, 53–54
sense-making and perceptual-induction activities, sequence of, 55–56
sense-making competencies, 52
sense-making support and perceptual-fluency support, combining, 54–55
Response-to-Text Assessment (RTA), 197
RFs, *see* Random forests
RL, *see* Reinforcement learning
RMSEA, *see* Root mean square error of approximation
RNNs, *see* Recurrent neural networks
Ronny McStat, 67, *68*
Root mean square error of approximation (RMSEA), 26
R package MultiLCIRT, 66
RTA, *see* Response-to-Text Assessment

SbLA techniques, *see* Sensor-based Learning Analytics techniques
Scaffolded science inquiry, 45–46
School leaders, values and goals held by, 202–203
Science (case study), 232
multi-faceted engagement, 233
self-regulation, 232–233
transformative and non-transformative discourse, 232
Science article preprocessing and vectorization, 114–115
Science article recommendation
based on topic similarity, 116
and item generation, 118–120
Science article topic analysis, 115–117
Science performance, 309
computer-based science education, 310
current study, 310–311
data collected, 311
individual differences data, 312
pretest/posttest assessments, 311
verbal data, 312
full model, 313–314
language ability and science scores, 310
linguistic model, 313
linguistic variables, 312
SEANCE, 312
TAACO, 312
TAALES, 312
non-linguistic model, 313
participants, 311
procedure, 311
statistical analysis, 313
transcriptions, 312
Science Projects Integrating Computation and Engineering (SPICE), 18, *23*
learning trajectory and curriculum, 20–22
system design perspectives, 18
coherence across modeling representations, 19
domain-specific modeling languages (DSML), 19, *21*
evidence-centered design (ECD), 19
SCL, *see* Skin conductance level
SCR, *see* Skin conductance response
SEANCE, 312
SEANCE, *see* SEntiment ANalysis and Cognition Engine
Self-efficacy, 61–62
Self-paced online learning (SPOL) model, 77, 83, 84, 86, 87
removing barriers in, 77
adaptive formative assessments, 80–82
analytic dashboards, learning, 84–86
automatic analysis of students' engagement, 78–79
automatic detection of students' engagement, 82–84
formative assessments, automatic generation of, 78
future work, 86–87
intelligent learning dashboard framework, 79–80
intelligent learning dashboards, 77–78
Self-regulated learning (SRL) theory, 215, 217, 218
Self-regulation, 79, 232–233
Sense-making and perceptual-induction activities, sequence of, 55–56
Sense-making competencies, 52
Sense-making support and perceptual-fluency support, combining, 54–55
Sensor-based Learning Analytics (SbLA) techniques, 247
"Sensor-free" detectors, 297
Sensor-free student factor measures, 297–298
Sensor technology, leveraging the advances in, 246–247
SEntiment ANalysis and Cognition Engine (SEANCE), 312
Sequence analytics, 272
Sequential pattern mining, 272–273
Serious games and simulations, 245–246
"Shake" button, 42
Simulation-based learning environments, 245
SIS, *see* Student information systems
Skin conductance level (SCL), 214
Skin conductance response (SCR), 214
SmartRamps, 41, 47, *47*
SMART technology, **95**, 96–97, *102*, 106
evaluation, 101
SMD, *see* Standardized mean difference
Social cognitive theory, 268–269
Socially shared regulation of learning (SSRL), 214
Social network analytics, 273
Society of Learning Analytics Research (SOLAR), 364
SOLAR, *see* Society of Learning Analytics Research
Spearman's ρ correlation analysis, 28
SPICE, *see* Science Projects Integrating Computation and Engineering
SPOL model, *see* Self-paced online learning model
SRL theory, *see* Self-regulated learning theory
SSRL, *see* Socially shared regulation of learning
Standardized mean difference (SMD), 284
StartSimulation, 23, 30, 31
Statistical knowledge, hierarchical structure for, 63–64
Statistics anxiety, 61
Steering Committee, 133
Stemming, 114

Strategy use on learning, 30
computational modeling strategies, 30–33
engineering design strategies, 33–35
Student behavior and assessment, 133–134
Student information systems (SIS), 216
Student interactions, AI-enabled systems to support, 343
design, 344–347
evaluation, 347–349
issues of design, 351–352
issues of evaluation, 352–353
Subject knowledge graph, 267
Support vector machines, 83, 84
SUTVA, 278

TAACO, 312
TAALES, 312
TAs, *see* Teaching assistants
Task difficulty, 241
challenges, 247
confounding factors, 250–251
data ethics and student privacy, 249
task difficulty and its operationalization, 247–249
what data to collect and how, 249–250
learning analytics (LA) and AI in digital learning environments, 244
serious games and simulations, 245–246
virtual labs, 244–245
multifaceted aspects of, 242–243
operationalization, 247–249
opportunities
personalized adaptive learning, 246
real-time feedback and intervention, 246
sensor technology, leveraging the advances in, 246–247
Taylor's computational model, 29
Teacher agency *vs.* automated systems, 157
Teacher learning, 163
active AIEd technologies for, 167–168
AI-based education (AIEd) tools, design guidance for, 165–166
constructive AIEd technologies for, 168–169
interactive AIEd technologies for, 169
interactive-constructive-active-passive (ICAP) framework, 166–169, **168**
limitations, 171
passive AIEd technologies for, 167
perspectives, 164–165
technologies for, 165
AI-based teacher tools, 165
Teacher orchestration systems, AI supported, 151
artificial intelligence for classroom orchestration, 152
classroom context interviews, 154–155
classroom orchestration, 151
discussion and future research, 159
orchestration assistant design, 155–156
automated systems *vs.* teacher agency, 157
classroom management *vs.* ambitious learning practices, 158–159
during class *vs.* beyond class assistance, 156–158
detailed information *vs.* actionable information, 158
orchestration assistant design, role of AI in, 153
research context, 153
Teacher Responding Tool (TRT), 168–169
Teachers, values and goals held by, 203–204
Teaching assistants (TAs), 165
Teaching Teamwork simulation platform, 225, 230, 232
Technological pedagogical statistical knowledge (TPSK), 63
Technology-enhanced environments to support STEM education, 226
collaborative problem-solving, teaching teamwork for, 227–228
Energy3D, 228–229
Term frequency-inverse document frequency (TF-IDF), 347
TextRanking algorithm, 97
TF-IDF, *see* Term frequency-inverse document frequency
Theory-driven data analysis and visualization, 219–220
Theory-driven data collection, 219
Theory-driven interpretation of results, 220
Theory-driven learning analytics, 216–218
Thompson Sampling, 81
Tool for the Automatic Analysis of Lexical Sophistication, 312
Topic modeling, 112–113
"Total absorption – total rainfall" expression, 29
TPSK, *see* Technological pedagogical statistical knowledge
Transformative and non-transformative discourse, 232
Treatment assignment indicator, generating, 282–283
Treatment weighting, inverse probability of, 279
Trial-and-error strategy, 34
TRT, *see* Teacher Responding Tool
"Turin Residence" attribute, 136
21st-century K-12 classrooms, 17

UCB, *see* Upper Confidence Bound
Upper Confidence Bound (UCB), 81

Variable neighborhood search (VNS), 343
VAR model, *see* Vector autoregression model
Vector autoregression (VAR) model, 219, 232
Virtual inquiry learning scene, web page presenting, 264
Virtual labs, 244–245
Virtual reality (VR), 260
Virtual Reality Modeling Language (VRML), 261
Visualizations, 86
VNS, *see* Variable neighborhood search
VR, *see* Virtual reality
VRML, *see* Virtual Reality Modeling Language

Water Runoff Challenge (WRC), 18, **23**
WATSON technology, 95–96, **95**, *96*
Web-enabled Intelligent SQL Tutor, 5
WebGL technology, 262
Webhose.io, 113
Web3D based inquiry learning environment, 262
concrete realization of learning analysis
generation and presentation of learner profile, 274–275
knowledge graph construction of chemistry experiments, 273–274
inquiry learning, combination with, 263
learning autonomy, 263
personalization of learning, 263
rich cognitive tools, providing, 263
virtual inquiry learning scene, web page presenting, 264
Web3D technology, 260–261, *261*, *262*

implementation technologies of, 261–262
and learning analytics, 264
Wheel-spinning detector, 94, 99
WRC, *see* Water Runoff Challenge
Writing Pal intelligent tutoring system, *296*
xAPI learning behavior statement, 269
X3D standard, *see* Extensible 3D standard

Zone of Proximal Development (ZPD), 81, 82, *82*
ZPD, *see* Zone of Proximal Development

For Product Safety Concerns and Information please contact our EU
representative GPSR@taylorandfrancis.com
Taylor & Francis Verlag GmbH, Kaufingerstraße 24, 80331 München, Germany

www.ingramcontent.com/pod-product-compliance
Ingram Content Group UK Ltd.
Pitfield, Milton Keynes, MK11 3LW, UK
UKHW052055230926
475702UK00031B/2920
9781032019604